T0234454

Lecture Notes in Computer Science   9399

Commenced Publication in 1973
Founding and Former Series Editors:
Gerhard Goos, Juris Hartmanis, and Jan van Leeuwen

More information about this series at http://www.springer.com/series/7407

Martin Leucker · Camilo Rueda
Frank D. Valencia (Eds.)

Theoretical Aspects of Computing – ICTAC 2015

12th International Colloquium
Cali, Colombia, October 29–31, 2015
Proceedings

 Springer

Editors

Martin Leucker
University of Lübeck
Lübeck
Germany

Camilo Rueda
Pontificia Universidad Javeriana-Cali
Cali
Colombia

Frank D. Valencia
CNRS LIX
École Polytechnique de Paris
Palaiseau
France

and

Pontificia Universidad Javeriana-Cali
Cali
Colombia

ISSN 0302-9743   ISSN 1611-3349   (electronic)
Lecture Notes in Computer Science
ISBN 978-3-319-25149-3   ISBN 978-3-319-25150-9   (eBook)
DOI 10.1007/978-3-319-25150-9

Library of Congress Control Number: 2015950887

LNCS Sublibrary: SL1 – Theoretical Computer Science and General Issues

Springer Cham Heidelberg New York Dordrecht London

Printed on acid-free paper

Springer International Publishing AG Switzerland is part of Springer Science+Business Media
(www.springer.com)

Preface

This volume contains the papers presented at ICTAC 2015: The 12th International Colloquium on Theoretical Aspects of Computing held during October 29–31, 2015, in Cali, Colombia.

The International Colloquia on Theoretical Aspects of Computing (ICTAC) is a series of annual events founded in 2003 by the United Nations University International Institute for Software Technology. Its purpose is to bring together practitioners and researchers from academia, industry, and government to present research results and exchange experience and ideas. Beyond these scholarly goals, another main purpose is to promote cooperation in research and education between participants and their institutions from developing and industrial regions.

The city of Cali, where this year's ICTAC took place, is the third largest city of Colombia and the seat of six major universities of the country. The Universidad Javeriana-Cali, host of the colloquium, has built a reputation on theoretical computer science through the works of Avispa, a research team founded in the Cali-based universities of Javeriana and Universidad del Valle, with active members in the universities of Cork (Ireland), École Polytechnique-Paris (France), Oxford (UK), and Groningen (The Netherlands). The latter three institutions were co-organizers of this year's colloquium, which was also sponsored by Microsoft Research Center, Inria, CNRS, CLEI, and the Colombian Computation Society.

We were honored to have seven distinguished guests as invited speakers:

- Jean-Raymond Abrial (consultant, France)
- Volker Diekert (University of Stuttgart, Germany)
- César Muñoz (NASA Langley, USA)
- Catuscia Palamidessi (Inria and École Polytechnique, France)
- Davide Sangiorgi (Inria and University of Bologna, Italy)
- Moshe Vardi (Rice University, USA)
- Glynn Winskel (University of Cambridge, UK)

Jean-Raymond Abrial's talk undertook the study of a proof of a well-known theorem in planar graphs to motivate the new discipline of mathematical engineering. Volker Diekert discussed different monitor constructions for checking safety properties of complex systems. César A. Muñoz's talk concerned the application of formal methods to the safety analysis of air traffic management systems. He described the detect and avoid (DAA) capability to address the challenge of NASA's Unmanned Aircraft Systems Integration project. Catuscia Palamidessi's talk discussed the problem of protecting the privacy of the user when dealing with location-based services. Davide Sangiorgi presented his work on refinements of co-inductive proof methods for functional and process languages. He discussed the contraction technique that refines Milner's unique solution of equations to reason about bisimilarity. Glynn Winskel discussed his work on optimal probabilistic strategies for distributed games. Moshe

Vardi's talk described the rise and fall of mathematical logic in computer science and then analyzed the quiet revolution in logic that has given rise again to modern applications of logic to computing.

ICTAC 2015 received 93 submissions from 30 different countries. Each submission was reviewed by at least three members of the Program Committee, along with help from external reviewers. Out of these 93 submissions, 25 full-length papers were accepted. The committee also accepted two short papers and three tool papers. This corresponds approximately to a 1/3 acceptance ratio.

Apart from the paper presentations and invited talks, ICTAC 2015 continued the tradition of previous ICTAC conferences in holding a four-course school on three important topics in theoretical aspects of computing: formal methods and verification, formal models of concurrency, and security in concurrency. These courses were: "Formal Modeling" given by Jean-Raymond Abrial (France), "Formal Verification Techniques," by Martin Leucker (University of Lübeck, Germany), "Security and Information Flow," by Kostas Chatzikokolakis (CNRS-École Polytechnique, France), and "Models for Concurrency" by Pawel Sobocinski (University of Southampton, UK). In addition, co-located for the first time with ICTAC, we hosted the 11th International Workshop on Developments in Computational Models (DCM 2015) chaired by César A. Muñoz (NASA) and Jorge A. Pérez (University of Groningen).

We thank all the authors for submitting their papers to the conference, and the Program Committee members and external reviewers for their excellent work in the review, discussion, and selection process. We are indebted to all the members of the Organizing Committee for their hard work in all phases of the conference. We also acknowledge our gratitude to the Steering Committee for their constant support.

We are also indebted to EasyChair that greatly simplified the assignment and reviewing of the submissions as well as the production of the material for the proceedings. Finally, we thank Springer for their cooperation in publishing the proceedings.

August 2015   Martin Leucker
                                                                 Camilo Rueda
                                                              Frank D. Valencia

Organization

Program Committee

Nazareno Aguirre	Universidad Nacional de Río Cuarto - CONICET, Argentina
Gerard Assayag	Ircam
Mauricio Ayala-Rincon	Universidade de Brasilia, Brazil
Pablo Barceló	Universidad de Chile, Chile
Gustavo Betarte	InCo, Facultad de Ingeniería, Universidad de la República, Uruguay
Filippo Bonchi	University of Pisa, Italy
Marco Carbone	IT University of Copenhagen, Denmark
Ilaria Castellani	Inria Sophia Antipolis, France
Néstor Cataño	The University EAFIT, Colombia
Gabriel Ciobanu	Romanian Academy, Iasi, Romania
Silvia Crafa	Università di Padova, Italy
Pedro R. D'Argenio	Universidad Nacional de Córdoba - CONICET, Argentina
Nicolas D'Ippolito	Universidad de Buenos Aires, Argentina
Stefan Dantchev	Durham University, UK
Rocco De Nicola	IMT - Institute for Advanced Studies Lucca, Italy
Yuxin Deng	Shanghai Jiao Tong University, China
Gilles Dowek	Inria, France
Moreno Falaschi	University of Siena, Italy
José Luiz Fiadeiro	Royal Holloway, University of London, UK
Wan Fokkink	Vrije Universiteit Amsterdam, The Netherlands
Fabio Gadducci	Università di Pisa, Italy
Kim Guldstrand Larsen	Aalborg University, Denmark
Julian Gutierrez	University of Oxford, UK
Stefan Haar	Inria Saclay/LSV, ENS Cachan, France
Thomas Hildebrandt	IT University of Copenhagen, Denmark
Einar Broch Johnsen	University of Oslo, Norway
Bartek Klin	University of Warsaw, Poland
Marta Kwiatkowska	University of Oxford, UK
Martin Leucker	University of Lübeck, Germany
Etienne Lozes	LSV, ENS Cachan, CNRS, France
Larissa Meinicke	The University of Queensland, Australia
Hernan Melgratti	Universidad de Buenos Aires, Argentina
Dominique Mery	Université de Lorraine, LORIA, France
Matteo Mio	University of Cambridge, UK

Andrzej Murawski	University of Warwick, UK
Vivek Nigam	Universidade Federal da Paraíba, Brazil
Mauricio Osorio	UDLA, México
Prakash Panangaden	McGill University, Canada
Elaine Pimentel	UFMG, Brazil
Jorge A. Pérez	University of Groningen, The Netherlands
António Ravara	Universidade Nova de Lisboa, Portugal
Camilo Rocha	Escuela Colombiana de Ingeniería, Colombia
Camilo Rueda	Universidad Javeriana-Cali, Colombia
Augusto Sampaio	Federal University of Pernambuco, Brazil
Vijay Saraswat	IBM, USA
Vladimiro Sassone	University of Southampton, UK
Gerardo Schneider	Chalmers — University of Gothenburg, Sewden
Alexandra Silva	Radboud University Nijmegen, The Netherlands
Jiri Srba	Aalborg University, Denmark
Jean-Bernard Stefani	Inria, France
Perdita Stevens	University of Edinburgh, UK
Kohei Suenaga	Graduate School of Informatics, Japan
Vasco T. Vasconcelos	University of Lisbon, Portugal
Alwen Tiu	Nanyang Technological University, Singapore
Kazunori Ueda	Waseda University, Japan
Frank D. Valencia	CNRS LIX, École Polytechnique de Paris, France and Universidad Javeriana-Cali, Colombia
Kapil Vaswani	Microsoft Research, India
Björn Victor	Uppsala University, Sweden
Igor Walukiewicz	CNRS, LaBRI, France
Farn Wang	National Taiwan University, Taiwan
Alan Wassyng	McMaster University, Canada
Gianluigi Zavattaro	University of Bologna, Italy

Organizing Committee

Gloria Inés Álvarez	Universidad Javeriana-Cali, Colombia
Jesús Aranda	Universidad del Valle, Colombia
Antal Buss	Universidad Javeriana-Cali, Colombia
Juan Francisco Díaz	Universidad del Valle, Colombia
Juan Pablo García	Universidad Javeriana-Cali, Colombia
Michell Guzmán	Inria and LIX École Polytechnique, France
Julian Gutierrez	University of Oxford, UK
Juan Carlos Martínez	Universidad Javeriana-Cali, Colombia
Andrés Navarro	Universidad Javeriana-Cali, Colombia
María Constanza Pabón	Universidad Javeriana-Cali, Colombia
Salim Perchy	Inria and LIX École Polytechnique, France
Jorge A. Pérez	University of Groningen, The Netherlands
Luis Fernando Pino	Inria and LIX École Polytechnique, France
Luisa Fernanda Rincón	Universidad Javeriana-Cali, Colombia

Gerardo Sarria   Universidad Javeriana-Cali, Colombia
Camilo Rueda   Universidad Javeriana-Cali, Colombia
Alexander Valencia   Universidad Javeriana-Cali, Colombia
Frank D. Valencia   CNRS LIX, École Polytechnique de Paris, France
                          and Universidad Javeriana-Cali, Colombia

Additional Reviewers

Almeida Matos, Ana
Aman, Bogdan
Åman Pohjola, Johannes
Ballis, Demis
Basold, Henning
Bianchni, Monica
Blankenburg, Martin
Bohorquez, Jaime
Brodo, Linda
Buss, Antal
Campo, Juan Diego
Cassel, Sofia
Castellan, Simon
Castellanos Joo, José Abel
Chimento, Jesus Mauricio
Colombo, Christian
Dardha, Ornela
Decker, Normann
Dezani-Ciancaglini,
    Mariangiola
Di Giusto, Cinzia
Fauconnier, Hugues
Fontaine, Gaelle
Forget, Julien
Gibson, J. Paul
Groote, Jan Friso
Gunadi, Hendra
Gutkovas, Ramūnas
Heijltjes, Willem
Hoffmann, Guillaume

Horne, Ross
Huang, Mingzhang
Ishii, Daisuke
Iwasaki, Hideya
Jensen, Peter Gjøl
Jovanovic, Aleksandra
Knight, Sophia
Kumar, Sandeep
König, Barbara
Lanese, Ivan
Laursen, Simon
Lee, Matias David
Lehtinen, Karoliina
Lin, Anthony Widjaja
Long, Huan
Luna, Carlos
Marcial-Romero,
    Jose-Raymundo
Markin, Grigory
Marques, Eduardo R.B.
Mastroeni, Isabella
Melo de Sousa, Simão
Merz, Stephan
Mezzina, Claudio Antares
Montesi, Fabrizio
Mostrous, Dimitris
Nieves, Juan Carlos
Olarte, Carlos
Pagano, Miguel
Paraskevas, Evripidis

Pardo, Raúl
Perelli, Giuseppe
Pun, Ka I.
Rocha Oliveira,
    Ana Cristina
Roldan, Christian
Salamon, Andras Z.
Salvati, Sylvain
Santini, Francesco
Sasse, Ralf
Scherer, Gabriel
Schlatte, Rudolf
Schmitz, Malte
Singh, Neeraj
Sobocinski, Pawel
Soncco-Álvarez, José Luis
Song, Lei
Steffen, Martin
Tasson, Christine
Ter Beek, Maurice H.
Thoma, Daniel
Tiezzi, Francesco
Tzameret, Iddo
Van Raamsdonk, Femke
Ventura, Daniel Lima
Viera, Marcos
Wiltsche, Clemens
Winter, Joost
Zepeda Cortes, Claudia

Invited Talks Abstracts

An Exercise in Mathematical Engineering: Stating and Proving Kuratowski Theorem

Jean-Raymond Abrial

Marseille, France
jrabrial@neuf.fr

Abstract. This paper contains the informal presentation of a well known theorem on planar graphs: the theorem of Kuratowski (1930). This study is supposed to serve as an example for the proposed new discipline of *Mathematical Engineering*. The intend if this discipline is to show to informaticians, by means of examples, that there must exist important connections between rigorous mathematics and rigorous computer science. Moreover, in both cases, the mechanisation of proofs is becoming more and more fashionable these days. Such mechanisations cannot be performed without a clear understanding of the mathematical context that has to be rigorously described before engaging in the proof itself.

Location Privacy via Geo-Indistinguishability

Konstantinos Chatzikokolakis[1,2], Catuscia Palamidessi[2,3],
and Marco Stronati[2]

[1] CNRS, France
[2] LIX, École Polytechnique, France
[3] INRIA, France
catuscia@lix.polytechnique.fr

Abstract. In this paper we report on the ongoing research of our team Comète on location privacy. In particular, we focus on the problem of protecting the privacy of the user when dealing with location-based services. The starting point of our approach is the principle of geo-indistinguishability, a formal notion of privacy that protects the user's exact location, while allowing approximate information – typically needed to obtain a certain desired service – to be released. Then, we discuss the problem that raise in the case of traces, when the user makes consecutive uses of the location based system, while moving along a path: since the points of a trace are correlated, a simple repetition of the mechanism would cause a rapid decrease of the level of privacy. We then show a method to limit such degradation, based on the idea of predicting a point from previously reported points, instead of generating a new noisy point. Finally, we discuss a method to make our mechanism more flexible over space: we start from the observation that space is not uniform from the point of view of location hiding, and we propose an approach to adapt the level of privacy to each zone.

A Note on Monitors and Büchi automata

Volker Diekert[1] and Anca Muscholl[2] and Igor Walukiewicz[2]

[1] Universität Stuttgart, FMI, Germany
diekert@fmi.uni-stuttgart.de
[2] LaBRI, University of Bordeaux, France

Abstract. When a property needs to be checked against an unknown or very complex system, classical exploration techniques like model-checking are not applicable anymore. Sometimes a monitor can be used, that checks a given property on the underlying system at runtime. A monitor for a property L is a deterministic finite automaton $\mathcal{M}_L$ that after each finite execution tells whether (1) every possible extension of the execution is in L, or (2) every possible extension is in the complement of L, or neither (1) nor (2) holds. Moreover, L being monitorable means that it is always possible that in some future the monitor reaches (1) or (2). Classical examples for monitorable properties are safety and cosafety properties. On the other hand, deterministic liveness properties like "infinitely many a's" are not monitorable.

We discuss various monitor constructions with a focus on deterministic ω-regular languages. We locate a proper subclass of deterministic ω-regular languages but also strictly large than the subclass of languages which are deterministic and codeterministic; and for this subclass there exists a canonical monitor which also accepts the language itself.

We also address the problem to decide monitorability in comparison with deciding liveness. The state of the art is as follows. Given a Büchi automaton, it is PSPACE-complete to decide liveness or monitorability. Given an LTL formula, deciding liveness becomes EXPSPACE-complete, but the complexity to decide monitorability remains open.

Formal Methods in Air Traffic Management: The Case of Unmanned Aircraft Systems (Invited Lecture)

César A. Muñoz

NASA Langley Research Center, Hampton, Virginia 23681-2199

Abstract. As the technological and operational capabilities of unmanned aircraft systems (UAS) continue to grow, so too does the need to introduce these systems into civil airspace. Unmanned Aircraft Systems Integration in the National Airspace System is a NASA research project that addresses the integration of civil UAS into non-segregated airspace operations. One of the major challenges of this integration is the lack of an on-board pilot to comply with the legal requirement that pilots see and avoid other aircraft. The need to provide an equivalent to this requirement for UAS has motivated the development of a *detect and avoid* (DAA) capability to provide the appropriate situational awareness and maneuver guidance in avoiding and remaining well clear of traffic aircraft. Formal methods has played a fundamental role in the development of this capability. This talk reports on the formal methods work conducted under NASA's Safe Autonomous System Operations project in support of the development of DAA for UAS. This work includes specification of low-level and high-level functional requirements, formal verification of algorithms, and rigorous validation of software implementations. The talk also discusses technical challenges in formal methods research in the context of the development and safety analysis of advanced air traffic management concepts.

This invited lecture reports on research conducted at NASA Langley Research Center at the Safety-Critical Avionics Systems Branch by several individuals including, in addition to the author, Anthony Narkawicz, George Hagen, Jason Upchurch, and Aaron Dutle.

The Proof Technique of Unique Solutions of Contractions

Davide Sangiorgi

Università di Bologna and INRIA
davide.sangiorgi@gmaol.com

This extended abstract summarises work conducted with Adrien Durier and Daniel Hirschkoff (ENS Lyon), initially reported in [38].

Bisimilarity is employed to define behavioural equivalences and reason about them. Originated in concurrency theory, bisimilarity is now widely used also in other areas, as well as outside Computer Science. In this work, behavioural equivalences, hence also bisimilarity, are meant to be *weak* because they abstract from internal moves of terms, as opposed to the *strong* ones, which make no distinctions between the internal moves and the external ones (i.e., the interactions with the environment). Weak equivalences are, practically, the most relevant ones: e.g., two equal programs may produce the same result with different numbers of evaluation steps.

D. Sangiorgi—The authors are partially supported by the ANR project 12IS02001 PACE.

A Logical Revolution

Moshe Y. Vardi

Rice University, Department of Computer Science, Rice University,
Houston, TX 77251-1892, USA
vardi@cs.rice.edu,
http://www.cs.rice.edu/~vardi

Abstract. Mathematical logic was developed in an effort to provide formal foundations for mathematics. In this quest, which ultimately failed, logic begat computer science, yielding both computers and theoretical computer science. But then logic turned out to be a disappointment as foundations for computer science, as almost all decision problems in logic are either unsolvable or intractable. Starting from the mid 1970s, however, there has been a quiet revolution in logic in computer science, and problems that are theoretically undecidable or intractable were shown to be quite feasible in practice. This talk describes the rise, fall, and rise of logic in computer science, describing several modern applications of logic to computing, include databases, hardware design, and software engineering.

References

1. Clarke, E.M., Emerson, E.A., Sifakis, J.: Model checking: algorithmic verification and debugging. Commun. ACM **52**(11), 74–84 (2009)
2. Codd, E.F.: A relational model for large shared data banks. Commun. ACM **13**, 377–387 (1970)
3. Codd, E.F.: Relational completeness of data base sublanguages. In: Rustin, R. (ed.) Database Systems, pp. 33–64. Prentice-Hall (1972)
4. Cook, B., Podelski, A., Rybalchenko, A.: Proving program termination. Commun. ACM **54**(5), 88–98 (2011)
5. de Moura, L.M., Bjørner, N.: Satisfiability modulo theories: introduction and applications. Commun. ACM **54**(9), 69–77 (2011)
6. Malik, S., Zhang, L.: Boolean satisfiability from theoretical hardness to practical success. Commun. ACM **52**(8), 76–82 (2009)
7. Pnueli, A.: The temporal logic of programs. In: Proceedings of 18th IEEE Symposium on Foundations of Computer Science, pp. 46–57 (1977)
8. Vardi, M.Y., Wolper, P.: An automata-theoretic approach to automatic program verification. In: Proceedings of 1st IEEE Symposium on Logic in Computer Science, pp. 332–344 (1986)

On Probabilistic Distributed Strategies

Glynn Winskel

Computer Laboratory, University of Cambridge
gwl04@cam.ac.uk

Abstract. In a distributed game we imagine a team Player engaging a team Opponent in a distributed fashion. No longer can we assume that moves of Player and Opponent alternate. Rather the history of a play more naturally takes the form of a partial order of dependency between occurrences of moves. How are we to define strategies within such a game, and how are we to adjoin probability to such a broad class of strategies? The answer yields a surprisingly rich language of probabilistic distributed strategies and the possibility of programming (optimal) probabilistic strategies. Along the way we shall encounter solutions to: the need to mix probability and nondeterminism; the problem of parallel causes in which members of the same team can race to make the same move, and why this leads us to invent a new model for the semantics of distributed systems.

Contents

Concurrency

Constraints

Logic and Semantic

Software Architecture and Component-Based Design

Verification

Tool Papers

Short Papers

Invited Talks

An Exercise in Mathematical Engineering: Stating and Proving Kuratowski Theorem

Jean-Raymond Abrial[✉]

Marseille, France
`jrabrial@neuf.fr`

Abstract. This paper contains the informal presentation of a well known theorem on planar graphs: the theorem of Kuratowski (1930). This study is supposed to serve as an example for the proposed new discipline of *Mathematical Engineering*. The intend if this discipline is to show to informaticians, by means of examples, that there must exist important connections between rigorous mathematics and rigorous computer science. Moreover, in both cases, the mechanisation of proofs is becoming more and more fashionable these days. Such mechanisations cannot be performed without a clear understanding of the mathematical context that has to be rigorously described before engaging in the proof itself.

1 Introduction

The writing of this paper originated in the frustration I felt after reading some mathematical works. Let me explain why in this introduction. First of all, why am I interested in reading some mathematical works either in papers, in textbooks, or in some presentation material, mainly through internet? There are two reasons for this: (i) I am interested in formalising and mechanically prove some well known mathematical theorems, and (ii) I have always thought that mathematics should be a good cultural framework for informaticians. In this paper, I will concentrate on this second reason.

In order to link mathematics and computer science, the idea would be to present some specific material that would help students to understand the way mathematics is presented by professional mathematicians. Examples could be taken from the presentations of important, complicated, and well accepted theorems by the scientific community. Before proving these theorems, the mathematician has to write down many definitions and sometimes a large collection of intermediate results that are needed for writing the statement of the theorem but also, of course, for its proof: let us call this kind of presentation a *mathematical context*. Sometimes this context is very short, but sometimes it might also be quite elaborate.

Now, in my opinion, such contexts are not so different from what the programmer should do before writing a program. He (she) has to state (most of the time implicitly) the properties of the data his (her) program is handling. An important part of software engineering these days consists in abstracting and

M. Leucker et al. (Eds.): ICTAC 2015, LNCS 9399, pp. 3–27, 2015.
DOI: 10.1007/978-3-319-25150-9_1

formalising such contexts. So, I (naively) thought that we might take advantage of learning how professional mathematicians do the same in their work. This is indeed why I plunge myself into some mathematical works.

Coming back to my frustration, it originated in the fact that I was not very happy with what I read. Mathematical contexts are quite often badly structured, hard to understand, and also sometimes miss to give some important definitions that are used later in a rather cavalier fashion. One of the main complaints I have is that these presentations are quite often *not abstract enough*. That seems to be a very strange conclusion regarding mathematical works, since mathematics should, in principle, be the realm of abstraction "par excellence". So, my conclusion after reading these works is that mathematics, after all, might not be such a good example to follow and could even induce some bad behaviours among informaticians. In conclusion, I had no choice but trying to reconstruct by myself some of these presentations until I feel comfortable with them.

In investigating some "interesting" mathematical works, I came across the theorem of Kuratowski [1]. This theorem gives a simple characteristic property of a *planar graph*. As a reminder, a planar graph is one where there exists a drawing of the graph on a plane with edges all intersecting on vertices only. I had access to one book [2] and several papers [3–5] where authors presented, as they claim it, some "simple" proofs of this theorem. In other documents (slides), some authors even miss completely the proof because "it is too complicated". In reading these presentations, I was surprised that many authors used drawings as justifications of their proofs although clearly such drawings did not receive any solid definitions and formalisations. I have even seen somewhere the following horrifying fragment in a proof: "... as demonstrated in Fig. 7.3". Of course, I have nothing against drawings, but they should be used with care and only for informal illustrations (in what follows, I will use many illustrating drawings).

It took me some times to understand that this theorem of Kuratowski is not so much a theorem of graph theory than rather a theorem of topology. In fact, most of the authors develop their proof within the framework of graph theory only, which is *not abstract enough* in this case.

This paper contains the result of my own reconstruction of the Theorem of Kuratowski. I hope it can be a good example for informaticians. This work shows that this theorem needs a large mathematical context in order to be stated and proved in a significant and convincing fashion.

Here is the way how the rest of the paper is structured. In Sect. 2, I present a very abstract concept of *regions*, simply defined by an interior, an exterior, and a border. In Sect. 3, I define some *relationships between regions*: regions can be either external to each other or else tangent to each other. In Sect. 4, I consider the case where regions are all connected to each other. In Sect. 5, I refine the notion of region in order to define the concept of *graph*. I also give a formal definition of *planar graphs*. In Sect. 6, I propose some *cutting axioms* fomalising the notion of intersection between region borders. Finally, in Sect. 7, I state and prove the *theorem of Kuratowski*. As can be seen, a "mathematical context" of a significant size is presented before entering into the theorem itself.

With such a paper, the idea is to start developing a (new) discipline entirely devoted to the rewriting of some well known mathematical works: *Mathematical Engineering*.

2   Definition of Regions

In this section, rather than recalling immediately the formal definition of a graph (this will be done in Sect. 5 only), we start with a definition of, so called, *regions* and propose some properties of them. The reason we are interested by regions is that the theorem of Kuratowski is concerned with graphs whose plane images show the presence of parts that are entirely surrounded by edges and thus form some "regions" of the plane. This is illustrated in Fig. 1, where thirteen regions can be seen, thirteen because some regions might contain others (Sect. 3.3).

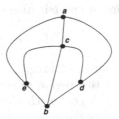

Fig. 1. A graph made of regions

2.1   Definition of a Region: Interior, Exterior, and Border

We suppose first that we are given a set P of *points*, and then we define a *finite set of regions*, R. Such regions are built on these points by means of three functions: interior, *int*, exterior, *ext*, and border, *brd*. Each of them defines a non-empty set of points:

$$
\begin{aligned}
int &\in R \to \mathbb{P}_1(P) \\
ext &\in R \to \mathbb{P}_1(P) \\
brd &\in R \to \mathbb{P}_1(P)
\end{aligned}
\tag{1}
$$

As an important property, the interior, exterior and border of a region r together *partition* the set of points:

$$
\begin{aligned}
int(r) \cap ext(r) &= \varnothing \\
int(r) \cap brd(r) &= \varnothing \\
ext(r) \cap bdr(r) &= \varnothing \\
int(r) \cup ext(r) \cup bdr(r) &= P
\end{aligned}
\tag{2}
$$

2.2   Decomposing the Border of a Region

Given two distinct points, we define the relation, reg, between these two points and the regions whose border contains these points:

$$reg \subseteq P \times P \times R \qquad (3)$$

We have thus the following property defining reg:

$$
\begin{aligned}
&(a \mapsto b \mapsto r) \in reg \\
&\Leftrightarrow \\
&a \neq b \\
&a \in brd(r) \\
&b \in brd(r)
\end{aligned}
\qquad (4)
$$

Notice that several regions might share these two points in their borders. Also note that the order of both points a and b is meaningless:

$$(a \mapsto b \mapsto r) \in reg \Leftrightarrow (b \mapsto a \mapsto r) \in reg \qquad (5)$$

Given two distinct points in the border of a region, we define the left, lft, and the right, rht, parts of the border of this region relative to these two points (look at Fig. 2). Don't take "left" and "right" with their usual meanings, these are just convenient names used here, we could have used "north" and "south", etc. instead:

$$
\begin{aligned}
lft &\in reg \to \mathbb{P}_1(P) \\
rht &\in reg \to \mathbb{P}_1(P)
\end{aligned}
\qquad (6)
$$

Given a region r and two distinct points a and b in its border, the "left" and "right" parts of the border of r, as determined by a and b, intersects in the set $\{a, b\}$ only and together exactly cover the border of r. Moreover, they each contain more points than just a and b. In order to enhance readability, we use the following abbreviations:

$$
\begin{aligned}
p &= lft(a \mapsto b \mapsto r) \\
q &= rht(a \mapsto b \mapsto r)
\end{aligned}
\qquad (7)
$$

We have then:

$$
\begin{aligned}
&(a \mapsto b \mapsto r) \in reg \\
&\Rightarrow \\
&p \cap q = \{a, b\} \\
&p \cup q = brd(r) \\
&p \setminus \{a, b\} \neq \varnothing \\
&q \setminus \{a, b\} \neq \varnothing
\end{aligned}
\qquad (8)
$$

All this is illustrated on Fig. 2. Again, the drawing in this figure is just a plane illustration. We might have different drawings within other contexts.

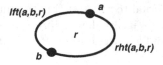

Fig. 2. A region border separated in two parts by two points a and b

3 Relations Between Regions

Given two *distinct* regions $r1$ and $r2$, we might consider four different relationships between them. They might be external to each other, one region might be internal to the other, they might be externally tangent to each other, or else one region might be internally tangent to the other. Let us give more information and properties about such relationships.

3.1 External Regions

Two regions $r1$ and $r2$ are *external to each other* when the interior of $r1$ together with its border are strictly included in the exterior of $r2$.

$$int(r1) \cup brd(r1) \subset ext(r2) \tag{9}$$

Notice that we automatically have a similar relationship between $r2$ and $r1$:

$$int(r2) \cup brd(r2) \subset ext(r1) \tag{10}$$

This is due to the partitioning Property (2).

3.2 Internal Regions

One region, say $r1$, is totally *inside another one*, $r2$, when the interior of $r1$ together with its border are strictly included in the interior of $r2$.

$$int(r1) \cup brd(r1) \subset int(r2) \tag{11}$$

As previously, we automatically have a reverse relationship between $r2$ and $r1$:

$$ext(r2) \cup brd(r2) \subset ext(r1) \tag{12}$$

This is due to the partitioning Property (2).

3.3 Externally Tangent Regions

The two regions $r1$ and $r2$ might be *externally tangent to each other*. In the rest of this paper, this will be the main relationship between regions we are interested in. By "externally tangent" we mean three things: (i) the two regions share some part of their borders, (ii) the part of the border of, say, $r1$ that is not shared with the border of $r2$ is external to $r2$, and (iii) the interior of both regions are incompatible. Have a look at Fig. 3 below. More precisely, we suppose that we have two distinct points a and b in the border of $r1$ and also in the border of $r2$, that is:

$$
\begin{aligned}
(a \mapsto b \mapsto r1) &\in reg \\
(a \mapsto b \mapsto r2) &\in reg
\end{aligned}
\tag{13}
$$

In order to enhance readability, we use the following abbreviations:

$$
\begin{aligned}
p1 &= lft(a \mapsto b \mapsto r1) \\
p2 &= lft(a \mapsto b \mapsto r2) \\
q1 &= rht(a \mapsto b \mapsto r1) \\
q2 &= rht(a \mapsto b \mapsto r2)
\end{aligned}
\tag{14}
$$

Moreover, the part of the borders in the left of a and b is the same for $r1$ and $r2$, and the other part of the border of each region together with its interior is strictly included in the exterior of the other region:

$$
\begin{aligned}
p1 &= p2 \\
int(r1) \cup (q1 \setminus \{a, b\}) &\subset ext(r2) \\
int(r2) \cup (q2 \setminus \{a, b\}) &\subset ext(r1)
\end{aligned}
\tag{15}
$$

Notice that our usage of lft is arbitrary here: we could have use rht or both as well. The external tangency of both regions $r1$ and $r2$ induces the existence of another region, $r3$, whose interior includes the interior of both regions $r1$ and $r2$ and the common part of their borders as well. Here are the characteristic elements of $r3$:

$$
\begin{aligned}
int(r3) &= int(r1) \cup int(r2) \cup (p1 \setminus \{a, b\}) \\
ext(r3) &= ext(r1) \setminus (int(r2) \cup q2) \\
brd(r3) &= q1 \cup q2
\end{aligned}
\tag{16}
$$

It is easy to prove that $r3$ is indeed a region with the partitioning property:

$$
\begin{aligned}
int(r3) \cap ext(r3) &= \varnothing \\
int(r3) \cap brd(r3) &= \varnothing \\
ext(r3) \cap brd(r3) &= \varnothing \\
int(r3) \cup ext(r3) \cup brd(r3) &= P
\end{aligned}
\tag{17}
$$

The main property of $r3$ with regards to both regions $r1$ and $r2$ is that the interior of $r1$ and that of $r2$ are *strictly included* in the interior of $r3$. We can indeed prove easily the following (this is illustrated in Fig. 3)

Lemma 1

$$
\begin{aligned}
int(r1) &\subset int(r3) \\
int(r2) &\subset int(r3)
\end{aligned}
\tag{18}
$$

Another very important property is the following: if two distinct points c and d both belong to the border of, say, region $r1$ then they also belong to the border

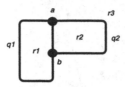

Fig. 3. Regions r1 and r2 are externally tangent to each other

of $r3$ provided they don't belong to the common part of the borders of $r1$ and $r2$ (except if they are the same as a or b):

Lemma 2

$$c \mapsto d \mapsto r1 \in reg$$
$$c \notin p1 \setminus \{a, b\}$$
$$d \notin p1 \setminus \{a, b\} \tag{19}$$
$$\Rightarrow$$
$$c \mapsto d \mapsto r3 \in reg$$

3.4 Internally Tangent Regions

One region, say $r1$, might be *internally tangent* to the region $r2$. This can be defined in a way that is very similar to what we have done in the previous subsection. This situation induces a region $r3$ whose interior is strictly included in that of $r2$.

Notice that other kinds of relationships between two regions are excluded. In particular, it is not possible for two regions to be tangent on a single point only of their respective borders. We consider by extension that such regions are either exterior to each other, or else one is inside the other.

3.5 Maximal Regions

In this section, we state an important theorem saying that there are some maximal regions whose border contains two given points.

Theorem 1: *Given a region $r1$ and two distinct points a and b in its border, then there exists a region $r2$ with a and b in its border, and which either cannot be made larger by means of further externally tangent regions, or else can be made larger by means of some further externally tangent regions but only by removing one or both points a and b from its border.*

In Fig. 4, one can see two maximal regions containing points a and b in their border. These regions have the following borders: $a-d-b-e-c$ and $a-e-b-d-c$. The proof of this theorem is a consequence of the two lemmas stated in Sect. 3.3 and of the finiteness hypothesis about regions (Sect. 2).

Fig. 4. a–d–b–e–c and a–e–b–d–c are both maximal regions containing a and b

4   Systems of Connected Regions and Definition of Faces

In the rest of the paper, we suppose that we have no regions that are entirely outside, or entirely inside other regions (Sects. 3.1 and 3.2). We have no isolated regions: they are all directly or indirectly *connected* by external tangency.

Among the various connected regions, we consider those having *no internally tangent regions*. Such regions are called *faces*. Let F be the set of such faces. We consider now the following relation, fr (for face relation), between faces: two faces are related in fr if they are externally tangent to each other (Sect. 3.3). Note that fr is obviously symmetric since it is the case that the external tangency relationship between faces (and more generally between regions) is indeed symmetric. Moreover, the relation fr is irreflexive (a region is not tangent to itself). Since we have no isolated regions then the relation fr is also connected. In Fig. 5, four faces $f1$, $f2$, $f3$, and $f4$ can be seen as well as the relation fr between them.

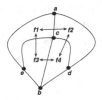

Fig. 5. The relation fr between faces $f1$, $f2$, $f3$ and $f4$

Finally, every path within the connected relation fr defines a region that is not a face. This is so because two tangent regions $r1$ and $r2$ lead to the formation of another region $r3$ including both regions $r1$ and $r2$ (again, Sect. 3.3). In Fig. 5, the path $f2$–$f4$–$f3$ corresponds to the region with border a–d–b–e–c.

We now state another important theorem saying that given two distinct points, situated in some borders (not apparently the border of the same face), then there always exists a region with these two points in its border.

Theorem 2: *Given two distinct points a and b, both situated in the borders of some regions, then there exists a region r, which is not necessary a face, with these two points being members of the border of r.*

Proof. We suppose that the point a belongs to the border of the region ra and that the point b belongs to the border of the region rb. If ra is the same as rb, then this common region is the solution. So, we suppose now that ra and rb are different. We have two cases:

Case 1: Suppose that the points a *belongs exclusively* to the borders of faces ra. Then the region r is the one corresponding to a path from ra to rb in the relation fr (remember that the relation fr is connected as stated in Sect. 4).

Case 2: Suppose the point a belongs to the common border of two tangent faces $ra1$ and $ra2$. If there is a path from $ra1$ to rb that is not involving $ra2$, then we can take the region corresponding to this path. If there is no such path, it means that there is a path from $ra2$ to rb not using $ra1$, then we can take the region corresponding to this path. ∎

Putting these two theorems together, we obtain the following theorem:

Theorem 3: *Given two distinct points, situated in some borders, then there exists a maximal region r with these two points being members of the border of r.*

5   Graphs and Planar Graphs

In this section we refine the concept of regions as stated in Sect. 2 (definition of regions), in Sect. 3 (relation between regions), and in Sect. 4 (definitions of connected regions and faces). This will allow us to define the concept of *graph*.

Notice that the definition of a "graph" we use in this section (and in this paper) is *less general* than the well known one that can be found everywhere in the graph theory literature and in textbooks. We use this restricted definition only because it is sufficient to state and prove the Theorem of Kuratowski.

5.1   Definitions of a Graph and of Graph Vertices

Proceeding now with our restricted graph definition, we distinguish a *finite set of points* in the borders of regions. We call this set of points the *set of vertices V*. A finite set of connected faces together with a finite set of vertices is called a *graph*. In Fig. 6, we suppose now that the points a, b, c, d, etc. are all vertices.

5.2   A Relation Between Vertices

There exists a symmetric, irreflexive, and connected binary relations, vr (for vertex relation), between distinct vertices. Two related vertices in vr belong to the border of the same face. Formally:

$$vr \subset \mathrm{dom}(reg) \tag{20}$$

In Fig. 6, we can see that the relation vr is as follows:

$$vr = \{a \mapsto d, c \mapsto d, d \mapsto b, \ldots\} \tag{21}$$

5.3   Definition of Edges

We now define another finite set called the *set of edges E* of the graph. An edge has a content, *edg*, that is a non-empty set of point:

$$edg \in E \to \mathbb{P}_1(P) \tag{22}$$

Moreover, edges are related to the previously defined binary relations *fr* and *vr*. This is done by means of the following two total functions *ve* and *fe* (*ve* is even a surjection):

$$\begin{aligned} ve &\in vr \twoheadrightarrow E \\ fe &\in fr \to E \end{aligned} \tag{23}$$

In Fig. 6, one can see edges *e1*, *e2*, etc. between vertices.

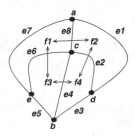

Fig. 6. A graph

These functions state that there corresponds an edge to each pair of related vertices and an edge to each pair of related faces. :

$$\begin{aligned} ve &= \{(a \mapsto d) \mapsto e1, (c \mapsto d) \mapsto e2, (d \mapsto b) \mapsto e3, \ldots\} \\ fe &= \{(f1 \mapsto f2) \mapsto e8, (f2 \mapsto f4) \mapsto e2, (f4 \mapsto f3) \mapsto e4, \ldots\} \end{aligned} \tag{24}$$

The following properties show that edges are *not oriented*:

$$\begin{aligned} v1 \mapsto v2 \in vr &\Rightarrow ve(v1 \mapsto v2) = ve(v2 \mapsto v1) \\ f1 \mapsto f2 \in fr &\Rightarrow fe(v1 \mapsto v2) = fe(v2 \mapsto v1) \end{aligned} \tag{25}$$

Finally, the contents of an edge is exactly the "left part" of the two vertices corresponding to that edge:

$$v1 \mapsto v2 \mapsto f \in reg \Rightarrow edg(ve(v1 \mapsto v2)) = lft(v1 \mapsto v2 \mapsto f) \tag{26}$$

5.4   Graphs with Loops

When in a graph, an edge links a vertex to itself, then the graph is said to *have a loop*. In this paper, we always have graphs without loops. This is because the relation *vr* is irreflexive .

5.5   Simple Graphs

When, in a graph, two vertices are linked with at most one edge, then the graph is said to be a *simple graph*. In this paper, we have simple graphs only. This is because the function *ve* links pairs of vertices to a single edge.

5.6   Chains in a Graph

Part of a region border linking two vertices a and b and possibly *containing other vertices* besides a and b is called a *chain* of the graph. Notice that a chain certainly contains some other non-vertex points. Also notice that a chain could be an edge. However a chain containing more than two vertices is not an edge.

5.7   Connected Graphs

When any two vertices in a graph are linked with a chain, then the graph is said to be a *connected graph*. In this paper, we have always connected graphs. This is because the relation *vr* is connected.

5.8   Subgraphs

A graph h is said to be a *subgraph* of a graph g when the sets of faces, vertices, and edges of h are subsets of corresponding sets in g.

5.9   Graph Drawings

Two graphs $g1$ and $g2$, where edges are linked by means of a bijective functions f, are said to be equivalent in the following circumstances: if $v1$ and $w1$ are two vertices of $g1$ linked by the edge $e1$, then $f(e1)$ links $v1$ to $w1$. We must make a clear distinction here between a *graph* and a *graph drawing*. The equivalence relation mentioned previously is rather one between various graph drawings of the same graph (more is explained in the next subsection). A graph then just appears to be the *equivalence class* of the mentioned graph drawing equivalence relation.

5.10   More on the Equivalence Relation Between Graph Drawings

We now give more information about the equivalence relation defined in the previous subsection. In fact, an edge between two vertices a and b can be drawn in the interior or in the exterior (left or right) of a region having a and b in its border. This distinction between the position of the edge characterises the equivalence relation. All this is illustrated in Fig. 7, where one can see four equivalent graph drawings. In fact, there are many more possible drawings since an external link could be positioned "on the left" or "on the right" (or equivalently "on the north" or "on the south").

Fig. 7. Graph drawing equivalences

5.11   Graph Extension

Given a graph g and the set sgd of all its corresponding graph drawings, we suppose to have two vertices a and b that are not connected by an edge in g. Let us connect them by means of a new edge thus forming a new graph h. The set shd of graph drawings of the graph h includes drawings of the set sgd where an edge linking a to b is drawn in all possible ways.

5.12   Chain Intersections in a Graph

When two chains of a graph drawing intersect each other on a single point, that point might be a vertex or not. In Fig. 7, the first and last drawings show chains intersecting in points that are not vertices. If two chains of a graph drawing intersect at a single point which is not a vertex, it is sometimes (but not always) possible to find an equivalent graph drawing where the corresponding chains are not intersecting at all. We can see this in the second and third graph drawings of Fig. 7.

5.13   Planar Graph

A connected, simple, and loop free graph g (as they are all in this paper) is said to be a *planar graph* if *there exists at least one graph drawing* of this graph where all chain intersections are vertices. If such a graph drawing does not exist, the graph g is said to be non-planar.

5.14   Example of a Non-planar Graph: The Graph K5

In Fig. 8, the first two graph drawings are intersecting on vertices only. So, the corresponding graph is indeed planar. By introducing an additional chain to these graph drawings (in the last two drawings) the corresponding graph becomes non-planar.

This non-planar graph is called K5. In the literature, the graph K5 is the graph where all connections are edge connections. In this paper, we suppose that the connections could be chain connections. Given a finite set V containing five vertices, the binary relation between chains corresponding to the graph K5 can be defined as follows:

$$\bigcup x \cdot x \in V \mid \{x\} \times (V \setminus \{x\})$$

The graph K5 is said to be the complete graph with five vertices.

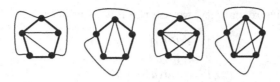

Fig. 8. The graph K5

5.15 Example of a Non-planar Graph: The Graph K3,3

In Fig. 9, the first two graph drawings are intersecting on vertices only. So, the corresponding graph is indeed planar. By introducing an additional chain to these graph drawings (in the last two drawings) the corresponding graph becomes non-planar.

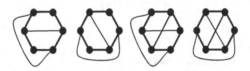

Fig. 9. The graph K3,3

This non-planar graph is called K3,3. In the literature, K3,3 is the graph where all connections are edge connections. Here we suppose that in this extension the connections could be chain connections. Given two finite and disjoint sets $V1$ and $V2$ containing three vertices each, the binary relation between chains corresponding to the graph K3,3 can be defined as follows:

$$(V1 \times V2) \cup (V2 \times V1)$$

6 Chain Intersection Axioms

In this section we precisely define in which circumstances two chains might intersect. More precisely, we are given a region r and four distinct vertices in the border of r: $a1$, $b1$, $a2$, and $b2$.

$$
\begin{aligned}
& a1 \neq b1 \\
& a2 \neq b2 \\
& \{a1, b1\} \cap \{a2, b2\} = \varnothing
\end{aligned}
\tag{27}
$$

We suppose that $a1$ and $b1$ are linked by a chain $e1$ which is distinct from the border of r except in $a1$ and $b1$. Likewise we suppose that $a2$ and $b2$ are linked by a chain $e2$ which is distinct from the border of r except in $a2$ and $b2$:

$$
\begin{aligned}
& e1 \cap brd(r) = \{a1, b1\} \\
& e2 \cap brd(r) = \{a2, b2\}
\end{aligned}
\tag{28}
$$

In order to enhance readability, we use the following abbreviations:

$$
\begin{aligned}
p1 &= lft(a1 \mapsto b1 \mapsto r) \\
p2 &= lft(a2 \mapsto b2 \mapsto r) \\
q1 &= rht(a1 \mapsto b1 \mapsto r) \\
q2 &= rht(a2 \mapsto b2 \mapsto r) \\
f1 &= e1 \setminus \{a1, b1\} \\
f2 &= e2 \setminus \{a2, b2\}
\end{aligned}
\tag{29}
$$

Here are one lemma and four axioms. When $e1$ links $a1$ to $b1$ by the interior of r whereas $e2$ links $a2$ to $b2$ by the exterior of r, then $e1$ and $e2$ do not intersect. Of course, this can be proved since the interior and the exterior of a region are disjoint:

Lemma 3

$$
\begin{aligned}
&f1 \subset int(r) \\
&f2 \subset ext(r) \\
&\Rightarrow \\
&e1 \cap e2 = \varnothing
\end{aligned}
\tag{30}
$$

Figure 10 illustrates this non intersecting case.

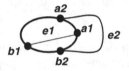

Fig. 10. Chains $e1$ and $e2$ do not intersect

We suppose now that $e1$ links $a1$ to $b1$ by the interior of r and likewise $e2$ links $a2$ to $b2$ by the interior of r. Moreover, if $a2$ and $b2$ are situated on the same side ($p1$ or $q1$) of the border of r, then $e1$ and $e2$ do not intersect:

Axiom 1

$$
\begin{aligned}
&f1 \subset int(r) \\
&f2 \subset int(r) \\
&\{a2, b2\} \subset p1 \ \lor \ \{a2, b2\} \subset q1 \\
&\Rightarrow \\
&e1 \cap e2 = \varnothing
\end{aligned}
\tag{31}
$$

Figure 11 illustrates this non intersecting case.

Fig. 11. Chains $e1$ and $e2$ do not intersect

We suppose now that $e1$ links $a1$ to $b1$ by the interior of r and likewise $e2$ links $a2$ to $b2$ by the interior of r. Moreover if $a2$ and $b2$ are situated on different sides ($p1$ or $q1$) of the border of r, then $e1$ and $e2$ do intersect:

Axiom 2

$$
\begin{aligned}
&f1 \subset int(r) \\
&f2 \subset int(r) \\
&\neg(\{a2, b2\} \subset p1 \ \vee \ \{a2, b2\} \subset q1) \\
&\Rightarrow \\
&e1 \cap e2 \neq \varnothing
\end{aligned}
\tag{32}
$$

Figure 12 illustrates this intersecting case.

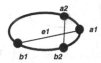

Fig. 12. Chains $e1$ and $e2$ do intersect

We suppose now that $e1$ links $a1$ to $b1$ by the exterior of r and likewise $e2$ links $a2$ to $b2$ by the exterior of r. Moreover, if $a2$ and $b2$ are situated on the same side ($p1$ or $q1$) of the border of r then $e1$ and $e2$ do not intersect:

Axiom 3

$$
\begin{aligned}
&f1 \subset ext(r) \\
&f2 \subset ext(r) \\
&\{a2, b2\} \subset p1 \ \vee \ \{a2, b2\} \subset q1 \\
&\Rightarrow \\
&e1 \cap e2 = \varnothing
\end{aligned}
\tag{33}
$$

Figure 13 illustrates this non intersecting case.

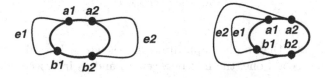

Fig. 13. Chains $e1$ and $e2$ do not intersect

We suppose finally that $e1$ links $a1$ to $b1$ by the exterior of r and likewise $e2$ links $a2$ to $b2$ by the exterior of r. Moreover, if $a2$ and $b2$ are situated on different sides ($p1$ or $q1$) of the border of r then $e1$ and $e2$ do intersect:

Axiom 4

$$f1 \subset ext(r)$$
$$f2 \subset ext(r)$$
$$\neg(\{a2, b2\} \subset p1 \;\vee\; \{a2, b2\} \subset q1) \tag{34}$$
$$\Rightarrow$$
$$e1 \cap e2 \neq \varnothing$$

Figure 14 illustrates this intersecting case.

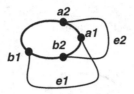

Fig. 14. Chains $e1$ and $e2$ do intersect

Here are some consequences of the cutting axioms presented in this section. They are given under the form of four exercises. We use these exercises in Sect. 7:

Exercise 1: Prove that both graph drawings shown in Fig. 15 are equivalent and that the link between a_2 and b_2 in the second one does not intersect the link between a_0 and b_0, nor with the link between a_1 and b_1.

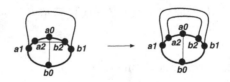

Fig. 15. Exercise 1

Exercise 2: Prove that both graph drawings shown in Fig. 16 are equivalent and that in the second one, the link between a_2 and b_2 intersects with the link between a_3 and b_3.

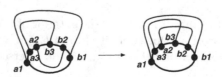

Fig. 16. Exercise 2

Exercise 3: From Exercises 1 and 2 prove that in the second drawing of Exercise 1, the link between a_2 and b_2 does not intersect any other link like a_3–b_3 provided the link between a_1 and b_1 does not contain any link like the one between a_3 and b_3 of Exercise 2. Moreover, if there is no chains between the link between a_1 and b_1 and the border of the region, then there is no intersection at all for the link between a_2 and b_2.

Exercise 4: Prove that both graph drawings shown in Fig. 17 are equivalent and that in the second one, the link between b_2' and v intersects with the border of the region.

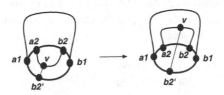

Fig. 17. Exercise 4

7   The Kuratowski Theorem

We have enough preliminary material to engage now into the proof of the Kuratowski Theorem. This theorem gives us a necessary and sufficient condition for a connected, simple, and loop free graph to be non-planar.

Kuratowski Theorem: *A graph, supposed to be connected, simple, and loop free, is non-planar if and only if it contains either K3,3 or K5 as a subgraph.*

Informal Proof of the "only if" Part of Kuratowski Theorem[1]: a non-planar graph contains either K3,3 or K5 as a subgraph.

We are given a connected, simple, loop free, and *planar* graph g. This means that there exists at least one drawing of g where all chains are intersecting with other chains on vertices only (Sect. 5.13). Let us call gd such a drawing.

We are given two distinct vertices a_0 and b_0 of g. We suppose that there is no edge already linking a_0 to b_0 in g. In other words, the graph g is simple and so remain after adding an edge linking a_0 to b_0. Notice that in g there exists a chain (which is not an edge) linking a_0 to b_0 since a_0 and b_0 are on the border of a region according to **Theorem 2** of Sect. 4. Now, we suppose that adding such an edge between a_0 and b_0 makes the obtained graph h being *non-planar*.

---
[1] This proof is inspired by that of Shimon Even in [2].

Our intention is now to prove the Kuratowski theorem on this graph h.

In the graph drawing gd of graph g, an edge linking a_0 to b_0 intersects with at least one chain in a point which is not a vertex. This is so because any graph drawing of g extended with the edge linking a_0 to b_0 is a graph drawing of h (Sect. 5.11) and, by definition, *all graph drawings* of a non-planar graph such as h, have this intersecting property (Sect. 5.13).

From the graph drawing gd, we can construct many graph drawings of h. This is done simply by drawing the edge linking a_0 to b_0 either inside or outside any of the regions having a_0 and b_0 in their borders (Sects. 5.9 and 5.10).

Among those new graph drawings of h, one of them is interesting: this is the one where the edge from a_0 to b_0 is *inside a maximal region r* having a_0 and b_0 in its border (such a region r exists according to **Theorem 1** of Sect. 3). Let us call hd this particular graph drawing of h. In order to enhance readability, we use the following abbreviations:

$$p_0 = lft(a_0 \mapsto b_0 \mapsto r)$$
$$q_0 = rht(a_0 \mapsto b_0 \mapsto r)$$

Now, there must exist a chain c_1 linking two vertices in the border of r and situated outside the maximal region r since, otherwise, the edge linking a_0 to b_0 could be moved outside r and thus h would be planar (remember **Lemma 3** and **Axiom 4** of Sect. 6). We suppose that this external chain c_1 links two vertices a_1 and b_1 of the border of region r. In order to enhance readability, we use the following:

$$p_1 = lft(a_1 \mapsto b_1 \mapsto r)$$
$$q_1 = rht(a_1 \mapsto b_1 \mapsto r)$$

The two vertices a_1 and b_1 cannot be both in p_0 or in q_0 since otherwise the region with border c_1 and, say, p_1 would be externally tangent to r and thus forms together with it a larger region with a_0 and b_0 in its border (Sect. 3.3). But this is not possible because r is supposed to be a *maximal region* with a_0 and b_0 in its border. So, a_1 and b_1 should be respectively in $p_0 \setminus \{a_0, b_0\}$ and $q_0 \setminus \{a_0, b_0\}$ (or the other way around). Moreover, we suppose that c_1 is the *smallest* such chain. In other words there is no chain like c_1 inside it. Also notice that there is no chain linking c_1 to p_1 since r is a maximal region, Thus c_1 follows the full assumptions of **Exercise 3** (Sect. 6). This is illustrated in Fig. 18.

Fig. 18. Edge linking a_0 to b_0 and chain linking a_1 to b_1 do not intersect

In order to enhance readability let us make precise which portions of the border of r, the chains p_0, q_0, p_1, and q_1 cover. In what follows, we use these denotations rather than the letters p_0, q_0, etc.

- p_0 is (a_0, a_1, b_0) denoted $[a_0, b_0]$

- q_0 is (a_0, b_1, b_0) denoted $[b_0, a_0]$

- p_1 is (a_1, a_0, b_1) denoted $[a_1, b_1]$

- q_1 is (a_1, b_0, b_1) denoted $[b_1, a_1]$

Notice some examples of the following denotations used when we remove some or all end points:

- $(a_0, a_1, b_0) \setminus \{b_0\}$ denoted $[a_0, b_0[$

- $(a_0, b_1, b_0) \setminus \{b_0\}$ denoted $]b_0, a_0]$

- $(a_1, a_0, b_1) \setminus \{a_1, b_1\}$ denoted $]a_1, b_1[$

- etc.

We consider now the offending chain c_2 crossing the new edge linking a_0 to b_0. This chain links two vertices a_2 and b_2 both situated on the border of region r. This is so because the graph g is supposed to be connected, so vertices in c_2 should be connected to vertices in the border of r. Notice that a_2 and b_2 are certainly different from a_0 and b_0 since c_2 intersects with the edge a_0–b_0 in a point that is not a vertex. We now have three cases:

Case 1: In this case (Fig. 19), we suppose for the moment that a_2 is in the chain $]a_0, a_1[$ (the chain linking a_0 to a_1 but neither in a_0 nor in a_1). The vertex b_2 must be in the chain $]b_0, a_0[$ (the chain linking b_0 to a_0 but neither in b_0 nor in a_0) since c_2 intersect with the edge linking a_0 to b_0 (**Axiom 2** of Sect. 6). But b_2 should also be in the chain $]b_1, a_1[$ (the chain linking b_1 to a_1 but neither in b_1 nor in a_1) since otherwise c_2 could be moved outside the region r without crossing any other chain (**Axiom 3** and **Exercise 3** of Sect. 6).

Fig. 19. Case 1

Thus b_2 is in the chain $]b_1, b_0[$ (the chain linking b_1 to b_0 but neither in b_1 nor in b_0). Again see Fig. 19. **Here K3,3 appears with vertices $V1 = \{a_0, a_1, b_2\}$**

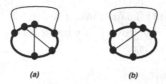

(a)   (b)

Fig. 20. Two sub-cases of Case 1

and $V2 = \{b_0, b_1, a_2\}$. Note that this **Case1** covers in fact two subcases as shown in Fig. 20 (a), where a_2 is in $]a_0, a_1[$ (studied above) and Fig. 20 (b), where a_2 is in $]b_1, a_0[$.

Case 2: In this case (Fig. 21), we do not modify the position of a_2 (it is still in $]a_0, a_1[$), but we investigate the possibility for b_2 to be in $]a_0, b_1]$ (the chain linking a_0 to b_1 but not in a_0).

Fig. 21. Case 2

Clearly, this is not possible as such because then there could exist a drawing of graph h where the chain c_2 could be moved outside region r (**Axiom 3** of Sect. 6) and thus not intersect the edge linking a_0 to b_0. This failing case is illustrated in Fig. 22 (see **Exercise 1** and **Exercise 3** in Sect. 6)

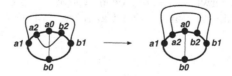

Fig. 22. Case 2: not possible. The chain $a2$–$b2$ can be moved outsid

In order to prevent this to happen we must suppose that we have a vertex v within c_2 that is connected to the border of r in a vertex b_2' (remember **Exercise 4** of Sect. 6). Now b_2' cannot be in $[a_1, b_1]$ (the chain linking a_1 to b_1 through a_0) since otherwise there could exist a drawing of graph h where $c2$ could again be moved outside r. This failing case is illustrated in Fig. 23.

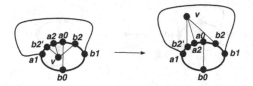

Fig. 23. Case 2: not possible. The chain $a2$–$b2$ and other chains can be moved outside

So, b'_2 is in $[b_1, a_1]$ (the chain linking a_1 to b_1 through b_0). But we can remove b'_2 from $[b_0, a_0]$ (the chain linking b_0 to a_0 through b_1) since then the connection between $a2$ and b'_2 would be the same as in **Case 1** (a) (Fig. 20).

In conclusion, b'_2 is in $]a_1, b_0]$ (the chain linking a_1 to b_0 but not in a_1). All this is shown in Fig. 21. **Here K3,3 appears with vertices** $V1 = \{a_0, a_1, v\}$ and $V2 = \{b_2, b'_2, a_2\}$. The link from a_0 to b'_2 is done through b_0 and that from a_1 to b_2 is done through b_1.

Notice that if b_2 and b'_2 are not common with b_1 and b_0 respectively then we have also K3,3 appearing with vertices $V1 = \{a_0, b_1, b'_2\}$ and $V2 = \{b_0, a_1, b_2\}$. In fact, this is just **Case 1** (b) as shown in Fig. 20. So, the genuine **Case 2** are those where b_2 is in b_1, or b'_2 is in b_0, or both b_2 and b'_2 in b_1 and b_0 respectively. This make 3 sub-cases and we have four other sub-cases by symmetry. All this is shown in Fig. 24.

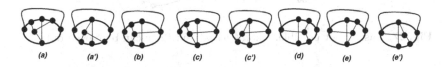

Fig. 24. Various sub-cases of Case 2

Case 3: In the two previous cases, a_2 was in the chain $]a_0, a_1[$ (or similar chains). In this third case, we now suppose now that a_2 is common with a_1. Remember that a_2 cannot be common with a_0 since the chain a_2–b_2 intersect with the edge a_0–b_0 in a non-vertex. As in previous cases, b_2 should be in the chain $]b_0, a_0[$ (the chain linking b_0 to a_0 but neither in b_0 nor in a_0).

We notice that b_2 cannot be in the chain $[b_1, b_0[$ (the chain linking b_1 to b_0 but not in b_0) since otherwise, the chain c_2 could be moved outside region r in another drawing and thus not cross the edge linking a_0 to b_0. This failing case is illustrated in Fig. 25.

In order to avoid this to happen, we must introduce a vertex v in c_2 and link it to the chain $]b_1, a_1[$ in a vertex b'_2. This technique is the same as that used in **Case 2**. Now b'_2 cannot be in the chain $]a_1, a_0[$ since this is **Case 1** (a), nor can $b2'$ be common with a_0 since it is **Case 2** (e'), nor can it be in the chain $[a_0, b_1[$ since it is **Case 2** (d). These already encountered cases are illustrated in Fig. 26.

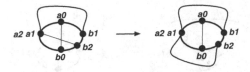

Fig. 25. Case 3: not possible. The chain $a1$–$b2$ can be moved outside

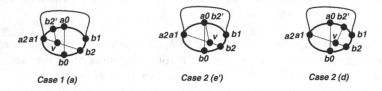

Case 1 (a)   Case 2 (e')   Case 2 (d)

Fig. 26. Case 3: already encountered cases

Finally $b2'$ cannot be in b_1 since then, again, c_2 could be moved outside region r in another drawing. This is illustrated in Fig. 27.

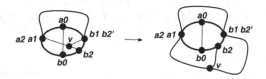

Fig. 27. Case 3: not possible. Chains $a1$–$b2$ and $b2'$–v can be moved outside

So, since all previous cases failed for $b2$ being in the chain $]b_1, b_0[$, we must suppose that b_2 is in the chain $]a_0, b_1[$. But then, the chain c_2 could be moved outside region r in another drawing and thus not cross the edge linking a_0 to b_0. This failing case is illustrated in Fig. 28 (see **Exercise 1** at the end of Sect. 6).

In order to avoid the chain c_2 to be moved outside in another drawing, we have to introduce a vertex v in the chain c_2 and link it to a vertex b_2' in the chain $]a_1, b_1[$. This vertex b_2' cannot be in the chain $]a_1, b_0[$, since this corresponds to **Case 1** (b), nor can it be common with b_0 since it corresponds to **Case 2** (e). Finally it cannot be in the chain $]b_0, b_1[$ since it corresponds to **Case 2** (d). This is illustrated in Fig. 29.

The only possibility then for b_2 is to be common with b_1. But then the chain c_2 can be moved outside. This is illustrated in Fig. 30 (see **Exercise 1** at the end of Sect. 6).

In order to avoid this to happen, we must introduce one or two vertices v and w in c_2 and connect them to two vertices b_2' and b_2'' situated in $]a_1, b_1[$ or in $]b_1, a_1[$ respectively. The only possibility for b_2' or b_2'' is to be common with a_0 and b_0 respectively, since otherwise we have situations already encountered in **Case 1** (a), **Case 2** (b), and **Case 2** (c). This is illustrated in Fig. 31.

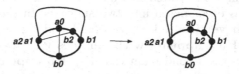

Fig. 28. Case 3: not possible. Chains $a1$–$b2$ can be moved outside

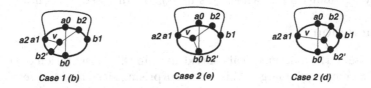

Case 1 (b)   Case 2 (e)   Case 2 (d)

Fig. 29. Case 3: already encountered cases

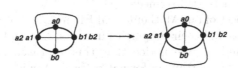

Fig. 30. Case 3: not possible. Chain $a_1 - b_1$ can be moved outside

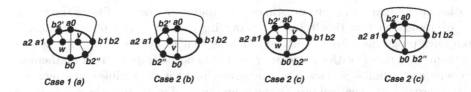

Case 1 (a)   Case 2 (b)   Case 2 (c)   Case 2 (c)

Fig. 31. Case 3: already encountered cases

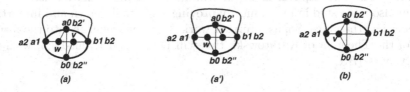

(a)   (a')   (b)

Fig. 32. Case 3

The final situation is illustrated in Fig. 32 where we have two main cases. The first main case is decomposed into two subcases. In the first subcase (a), **K3.3 appears with vertices** $V1 = \{b_0, b_1, w\}$ and $V2 = \{a_0, a_1, v\}$. In the second subcase (a'), **K3.3 appears with vertices** $V1 = \{a_0, b_1, w\}$ and $V2 = \{b_0, a_1, v\}$, Finally, in the second main case (b), **K5 appears with vertices** $V = \{a_0, b_0, a_1, b_1, v\}$.

Notice that we cover all cases. In **Case 1**, we have $a_2 \in]a_0, a_1[$ and $b_2 \in]b_0, b_1[$ (and similar cases). In **Case 2**, we have $a_2 \in]a_0, a_1[$ and $b_2 \in]a_0, b_1]$ (and similar cases). So, in **Case 1** and **Case 2** together, $b2$ is in $]a_0, b_0[$, which is the only possibility for b_2 when a_2 is in $]a_0, a_1[$. In **Case 3**, we have $a_2 = a_1$.

8 Conclusion

As already said, the mathematical context used in this presentation of the Kuratowski Theorem is quite large. This is not surprising since this theorem involves many properties of the concerned graphs: they should be simple, loop free, and connected. Another important property is the existence of a binary relation between externally tangent faces and the single connection between related faces and edges (function *fe* introduced in Sect. 5.3). Consequently, vertices situated on internal edges are at least 3-connected.

An important aspect of our Mathematical Engineering approach is the usage of abstraction (regions and their properties in Sects. 2, 3, and 4) and refinement (graphs in Sects. 5 and 6). We notice that this approach of abstraction and refinement is also very important in formal Software Engineering.

Another peculiarity we encounter, this time in the proof, is the rather *large number of cases*. Here we touch one of the main problems of mathematical presentations. Usually mathematicians *omit to cover all cases*, thus putting the reader in a difficult position, oscillating between a poor comprehension of the mathematical text and even sometimes a doubt about the validity of the proof.

We remember that this lack of case covering is at the origin of many bugs in computer programs. However, the problem is that the consequence of a wrong mathematical proof (with some cases missing) is not so important in mathematics, whereas a similar situation in a program might have terrible consequences. In the case of mathematical proofs, the community of mathematicians is eager to check that a new proposed proof is correct (it might take several years). Whereas in informatics, the non-correctness of a program is usually discovered by the consequence of a bug only (but then it is sometimes too late). In the programming discipline we have no equivalent to the community of mathematicians.

I am also interested in mechanically proving some well known and important mathematical theorems. So, as a further work, I will now engage into the formal proof of the Theorem of Kuratowski: this will be done in Event-B [6] with the Rodin toolset [7].

Acknowledgements. I would like to thank very much Dominique Cansell for his reading of earlier drafts of this work. He gave me many useful suggestions able to improve the paper.

References

1. Kuratowski, C.: Sur le problème des courbes gauche en topologie. Fund. Math. **15**, 271–283 (1930)
2. Even, S.: Graph Algorithms, 2nd edn. Cambridge University Press, New York (2012)
3. Fournier, J.C.: Démonstration simple du théorème de Kuratowski et de sa forme duale. Discrete Math. **31**, 329–332 (1980)
4. Klotz, W.: A constructive proof of kuratowski theorem. Ars Combinatoria **28**, 51–54 (1989)
5. Thomassen, C.: A refinement of kuratowski theorem. J. Comb. Theory **37**, 245–253 (1984)
6. Abrial, J.R.: Modeling in Event-B: System and Software Engineering. Cambridge University Press, New York (2010)
7. http://www.event-b.org

Location Privacy via Geo-Indistinguishability

Konstantinos Chatzikokolakis[1,2], Catuscia Palamidessi[2,3](✉),
and Marco Stronati[2]

[1] CNRS, Paris, France
[2] LIX, École Polytechnique, Rocquencourt, France
[3] INRIA, Paris, France
`catuscia@lix.polytechnique.fr`

Abstract. In this paper we report on the ongoing research of our team Comète on location privacy. In particular, we focus on the problem of protecting the privacy of the user when dealing with location-based services. The starting point of our approach is the principle of geo-indistinguishability, a formal notion of privacy that protects the user's exact location, while allowing approximate information – typically needed to obtain a certain desired service – to be released. Then, we discuss the problem that raise in the case of traces, when the user makes consecutive uses of the location based system, while moving along a path: since the points of a trace are correlated, a simple repetition of the mechanism would cause a rapid decrease of the level of privacy. We then show a method to limit such degradation, based on the idea of predicting a point from previously reported points, instead of generating a new noisy point. Finally, we discuss a method to make our mechanism more flexible over space: we start from the observation that space is not uniform from the point of view of location hiding, and we propose an approach to adapt the level of privacy to each zone.

1 Introduction

In recent years, the increasing availability of location information about individuals has led to a growing use of systems that record and process location data, generally referred to as "location-based systems". Examples of these systems include Location Based Services (LBSs), location-data mining algorithms to determine points of interest, and location-based machine learning algorithms to predict traffic patterns.

While location-based systems have demonstrated to provide enormous benefits to individuals and society, the growing exposure of users' location information raises important privacy issues. First of all, location information itself may be considered as sensitive. Furthermore, it can be easily linked to a variety of other information that an individual usually wishes to protect: by collecting and processing accurate location data on a regular basis, it is possible to infer an individual's home or work location, sexual preferences, political views, religious inclinations, etc.

M. Leucker et al. (Eds.): ICTAC 2015, LNCS 9399, pp. 28–38, 2015.
DOI: 10.1007/978-3-319-25150-9_2

It is therefore important to design and implement methods for protecting the user's privacy while preserving the utility and the dependability of location data for their use in location-based systems. In this paper, we report on the research of the INRIA Comète team on this field.

A characteristics of our approach is that we focus on the problem of *protecting the user's location, rather than the user's anonymity*. The latter is based on the idea of hiding the association between the user's location data and his name. However, there have been several examples of attacks showing that anonymity is not sufficient to protect the user: in the large majority of cases, location data can be re-identified by using correlated information.

Furthermore, we focus on methods that provide privacy guarantees which are (a) *based on solid mathematical basis*, (b) *independent from the adversary side information*, and (c) *robust with respect to composition of attacks*.

Our approach is based on the notion of *geo-indistinguishability*, which is a property similar to that of *differential privacy* [8]. Basically, the idea is to obfuscate the real location by reporting an approximate one, using some random noise. The idea is that from the reported location, the attacker may be able to make a good guess of the area where the user is actually located, but it should not be able to make a good guess of the exact location of the user within this area. This meachanism can be implemented by using a noise with a Laplacian distribution, that is a negative exponential with respect to the distance from the real location, like in the case of differential privacy. This method provides a good level of robustness with respect to composition of attacks, in that the level of privacy decreases in a controlled way (linearly).

When the user makes several repeated applications of the mechanism from related points (typically in the case of a trace), however, even a linear decrease of the level of privacy poses a tall too high to the privacy level. To address this problem, we propose a *predictive mechanism*, which avoids the application of the mechanism when a new (noisy) point can be derived from the previous ones.

Finally, we consider the problem that raises when the space is not uniform with respect to the hiding value: the point is that in different zones the number of locations where the user could be located may vary a lot, and as a consequence these zones should have a different privacy parameter. We address this problem by proposing an *elastic mechanism*, which is based on a notion of distance adapted to the different zones.

1.1 Related Work

Most location privacy mechanisms proposed in the literature involve obfuscation of the real location. The simplest methods are those based on variants of the *cloaking technique*, which consists in hiding the real location within *a region of possible locations*, for instance by reporting the area around the real location, or by using dummy locations [2,6,7,11,14,17]. Unfortunately, cloacking methods are not robust with respect to composition. For instance, reporting the area is subject to triangulation attacks. Furthermore, they require assumptions about the attacker's side information. For example, dummy locations are only useful if they look equally likely to be the real location from the point of view of the attacker.

A second class of location obfuscation mechanisms involve the generation of controlled noise for Bayesian adversaries. We mention in particular [10] and [15]: The first obtains a perturbation mechanism by crossing paths of individual users, thus rendering the task of tracking individual paths challenging. The second obtains an optimal mechanism (i.e., achieving maximum level of privacy for the user) by solving a linear program in which the constraints are determined by the quality of service and by the user's profile.

1.2   Plan of the Paper

In the next section we present our basic approach to location privacy, based o the notion of geo-indistinguishability. In Sect. 3 we then discuss the problems that raise when we repeatedly use the mechanism along a trace, and when the space is not uniform from the point of view of location hiding, and we illustrate our approach to address these problems. Finally, Sect. 4 presents some future work.

2   Geo-Indistinguishability

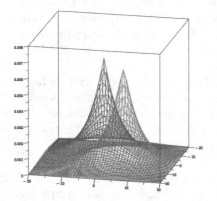

Our approach is based on the property of *geo-indistinguishability* [1], which guarantees that the user's location is protected, within a radius r, with a level of noise that decreases with r, at a rate that depends on the desired level of privacy. Intuitively, this means that the real location is highly indistinguishable from the locations that are close, and gradually more distinguishable from those that are far away. This characteristics allows us to obtain a good level of privacy without significant loss of utility.

Fig. 1. The prob. density functions of two planar Laplacians, centered on the (real) locations $(-2, -4)$ and $(5, 3)$ respectively.

From a technical point of view, geo-indistinguishability is a particular instance of d-privacy [4], an extension of *differential privacy* [8] to arbitrary metric domains, obtained by replacing the Hamming distance, implicit in the definition of differential privacy, with the intended distance – namely the geographical distance in our case. Like differential privacy, geo-indistinguishability is independent from the side knowledge of the adversary and robust with respect to composition of attacks.

We have implemented geo-indistinguishability by adding random noise drawn from a planar Laplace distribution, see Fig. 1. In [1] we have compared this mechanism with the representatives of the other methods proposed in the literature (the cloaking and the linear programming mechanisms), using the privacy metric proposed in [15]. It turns out that our mechanism offers the best privacy guarantees, for the same utility, among all those which do not depend on the prior knowledge of the adversary. The advantages of the independence from the prior are obvious: first, the mechanism is designed once and for all (i.e. it does not need to be recomputed every time the adversary changes, it works also in simultaneous presence of different adversaries, etc.). Second, and even more important, it is applicable also when we do not know the prior.

Our technique can be used to enhance any application for location-based services with privacy guarantees, and can be implemented on the client side of the application. To this purpose, we are developing a tool, called Location Guard.

2.1 Location Guard

Location Guard [https://github.com/chatziko/location-guard] is an open source web browser extension based on geo-indistinguishability, that provides location privacy when using the HTML5 geolocation API (Fig. 2).

Fig. 2. Privacy level config-uration on Android, r_u in purple and r_p in pink.

When a page is loaded and before any other code is executed, Location Guard injects a small snippet of JavaScript that redefines `geolocation.getCurrentPosition`, the main function provided by the Geolocation API to retrieve the current position. When the rest of the page code runs and tries to access this function, it gets intercepted by Location Guard, which in turn obtains the real location from the browser, sanitizes it and returns it to the page.

The location is sanitized through the use of random noise drawn from a Planar Laplace distribution. The amount of noise added can be configured easily with a single parameter, the privacy *level*. Location guard provides three predefined levels {high,medium,low} and the user is also free to pick any other value. Additionally the privacy level can be adjusted per domain, so that different protection can be applied to different services: a larger amount of noise can be added to a weather service as opposed to a point of interest search engine.

An advantage of geo-indistinguishability is that it is relatively intuitive to explain to the user the effect of changing the levels on privacy and utility. For a certain privacy level we can compute two radiuses r_p and r_u, respectively the radius of privacy protection and of utility. r_p is the area of locations highly indistinguishable from the actual one, i.e. all locations producing the same sanitized one with similar probabilities. r_u is the area in which the reported location lies with high probability, thus giving an idea of the utility that the user can expect.

Both these radiuses can be easily plotted on a map to give the user a direct impression of privacy and utility, according to the level of protection chosen.

Location Guard has reached considerable popularity since its release in Fall 2014, covering Chrome, Firefox and Opera browsers, and more recently moving to mobile devices with Firefox for Android. As of June 2015 Location Guard counts 9,800 active users in Google Chrome, 29,400 in Mozilla Firefox (including Android) and 5,000 downloads in Opera. Adoption has been mainly through the browser extension stores, as well as through technology blogs covering Location Guard [3,13]. In June 2015 it was chosen as "Pick of the Month" in Mozilla Add-ons Blog [16].

3   Making Geo-Indistinguishability Flexible Over Time and Space.

Geo-indistinguishability and its current implementation Location Guard are just a preliminary approach to location privacy, and they present two main limitations. First, when used repeatedly, there is a linear degradation of the user's privacy that limits the use of the mechanism over time. Second, the level of noise of the Laplacian mechanism has to be fixed in advance independently of the movements of the user, providing the same protection in areas with very different privacy characteristic, like a dense city or a sparse countryside. This limits the flexibility of the mechanism over space.

In this section we present two extensions that we developed to overcome these issues as well as future challenges that we plan to tackle. Many of techniques presented are currently being introduced into Location Guard, in order to extend its range of applications and at the same time provide a realistic experimentation platform to evaluate them.

3.1   Repeated Use Over Time

The main limitation of Location Guard is that, so far, it works well when used sporadically, to protect a single location, for instance when querying an LBS to find some point of interest (restaurants, cinemas,...) in the vicinity.

We aim at extending the range of applications by handling traces (sequences of location points). This is a very challenging task. Note, in fact, that the naive approach of applying the noise at every step would cause a dramatic privacy degradation, due to the large number of points. Intuitively, in the extreme case when the user never moves (which corresponds to maximum correlation), the reported locations would be centered around the real one, thus revealing it more and more precisely as the number of queries increases. Technically, the independent mechanism applying ϵ-geo-indistinguishable noise (where ϵ is the privacy parameter) to n locations can be shown to satisfy $n\epsilon$-geo-indistinguishability. This is a typical phenomenon in the framework of differential privacy, and consequently $n\epsilon$ is thought as a privacy *budget*, consumed by each query. This linear increase makes the mechanism applicable only when the number of queries remains small.

Fig. 3. Original trace (red), sampled trace (light blue) and reported trace (yellow) (Color figure online).

In [5] we explore a *trace obfuscation* mechanism with a smaller *budget consumption rate* than the one produced by applying independent noise. We show that correlation in the trace can be in fact exploited through a *prediction function* that tries to guess the new location based on the previously reported locations. Predicted points are safe to report directly (the adversary would have guessed them in any case) and thus have a smaller footprint on the privacy budget, because they reduce the need of applying the noise at every step. However the inclusion of the prediction function in a privacy mechanism has to be private itself, leading to additional costs for the privacy budget of the user. If there is considerable correlation in the input trace, our carefully designed *budget managers* handle this balance of costs, producing a more efficient *predictive mechanism.*

The mechanism is evaluated using the Geolife and T-Drive datasets, containing traces of thousands of users in the Beijing area. The users are modeled as accessing a location-based service while moving around the city. The prediction function used is simply behaving like a cache: It predicts that the user doesn't move and that the next location will be the same as the last one. This prediction function has the advantages of being trivial to implement, independent of the user profile and proved to be very effective in our evaluation.

Example of Sanitized Trace. Fig. 3 displays one of Geolife trajectories sanitized with fixed utility. The original trace, in red, starts south with low speed, moves north on a high speed road and then turns around Tsinghua University for some time. In order to model a user's sporadic behavior we sample the trace obtaining the 9 light blue dots, which are locations where the user queries the LBS. Finally in yellow we have the reported trace, sanitized by the predictive mechanism, with only 3 locations. The first used once for the point at the bottom, the second 7 times for the one in the middle and the third twice for point in the top. In this example the mechanism needed to sanitize with noise only 3 locations, using them as prediction for the other 6.

3.2 Highly Recurrent Locations

Even with the budget savings of the predictive mechanism, the user's privacy is bound to be breached in the long run in those locations that are *highly recurrent*, such as home and work. We propose a simple construction to model "geographic fences": Areas around highly recurrent locations where the mechanism reports uniformly, effectively stopping the privacy erosion. On one side the user has to

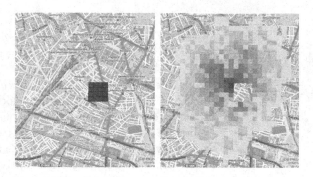

Fig. 4. Probability distribution of reported location inside and outside the fence. Darker colors indicate more likely values (Color figure online).

release publicly the position of her fences but on the other the budget cost when reporting from inside them is zero, leading to a practical solution that can be used in combination with the predictive mechanism.

In Fig. 4 we can see an example of fence introduced in an elastic metric. On the left we have the distribution of reported locations inside the fence, that is perfectly uniform, covering a few blocks and proving an adequate level of privacy while costing zero on the budget. On the right we can see the distribution of reported locations of a point right outside, the fence is clearly visible and the mechanism reports right around it.

3.3   Flexible Behavior Over Space

Another shortcoming of standard geo-indistinguishability is that the privacy level has to be fixed independently of the user location. For example, once set to have a protection in a radius of 200m, that is sufficient in a dense urban environment, the same protection will be provided when the user moves outside the city, possibly in sparsely populated area. The problem is described in more depth in [12], where we propose an *elastic mechanism* that adapts the level of noise to the semantic characteristics of each location, such as population and presence of POIs. We perform an extensive evaluation of our technique by building an elastic mechanism for Paris' wide metropolitan area, using semantic information from the OpenStreetMap database.

The resulting privacy *mass* of each location is shown in Fig. 5a, where white color indicates a small mass while yellow, red and black indicate increasingly greater mass. The figure is just a small extract of the whole grid depicting the two smaller areas used in the evaluation: central Paris and the nearby suburb of Nanterre. Note that the colors alone depict a fairly clear picture of the city: in white we can see the river traversing horizontally, the main ring-road and several spots mark parks and gardens. In yellow colors we find low density areas as well as roads and railways while red colors are present in residential areas. Finally dark colors indicate densely populated areas with presence of POIs.

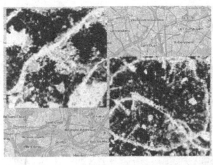 

(a) Privacy mass of each location   (b) Expected error at each location

Fig. 5. Paris' center (right) and the nearby suburb of Nanterre (left)

Figure 5b shows our utility per location, computed as the expected distance between the real and the reported location. Compared to Fig. 5a it is clear that areas with higher privacy mass result to less noise. Populated areas present a good and uniform error that starts to increase on the river and ring-road. On the other hand, the large low-density areas, especially in the Nanterre suburb, have a higher error because they need to report over larger areas to reach the needed amount of privacy.

We compare the resulting mechanism against the Planar Laplace mechanism satisfying standard geo-indistinguishability, using two real-world datasets from the Gowalla and Brightkite location-based social networks. The results show that the elastic mechanism adapts well to the semantics of each area, adjusting the noise as we move outside the city center, hence offering better overall privacy.

3.4 A Tiled Mechanism

The extreme flexibility of the elastic mechanism, that can change its behavior for locations just 100 meters apart, comes with the cost of a heavy phase of pre-processing to build its semantic map, which is not suitable for Location Guard.

For this reason we propose a lighter version of the *elastic mechanism*, that requires no pre-computation of the metric, and is thus suitable for lower end devices and for an easier inclusion in existing systems. Of course this *tiled mechanism* provides less flexibility: Instead of adapting the noise differently in locations tens of meters apart, it can only adapt to large areas of a city, covering tens of square kilometers. These areas, that we call tiles, area small enough to distinguish a park from a residential area, but still easily computable. In order to build the set of tiles, we query two online geographical services, `overpass-turbo` and `dbpedia` to obtain a set of polygons together with a quantitative description of the amount of privacy they provide. This dataset should cover an area large enough to contain most of the user usual movement and it can easily reach a few tens of kilometers while retaining a small size. Once this small dataset is build, we would have a mapping from tiles to their privacy mass, and we would

Fig. 6. Polygons computed for New York and Paris

use it to define a function ℓ that, for each location, finds the containing polygon and returns a privacy level adapted to the privacy mass provided by the tile. Examples of the kind of maps that we aim at obtaining with this method are shown in Fig. 6.

The mechanism described above, despite achieving the flexible behavior we needed, would not satisfy geo-indistinguishability. It is enough to notice that the level of protection, a public information of the mechanism, depends on the current location of the user, which is sensitive. In order to solve this problem we would need to make ℓ itself differentially private. A simple way to do it could be to first sanitize the current location with a fixed privacy level and then feed it to ℓ. Post processing a sanitized location does not pose any threat to privacy and would allow the mechanism to reduce sharply the amount of noise added to location in very private area.

4   Future Work

Regarding the geographic fences we are currently evaluating how to automatically configure their position and size. The user input would be the best option, however they could also be inferred and suggested automatically. In [9] the authors developed an attack to identify POI of a specific user, from a set of mobility traces. A similar technique could be employed on the user's phone, over a training period, to collect and analyze her movements for a few days. The mechanism would then automatically detect recurrent locations and suggest the user to fence them, possibly detecting more than just home/work locations.

With the use of geolocated queries, such as those used to extract privacy mass of the elastic mechanism, we could determine the size of the fence so to include a reasonable amount of buildings for home and other POIs for work.

Concerning the elastic mechanism in some cases we might want to tailor our mechanism to a specific group of users, to increase the performance in terms of both privacy and utility. In this case, given a *prior* probability distribution over the grid of locations, we can use it to influence the privacy mass of each cell. For instance, if we know that our users never cross some locations or certain kind of POIs, we can reduce their privacy mass.

Moreover, we are interested in queries that reward variety other that richness e.g. a location with 50 restaurants should be considered less private than one with 25 restaurant and 25 shops.

Finally, different grids could be computed for certain periods of the day or of the year. For instance, our user could use the map described above during the day, feeling private in a road with shops, but in the evening only a subset of the tags should be used as many activities are closed, making a road with many restaurants a much better choice. The same could be applied to seasons, imagine for example how snow affects human activities in many regions.

Additionally we are also actively working on the tiled mechanism in order to provide both a formal proof of privacy as well as an efficient implementation to include in Location Guard.

References

1. Andrés, M.E., Bordenabe, N.E., Chatzikokolakis, K., Palamidessi, C.: Geo-indistinguishability: differential privacy for location-based systems. In: Proceedings of CCS, pp. 901–914. ACM (2013)
2. Bamba, B., Liu, L., Pesti, P., Wang, T.: Supporting anonymous location queries in mobile environments with privacygrid. In: Proceedings of WWW, pp. 237–246. ACM (2008)
3. Brinkmann, M.: Change your location in firefox using location guard. Ghacks.net (2014)
4. Chatzikokolakis, K., Andrés, M.E., Bordenabe, N.E., Palamidessi, C.: Broadening the scope of differential privacy using metrics. In: De Cristofaro, E., Wright, M. (eds.) PETS 2013. LNCS, vol. 7981, pp. 82–102. Springer, Heidelberg (2013)
5. Chatzikokolakis, K., Palamidessi, C., Stronati, M.: A predictive differentially-private mechanism for mobility traces. In: De Cristofaro, E., Murdoch, S.J. (eds.) PETS 2014. LNCS, vol. 8555, pp. 21–41. Springer, Heidelberg (2014)
6. Cheng, R., Zhang, Y., Bertino, E., Prabhakar, S.: Preserving user location privacy in mobile data management infrastructures. In: Danezis, G., Golle, P. (eds.) PET 2006. LNCS, vol. 4258, pp. 393–412. Springer, Heidelberg (2006)
7. Duckham, M., Kulik, L.: A formal model of obfuscation and negotiation for location privacy. In: Gellersen, H.-W., Want, R., Schmidt, A. (eds.) PERVASIVE 2005. LNCS, vol. 3468, pp. 152–170. Springer, Heidelberg (2005)
8. Dwork, C.: Differential privacy. In: Bugliesi, M., Preneel, B., Sassone, V., Wegener, I. (eds.) ICALP 2006. LNCS, vol. 4052, pp. 1–12. Springer, Heidelberg (2006)
9. Gambs, S., Killijian, M.O., del Prado Cortez, M.N.: Show me how you move and i will tell you who you are. Trans. Data Priv. 4(2), 103–126 (2011)
10. Hoh, B., Gruteser, M.: Protecting location privacy through path confusion. In: Proceedings of SecureComm, pp. 194–205. IEEE (2005)
11. Kido, H., Yanagisawa, Y., Satoh, T.: Protection of location privacy using dummies for location-based services. In: Proceedings of ICDE Workshops, p. 1248 (2005)
12. Stronati, M., Chatzikokolakis, K., Palamidessi, C.: Constructing elastic distinguishability metrics for location privacy. In: Proceedings of PETS (2015). To appear

13. Korben, D.: Géolocalisation - restez maître de votre situation (2015). http://korben.info/geolocalisation-restez-maitre-de-votre-situation.html
14. Shankar, P., Ganapathy, V., Iftode, L.: Privately querying location-based services with SybilQuery. In: Proceedings of UbiComp, pp. 31–40. ACM (2009)
15. Shokri, R., Theodorakopoulos, G., Troncoso, C., Hubaux, J.P., Le Boudec, J.Y.: Protecting location privacy: optimal strategy against localization attacks. In: Proceedings of CCS, pp. 617–627. ACM (2012)
16. Tsay, A.: Mozilla add-ons blog, June 2015. https://blog.mozilla.org/addons/2015/06/01/june-2015-featured-add-ons/
17. Xue, M., Kalnis, P., Pung, H.K.: Location diversity: enhanced privacy protection in location based services. In: Choudhury, T., Quigley, A., Strang, T., Suginuma, K. (eds.) LoCA 2009. LNCS, vol. 5561, pp. 70–87. Springer, Heidelberg (2009)

A Note on Monitors and Büchi Automata

Volker Diekert[1]([⊠]), Anca Muscholl[2], and Igor Walukiewicz[2]

[1] FMI, Universität Stuttgart, Stuttgart, Germany
diekert@fmi.uni-stuttgart.de
[2] LaBRI, University of Bordeaux, Talence, France

Abstract. When a property needs to be checked against an unknown or very complex system, classical exploration techniques like model-checking are not applicable anymore. Sometimes a monitor can be used, that checks a given property on the underlying system at runtime. A monitor for a property L is a deterministic finite automaton $\mathcal{M}_L$ that after each finite execution tells whether (1) every possible extension of the execution is in L, or (2) every possible extension is in the complement of L, or neither (1) nor (2) holds. Moreover, L being monitorable means that it is always possible that in some future the monitor reaches (1) or (2). Classical examples for monitorable properties are safety and cosafety properties. On the other hand, deterministic liveness properties like "infinitely many a's" are not monitorable.

We discuss various monitor constructions with a focus on deterministic ω-regular languages. We locate a proper subclass of deterministic ω-regular languages but also strictly larger than the subclass of languages which are deterministic and codeterministic; and for this subclass there exist canonical monitors which also accept the language itself.

We also address the problem to decide monitorability in comparison with deciding liveness. The state of the art is as follows. Given a Büchi automaton, it is PSPACE-complete to decide liveness or monitorability. Given an LTL formula, deciding liveness becomes EXPSPACE-complete, but the complexity to decide monitorability remains open.

Introduction

Automata theoretic verification has its mathematical foundation in classical papers written in the 1950's and 1960's by Büchi, Rabin and others. Over the past few decades it became a success story with large scale industrial applications. However, frequently properties need to be checked against an unknown or very complex system. In such a situation classical exploration techniques like model-checking might fail. The model-checking problem asks whether all runs satisfy a given specification. If the specification is written in monadic second-order logic, then all runs obeying the specification can be expressed effectively by some Büchi automaton (BA for short). If the abstract model of the system is given by some finite transition system, then the model-checking problem becomes an inclusion problem on ω-regular languages: all runs of the transition system must be accepted by the BA for the specification, too. In formal terms

© Springer International Publishing Switzerland 2015
M. Leucker et al. (Eds.): ICTAC 2015, LNCS 9399, pp. 39–57, 2015.
DOI: 10.1007/978-3-319-25150-9_3

we wish to check $L(\mathcal{A}) \subseteq L(\varphi)$ where $\mathcal{A}$ is the transition system of the system and φ is a formula for the specification. Typically testing inclusion is expensive, hence it might be better to check the equivalent assertion $L(\mathcal{A}) \cap L(\neg\varphi) = \emptyset$. This is a key fact, because then the verification problem becomes a reachability problem in finite graphs.

Whereas the formulas are typically rather small, so we might be able to construct the Büchi automaton for $L(\neg\varphi)$, the transition systems tend to be very large. Thus, "state explosion" on the system side might force us to use weaker concepts. The idea is to construct a "monitor" for a given specification. A monitor observes the system during runtime. It is a finite deterministic automaton with at most two distinguished states $\bot$ and $\top$. If it reaches the state $\bot$, the monitor stops and raises an "alarm" that no continuation of the so far observed run will satisfy the specification. If it reaches $\top$, the monitor stops because all continuations will satisfy the specification. Usually, this means we must switch to a finer monitor. Finally, we say that a language is monitorable, if in every state of the monitor it is possible to reach either $\bot$ or $\top$ or both.

The formal definition of monitorable properties has been given in [18] by Pnueli and Zaks. It generalizes the notion of a *safety property* because for a safety property some deterministic finite automaton can raise an alarm $\bot$ by observing a finite "bad prefix", once the property is violated. The extension to the more general notion of monitorability is that a monitorable property gives also a positive feedback $\top$, if all extensions of a finite prefix obey the specification. Monitors are sometimes easy to implement and have a wide range of applications. See for example [13] and the references therein. Extensions of monitors have been proposed in more complex settings such as for stochastic automata [7,20] and for properties expressed in metric first-order temporal logic [2]. For practical use of monitors, various parameters may be relevant, in particular the size of the monitor or the runtime overhead generated by the monitor (see also the discussion in [24]).

In the present paper we discuss various monitor constructions. A monitor for a safety property L can have much less states than the smallest DBA accepting L. For example, let $\Sigma = \{a, b\}$ and $n \in \mathbb{N}$. Consider the language $L = a^n b a \Sigma^\omega \setminus \Sigma^* b b \Sigma^\omega$. The reader is invited to check that L is a safety property and every DBA accepting L has more than n states. But there is a monitor with three states, only. The monitor patiently waits to see an occurrence of a factor bb and then switches to $\bot$. Hence, there is no bound between a minimal size of an accepting DBA and the minimal size of a possible monitor. This option, that a monitor might be much smaller than any accepting DBA, has been one of the main motivations for the use of monitors.

There are many deterministic languages which are far away from being monitorable. Consider again $\Sigma = \{a, b\}$ and let L be the deterministic language of "infinitely many a's". It is shown in [5] that L cannot be written as any countable union of monitorable languages. On the other hand, if L is monitorable and also accepted by some DBA with n states and a single initial state, then there is some monitor accepting L with at most n states.

In the last section of this paper we discuss the question how to decide whether a language is monitorable and its complexity. If the input is a Büchi automaton, then deciding safety, liveness, or monitorability is PSPACE-complete. If the input is an LTL formula, then deciding safety remains PSPACE-complete. It becomes surprisingly difficult for liveness: EXPSPACE-complete. For monitorability the complexity is wide open: we only know that it is PSPACE-hard and that monitorability can be solved in EXPSPACE.

1 Preliminaries

We assume that the reader is familiar with the basic facts about automata theory for infinite words as it is exposed in the survey [25]. In our paper Σ denotes a finite nonempty alphabet. We let Σ^* (resp. Σ^ω) be the set of finite (resp. infinite) words over Σ. Usually, lower case letters like a, b, c denote letters in Σ, $u, \ldots, z$ denote finite words, 1 is the empty word, and α, β, γ denote infinite words. By language we mean a subset $L \subseteq \Sigma^\omega$. The complement of L w.r.t. Σ^ω is denoted by L^{co}. Thus, $L^{co} = \Sigma^\omega \setminus L$.

A *Büchi automaton* (*BA* for short) is a tuple $\mathcal{A} = (Q, \Sigma, \delta, I, F)$ where Q is the nonempty finite set of states, $I \subseteq Q$ is the set of initial states, $F \subseteq Q$ is the set of final states, and $\delta \subseteq Q \times \Sigma \times Q$ is the transition relation. The accepted language $L(\mathcal{A})$ is the set of infinite words $\alpha \in \Sigma^\omega$ which label an infinite path in $\mathcal{A}$ which begins at some state in I and visits some state in F infinitely often. Languages of type $L(\mathcal{A})$ are called ω-*regular*.

If for each $p \in Q$ and $a \in \Sigma$ there is at most one $q \in Q$ with $(p, a, q) \in \delta$, then $\mathcal{A}$ is called *deterministic*. We write *DBA* for deterministic Büchi automaton. In a DBA we view δ as a partially defined function and we also write $p \cdot a = q$ instead of $(p, a, q) \in \delta$. Frequently it is asked that a DBA has a unique initial state. This is not essential, but in order to follow the standard notation $(Q, \Sigma, \delta, q_0, F)$ refers to a BA where I is the singleton $\{q_0\}$.

A *deterministic weak Büchi automaton* (*DWA* for short) is a DBA where all states in a strongly connected component are either final or not final. Note that a strongly connected component may have a single state because the underlying directed graph may have self-loops. A language is accepted by some DWA if and only if it is deterministic and simultaneously codeterministic. The result is in [22] which in turn is based on previous papers by Staiger and Wagner [23] and Wagner [27].

According to [18] a *monitor* is a finite deterministic transition system $\mathcal{M}$ with at most two distinguished states $\bot$ and $\top$ such that for all states p either there exist a path from p to $\bot$, or to $\top$, or to both. It is a *monitor for an* ω-*language* $L \subseteq \Sigma^\omega$ if the following additional properties are satisfied:

- If u denotes the label of a path from an initial state to $\bot$, then $u\Sigma^\omega \cap L = \emptyset$.
- If u denotes the label of a path from an initial state to $\top$, then $u\Sigma^\omega \subseteq L$.

A language $L \subseteq \Sigma^\omega$ is called *monitorable* if there exists a monitor for L. Thus, even non regular languages might be monitorable. If a property is monitorable, then the following holds:

$$\forall x \, \exists w : xw\Sigma^\omega \subseteq L \vee xw\Sigma^\omega \cap L = \emptyset . \tag{1}$$

The condition in (1) is not sufficient for non-regular languages: indeed consider $L = \{a^n b^n a \mid n \in \mathbb{N}\}\Sigma^\omega$. There is no finite state monitor for this language. In the present paper, the focus is on monitorable ω-regular languages. For ω-regular languages (1) is also sufficient; and Remark 2 below shows an equivalent condition for monitorability (although stronger for non-regular languages).

The common theme in "automata on infinite words" is that finite state devices serve to classify ω-regular properties. The most prominent classes are:

- *Deterministic properties:* there exists a DBA.
- *Deterministic properties which are simultaneously codeterministic:* there exists a DWA.
- *Safety properties:* there exists a DBA where all states are final.
- *Cosafety properties:* the complement is a safety property.
- *Liveness properties:* there exists a BA where from all states there is a path to some final state lying in a strongly connected component.
- *Monitorable properties:* there exists a monitor.

According to our definition of a monitor, not both states $\bot$ and $\top$ need to be defined. Sometimes it is enough to see $\bot$ or $\top$. For example, let $\emptyset \neq L \neq \Sigma^\omega$ be a safety property and $\mathcal{A} = (Q, \Sigma, \delta, I, Q)$ be a DBA accepting L where all states are final. Since $\emptyset \neq L$ we have $I \neq \emptyset$. Since $L \neq \Sigma^\omega$, the partially defined transition function δ is not defined everywhere. Adding a state $\bot$ as explained above turns $\mathcal{A}$ into a monitor $\mathcal{M}$ for L where the state space is $Q \cup \{\bot\}$. There is no need for any state $\top$. The monitor $\mathcal{M}$ also accepts L. This is however not the general case.

2   Topological Properties

A topological space is a pair $(X, \mathcal{O})$ where X is a set and $\mathcal{O}$ is collection of subsets of X which is closed under arbitrary unions and finite intersections. In particular, $\emptyset, X \in \mathcal{O}$. A subset $L \in \mathcal{O}$ is called *open*; and its complement $X \setminus L$ is called *closed*.

For $L \subseteq X$ we denote by $\overline{L}$ the intersection over all closed subsets K such that $L \subseteq K \subseteq X$. It is the *closure* of L. The complement $X \setminus L$ is denoted by L^{co}.

A subset $L \subseteq X$ is called *nowhere dense* if its closure $\overline{L}$ does not contain any open subset. The classical example of the uncountable Cantor set C inside the closed interval $[0, 1]$ is nowhere dense. It is closed and does not have any open subset. On the other hand, the subset of rationals $\mathbb{Q}$ inside $\mathbb{R}$ (with the usual topology) satisfies $\overline{\mathbb{Q}} = \mathbb{R}$. Hence, $\mathbb{Q}$ is "dense everywhere" although $\mathbb{Q}$ itself does not have any open subset.

The *boundary* of L is sometimes denoted as $\delta(L)$; it is defined by

$$\delta(L) = \overline{L} \cap \overline{L^{co}}.$$

In a metric space $B(x, 1/n)$ denotes the *ball of radius* $1/n$. It is the set of y where the distance between x and y is less than $1/n$. A set is open if and only if it is some union of balls, and the closure of L can be written as

$$\overline{L} = \bigcap_{n \geq 1} \bigcup_{x \in L} B(x, 1/n).$$

In particular, every closed set is a countable intersection of open sets. Following the traditional notation we let F be the family of closed subsets and G be the family of open subsets. Then F_σ denotes the family of countable unions of closed subsets and G_δ denotes the family of countable intersections of open subsets. We have just seen $F \subseteq G_\delta$, and we obtain $G \subseteq F_\sigma$ by duality. Since G_δ is closed under finite union, $G_\delta \cap F_\sigma$ is Boolean algebra which contains all open and all closed sets.

In this paper we deal mainly with ω-regular sets. These are subsets of Σ^ω; and Σ^ω is endowed with a natural topology where the open sets are defined by the sets of the form $W\Sigma^\omega$ where $W \subseteq \Sigma^*$. It is called the *Cantor topology*. The Cantor topology corresponds to a complete ultra metric space: for example, we let $d(\alpha, \beta) = 1/n$ for $\alpha, \beta \in \Sigma^\omega$ where $n - 1 \in \mathbb{N}$ is the length of a maximal common prefix of α and β. (The convention is $0 = 1/\infty$.)

The following dictionary translates notation about ω-regular sets into its topological counterpart.

- Safety = closed sets = F.
- Cosafety = open sets = G.
- Liveness = *dense* = closure is Σ^ω.
- Deterministic = G_δ, see [12].
- Codeterministic = F_σ, by definition and the previous line.
- Deterministic and simultaneously codeterministic = $G_\delta \cap F_\sigma$, by definition.
- Monitorable = the boundary is nowhere dense, see [5].

Monitorability depends on the ambient space X. Imagine we embed $\mathbb{R}$ into the plane $\mathbb{R}^2$ in a standard way. Then $\mathbb{R}$ is a line which is nowhere dense in $\mathbb{R}^2$. As a consequence every subset $L \subseteq \mathbb{R}$ is monitorable in $\mathbb{R}^2$. The same phenomenon happens for ω-regular languages. Consider the embedding of $\{a, b\}^\omega$ into $\{a, b, c\}^\omega$ by choosing a third letter c. Then $\{a, b\}^\omega$ is nowhere dense in $\{a, b, c\}^\omega$ and hence, every subset $L \subseteq \{a, b\}^\omega$ is monitorable in $\{a, b, c\}^\omega$. The monitor has 3 states. One state is initial and by reading c we switch into the state $\bot$. The state $\top$ can never be reached. In some sense this 3-state minimalistic monitor is useless: it tells us almost nothing about the language. Therefore the smallest possible monitor is rarely the best one.

Remark 1. In our setting many languages are monitorable because there exists a "forbidden factor", for example a letter c in the alphabet which is never used.

More precisely, let $L \subseteq \Sigma^\omega$ be any subset and assume that there exists a finite word $f \in \Sigma^*$ such that either $\Sigma^* f \Sigma^\omega \subseteq L$ or $\Sigma^* f \Sigma^\omega \cap L = \emptyset$. Then L is monitorable. Indeed, the monitor just tries to recognize $\Sigma^* f \Sigma^\omega$. Its size is $|f| + 2$ and can be constructed in linear time from f by algorithms of Matiyasevich [16] or Knuth-Morris-Pratt [10].

3   Constructions of Monitors

Remark 1 emphasizes that one should not try simply to minimize monitors. The challenge is to construct "useful" monitors. In the extreme, think that we encode a language L in printable ASCII code, hence it is a subset of $\{0, 1\}^*$. But even in using a 7-bit encoding there were 33 non-printable characters. A monitor can choose any of them and then waits patiently whether this very special encoding error ever happens. This might be a small monitor, but it is of little interest. It does not even check all basic syntax errors.

3.1   Monitors for ω-regular Languages in $G_\delta \cap F_\sigma$

The ω-regular languages in $G_\delta \cap F_\sigma$ are those which are deterministic and simultaneously codeterministic. In every complete metric space (as for example the Cantor space Σ^ω) all sets in $G_\delta \cap F_\sigma$ have a boundary which is nowhere dense. Thus, deterministic and simultaneously codeterministic languages are monitorable by a purely topological observation, see [5].

Recall that there is another characterization of ω-regular languages in $G_\delta \cap F_\sigma$ due to Staiger, [22]. It says that these are the languages which are accepted by some DWA, thus by some DBA where in every strongly connected component either all states are final or none is final.

In every finite directed graph there is at least one strongly connected component which cannot be left anymore. In the minimal DWA (which exists and which is unique and where, without restriction, the transition function is totally defined) these end-components consist of a single state which can be identified either with $\bot$ or with $\top$. Thus, the DWA is itself a monitor. Here we face the problem that this DWA might be very large and also too complicated for useful monitoring.

3.2   General Constructions

Let $w \in \Sigma^*$ be any word. Then the language $L = w \Sigma^\omega$ is *clopen* meaning simultaneously open and closed. The minimal monitor for $w \Sigma^\omega$ must read the whole word w before it can make a decision; and the minimal monitor has exactly $|w| + 2$ states. On the other hand, its boundary, $\overline{L} \cap \overline{L^{co}}$ is empty and therefore nowhere dense. This suggests that deciding monitorability might be much simpler than constructing a monitor. For deciding we just need any DBA accepting the safety property $\overline{L} \cap \overline{L^{co}}$. Then we can see on that particular DBA whether L is monitorable, although this particular DBA might be of no help for monitoring.

Phrased differently, there is no bound between the size of a DBA certifying that L is monitorable and the size of an actual monitor for L.

Indeed, the standard construction for a monitor $\mathcal{M}_L$ is quite different from a direct construction of the DBA for the boundary, see for example [5]. The construction for the monitor $\mathcal{M}_L$ is as follows. Let $L \subseteq \Sigma^\omega$ be monitorable and given by some BA. First, we construct two DBAs: one DBA with state set Q_1, for the closure $\overline{L}$ and another one with state set Q_2 for the closure of the complement $\overline{L^{co}}$. We may assume that in both DBAs all states are final and reachable from a unique initial state q_{01} and q_{02}, respectively. Second, let $Q' = Q_1 \times Q_2$. Now, if we are in a state $(p, q) \in Q'$ and we want to read a letter $a \in \Sigma$, then exactly one out of the three possibilities can happen.

1. The states $p \cdot a$ and $q \cdot a$ are defined, in which case we let $(p, q) \cdot a = (p \cdot a, q \cdot a)$.
2. The state $p \cdot a$ is not defined, in which case we let $(p, q) \cdot a = \bot$.
3. The state $q \cdot a$ is not defined, in which case we let $(p, q) \cdot a = \top$.

Here $\bot$ and $\top$ are new states. Moreover, we let $q \cdot a = q$ for $q \in \{\bot, \top\}$ and $a \in \Sigma$. Hence, the transition function is totally defined. Finally, we let $Q \subseteq Q' \cup \{\bot, \top\}$ be the subset which is reachable from the initial state (q_{01}, q_{02}). Since L is monitorable, $Q \cap \{\bot, \top\} \neq \emptyset$; and Q defines a set of a monitor $\mathcal{M}_L$. Henceforth, the monitor $\mathcal{M}_L$ above is called a *standard monitor for L*. The monitor has exactly one initial state. From now on, for simplicity, we assume that every monitor $\mathcal{M}$ has exactly one initial state and that the transition function is totally defined. Thus, we can denote a monitor $\mathcal{M}$ as a tuple

$$\mathcal{M} = (Q, \Sigma, \delta, q_0, \bot, \top). \tag{2}$$

Here, $\delta : Q \times \Sigma \to Q$, $(p, a) \mapsto p \cdot a$ is the transition function, q_0 is the unique initial state, $\bot$ and $\top$ are distinguished states with $Q \cap \{\bot, \top\} \neq \emptyset$.

Definition 1. *Let $\mathcal{M} = (Q, \Sigma, \delta, q_0, \bot, \top)$, $\mathcal{M}' = (Q', \Sigma, \delta', q_0', \bot, \top)$ be monitors. A morphism between $\mathcal{M}$ and $\mathcal{M}'$ is mapping $\varphi : Q \cup \{\bot, \top\} \to Q' \cup \{\bot, \top\}$ such that $\varphi(q_0) = q_0'$, $\varphi(\bot) = \bot$, $\varphi(\top) = \top$, and $\varphi(p \cdot a) = \varphi(p) \cdot a$ for all $p \in Q$ and $a \in \Sigma$.*

If φ is surjective, then φ is called an epimorphism.

Another canonical monitor construction uses the classical notion of right-congruence. A *right-congruence* for the monoid Σ^* is an equivalence relation $\sim$ such that $x \sim y$ implies $xz \sim yz$ for all $x, y, z \in \Sigma^*$. There is a canonical right-congruence $\sim_L$ associated with every ω-language $L \subseteq \Sigma^\omega$: for $x \in \Sigma^*$ denote by $L(x) = \{\alpha \in \Sigma^\omega \mid x\alpha \in L\}$ the *quotient* of L by x. Then defining $\sim_L$ by $x \sim_L y \iff L(x) = L(y)$ yields a right-congruence. More precisely, Σ^* acts on the set of quotients $Q_L = \{L(x) \mid x \in \Sigma^*\}$ on the right, and the formula for the action becomes $L(x) \cdot z = L(xz)$. Note that this is well-defined. This yields the *associated automaton* [22, Sect. 2]. It the finite deterministic transition system with state set Q_L and arcs $(L(x), a, L(xa))$ where $x \in \Sigma^*$ and $a \in \Sigma$.

There is a canonical initial state $L = L(1)$, but unlike in the case of regular sets over finite words there is no good notion of final states in Q_L for infinite words. The right congruence is far too coarse to recognize L, in general. For example, consider the deterministic language L of "infinitely many a's" in $\{a, b\}^\omega$. For all x we have $L = L(x)$, but in order to recognize L we need two states.

It is classical that if L is ω-regular, then the set Q_L is finite, but the converse fails badly [22, Sect. 2]: there are uncountably many languages where $|Q_L| = 1$. To see this define for each $\alpha \in \Sigma^\omega$ a set

$$L_\alpha = \{\beta \in \Sigma^\omega \mid \alpha \text{ and} \beta \text{ share an infinite suffix}\}.$$

All L_α are countable, but the union $\{L_\alpha \mid \alpha \in \Sigma^\omega\}$ covers the uncountable Cantor space Σ^ω. Hence, there are uncountably many L_α. However, $|Q_{L_\alpha}| = 1$ since $L_\alpha(x) = L_\alpha$ for all x.

Recall that a monitor is a DBA where the monitoring property is not defined using final states, but it is defined using the states $\bot$ and $\top$. Thus, a DBA with an empty set of final states can be used as a monitor as long as $\bot$ and $\top$ have been assigned and the required properties for a monitor are satisfied.

Proposition 1. *Let $L \subseteq \Sigma^\omega$ be ω-regular and monitorable. Assume that L is accepted by some BA with n states. As above let $Q_L = \{L(x) \mid x \in \Sigma^*\}$ and denote $\bot = \emptyset$ and $\top = \Sigma^\omega$. Then $|Q_L| \leq 2^n$ and $Q_L \cup \{\top, \bot\}$ is the set of states for a monitor for L. At least one of the states in $\{\top, \bot\}$ is reachable from the initial state $L = L(1)$.*

The monitor in Proposition 1 with state space Q_L is denoted by $\mathcal{A}_L$ henceforth. We say that $\mathcal{A}_L$ is the *right-congruential* monitor for L.

Proposition 2. *Let $\mathcal{A}$ be the right-congruential monitor for L. Then the mapping*

$$L(x) \mapsto \varphi(L(x)) = (\overline{L}(x), \overline{L^{co}}(x))$$

induces a canonical epimorphism from $\mathcal{A}_L$ onto some standard monitor $\mathcal{M}_L$.

Proof. Observe that $\overline{L}(x) = \overline{L(x)}$ and $L^{co}(x) = L(x)^{co}$. Hence, $(\overline{L}(x), \overline{L^{co}}(x)) = (\overline{L(x)}, \overline{L(x)^{co}})$ and $\varphi(L(x))$ is well-defined. Now, if $\overline{L}(x) \neq \emptyset$ and $\overline{L^{co}}(x) \neq \emptyset$, then $\varphi(L(x)) \in Q$ where Q is the state space of the standard monitor $\mathcal{M}$. If $\overline{L}(x) = \emptyset$ then we can think that all $(\emptyset, \overline{L^{co}}(x))$ denote the state $\bot$; and if $\overline{L^{co}}(x) = \emptyset$ then we can think that all $(\overline{L}(x), \emptyset)$ denote the state $\top$. $\square$

Corollary 1. *Let $L \subseteq \Sigma^\omega$ be monitorable and given by some BA with n states. Then some standard monitor $\mathcal{M}_L$ for L has at most 2^n states.*

Proof. Without restriction we may assume that in the BA $(Q, \Sigma, \delta, I, F)$ accepting L every state $q \in Q$ leads to some final state. The usual subset construction leads first to a DBA accepting $\overline{L}$, where all states are final and the states of this DBA are the nonempty subsets of Q. Thus, these are $2^n - 1$ states. Adding the empty set $\emptyset = \bot$ we obtain a DBA with 2^n states where the transition function is defined everywhere. If the complement L^{co} is dense, this yields a standard monitor. In the other case we can use the subset construction also for a DBA accepting $\overline{L^{co}}$. In this case we remove all subsets $P \subseteq Q$ where $L(Q, \Sigma, \delta, P, F) = \Sigma^\omega$. (Note, for all $a \in \Sigma$ we have: if $L(Q, \Sigma, \delta, P, F) = \Sigma^\omega$ and $P' = \{q \in Q \mid \exists p \in P : (p, a, q) \in \delta\}$, then $L(Q, \Sigma, \delta, P', F) = \Sigma^\omega$, too.) Thus, if L^{co} is not dense, then the construction for a standard monitor has at most $2^n - 2$ states of the form (P, P) where $\emptyset \neq P$ and $L(Q, \Sigma, \delta, P, F) \neq \Sigma^\omega$. In addition there exists the reachable state $\top$ and possibly the state $\bot$.   □

Proposition 2 leads to the question of a canonical minimal monitor, at least for a safety language where a minimal accepting DBA exists. The answer is "no" as we will see in Example 1 later.

Let us finish the section with a result on arbitrary monitorable subsets of Σ^ω which is closely related to [21, Lemma 2]. Consider any subset $L \subseteq \Sigma^\omega$ where the set of quotients $Q_L = \{L(x) \mid x \in \Sigma^*\}$ is finite (="zustandsendlich" or "finite state"in the terminology of [21]). If Q_L is finite, then L is monitorable if and only if the boundary is nowhere dense. In every topological space this latter condition is equivalent to the condition that the interior of L is dense in its closure $\overline{L}$. Translating Staiger's result in [21] to the notion of monitorability we obtain the following fact.

Proposition 3. *Let $L \subseteq \Sigma^\omega$ be any monitorable language and let $\mathcal{M}$ be a monitor for L with n states. Then there exists a finite word w of length at most $(n-1)^2$ such that for all $x \in \Sigma^*$ we have either $xw\Sigma^\omega \subseteq L$ or $xw\Sigma^\omega \cap L = \emptyset$.*

Proof. We may assume that $n \geq 1$ and that the state space of $\mathcal{M}$ is included in $\{1, \ldots, n-1, \bot, \top\}$. Merging $\top$ and $\bot$ into a single state 0 we claim that there is a word w of length at most $(n-1)^2$ such that $q \cdot w = 0$ for all $0 \leq q \leq n-1$. Since L is monitorable, there is for each $q \in \{0, \ldots, n-1\}$ a finite word v_q of length at most $n-1$ such that $q \cdot v_q = 0$. By induction on k we may assume that there is a word w_k of length at most $k(n-1)$ such that for each $q \in \{0, \ldots, k\}$ we have $q \cdot w_k = 0$. (Note that the assertion trivially holds for $k = 0$.) If $k \geq n-1$ we are done: $w = w_{n-1}$. Otherwise consider the state $q = k+1$ and the state $p = q \cdot w_k$. Define the word w_{k+1} by $w_{k+1} = w_k v_p$. Then the length of w_{k+1} is at most $(k+1)(n-1)$. Since w_k is a prefix of w_{k+1} and since $0 \cdot v = v$ for all v, we have $q \cdot w_{k+1} = 0$ for all $0 \leq q \leq k+1$.   □

Remark 2. The interest in Proposition 3 is that monitorability can be characterized by a single alternation of quantifiers. Instead of saying that

$$\forall x \, \exists w \, (\forall \alpha : xw\alpha \in L) \vee (\forall \alpha : xw\alpha \notin L)$$

it is enough to say

$$\exists w \, \forall x \, (\forall \alpha : xw\alpha \in L) \vee (\forall \alpha : xw\alpha \notin L).$$

The length bound $(n-1)^2$ is not surprising. It confirms Černý's Conjecture in the case of monitors. (See [26] for a survey on Černý's Conjecture.) Actually, in the case of monitors with more than 3 states the estimation of the length of the "reset word" is not optimal. For example in the proof of Proposition 3 we can choose the word v_1 to be a letter, because there must be a state with distance at most one to 0. The precise bound is $\binom{n+1}{2} = (n+1)n/2$ if the alphabet is allowed to grow with n [19, Theorem 6.1]. If the alphabet is fixed, then the lower bound for the length of w is still in $n^2/4 + \Omega(n)$ [15].

4 Monitorable Deterministic Languages

The class of monitorable languages form a Boolean algebra and every ω-regular set L can be written as a finite union $L = \bigcup_{i=1}^{n} L_i \setminus K_i$ where the L_i and K_i are deterministic ω-regular, [25]. Thus, if L is not monitorable, then one of the deterministic L_i or K_i is not monitorable. This motivates to study monitorable deterministic languages more closely.

Definition 2. *Let $L \subseteq \Sigma^\omega$ be deterministic ω-regular. A deterministic Büchi monitor (DBM for short) for L is a tuple*

$$\mathcal{B} = (Q, \Sigma, \delta, q_0, F, \bot, \top)$$

where $\mathcal{A} = (Q, \Sigma, \delta, q_0, F)$ is a DBA with $L = L(\mathcal{A})$ and where $(Q, \Sigma, \delta, q_0, \bot, \top)$ is a monitor in the sense of Eq. (2) for L.

The next proposition justifies the definition.

Proposition 4. *Let $L \subseteq \Sigma^\omega$ be any subset. Then L is a monitorable deterministic ω-regular language if and only if there exists a DBM for L.*

Proof. The direction from right to left is trivial. Thus, let L be monitorable and let $L = L(\mathcal{A})$ for some DBA $\mathcal{A} = (Q, \Sigma, \delta, q_0, F)$ where all states are reachable from the initial state q_0. For a state $p \in Q$ let $L(p) = L(Q, \Sigma, \delta, p, F)$. If $L(p) = \emptyset$, then $L(p \cdot a) = \emptyset$; and if $L(p) = \Sigma^\omega$, then $L(p \cdot a) = \Sigma^\omega$. Thus, we can merge all states p with $L(p) = \emptyset$ into a single non-final state $\bot$; and we can merge all all states p with $L(p) = \Sigma^\omega$ into a single final state $\top$ without changing the accepted language. All states are of the form $q_0 \cdot x$ for some $x \in \Sigma^*$; and, since L is monitorable, for each x either there is some y with $xy\Sigma^\omega \cap L = \emptyset$ or there is some y with $xy\Sigma^\omega \subseteq L$ (or both). In the former case we have $q_0 \cdot xy = \bot$ and in the latter case we have $q_0 \cdot xy = \top$. □

Corollary 2. *Let $L \subseteq \Sigma^\omega$ be a monitorable deterministic ω-regular language and $\mathcal{A}$ be a DBA with n states accepting L. Let $\mathcal{B}$ be a DBM for L with state set $Q_\mathcal{B}$ where the size of $Q_\mathcal{B}$ is as small as possible. Let further $Q_\mathcal{R}$ (resp. $Q_\mathcal{M}$) be the state set of the congruential (resp. smallest standard) monitor for L. Then we have*

$$n \geq |Q_\mathcal{B}| \geq |Q_\mathcal{R}| \geq |Q_\mathcal{M}|.$$

Example 1. Let $\Sigma = \{a, b\}$ and $\Gamma = \{a, b, c, d\}$.

1. For $n \in \mathbb{N}$ consider $L = a^n b \Sigma^\omega \setminus \Sigma^* bb \Sigma^\omega$. It is a safety property. Hence, we have $\overline{L} = L$. Moreover, $\Sigma^* bb \Sigma^\omega$ is a liveness property (i.e., dense). Hence $\overline{L^{co}} = \Sigma^\omega$. It follows that the standard monitor is just the minimal DBA for L augmented by the state $\bot$. There are exactly $n + 4$ right-congruence classes defined by prefixes of the words $a^n ba$ and $a^n b^2$. We have $L(a^n b^2) = \emptyset$. Hence reading $a^n b^2$ leads to the state $\bot$. This, shows that the inequalities in Corollary 2 become equalities in that example. On the other hand b^2 is a forbidden factor for L. Hence there is a 3 state monitor for L. Still there is no epimorphism from the standard monitor onto that monitor, since in the standard monitor we have $L(a^{n+1}) = \emptyset$ but in the 3-state monitor $\bot$ has not an incoming arc labeled by a.

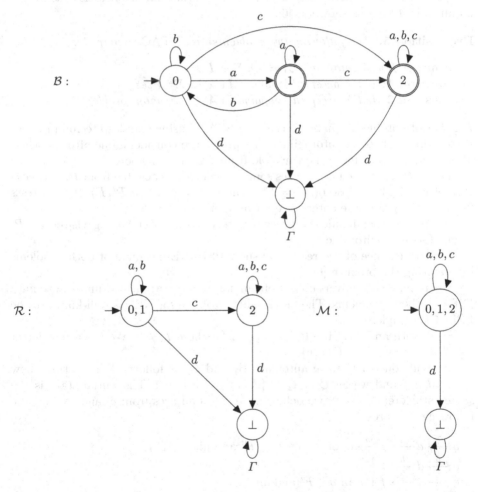

Fig. 1. Monitors $\mathcal{B}$, $\mathcal{R}$, $\mathcal{M}$ for $L = L(\mathcal{B})$.

2. Every monitor for the language $\Sigma^*(bab \cup b^3)\Sigma^\omega$ has at least 4 states. There are three monitors with 4 states which are pairwise non-isomorphic.
3. Let $L = (b^*a)^\omega \cup \{a, b\}^* c\{a, b, c\}^\omega \subseteq \Gamma^\omega$. Then L is monitorable and deterministic, but not codeterministic. Its minimal DBM has 4 states, but the congruential monitor $Q_\mathcal{R}$ has 3 states, only. We have $\overline{L} = \{a, b, c\}^\omega$ and $\overline{L}^{\mathrm{co}} = \Gamma^\omega$. Hence, the smallest standard monitor has two states. In particular, we have $|Q_\mathcal{B}| > |Q_\mathcal{R}| > |Q_\mathcal{M}|$, see also Fig. 1.

5   Deciding Liveness and Monitorability

5.1   Decidability for Büchi Automata

It is well-known that decidability of liveness (monitorability resp.) is PSPACE-complete for Büchi automata. The following result for liveness is classic, for monitorability it was shown in [6].

Proposition 5. *The following two problems are PSPACE-complete:*

- **Input:** *A Büchi automaton $\mathcal{A} = (Q, \Sigma, \delta, I, F)$.*
- **Question 1:** *Is the accepted language $L(\mathcal{A}) \subseteq \Sigma^\omega$ live?*
- **Question 2:** *Is the accepted language $L(\mathcal{A}) \subseteq \Sigma^\omega$ monitorable?*

Proof. Both problems can be checked in PSPACE using standard techniques. We sketch this part for monitorability. The procedure considers, one after another, all subsets P such that P is reachable from I by reading some input word. For each such P the procedure guesses some P' which is reachable from P. It checks that either $L(\mathcal{A}') = \emptyset$ or $L(\mathcal{A}') = \Sigma^\omega$, where $\mathcal{A}' = (Q, \Sigma, \delta, P', F)$. If both tests fail then the procedure enters a rejecting loop.

If, on the other hand, the procedure terminates after having visited all P, then $L(\mathcal{A})$ is monitorable.

For convenience of the reader we show PSPACE-hardness of both problems by adapting the proof in [6].

We reduce the universality problem for non-deterministic finite automata (NFA) to both problems. The universality problem for NFA is well-known to be PSPACE-complete.

Start with an NFA $\mathcal{A} = (Q', \Gamma, \delta', q_0, F')$ where $\Gamma \neq \emptyset$. We use a new letter $b \notin \Gamma$ and we let $\Sigma = \Gamma \cup \{b\}$.

We will construct Büchi automata $\mathcal{B}_1$ and $\mathcal{B}_2$ as follows. We use three new states d, e, f and we let $Q = Q' \cup \{d, e, f\}$, see Fig. 2. The initial state is the same as before: q_0. Next, we define δ. We keep all arcs from δ' and we add the following new arcs.

- $q \xrightarrow{b} d \xrightarrow{a} e \xrightarrow{a} e$ for all $q \in Q' \setminus F'$ and all $a \in \Gamma$.
- $e \xrightarrow{b} d \xrightarrow{b} d$
- $q \xrightarrow{b} f \xrightarrow{c} f$ for all $q \in F'$ and all $c \in \Sigma$.

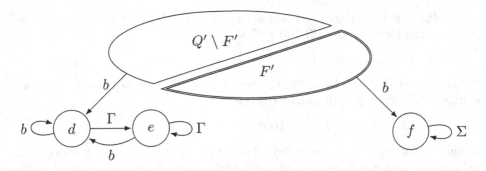

Fig. 2. PSPACE-hardness for liveness and monitorability for Büchi automata.

Let us define two final sets of states: $F_1 = \{f\}$ and $F_2 = \{d, f\}$. Thus, we have constructed Büchi automata $\mathcal{B}_1$ and $\mathcal{B}_2$ where

$$\mathcal{B}_i = (Q, \Gamma, \delta, q_0, F_i) \text{ for } i = 1, 2.$$

For the proof of the proposition it is enough to verify the following two claims which are actually more precise than needed.

1. The language $L(\mathcal{B}_1)$ is monitorable. It is live if and only if $L(\mathcal{A}) = \Gamma^*$.
2. The language $L(\mathcal{B}_2)$ is live. It is monitorable if and only if $L(\mathcal{A}) = \Gamma^*$.

If $L(\mathcal{A}) = \Gamma^*$, then we have $L(\mathcal{B}_1) = L(\mathcal{B}_2) = \Sigma^\omega$, so both languages are live and monitorable.

If $L(\mathcal{A}) \neq \Gamma^*$, then there exists some word $u \notin L(\mathcal{A})$ and hence reading ub we are necessarily in state d. It follows that $ub\Sigma^\omega \cap L(\mathcal{B}_1) = \emptyset$ and $L(\mathcal{B}_1)$ is not live. Still, $L(\mathcal{B}_1)$ is monitorable. Now, for all $w \in \Sigma^*$ we have $wb^\omega \in L(\mathcal{B}_2)$. Hence, $L(\mathcal{B}_2)$ is live. However, if $u \notin L(\mathcal{A})$, then after reading ub we are in state d. Now, choose some letter $a \in \Gamma$. For all $v \in \Sigma^*$ we have $ubva^\omega \notin L(\mathcal{B}_2)$, but $ubvb^\omega \in L(\mathcal{B}_2)$. Hence, if $L(\mathcal{A}) \neq \Gamma^*$, then $L(\mathcal{B}_2)$ is not monitorable.   □

5.2   Decidability for LTL

We use the standard syntax and semantics of the linear temporal logic LTL for infinite words over some finite nonempty alphabet Σ. We restrict ourselves the pure future fragment and the syntax of $\text{LTL}_\Sigma[\textsf{XU}]$ is given as follows.

$$\varphi ::= \top \mid a \mid \neg\varphi \mid \varphi \vee \varphi \mid \varphi \,\textsf{XU}\, \varphi,$$

where a ranges over Σ. The binary operator $\textsf{XU}$ is called the *next-until* modality.

In order to give the semantics we identify each $\varphi \in \text{LTL}_\Sigma$ with some first-order formula $\varphi(x)$ in at most one free variable. The identification is done as usual by structural induction. The formula a becomes $a(x) = P_a(x)$, where $P_a(x)$ is the unary predicate saying that the label of position x is the letter a. The formula "φ *neXt-Until* ψ" is defined by:

$$(\varphi \, \mathsf{XU} \, \psi)(x) = \exists z : \ (x < z \land \psi(z) \land \forall y : \ \varphi(y) \lor y \leq x \lor z \leq y).$$

Finally let $\alpha \in \Sigma^\omega$ be an infinite word with the first position 0, then we define $\alpha \models \varphi$ by $\alpha \models \varphi(0)$; and we define

$$L(\varphi) = \{\alpha \in \Sigma^\omega \mid \alpha \models \varphi\}.$$

Languages of type $L(\varphi)$ are called LTL *definable*, It is clear that every LTL definable language is first-order definable; and Kamp's famous theorem [9] states the converse. In particular, given $L(\varphi)$ there exists a BA $\mathcal{A}$ such that $L(\varphi) = L(\mathcal{A})$. There are examples where the size of the formula φ is exponentially smaller than the size of any corresponding BA $\mathcal{A}$.

For a survey on first-order definable languages we refer to [4]. By LTL decidability of a property $\mathcal{P}$ we mean that the input is a formula $\varphi \in \mathrm{LTL}_\Sigma$ and we ask whether property $\mathcal{P}$ holds for $L(\varphi)$. By Proposition 5 we obtain straightforwardly the following lower and upper bounds for the LTL decidability of monitorability and liveness.

Remark 3. The following two problems are PSPACE-hard and can be solved in EXPSPACE:

- **Input:** A formula $\varphi \in \mathrm{LTL}_\Sigma$.
- **Question 1:** Is the accepted language $L(\varphi) \subseteq \Sigma^\omega$ live?
- **Question 2:** Is the accepted language $L(\varphi) \subseteq \Sigma^\omega$ monitorable?

Remark 3 is far from satisfactory since there is huge gap between PSPACE-hardness and containment in EXPSPACE. Very unfortunately, we were not able to make the gap any smaller for monitorability. There was some belief in the literature that, at least, LTL liveness can be tested in PSPACE, see for example [17]. But surprisingly this last assertion is wrong: testing LTL liveness is EXPSPACE-complete!

Proposition 6. *Deciding* LTL *liveness is EXPSPACE-complete:*

- **Input:** *A formula* $\varphi \in \mathrm{LTL}_\Sigma$.
- **Question** *Is the accepted language* $L(\varphi) \subseteq \Sigma^\omega$ *live?*

EXPSPACE-completeness of liveness was proved by Muscholl and Walukiewicz in 2012, but never published. Independently, it was proved by Orna Kupferman and Gal Vardi in [11].

We give a proof of Proposition 6 in Sects. 5.3 and 5.4 below. We also point out why the proof technique fails to say anything about the hardness to decide monitorability. Our proof for Proposition 6 is generic. This means that we start with a Turing machine M which accepts a language $L(M) \subseteq \Gamma^*$ in EXPSPACE. We show that we can construct in polynomial time a formula $\varphi(w) \in \mathrm{LTL}_\Sigma$ such that

$$w \in L(M) \iff L(\varphi(w)) \subseteq \Sigma^\omega \text{is not live.}$$

5.3   Encoding EXPSPACE Computations

For the definition of Turing machines we use standard conventions, very closely
to the notation e.g. in [8]. Let $L = L(M)$ be accepted by a deterministic Turing
machine M, where M has set of states Q and the tape alphabet is Γ containing
a "blank" symbol B. We assume that for some fixed polynomial $p(n) \geq n+2$ the
machine M uses on an input word $w \in (\Gamma \setminus \{B\})^*$ of length n strictly less space
than $2^N - 2$, where $N = p(n)$. (It does not really matter that M is deterministic.)
Configurations are words from $\Gamma^*(Q \times \Gamma)\Gamma^*$ of length precisely 2^N, where the
head position corresponds to the symbol from $Q \times \Gamma$. For technical reasons we
will assume that the first and the last symbol in each configuration is B. Let
$A = \Gamma \cup (Q \times \Gamma)$.

If the input is nonempty word $w = a_1 \cdots a_n$ where the a_i are letters, then
the initial configuration is defined here as

$$C_0 = B(q_0, a_1)a_2 \cdots a_n \underbrace{BBBBB \cdots B}_{2^N - n - 1 \text{ times}}.$$

For $t \geq 0$ let C_t be configuration of M at time t during the computation
starting with the initial configuration C_0 on input w. We may assume that the
computation is successful if and only if there is some t such that a special symbol,
say q_f, appears in C_t. Thus, we can write each C_t as a word $C_t = a_{0,t} \cdots a_{m,t}$
with $m = 2^N - 1$; and we have $w \in L(M)$ if and only if there are some $i \geq 1$
and $t \geq 1$ such that $a_{i,t} = q_f$.

In order to check that a sequence $C_0, C_1, \ldots$ is a valid computation we may
assume that the Turing machine comes with a table $\Delta \subseteq A^4$ such that the
following formula holds:

$$\forall t > 0 \, \forall 1 \leq i < 2^N - 1 : (a_{i-1,t-1}, a_{i,t-1}, a_{i+1,t-1}, a_{i,t}) \in \Delta.$$

Without restriction we have $(B, B, B, B) \in \Delta$, because otherwise M would
accept only finitely many words.

We may express that we can reach a final configuration C_t by saying:

$$\exists t \geq 1 \, \exists 1 \leq i < 2^N : a_{i,t} = q_f.$$

As in many EXPSPACE-hardness proofs, for comparing successive configurations we need to switch to a slightly different encoding, by adding the tape
position after each symbol from A. To do so, we enlarge the alphabet A by
new symbols $0, 1, \$, \#, k_1, \ldots k_N$ which are not used in any C_t so far. Hence,
$\Sigma = A \cup \{0, 1, \$, \#, k_1, \ldots k_N\}$. We encode a position $0 \leq i < 2^N$ by using
its binary representation with exactly N bits. Thus, each i is written as a
word $\text{bin}(i) = b_1 \cdots b_N$ where each $b_p \in \{0, 1\}$. In particular, $\text{bin}(0) = 0^N$,
$\text{bin}(1) = 0^{N-1}1, \ldots, \text{bin}(2^N - 1) = 1^N$.

Henceforth, a configuration $C_t = a_{0,t} \cdots a_{m,t}$ with $m = 2^N - 1$ is encoded as
a word

$$c_t = a_{0,t} \, \text{bin}(0) \cdots a_{m,t} \, \text{bin}(m)\$.$$

Words of this form are called *stamps* in the following. Each stamp has length $2^N \cdot N + 1$. If a factor $\text{bin}(i)$ occurs, then either $i = m$ (i.e., $\text{bin}(i) = 1^N$) and the next letter is \$ or $i < m$ and the next letter is some letter from the original alphabet A followed by the word $\text{bin}(i+1)$.

Now we are ready to define a language $L = L(w)$ which has the property that L is not live if and only if $w \in L(M)$. We describe the words $\alpha \in \Sigma^\omega$ which belong to L as follows.

1. Assume that α does not start with a prefix of the form $c_0 \cdots c_\ell \#$, where c_0 corresponds to the initial configuration w.r.t. w, each c_t is a stamp and in the stamp c_ℓ the symbol q_f occurs. Then α belongs to L.
2. Assume now that α starts with a prefix $c_0 \cdots c_\ell \#$ as above. Then we let α belong to L if and only if the set of letters occurring infinitely often in α witness that the prefix $c_0 \cdots c_\ell$ of stamps is **not** a valid computation. Thus, we must point to some $t \geq 1$ and some position $1 \leq i < m$ such that $(a_{i-1,t-1}, a_{i,t-1}, a_{i+1,t-1}, a_{i,t}) \notin \Delta$. The position i is given as $\text{bin}(i) = b_1 \cdots b_N \in \{0,1\}^N$. The string $\text{bin}(i)$ defines a subset of Σ:

$$I(i) = \{k_p \in \{k_1, \ldots, k_N\} \mid b_p = 1\}.$$

The condition for α to be in L is that for some t the mistake from c_{t-1} to c_t is reported by $(a_{i-1,t-1}, a_{i,t-1}, a_{i+1,t-1}, a_{i,t}) \notin \Delta$ and the position i such that $I(i)$ equals the set of letters k_p which appear infinitely often in α. Note that since we excluded mistakes at positition $i = 0$ (because of the leftmost B), the set $I(i)$ is non-empty.

Lemma 1. *The language $L = L(w)$ is not live if and only if $w \in L(M)$.*

Proof. First, let $w \in L(M)$. Then we claim that L is not live. To see this let $u = c_0 \cdots c_\ell \#$, where the prefix $c_0 \cdots c_\ell$ is a valid accepting computation of M. There is no mistake in $c_0 \cdots c_\ell$. Thus we have $u\Sigma^\omega \cap L = \emptyset$, so indeed, L is not live.

Second, let $w \notin L(M)$. We claim that L is live. Consider any $u \in \Sigma^*$. Assume first that u does not start with a prefix of the form $c_0 \cdots c_\ell \#$, where c_0 corresponds to the initial configuration w.r.t. w, each c_t is a stamp and in the stamp c_ℓ the symbol q_f occurs. Then we have $u\Sigma^\omega \subseteq L$.

Otherwise, assume that $c_0 \cdots c_\ell \#$ is a prefix of u and that all c_t's are stamps, with c_0 initial and c_ℓ containing q_f. There must be some mistake in $c_0 \cdots c_\ell \#$, say for some i and t. Let $I(i)$ be as defined a above. As $i \geq 1$ we have $I(i) \neq \emptyset$. Therefore we let β be any infinite word where the set of letters appearing infinitely often is exactly the set $I(i)$. By definition of L we have $u\beta \in L$. Hence, L is live. $\square$

There are other ways to encode EXPSPACE computations which may serve to prove Proposition 6, see for example [11]. However, these proofs do not reveal any hardness for LTL monitorability. In particular, they do not reveal EXPSPACE or EXPTIME hardness. For our encoding this is explained in Remark 4.

Remark 4. Since are interested in EXPSPACE-hardness, we may assume that there infinitely many w with $w \notin L(M)$. Let n be large enough, say $n \geq 3$ and $w \notin L(M)$, then $(B, (q_0, a_1), a_2, q_f) \notin \Delta$, where $w = a_1 a_2 \cdots$ because otherwise $w \in L(M)$. Define c_1 just as the initial stamp c_0 with the only difference that the letter (q_0, a_1) is replaced by the symbol q_f. Let $u = c_0 c_1 \#$, then for every $v \in \Sigma^*$ we have that $uv(k_N)^\omega \in L$ (i.e., there is a mistake at position 1), but $uv(k_1 k_2 \cdots k_N)^\omega \cap L = \emptyset$ (i.e., there is no mistake at position $2^N - 1$) because $(B, B, B, B) \in \Delta$. Thus, L is not monitorable.

5.4 Proof of Proposition 6

LTL liveness is in EXPSPACE by Remark 3. The main ideas for the proof are in the previous subsection. We show that we can construct in polynomial time on input w some $\varphi \in \text{LTL}_\Sigma$ such that $L(\varphi) = L(w)$. This can be viewed as a standard exercise in LTL. The solution is a little bit tedious and leads to a formula of at most quadratic size in n. The final step in the proof is to apply Lemma 1. □

6 Conclusion and Outlook

In the paper we studied monitorable languages from the perspective of what is a "good monitor". In some sense we showed that there is no final answer yet, but monitorability is a field where various interesting questions remain to be answered.

Given an LTL formula for a monitorable property one can construct monitors of at most doubly exponential size; and there is some indication that this is the best we can hope for, see [3]. Still, we were not able to prove any hardness for LTL monitorability beyond PSPACE. This does not mean anything, but at least in theory, it could be that LTL monitorability cannot be tested in EXPTIME, but nevertheless it is not EXPTIME-hard.

There is also another possibility. Deciding monitorability might be easier than constructing a monitor. Remember that deciding monitorability means to test that the boundary is nowhere dense. However we have argued that a DBA for the boundary does not give necessarily any information about a possible monitor, see the discussion at the beginning of Sect. 3.2.

A more fundamental question is about the notion of monitorability. The definition is not robust in the sense that every language becomes monitorable simply by embedding the language into a larger alphabet. This is somewhat puzzling, so the question is whether a more robust and still useful notion of monitorability exist.

Finally, there is an interesting connection to learning. In spite of recent progress to learn general ω-regular languages by [1] it not known how to learn a DBA for deterministic ω-regular languages in polynomial time. The best result is still due to Maler and Pnueli in [14]. They show that it is possible to learn a DWA for a ω-regular language L in $G_\delta \cap F_\sigma$ in polynomial time. The queries to the oracle are membership question "$uv^\omega \in L$?" where u and v are finite words and the query whether a proposed DWA is correct. If not, the oracle provides a shortest counterexample of the form uv^ω.

Since a DWA serves also as a monitor we can learn a monitor the very same way, but beyond $G_\delta \cap F_\sigma$ it is not known that membership queries to L and queries whether a proposed monitor is correct suffice. As a first step one might try find out how to learn a deterministic Büchi monitor in case it exists. This is a natural class beyond $G_\delta \cap F_\sigma$ because canonical minimal DBA for these languages exist. Moreover, just as for DWA this minimal DBA is an DBM, too.

Another interesting branch of research is monitorability in a distributed setting. A step in this direction for infinite Mazurkiewicz traces was outlined in [6].

Acknowledgment. The work was done while the first author was visiting LaBRI in the framework of the IdEx Bordeaux Visiting Professors Programme in June 2015. The hospitality of LaBRI and their members is greatly acknowledged.

The authors thank Andreas Bauer who communicated to us (in June 2012) that the complexity of LTL-liveness should be regarded as open because published proofs stating PSPACE-completeness were not convincing. We also thank Ludwig Staiger, Gal Vardi, and Mikhail Volkov for helpful comments.

References

1. Angluin, D., Fisman, D.: Learning regular omega languages. In: Auer, P., Clark, A., Zeugmann, T., Zilles, S. (eds.) ALT 2014. LNCS, vol. 8776, pp. 125–139. Springer, Heidelberg (2014)
2. Basin, D., Klaedtke, F., Müller, S., Zălinescu, E.: Monitoring metric first-order temporal properties. J. ACM **62**, 15:1–15:45 (2015)
3. Bauer, A., Leucker, M., Schallhart, C.: Monitoring of real-time properties. In: Arun-Kumar, S., Garg, N. (eds.) FSTTCS 2006. LNCS, vol. 4337, pp. 260–272. Springer, Heidelberg (2006)
4. Diekert, V., Gastin, P.: First-order definable languages. In: Flum, J., Grädel, E., Wilke, Th. (eds.) Logic and Automata: History and Perspectives, Texts in Logic and Games, pp. 261–306. Amsterdam University Press (2008)
5. Diekert, V., Leucker, M.: Topology, monitorable properties and runtime verification. Theor. Comput. Sci. **537**, 29–41 (2014). Special Issue of ICTAC 2012
6. Diekert, V., Muscholl, A.: On distributed monitoring of asynchronous systems. In: Ong, L., de Queiroz, R. (eds.) WoLLIC 2012. LNCS, vol. 7456, pp. 70–84. Springer, Heidelberg (2012)
7. Gondi, K., Patel, Y., Sistla, A.P.: Monitoring the full range of ω-regular properties of stochastic systems. In: Jones, N.D., Müller-Olm, M. (eds.) VMCAI 2009. LNCS, vol. 5403, pp. 105–119. Springer, Heidelberg (2009)
8. Hopcroft, J.E., Ulman, J.D.: Introduction to Automata Theory, Languages and Computation. Addison-Wesley, Reading (1979)
9. Kamp, H.: Tense logic and the theory of linear order. Ph.D. thesis, University of California (1968)
10. Knuth, D., Morris, J.H., Pratt, V.: Fast pattern matching in strings. SIAM J. Comput. **6**, 323–350 (1977)

11. Kupferman, O., Vardi, G.: On relative and probabilistic finite counterabilty. In: Kreutzer, S. (ed.) Proceedings 24th EACSL Annual Conference on Computer Science Logic (CSL 2015). LIPIcs, vol. 41, Dagstuhl, Germany, pp. 175–192. Schloss Dagstuhl – Leibniz-Zentrum für Informatik (2015)

12. Landweber, L.H.: Decision problems for ω-automata. Math. Syst. Theory **3**(4), 376–384 (1969)

13. Leucker, M., Schallhart, C.: A brief account of runtime verification. J. Logic Algebraic Program. **78**(5), 293–303 (2009)

14. Maler, O., Pnueli, A.: On the learnability of infinitary regular sets. Inf. Comput. **118**, 316–326 (1995)

15. Martugin, P.V.: A series of slowly synchronizing automata with a zero state over a small alphabet. Inf. Comput. **206**, 1197–1203 (2008)

16. Matiyasevich, Y.: Real-time recognition of the inclusion relation. J. Sov. Math. **1**, 64–70 (1973). Translated from Zapiski Nauchnykh Seminarov Leningradskogo Otdeleniya Matematicheskogo Instituta im. V. A. Steklova Akademii Nauk SSSR, vol. 20, pp. 104–114 (1971)

17. Nitsche, U., Wolper, P.: Relative liveness and behavior abstraction (extended abstract). In: Burns, J.E., Attiya, H. (eds.) Proceedings of the Sixteenth Annual ACM Symposium on Principles of Distributed Computing (PODS 1997), Santa Barbara, California, USA, 21–24 August 1997, pp. 45–52. ACM (1997)

18. Pnueli, A., Zaks, A.: PSL model checking and run-time verification via testers. In: Misra, J., Nipkow, T., Sekerinski, E. (eds.) FM 2006. LNCS, vol. 4085, pp. 573–586. Springer, Heidelberg (2006)

19. Rystsov, I.: Reset words for commutative and solvable automata. Theor. Comput. Sci. **172**, 273–279 (1997)

20. Sistla, A.P., Žefran, M., Feng, Y.: Monitorability of stochastic dynamical systems. In: Gopalakrishnan, G., Qadeer, S. (eds.) CAV 2011. LNCS, vol. 6806, pp. 720–736. Springer, Heidelberg (2011)

21. Staiger, L.: Reguläre Nullmengen. Elektronische Informationsverarbeitung und Kybernetik **12**(6), 307–311 (1976)

22. Staiger, L.: Finite-state ω-languages. J. Comput. Syst. Sci. **27**, 434–448 (1983)

23. Staiger, L., Wagner, K.W.: Automatentheoretische und automatenfreie Charakterisierungen topologischer Klassen regulärer Folgenmengen. Elektronische Informationsverarbeitung und Kybernetik **10**, 379–392 (1974)

24. Tabakov, D., Rozier, K.Y., Vardi, M.Y.: Optimized temporal monitors for SystemC. Formal Methods Syst. Des. **41**, 236–268 (2012)

25. Thomas, W.: Automata on infinite objects. In: van Leeuwen, J. (ed.) Handbook of Theoretical Computer Science, chap. 4, pp. 133–191. Elsevier Science Publishers B.V., Amsterdam (1990)

26. Volkov, M.V.: Synchronizing automata and the Černý conjecture. In: Martín-Vide, C., Otto, F., Fernau, H. (eds.) LATA 2008. LNCS, vol. 5196, pp. 11–27. Springer, Heidelberg (2008)

27. Wagner, K.W.: On omega-regular sets. Inf. Control **43**, 123–177 (1979)

Formal Methods in Air Traffic Management: The Case of Unmanned Aircraft Systems (Invited Lecture)

César A. Muñoz[✉]

NASA Langley Research Center, Hampton, VA 23681-2199, USA
cesar.a.munoz@nasa.gov

Abstract. As the technological and operational capabilities of unmanned aircraft systems (UAS) continue to grow, so too does the need to introduce these systems into civil airspace. Unmanned Aircraft Systems Integration in the National Airspace System is a NASA research project that addresses the integration of civil UAS into non-segregated airspace operations. One of the major challenges of this integration is the lack of an on-board pilot to comply with the legal requirement that pilots see and avoid other aircraft. The need to provide an equivalent to this requirement for UAS has motivated the development of a *detect and avoid* (DAA) capability to provide the appropriate situational awareness and maneuver guidance in avoiding and remaining well clear of traffic aircraft. Formal methods has played a fundamental role in the development of this capability. This talk reports on the formal methods work conducted under NASA's Safe Autonomous System Operations project in support of the development of DAA for UAS. This work includes specification of low-level and high-level functional requirements, formal verification of algorithms, and rigorous validation of software implementations. The talk also discusses technical challenges in formal methods research in the context of the development and safety analysis of advanced air traffic management concepts.

Extended Abstract

The unmanned aircraft industry represents a potential source of significant increase in economic developments and safety capabilities. According to the 2013 economic report by the Association for Unmanned Vehicle Systems International (AUVSI) [6], the cumulative impact between 2015 and 2025 to the US economy resulting from the integration of Unmanned Aircraft Systems (UAS) into the National Airspace System (NAS) will be more than US $80 billions and will

---

This invited lecture reports on research conducted at NASA Langley Research Center at the Safety-Critical Avionics Systems Branch by several individuals including, in addition to the author, Anthony Narkawicz, George Hagen, Jason Upchurch, and Aaron Dutle.

M. Leucker et al. (Eds.): ICTAC 2015, LNCS 9399, pp. 58–62, 2015.
DOI: 10.1007/978-3-319-25150-9_4

generate more than 100 thousand jobs. The report identifies precision agriculture and public safety as the two main potential markets for UAS in the US.

As the availability and applications of UAS grow, these systems will inevitably become part of standard airspace operations. A fundamental challenge for the integration of UAS into the NAS is the lack of an on-board pilot to comply with the legal requirement identified in the US Code of Federal Regulations to see and avoid traffic aircraft. As a means of compliance with this legal requirement, the final report of the FAA-sponsored Sense and Avoid (SAA) Workshop [4] defines the concept of *sense and avoid* for remote pilots as "the capability of a UAS to remain well clear from and avoid collisions with other airborne traffic."

NASA's Unmanned Aircraft Systems Integration in the National Airspace System project aims to develop key capabilities to enable routine and safe access for public and civil use of UAS in non-segregated airspace operations. As part of this project, NASA has developed a *detect and avoid* (DAA) concept for UAS [1] that implements the sense and avoid concept outlined by the SAA Workshop. The NASA DAA concept defines a volume representing a well-clear boundary where aircraft inside this volume are considered to be in well-clear violation. This volume is intended to be large enough to avoid safety concerns for controllers and see-and-avoid pilots. It shall also be small enough to avoid disruptions to traffic flow. Formally, this volume is defined by a boolean predicate on the states of two aircraft, i.e., their position and velocity vectors at current time. The predicate states that two aircraft are *well clear* of each other if appropriate distance and time variables determined by the relative aircraft states remain outside a set of predefined threshold values. These distance and time variables are closely related to variables used in the Resolution Advisory (RA) logic of the Traffic Alerting and Collision Avoidance System (TCAS).

TCAS is a family of airborne devices that are designed to reduce the risk of mid-air collisions between aircraft equipped with operating transponders. TCAS II [16], the current generation of TCAS devices, is mandated in the US for aircraft with greater than 30 seats or a maximum takeoff weight greater than 33,000 pounds. Although it is not required, TCAS II is also installed on many turbine-powered general aviation aircraft. An important characteristic of the well-clear violation volume is that it conservatively extends the volume defined by TCAS II, i.e., for an appropriate choice of threshold values, the TCAS II RA volume is strictly contained within the well-clear violation volume [10]. Hence, aircraft are declared to be in a well-clear violation before an RA is issued. This relation between the well-clear violation volume and the TCAS II volume guarantees that software capabilities supporting the DAA concept safely interact well with standard collision avoidance systems for commercial aircraft.

The well-clear definition proposed by NASA satisfies several geometric and operational properties [11]. For example, it is *symmetric*, i.e., in a pair-wise scenario, both aircraft make the same determination of being well-clear or not. Furthermore, the well-clear violation volume is *locally convex*, i.e., in a non-maneuvering pair-wise scenario, there is at most one time interval in which the aircraft are not well clear. Symmetry and local convexity represent fundamental safety properties of the DAA concept. In particular, symmetry ensures that all aircraft are simultaneously aware of a well-clear violation. Local convexity states that in a non-maneuvering scenario, a predicted well-clear violation is continuously alerted until it disappears. Once the alert disappears, it does not reappear unless the aircraft change their trajectories.

The NASA DAA concept also includes self-separation and alerting algorithms intended to provide remote pilots appropriate situational awareness of proximity to other aircraft in the airspace. These algorithms are implemented in a software library called DAIDALUS (Detect & Avoid Alerting Logic for Unmanned Systems) [12]. DAIDALUS consists of algorithms for determining the current well-clear status between two aircraft and for predicting a well-clear violation within a lookahead time, assuming non-maneuvering trajectories. In the case of a predicted well-clear violation, DAIDALUS also provides an algorithm that computes the time interval of well-clear violation. Furthermore, DAIDALUS implements algorithms for computing prevention bands, assuming a simple kinematic trajectory model. Prevention bands are ranges of track, ground speed, and vertical speed maneuvers that are predicted to be in well-clear violation within a given lookahead time. These bands provide awareness information to remote pilots and assist them in avoiding certain areas in the airspace. When aircraft are not well clear, or when a well-clear violation is unavoidable, the DAIDALUS algorithms compute well-clear recovery bands. Recovery bands are ranges of horizontal and vertical maneuvers that assist pilots in regaining well-clear status within the minimum possible time. Recovery bands are designed so that they do not conflict with resolution advisory maneuvers generated by systems such as TCAS II. DAIDALUS implements two alternative alerting schemas. One schema is based on the prediction of well-clear violations for different sets of increasingly conservative threshold values. The second schema is based on the types of bands, which can be either preventive or corrective, computed for a single set of threshold values. A band is preventive if it does not include the current trajectory. Otherwise, it is corrective. Recovery bands, by definition, are always corrective. In general, both schemas yield alert levels that increase in severity as a potential pair-wise conflict scenario evolves. The DAIDALUS library is written in both C++ and Java and the code is available under NASA's Open Source Agreement. DAIDALUS is currently under consideration for inclusion as DAA reference implementation of the RTCA Special Committee 228 Minimum Operational Performance Standards (MOPS) for Unmanned Aircraft Systems.

Given the safety-critical nature of the UAS in the NAS project, formal methods research has been conducted under NASA's Safe Autonomous System Operations project in support of the development of the DAA concept for UAS. The use of formal methods includes a formal definition of the well-clear violation volume, formal proofs of its properties, formal specification and verification of all DAIDALUS algorithms, and the rigorous validation of the software implementation of DAIDALUS algorithms against their formal specifications. All formal specifications and proofs supporting this work are written and mechanically verified in the Prototype Verification System (PVS) [15]. The tool PVSio [8] is used to animate PVS functional specifications.

The application of formal methods to the safety analysis of air traffic management systems faces technical challenges common to complex cyber-physical systems (CPS). Chief among those challenges is the interaction of CPS with the physical environment that yields mathematical models with both continuous and discrete behaviors. Formally proving properties involving continuous mathematics, and in particular, non-linear arithmetic is a well-known problem in automated deduction. As part of this research effort, several automated decision and semi-decision procedures for dealing with different kinds of non-linear real arithmetic problems have been developed [2,7,9,13,14]. Most of these procedures are formally verified and are available as proof-producing automated strategies in the PVS theorem prover.

The formal verification of software implementations of a CPS is a major endeavor even when the algorithms that are implemented have been formally verified. First, there is a large semantic gap between modern programming languages and the functional notation used in formal tools such as PVS. However, the main difficulty arises from the fact that modern programming languages utilize floating point arithmetic while formal verification is usually performed over the real numbers. An idea for lifting functional correctness properties from algorithms that use real numbers to algorithms that use floating-point numbers is discussed in [5]. However, this research area is still in an early stage. In [3], a practical approach to the validation of numerical software is proposed. The approach, which is called *model animation*, compares computations performed in the software implementations against those symbolically evaluated to an arbitrary precision on the corresponding formal models. While model animation does not provide an absolute guarantee that the software is correct, it increases the confidence that the formal models are faithfully implemented in code. Model animation has been used to validate in a rigorous way the software implementation of DAIDALUS algorithms against their formal specifications.

Finally, air traffic management systems are unique in some aspects. For instance, these systems involve human and automated elements and these elements are often subject to strict operational (and sometimes legal) requirements. These requirements restrict the design space of operational concepts, such as detect and avoid for UAS. More importantly, new concepts and algorithms have to support an incremental evolution of the air space system at a global scale. All these requirements and restrictions may result in solutions that are non-optimal from a theoretical point of view or that have complex verification issues due to legacy systems.

References

1. Consiglio, M., Chamberlain, J., Muñoz, C., Hoffler, K.: Concept of integration for UAS operations in the NAS. In: Proceedings of 28th International Congress of the Aeronautical Sciences, ICAS 2012, Brisbane, Australia (2012)

2. Denman, W., Muñoz, C.: Automated real proving in PVS via MetiTarski. In: Jones, C., Pihlajasaari, P., Sun, J. (eds.) FM 2014. LNCS, vol. 8442, pp. 194–199. Springer, Heidelberg (2014)

3. Dutle, A.M., Muñoz, C.A., Narkawicz, A.J., Butler, R.W.: Software validation via model animation. In: Blanchette, J.C., Kosmatov, N. (eds.) TAP 2015. LNCS, vol. 9154, pp. 92–108. Springer, Heidelberg (2015)

4. FAA Sponsored Sense and Avoid Workshop. Sense and avoid (SAA) for Unmanned Aircraft Systems (UAS), October 2009

5. Goodloe, A.E., Muñoz, C., Kirchner, F., Correnson, L.: Verification of numerical programs: from real numbers to floating point numbers. In: Brat, G., Rungta, N., Venet, A. (eds.) NFM 2013. LNCS, vol. 7871, pp. 441–446. Springer, Heidelberg (2013)

6. Jenkins, D., Vasigh, B.: The economic impact of Unmanned Aircraft Systems integration in the United States. Economic report of the Association For Unmanned Vehicle Systems International (AUVSI), March 2013

7. Mariano Moscato, César Muñoz, and Andrew Smith. Affine arithmetic and applications to real-number proving. In: Urban, C., Zhang, X. (ed.), Proceedings of the 6th International Conference on Interactive Theorem Proving (ITP 2015), vol. 9236 of Lecture Notes in Computer Science, Nanjing, China, Springer, Heidelberg, August 2015

8. Muñoz, C.: Rapid prototyping in PVS. Contractor Report NASA/CR-2003-212418, NASA, Langley Research Center, Hampton VA 23681–2199, USA, May 2003

9. Muñoz, C., Narkawicz, A.: Formalization of a representation of Bernstein polynomials and applications to global optimization. J. Autom. Reason. **51**(2), 151–196 (2013)

10. Muñoz, C., Narkawicz, A., Chamberlain, J.: A TCAS-II resolution advisory detection algorithm. In: Proceedings of the AIAA Guidance Navigation, and Control Conference and Exhibit 2013, number AIAA-2013-4622, Boston, Massachusetts, August 2013

11. Muñoz, C., Narkawicz, A., Chamberlain, J., Consiglio, M., Upchurch, J.: A family of well-clear boundary models for the integration of UAS in the NAS. In: Proceedings of the 14th AIAA Aviation Technology, Integration, and Operations (ATIO) Conference, number AIAA-2014-2412, Georgia, Atlanta, USA, June 2014

12. Muñoz, C., Narkawicz, A., Consiglio, G.: DAIDALUS: detect and avoid alerting logic for unmanned systems. In: Proceedings of the 34th Digital Avionics Systems Conference (DASC 2015), Prague, Czech Republic, September 2015

13. Narkawicz, A., Muñoz, C.: A formally verified generic branching algorithm for global optimization. In: Cohen, E., Rybalchenko, A. (eds.) VSTTE 2013. LNCS, vol. 8164, pp. 326–343. Springer, Heidelberg (2014)

14. Narkawicz, A., Muñoz, C., Dutle, A.: Formally-verified decision procedures for univariate polynomial computation based on Sturm's and Tarski's theorems. J. Autom. Reason. **54**(4), 285–326 (2015)

15. Owre, S., Rushby, J., Shankar, N.: PVS: A prototype verification system. In: Kapur, D. (ed.) Automated Deduction–CADE-11. LNAI, vol. 607, pp. 748–752. Springer, Heidelberg (1992)

16. RTCA SC-147. RTCA-DO-185B, Minimum operational performance standards for traffic alert and collision avoidance system II (TCAS II), July 2009

The Proof Technique of Unique Solutions of Contractions

Davide Sangiorgi[✉]

Università di Bologna & INRIA, Bologna, Italy
davide.sangiorgi@gmaol.com

This extended abstract summarises work conducted with Adrien Durier and Daniel Hirschkoff (ENS Lyon), initially reported in [38].

Bisimilarity is employed to define behavioural equivalences and reason about them. Originated in concurrency theory, bisimilarity is now widely used also in other areas, as well as outside Computer Science. In this work, behavioural equivalences, hence also bisimilarity, are meant to be *weak* because they abstract from internal moves of terms, as opposed to the *strong* ones, which make no distinctions between the internal moves and the external ones (i.e., the interactions with the environment). Weak equivalences are, practically, the most relevant ones: e.g., two equal programs may produce the same result with different numbers of evaluation steps.

In proofs of bisimilarity results, the bisimulation proof method has become predominant, particularly with the enhancements of the method provided by the so called 'up-to techniques' [29]. Among these, one of the most powerful ones is 'up-to expansion and context', whereby the derivatives of two terms can be rewritten using expansion and bisimilarity and then a common context can be erased. Forms of 'bisimulations up-to context' have been shown to be effective in various fields, including process calculi [27,29,39], λ-calculi [16,18,19,40], and automata [7,34].

The landmark document for bisimilarity is Milner's CCS book [21]. In the book, Milner carefully explains that the bisimulation proof method is not supposed to be the only method for reasoning about bisimilarity. Indeed, various interesting examples in the book are handled using other techniques, notably *unique solution of equations*, whereby two tuples of processes are componentwise bisimilar if they are solutions of the same system of equations. This method is important in verification techniques and tools based on algebraic reasoning [2,32,33].

Milner's theorem that guarantees unique solutions [21] has however limitations: the equations must be 'guarded and sequential', that is, the variables of the equations may only be used underneath a visible prefix and preceded, in the syntax tree, only by the sum and prefix operators. This limits the expressiveness of the technique (since occurrences of other operators above the variables, such as parallel composition and restriction, in general cannot be removed), and its transport onto other languages (e.g., languages for distributed systems or higher-order languages, which usually do not include the sum operator).

The authors are partially supported by the ANR project 12IS02001 PACE.

M. Leucker et al. (Eds.): ICTAC 2015, LNCS 9399, pp. 63–68, 2015.
DOI: 10.1007/978-3-319-25150-9_5

We propose a refinement of Milner's technique in which equations are replaced by special inequations called *contractions*. Intuitively, for a behavioural equivalence $\asymp$, its contraction $\succeq_\asymp$ is a preorder in which $P \succeq_\asymp Q$ holds if $P \asymp Q$ and, in addition, Q has the *possibility* of being as efficient as P. That is, Q is capable of simulating P by performing less internal work. It is sufficient that Q has one 'efficient' path; Q could also have other paths, that are slower than any path in P. Uniqueness of the solution of a system of contractions is defined as with systems of equations: any two solutions must be equivalent with respect to $\asymp$. The difference with equations is in the meaning of solution: in the case of contractions the solution is evaluated with respect to the preorder $\succeq_\asymp$, rather than the equivalence $\asymp$.

If a system of equations has a unique solution, then the corresponding system of contractions, obtained by replacing the equation symbol with the contraction symbol, has a unique solution too. The converse however is false: it may be that only the system of contractions has a unique solution. More important, the condition that guarantees a unique solution in Milner's theorem about equations can be relaxed:'sequentiality' is not required, and 'guardedness' can be replaced by 'weak guardedness', that is, the variables of the contractions can be underneath *any* prefix, including a prefix representing internal work. (This is the same constraint in Milner's 'unique solution of equations' theorem for *strong* bisimilarity; the constraint is unsound for equations on weak bisimilarity.)

Milner's theorem is not complete for *pure* equations (equations in which recursion is only expressible through the variables of the equations, without using the recursion construct of the process language): there are bisimilar processes that cannot be solutions to the same system of guarded and sequential pure equations.In contrast, completeness holds for weakly-guarded pure contractions. The contraction technique is als *computationally* complete: any bisimulation $\mathcal{R}$ can be transformed into an equivalent system of weakly-guarded contractions that has the same size of $\mathcal{R}$ (where the size of a relation is the number of its pairs, and the size of a system of contractions is the number of its contractions). An analogous result also holds with respect to bisimulation enhancements such as 'bisimulation up-to expansion and context'.The contraction technique is in fact computationally equivalent to the 'bisimulation up-to contraction and context' technique — a refinement of 'bisimulation up-to expansion and context'.

The contraction technique can be generalised to languages whose syntax is the term algebra derived from some signature, and whose semantics is given as an LTS. In this generalisation the weak-guardedness condition for contractions becomes a requirement of *autonomy*, essentially saying that the processes that replace the variables of a contraction do not contribute to the initial action of the resulting expression. The technique can also be transported onto other equivalences, including contextually-defined equivalences such as barbed congruence, and non-coinductive equivalences such as contextual equivalence (i.e., may testing) and trace equivalence [9,10,24]. For each equivalence, one defines its contraction preorder by controlling the amount of internal work performed.

Further, a contraction preorder can be injected into the bisimulation game. That is, given an equivalence $\asymp$ and its contraction preorder $\succeq_\asymp$, one can define the technique of 'bisimulation up-to $\succeq_\asymp$ and context' whereby, in the bisimulation game, the derivatives of the two processes can be manipulated with $\succeq_\asymp$ and $\asymp$ (similarly to the manipulations that are possible in the standard 'bisimulation up-to expansion and context' using the expansion relation and bisimilarity) and a common context can then be erased. The resulting 'bisimulation up-to $\succeq_\asymp$ and context' is sound for $\asymp$. This technique allows us to derive results for $\asymp$ using the (enhanced) bisimulation proof method, thus transferring 'up-to context' forms of reasoning, originally proposed for labeled bisimilarities and their proof method, onto equivalences that are contextual or non-coinductive.

The contraction technique cannot however be transported onto all (weak) behavioural equivalences. For instance, it does not work in the setting of infinitary trace equivalence (whereby two processes are equal if they have the same finite and infinite traces) [10,11] and must testing [9]. A discussion on this point is deferred to the concluding section.

An example of application of contractions to a higher-order language, which exploits the autonomy condition, is also reported in [38]

Milner's theorem about unique solution of equations stems from an axiomatisation of bisimulation on finite-state processes [23]. Indeed, in axiomatisations of behavioural equivalences [2,21], the corresponding rule plays a key role and is called *fixed-point rule*, or *recursive specification principle*; see also [30], for trace equivalence. The possible shapes of the solutions of systems of equations, in connection with conditions on the guardedness of the equations, is studied by Baeten and Luttik [4].

Unique solution of equations has been considered in various settings, including languages, algebraic power series and pushdown automata (see the surveys [17,26]), as well as in coalgebras (e.g., [20]). These models, however, do not have the analogous of 'internal step', around which all the theory of contractions is built. In functional languages, unique solution of equations is sometimes called 'unique fixed-point induction principle'. See for instance [35], in which the conditions resembles Milner's conditions for CCS, and [15], which studies equations on streams advocating a condition based on the notion of 'contractive function' (the word 'contraction' here is unrelated to its use in our paper).

A tutorial on bisimulation enhancements is [29]. 'Up-to context' techniques have been formalised in a coalgebraic setting, and adapted to languages whose LTS semantics adheres to the GSOS format [5]; see for instance [6], which uses lambda-bialgebras, a generalisation of GSOS to the categorical framework.

Our transporting of the bisimulation proof method and some of its enhancements onto non-coinductive equivalences reminds us of techniques for reducing non-coinductive equivalences to bisimilarity. For instance, trace equivalence on nondeterministic processes can be reduced to bisimilarity on deterministic processes, following the powerset construction for automata [14]; a similar reduction can be made for testing equivalence [8]. These results rely on transformations of transitions systems, which modify the nondeterminism and the set of states, in such a way that a given equivalence on the original systems corresponds to bisimilarity on the altered systems. In contrast, in the techniques based on contractions the transformation of processes is performed dynamically, alongside the bisimulation game: two processes are manipulated only when necessary, i.e., when their immediate transitions would break the bisimulation game.

In CSP [12], some beautiful results have been obtained in which systems of equations have unique solutions provided their least fixed point (intuitively obtained by infinite unfolding of the equations) does not contain divergent states; see [32,33]. In CSP the semantics has usually a denotational flavour and, most important, the reference behavioural equivalence, failure equivalence, is divergent sensitive. We are currently trying to compare this kind of techniques, based on divergence, with those based on contractions. We just note here that unique solution of contractions holds in cases where the infinite unfolding of the contractions would introduce divergence.

As for the technique based on equations, so the technique based on contractions is meant to be used in combination with algebraic reasoning, on terms whose behaviour is not finite or finite-state: the recursion on the contraction variables captures the infinite behaviour of terms, and the proof that certain processes are solutions is carried out with pure algebraic reasoning. In comparison with equations, a drawback of unique solution of contractions for an equivalence $\asymp$ is that the solutions are not $\asymp$-interchangeable: it may be that P is solution and Q is not, even though $P \asymp Q$.

The proof of completeness of the 'unique solution of contractions' method with respect to the bisimulation proof method uses the sum operator to express the possible initial actions of a process. We are currently exploring how completeness could be recovered in languages in which the sum operator is missing.

We also plan to explore more in depth the contraction techniques in higher-order languages. Such study may shed light on the applicability of up-to context techniques to higher-order languages. In a higher-order language, while there are well-developed techniques for proving that a bisimulation is a congruence [28], up-to context is still poorly understood [16,18,19,27,40]. For instance, for pure λ-calculi and applicative bisimilarity, the soundness of the full up-to context technique (allowing one to remove any context, possibly binding variables of the enclosed terms) still represents an open problem.

Another setting in which up-to context techniques have been recently applied is that of language equivalence for automata, see e.g., [7,34]. Our techniques are however for languages with internal moves. In the case of automata, a τ-action could correspond to the empty word, which is absorbed in concatenations of words, in the same way as τ-actions are absorbed in concatenation of traces. Even taking into account the way the empty word (or the empty language) and τ-steps are used, the analogy seems light. It is unclear whether contractions could be useful on automata.

Our original motivation for studying contractions was to better understand 'up-to context' enhancements of the bisimulation proof method and their soundness. More broadly, the goal of the line of work reported is to improve our understanding of bisimilarity and the proof techniques for it, including the possibility of exporting the techniques onto other equivalences.

References

1. Arun-Kumar, S., Hennessy, M.: An efficiency preorder for processes. Acta Inform. **29**, 737–760 (1992)
2. Baeten, J.C.M., Basten, T., Reniers, M.A.: Process Algebra: Equational Theories of Communicating Processes. Cambridge University Press, Cambridge (2010)
3. Baeten, J.C.M., Bergstra, J.A., Klop, J.W.: Ready-trace semantics for concrete process algebra with the priority operator. Comput. J. **30**(6), 498–506 (1987)
4. Baeten, J.C.M., Luttik, B.: Unguardedness mostly means many solutions. Theor. Comput. Sci. **412**(28), 3090–3100 (2011)
5. Bloom, B., Istrail, S., Meyer, A.R.: Bisimulation can't be traced. J. ACM **42**(1), 232–268 (1995)
6. Bonchi, F., Petrisan, D., Pous, D., Rot, J.: Coinduction up to in a fibrational setting. In: Proceedings of CSL-LICS 2014. ACM (2014)
7. Bonchi, F., Pous, D.: Checking nfa equivalence with bisimulations up to congruence. In: Proceedings of POPL 2013, pp. 457–468. ACM (2013)
8. Cleaveland, R., Hennessy, M.: Testing equivalence as a bisimulation equivalence. Formal Asp. Comput. **5**(1), 1–20 (1993)
9. De Nicola, R., Hennessy, R.: Testing equivalences for processes. Theor. Comput. Sci. **34**, 83–133 (1984)
10. van Glabbeek, R.J.: The linear time–branching time spectrum II. In: Best, E. (ed.) CONCUR 1993. LNCS, vol. 715. Springer, Heidelberg (1993)
11. van Glabbeek, R.J.: The linear time–branching time spectrum I. In: Handbook of Process Algebra, pp. 3–99. Elsevier (2001)
12. Hoare, C.A.R.: Communicating Sequential Processes. Prentice Hall, Upper Saddle River (1985)
13. Honda, K., Yoshida, N.: On reduction-based process semantics. Theor. Comput. Sci. **152**(2), 437–486 (1995)
14. Hopcroft, J.E., Motwani, R., Ullman, J.D.: Introduction to Automata Theory, Languages, and Computation. Addison-Wesley, Boston (2006)
15. Hutton, G., Jaskelioff, M.: Representing Contractive Functions on Streams (2011)
16. Koutavas, V., Wand, M.: Small bisimulations for reasoning about higher-order imperative programs. In: Proceedings of POPL 2006, pp. 141–152. ACM (2006)
17. Kunc, M.: Simple language equations. Bull. EATCS **85**, 81–102 (2005)
18. Lassen, S.B.: Relational reasoning about contexts. In: Higher-order operational techniques in semantics, pp. 91–135. Cambridge University Press (1998)
19. Lassen, S.B.: Bisimulation in untyped lambda calculus: Böhm trees and bisimulation up to context. Electr. Notes Theor. Comput. Sci. **20**, 346–374 (1999)
20. Milius, S., Moss, L.S., Schwencke, D.: Abstract GSOS rules and a modular treatment of recursive definitions. Log. Methods Comput. Sci. 9(3) (2013)
21. Milner, R.: Communication and Concurrency. Prentice Hall, Upper Saddle River (1989)
22. Milner, R., Sangiorgi, D.: Barbed bisimulation. In: Kuich, W. (ed.) ICALP 1992. LNCS, vol. 623. Springer, Heidelberg (1992)
23. Milner, R.: A complete axiomatisation for observational congruence of finite-state behaviors. Inf. Comput. **81**(2), 227–247 (1989)
24. Morris, J.H.: Lambda-Calculus Models of Programming Languages. Ph.D. thesis MAC-TR-57, M.I.T., project MAC, December 1968
25. Natarajan, V., Cleaveland, R.: Divergence and fair testing. In: Fülöp, Z. (ed.) ICALP 1995. LNCS, vol. 944, pp. 648–659. Springer, Heidelberg (1995)

26. Petre, I., Salomaa, A.: Algebraic systems and pushdown automata. In: Droste, M., Kuich, W., Vogler, H. (eds.) Handbook of Weighted Automata. EATCS Series, pp. 257–289. Springer, Heidelberg (2009)

27. Piérard, A., Sumii, E.: Sound bisimulations for higher-order distributed process calculus. In: Hofmann, M. (ed.) FOSSACS 2011. LNCS, vol. 6604, pp. 123–137. Springer, Heidelberg (2011)

28. Pitts, A.: Howe's method. In: Advanced Topics in Bisimulation and Coinduction. ambridge University Press (2012)

29. Pous, D., Sangiorgi, D.: Enhancements of the bisimulation proof method. In: Advanced Topics in Bisimulation and Coinduction. Cambridge University Press (2012)

30. Rabinovich, A.M.: A complete axiomatisation for trace congruence of finite state behaviors. In: Main, M.G., Melton, A.C., Mislove, M.W., Schmidt, D., Brookes, S.D. (eds.) MFPS 1993. LNCS, vol. 802, pp. 530–543. Springer, Heidelberg (1994)

31. Rensink, A., Volger, W.: Fair testing. Inf. Comput. **205**, 125–198 (2007)

32. Roscoe, A.W.: The Theory and Practice of Concurrency. Prentice Hall, Upper Saddle River (1998)

33. Roscoe, A.W.: Understanding Concurrent Systems. Springer, London (2010)

34. Rot, J., Bonsangue, M., Rutten, J.: Coinductive proof techniques for language equivalence. In: Dediu, A.-H., Martín-Vide, C., Truthe, B. (eds.) LATA 2013. LNCS, vol. 7810, pp. 480–492. Springer, Heidelberg (2013)

35. Sands, D.: Computing with contexts: a simple approach. ENTCS **10**, 134–149 (1998)

36. Sangiorgi, D., Milner, R.: The problem of "weak bisimulation up to". In: Cleaveland, W.R. (ed.) CONCUR 1992. LNCS, vol. 630, pp. 32–46. Springer, Heidelberg (1992)

37. Sangiorgi, D.: Locality and true-concurrency in calculi for mobile processes. In: Hagiya, M., Mitchell, J.C. (eds.) TACS 1994. LNCS, vol. 789, pp. 405–424. Springer, Heidelberg (1994)

38. Sangiorgi, D.: Equations, contractions, and unique solutions. In: Proceedings of POPL 15, pp. 421–432. ACM (2015)

39. Sangiorgi, D., Walker, D.: The π-calculus: A Theory of Mobile Processes. Cambridge University Press, Upper Saddle River (2001)

40. Sangiorgi, D., Kobayashi, N., Sumii, E.: Environmental bisimulations for higher-order languages. ACM Trans. Program. Lang. Syst. **33**(1), 5 (2011)

On Probabilistic Distributed Strategies

Glynn Winskel[✉]

Computer Laboratory, University of Cambridge, Cambridge, UK
gw104@cam.ac.uk

Abstract. In a distributed game we imagine a team Player engaging a team Opponent in a distributed fashion. No longer can we assume that moves of Player and Opponent alternate. Rather the history of a play more naturally takes the form of a partial order of dependency between occurrences of moves. How are we to define strategies within such a game, and how are we to adjoin probability to such a broad class of strategies? The answer yields a surprisingly rich language of probabilistic distributed strategies and the possibility of programming (optimal) probabilistic strategies. Along the way we shall encounter solutions to: the need to mix probability and nondeterminism; the problem of parallel causes in which members of the same team can race to make the same move, and why this leads us to invent a new model for the semantics of distributed systems.

1 Introduction

I am working on a theory of distributed games and strategies. The games are distributed in the sense that they involve a team Player in competition with a team Opponent in widely-spread, possibly varying locations. It is no longer sensible to regard the history of the play as a sequence of alternating moves, the case in traditional games. Rather at a reasonable level of abstraction it is sensible to view a history as a partial order showing the dependency of moves on earlier moves. Of course the terms Player and Opponent are open to a variety of interpretations so the intended application areas are very broad.

My own original motivation comes from the wish to generalise domain theory as a basis for denotational semantics. While domain theory provides a beautiful paradigm for formalising and analysing computation it has been increasingly falling short in the burgeoning world of distributed, nondeterministic and probabilistic computation we live in today. In brief, with the wisdom of hindsight, domain theory abstracted from operational concerns too early. So one aim is to repair the "little divide" between operational and denotational semantics. There is also some hope that the common vocabulary and techniques games provide will help bridge the "big divide" in theoretical computer science between the fields of semantics and algorithmics.

One could summarise the enterprise as redoing traditional of games and strategies as a theory based on histories as partial orders of moves. However, the move from sequences to partial orders brings in its wake a lot of technical

© Springer International Publishing Switzerland 2015
M. Leucker et al. (Eds.): ICTAC 2015, LNCS 9399, pp. 69–88, 2015.
DOI: 10.1007/978-3-319-25150-9_6

difficulty and potential for undue complexity unless it's done artfully. Here we have been in a good position to take advantage techniques from the early 1980's on a mathematical foundation for work of Hoare and Milner on synchronising processes in categories of models for concurrency [1] and in particular the model and techniques of event structures—an analogue of trees where branches have the form of partial orders [2,3]. The work on distributed strategies described here could have been done then.

One surprise has been how adeptly distributed strategies fit with probability, at least once a general enough definition of probabilistic event structures was discovered and characterised. It was certainly an advantage to have started with *nondeterministic* strategies [4]. But as we shall see in the move from nondeterministic to probabilistic strategies new phenomena and an unexpected limitation appear.

It has become clear recently that there is a built-in limitation in basing strategies on traditional event structures. Sometimes a distributed strategy can rely on certain "benign races" where, intuitively, several members of team Player may race each other to make a common move. If we are to support benign races in strategies there is a need to work with mathematical structures which support parallel causes—in which an event can be enabled in several compatible ways. This extension seems not to be needed for nondeterministic strategies. It was only revealed in the extension to probabilistic strategies when it was realised that certain intuitively natural probabilistic strategies could not be expressed, with the event structures we were working with. Why, will be explained later.

Though event structures allowing parallel causes have been studied existing structures do not support an operation of hiding central to the composition of strategies. So to some extent we have had go back to the drawing board and invent appropriate structures to support parallel causes and simultaneously a hiding operation. We now know ways to do this. Fortunately the new structures are not so removed from traditional event structures. They involve the objectification of cause, so that one can express *e.g.* that one cause is in parallel with another or in conflict with another, and assign probabilities to causes—see the final section which sketches recent work with Marc de Visme.

2   Event Structures [3]

The behaviour of distributed games is based on event structures, rather than trees. Instead of regarding a play in a game as a sequence of Player and Opponent moves it is given the structure of a partial order of occurrences of moves.

Event structures describe a process, or system, in terms of its possible event occurrences, their causal dependency and consistency. Just as it can be helpful to understand the behaviour of a state-transition diagram in terms of its unfolding to a tree, more detailed models, such as Petri nets, which make explicit the local nature of events and their changes on state, unfold to an event structure [5]. In this sense event structures are a concurrent, or distributed, analogue of trees; though in an event structure the individual 'branches' are no longer necessarily sequences but have the shape of a partial order of events.

An *event structure* comprises $(E, \leq, \mathrm{Con})$, consisting of a set E of *events* (really event occurrences) which are partially ordered by $\leq$, the *causal dependency relation*, and a nonempty *consistency relation* Con consisting of finite subsets of E. The relation $e' \leq e$ expresses that event e causally depends on the previous occurrence of event e'. That a finite subset of events is consistent conveys that its events can occur together by some stage in the evolution of the process. Together the relations satisfy several axioms:

$$\{e' \mid e' \leq e\} \text{ is finite for all } e \in E,$$
$$\{e\} \in \mathrm{Con} \text{ for all } e \in E,$$
$$Y \subseteq X \in \mathrm{Con} \text{ implies } Y \in \mathrm{Con}, \quad \text{and}$$
$$X \in \mathrm{Con} \ \& \ e \leq e' \in X \text{ implies } X \cup \{e\} \in \mathrm{Con}.$$

The first axiom says that an event causally depends on only a finite number of events, the second that there are no redundant events, which are in themselves inconsistent. The third axiom expresses the reasonable property that a subset of consistent events is consistent, while the final axiom entails that the $\leq$-down-closure of any consistent set of events is also consistent. Two events e, e' are considered to be *concurrent* if the set $\{e, e'\}$ is in Con and neither event is causally dependent on the other.

It is sometimes convenient to draw event structures. For example,

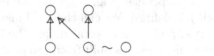

illustrates an event structure consisting of five events where, in particular, the top event on the left causally depends on the previous occurrences of two concurrent events—the arrows express the causal dependency—one of which is inconsistent with the event on the far right—we have indicated the inconsistency between the two events by a wiggly line.

Given this understanding of an event structure, there is an accompanying notion of state, or history, those events that may occur up to some stage in the behaviour of the process described. A *configuration* is a, possibly infinite, set of events $x \subseteq E$ which is both consistent and down-closed w.r.t. causal dependency:

Consistent: $X \subseteq x$ and X is finite implies $X \in \mathrm{Con}$, and
Down-closed: $e' \leq e \in x$ implies $e' \in x$.

An individual configuration inherits a partial order from the ambient event structure, and represents one possible partial-order history.

It will be very useful to relate event structures by maps. A *map* of event structures $f : E \to E'$ is a partial function f from E to E' such that the image of a configuration x is a configuration fx and any event of fx arises as the image of a unique event of x. In particular, when f is a total function it restricts to a bijection $x \cong fx$ between any configuration x and its image fx.

A map $f : E \to E'$ preserves concurrency: if two events in E are concurrent, then their images if defined are also concurrent. The map also reflects causal dependency locally, in the sense that if e, e' are events in a configuration x of E for which $f(e') \le f(e)$ in E', then $e' \le e$ also in E; the event structure E inherits causal dependencies from the event structure E' via the map f. In general a map of event structures need not preserve causal dependency; when it does we say it is *rigid*.

In describing distributed games and strategies we shall rely on two properties of maps. Firstly, any map of event structures $f : E \to E'$ factors into the composition of a partial map of event structures followed by a total map of event structures

$$E \xrightarrow{p} E_0 \xrightarrow{t} E'$$

in such a way that for any other factorisation $E \xrightarrow{p_1} E_1 \xrightarrow{t_1} E'$ with p_1 partial and t_1 total, there is a unique (necessarily total) map $h : E_0 \to E_1$ such that

$$
\begin{array}{ccc}
 & E_1 & \\
\begin{smallmatrix}p_1\end{smallmatrix}\nearrow & \uparrow h & \nwarrow \begin{smallmatrix}t_1\end{smallmatrix} \\
E \xrightarrow{p} & E_0 & \xrightarrow{t} E'
\end{array}
$$

commutes. The event structure E_0 is obtained as the "projection," or restriction, of the relations of causal dependency and consistency of the event structure E to the events on which f is defined. We call the total map t *the defined part* of f.

Secondly we shall use pullbacks of total maps. Pullbacks are an important construction in representing a process built from two processes sharing a common interface. Maps $f : A \to C$ and $g : B \to C$ always have pullbacks in the category of event structures, but they are more simple to describe in the case where f and g are total, and this is all we shall need:

$$
\begin{array}{ccc}
 & A & \\
\begin{smallmatrix}\pi_1\end{smallmatrix}\nearrow & & \searrow \begin{smallmatrix}f\end{smallmatrix} \\
P \; > & & C \\
\begin{smallmatrix}\pi_2\end{smallmatrix}\searrow & & \nearrow \begin{smallmatrix}g\end{smallmatrix} \\
 & B &
\end{array}
$$

Roughly, configurations of the pullback P are matches between configurations of A and B which satisfy the causal constraints of both. Precisely, finite configurations of P correspond to composite bijections

$$\theta : x \cong fx = gy \cong y$$

between finite configurations x of A and y of B such that $fx = gy$, for which the transitive relation generated on θ by $(a, b) \le (a', b')$ if $a \le_A a'$ or $b \le_B b'$ has no non-trivial causal loops, and so forms a partial order.

3   Distributed Games and Strategies—the Definitions [4]

Often the behaviour of a game is represented by a tree in which the arcs correspond to occurrences of moves by Player or Opponent. Instead we can represent the behaviour of a distributed game more accurately by an event structure together with a polarity function from its events to $+$ or $-$ to signify whether they are move occurrences of Player or Opponent, respectively.

A game might generally have winning conditions, a subset of configurations at which Player is deemed to have won, or more generally a payoff function from configurations to the reals.

There are two fundamentally important operations on two-party games. One is that of forming the dual game in which the moves of Player and Opponent are reversed. On an event structure with polarity A this amounts to reversing the polarities of events to produce the dual $A^\perp$. By a strategy in a game we will mean a strategy for Player. A strategy for Opponent, or a counter-strategy, in a game A will be identified with a strategy in $A^\perp$. The other operation is a simple parallel composition of games, achieved on event structures with polarity A and B very directly by simply juxtaposing them, ensuring a finite subset of events is consistent if its overlaps with the two games are individually consistent, to form $A\|B$.

As an example of a strategy in a game consider the *copy-cat* strategy for a game A. This is a strategy in the game $A^\perp\|A$ which, following the spirit of a copy-cat, has Player moves copy the corresponding Opponent moves in the other component. In more detail, the copy-cat strategy $\mathbb{CC}_A$ is obtained by adding extra causal dependencies to $A^\perp\|A$ so that any Player move in either component causally depends on its copy, an Opponent move, in the other component. It can be checked that this generates a partial order of causal dependency. A finite set is taken to be consistent if its down-closure w.r.t. the order generated is consistent in $A^\perp\|A$. We illustrate the construction on the simple game comprising a Player move causally dependent on a single Opponent move:

$$
\begin{array}{ccc}
 & \ominus\; \text{-}\text{-}\text{-}\text{-}\!\twoheadrightarrow\oplus & \\
 & \uparrow\qquad\qquad\uparrow & \\
A^\perp & \quad\mathbb{CC}_A\quad & A \\
 & \downarrow\qquad\qquad\downarrow & \\
 & \oplus\!\twoheadleftarrow\; \text{-}\text{-}\text{-}\ominus &
\end{array}
$$

In characterising the configurations of the copy-cat strategy an important partial order on configurations is revealed. Clearly configurations of a game A are ordered by inclusion $\subseteq$. For configurations x and y, write $x \subseteq^- y$ and $x \subseteq^+ y$ when all the additional events of the inclusion are purely Opponent, respectively, Player moves. A configuration x of $\mathbb{CC}_A$ is also a configuration of $A^\perp\|A$ and as such splits into two configurations x_1 on the left and x_2 on the right. The extra causal constraints of copy-cat ensure that the configurations of $\mathbb{CC}_A$ are precisely those configurations of $A^\perp\|A$ for which it holds that

$$x_2 \sqsubseteq_A x_1, \text{ defined as } x_2 \supseteq^- y \subseteq^+ x_1,$$

for some configuration y (necessarily $x_1 \cap x_2$). The relation $\sqsubseteq_A$ is in fact a partial order on configurations. Increasing in the order $\sqsubseteq_A$ involves losing Opponent moves and gaining Player moves. Because it generalises the pointwise order of domain theory, initiated by Dana Scott, we call the order $\sqsubseteq_A$ the *Scott order*.

Strategies in a game are not always obtained by simply adding extra causal dependencies to the game. For example, consider the game comprising two Opponent moves in parallel with a Player move and the strategy (for Player) in which Player make their move if Opponent makes one of theirs. Here the strategy is represented by

We are forced to split the Player move of the game into two moves, each dependent on different Opponent moves, and mutually inconsistent indicated by the wiggly line. For reasons such as this we are led to separate the actual moves of the strategy into an event structure with polarity S and, in order to track how actual moves correspond to moves in the game, formalise a strategy in a game A as a total map of event structures

$$\sigma : S \to A$$

which preserves polarity. (We have met a very special case of this in the copy-cat strategy where the role of S is taken by $\mathbb{CC}_A$ and σ acts as the identity function on events.) The event structure S describes the possibly nondeterministic plays of the strategy. Automatically a state of play of the strategy, represented by a configuration x of S, determines a position of the game, a configuration σx of A. Directly from the fact that σ is a map, we know that any move in σx is due to the play of a unique move in x. The total map $\sigma : S \to A$ really just expresses that S represents a nondeterministic play in the game A. More is expected of a strategy. For example, consider the game consisting of a Player move concurrent with a move of Opponent and the two total maps indicated:

(i) $\quad S \quad \ominus \dashrightarrow \oplus$
$\qquad \sigma{\downarrow} \quad {\vdots} \quad {\vdots}$
$\qquad A \quad \ominus \qquad \oplus$

(ii) $\quad S \quad \oplus \dashrightarrow \ominus$
$\qquad \sigma{\downarrow} \quad {\vdots} \quad {\vdots}$
$\qquad A \quad \oplus \qquad \ominus$

The first map (i) seems reasonable as a strategy; Player awaits the move of Opponent and then makes a move. However, the second map (ii) seems dubious; Player forces Opponent to wait until they have made their move, inappropriate in a distributed strategy.

Instead of guessing, we seek a principled way to determine what further properties a strategy should satisfy. In fact, the further conditions we shall impose on strategies will be precisely those needed to ensure that the copy-cat strategy behaves as an identity w.r.t. the composition of strategies.[1] To do so we adapt an important idea of Conway followed up by Joyal, explaining how to extend the notion of strategy *in* a game to that of a strategy *between* games [6,7]. The operations of dual and simple parallel composition of games are the key.

A strategy *from* a game A *to* a game B is a strategy in the compound game $A^\perp \| B$. In particular, copy-cat of a game A is now seen as a strategy from A to A.

In composing two strategies one σ in $A^\perp \| B$ and another τ in $B^\perp \| C$ one firstly instantiates the Opponent moves in component B by Player moves in $B^\perp$ and *vice versa*, and then secondly hides the resulting internal moves over B. The first step is achieved efficiently via pullback. Temporarily ignoring polarities, the pullback

$$
\begin{array}{ccc}
 & A \| T & \\
{\scriptstyle \pi_2} \nearrow & & \searrow {\scriptstyle A\|\tau} \\
T \circledast S > & & A \| B \| C \\
{\scriptstyle \pi_1} \searrow & & \nearrow {\scriptstyle \sigma\|C} \\
 & S \| C &
\end{array}
$$

"synchronises" matching moves of S and T over the game B. But we require a strategy over the game $A^\perp \| C$ and the pullback $T \circledast S$ has internal moves over the game B. We achieve this via the projection of $T \circledast S$ to its moves over A and C. We make use of the partial map from $A\|B\|C$ to $A\|C$ which acts as the identity function on A and C and is undefined on B. The composite partial map

$$
\begin{array}{ccc}
 & A \| T & \\
{\scriptstyle \pi_2} \nearrow & & \searrow {\scriptstyle A\|\tau} \\
T \circledast S > & & A \| B \| C \rightharpoonup A \| C \\
{\scriptstyle \pi_1} \searrow & & \nearrow {\scriptstyle \sigma\|C} \\
 & S \| C &
\end{array}
$$

has defined part, yielding the composition

$$
\tau \odot \sigma : T \odot S \to A^\perp \| C
$$

once we reinstate polarities. The composition of strategies $\tau \odot \sigma$ is a form of synchronised composition of processes followed by the hiding of internal moves, a view promulgated by Abramsky within traditional game semantics of programs.

Two further conditions, *receptivity* and *innocence*, are demanded of strategies. The conditions are necessary and sufficient to ensure that copy-cat strategies behave as identities w.r.t. composition [4]. Receptivity expresses that any Opponent move allowed from a reachable position of the game is present as a move in the strategy. In more detail, $\sigma : S \to A$ is receptive when for any configurations x of S if σx extends purely by Opponent events to a configuration y

---

[1] We consider two strategies $\sigma : S \to A$ and $\sigma' : S' \to A$ to be essentially the same if there is an isomorphism $f : S \cong S'$ of event structures respecting polarity such that $\sigma = \sigma' f$.

then there is a unique extension of x to a configuration x' of S such that $\sigma x' = y$. Innocence says a strategy can only adjoin new causal dependencies of the form $\ominus \rightarrowtail \oplus$, where Player awaits moves of Opponent, beyond those already inherited from the game.

The literature is often concerned with deterministic strategies, in which Player has at most one consistent response to Opponent. We can broaden the concept of deterministic strategy to distributed strategies by taking such a strategy to be *deterministic* if consistent moves of Opponent entail consistent moves of Player—see [4,8]. Formally, we say an event structure with polarity is *deterministic* if any finite down-closed subset is consistent when its Opponent events form a consistent subset. In general the copy-cat strategy for a game need not be deterministic. Copy-cat is however deterministic precisely for games which are *race-free*, *i.e.* such that at any configuration, if both a move of Player and a move of Opponent are possible then they may occur together: if whenever x, $x \cup \{\oplus\}$ and $x \cup \{\ominus\}$ are configurations of A, where the events $\oplus$ and $\ominus$ have the opposing polarities indicated, then $x \cup \{\oplus, \ominus\}$ is a configuration. Deterministic distributed strategies coincide with the *receptive* ingenuous strategies of Melliès and Mimram [9].

Just as strategies generalise relations, deterministic strategies generalise functions. In fact, multirelations and functions are recovered as strategies, respectively deterministic strategies, in the special case where the games are composed solely of Player moves with trivial causal dependency and where only the empty set and singletons are consistent.

As would be hoped the concepts of strategy and deterministic strategy espoused here reduce to the expected traditional notions on traditional games. There have also been pleasant surprises. In the extreme case where games comprise purely Player moves, strategies correspond precisely to the 'stable spans' used in giving semantics to nondeterministic dataflow [10], and in the deterministic subcase one recovers exactly the *stable domain theory* of Gérard Berry [11].

We now turn to how a strategy might be made probabilistic. We first address an appropriately general way to adjoin probability to event structures.

4   Probabilistic Event Structures [12]

The extension of distributed strategies to probabilistic strategies required a new general definition of probabilistic event structure. A probabilistic event structure essentially comprises an event structure together with a continuous valuation on the Scott-open sets of its domain of configurations.[2] The continuous valuation

---

[2] A *Scott-open* subset of configurations is upwards-closed w.r.t. inclusion and such that if it contains the union of a directed subset S of configurations then it contains an element of S. A *continuous valuation* is a function w from the Scott-open subsets of $\mathcal{C}^\infty(E)$ to $[0,1]$ which is *(normalized)* $w(\mathcal{C}^\infty(E)) = 1$; *(strict)* $w(\varnothing) = 0$; *(monotone)* $U \subseteq V \implies w(U) \le w(V)$; *(modular)* $w(U \cup V) + w(U \cap V) = w(U) + w(V)$; and *(continuous)* $w(\bigcup_{i \in I} U_i) = \sup_{i \in I} w(U_i)$, for *directed* unions. The idea: $w(U)$ is the probability of a result in open set U.

assigns a probability to each open set and can then be extended to a probability measure on the Borel sets [13]. However open sets are several levels removed from the events of an event structure, and an equivalent but more workable definition is obtained by considering the probabilities of basic open sets, generated by single finite configurations; for each finite configuration this specifies the probability of obtaining a result which extends the finite configuration. Such valuations on configurations determine the continuous valuations from which they arise, and can be characterised through the device of "drop functions" which measure the drop in probability across certain generalised intervals. The characterisation yields a workable general definition of probabilistic event structure as event structures with *configuration valuations*, viz. functions from finite configurations to the unit interval for which the drop functions are always nonnegative.

In detail, a *probabilistic event structure* comprises an event structure E with a *configuration valuation*, a function v from the finite configurations of E to the unit interval which is

(normalized) $v(\varnothing) = 1$ and has
(non−ve drop) $d_v[y; x_1, \cdots, x_n] \geq 0$ when $y \subseteq x_1, \cdots, x_n$ for finite configurations $y, x_1, \cdots, x_n$ of E,

where the "drop" across the generalized interval starting at y and ending at one of the $x_1, \cdots, x_n$ is given by

$$d_v[y; x_1, \cdots, x_n] =_{\text{def}} v(y) - \sum_I (-1)^{|I|+1} v(\bigcup_{i \in I} x_i)$$

—the index I ranges over nonempty $I \subseteq \{1, \cdots, n\}$ such that the union $\bigcup_{i \in I} x_i$ is a configuration. The "drop" $d_v[y; x_1, \cdots, x_n]$ gives the probability of the result being a configuration which includes the configuration y and does not include any of the configurations $x_1, \cdots, x_n$.[3]

5   Probabilistic Strategies [15]

The above has prepared the ground for a general definition of distributed probabilistic strategies, based on event structures. One hurdle is that in a strategy it is impossible to know the probabilities assigned by Opponent. We need to address the problem—notorious in domain theory—of how to mix probability (which Player attributes to their moves) and nondeterminism (ensuing from Player's ignorance of the probabilities assigned to Opponent moves). A probabilistic strategy in a game A, presented as a race-free event structure with polarity, is a strategy $\sigma : S \to A$ in which we endow S with probability, while taking

---

[3] Samy Abbes has pointed out that the same "drop condition" appears in early work of the Russian mathematician V.A.Rohlin [14](as relation (6) of Sect. 3, p.7). Its rediscovery in the context of event structures was motivated by the need to tie probability to the occurrences of events; it is sufficient to check the 'drop condition' for $y \mathbin{-\!\subset} x_1, \cdots, x_n$, in which the configurations x_i extend y with a single event.

account of the fact that in a strategy Player can't be aware of the probabilities assigned by Opponent. We do this by extending the notion of configuration valuation so that: causal independence between Player and Opponent moves entails their probabilistic independence, or equivalently, so probabilistic dependence of Player on Opponent moves will presuppose their causal dependence (the effect of the condition of "±-independence" below); the "drop condition" only applies to moves of Player. Precisely, a *configuration valuation* is now a function v, from finite configurations of S to the unit interval, which is

(normalized) $v(\varnothing) = 1$, has
(±-independence) $v(x) = v(y)$ when $x \subseteq^- y$ for finite configurations x, y of S, and satisfies the
(+ve drop condition) $d_v[y; x_1, \cdots, x_n] \geq 0$ when $y \subseteq^+ x_1, \cdots, x_n$ for finite configurations of S.

One can think of the value $v(x)$, where x is a finite configurations of S, as the probability of obtaining a result which extends x conditional on the Opponent moves in x.

We return to the point that "±-independence" expresses that causal independence between Player and Opponent moves entails their probabilistic independence. Consider two moves, $\oplus$ of Player and $\ominus$ of Opponent able to occur independently, *i.e.* concurrently, at some finite configuration x, taking it to the configuration $x \cup \{\oplus, \ominus\}$. There are intermediate configurations $x \cup \{\oplus\}$ and $x \cup \{\ominus\}$ associated with just one additional move. The condition of "±-independence" ensures $v(x \cup \{\oplus, \ominus\}) = v(x \cup \{\oplus\})$, *i.e.* the probability of $\oplus$ with $\ominus$ is the same as the probability of $\oplus$ at configuration x. At x the probability of the Player move conditional on the Opponent move equals the probability of the Player move—the moves are probabilistically independent.

5.1   A Bicategory of Probabilistic Strategies

Probabilistic strategies compose. Assume probabilistic strategies $\sigma : S \to A^\perp \| B$ with configuration valuation v_S and $\tau : T \to B^\perp \| C$ with configuration valuation v_T. Recall how the composition $\tau \odot \sigma$ is obtained via pullback, to synchronise the strategies over common moves, followed by projection, to hide the synchronisations.

Given z a finite configuration of the pullback $T \circledast S$ its image $\pi_1 z$ under the projection π_1 is a finite configuration of $S \| C$; taking its left component we obtain $(\pi_1 z)_1$, a finite configuration of S. Similarly, taking the right component of the image $\pi_2 z$ we obtain a finite configuration $(\pi_2 z)_2$ of T. It can be shown that defining $v(z) = v_S((\pi_1 z)_1) \times v_T((\pi_2 z)_2)$ for z a finite configuration of $T \circledast S$ satisfies the conditions of a configuration valuation (with the proviso that we treat synchronisation and Player events alike in the drop condition). In the proof 'drop functions' come into their own. A finite configuration x of $T \odot S$, after hiding, is a subset of $T \circledast S$ so we can form its down-closure there to obtain $[x]$, a finite configuration of $T \circledast S$. The assignment of value $v([x])$ to x a finite configuration of $T \odot S$ yields a configuration valuation to associate with the composition $\tau \odot \sigma$.

Above, notice in the special case where $\sigma : S \to B$ and $\tau : T \to B^{\perp}$, *i.e.* of a strategy and a counter-strategy in the game B, that the resulting probabilistic play is captured by $T \circledast S$, which is now a probabilistic event structure.[4]

Because we restrict to race-free games, copy-cat strategies are deterministic ensuring that the assignment of one to each finite configuration of copy-cat is a configuration valuation; this provides us with identities w.r.t. composition.

We don't have a category however, as the laws for categories are only true up to isomorphism. Technically we have a bicategory of games and probabilistic strategies in which the objects are race-free games and the arrows are probabilistic strategies. The 2-cells, the maps between strategies, require some explanation.[5] Without the presence of probability it is sensible to take a 2-cell between two strategies $\sigma : S \to A^{\perp} \| B$ and $\sigma' : S' \to A^{\perp} \| B$ to be a map $f : S \to S'$ making

commute. However, in the situation where the strategies are probabilistic, when σ is accompanied by a configuration valuation v and σ' by configuration valuation v', we need a further constraint to relate probabilities. Normally probability distributions can be "pushed forward" across measurable functions. But configuration valuations don't correspond to probability distributions in the presence of Opponent moves and in general we can't push forward the configuration valuation v of S to a configuration valuation fv of S'. We can however do so when f is rigid: then defining

$$(fv)(y) =_{\text{def}} \sum \{v(x) \mid fx = y\},$$

for $y \in \mathcal{C}(S')$, yields a configuration valuation fv of S' —the *push-forward* of v. So finally we constrain 2-cells between probabilistic strategies, from σ with v to σ' with v', to those rigid maps f for which $\sigma = \sigma' f$ and the push-forward fv is pointwise less than or equal to v'.

The vertical composition of 2-cells is the usual composition of maps. Horizontal composition is given by the composition of strategies $\odot$ (which extends to a functor on 2-cells via the universal properties of pullback and factorisation used in its definition).

2-cells include rigid embeddings preserving the value assigned by configuration valuations.[6] Amongst these are those 2-cells in which the rigid embedding is an inclusion—providing a very useful order for defining probabilistic strategies

---

[4] The use of "schedulers to resolve the probability or nondeterminism" in earlier work is subsumed by that of probabilistic and deterministic counter-strategies. Deterministic strategies coincide with those with assignment one to each finite configuration.

[5] Their treatment in [15] is slapdash.

[6] One way to define a rigid embedding is as a rigid map whose function is injective and reflects consistency.

recursively. Let $\sigma : S \to A^\perp \| B$ with configuration valuation v and $\sigma' : S' \to A^\perp \| B$ with v' be two probabilistic strategies. Define $\sigma \trianglelefteq \sigma'$ when $S \subseteq S'$ and the associated inclusion map is a rigid embedding and a 2-cell for which $v(x) = v'(x)$ for all finite configurations of S. This enables us to exploit old techniques to define strategies recursively: the substructure order on event structures, of which we have an example, forms a "large complete partial order" on which continuous operations possess least fixed points—see [2, 3].

5.2   Extensions: Payoff and Imperfect Information

We can add *payoff* to a game as a function from its configurations to the real numbers [15, 16]. For such quantitative games, determinacy is expressed in terms of the game possessing a *value*, a form of minimax property. The interest is now focussed on *optimal* strategies which achieve the value of the game. In games of *imperfect information* some moves are masked, or inaccessible, and strategies with dependencies on unseen moves are ruled out. It is straightforward to extend probabilistic distributed games with payoff and imperfect information in way that respects the operations of distributed games and strategies [17]. *Blackwell games* [18], of central importance in logic and computer science, become a special case of probabilistic distributed games of imperfect information with payoff [15].

6   Constructions on Probabilistic Strategies [19]

There is a richness of constructions in the world of distributed strategies and games. The language of games and strategies that ensues is largely stable under the addition of probability and extra features such as imperfect information and payoff. Though for instance we shall need to restrict to race-free games in order to have identities w.r.t. the composition of probabilistic strategies.

In the language for probabilistic strategies, race-free games $A, B, C, \cdots$ will play the role of types. There are operations on games of forming the dual $A^\perp$, simple parallel composition $A \| B$, sum $\Sigma_{i \in I} A_i$ as well as recursively-defined games —the latter rest on well-established techniques [2] and will be ignored here. The operation of sum of games is similar to that of simple parallel composition but where now moves in different components are made inconsistent; we restrict its use to those cases in which it results in a game which is race-free.

Terms have typing judgements:

$$x_1 : A_1, \cdots, x_m : A_m \vdash t \dashv y_1 : B_1, \cdots, y_n : B_n \ ,$$

where all the variables are distinct, interpreted as a probabilistic strategy from the game $\vec{A} = A_1 \| \cdots \| A_m$ to the game $\vec{B} = B_1 \| \cdots \| B_n$. We can think of the term t as a box with input and output wires for the variables:

The idea is that t denotes a probabilistic strategy $S \rightarrow \vec{A}^\perp \| \vec{B}$ with configuration valuation v. The term t describes witnesses, finite configurations of S, to a relation between finite configurations $\vec{x}$ of $\vec{A}$ and $\vec{y}$ of $\vec{B}$, together with their probability conditional on the Opponent moves involved.

Duality. The duality, that a probabilistic strategy from A to B can equally well be seen as a probabilistic strategy from $B^\perp$ to $A^\perp$, is caught by the rules:

$$\frac{\Gamma, x : A \vdash t \dashv \Delta}{\Gamma \vdash t \dashv x : A^\perp, \Delta} \qquad \frac{\Gamma \vdash t \dashv x : A, \Delta}{\Gamma, x : A^\perp \vdash t \dashv \Delta}$$

Composition. The composition of probabilistic strategies is described in the rule

$$\frac{\Gamma \vdash t \dashv \Delta \qquad \Delta \vdash u \dashv H}{\Gamma \vdash \exists \Delta . [t \| u] \dashv H}$$

which, in the picture of strategies as boxes, joins the output wires of one strategy to input wires of the other.

Probabilistic Sum. For I countable and a sub-probability distribution $p_i, i \in I$, we can form the probabilistic sum of strategies of the same type:

$$\frac{\Gamma \vdash t_i \dashv \Delta \qquad i \in I}{\Gamma \vdash \Sigma_{i \in I} p_i t_i \dashv \Delta .}$$

In the probabilistic sum of strategies, of the same type, the strategies are glued together on their initial Opponent moves (to maintain receptivity) and only commit to a component with the occurrence of a Player move, from which component being determined by the distribution $p_i, i \in I$. We use $\perp$ for the empty probabilistic sum, when the rule above specialises to

$$\Gamma \vdash \perp \dashv \Delta ,$$

which denotes the minimum strategy in the game $\Gamma^\perp \| \Delta$ — it comprises the initial segment of the game $\Gamma^\perp \| \Delta$ consisting of its initial Opponent events.

Conjoining Two Strategies. The pullback of a strategy across a map of event structures is itself a strategy [15]. We can use the pullback of one strategy against another to conjoin two probabilistic strategies of the same type:

$$\frac{\Gamma \vdash t_1 \dashv \Delta \quad \Gamma \vdash t_2 \dashv \Delta}{\Gamma \vdash t_1 \wedge t_2 \dashv \Delta}$$

Such a strategy acts as the two component strategies agree to act jointly. In the case where t_1 and t_2 denote the probabilistic strategies $\sigma_1 : S_1 \rightarrow \Gamma^\perp \| \Delta$ with configuration valuation v_1 and $\sigma_2 : S_2 \rightarrow \Gamma^\perp \| \Delta$ with v_2 the strategy $t_1 \wedge t_2$ denotes the pullback

$$\begin{array}{ccc}
 & S_1 \wedge S_2 & \\
 \pi_1 \swarrow & & \searrow \pi_2 \\
 S_1 & \vdots \sigma_1 \wedge \sigma_2 & S_2 \\
 \sigma_1 \searrow & \downarrow & \swarrow \sigma_2 \\
 & \Gamma^\perp \| \Delta &
\end{array}$$

with configuration valuation $x \mapsto v_1(\pi_1 x) \times v_2(\pi_2 x)$ for $x \in \mathcal{C}(S_1 \wedge S_2)$.

Copy-cat Terms. Copy-cat terms are a powerful way to lift maps or relations expressed in terms of maps to strategies. Along with duplication they introduce new "causal wiring." Copy-cat terms have the form

$$x : A \vdash gy \sqsubseteq_C fx \dashv y : B\,,$$

where $f : A \to C$ and $g : B \to C$ are maps of event structures preserving polarity. (In fact, f and g may even be "affine" maps, which don't necessarily preserve empty configurations, provided $g\varnothing \sqsubseteq_C f\varnothing$—see [19].) This denotes a deterministic strategy—so a probabilistic strategy with configuration valuation constantly one—provided f reflects $-$-compatibility and g reflects $+$-compatibility. The map g reflects $+$-compatibility if whenever $x \subseteq^+ x_1$ and $x \subseteq^+ x_2$ in the configurations of B and $fx_1 \cup fx_2$ is consistent, so a configuration, then so is $x_1 \cup x_2$. The meaning of f reflecting $-$-compatibility is defined analogously.

A term for copy-cat arises as a special case,

$$x : A \vdash y \sqsubseteq_A x \dashv y : A\,,$$

as do terms for the jth injection into and jth projection out of a sum $\Sigma_{i \in I} A_i$ w.r.t. its component A_j,

$$x : A_j \vdash y \sqsubseteq_{\Sigma_{i \in I} A_i} jx \dashv y : \Sigma_{i \in I} A_i$$

and

$$x : \Sigma_{i \in I} A_i \vdash jy \sqsubseteq_{\Sigma_{i \in I} A_i} x \dashv y : A_j\,,$$

as well as terms which split or join 'wires' to or from a game $A \| B$.

In particular, a map $f : A \to B$ of games which reflects $-$-compatibility lifts to a deterministic strategy $f_! : A \longrightarrow B$:

$$x : A \vdash y \sqsubseteq_B fx \dashv y : B\,.$$

A map $f : A \to B$ which reflects $+$-compatibility lifts to a deterministic strategy $f^* : B \longrightarrow A$:

$$y : B \vdash fx \sqsubseteq_B y \dashv x : A\,.$$

The construction $f^* \odot t$ denotes the pullback of a strategy t in B across the map $f : A \to B$. It can introduce extra events and dependencies in the strategy. It subsumes the operations of prefixing by an initial Player or Opponent move on games and strategies.

Trace. A probabilistic *trace*, or feedback, operation is another consequence of such "wiring." Given a probabilistic strategy $\Gamma, x : A \vdash t \dashv y : A, \Delta$ represented by the diagram

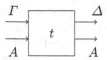

we obtain

$$\Gamma, \Delta^\perp \vdash t \dashv x : A^\perp, y : A$$

which post-composed with the term

$$x : A^\perp, y : A \vdash x \sqsubseteq_A y \dashv,$$

denoting the copy-cat strategy $\gamma_{A^\perp}$, yields

$$\Gamma \vdash \exists x : A^\perp, y : A . [\, t \parallel x \sqsubseteq_A y \,] \dashv \Delta,$$

representing its trace:

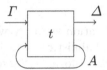

The composition introduces causal links from the Player moves of $y : A$ to the Opponent moves of $x : A$, and from the Player moves of $x : A$ to the Opponent moves of $y : A$—these are the usual links of copy-cat $\gamma_{A^\perp}$ as seen from the left of the turnstile. If we ignore probabilities, this trace coincides with the feedback operation which has been used in the semantics of nondeterministic dataflow (where only games comprising solely Player moves are needed) [10].

Duplication. Duplications of arguments is essential if we are to support the recursive definition of strategies. We duplicate arguments through a probabilistic strategy $\delta_A : A \rightarrowtail A \| A$. Intuitively it behaves like the copy-cat strategy but where a Player move in the left component may choose to copy from either of the two components on the right. In general the technical definition is involved, even without probability—see [19]. The introduction of probability begins to reveal a limitation within probabilistic strategies as we have defined them, a point we will follow up on in the next section. We can see the issue in the second of two simple examples. The first is that of δ_A in the case where the game A consists of a single Player move $\oplus$. Then, δ_A is the deterministic strategy

in which the configuration valuation assigns one to all finite configurations —we have omitted the obvious map to the game $A^\perp \| A \| A$. In the second example, assume A consists of a single Opponent move $\ominus$. Now δ_A is no longer deterministic and takes the form and the strategy is forced to choose probabilistically between reacting to the upper or lower move of Opponent in order to satisfy the drop condition of its configuration valuation. Given the symmetry of the situation, in this case any configuration containing a Player move is assigned value a half by the configuration valuation associated with δ_A. (In the definition of the probabilistic duplication for general A the configuration valuation is distributed uniformly over the different ways Player can copy Opponent moves.) But this is odd: in the second example, if the Opponent makes only one move there is a 50 % chance that Player will not react to it! There are mathematical consequences too. In the absence of probability δ_A forms a comonoid with counit $\perp : A \longrightarrow \varnothing$. However, as a probabilistic strategy δ_A is no longer a comonoid—it fails associativity. It is hard to see an alternative definition of a probabilistic duplication strategy within the limitations of the event structures we have been using. We shall return to duplication, and a simpler treatment through a broadening of event structures in the next section.

Recursion. Once we have duplication strategy we can treat recursion. Recall that 2-cells, the maps between probabilistic strategies, include the approximation order $\trianglelefteq$ between strategies. The order forms a 'large complete partial order' with a bottom element the minimum strategy $\perp$. Given $x : A, \Gamma \vdash t \dashv y : A$, the term $\Gamma \vdash \mu x : A. t \dashv y : A$ denotes the $\trianglelefteq$-least fixed point amongst probabilistic strategies $X : \Gamma \longrightarrow A$ of the $\trianglelefteq$-continuous operation $F(X) = t \odot (\mathrm{id}_\Gamma \| X) \odot \delta_\Gamma$. (With one exception, F is built out of operations which it's been shown can be be defined concretely in such a way that they are $\trianglelefteq$-continuous; the one exception which requires separate treatment is the 'new' operation of projection, used to hide synchronisations.) With probability, as δ_Γ is no longer a comonoid not all the "usual" laws of recursion will hold, though the unfolding law will hold by definition.

7   A Limitation

One limitation that is not seen when working with purely nondeterministic strategies has revealed itself when strategies are made probabilistic. The simple event structures on which we have based games and strategies do not support "parallel causes" and this has the consequence that certain informal but intuitively convincing strategies are not expressible. We met this in the previous section in our treatment of a probabilistic duplication strategy $\delta_A : A \longrightarrow A \| A$.

Probabilistic strategies, as presented, do not cope with stochastic behaviour such as races as in the game

$$\ominus \sim \oplus.$$

To do such we would expect to have to equip events in the strategy with stochastic rates (which isn't hard to do if synchronisation events are not hidden). So this is to be expected. But at present probabilistic strategies do not cope with benign Player-Player races either! Consider the game

where Player would like a strategy in which they play a move iff Opponent plays one of theirs. We might stipulate that Player wins if a play of any ⊖ is accompanied by the play of ⊕ and *vice versa*. Intuitively a winning strategy would be got by assigning watchers (in the team Player) for each ⊖ who on seeing their ⊖ race to play ⊕. This strategy should win with certainty against any counter-strategy: no matter how Opponent plays one or both of their moves at least one of the watchers will report this with the Player move. But we cannot express this with event structures. The best we can do is a probabilistic strategy

with configuration valuation assigning $1/2$ to configurations containing either Player move and 1 otherwise. Against a counter-strategy with Opponent playing one of their two moves with probability $1/2$ this strategy only wins half the time. In fact, the strategy together with the counter-strategy form a Nash equilibrium when a winning configuration for Player is assigned payoff $+1$ and a loss -1. This strategy really is the best we can do presently in that it is optimal amongst those expressible using the simple event structures of Sect. 2.

8 A Solution

If we are to be able to express the intuitive strategy which wins with certainty we need to develop distributed probabilistic strategies which allow such parallel causes as in 'general event structures' $(E, \vdash, \mathrm{Con})$ which permit *e.g.* two distinct compatible causes $X \vdash e$ and $Y \vdash e$ (see [3]). In the informal strategy described in the previous section both Opponent moves would individually enable the Player move, with all events being consistent. But it can be shown that general event structures do not support an appropriate operation of hiding. Nor is it clear how within general event structures one could express a variant of the strategy above, in which the two watchers succeed in reporting the Player move with different probabilities.

It is informative to see why general event structures are not closed under hiding. The following describes a general event structure.

Events: a, b, c, d and e.
Enablings: $b, c \vdash e$ and $d \vdash e$, with all events other than e being enabled by the empty set.
Consistency: all subsets are consistent unless they contain the events a and b; in other words, the events a and b are in conflict.

Any configuration will satisfy the assertion

$$(a \wedge e) \implies d$$

because if e has occurred it has to have been enabled by (1) or (2) and if a has occurred its conflict with b has prevented the enabling (1), so e can only have occurred via enabling (2).

Now imagine the event b is hidden, so allowed to occur invisibly in the background. The "configurations after hiding" are those obtained by hiding (*i.e.* removing) the invisible event b from the configurations of the original event structure. The assertion above will still hold of the configurations after hiding. There isn't a general event structure with events a, c, d and e, and configurations those which result when we hide (or remove) b from the configurations of the original event structure. One way to see this is to observe that amongst the configurations after hiding we have $\{c\} \subseteq \{c, e\}$ and $\{c\} \subseteq \{a, c\}$ where both $\{c, e\}$ and $\{a, c\}$ have upper bound $\{a, c, d, e\}$, and yet $\{a, c, e\}$ is not a configuration after hiding as it fails to satisfy the assertion. (In a general event structure it would have to also be a configuration.)

The first general event structure can be built out of the composition *without hiding* of strategies described by general event structures, one from a game A to a game B and the other from B to C; the second structure, not a general event structure, would arise when hiding the events over the intermediate game B.

To obtain a bicategory of strategies with disjunctive causes we need to support hiding. We need to look for structures more general than general event structures. The example above gives a clue: the inconsistency is one of inconsistency between complete enablings rather than events.

Marc de Visme and I have explored the space of possibilities and discovered a refinement of event structures into which general event structures embed, which supports hiding, and provides a basis on which to develop probabilistic distributed strategies with disjunctive and parallel causes. One is led to introduce structures in which we *objectify* cause : a minimal causal enabling is no longer an instance of a relation but an object that realises that instance (*cf.* a proof in contrast to a judgement of theorem-hood). This is in order to express inconsistency between complete enablings, inexpressible as inconsistencies on events, that can arise when hiding.

An event structure with disjunctive causes (an edc) is a structure

$$(P, \leq, \mathrm{Con}_P, \equiv)$$

where $(P, \leq, \mathrm{Con}_P)$ satisfies the axioms of a event structure and $\equiv$ is an equivalence relation on P such that

$$\forall p_1, p_2 \leq p.\ p_1 \equiv p_2 \implies p_1 = p_2.$$

The events of P represent *prime causes* while the $\equiv$-equivalence classes of P represent *disjunctive events*: p in P is a prime cause of the event $\{p\}_\equiv$. Notice there may be several prime causes of the same event and that these may be parallel causes in the sense that they are consistent with each other and causally independent. A *configuration* of the edc is a configuration of $(P, \leq, \text{Con}_P)$. An edc dissociates the two roles of enabling and atomic action conflated in the events of an event structures. The elements of P are to be thought of as complete minimal enablings and the equivalence classes as atomic actions representing the occurrence of at least one prime cause.

When the equivalence relation $\equiv$ of an edc is the identity it is essentially an event structure. This view is reinforced in our choice of maps. A map from $(P, \leq_P, \text{Con}_P, \equiv_P)$ to $(Q, \leq_Q, \text{Con}_Q, \equiv_Q)$ is a partial function $f : P \rightharpoonup Q$ which preserves $\equiv$ (*i.e.* if $p_1 \equiv_P p_2$ then either both $f(p_1)$ and $f(p_2)$ are undefined or both defined with $f(p_1) \equiv_Q f(p_2)$) such that for all $x \in \mathcal{C}(P)$

(i) the direct image $fx \in \mathcal{C}(Q)$, and
(ii) $\forall p_1, p_2 \in x.\ f(p_1) \equiv_Q f(p_2) \implies p_1 \equiv_P p_2$.

Edc's support a hiding operation along the same lines as event structures. There is an adjunction expressing the sense in which general event structures embed within edc's. There is also an adjunction (in fact a coreflection) from event structures to edc's which helps give a tight connection between strategies based on event structures and their generalisation to edc's. Probability extends straightforwardly to strategies based on edc's. The work is recent and in the process of being written up [20]. We conclude by presenting the *deterministic* strategy in the game

$$\oplus$$
$$\ominus \qquad \ominus$$

in which Player makes a move iff Opponent does:

$$
\begin{array}{ccc}
\oplus & \equiv & \oplus \\
\uparrow & & \uparrow \\
\ominus & & \ominus
\end{array}
$$

(In the deterministic case each watcher succeeds with certainty. We can also represent the situation where one watcher succeeds with probability $p \in [0, 1]$ and the other with probability $q \in [0, 1]$ through the obvious configuration valuation.) The same strategy serves as the duplication strategy for the game comprising a single Opponent event. This indicates how within the broader framework of edc's there are deterministic duplication strategies $\delta_A : A \dashrightarrow A \| A$ in which a Player move in the left component is alerted in parallel by a corresponding Opponent move in either of the two components on the right. The fact that the duplication strategies are now deterministic obviates the difficulties we encountered earlier: duplication now forms a comonoid and we recover the usual laws for recursive definitions of strategies.

Acknowledgements. Thanks to Samy Abbes, Nathan Bowler, Simon Castellan, Pierre Clairambault, Marcelo Fiore, Mai Gehrke, Julian Gutierrez, Jonathan Hayman, Martin Hyland, Marc Lasson, Silvain Rideau, Daniele Varacca and Marc de Visme for helpful discussions. The concluding section is based on recent joint work with Marc de Visme while on his internship from ENS Paris. The support of Advanced Grant ECSYM of the European Research Council is acknowledged with gratitude.

References

1. Winskel, G., Nielsen, M.: Handbook of Logic in Computer Science. Oxford University Press, New York (1995). ch. Models for Concurrency
2. Winskel, G.: Event structure semantics for CCS and related languages. In: Schmidt, E.M., Nielsen, M. (eds.) Automata, Languages and Programming. LNCS, vol. 140, pp. 561–576. Springer, Heidelberg (1982)
3. —, "Event structures," in Advances in Petri Nets, ser. Lecture Notes in Computer Science, vol. 255. Springer, pp. 325–392 (1986)
4. Rideau, S., Winskel, G.: Concurrent strategies. In: LICS 2011. IEEE Computer Society (2011)
5. Nielsen, M., Plotkin, G., Winskel, G.: Petri nets, event structures and domains. Theor. Comput. Sci. **13**, 85–108 (1981)
6. Conway, J.: On Numbers and Games. A K Peters, Wellesley (2000)
7. Joyal, A.: Remarques sur la théorie des jeux à deux personnes. Gaz. des Sci. Mathématiques du Que **1**(4), 46–52 (1997)
8. Winskel, G.: Deterministic concurrent strategies. Formal Asp. Comput. **24**(4-6), 647–660 (2012)
9. Melliès, P.-A., Mimram, S.: Asynchronous games: innocence without alternation. In: Caires, L., Vasconcelos, V.T. (eds.) CONCUR 2007. LNCS, vol. 4703, pp. 395–411. Springer, Heidelberg (2007)
10. Saunders-Evans, L., Winskel, G.: Event structure spans for nondeterministic dataflow. Electr. Notes Theor. Comput. Sci. **175**(3), 109–129 (2007)
11. Berry, G.: Stable models of typed lambda-calculi. In: ICALP, ser. Lecture Notes in Computer Science, vol. 62. Springer, pp. 72–89 (1978)
12. Winskel, G.: Probabilistic and quantum event structures. In: van Breugel, F., Kashefi, E., Palamidessi, C., Rutten, J. (eds.) Horizons of the Mind. LNCS, vol. 8464, pp. 476–497. Springer, Heidelberg (2014)
13. Jones, C., Plotkin, G.: A probabilistic powerdomain of valuations. In: LICS 1989. IEEE Computer Society (1989)
14. Rohlin, V.A.: On the fundamental ideas of measure theory. Amer. Math. Soc. Translation. **71**, 55 (1952)
15. Winskel, G.: Distributed probabilistic and quantum strategies. Electr. Notes Theor. Comput. Sci. **298**, 403–425 (2013)
16. Clairambault, P., Winskel, G.: On concurrent games with payoff. Electr. Notes Theor. Comput. Sci. **298**, 71–92 (2013)
17. Winskel, G.: Winning, losing and drawing in concurrent games with perfect or imperfect information. In: Constable, R.L., Silva, A. (eds.) Logic and Program Semantics, Kozen Festschrift. LNCS, vol. 7230, pp. 298–317. Springer, Heidelberg (2012)
18. Peters, J.F.: In: Peters, J.F. (ed.). ISRL, vol. 63, pp. 1–76. Springer, Heidelberg (2014)
19. Castellan, S., Hayman, J., Lasson, M., Winskel, G.: Strategies as concurrent processes. Electr. Notes Theor. Comput. Sci. **308**, 87–107 (2014)
20. de Visme, M., Winskel, G.: Strategies with Parallel Causes. Draft (2015)

Algebra and Category Theory

Newton Series, Coinductively

Henning Basold[1], Helle Hvid Hansen[2]([✉]), Jean-Éric Pin[3], and Jan Rutten[1,4]

[1] Radboud University Nijmegen, Nijmegen, Netherlands
[2] Delft University of Technology, Delft, Netherlands
h.h.hansen@tudelft.nl
[3] LIAFA Université Paris VII and CNRS, Paris, France
[4] CWI Amsterdam, Amsterdam, Netherlands

Abstract. We present a comparative study of four product operators on weighted languages: (i) the *convolution*, (ii) the *shuffle*, (iii) the *infiltration*, and (iv) the *Hadamard* product. Exploiting the fact that the set of weighted languages is a final coalgebra, we use coinduction to prove that an operator of the classical difference calculus, the *Newton transform*, generalises (from infinite sequences) to weighted languages. We show that the Newton transform is an isomorphism of rings that transforms the Hadamard product of two weighted languages into their infiltration product, and we develop various representations for the Newton transform of a language, together with concrete calculation rules for computing them.

1 Introduction

Formal languages [8] are a well-established formalism for the modelling of the behaviour of systems, typically represented by automata. *Weighted* languages – aka formal power series [3] – are a common generalisation of both formal languages (sets of words) and streams (infinite sequences). Formally, a weighted language is an assignment from words over an alphabet A to values in a set k of *weights*. Such weights can represent various things such as the multiplicity of the occurrence of a word, or its duration, or probability etc. In order to be able to add and multiply, and even subtract such weights, k is typically assumed to be a semi-ring (e.g., the Booleans) or a ring (e.g., the integers).

We present a comparative study of four product operators on weighted languages, which give us four different ways of composing the behaviour of systems. The operators under study are (i) the *convolution*, (ii) the *shuffle*, (iii) the *infiltration*, and (iv) the *Hadamard* product, representing, respectively: (i) the concatenation or sequential composition, (ii) the interleaving without synchronisation, (iii) the interleaving *with* synchronisation, and (iv) the fully synchronised interleaving, of systems. The set of weighted languages, together with the operation of *sum* and combined with any of these four product operators, is a ring

Supported by NWO project 612.001.021.

Supported by NWO Veni grant 639.021.231.

itself, assuming that k is a ring. This means that in all four cases, we have a well-behaved calculus of behaviours.

Main Contributions: (1) We show that a classical operator from difference calculus in mathematics: the *Newton transform*, generalises (from infinite sequences) to weighted languages, and we characterise it in terms of the shuffle product. (2) Next we show that the Newton transform is an isomorphism of rings that transforms the Hadamard product of two weighted languages into an infiltration product. This allows us to switch back and forth between a fully synchronised composition of behaviours, and a shuffled, partially synchronised one. (3) We develop various representations for the Newton transform of a language, together with concrete calculation rules for computing them.

Approach: We exploit the fact that the set of weighted languages is a *final coalgebra* [16,17]. This allows us to use *coinduction* as the guiding methodology for both our definitions and proofs. More specifically, we define our operators in terms of *behavioural differential equations*, which yields, for instance, a uniform and thereby easily comparable presentation of all four product operators. Moreover, we construct *bisimulation* relations in order to prove various identities.

As the set of weighted languages over a one-letter alphabet is isomorphic to the set of streams, it turns out to be convenient to prove our results first for the special case of streams and then to generalise them to weighted languages.

Related Work: The present paper fits in the coalgebraic outlook on systems behaviour, as in, for instance, [1,17]. The definition of Newton series for weighted languages was introduced in [14], where Mahler's theorem (which is a p-adic version of the classical Stone-Weierstrass theorem) is generalised to weighted languages. The Newton transform for streams already occurs in [12] (where it is called the discrete Taylor transform), but not its characterisation using the shuffle product, which for streams goes back to [18], and which for weighted languages is new. Related to that, we present elimination rules for (certain uses of) the shuffle product, which were known for streams [18] and are new for languages. The proof that the Newton transform for weighted languages is a ring *isomorphism* that exchanges the Hadamard product into the infiltration product, is new. In [11, Chap. 6], an operation was defined that does the reverse; it follows from our work that this operation is the inverse of the Newton transform. The infiltration product was introduced in [6]; as we already mentioned, [11, Chap. 6] studies some of its properties, using a notion of binomial coefficients for words that generalises the classical notions for numbers. The present paper introduces a new notion of binomial coefficients for words, which refines the definition of [11, Chap. 6].

2   Preliminaries: Stream Calculus

We present basic facts from coinductive stream calculus [18]. In the following, we assume k to be a ring, unless stated otherwise. Let then the set of streams over k be given by $k^\omega = \{\, \sigma \mid \sigma : \mathbb{N} \to k \,\}$. We define the *initial value* of a stream σ by $\sigma(0)$ and its *stream derivative* by $\sigma' = (\sigma(1), \sigma(2), \sigma(3), \ldots)$. In order to conclude that two streams σ and τ are equal, it suffices to prove $\sigma(n) = \tau(n)$, for all $n \geq 0$. Sometimes this can be proved by *induction* on the natural number n but, more often than not, we will not have a succinct description or formula for $\sigma(n)$ and $\tau(n)$, and induction will be of no help. Instead, we take here a coalgebraic perspective on k^ω, and most of our proofs will use the proof principle of *coinduction*, which is based on the following notion.

A relation $R \subseteq k^\omega \times k^\omega$ is a *(stream) bisimulation* if for all $(\sigma, \tau) \in R$,

$$\sigma(0) = \tau(0) \qquad \text{and} \qquad (\sigma', \tau') \in R. \tag{1}$$

Theorem 1 (Coinduction Proof Principle). *If there exists a bisimulation relation containing (σ, τ), then $\sigma = \tau$.*

Coinductive *definitions* are phrased in terms of stream derivatives and initial values, and are called *stream differential equations*; see [10,17,18] for examples and details.

Definition 2 (Basic Operators). *The following system of* stream differential equations *defines our first set of constants and operators:*

Derivative	Initial value	Name
$[r]' = [0]$	$[r](0) = r$	$r \in k$
$X' = [1]$	$X(0) = 0$	
$(\sigma + \tau)' = \sigma' + \tau'$	$(\sigma + \tau)(0) = \sigma(0) + \tau(0)$	*Sum*
$(\Sigma_{i \in I} \sigma_i)' = \Sigma_{i \in I} \sigma_i'$	$(\Sigma_{i \in I} \sigma_i)(0) = \sum_{i \in I} \sigma_i(0)$	*Infinite sum*
$(-\sigma)' = -(\sigma')$	$(-\sigma)(0) = -\sigma(0)$	*Minus*
$(\sigma \times \tau)' = (\sigma' \times \tau) + ([\sigma(0)] \times \tau')$	$(\sigma \times \tau)(0) = \sigma(0)\tau(0)$	*Convolution product*
$(\sigma^{-1})' = -[\sigma(0)^{-1}] \times \sigma' \times \sigma^{-1}$	$(\sigma^{-1})(0) = \sigma(0)^{-1}$	*Convolution inverse*

The unique existence of constants and operators satisfying the equations above is ultimately due to the fact that k^ω, together with the operations of initial value and stream derivative, is a final coalgebra.

For $r \in k$, we have the constant stream $[r] = (r, 0, 0, 0, \ldots)$ which we often denote again by r. Then we have the constant stream $X = (0, 1, 0, 0, 0, \ldots)$. We define $X^0 = [1]$ and $X^{i+1} = X \times X^i$. The infinite sum $\Sigma_{i \in I} \sigma_i$ is defined only when the family $\{\sigma_i\}_{i \in I}$ is *summable*, that is, if for all $n \in \mathbb{N}$ the set $\{i \in I \mid \sigma_i(n) \neq 0\}$ is finite. If $I = \mathbb{N}$, we denote $\Sigma_{i \in I} \sigma_i$ by $\sum_{i=0}^{\infty} \sigma_i$. Note that $(\tau_i \times X^i)_i$ is summable for any sequence of streams $(\tau_i)_i$. Minus is defined only if k is a ring. In spite of its non-symmetrical definition, convolution product on streams is commutative (assuming that k is). Convolution inverse is defined for those streams σ for which the initial value $\sigma(0)$ is invertible. We will often write $r\sigma$ for $[r] \times \sigma$, and $1/\sigma$ for σ^{-1} and τ/σ for $\tau \times (1/\sigma)$, which – for streams – is equal to $(1/\sigma) \times \tau$.

The following analogue of the fundamental theorem of calculus, tells us how to compute a stream σ from its initial value $\sigma(0)$ and derivative σ'.

Theorem 3. *We have* $\sigma = \sigma(0) + (X \times \sigma')$, *for every* $\sigma \in k^\omega$.   □

We will also use coinduction-up-to [15,18], a strengthening of the coinduction proof principle. Let us first introduce some convenient notation.

Given a relation R on k^ω, we denote by $\bar{R}$ the smallest reflexive relation on k^ω containing R and closed under the element-wise application of the operators in Definition 2. For instance, if $(\alpha, \beta), (\gamma, \delta) \in \bar{R}$ then $(\alpha + \gamma, \beta + \delta) \in \bar{R}$, etc. A relation $R \subseteq k^\omega \times k^\omega$ is a *(stream) bisimulation-up-to* if, for all $(\sigma, \tau) \in R$,

$$\sigma(0) = \tau(0) \qquad \text{and} \qquad (\sigma', \tau') \in \bar{R}. \tag{2}$$

Theorem 4 (Coinduction-up-to). *If* R *is a bisimulation-up-to and* $(\sigma, \tau) \in R$, *then* $\sigma = \tau$.

Proof. If R is a bisimulation-up-to then $\bar{R}$ can be shown to be a bisimulation relation, by structural induction on its definition. Now apply Theorem 1.

Using coinduction (up-to), one can easily prove the following.

Proposition 5 (Semiring of Streams – with Convolution Product). *If* k *is a semiring then the set of streams with sum and convolution product forms a semiring as well:* $(k^\omega, +, [0], \times, [1])$. *If* k *is commutative then so is* k^ω.   □

Polynomial and rational streams are defined as usual, cf. [17].

Definition 6 (Polynomial, Rational Streams). *We call a stream* $\sigma \in k^\omega$ *polynomial if it is of the form* $\sigma = a_0 + a_1X + a_2X^2 + \cdots + a_nX^n$, *for* $n \geq 0$ *and* $a_i \in k$. *We call* σ *rational if it is of the form*

$$\sigma = \frac{a_0 + a_1X + a_2X^2 + \cdots + a_nX^n}{b_0 + b_1X + b_2X^2 + \cdots + b_mX^m}$$

with $n, m \geq 0$, $a_i, b_j \in k$, *and* b_0 *is invertible.*

Example 7. Here are a few concrete examples of streams (over the natural numbers): $1 + 2X + 3X^2 = (1, 2, 3, 0, 0, 0, \ldots)$, $\frac{1}{1-2X} = (2^0, 2^1, 2^2, \ldots)$, $\frac{1}{(1-X)^2} = (1, 2, 3, \ldots)$, $\frac{X}{1-X-X^2} = (0, 1, 1, 2, 3, 5, 8, \ldots)$. We note that convolution product behaves naturally, as in the following example: $(1 + 2X^2) \times (3 - X) = 3 - X + 6X^2 - 2X^3$.   □

We shall be using yet another operation on streams.

Definition 8 (Stream Composition). *We define the composition of streams by the following stream differential equation:*

Derivative	Initial value	Name
$(\sigma \circ \tau)' = \tau' \times (\sigma' \circ \tau)$	$(\sigma \circ \tau)(0) = \sigma(0)$	*Stream composition*

We will consider the composition of streams σ with τ if $\tau(0) = 0$, in which case composition enjoys the following properties.

Proposition 9 (Properties of Composition). *For all ρ, σ, τ with $\tau(0) = 0$, we have $[r] \circ \tau = [r]$, $X \circ \tau = \tau$, and*

$$(\rho + \sigma) \circ \tau = (\rho \circ \tau) + (\sigma \circ \tau), \quad (\rho \times \sigma) \circ \tau = (\rho \circ \tau) \times (\sigma \circ \tau), \quad \sigma^{-1} \circ \tau = (\sigma \circ \tau)^{-1}$$

and similarly for infinite sum.

Example 10. As a consequence, for rational σ, τ, the composition $\sigma \circ \tau$ amounts to replacing every X in σ by τ. For instance, $\frac{X}{1-X-X^2} \circ \frac{X}{1+X} = \frac{X(1+X)}{1+X-X^2}$. $\square$

Defining $\sigma^{(0)} = \sigma$ and $\sigma^{(n+1)} = (\sigma^{(n)})'$, for any stream $\sigma \in k^\omega$, we have $\sigma^{(n)}(0) = \sigma(n)$. Thus $\sigma = (\sigma(0), \sigma(1), \sigma(2), \ldots) = (\sigma^{(0)}(0), \sigma^{(1)}(0), \sigma^{(2)}(0), \ldots)$. Hence every stream is equal to the stream of its *Taylor coefficients* (with respect to stream derivation). There is also the corresponding *Taylor series* representation for streams.

Theorem 11 (Taylor Series). *For every $\sigma \in k^\omega$,*

$$\sigma = \sum_{i=0}^{\infty} [\sigma^{(i)}(0)] \times X^i = \sum_{i=0}^{\infty} [\sigma(i)] \times X^i$$

For some of the operations on streams, we have explicit formulae for the n-th Taylor coefficient, that is, for their value in n.

Proposition 12. *For all $\sigma, \tau \in k^\omega$, for all $n \geq 0$,*

$$(\sigma + \tau)(n) = \sigma(n) + \tau(n), \quad (-\sigma)(n) = -\sigma(n), \quad (\sigma \times \tau)(n) = \sum_{k=0}^{n} \sigma(k)\tau(n-k)$$

3   Four Product Operators

In addition to convolution product, we shall discuss also the following product operators (repeating below the definitions of convolution product and inverse).

Definition 13 (Product Operators). *We define four product operators by the following system of stream differential equations:*

Derivative	Initial value	Name
$(\sigma \times \tau)' = (\sigma' \times \tau) + ([\sigma(0)] \times \tau')$	$(\sigma \times \tau)(0) = \sigma(0)\tau(0)$	*Convolution*
$(\sigma \otimes \tau)' = (\sigma' \otimes \tau) + (\sigma \otimes \tau')$	$(\sigma \otimes \tau)(0) = \sigma(0)\tau(0)$	*Shuffle*
$(\sigma \odot \tau)' = \sigma' \odot \tau'$	$(\sigma \odot \tau)(0) = \sigma(0)\tau(0)$	*Hadamard*
$(\sigma \uparrow \tau)' = (\sigma' \uparrow \tau) + (\sigma \uparrow \tau') + (\sigma' \uparrow \tau')$	$(\sigma \uparrow \tau)(0) = \sigma(0)\tau(0)$	*Infiltration*

For streams σ with invertible initial value $\sigma(0)$, we can define both convolution and shuffle inverse, as follows:

Derivative	Initial value	Name
$(\sigma^{-1})' = -[\sigma(0)^{-1}] \times \sigma' \times \sigma^{-1}$	$(\sigma^{-1})(0) = \sigma(0)^{-1}$	*Convolution inverse*
$(\sigma^{-\frac{1}{}})' = -\sigma' \otimes \sigma^{-\frac{1}{}} \otimes \sigma^{-\frac{1}{}}$	$(\sigma^{-\frac{1}{}})(0) = \sigma(0)^{-1}$	*Shuffle inverse*

(We will not need the inverse of the other two products.) Convolution and Hadamard product are standard operators in mathematics. Shuffle and infiltration product are, for streams, less well-known, and are better explained and understood when generalised to weighted languages, which we shall do in Sect. 7. Closed forms for shuffle and Hadamard are given in Proposition 15 below. In the present section and the next, we shall relate convolution product and Hadamard product to, respectively, shuffle product and infiltration product, using the so-called *Laplace* and the *Newton* transforms.

Example 14. Here are a few simple examples of streams (over the natural numbers), illustrating the differences between these four products.

$$\frac{1}{1-X} \times \frac{1}{1-X} = \frac{1}{(1-X)^2} = (1, 2, 3, \ldots)$$

$$\frac{1}{1-X} \otimes \frac{1}{1-X} = \frac{1}{1-2X} = (2^0, 2^1, 2^2, \ldots)$$

$$\frac{1}{1-X} \odot \frac{1}{1-X} = \frac{1}{1-X}$$

$$\frac{1}{1-X} \uparrow \frac{1}{1-X} = \frac{1}{1-3X} = (3^0, 3^1, 3^2, \ldots)$$

$$(1-X)^{-\frac{1}{}} = (0!, 1!, 2!, \ldots) \tag{3}$$

We have the following closed formulae for the shuffle and Hadamard product. Recall Proposition 12 for the closed form of convolution product. In Proposition 23 below, we derive a closed formula for the infiltration product as well.

Proposition 15.

$$(\sigma \otimes \tau)(n) = \sum_{i=0}^{n} \binom{n}{i} \sigma(i)\tau(n-i) \tag{4}$$

$$(\sigma \odot \tau)(n) = \sigma(n)\tau(n) \tag{5}$$

$\square$

Next we consider the set of streams k^ω together with sum and, respectively, each of the four product operators.

Proposition 16 (Four (semi-)rings of Streams). *If k is a (semi-)ring then each of the four product operators defines a corresponding (semi-)ring structure on k^ω, as follows:*

$$\mathcal{R}_c = (k^\omega, +, [0], \times, [1]), \qquad \mathcal{R}_s = (k^\omega, +, [0], \otimes, [1])$$
$$\mathcal{R}_H = (k^\omega, +, [0], \odot, \text{ones}), \qquad \mathcal{R}_i = (k^\omega, +, [0], \uparrow, [1])$$

where ones *denotes* $(1, 1, 1, \ldots)$. □

We recall from [12] and [18, Theorem 10.1] the following ring isomorphism between $\mathcal{R}_c$ and $\mathcal{R}_s$.

Theorem 17 (Laplace for Streams, [12,18]). *Let the Laplace transform Λ : $k^\omega \to k^\omega$ be given by the following stream differential equation:*

Derivative	Initial value	Name
$(\Lambda(\sigma))' = \Lambda(d/dX(\sigma))$	$\Lambda(\sigma)(0) = \sigma(0)$	*Laplace*

where $d/dX(\sigma) = (X \otimes \sigma')' = (\sigma(1), 2\sigma(2), 3\sigma(3), \ldots)$. Then $\Lambda : \mathcal{R}_c \to \mathcal{R}_s$ is an isomorphism of rings; notably, for all $\sigma, \tau \in k^\omega$, $\Lambda(\sigma \times \tau) = \Lambda(\sigma) \otimes \Lambda(\tau)$. □

(The Laplace transform is also known as the Laplace-Carson transform.) One readily shows that $\Lambda(\sigma) = (0!\sigma(0), 1!\sigma(1), 2!\sigma(2), \ldots)$, from which it follows that Λ is bijective. Coalgebraically, Λ arises as the unique final coalgebra homomorphism between two different coalgebra structures on k^ω:

$$
\begin{array}{ccc}
k^\omega & \xrightarrow{\ \Lambda\ } & k^\omega \\
{\scriptstyle \langle(-)(0),\, d/dX\rangle} \downarrow & & \downarrow {\scriptstyle \langle(-)(0),\, (-)'\rangle} \\
k \times k^\omega & \xrightarrow{\ 1 \times \Lambda\ } & k \times k^\omega
\end{array}
$$

On the right, we have the standard (final) coalgebra structure on streams, given by: $\sigma \mapsto (\sigma(0), \sigma')$, whereas on the left, the operator d/dX is used instead of stream derivative: $\sigma \mapsto (\sigma(0), d/dX(\sigma))$. The commutativity of the diagram above is precisely expressed by the stream differential equation defining Λ above. It is this definition, in terms of stream derivatives, that enables us to give an easy proof of Theorem 17, by coinduction-up-to.

As we shall see, there exists also a ring isomorphism between $\mathcal{R}_H$ and $\mathcal{R}_i$. It will be given by the *Newton* transform, which we will consider next.

4 Newton Transform

Assuming that k is a ring, let the *difference operator* on a stream $\sigma \in k^\omega$ be defined by $\Delta\sigma = \sigma' - \sigma = (\sigma(1) - \sigma(0), \sigma(2) - \sigma(1), \sigma(3) - \sigma(2), \ldots)$.

Definition 18 (Newton Transform). *We define the* Newton transform $\mathcal{N}$: $k^\omega \to k^\omega$ *by the following stream differential equation:*

Derivative	Initial value	Name
$(\mathcal{N}(\sigma))' = \mathcal{N}(\Delta\sigma)$	$\mathcal{N}(\sigma)(0) = \sigma(0)$	*Newton transform*

It follows that $\mathcal{N}(\sigma) = ((\Delta^0\sigma)(0), (\Delta^1\sigma)(0), (\Delta^2\sigma)(0), \dots)$, where $\Delta^0\sigma = \sigma$ and $\Delta^{n+1}\sigma = \Delta(\Delta^n\sigma)$. We call $\mathcal{N}(\sigma)$ the stream of the *Newton coefficients* of σ. Coalgebraically, $\mathcal{N}$ arises as the unique mediating homomorphism – in fact, as we shall see below, an isomorphism – between the following two coalgebras:

$$
\begin{array}{ccc}
k^\omega & \xrightarrow{\;\;\mathcal{N}\;\;} & k^\omega \\
{\scriptstyle \langle(-)(0), \Delta\rangle}\big\downarrow & & \big\downarrow{\scriptstyle \langle(-)(0), (-)'\rangle} \\
k \times k^\omega & \xrightarrow{\;\;1\times\mathcal{N}\;\;} & k \times k^\omega
\end{array}
$$

On the right, we have as before the standard (final) coalgebra structure on streams, whereas on the left, the difference operator is used instead: $\sigma \mapsto (\sigma(0), \Delta\sigma)$. We note that the term Newton transform is used in mathematical analysis [5] for an operational method for the transformation of differentiable functions. In [12], where the diagram above is discussed, our present Newton transform $\mathcal{N}$ is called the discrete Taylor transformation.

The fact that $\mathcal{N}$ is bijective follows from Theorem 20 below, which characterises $\mathcal{N}$ in terms of the shuffle product. Its proof uses the following lemma.

Lemma 19. $\frac{1}{1-X} \otimes \frac{1}{1+X} = 1.$

Note that this formula combines the convolution inverse with the shuffle product. The function $\mathcal{N}$, and its inverse, can be characterised by the following formulae.

Theorem 20 ([18]). *The function $\mathcal{N}$ is bijective and satisfies, for all $\sigma \in k^\omega$,*

$$
\mathcal{N}(\sigma) = \frac{1}{1+X} \otimes \sigma, \qquad \mathcal{N}^{-1}(\sigma) = \frac{1}{1-X} \otimes \sigma.
$$

At this point, we observe the following structural parallel between the Laplace transform from Theorem 17 and the Newton transform: for all $\sigma \in k^\omega$,

$$
\Lambda(\sigma) = (1-X)^{-1} \odot \sigma \tag{6}
$$
$$
\mathcal{N}(\sigma) = (1+X)^{-1} \otimes \sigma \tag{7}
$$

The first equality is immediate from the observation that $(1-X)^{-1} = (0!, 1!, 2!, \dots)$. The second equality is Theorem 20.

The Newton transform is also an isomorphism of *rings*, as follows.

Theorem 21 (Newton Transform as Ring Isomorphism). *We have that* $\mathcal{N} : \mathcal{R}_H \to \mathcal{R}_i$ *is an isomorphism of rings; notably,* $\mathcal{N}(\sigma \odot \tau) = \mathcal{N}(\sigma) \uparrow \mathcal{N}(\tau)$, *for all* $\sigma, \tau \in k^\omega$.

Expanding the definition of the shuffle product in Theorem 20, we obtain the following closed formulae.

Proposition 22. *For all* $\sigma \in k^\omega$ *and* $n \geq 0$,

$$\mathcal{N}(\sigma)(n) = \sum_{i=0}^{n} \binom{n}{i}(-1)^{n-i}\sigma(i), \qquad \mathcal{N}^{-1}(\sigma)(n) = \sum_{i=0}^{n} \binom{n}{i}\sigma(i)$$

From these, we can derive the announced closed formula for the infiltration product.

Proposition 23. *For all* $\sigma, \tau \in k^\omega$,

$$(\sigma \uparrow \tau)(n) = \sum_{i=0}^{n} \binom{n}{i}(-1)^{n-i} \left(\sum_{j=0}^{i} \binom{i}{j}\sigma(j) \right) \left(\sum_{l=0}^{i} \binom{i}{l}\tau(l) \right)$$

5 Calculating Newton Coefficients

The Newton coefficients of a stream can be computed using the following theorem [18, Theorem 10.2(68)]. Note that the righthand side of (8) below no longer contains the shuffle product.

Theorem 24 (Shuffle Product Elimination). *For all* $\sigma \in k^\omega$, $r \in k$,

$$\frac{1}{1 - rX} \otimes \sigma = \frac{1}{1 - rX} \times \left(\sigma \circ \frac{X}{1 - rX} \right) \tag{8}$$

Example 25. For the Fibonacci numbers, we have

$$\mathcal{N}(0, 1, 1, 2, 3, 5, 8, \ldots) = \mathcal{N}\left(\frac{X}{1 - X - X^2} \right) = \frac{X}{1 + X - X^2}$$

It is immediate by Theorems 20 and 24 and Example 10 that the Newton transform preserves rationality.

Corollary 26. *A stream* $\sigma \in k^\omega$ *is rational iff its Newton transform* $\mathcal{N}(\sigma)$ *is rational.*   □

6 Newton Series

Theorem 20 tells us how to compute for a given stream σ the stream of its Newton coefficients $\mathcal{N}(\sigma)$, using the shuffle product. Conversely, there is the following Newton series representation, which tells us how to express a stream σ in terms of its Newton coefficients.

Theorem 27 (Newton Series for Streams, 1st). *For all $\sigma \in k^\omega$, $n \geq 0$,*

$$\sigma(n) = \sum_{i=0}^{n} (\Delta^i \sigma)(0) \binom{n}{i}$$

Using $\binom{n}{i} = n!/i!(n-i)!$ and writing $n^{\underline{i}} = n(n-1)(n-2)\cdots(n-i+1)$ (not to be confused with our notation for the shuffle inverse), Newton series are sometimes (cf. [9, Eq. (5.45)]) also denoted as

$$\sigma(n) = \sum_{i=0}^{n} \frac{(\Delta^i \sigma)(0)}{i!} n^{\underline{i}}$$

thus emphasizing the structural analogy with Taylor series.

Combining Theorem 20 with Theorem 24 leads to yet another, and less familiar expansion theorem (see [19] for a *finitary* version thereof).

Theorem 28 (Newton Series for Streams, 2nd; Euler Expansion). *For all $\sigma \in k^\omega$,*

$$\sigma = \sum_{i=0}^{\infty} (\Delta^i \sigma)(0) \times \frac{X^i}{(1-X)^{i+1}}$$

Example 29. Theorem 28 leads, for instance, to an easy derivation of a rational expression for the stream of cubes, namely

$$(1^3, 2^3, 3^3, \ldots) = \frac{1 + 4X + X^2}{(1-X)^4}$$

$\square$

7   Weighted Languages

Let k again be a ring or semiring and let A be a set. We consider the elements of A as *letters* and call A the *alphabet*. Let A^* denote the set of all finite sequences or *words* over A. We define the set of *languages over A with weights in k* by

$$k^{A^*} = \{ \sigma \mid \sigma : A^* \to k \}$$

Weighted languages are also known as *formal power series* (over A with coefficients in k), cf. [3]. If k is the Boolean semiring $\{0, 1\}$, then weighted languages are just sets of words. If k is arbitrary again, but we restrict our alphabet to a singleton set $A = \{X\}$, then $k^{A^*} \cong k^\omega$, the set of streams with values in k. In other words, by moving from a one-letter alphabet to an arbitrary one, streams generalise to weighted languages.

From a coalgebraic perspective, much about streams holds for weighted languages as well, and typically with an almost identical formulation. This structural similarity between streams and weighted languages is due to the fact that weighted languages carry a final coalgebra structure that is very similar to that of streams, as follows. We define the *initial value* of a (weighted) language σ by $\sigma(\varepsilon)$, that is, σ applied to the *empty word* ε. Next, we define for every $a \in A$ the *a-derivative* of σ by $\sigma_a(w) = \sigma(a \cdot w)$, for every $w \in A^*$. Initial value and derivatives together define a final coalgebra structure on weighted languages, given by

$$k^{A^*} \to k \times (k^{A^*})^A \qquad \sigma \mapsto (\sigma(\varepsilon), \lambda a \in A. \sigma_a)$$

(where $(k^{A^*})^A = \{f \mid f : A \to k^{A^*}\}$). For the case that $A = \{X\}$, the coalgebra structure on the set of streams is a special case of the one above, since under the isomorphism $k^{A^*} \cong k^\omega$, we have that $\sigma(\varepsilon)$ corresponds to $\sigma(0)$, and σ_X corresponds to σ'.

We can now summarize the remainder of this paper, roughly and succinctly, as follows: if we replace in the previous sections $\sigma(0)$ by $\sigma(\varepsilon)$, and σ' by σ_a (for $a \in A$), everywhere, then most of the previous definitions and properties for streams generalise to weighted languages. Notably, we will again have a set of basic operators for weighted languages, four different product operators, four corresponding ring stuctures, and the Newton transform between the rings of Hadamard and infiltration product. (An exception to this optimistic program of translation sketched above, however, is the Laplace transform: there does not seem to exist an obvious generalisation of the Laplace transform for streams – transforming the convolution product into the shuffle product – to the corresponding rings of weighted languages.)

Let us now be more precise and discuss all of this in some detail. For a start, there is again the proof principle of coinduction, now for weighted languages. A relation $R \subseteq k^{A^*} \times k^{A^*}$ is a *(language) bisimulation* if for all $(\sigma, \tau) \in R$:

$$\sigma(\varepsilon) = \tau(\varepsilon) \qquad \text{and} \qquad (\sigma_a, \tau_a) \in R, \text{ for all } a \in A. \tag{9}$$

We have the following *coinduction proof principle*, similar to Theorem 1:

Theorem 30 (Coinduction for Languages). *If there exists a (language) bisimulation relation containing (σ, τ), then $\sigma = \tau$.*

Coinductive *definitions* are given again by differential equations, now called *behavioural differential equations* [17,18].

Definition 31 (Basic Operators for Languages). *The following system of* behavioural differential equations *defines the basic constants and operators for* languages:

Derivative	Initial value	Name
$[r]_a = [0]$	$[r](\varepsilon) = r$	$r \in k$
$b_a = [0]$	$b(\varepsilon) = 0$	$b \in A,\ b \neq a$
$b_a = [1]$	$b(\varepsilon) = 0$	$b \in A,\ b = a$
$(\sigma + \tau)_a = (\sigma_a + \tau_a)$	$(\sigma + \tau)(\varepsilon) = \sigma(\varepsilon) + \tau(\varepsilon)$	Sum
$(\Sigma_{i \in I} \sigma_i)_a = \Sigma_{i \in I} (\sigma_i)_a$	$(\Sigma_{i \in I} \sigma_i)(\varepsilon) = \sum_{i \in I} \sigma_i(\varepsilon)$	Infinite sum
$(-\sigma)_a = -(\sigma_a)$	$(-\sigma)(\varepsilon) = -\sigma(\varepsilon)$	Minus
$(\sigma \times \tau)_a = (\sigma_a \times \tau) + ([\sigma(\varepsilon)] \times \tau_a)$	$(\sigma \times \tau)(\varepsilon) = \sigma(\varepsilon)\tau(\varepsilon)$	Convolution product
$(\sigma^{-1})_a = -[\sigma(\varepsilon)^{-1}] \times \sigma_a \times \sigma^{-1}$	$(\sigma^{-1})(\varepsilon) = \sigma(\varepsilon)^{-1}$	Convolution inverse

The convolution inverse is again defined only for σ with $\sigma(\varepsilon)$ invertible in k. We will write a both for an element of A and for the corresponding constant weighted language. We shall often use shorthands like $ab = a \times b$, where the context will determine whether a word or a language is intended. Also, we will sometimes write A for $\Sigma_{a \in A} a$. The infinite sum $\Sigma_{i \in I} \sigma_i$ is, again, only defined if the family $\{\sigma_i\}_{i \in I}$ is *summable*, i.e., if for all $w \in A^*$ the set $\{i \in I \mid \sigma_i(w) \neq 0\}$ is finite. As before, we shall often write $1/\sigma$ for σ^{-1}. Note that convolution product is weighted concatenation and is no longer commutative. As a consequence, τ/σ is now generally ambiguous as it could mean either $\tau \times \sigma^{-1}$ or $\sigma^{-1} \times \tau$. Only when the latter are equal, we shall sometimes write τ/σ. An example is $A/(1 - A)$, which is A^+, the set of all non-empty words.

Theorem 32 (Fundamental Theorem, for Languages). *For every* $\sigma \in k^{A^*}$, $\sigma = \sigma(\varepsilon) + \sum_{a \in A} a \times \sigma_a$ *(cf. [7,17]).*   □

We can now extend Theorem 4 to languages. Given a relation R on k^{A^*}, we denote by $\bar{R}$ the smallest reflexive relation on k^{A^*} containing R and is closed under the element-wise application of the operators in Definition 31. For instance, if $(\alpha, \beta), (\gamma, \delta) \in \bar{R}$ then $(\alpha + \gamma, \beta + \delta) \in \bar{R}$, etc.

A relation $R \subseteq k^{A^*} \times k^{A^*}$ is a *(weighted language) bisimulation-up-to* if for all $(\sigma, \tau) \in R$:

$$\sigma(\varepsilon) = \tau(\varepsilon), \qquad \text{and} \qquad \text{for all } a \in A : (\sigma_a, \tau_a) \in \bar{R}. \qquad (10)$$

Theorem 33 (Coinduction-up-to for Languages). *If* $(\sigma, \tau) \in R$ *for some bisimulation-up-to, then* $\sigma = \tau$.   □

Composition of languages is defined by the following differential equation:

Derivative	Initial value	Name
$(\sigma \circ \tau)_a = \tau_a \times (\sigma_a \circ \tau)$	$(\sigma \circ \tau)(\varepsilon) = \sigma(\varepsilon)$	Composition

Language composition $\sigma \circ \tau$ is well-behaved, for arbitrary σ and τ with $\tau(\varepsilon) = 0$.

Proposition 34 (Composition of Languages). *For* $\tau \in k^{A^*}$ *with* $\tau(\varepsilon) = 0,$

$$[r] \circ \tau = [r], \qquad a \circ \tau = a \times \tau_a, \qquad A \circ \tau = \tau, \qquad \sigma^{-1} \circ \tau = (\sigma \circ \tau)^{-1}$$

$$(\rho + \sigma) \circ \tau = (\rho \circ \tau) + (\sigma \circ \tau), \qquad (\rho \times \sigma) \circ \tau = (\rho \circ \tau) \times (\sigma \circ \tau)$$

Definition 35 (Polynomial, Rational Languages). *We call* $\sigma \in k^{A^*}$ *poly-nomial if it can be constructed using constants* ($r \in k$ *and* $a \in A$) *and the operations of finite sum and convolution product. We call* $\sigma \in k^{A^*}$ *rational if it can be constructed using constants and the operations of finite sum, convolution product and convolution inverse.* $\square$

As a consequence of Proposition 34, for every rational σ, $\sigma \circ \tau$ is obtained by replacing every occurrence of a in σ by $a \times \tau_a$, for every $a \in A$.

Defining $\sigma_\varepsilon = \sigma$ and $\sigma_{w \cdot a} = (\sigma_w)_a$, for any language $\sigma \in k^{A^*}$, we have $\sigma_w(\varepsilon) = \sigma(w)$. This leads to a *Taylor series* representation for languages.

Theorem 36 (Taylor Series, for Languages). *For every* $\sigma \in k^{A^*}$,

$$\sigma = \sum_{w \in A^*} \sigma_w(\varepsilon) \times w = \sum_{w \in A^*} \sigma(w) \times w$$

Example 37. Here are a few concrete examples of weighted languages:

$$\frac{1}{1 - A} = \sum_{w \in A^*} w = A^*$$

$$\frac{1}{1 + A} = \sum_{w \in A^*} (-1)^{|w|} \times w, \qquad \frac{1}{1 - 2ab} = \sum_{i \geq 0} 2^i \times (ab)^i$$

8   Four Rings of Weighted Languages

The definitions of the four product operators for streams generalise straightfor-wardly to languages, giving rise to four different ring structures on languages.

Definition 38 (Product Operators for Languages). *We define four prod-uct operators by the following system of behavioural differential equations:*

Derivative	Initial value	Name
$(\sigma \times \tau)_a = (\sigma_a \times \tau) + ([\sigma(\varepsilon)] \times \tau_a)$	$(\sigma \times \tau)(\varepsilon) = \sigma(\varepsilon)\tau(\varepsilon)$	*Convolution*
$(\sigma \otimes \tau)_a = (\sigma_a \otimes \tau) + (\sigma \otimes \tau_a)$	$(\sigma \otimes \tau)(\varepsilon) = \sigma(\varepsilon)\tau(\varepsilon)$	*Shuffle*
$(\sigma \odot \tau)_a = \sigma_a \odot \tau_a$	$(\sigma \odot \tau)(\varepsilon) = \sigma(\varepsilon)\tau(\varepsilon)$	*Hadamard*
$(\sigma \uparrow \tau)_a = (\sigma_a \uparrow \tau) + (\sigma \uparrow \tau_a) + (\sigma_a \uparrow \tau_a)$	$(\sigma \uparrow \tau)(\varepsilon) = \sigma(\varepsilon)\tau(\varepsilon)$	*Infiltration*

For languages σ *with invertible initial value* $\sigma(\varepsilon)$, *we can define both convolution and shuffle inverse, as follows:*

Derivative	Initial value	Name
$(\sigma^{-1})_a = -[\sigma(0)^{-1}] \times \sigma_a \times \sigma^{-1}$	$(\sigma^{-1})(0) = \sigma(0)^{-1}$	*Convolution inverse*
$(\sigma^{-\underline{1}})_a = -\sigma_a \otimes \sigma^{-\underline{1}} \otimes \sigma^{-\underline{1}}$	$(\sigma^{-\underline{1}})(0) = \sigma(0)^{-1}$	*Shuffle inverse*

Convolution product is concatenation of (weighted) languages and Hadamard product is the fully synchronised product, which corresponds to the intersection of weighted languages. The shuffle product generalises the definition of the shuffle operator on classical languages (over the Boolean semiring), and can be, equivalently, defined by induction. The following definition is from [11, p. 126] (where shuffle product is denoted by the symbol ○): for all $v, w \in A^*$, $\sigma, \tau \in k^{A^*}$,

$$v \otimes \varepsilon = \varepsilon \otimes v = v$$
$$va \otimes wb = (v \otimes wb)a + (va \otimes w)b \tag{11}$$
$$\sigma \otimes \tau = \sum_{v, w \in A^*} \sigma(v) \times \tau(w) \times (v \otimes w) \tag{12}$$

The infiltration product, originally introduced in [6], can be considered as a variation on the shuffle product that not only interleaves words but als synchronizes them on identical letters. In the differential equation for the infiltration product above, this is apparent from the presence of the additional term $\sigma_a \uparrow \tau_a$. There is also an inductive definition of the infiltration product, in [11, p. 128]. It is a variant of (11) above that for the case that $a = b$ looks like

$$va \uparrow wa = (v \uparrow wa)a + (va \uparrow w)a + (v \uparrow w)a$$

However, we shall be using the coinductive definitions, as these allow us to give proofs by coinduction.

Example 39. Here are a few simple examples of weighted languages, illustrating the differences between these four products. Recall that $1/1 - A = A^*$, that is, $(1/1 - A)(w) = 1$, for all $w \in A^*$. Indicating the length of a word $w \in A^*$ by $|w|$, we have the following identities:

$$\left(\frac{1}{1-A} \times \frac{1}{1-A}\right)(w) = |w| + 1, \qquad \left(\frac{1}{1-A} \otimes \frac{1}{1-A}\right)(w) = 2^{|w|}$$

$$\frac{1}{1-A} \odot \frac{1}{1-A} = \frac{1}{1-A}, \qquad \left(\frac{1}{1-A} \uparrow \frac{1}{1-A}\right)(w) = 3^{|w|}$$

$$\left((1-A)^{-\underline{1}}\right)(w) = |w|! \tag{13}$$

If we restrict the above identities to streams, that is, if the alphabet $A = \{X\}$, then we obtain the identities on streams from Example 14. □

Next we consider the set of weighted languages together with sum and each of the four product operators.

Proposition 40 (Four Rings of Weighted Languages). *If k is a (semi-)ring then each of the four product operators defines a corresponding (semi-)ring structure on k^{A^*}, as follows:*

$$\mathcal{L}_c = \left(k^{A^*}, +, [0], \times, [1] \right), \qquad \mathcal{L}_s = \left(k^{A^*}, +, [0], \otimes, [1] \right)$$

$$\mathcal{L}_H = \left(k^{A^*}, +, [0], \odot, \frac{1}{1-A} \right), \qquad \mathcal{L}_i = \left(k^{A^*}, +, [0], \uparrow, [1] \right)$$

Proof. A proof is again straightforward by coinduction-up-to, once we have adapted Theorem 33 by requiring $\bar{R}$ to be also closed under the element-wise application of all four product operators above.

We conclude the present section with closed formulae for the Taylor coefficients of the above product operators, thus generalising Propositions 12 and 15 to languages. We first introduce the following notion.

Definition 41 (Binomial Coefficients on Words). *For all $u, v, w \in A^*$, we define $\binom{w}{u \mid v}$ as the number of different ways in which u can be taken out of w as a subword, leaving v; or equivalently – and more formally – as the number of ways in which w can be obtained by shuffling u and v; that is,*

$$\binom{w}{u \mid v} = (u \otimes v)(w) \tag{14}$$

The above definition generalises the notion of binomial coefficient for words from [11, p. 121], where one defines $\binom{w}{u}$ as the number of ways in which u can be taken as a subword of w. The two notions of binomial coefficient are related by the following formula:

$$\binom{w}{u} = \sum_{v \in A^*} \binom{w}{u \mid v} \tag{15}$$

As an immediate consequence of the defining equation (14), we find the following recurrence.

Proposition 42. *For all $a \in A$ and $u, v, w \in A^*$,*

$$\binom{aw}{u \mid v} = \binom{w}{u_a \mid v} + \binom{w}{u \mid v_a} \tag{16}$$

Note that for the case of streams, (16) gives us Pascal's formula for classical binomial coefficients (by taking $a = X$, $w = X^n$, $u = X^k$ and $v = X^{n+1-k}$):

$$\binom{n+1}{k} = \binom{n}{k-1} + \binom{n}{k}$$

Proposition 43 gives another property, the easy proof of which illustrates the convenience of the new definition of binomial coefficient. (It is also given in [11, Proposition 6.3.13], where $1/1 - A$ is written as A^* and convolution product as $\circ$.)

Proposition 43. *For all* $u, w \in A^*$, $\left(u \otimes \frac{1}{1-A}\right)(w) = \binom{w}{u}$.

Example 44. $\binom{abab}{ab} = \binom{abab}{ab|ab} + \binom{abab}{ab|ba} = 2 + 1 = 3$ $\qquad\qquad$ □

We have the following closed formulae for three of our product operators.

Proposition 45. *For all* $\sigma, \tau \in k^{A^*}$, $w \in A^*$,

$$(\sigma \times \tau)(w) = \sum_{u,v \in A^* \ s.t. \ u \cdot v = w} \sigma(u)\tau(v)$$

$$(\sigma \otimes \tau)(w) = \sum_{u,v \in A^*} \binom{w}{u \mid v} \sigma(u)\tau(v) \tag{17}$$

$$(\sigma \odot \tau)(w) = \sigma(w)\tau(w) \tag{18}$$

A closed formula for the infiltration product can be derived later, once we have introduced the Newton transform for weighted languages. $\qquad\qquad$ □

9   Newton Transform for Languages

Assuming again that k is a ring, we define the *difference operator* (with respect to $a \in A$) by $\Delta^a \sigma(w) = \sigma_a(w) - \sigma(w) = \sigma(a \cdot w) - \sigma(w)$, for $\sigma \in k^{A^*}$.

Definition 46 (Newton Transform for Languages). *We define the* Newton transform $\mathcal{N} : k^{A^*} \to k^{A^*}$ *by the following behavioural differential equation:*

Derivative	Initial value	Name
$(\mathcal{N}(\sigma))_a = \mathcal{N}(\Delta^a \sigma)$	$\mathcal{N}(\sigma)(\varepsilon) = \sigma(\varepsilon)$	*Newton transform*

(using again the symbol $\mathcal{N}$, now for weighted languages instead of streams). □

It follows that $\mathcal{N}(\sigma)(w) = (\Delta^w \sigma)(\varepsilon)$, for all $w \in A^*$, where $\Delta^\varepsilon \sigma = \sigma$ and $\Delta^{w \cdot a} \sigma = \Delta^a(\Delta^w \sigma)$. Coalgebraically, $\mathcal{N}$ arises again as a unique mediating isomorphism between two final coalgebras:

$$
\begin{array}{ccc}
k^{A^*} & \xrightarrow{\ \ \mathcal{N}\ \ } & k^{A^*} \\
{\scriptstyle \langle(-)(\varepsilon),\lambda a.\Delta^a\rangle} \big\downarrow & & \big\downarrow {\scriptstyle \langle(-)(\varepsilon),\lambda a.(-)_a\rangle} \\
k \times (k^{A^*})^A & \xrightarrow{\ 1 \times \mathcal{N}\ } & k \times (k^{A^*})^A
\end{array}
$$

On the right, we have the standard (final) coalgebra structure on weighted languages, given by: $\sigma \mapsto (\sigma(\varepsilon), \lambda a \in A. \sigma_a)$, whereas on the left, the difference operator is used instead of the stream derivative: $\sigma \mapsto (\sigma(\varepsilon), \lambda a \in A. \Delta^a \sigma)$.

Theorem 47. *The function $\mathcal{N}$ is bijective and satisfies, for all $\sigma \in k^{A^*}$,*

$$\mathcal{N}(\sigma) = \frac{1}{1+A} \otimes \sigma, \qquad \mathcal{N}^{-1}(\sigma) = \frac{1}{1-A} \otimes \sigma$$

(Note again that these formulae combine the convolution inverse with the shuffle product.) The Newton transform is again an isomorphism of *rings*.

Theorem 48 (Newton Transform as Ring Isomorphism for Languages).
The Newton transform $\mathcal{N} : \mathcal{L}_H \to \mathcal{L}_i$ is an isomorphism of rings; notably,
$\mathcal{N}(\sigma \odot \tau) = \mathcal{N}(\sigma) \uparrow \mathcal{N}(\tau)$, *for all $\sigma, \tau \in k^\omega$.*

Noting that $\mathcal{N}(\frac{1}{1-A}) = [1]$, a proof of the theorem by coinduction-up to is straightforward. Part of this theorem is already known in the literature: [11, Theorem 6.3.18] expresses (for the case that $k = \mathbb{Z}$) that $\frac{1}{1-A} \otimes (-)$ transforms the infiltration product of two words into a Hadamard product.

Propositions 22 and 23 for streams straightforwardly generalise to weighted languages. Also Theorem 24 generalises to weighted languages, as follows.

Theorem 49 (Shuffle Product Elimination for Languages). *For all $\sigma \in k^{A^*}$, $r \in k$,*

$$\frac{1}{1-(r \times A)} \otimes \sigma = \frac{1}{1-(r \times A)} \times \left(\sigma \circ \frac{A}{1-(r \times A)} \right) \tag{19}$$

Corollary 50. *For all $\sigma \in k^{A^*}$, σ is rational iff $\mathcal{N}(\sigma)$ is rational. For all $\sigma, \tau \in k^{A^*}$, if both $\mathcal{N}(\sigma)$ and $\mathcal{N}(\tau)$ are polynomial resp. rational, then so is $\mathcal{N}(\sigma \odot \tau)$.*

Example 51. We illustrate the use of Theorem 49 in the calculation of the Newton transform with an example, stemming from [14, Example 2.1]. Let $A = \{\hat{0}, \hat{1}\}$, where we use the little festive hats to distinguish these alphabet symbols from $0, 1 \in k$. We define $\beta \in k^{A^*}$ by the following behavioural differential equation: $\beta_{\hat{0}} = 2 \times \beta$, $\beta_{\hat{1}} = (2 \times \beta) + \frac{1}{1-A}$, $\beta(\varepsilon) = 0$. Using Theorem 32, we can solve the differential equation above, and obtain the following expression: $\beta = \frac{1}{1-2A} \times \hat{1} \times \frac{1}{1-A}$. We have, for instance, that $\beta(\hat{0}\hat{1}\hat{1}) = \beta_{\hat{0}\hat{1}\hat{1}}(\varepsilon) = \left((8 \times \beta) + \frac{6}{1-A} \right)(\varepsilon) = 6$. More generally, β assigns to each word in A^* its value as a binary number (least significant digit first). By an easy computation, we find: $\mathcal{N}(\beta) = \frac{1}{1-A} \times \hat{1}$; in other words, $\mathcal{N}(\beta)(w) = 1$, for all w ending in $\hat{1}$. $\qquad \square$

10   Newton Series for Languages

Theorem 27 generalises to weighted languages as follows.

Theorem 52 (Newton Series for Languages, 1st). *For all $\sigma \in k^{A^*}$, $w \in A^*$,*

$$\sigma(w) = \sum_u \binom{w}{u} (\Delta^u \sigma)(\varepsilon)$$

Also Theorem 28 generalises to weighted languages.

Theorem 53 (Newton Series for Languages, 2nd; Euler Expansion).
For all $\sigma \in k^{A^*}$,

$$\sigma = \sum_{a_1 \cdots a_n \in A^*} (\Delta^{a_1 \cdots a_n} \sigma)(\varepsilon) \times \frac{1}{1-A} \times a_1 \times \frac{1}{1-A} \times \cdots \times a_n \times \frac{1}{1-A}$$

where we understand this sum to include $\sigma(\varepsilon) \times \frac{1}{1-A}$, corresponding to $\varepsilon \in A^*$.

11 Discussion

All our definitions are coinductive, given by behavioural differential equations, allowing all our proofs to be coinductive as well, that is, based on constructions of bisimulation (up-to) relations. This makes all proofs uniform and transparent. Moreover, coinductive proofs can be easily automated and often lead to efficient algorithms, for instance, as in [4]. There are several topics for further research: (i) *Theorems* 52 *and* 53 are pretty but are they also useful? We should like to investigate possible applications. (ii) The *infiltration product* deserves further study (including its restriction to streams, which seems to be new). It is reminiscent of certain versions of synchronised merge in process algebra (cf. [2]), but it does not seem to have ever been studied there. (iii) *Theorem* 47 characterises the Newton transform in terms of the shuffle product, from which many subsequent results follow. Recently [13], Newton series have been defined for functions from words to words. We are interested to see whether our present approach could be extended to those as well. (iv) *Behavioural differential equations* give rise to weighted automata (by what could be called the 'splitting' of derivatives into their summands, cf. [10]). We should like to investigate whether our representation results for Newton series could be made relevant for weighted automata as well. (v) Our new *Definition* 41 of binomial coefficients for words, which seems to offer a precise generalisation of the standard notion for numbers and, e.g., Pascal's formula, deserves further study.

Acknowledgments. We thank the anonymous referees for their constructive comments.

References

1. Barbosa, L.: Components as coalgebras. Ph.D. thesis, Universidade do Minho, Braga, Portugal (2001)
2. Bergstra, J., Klop, J.W.: Process algebra for synchronous communication. Inf. Control **60**(1), 109–137 (1984)
3. Berstel, J., Reutenauer, C.: Rational Series and Their Languages. EATCS Monographs on Theoretical Computer Science, vol. 12. Springer, Heidelberg (1988)
4. Bonchi, F., Pous, D.: Hacking nondeterminism with induction and coinduction. Commun. ACM **58**(2), 87–95 (2015)

5. Burns, S.A., Palmore, J.I.: The newton transform: an operational method for constructing integral of dynamical systems. Physica D Nonlinear Phenom. **37**(1–3), 83–90 (1989)
6. Chen, K., Fox, R., Lyndon, R.: Free differential calculus, IV - the quotient groups of the lower series. Ann. Math. Second Ser. **68**(1), 81–95 (1958)
7. Conway, J.: Regular Algebra and Finite Machines. Chapman and Hall, London (1971)
8. Eilenberg, S.: Automata, Languages and Machines. Pure and Applied Mathematics, vol. A. Academic Press, London (1974)
9. Graham, R., Knuth, D., Patashnik, O.: Concrete Mathematics, 2nd edn. Addison-Wesley, Reading (1994)
10. Hansen, H.H., Kupke, C., Rutten, J.: Stream differential equations: specification formats and solution methods. Report FM-1404, CWI (2014). www.cwi.nl
11. Lothaire, M.: Combinatorics on Words. Cambridge Mathematical Library. Cambridge University Press, Cambridge (1997)
12. Pavlović, D., Escardó, M.: Calculus in coinductive form. In: Proceedings of the 13th Annual IEEE Symposium on Logic in Computer Science, pp. 408–417. IEEE Computer Society Press (1998)
13. Pin, J.É.: Newton's forward difference equation for functions from words to words. In: Beckmann, A., Mitrana, V., Soskova, M. (eds.) CiE 2015. LNCS, vol. 9136, pp. 71–82. Springer, Heidelberg (2015)
14. Pin, J.E., Silva, P.V.: A noncommutative extension of Mahler's theorem on interpolation series. Eur. J. Comb. **36**, 564–578 (2014)
15. Rot, J., Bonsangue, M., Rutten, J.: Coalgebraic bisimulation-Up-To. In: van Emde Boas, P., Groen, F.C.A., Italiano, G.F., Nawrocki, J., Sack, H. (eds.) SOFSEM 2013. LNCS, vol. 7741, pp. 369–381. Springer, Heidelberg (2013)
16. Rutten, J.: Universal coalgebra: a theory of systems. Theor. Comput. Sci. **249**(1), 3–80 (2000). Fundamental Study
17. Rutten, J.: Behavioural differential equations: a coinductive calculus of streams, automata, and power series. Theor. Comput. Sci. **308**(1), 1–53 (2003). Fundamental Study
18. Rutten, J.: A coinductive calculus of streams. Math. Struct. Comput. Sci. **15**, 93–147 (2005)
19. Scheid, F.: Theory and Problems Of Numerical Analysis. Schaum's outline series. McGraw-Hill, New York (1968)

Quotienting the Delay Monad
by Weak Bisimilarity

James Chapman, Tarmo Uustalu, and Niccolò Veltri[✉]

Institute of Cybernetics, Tallinn University of Technology,
Akadeemia tee 21, 12618 Tallinn, Estonia
{james,tarmo,niccolo}@cs.ioc.ee

Abstract. The delay datatype was introduced by Capretta [3] as a
means to deal with partial functions (as in computability theory) in
Martin-Löf type theory. It is a monad and it constitutes a constructive alternative to the maybe monad. It is often desirable to consider
two delayed computations equal, if they terminate with equal values,
whenever one of them terminates. The equivalence relation underlying
this identification is called weak bisimilarity. In type theory, one commonly replaces quotients with setoids. In this approach, the delay monad
quotiented by weak bisimilarity is still a monad. In this paper, we consider Hofmann's alternative approach [6] of extending type theory with
inductive-like quotient types. In this setting, it is difficult to define the
intended monad multiplication for the quotiented datatype. We give a
solution where we postulate some principles, crucially proposition extensionality and the (semi-classical) axiom of countable choice. We have
fully formalized our results in the Agda dependently typed programming
language.

1 Introduction

The delay datatype was introduced by Capretta [3] as a means to deal with partial functions (as in computability theory) in Martin-Löf type theory. It is used
in this setting to cope with possible non-termination of computations (as, e.g.,
in the unbounded search of minimalization). Inhabitants of the delay datatype
are delayed values, that we call computations throughout this paper. Crucially
computations can be non-terminating and not return a value at all. The delay
datatype constitutes a (strong) monad, which makes it possible to deal with possibly non-terminating computations just like any other flavor of effectful computations following Moggi's general monad-based method [12]. Often, one is only
interested in termination of computations and not the exact computation time.
Identifying computations that only differ by finite amounts of delay corresponds
to quotienting the delay datatype by weak bisimilarity. The quotient datatype
is used as a constructive alternative to the maybe datatype (see, e.g., [2]) and
should also be a (strong) monad.

Martin-Löf type theory does not have built-in quotient types. The most common approach to compensate for this is to mimic them by working with setoids.

© Springer International Publishing Switzerland 2015
M. Leucker et al. (Eds.): ICTAC 2015, LNCS 9399, pp. 110–125, 2015.
DOI: 10.1007/978-3-319-25150-9_8

But this approach has some obvious shortcomings as well, for example, the concept of a function type is changed (every function has to come with a compatibility proof) etc. An alternative approach, which we pursue here, consists in extending the theory by postulating the existence of inductive-like quotient types à la Hofmann [6]. These quotient types are ordinary types rather than setoids.

In this paper, we ask the question: is the monad structure of the delay datatype preserved under quotienting by weak bisimilarity? Morally, this ought to be the case. In the setoid approach, this works out unproblematically indeed. But with inductive-like quotient types, one meets a difficulty when attempting to reproduce the monad structure on the quotiented datatype. Specifically, one cannot define the multiplication. The difficulty has to do with the interplay of the coinductive nature of the delay datatype, or more precisely the infinity involved, and quotient types. We discuss the general phenomenon behind this issue and provide a solution where we postulate some principles, the crucial ones being proposition extensionality (accepted in particular in homotopy type theory) and the (semi-classical) axiom of countable choice. It is very important here to be careful and not postulate too much: in the presence of proposition extensionality, the full axiom of choice implies the law of excluded middle.

As an aside, we also look at the (strong) arrow structure (in the sense of Hughes [7]) on the Kleisli function type for the delay datatype and ask whether this survives quotienting by pointwise weak bisimilarity. Curiously, here the answer is unconditionally positive also for inductive-like quotient types.

This paper is organized as follows. In Sect. 2, we give an overview of the type theory we are working in. In Sect. 3, we introduce the delay datatype and weak bisimilarity. In Sect. 4, we extend type theory with quotients à la Hofmann. In Sect. 5, we analyze why a multiplication for the quotiented delay type is impossible to define. We notice that the problem is of a more general nature, and a larger class of types, namely non-wellfounded and non-finitely branching trees, suffer from it. In Sect. 6, we introduce the axiom of countable choice and derive some important consequences from postulating it. In Sect. 7, using the results of Sect. 6, we define multiplication for the delay type quotiented by weak bisimilarity (we omit the proof of the monad laws, which is the easy part—essentially the proofs for the unquotiented delay datatype carry over). In Sect. 8, we quotient the arrow corresponding to the monad by pointwise weak bisimilarity. Finally, in Sect. 9, we draw some conclusions and discuss future work.

We have fully formalized the results of this paper in the dependently typed programming language Agda [13]. The formalization is available at http://cs.ioc.ee/~niccolo/delay/.

2   The Type Theory Under Consideration

We consider Martin-Löf type theory with inductive and coinductive types and a cumulative hierarchy of universes $\mathcal{U}_k$. To define functions from inductive types or to coinductive types, we use guarded (co)recursion. The first universe is simply denoted $\mathcal{U}$ and when we write statements like "X is a type", we mean

$X : \mathcal{U}$ unless otherwise specified. We allow dependent functions to have implicit arguments and indicated implicit argument positions with curly brackets (as in Agda). We write $\equiv$ for propositional equality (identity types) and $=$ for judgmental (definitional) equality. Reflexivity, transitivity and substitutivity of $\equiv$ are named refl, trans and subst, respectively.

We assume the principle of *function extensionality*, expressing that pointwise equal functions are equal, i.e., the inhabitedness of

$$ \mathsf{FunExt} = \prod_{\{X,Y:\mathcal{U}\}} \prod_{\{f_1,f_2:X\to Y\}} \left(\prod_{x:X} f_1\,x \equiv f_2\,x \right) \to f_1 \equiv f_2 $$

Likewise we will assume analogous extensionality principles stating that strongly bisimilar coinductive data and proofs are equal for the relevant coinductive types and predicates, namely, the delay datatype and weak bisimilarity (check DExt, $\approx$Ext below in Sects. 3 and 4).

We also assume *uniqueness of identity proofs* for all types,[1] i.e., an inhabitant for

$$ \mathsf{UIP} = \prod_{\{X:\mathcal{U}\}} \prod_{\{x_1,x_2:X\}} \prod_{p_1,p_2:x_1\equiv x_2} p_1 \equiv p_2. $$

A type X is said to be a *proposition*, if it has at most one inhabitant, i.e., if the type

$$ \mathsf{isProp}\,X = \prod_{x_1,x_2:X} x_1 \equiv x_2 $$

is inhabited.

For propositions, we postulate a further and less standard principle of *proposition extensionality*, stating that logically equivalent propositions are equal:[2]

$$ \mathsf{PropExt} = \prod_{\{X,Y:\mathcal{U}\}} \mathsf{isProp}\,X \to \mathsf{isProp}\,Y \to X \leftrightarrow Y \to X \equiv Y $$

Here $X \leftrightarrow Y = (X \to Y) \times (Y \to X)$.

3   Delay Monad

For a given type X, each element of $\mathsf{D}\,X$ is a possibly infinite computation that returns a value of X, if it terminates. We define $\mathsf{D}\,X$ as a coinductive type by the rules

$$ \frac{}{\mathsf{now}\,x : \mathsf{D}\,X} \qquad \frac{c : \mathsf{D}\,X}{\mathsf{later}\,c : \mathsf{D}\,X} $$

---

[1] Working in homotopy type theory [15], we would assume this principle only for 0-types, i.e., sets, and that would also be enough for our purposes.

[2] Propositions are (-1)-types and proposition extensionality is univalence for (-1)-types.

Let R be an equivalence relation on a type X. The relation lifts to an equivalence relation $\sim_R$ on $\mathsf{D}\,X$ that we call *strong R-bisimilarity*. The relation is coinductively defined by the rules

$$\frac{p : x_1 R x_2}{\mathsf{now}_\sim p : \mathsf{now}\, x_1 \sim_R \mathsf{now}\, x_2} \qquad \frac{p : c_1 \sim_R c_2}{\mathsf{later}_\sim p : \mathsf{later}\, c_1 \sim_R \mathsf{later}\, c_2}$$

We alternatively denote the relation $\sim_R$ with $\mathsf{D}\,R$, since strong R-bisimilarity is the functorial lifting of the relation R to $\mathsf{D}\,X$. Strong $\equiv$-bisimilarity is simply called strong bisimilarity and denoted $\sim$. While it ought to be the case morally, one cannot prove that strongly bisimilar computations are equal in Martin-Löf type theory. Therefore we postulate an inhabitant for

$$\mathsf{DExt} = \prod_{\{X:\mathcal{U}\}} \prod_{\{c_1,c_2 : \mathsf{D}\,X\}} c_1 \sim c_2 \to c_1 \equiv c_2$$

We take into account another equivalence relation $\approx_R$ on $\mathsf{D}\,X$ called *weak R-bisimilarity*, which is in turn defined in terms of *convergence*. The latter is a binary relation between $\mathsf{D}\,X$ and X relating terminating computations to their values. It is inductively defined by the rules

$$\frac{p : x_1 \equiv x_2}{\mathsf{now}_\downarrow p : \mathsf{now}\, x_1 \downarrow x_2} \qquad \frac{p : c \downarrow x}{\mathsf{later}_\downarrow p : \mathsf{later}\, c \downarrow x}$$

Two computations are considered weakly R-bisimilar, if they differ by a finite number of applications of the constructor later (from where it follows classically that they either converge to R-related values or diverge). Weak R-bisimilarity is defined coinductively by the rules

$$\frac{p_1 : c_1 \downarrow x_1 \quad p_2 : x_1 R x_2 \quad p_3 : c_2 \downarrow x_2}{\downarrow_\approx p_1\, p_2\, p_3 : c_1 \approx_R c_2} \qquad \frac{p : c_1 \approx_R c_2}{\mathsf{later}_\approx p : \mathsf{later}\, c_1 \approx_R \mathsf{later}\, c_2}$$

Weak $\equiv$-bisimilarity is called just weak bisimilarity and denoted $\approx$. In this case, we modify the first constructor for simplicity:

$$\frac{p_1 : c_1 \downarrow x \quad p_2 : c_2 \downarrow x}{\downarrow_\approx p_1\, p_2 : c_1 \approx c_2}$$

The delay datatype D is a (strong) monad. The unit η is the constructor now while the multiplication μ is "concatenation" of laters:

$$\mu : \mathsf{D}\,(\mathsf{D}\,X) \to \mathsf{D}\,X$$
$$\mu\,(\mathsf{now}\, c) = c$$
$$\mu\,(\mathsf{later}\, c) = \mathsf{later}\,(\mu\, c)$$

In the quotients-as-setoids approach, it is trivial to define the corresponding (strong) monad structure on the quotient of D by $\approx$. The role of the quotiented datatype is played by the setoid functor $\hat{\mathsf{D}}$, defined by $\hat{\mathsf{D}}(X, R) = (\mathsf{D}\,X, \approx_R)$. The unit $\hat{\eta}$ and multiplication $\hat{\mu}$ are just η and μ together with proofs of that the appropriate equivalences are preserved. The unit $\hat{\eta}$ is a setoid morphism from (X, R) to $(D\,X, \approx_R)$, as $x_1\,R\,x_2 \to \mathsf{now}\,x_1 \approx_R \mathsf{now}\,x_2$ by definition of $\approx_R$. The multiplication $\hat{\mu}$ is a setoid morphism from $(\mathsf{D}\,(\mathsf{D}\,X), \approx_{\approx_R})$ to $(\mathsf{D}\,X, \approx_R)$, since $c_1 \approx_{\approx_R} c_2 \to \mu\,c_1 \approx_R \mu\,c_2$ for all $c_1, c_2 : \mathsf{D}\,(\mathsf{D}\,X)$. The monad laws hold up to $\approx_R$, since they hold up to $\sim_R$.

In this paper, our goal is to establish that the delay datatype quotiented by weak bisimilarity is a monad also in Hofmann's setting [6], where the quotient type of a given type has its propositional equality given by the equivalence relation. We discuss such quotient types in the next section.

4   Inductive-Like Quotients

In this section, we describe quotient types as particular inductive-like types introduced by M. Hofmann in his PhD thesis [6]. Let X be a type and R an equivalence relation on X. For any type Y and function $f : X \to Y$, we say that f is R-*compatible* (or simply *compatible*, when the intended equivalence relation is clear from the context), if the type

$$\mathsf{compat}\,f = \prod_{\{x_1, x_2 : X\}} x_1 R x_2 \to f\,x_1 \equiv f\,x_2$$

is inhabited. The quotient of X by the relation R is described by the following data:

(i) a carrier type X/R;
(ii) a constructor $[\_] : X \to X/R$ together with a proof $\mathsf{sound} : \mathsf{compat}\,[\_]$;
(iii) a dependent eliminator: for every family of types $Y : X/R \to \mathcal{U}_k$ and function $f : \prod_{x:X} Y\,[x]$ with $p : \mathsf{dcompat}\,f$, there exists a function $\mathsf{lift}\,f\,p : \prod_{q:X/R} Y\,q$ together with a computation rule

$$\mathsf{lift}_\beta\,f\,p\,x : \mathsf{lift}\,f\,p\,[x] \equiv f\,x$$

for all $x : X$.

The predicate $\mathsf{dcompat}$ is compatibility for dependent functions $f : \prod_{x:X} Y\,[x]$:

$$\mathsf{dcompat}\,f = \prod_{\{x_1, x_2 : X\}} \prod_{r : x_1 R x_2} \mathsf{subst}\,Y\,(\mathsf{sound}\,r)\,(f\,x_1) \equiv f\,x_2.$$

We postulate the existence of data (i)–(iii) for all types X and equivalence relations R on X. Notice that the predicate $\mathsf{dcompat}$ depends of the availability of sound. Also notice that, in (iii), we allow elimination on every universe $\mathcal{U}_k$. In our development, we actually eliminate only on $\mathcal{U}$ and once on $\mathcal{U}_1$ (Proposition 2).

The *propositional truncation* (or *squash*) $\|X\|$ of a type X is the quotient of X by the total relation $\lambda\,x_1\,x_2.\,\top$. We write $|\_|$ instead of $[\_]$ for the constructor of $\|X\|$. The non-dependent version of the elimination principle of $\|X\|$ is employed several times in this paper, so we spell it out: in order to construct a function of type $\|X\| \to Y$, one has to construct a constant function of type $X \to Y$.

Informally an inhabitant of $\|X\|$ corresponds to an "uninformative" proof of inhabitedness of X. For example, an inhabitant of $\|\sum_{x:X} P x\|$ can be thought of as a proof of there existing an element of X that satisfies P that has forgotten the information of which element satisfies the predicate P. Propositional truncation and other notions of weak or anonymous existence have been thoroughly studied in type theory [9].

We call a function $f : X \rightarrow Y$ *surjective*, if the type $\prod_{y:Y} \|\sum_{x:X} f x \equiv y\|$ is inhabited, and a *split epimorphism*, if the type $\|\sum_{g:Y \rightarrow X} \prod_{y:Y} f(g y) \equiv y\|$ is inhabited. We say that f is a *retraction*, if the type $\sum_{g:Y \rightarrow X} \prod_{y:Y} f(g y) \equiv y$ is inhabited. Every retraction is a split epimorphism, and every split epimorphism is surjective.

Proposition 1. *The constructor [_] is surjective for all quotients.*

Proof. Given a type X and an equivalence relation R on X, we define:

$$[\_]\mathsf{surj} : \prod_{q:X/R} \left\| \sum_{x:X} [x] \equiv q \right\|$$

$$[\_]\mathsf{surj} = \mathsf{lift}\,(\lambda x.\,|x, \mathsf{refl}|)\,p$$

The compatibility proof p is trivial, since $|x_1, \mathsf{refl}| \equiv |x_2, \mathsf{refl}|$ for all $x_1, x_2 : X$. $\square$

A quotient X/R is said to be *effective*, if the type $\prod_{x_1,x_2:X} [x_1] \equiv [x_2] \rightarrow x_1 R x_2$ is inhabited. In general, effectiveness does not hold for all quotients. But we can prove that all quotients satisfy a weaker property. We say that a quotient X/R is *weakly effective*, if the type $\prod_{x_1,x_2:X} [x_1] \equiv [x_2] \rightarrow \|x_1 R x_2\|$ is inhabited.

Proposition 2. *All quotients are weakly effective.*

Proof. Let X be a type, R an equivalence relation on X and $x : X$. Consider the function $\|x R \_\| : X \rightarrow \mathcal{U}$, $\|x R \_\| = \lambda x'.\|x R x'\|$. We show that $\|x R \_\|$ is R-compatible. Let $x_1, x_2 : X$ with $x_1 R x_2$. We have $x R x_1 \leftrightarrow x R x_2$ and therefore $\|x R x_1\| \leftrightarrow \|x R x_2\|$. Since propositional truncations are propositions, using proposition extensionality, we conclude $\|xRx_1\| \equiv \|xRx_2\|$. We have constructed a term $p_x : \mathsf{compat}\,\|x R \_\|$, and therefore a function $\mathsf{lift}\,\|x R \_\|\,p_x : X/R \rightarrow \mathcal{U}$ (large elimination is fundamental in order to apply lift, since $\|x R \_\| : X \rightarrow \mathcal{U}$ and $X \rightarrow \mathcal{U} : \mathcal{U}_1$). Moreover, $\mathsf{lift}\,\|x R \_\|\,p_x\,[y] \equiv \|x R y\|$ by its computation rule.

Let $[x_1] \equiv [x_2]$ for some $x_1, x_2 : X$. We have:

$$\|x_1 R x_2\| \equiv \mathsf{lift}\,\|x_1 R \_\|\,p_{x_1}\,[x_2] \equiv \mathsf{lift}\,\|x_1 R \_\|\,p_{x_1}\,[x_1] \equiv \|x_1 R x_1\|$$

and $x_1 R x_1$ holds, since R is reflexive. $\square$

Notice that the constructor [_] is not a split epimorphism for all quotients. The existence of a choice of representative for each equivalence class is a non-constructive principle, since it implies the *law of excluded middle*, i.e., the inhabitedness of the following type:

$$\mathsf{LEM} = \prod_{\{X:\mathcal{U}\}} \mathsf{isProp}\,X \rightarrow X + \neg X$$

where $\neg X = X \rightarrow \bot$.

Proposition 3. *Suppose that* [_] *is a split epimorphism for all quotients. Then* LEM *is inhabited.*

Proof. Let X be a type together with a proof of isProp X. We consider the equivalence relation R on Bool, $x_1 R x_2 = x_1 \equiv x_2 + X$. By hypothesis we obtain $\| \sum_{\text{rep:Bool}/R \to \text{Bool}} \prod_{q:\text{Bool}/R} [\text{rep } q] \equiv q \|$. Using the elimination principle of propositional truncation, it is sufficient to construct a constant function of type:

$$\sum_{\text{rep:Bool}/R \to \text{Bool}} \prod_{q:\text{Bool}/R} [\text{rep } q] \equiv q \ \to \ X + \neg X$$

Let rep : Bool$/R \to$ Bool with $[\text{rep } q] \equiv q$ for all q : Bool$/R$. We have $[\text{rep } [x]] \equiv [x]$ for all x : Bool, which by Proposition 2 implies $\|\text{rep } [x] R x\|$.

Note now that the following implication (a particular form of axiom of choice on Bool) holds:

$$\text{acBool} : \prod_{x:\text{Bool}} \|\text{rep } [x] R x\| \ \to \ \left\| \prod_{x:\text{Bool}} \text{rep } [x] R x \right\|$$

$$\text{acBool } r = \text{lift}_2 \left(\lambda r_1 r_2. |d\, r_1 r_2| \right) p \, (r\,\text{true}) \, (r\,\text{false})$$

where $d\, r_1 r_2 \,\text{true} = r_1$ and $d\, r_1 r_2 \,\text{false} = r_2$, and lift_2 is the two-argument version of lift. The compatibility proof p is immediate, since the return type is a proposition.

We now construct a function of type $\|\prod_{x:\text{Bool}} \text{rep } [x] R x\| \to X + \neg X$. It is sufficient to define a function $\prod_{x:\text{Bool}} \text{rep } [x] R x \to X + \neg X$ (it will be constant, since the type $X + \neg X$ is a proposition, if X is a proposition), so we suppose rep $[x] R x$ for all x : Bool. We analyze the (decidable) equality rep $[\text{true}] \equiv$ rep $[\text{false}]$ on Bool. If it holds, then we have true R false and therefore an inhabitant of X. If it does not hold, we have an inhabitant of $\neg X$: let x : X, therefore true R false, and this implies rep $[\text{true}] \equiv$ rep $[\text{false}]$ holds, which contradicts the hypothesis. □

We already noted that not all quotients are effective. In fact, postulating effectiveness for all quotients implies LEM [10]. But the quotient we are considering in this paper, namely $D\,X/\approx$ for a type X, is indeed effective. Notice that, by Proposition 2, it suffices to prove that $\|c_1 \approx c_2\| \to c_1 \approx c_2$ for all c_1, c_2 : $D\,X$.

Lemma 1. *For all types X and c_1, c_2 : $D\,X$, there exists a constant endofunction on $c_1 \approx c_2$. Therefore, the type $\|c_1 \approx c_2\| \to c_1 \approx c_2$ is inhabited.*

Proof. Let X be a type and c_1, c_2 : $D\,X$. We consider the following function.

$$\text{canon}\approx \,:\, c_1 \approx c_2 \to c_2 \approx c_2$$
$$\text{canon}\approx \,(\downarrow_\approx (\text{now}_\downarrow p_1)\, p_2) = \downarrow_\approx (\text{now}_\downarrow p_1)\, p_2$$
$$\text{canon}\approx \,(\downarrow_\approx (\text{later}_\downarrow p_1)\, (\text{now}_\downarrow p_2)) = \downarrow_\approx (\text{later}_\downarrow p_1)\, (\text{now}_\downarrow p_2)$$
$$\text{canon}\approx \,(\downarrow_\approx (\text{later}_\downarrow p_1)\, (\text{later}_\downarrow p_2)) = \text{later}_\approx (\text{canon}\approx (\downarrow_\approx p_1 p_2))$$
$$\text{canon}\approx \,(\text{later}_\approx p) = \text{later}_\approx (\text{canon}\approx p)$$

The function canon$\approx$ canonizes a given weak bisimilarity proof by maximizing the number of applications of the constructor later$_\approx$. This function is indeed constant, i.e., one can prove $\prod_{p_1,p_2:c_1\approx c_2} p_1 \cong p_2$ for all $c_1, c_2 : D\,X$, where the relation $\cong$ is strong bisimilarity on proofs of $c_1 \approx c_2$, coinductively defined by the rules:

$$\frac{}{\downarrow_\approx p_1\,p_2 \cong \downarrow_\approx p_1\,p_2} \qquad \frac{p_1 \cong p_2}{\mathsf{later}_\approx p_1 \cong \mathsf{later}_\approx p_2}$$

Similarly to extensionality of delayed computations, we assume that strongly bisimilar weak bisimilarity proofs are equal, i.e., that we have an inhabitant for

$$\approx\!\mathsf{Ext} = \prod_{\{X:\mathcal{U}\}} \prod_{\{c_1,c_2:D\,X\}} \prod_{p_1,p_2:c_1\approx c_2} p_1 \cong p_2 \rightarrow p_1 \equiv p_2 \qquad\qquad \Box$$

5 Multiplication: What Goes Wrong?

Consider now the type functor $\bar{D}$, defined by $\bar{D}\,X = D\,X/\approx$. Let us try to equip it with a monad structure. Let X be a type. As the unit $\bar{\eta} : X \rightarrow D\,X/\approx$, we can take $[\_]\circ\mathsf{now}$. But when we try to construct a multiplication $\bar{\mu} : D\,(D\,X/\approx)/\approx \rightarrow D\,X/\approx$, we get stuck immediately. Indeed, $\bar{\mu}$ must be of the form $\mathsf{lift}\,\bar{\mu}'\,p$ for some $\bar{\mu}' : D\,(D\,X/\approx) \rightarrow D\,X/\approx$ with $p : \mathsf{compat}\,\bar{\mu}'$, but we cannot define such $\bar{\mu}'$ and p. The problem lies in the coinductive nature of the delay datatype. A function of type $D\,(D\,X/\approx) \rightarrow D\,X/\approx$ should send a converging computation to its converging value and a non-terminating one to the equivalence class of non-termination. This discontinuity makes constructing such a function problematic. Moreover, one can show that a right inverse of $[\_] : D\,X \rightarrow D\,X/\approx$, i.e., a canonical choice of representative for each equivalence class in $D\,X/\approx$, is not definable. Therefore, we cannot even construct $\bar{\mu}'$ as a composition $[\_] \circ \bar{\mu}''$ with $\bar{\mu}'' : D\,(D\,X/\approx) \rightarrow D\,X$, since we do not know how to define $\bar{\mu}''(\mathsf{now}\,q)$ for $q : D\,X/\approx$.

A function $\bar{\mu}'$ would be constructable, if the type $D\,(D\,X/\approx)$ were a quotient of $D\,(D\,X)$ by the equivalence relation $D\approx$ (remember that $D\approx$ is a synonym of $\sim_\approx$, the functorial lifting of $\approx$ from $D\,X$ to $D\,(D\,X)$). In fact, the function $[\_] \circ \mu : D\,(D\,X) \rightarrow D\,X/\approx$ is $D\approx$-compatible, since $x_1(D\approx)x_2 \rightarrow \mu x_1 \approx \mu x_2$, and therefore the elimination principle would do the job. But how "different" are $D\,(D\,X/\approx)$ and the quotient $D\,(D\,X)/D\approx$? More generally, how "different" are $D\,(X/R)$ and the quotient $D\,X/D\,R$, for a given type X and equivalence relation R on X?

A function $\theta^D : D\,X/D\,R \rightarrow D\,(X/R)$ always exists, $\theta^D = \mathsf{lift}\,(D\,[\_])\,p$. The compatibility proof p follows directly from $c_1(D\,R)c_2 \rightarrow D\,[\_]\,c_1 \sim D\,[\_]\,c_2$. But an inverse function $\psi^D : D\,(X/R) \rightarrow D\,X/D\,R$ is not definable. This phenomenon can be spotted more generally in non-wellfounded trees, i.e., the canonical function $\theta^T : T\,X/T\,R \rightarrow T\,(X/R)$ does not have an inverse, if $T\,X$ is coinductively defined, where $T\,R$ is the functorial lifting of R to $T\,X$. On the other hand, a large class of purely inductive types, namely, the datatypes of wellfounded trees where branching is finite, is free of this problem. As an example, for binary trees the inverse $\psi^{\mathsf{BTree}} : \mathsf{BTree}\,(X/R) \to \mathsf{BTree}\,X/\mathsf{BTree}\,R$ of $\theta^{\mathsf{BTree}} : \mathsf{BTree}\,X/\mathsf{BTree}\,R \to \mathsf{BTree}\,(X/R)$ is defined as follows:

$$\psi^{\mathsf{BTree}} : \mathsf{BTree}\,(X/R) \to \mathsf{BTree}\,X/\mathsf{BTree}\,R$$

$$\psi^{\mathsf{BTree}}\,(\mathsf{leaf}\,q) = \mathsf{lift}\,(\lambda\,x.\,[\mathsf{leaf}\,x])\,p_{\mathsf{leaf}}\,q$$

$$\psi^{\mathsf{BTree}}\,(\mathsf{node}\,t_1\,t_2) = \mathsf{lift}_2\,(\lambda\,s_1\,s_2.\,[\mathsf{node}\,s_1\,s_2])\,p_{\mathsf{node}}\,(\psi^{\mathsf{BTree}}\,t_1)\,(\psi^{\mathsf{BTree}}\,t_2)$$

where lift_2 is the two-argument version of lift. The simple compatibility proofs p_{leaf} and p_{node} are omitted. Wellfounded non-finitely branching trees are affected by the same issues that non-wellfounded trees have. And in general, for a W-type T, the function $\theta^T : T\,X/T\,R \to T\,(X/R)$ is not invertible, since for function spaces the function $\theta^{\to} : (Y \to X)/(Y \to R) \to (Y \to X/R)$ is not invertible. Invertibility of the function $\theta^{\to} : (Y \to X)/(Y \to R) \to (Y \to X/R)$, for all types Y, X and equivalence relation R on X, has been analyzed in the Calculus of Inductive Constructions [4]. It turns out that surjectivity of $\theta^{\to}$ is logically equivalent to the full axiom of choice (AC)[3], i.e., the following type is inhabited:

$$\mathsf{AC} = \prod_{\{X,Y:\mathcal{U}\}}\;\prod_{P:X\to Y\to\mathcal{U}}\left(\prod_{x:X}\left\|\sum_{y:Y} P\,x\,y\right\|\right) \to \left\|\sum_{f:X\to Y}\prod_{x:X} P\,x\,(f\,x)\right\|$$

Together with weak effectiveness (Proposition 2), AC not only implies surjectivity of $\theta^{\to}$, but also the existence of an inverse $\psi^{\to} : (Y \to X/R) \to (Y \to X)/(Y \to R)$. We refrain from proving these facts, but we prove Lemma 2 and Proposition 5, which are weaker statements, but have analogous proofs.

The existence of an inverse $\psi^{\to}$ of $\theta^{\to}$ would immediately allow us to define the bind operation for $\bar{\mathsf{D}}$. Let us consider the case where X is $\mathsf{D}\,X$ and R is weak bisimilarity, so $\psi^{\to} : (Y \to \mathsf{D}\,X/\approx) \to (Y \to \mathsf{D}\,X)/(Y \to \approx)$. We define

$$\overline{\mathsf{bind}} : (Y \to \mathsf{D}\,X/\approx) \to \mathsf{D}\,Y/\approx \to \mathsf{D}\,X/\approx$$

$$\overline{\mathsf{bind}}\,f\,q = \mathsf{lift}_2\,(\lambda\,g\,c.\,[\mathsf{bind}\,g\,c])\,p\,(\psi^{\to}\,f)\,q$$

where bind is the bind operation of the unquotiented delay monad. The compatibility proof p is obtained from the fact that $\mathsf{bind}\,g_1\,c_1 \approx \mathsf{bind}\,g_2\,c_2$ if $c_1 \approx c_2$ and $g_1\,y \approx g_2\,y$ for all $y : Y$.

AC is a controversial semi-classical axiom, generally not accepted in constructive systems [11]. We reject it too, since in our system the axiom of choice implies the law of excluded middle.

---

[3] Notice that AC is fundamentally different from the *type-theoretic* axiom of choice:

$$\prod_{\{X,Y:\mathcal{U}\}}\;\prod_{P:X\to Y\to\mathcal{U}}\left(\prod_{x:X}\sum_{y:Y} P\,x\,y\right) \to \sum_{f:X\to Y}\prod_{x:X} P\,x\,(f\,x)$$

which is provable in type theory.

Proposition 4. AC *implies* LEM.

Proof. Assume AC. With a proof analogous to Lemma 2, we can prove that the function $\lambda f. [\_] \circ f : (X \to Y) \to (X \to Y/R)$ is surjective, for any types X, Y and equivalence relation R on Y. In particular, given a type X and an equivalence relation R on X, we have that the type $\prod_{g:X/R \to X/R} \left\| \sum_{f:X/R \to X} [\_] \circ f \equiv g \right\|$ is inhabited. Instantiating g with the identity function on X/R, we obtain $\left\| \sum_{f:X/R \to X} \prod_{q:X/R} [f\, q] \equiv q \right\|$, i.e., the constructor $[\_]$ is a split epimorphism for all quotients X/R. By Proposition 3, this implies LEM. $\square$

In the following sections, we show that the weaker axiom of countable choice is already enough for constructing a multiplication for $\bar{\mathsf{D}}$. Countable choice does not imply excluded middle and constructive mathematicians like it more [14, Ch. 4].

6   Axiom of Countable Choice and Streams of Quotients

The axiom of countable choice ($\mathsf{AC}\omega$) is a specific instance of AC where the binary predicate P has its first argument in $\mathbb{N}$:

$$\mathsf{AC}\omega = \prod_{\{X:\mathcal{U}\}} \prod_{P:\mathbb{N} \to X \to \mathcal{U}} \left(\prod_{n:\mathbb{N}} \left\| \sum_{x:X} P\,n\,x \right\| \right) \to \left\| \sum_{f:\mathbb{N} \to X} \prod_{n:\mathbb{N}} P\,n\,(f\,n) \right\|$$

We also introduce a logically equivalent formulation of $\mathsf{AC}\omega$ that will be used in Proposition 5:

$$\mathsf{AC}\omega_2 = \prod_{P:\mathbb{N} \to \mathcal{U}} \left(\prod_{n:\mathbb{N}} \|P\,n\| \right) \to \left\| \prod_{n:\mathbb{N}} P\,n \right\|$$

Let X be a type and R an equivalence relation on it. We show that $\mathsf{AC}\omega$ implies the surjectivity of the function $[\_]^{\mathbb{N}} : (\mathbb{N} \to X) \to (\mathbb{N} \to X/R)$, $[f]^{\mathbb{N}} n = [f\,n]$. This in turn implies the definability of a function $\psi^{\mathbb{N}} : (\mathbb{N} \to X/R) \to (\mathbb{N} \to X)/(\mathbb{N} \to R)$ that is inverse of the canonical function $\theta^{\mathbb{N}} = \mathsf{lift}\,[\_]^{\mathbb{N}}\,\mathsf{sound}^{\mathbb{N}}$, where

$$\mathsf{sound}^{\mathbb{N}} : \mathsf{compat}\,[\_]^{\mathbb{N}}$$

$$\mathsf{sound}^{\mathbb{N}}\,r = \mathsf{funext}\,(\lambda n.\,\mathsf{sound}\,(r\,n)).$$

using $\mathsf{funext} : \mathsf{FunExt}$.

Lemma 2. *Assume* $\mathsf{ac}\omega : \mathsf{AC}\omega$. *Then* $[\_]^{\mathbb{N}}$ *is surjective.*

Proof. Given any $g : \mathbb{N} \to X/R$, we construct a term $e_g : \left\| \sum_{f:\mathbb{N} \to X} [f]^{\mathbb{N}} \equiv g \right\|$. Since we are assuming the principle of function extensionality, it is sufficient to find a term $e'_g : \left\| \sum_{f:\mathbb{N} \to X} \prod_{n:\mathbb{N}} [f\,n] \equiv g\,n \right\|$. Define $P : \mathbb{N} \to X \to \mathcal{U}$ by $P\,n\,x = [x] \equiv g\,n$. We take $e'_g = \mathsf{ac}\omega\,P\,(\lambda n.\,[\_]\mathsf{surj}\,(g\,n))$, with $[\_]\mathsf{surj}$ introduced in Proposition 1. $\square$

Proposition 5. *Assume* $\mathsf{AC}\omega$. *Then* $\theta^{\mathbb{N}} : (\mathbb{N} \to X)/(\mathbb{N} \to R) \to (\mathbb{N} \to X/R)$ *is invertible.*

Proof. We construct a term

$$r : \sum_{\psi^{\mathbb{N}} : (\mathbb{N} \to X/R) \to (\mathbb{N} \to X)/(\mathbb{N} \to R)} \prod_{g : \mathbb{N} \to X/R} \theta^{\mathbb{N}} (\psi^{\mathbb{N}} g) \equiv g$$

Given any $g : \mathbb{N} \to X/R$, we define:

$$h'_g : \left(\sum_{f : \mathbb{N} \to X} [f]^{\mathbb{N}} \equiv g \right) \to \sum_{q : (\mathbb{N} \to X)/(\mathbb{N} \to R)} \theta^{\mathbb{N}} q \equiv g$$

$$h'_g (f, p) = \left([f], \mathsf{trans} \left(\mathsf{lift}_\beta [\_]^{\mathbb{N}} \mathsf{sound}^{\mathbb{N}} f \right) p \right)$$

The function h'_g is constant. Indeed, let $f_1, f_2 : \mathbb{N} \to X$ with $p_1 : [f_1]^{\mathbb{N}} \equiv g$ and $p_2 : [f_2]^{\mathbb{N}} \equiv g$. By uniqueness of identity proofs, it is sufficient to show $[f_1] \equiv [f_2]$. By symmetry and transitivity, we get $[f_1]^{\mathbb{N}} \equiv [f_2]^{\mathbb{N}}$. We construct the following series of implications:

$$[f_1]^{\mathbb{N}} \equiv [f_2]^{\mathbb{N}} \to \prod_{n : \mathbb{N}} [f_1 \, n] \equiv [f_2 \, n]$$

$$\to \prod_{n : \mathbb{N}} \|(f_1 \, n) \, R \, (f_2 \, n)\| \qquad \text{(by weak effectiveness)}$$

$$\to \left\| \prod_{n : \mathbb{N}} (f_1 \, n) \, R \, (f_2 \, n) \right\| \qquad \text{(by $\mathsf{AC}\omega$ and $\mathsf{AC}\omega \to \mathsf{AC}\omega_2$)}$$

$$= \| f_1 \, (\mathbb{N} \to R) \, f_2 \|$$

$$\to [f_1] \equiv [f_2]$$

The last implication is given by the elimination principle of propositional truncation applied to sound, which is a constant function by uniqueness of identity proofs. Therefore h'_g is constant and we obtain a function

$$h_g : \left\| \sum_{f : \mathbb{N} \to X} [f]^{\mathbb{N}} \equiv g \right\| \to \sum_{q : (\mathbb{N} \to X)/(\mathbb{N} \to R)} \theta^{\mathbb{N}} q \equiv g$$

We get $h_g \, e_g : \sum_{q : (\mathbb{N} \to X)/(\mathbb{N} \to R)} \theta^{\mathbb{N}} q \equiv g$, with e_g constructed in Lemma 2. We take $r = (\lambda g. \mathsf{fst} \, (h_g \, e_g), \lambda g. \mathsf{snd} \, (h_g \, e_g))$ and $\psi^{\mathbb{N}} = \mathsf{fst} \, r$.

We now prove that $\psi^{\mathbb{N}} (\theta^{\mathbb{N}} q) \equiv q$ for all $q : (\mathbb{N} \to X)/(\mathbb{N} \to R)$. It is sufficient to prove this equality for $q = [f]$ with $f : \mathbb{N} \to X$. By the computation rule of quotients, we have to show $\psi^{\mathbb{N}} [f]^{\mathbb{N}} \equiv [f]$. This is true, since

$$\psi^{\mathbb{N}} [f]^{\mathbb{N}} = \mathsf{fst} \, (h_{[f]^{\mathbb{N}}} \, e_{[f]^{\mathbb{N}}}) \equiv \mathsf{fst} \, (h_{[f]^{\mathbb{N}}} \, |f, \mathsf{refl}|) \equiv \mathsf{fst} \, (h'_{[f]^{\mathbb{N}}} (f, \mathsf{refl})) = [f] \qquad \square$$

Corollary 1. *Assume* ACω. *The type* $\mathbb{N} \to X/R$ *is the carrier of a quotient of* $\mathbb{N} \to X$ *by the equivalence relation* $\mathbb{N} \to R$. *The constructor is* $[\_]^{\mathbb{N}}$ *and we have the following dependent eliminator and computation rule: for every family of types* $Y : (\mathbb{N} \to X/R) \to \mathcal{U}_k$ *and function* $h : \prod_{f:\mathbb{N}\to X} Y [f]^{\mathbb{N}}$ *with* $p : \mathsf{dcompat}^{\mathbb{N}} h$, *there exists a function* $\mathsf{lift}^{\mathbb{N}} h p : \prod_{g:\mathbb{N}\to X/R} Y g$ *with the property that* $\mathsf{lift}^{\mathbb{N}} h p [f]^{\mathbb{N}} \equiv h f$ *for all* $f : \mathbb{N} \to X$, *where*

$$\mathsf{dcompat}^{\mathbb{N}} h = \prod_{\{f_1, f_2 : \mathbb{N}\to X\}} \prod_{r : f_1 (\mathbb{N}\to R) f_2} \mathsf{subst}\, Y \,(\mathsf{sound}^{\mathbb{N}} r)\,(h\, f_1) \equiv h\, f_2$$

7   Multiplication: A Solution Using ACω

We can now build the desired monad structure on $\bar{\mathsf{D}}$ using the results proved in Sect. 6. In particular, we can define $\bar{\mu} : \mathsf{D}\,(\mathsf{D}\,X/\approx)/\approx \to \mathsf{D}\,X/\approx$. We rely on ACω.

7.1   Delayed Computations as Streams

In order to use the results of Sect. 6, we think of possibly non-terminating computations as streams. More precisely, let X be a type and $c : \mathsf{D}\,X$. Now c can be thought of as a stream $\varepsilon\, c : \mathbb{N} \to X + 1$ with at most one value element in the left summand X.

$$\varepsilon : \mathsf{D}\,X \to \mathbb{N} \to X + 1$$
$$\varepsilon\,(\mathsf{now}\,x)\,\mathsf{zero} = \mathsf{inl}\,x$$
$$\varepsilon\,(\mathsf{later}\,c)\,\mathsf{zero} = \mathsf{inr}\star$$
$$\varepsilon\,(\mathsf{now}\,x)\,(\mathsf{suc}\,n) = \mathsf{inr}\star$$
$$\varepsilon\,(\mathsf{later}\,c)\,(\mathsf{suc}\,n) = \varepsilon\,c\,n$$

Conversely, from a stream $f : \mathbb{N} \to X + 1$, one can construct a computation $\pi\, f : \mathsf{D}\,X$. This computation corresponds to the "truncation" of the stream to its first value in X.

$$\pi : (\mathbb{N} \to X + 1) \to \mathsf{D}\,X$$
$$\pi\, f = \mathsf{case}\; f\, \mathsf{zero}\; \mathsf{of}$$
$$\mathsf{inl}\,x \mapsto \mathsf{now}\,x$$
$$\mathsf{inr}\star \mapsto \mathsf{later}\,(\pi\,(f \circ \mathsf{suc}))$$

We see that $\mathsf{D}\,X$ is a subset of $\mathbb{N} \to X + 1$ in the sense that, for all $c : \mathsf{D}\,X$, $\pi\,(\varepsilon\, c) \sim c$, and therefore $\pi(\varepsilon\, c) \equiv c$ by delayed computation extensionality.

Now let R be an equivalence relation on X. The canonical function $\theta^{+1} : (X+1)/(R+1) \to X/R+1$ has an inverse ψ^{+1} whose construction is similar to the construction of ψ^{BTree} for binary trees in Sect. 5. Therefore, for all $q : \mathsf{D}\,(X/R)$, we have $\pi\,(\theta^{+1} \circ (\psi^{+1} \circ \varepsilon\, q)) \equiv q$.

We define $[\_]^{\mathsf{D}} : \mathsf{D}\,X \to \mathsf{D}\,(X/R)$ by $[\_]^{\mathsf{D}} = \mathsf{D}\,[\_]$. This function is compatible with the relation $\mathsf{D}\,R$, i.e., there exists a term $\mathsf{sound}^{\mathsf{D}} : \mathsf{compat}\,[\_]^{\mathsf{D}}$.

Theorem 1. *The type* $D(X/R)$ *is the carrier of a quotient of* DX *by the equivalence relation* DR. *The constructor is* $[\_]^D$ *and we have the following dependent eliminator and computation rule: for every family of types* $Y : D(X/R) \to \mathcal{U}_k$ *and function* $h : \prod_{c:DX} Y[c]^D$ *with* $p : \mathrm{dcompat}^D h$, *there exists a function* $\mathrm{lift}^D h p : \prod_{q:D(X/R)} Y q$ *such that* $\mathrm{lift}^D h p [c]^D \equiv h c$ *for all* $c : DX$, *where*

$$\mathrm{dcompat}^D h = \prod_{\{c_1,c_2:DX\}} \prod_{r:c_1(DR)c_2} \mathrm{subst}\, Y (\mathrm{sound}^D r)(h c_1) \equiv h c_2$$

Proof. We only define the dependent eliminator. Let $h : \prod_{x:DX} Y[x]^D$ with $p : \mathrm{dcompat}^D h$, and $q : D(X/R)$. Let $g : \mathbb{N} \to (X+1)/(R+1)$, $g = \psi^{+1} \circ \epsilon q$ so $\pi(\theta^{+1} \circ g) \equiv q$.

We prove $Y(\pi(\theta^{+1} \circ g))$. By Corollary 1, it suffices to construct a function $h' : \prod_{f:\mathbb{N}\to X+1} Y(\pi(\theta^{+1} \circ [f]^{\mathbb{N}}))$ together with a proof $r : \mathrm{dcompat}^{\mathbb{N}} h'$. One can easily construct a proof $s : [\pi f]^D \equiv \pi(\theta^{+1} \circ [f]^{\mathbb{N}})$, so we take $h' f = \mathrm{subst}\, Y s (h(\pi f))$. A proof $r : \mathrm{dcompat}^{\mathbb{N}} h'$ can be constructed by observing that, for all $f_1, f_2 : \mathbb{N} \to X+1$ satisfying $f_1(\mathbb{N} \to R+1) f_2$, one can prove $\pi f_1(DR)\pi f_2$.   □

7.2   Construction of $\bar{\mu}$

Using the elimination rule of the quotient $D(X/R)$ defined in Theorem 1, we can finally define the multiplication $\bar{\mu}$ of $\bar{D}$.

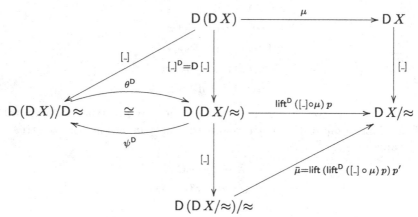

The above diagram makes sense only, if one constructs two compatibility proofs $p : \mathrm{compat}^D ([\_] \circ \mu)$ and $p' : \mathrm{compat}(\mathrm{lift}^D ([\_] \circ \mu) p)$, where compat^D is the non-dependent version of $\mathrm{dcompat}^D$.

The first proof is easy, since $c_1(D\approx)c_2 \to \mu c_1 \approx \mu c_2$ for all $c_1, c_2 : D(DX)$.

It is more complicated to prove compatibility of the second function. Let $q_1, q_2 : D(DX/\approx)$. We have to show $q_1 \approx q_2 \to \mathrm{lift}^D([\_] \circ \mu) p q_1 \equiv \mathrm{lift}^D([\_] \circ \mu) p q_2$. By the elimination principle of the quotient $D(DX/\approx)$, described in Theorem 1, it is sufficient to prove $[x_1]^D \approx [x_2]^D \to \mathrm{lift}^D([\_] \circ \mu) p [c_1]^D \equiv \mathrm{lift}^D([\_] \circ \mu) p [c_2]^D$ for some $c_1, c_2 : D(DX)$. Applying the computation rule of the quotient $D(DX/{\approx})$ and spelling out the definition of the constructor $[\_]^D$, it remains to show $D[\_]c_1 \approx D[\_]c_2 \to [\mu c_1] \equiv [\mu c_2]$, which holds, if one can prove $D[\_]c_1 \approx D[\_]c_2 \to \mu c_1 \approx \mu c_2$. This is provable thanks to Lemma 1. It is easy to see why Lemma 1 is important for completing the compatibility proof of $\mathsf{lift}^D([\_] \circ \mu)p$. The difficult case in the proof of $D[\_]c_1 \approx D[\_]c_2 \to \mu c_1 \approx \mu c_2$ is the case where $c_1 = \mathsf{now}\,y_1$ and $c_2 = \mathsf{now}\,y_2$, so we are given an assumption of type $[y_1] \equiv [y_2]$. From this, by Lemma 1, we obtain $\mu(\mathsf{now}\,y_1) = y_1 \approx y_2 = \mu(\mathsf{now}\,y_2)$.

8 A Monad or an Arrow?

Hughes [7] has proposed arrows as a generalization of monads. Jacobs et al. [8] have sorted out their mathematical theory.

We have seen that it takes a semi-classical principle to show that quotienting the functor D by weak bisimilarity preserves its monad structure. In contrast, quotienting the corresponding profunctor KD, defined by $KD\,X\,Y = X \to DY$, by pointwise weak bisimilarity can easily be shown to preserve its (strong) arrow structure (whose Freyd category is isomorphic to the Kleisli category of the monad) without invoking such principles.

Indeed, the arrow structure on KD is given by $\mathsf{pure}: (X \to Y) \to KD\,X\,Y$, $\mathsf{pure}\,f = \eta \circ f$ and $\lll: KD\,Y\,Z \to KD\,X\,Y \to KD\,X\,Z$, $\ell \lll k = \mathsf{bind}\,\ell \circ k$.

Now, define the quotiented profunctor by $\overline{KD}\,X\,Y = (X \to DY)/(X \to {\approx})$. We can define $\overline{\mathsf{pure}}: (X \to Y) \to \overline{KD}\,X\,Y$ straightforwardly by $\overline{\mathsf{pure}}\,f = [\mathsf{pure}f]$. But we can also construct $\overline{\lll}: \overline{KD}\,Y\,Z \to \overline{KD}\,X\,Y \to \overline{KD}\,X\,Z$ as $\ell \overline{\lll} k = \mathsf{lift}_2(\lll)\,p\,\ell\,k$, where p is an easy proof of $\ell_1(Y \to {\approx})\ell_2 \to k_1(X \to {\approx})k_2 \to (\ell_1 \lll k_1)(X \to {\approx})(\ell_2 \lll k_2)$.

This works entirely painlessly, as there is no need in this construction for a coercion $(X \to Y/{\approx}) \to (X \to Y)/(X \to {\approx})$ (cf. the discussion above in Sect. 5). From the beginning, we quotient the relevant function types here rather than their codomains.

There are some further indications that quotienting the arrow may be a righter thing to do than quotienting the monad. In particular, the work by Cockett et al. [5] suggests that working with finer quotients of the arrow considered here may yield a setting for dealing with computational complexity rather computability constructively.

9 Conclusions

In this paper, studied the question of whether the delay datatype quotiented by weak bisimilarity is still a monad? As we saw, different approaches to quotients in type theory result in different answers. In the quotients-as-setoids, the answer is immediately positive. We focussed on the more interesting and (as it turned out) more difficult case of the quotient types à la Hofmann. The main issue in this case, highlighted in Sect. 5, is that quotient types interact badly with infinite datatypes, such as datatypes of non-wellfounded or non-finitely branching trees; such datatypes do not commute with quotienting. For the delay datatype, and more generally for types that can be injectively embedded into streams or countably branching trees, a solution is possible assuming the axiom of countable choice.

In the type theory that we are considering, the employment of semi-classical principles, such as countable choice, is unavoidable. In homotopy type theory with higher inductive types [15, Ch. 6], the problem may have a different solution. One might be able to implement the delay type quotiented by weak bisimilarity as an higher inductive type, proceeding similarly to the construction of Cauchy reals in [15, Sect. 11.3], mutually defining the type and the equivalence relation, and adding a 1-constructor stating that the equivalence has to be read as equality. Note that this technique is not immediately applicable, since the delay datatype is coinductive and weak bisimilarity is mixed inductive-coinductive. One would have to come up with a different construction. We think that the idea should be to construct the intended monad as a datatype delivering free completely Elgot algebras [1]. Notice that this would be analogous to the already mentioned implementation of Cauchy reals, which are constructed as the free completion of the rational numbers.

Acknowledgement. We thank Thorsten Altenkirch, Andrej Bauer, Bas Spitters and our anonymous referees for comments.

This research was supported by the ERDF funded Estonian CoE project EXCS and ICT national programme project "Coinduction", the Estonian Science Foundation grants No. 9219 and 9475 and the Estonian Ministry of Education and Research institutional research grant IUT33-13.

References

1. Adámek, J., Milius, S., Velebil, J.: Elgot algebras. Log. Methods Comput. Sci. **2**(5:4), 1–31 (2006)
2. Benton, N., Kennedy, A., Varming, C.: Some domain theory and denotational semantics in Coq. In: Berghofer, S., Nipkow, T., Urban, C., Wenzel, M. (eds.) TPHOLs 2009. LNCS, vol. 5674, pp. 115–130. Springer, Heidelberg (2009)
3. Capretta, V.: General recursion via coinductive types. Log. Methods Comput. Sci. **1**(2:1), 1–28 (2005)
4. Chicli, L., Pottier, L., Simpson, D.: Mathematical quotients and quotient types in Coq. In: Geuvers, H., Wiedijk, F. (eds.) TYPES 2002. LNCS, vol. 2646, pp. 95–107. Springer, Heidelberg (2003)
5. Cockett, R., Díaz-Boïls, J., Gallagher, J., Hrubes, P.: Timed sets, complexity, and computability. In: Berger, U., Mislove, M. (eds.) Proceedings of 28th Conference on the Mathematical Foundations of Program Semantics, MFPS XXVIII. Electron. Notes in Theor. Comput. Sci., vol. 286, pp. 117–137. Elsevier, Amsterdam (2012)
6. Hofmann, M.: Extensional Constructs in Intensional Type Theory. CPHS/BCS Distinguished Dissertations. Springer, London (1997)
7. Hughes, J.: Generalising monads to arrows. Sci. Comput. Program. **37**(1–3), 67–111 (2000)

8. Jacobs, B., Heunen, C., Hasuo, I.: Categorical semantics for arrows. J. Funct. Program. **19**(3–4), 403–438 (2009)
9. Kraus, N., Escardó, M., Coquand, T., Altenkirch, T.: Notions of anonymous existence in Martin-Löf type theory. Manuscript (2014)
10. Maietti, M.E.: About effective quotients in constructive type theory. In: Altenkirch, T., Naraschewski, W., Reus, B. (eds.) TYPES 1998. LNCS, vol. 1657, pp. 166–178. Springer, Heidelberg (1999)
11. Martin-Löf, P.: 100 years of Zermelo's axiom of choice: what was the problem with it? Comput. J. **49**(3), 345–350 (2006)
12. Moggi, E.: Notions of computation and monads. Inf. Comput. **93**(1), 55–92 (1991)
13. Norell, U.: Dependently typed programming in Agda. In: Koopman, P., Plasmeijer, R., Swierstra, D. (eds.) AFP 2008. LNCS, vol. 5832, pp. 230–266. Springer, Heidelberg (2009)
14. Troelstra, A.S., Van Dalen, D.: Constructivism in Mathematics: An Introduction, v. I. Studies in Logic and the Foundations of Mathematics, vol. 121. North-Holland, Amsterdam (1988)
15. The Univalent Foundations Program: Homotopy Type Theory: Univalent Foundations of Mathematics. Institute for Advanced Study, Princeton, NY (2013). http://homotopytypetheory.org/book

Inverse Monoids of Higher-Dimensional Strings

David Janin[✉]

LaBRI CNRS UMR 5800, Bordeaux INP, Université de Bordeaux,
INRIA Bordeaux Sud-Ouest, 33405 Talence, France
janin@labri.fr

Abstract. Halfway between graph transformation theory and inverse semigroup theory, we define higher dimensional strings as bi-deterministic graphs with distinguished sets of input roots and output roots. We show that these generalized strings can be equipped with an associative product so that the resulting algebraic structure is an inverse semigroup. Its natural order is shown to capture existence of root preserving graph morphism. A simple set of generators is characterized. As a subsemigroup example, we show how all finite grids are finitely generated. Finally, simple additional restrictions on products lead to the definition of subclasses with decidable Monadic Second Order (MSO) language theory.

1 Introduction

A never-ending challenge faced by computer science is to provide modeling concepts and tools that, on the one hand, allows for representing data and computations in a more and more abstract and richly structured way, but, on the other hand, remains simple enough to be taught to and used by application designers and software engineers [33].

A possible approach to this goal consists in generalizing to graphs the techniques that have already been developed for strings or trees such as the notion of recognizable languages and the associated notion of recognizers. In these directions, an enormous amount of techniques and works has been developed ranging from Lewis' graph composition techniques [27] and Courcelle's developments of recognizability to graph languages [8] (see also [9]) up to more recent advances based on category theoretical development (see [6,13] to name but a few).

Despite numerous achievements in theoretical computer science, there is still room for polishing these techniques towards applications to computer engineering. The ideal balance to achieve between usage simplicity and mathematical coherence is a long-term goal [33]. While the underlying frameworks (the back end) of application tools to be designed can (and probably should) be based on robust mathematics, the interface (the front end) of these tools must be kept simple enough to be taught ad used.

Keeping in mind that strings, free monoids and related automata techniques are among the simplest and the most robust available models and are already and successfully put in practice in system modeling methods like *event B* [2], we develop in this paper a notion of generalized strings, called *higher dimensional strings*, in such a way that:

M. Leucker et al. (Eds.): ICTAC 2015, LNCS 9399, pp. 126–143, 2015.
DOI: 10.1007/978-3-319-25150-9_9

1. higher dimensional strings are simple: they are finitely generated from elementary graphs composed via a single and associative product that generalizes string concatenation in free monoids (Theorem 4.14),
2. the resulting classes of generalized strings include large classes of finite graphs such as, in particular, hypercubes, hence the name higher dimensional (Sect. 5 for the case of grids),
3. the resulting semigroups are inverse semigroups (Theorems 4.6 and 4.8) henceforth mathematically rich enough to provide algebraic characterization of graph-based concepts such as, for instance, existence of graph morphisms characterized by natural order (Theorem 4.12) or acyclicity defined by a quotient with an adequate ideal (Lemma 6.1),
4. some well-defined and rich subclasses of these generalized strings still has efficient, expressive and decidable language theory (Theorem 5.5).

Technically, following the lines already sketched in [18], we use and generalize the concept of birooted graphs (with single input and output roots) defined and used in [31] into the notion of higher dimensional strings (with sets of input and output roots). This provides a better measure of the amount of overlaps that occurs in birooted graphs products can be better measured. Thus we can extend the notion of disjoint product [15,17] and the applicable partial algebra techniques [5]). This yields to our main decidability result (Theorem 5.5).

In some sense, our proposal amounts to combining concepts and results arising from the theory of inverse semigroups [26,29] with graph transformation approaches [6,9,13,27].

Of course, various research developments have already shown that inverse semigroup theory is applicable to computer science, be it for data, computation, language or system modeling. Concerning data modeling, experiments in theoretical physics have already shown that structured data as complex as quasi-crystals can be described by means of some notion of (inverse) tiling semigroup [23–25]. Inverse semigroup theory has also been used to study reversible computations [1,10]. More recently, various modeling experiments have been conducted in computational music [3,21]. These last experiments also led to the definition of a Domain Specific (Programing) Language (DSL) which semantics is based on concepts arising from inverse semigroup theory [14,22].

2   Preliminaries

Let $A = \{a, b, c, \cdots\}$ be a finite alphabet of graph edge labels. Every concept defined in the sequel could be extended to hypergraphs, that is, graphs with edges that possibly relate more than two vertices (see Footnote 1). However, restricting our presentation to standard (binary) graph structures allows us to keep statements (and proofs) simpler.

Relational Graphs. A (relational) graph on the (binary symbols) alphabet A, simply called A-*graph* or even *graph* when A is clear from the context, is a pair $G = \langle V, \{E_a\}_{a \in A} \rangle$ with set of vertices V and a-labeled edge relation $E_a \subseteq V \times V$ for every $a \in A$.

Back and Forth Path Labels. Let $\bar{A} = \{\bar{a}, \bar{b}, \bar{c}, \cdots\}$ be a disjoint copy of the alphabet A. A back and forth path label (or simply path label) is a word from the free monoid $(A + \bar{A})^*$ on the alphabet $A + \bar{A}$, with empty word denoted by 1 and the product of two words u and $v \in (A + \bar{A})^*$ denoted by $u \cdot v$ or simply uv. Then, the reverse mapping $w \mapsto \bar{w}$ from $(A + \bar{A})^*$ into itself is inductively defined by $\bar{1} = 1$, $\overline{a \cdot v} = \bar{v} \cdot \bar{a}$ and $\overline{\bar{a} \cdot v} = \bar{v} \cdot a$ for every $a \in A$, $x \in A + \bar{A}$ and $v \in (A + \bar{A})^*$. It is an easy observation that the reverse mapping is an involutive monoid anti-isomorphism, that is, we have $\overline{u \cdot v} = \bar{v} \cdot \bar{u}$ and $\overline{\overline{w}} = w$ for every $u, v, w \in (A + \bar{A})^*$.

Back and Forth Path Actions. For every $X \subseteq V$ and $w \in (A + \bar{A})^*$, the set $X \cdot w \subseteq V$ of vertices reachable from X following w is inductively defined by $X \cdot 1 = X$, $X \cdot aw = \{y \in V : \exists x \in X, (x, y) \in E_a\} \cdot w$ and $X \cdot \bar{a}w = \{y \in V : \exists x \in X, (y, x) \in E_a\} \cdot w$, for every letter $a \in A$ and every string $v \in (A + \bar{A})^*$. In other words, $X \cdot w$ is the set of vertices that can be reached from a vertex in X along a path labeled by w, where a (resp. $\bar{a}$) denotes the forward (resp. backward) traversal of an a-labeled edge in the graph G.

One can check that $X \cdot 1 = X$ and $X \cdot (u \cdot v) = (X \cdot u) \cdot v$ for every $X \subseteq V$ and every string $u, v \in (A + \bar{A})^*$. Rephrased in semigroup theoretical term, the edge relations of the graph G induce an *action* of the monoid $(A + \bar{A})^*$ on the powerset of the set of vertices of the graph G. It follows that parentheses can be removed without ambiguity in expressions like $(X \cdot u) \cdot v$..

Notation for the Singleton Case. When X is a singleton $\{x\}$, we may simply write $x \cdot w$ instead of $\{x\} \cdot w$. Similarly, when $x \cdot w$ itself is a singleton we may also treat it just as the element it contains. In other words, we may simply write $x \cdot w = y$ instead of $\{x\} \cdot w = \{y\}$, to denote both the fact that there exists a (back and forth) path from vertex x to vertex y labeled by w and the fact that this path is unique. Similarly, we may say that $x \cdot w$ is undefined (as a vertex) in the case $x \cdot w = \emptyset$ (as a set).

Graph Morphism. The usual notion of graph morphism can then be (re)defined via path actions as follows. Let $G = \langle V, \{E_a\}_{a \in A} \rangle$ and $G' = \langle V', \{E'_a\}_{a \in A} \rangle$ be two graphs on the alphabet A. A morphism f from G to G', denoted by $f : G \to G'$, is a mapping $f : V \to V'$ such that we have $f(x \cdot a) \subseteq f(x) \cdot a$ and $f(x \cdot \bar{a}) \subseteq f(x) \cdot \bar{a}$ for every $x \in V$ and every $a \in A$. Then, by induction, we can easily prove that $f(x \cdot w) \subseteq f(x) \cdot w$ for every $x \in V$ and every $w \in (A + \bar{A})^*$.

Graph Quotient. Let $G = \langle V, \{E_a\}_{a \in A} \rangle$ be a graph. Let $\simeq$ be an equivalence relation over the set V, that is, a reflexive, symmetric and transitive relation. Let $V/\simeq$ be the set of equivalence classes $\{[x]_\simeq \subseteq V : x \in V\}$ where $[x]_\simeq = \{x' \in V : x \simeq x'\}$. Then, the quotient of the graph G by the equivalence $\simeq$ is defined to be the graph $G/\simeq = \langle V', \{E'_a\}_{a \in A} \rangle$ with set of vertices $V' = V/\simeq_G$ and set of edges $E'_a = \{([x], [y]) \in V' \times V' : ([x] \times [y]) \cap E_a \neq \emptyset\}$. The mapping $\eta_\simeq : V \to V/\simeq$ defined by $\eta_\simeq(x) = [x]_\simeq$ for every $x \in V$ is a surjective morphism called the canonical morphism from the graph G onto the quotient graph $G/\simeq$.

3   Unambiguous Graphs and Connecting Morphisms

We define and study in this section the category of unambiguous graphs and connecting morphisms. Though fairly simple, this study is quite detailed for it constitutes the foundation of the notion of birooted graphs defined in the next section.

Definition 1. (Unambiguous Graphs). A graph $G = \langle V, \{E_a\}_{a \in A} \rangle$ is *unambiguous*[1] when, for every vertex $x \in V$, for every path $w \in (A + \bar{A})^*$, there is at most one vertex y such that $x \cdot w = \{y\}$.

Clearly, by simple inductive argument, G is unambiguous as soon as the above condition is satisfied for every one letter path.

Examples. Graphs examples are depicted in Fig. 1 with ambiguous graph G_1 and unambiguous graphs I_2 and G_2. In this figure, vertices are named only for illustrative purposes. These vertex names should not be understood as labels. Only edges are labeled in relational graphs.

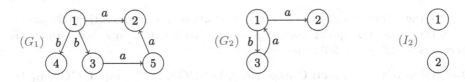

Fig. 1. Ambiguous graph G_1 and unambiguous graph G_2.

One can observe that graph G_1 is ambiguous for two reasons. First, the upper left vertex 1 is the source of two edges labeled by b. Second, the upper right vertex 2 is the target of two edges labeled by a.

Remark. Observe that when a graph G is seen as a graph automaton on the alphabet A, it is unambiguous when it is both deterministic and co-deterministic. In the connected case, these unambiguous graphs are the Schützenberger graphs studied and used in [31].

Definition 2. (Connecting Morphisms). Let $f : G \to G'$ be a graph morphism between two graphs $G = \langle V, \{E_a\}_{a \in A} \rangle$ and let $G' = \langle V', \{E'_a\}_{a \in A} \rangle$. The morphism f is a *connecting morphism* when for every $x' \in V'$ there exist $x \in V$ and $w \in (A + \bar{A})^*$ such that $x' \in f(x) \cdot w$.

In other words, a morphism $f : G \to G'$ is a connecting morphism when every vertex of graph G' is connected to the image of a vertex of G in graph G'.

Examples. Clearly, every surjective (i.e. onto) morphism is a connecting morphism. Another example of (non surjective) connecting morphism $f : I_2 \to G$ is depicted in Fig. 2.

---

[1] unambiguity can be generalized to hypergraphs by viewing every binary relation of the form $\exists z_1 z_2 z_3 \, a(z_1, x, z_2, y, z_3)$ with tuples of FO-variables z_1, z_2 and z_3 of adequate lengths as a primitive binary relation.

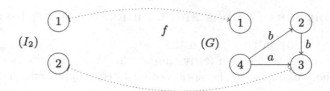

Fig. 2. A connecting morphism $\varphi : I_2 \rightarrow G$ with $\varphi(1) = 1$ and $\varphi(2) = 3$.

Remark. Observe that when both G and G' are unambiguous, then, for every $x \in V$, every $w \in (A + \bar{A})^*$, if $x \cdot w$ is not empty then so is $f(x) \cdot w$ and we have $f(x \cdot w) = f(x) \cdot w$. This leads us to the following Lemma.

Lemma 3.3. (Unique Morphism Completion). *Let G, G_1 and G_2 be three graphs. Let $f_1 : G \rightarrow G_1$ and $f_2 : G \rightarrow G_2$ be two graph morphisms. Assume that f_1 is connecting and that both G_1 and G_2 are unambiguous. Then there exists at most one morphism $g : G_1 \rightarrow G_2$ such that $g \circ f_1 = f_2$. Moreover, if f_2 is connecting, then so is g.*

Clearly, the composition of two connecting morphisms is a connecting morphism. Since the identity mapping over a graph is also a connecting morphism, this allows us to define the following categories.

Definition 3.4. (Induced Categories). Let **CGrph**(A) (resp. **UCGrph**(A)) be the category defined by finite graphs (resp. by finite unambiguous graphs) as objects and connecting morphisms as arrows.

We aim now at studying the properties of both category **CGrph**(A) and category **UCGrph**(A) and, especially, the way they are related. The notion of unambiguous congruence defined below allows us to transform any graph into its greatest unambiguous image. In group theory, this generalizes the notion of Stallings foldings [29].

Definition 3.5. (Unambiguous Congruence). Let $G = \langle V, \{E_a\}_{a \in A} \rangle$ be a graph on the alphabet A. A relation $\simeq \subseteq V \times V$ over the vertices of G is an *unambiguous congruence* when it is an equivalence relation such that, for every $a \in A$, for every $x, y \in V$, if $x \simeq y$ then we have both $x \cdot a \times y \cdot a \subseteq \simeq$ and $x \cdot \bar{a} \times y \cdot \bar{a} \subseteq \simeq$.

The existence of a least congruence is stated in Lemma 3.6 and the associated universality property is stated in Lemma 3.7.

Lemma 3.6. (Least Unambiguous Congruence). *Let G be a graph, possibly ambiguous. Then there exists a least unambiguous congruence $\simeq_G$ over G. Moreover, in the case G is unambiguous, then $\simeq_G$ is the identity relation.*

The graph $G/ \simeq_G$ is called the *greatest unambiguous graph image* of the graph G. Its maximality is to be understood in the following sense.

Lemma 3.7. (Maximal Unambiguous Image). *Let G be a graph. Let $\simeq_G$ be its least unambiguous congruence. Then, for every graph morphism $f : G \to H$ with unambiguous graph H, there exists a unique morphism $g : G/\simeq_G \to H$ such that $f = g \circ \eta_{\simeq_G}$. Moreover, if f is connecting then so is g.*

Example. An example of maximal graph image is provided by the graphs already depicted in Fig. 1 where G_2 has not been chosen at random since $G_2 = G_1/\simeq_{G_1}$.

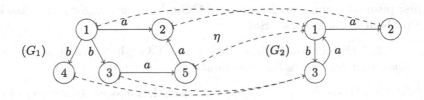

Fig. 3. Graph G_2 is the maximal unambiguous image of graph G_2.

The canonical onto morphism $\eta : G_1 \to G_1/\simeq_{G_1} = G_2$ is depicted in Fig. 3, encoding the least unambiguous congruence on G_1 that glues 1 with 5, and 3 with 4.

Remark. The construction described above is a generalization of what is known in algebra as Stallings folding [29]. Observe that with $G = \langle V, \{E_a\}_{a \in V}\rangle$, the least unambiguous congruence $\simeq_G$ equals the least fixpoint of the mapping $F : V \times V \to V \times V$ defined by

$$F(R) = R \cup \bigcup \{(x \cdot a) \times (y \cdot a) \cup (x \cdot \bar{a}) \times (y \cdot \bar{a}) : (x, y) \in R, a \in A\}$$

that contains the equality. It follows, by applying classical fixpoint techniques, that $\simeq_G = \bigcup_{n \geq 0} F^n(=)$, henceforth it can be computed in quasi linear time. In other words, computing the maximal unambiguous image $G/\simeq_G$ of the graph G can be done in time quasi linear in the size of the graph G.

Clearly, the category $\mathbf{UCGrph}(A)$ is a subcategory of $\mathbf{CGrph}(A)$. The next lemma shows that maximal graph images extend to morphisms henceforth defining a projection functor from $\mathbf{CGrph}(A)$ into $\mathbf{UCGrph}(A)$.

Lemma 3.8. (Projected Morphisms). *Let G and H be two graphs with a connecting morphism $f : G \to H$. Let $\eta_G : G \to G/\simeq_G$ and $\eta_H : H \to H/\simeq_H$ be the related canonical onto morphisms. Then there exists a unique connecting morphism $\varphi(f) : G/\simeq_G \to H/\simeq_H$ such that $\varphi(f) \circ \eta_G = \eta_H \circ f$.*

In other words, we can define the functor $\varphi : \mathbf{CGrph}(A) \to \mathbf{UCGrph}(A)$ by $\varphi(G) = G/\simeq_G$ for every graph G and by $\varphi(f)$ as given by Lemma 3.8 for every connecting morphism f. Then, we have $\varphi(G) = G$ for every unambiguous graph G and $\varphi(f) = f$ for every connecting graph morphism f between unambiguous graphs. In other words, φ is a projection from $CG(A)$ into $\mathbf{UCGrph}(A)$ henceforth a left inverse of the inclusion functor from $\mathbf{UCGrph}(A)$ to $\mathbf{CGrph}(A)$.

We study a bit further the morphisms in these categories showing that they both admit pushouts. The following definition, classical in category theory, is given here for the sake of completeness.

Definition 3.9. (Pushouts). Let $\langle f_1 : G \to G_1, f_2 : G \to G_2 \rangle$ be a pair of morphisms. A pair of morphisms $\langle g_1 : G_1 \to H, g_2 : G_2 \to H \rangle$ is a pushout of the pair $\langle f_1, f_2 \rangle$ when $f_1 \circ g_1 = f_2 \circ g_2$, and, for every other pair of morphisms $\langle g_1' : G_1 \to H', g_2' : G_2 \to H' \rangle$, if $f_1 \circ g_1' = f_2 \circ g_2'$ then there exists a unique morphism $h : H \to H'$ such that $g_1' = h \circ g_1$ and $g_2' = h \circ g_2$.

The first pushout lemma, in the category $\mathbf{CGrph}(A)$, is a slight generalization of the pushout in the category $\mathbf{Set}$.

Lemma 3.10. (Synchronization). *In category $\mathbf{CGrph}(A)$, every pair of morphisms with common source has a pushout.*

Proof (sketch of). Let $\equiv_{f_1,f_2}$ be the equivalence relation over the vertices of the disjoint sum $G_1 + G_2$ induced by $f_1(x) \equiv_{f_1,f_2} f_2(x)$ for every vertex x of G. Let $H = G_1 + G_2 / \equiv_{f_1,f_2}$. Then, the pair $\langle \eta_{\equiv_{f_1,f_2}} \circ i_1, \eta_{\equiv_{f_1,f_2}} \circ i_2 \rangle$ with canonical injection i_1 (resp. i_2) of G_1 (resp. G_2) into $G_1 + G_2$ is a pushout of $\langle f_1, f_2 \rangle$ in category $\mathbf{CGrph}(A)$. □

Example. An example of such a pushout in the category $\mathbf{CGrph}(A)$ is depicted in Fig. 4.

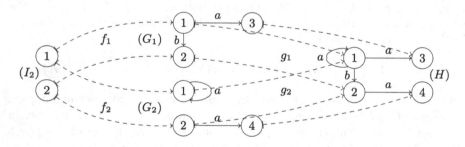

Fig. 4. A "synchronization" pushout example.

Remark. Existence of pushouts in $\mathbf{CGrph}(A)$ essentially follows from the existence of pushouts in the category $\mathbf{Set}$. These pushouts are called synchronization (or glueing) pushouts since, the pushout of $\langle f_1 : G \to G_1, f_2 : G \to G_2 \rangle$ essentially glues the vertices of G_1 and G_2 that have common ancestors in G either via f_1 or via f_2.

The second pushout lemma, in the category $\mathbf{UCGrph}(A)$, is completed by a fusion phase (or glueing propagation) defined by taking the maximal unambiguous image of the graph resulting from the pushout in $\mathbf{CGrph}(A)$.

Lemma 3.11. (Synchronization and Fusion). *In category $\mathbf{UCGrph}(A)$, every pair of morphisms with common source has a pushout.*

Proof (sketch of). Take $H = G_1 + G_2 / \simeq_{f_1,f_2}$ as for Lemma 3.10 with pushout $\langle g_1, g_2 \rangle$. Then, take $U = H / \simeq_H$ the greatest unambiguous image of H. The pair $\langle \eta_H \circ g_1, \eta_H \circ g_2 \rangle$ is a pushout of $\langle f_1, f_2 \rangle$ in **UCGrph**(A). $\square$

Example. An example of a synchronization + fusion is depicted in Fig. 5.

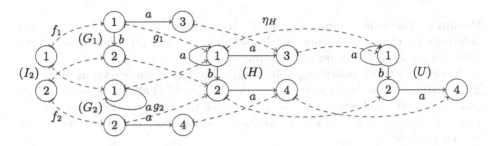

Fig. 5. A "synchronization + fusion" pushout example.

4   The Inverse Monoid of Birooted Graphs

We are now ready to define birooted graphs as certain cospans in the category **UCGrph**(A). For such a purpose, for every integer $k > 0$, let I_k be the unambiguous defined by k distinct vertices $\{1, 2, \cdots, k\}$ and empty edge relations, and let $id_k : I_k \to I_k$ be the identity isomorphism.

Definition 4.1. (Birooted Graphs). A birooted graph B is a pair of connecting morphisms

$$B = \langle in : I_p \to G, out : I_q \to G \rangle$$

from two trivial graphs I_p and I_q to a common unambiguous graph G.

The morphism *in* is called the *input root morphism*, or, more simply, the *input root* of the birooted graph B. The morphism *out* is called the *output root morphism*, or, more simply, the *output root* of the birooted graph B.

The pair of positive integers (p, q) that defines the domains of root morphisms is called the *type* of the birooted graph. It is denoted by $dom(B)$. The underlying graph G is the codomain of the input and output morphisms. It is called the *graph* of B and it is also denoted by $cod(B)$.

Remark. A birooted graph of type (p, q) can simply be seen as a unambiguous graph $G = \langle V, \{E_a\}_{a \in A} \rangle$ enriched with two tuples of distinguished vertices $(x_1, x_2, \cdots, x_p) \in V^p$ and $(y_1, y_2, \cdots, y_q) \in V^q$ that label the vertices marked by the input and the output roots of the birooted graph.

This point of view is depicted in Fig. 6 with two birooted graphs B_1 and B_2 of type $(2, 2)$. In such a figure, vertices of input roots are marked by dangling input arrows, and vertices of output roots are marked by dangling output arrows.

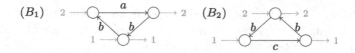

Fig. 6. Examples of $(2, 2)$-birooted graphs.

Remark. The name "birooted graphs" is borrowed from [31]. However, our definition is a clear generalization of the definition given in [31]. Indeed, Stephen's birooted graphs are only birooted graphs of type $(1, 1)$.

In category theoretical term, a birooted graph is a cospan (see for instance [4]). The existence of pushouts in the category **UCGrph**(A) allows us to define the product of birooted graphs as the product of their cospan. However, such a product is (so far) not uniquely determined since, a priori, it may depend on the chosen pushout.

Definition 4.2. (Birooted Graph Product Instance). Let $B_1 = \langle in_1, out_1 \rangle$ and let $B_2 = \langle in_2, out_2 \rangle$ be two birooted graphs. Assume that B_1 is of type (p, q) and that B_2 is of type (q, r). Let $\langle h_1, h_2 \rangle$ be a pushout of the pair $\langle out_1, in_2 \rangle$. Then, the *product instance* of birooted graphs *via the pushout* $\langle h_1, h_2 \rangle$ is defined to be the birooted graphs $\langle h_1 \circ in_1, h_2 \circ out_2 \rangle$, and it is denoted by $B_1 \cdot_{h_1, h_2} B_2$.

A concrete example of a product instance built from the $(2, 2)$-birooted graphs given in Fig. 6 is depicted in Fig. 7.

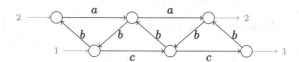

Fig. 7. A product instance of $B_1 \cdot B_2 \cdot B_1 \cdot B_2$.

We aim now at defining products of birooted graphs up to some adequate notion of birooted graph equivalence. This is done via the notion of birooted graph morphisms (Definition 4.3) and the proof that birooted graph product instances are stable under birooted graph morphisms (Lemma 4.4).

Definition 4.3. (Birooted Graph Morphisms). Let $B_1 = \langle in_1, out_1 \rangle$ and $B_2 = \langle in_2, out_2 \rangle$ be two birooted graphs. A birooted graph morphism from B_1 to B_2 is defined as root preserving graph morphism of their codomain, that is, a graph morphism $h : cod(B_1) \rightarrow cod(B_2)$ such that $in_2 = h \circ in_1$ and $out_2 = h \circ out_1$. Such a morphism is denoted by $h : B_1 \Rightarrow B_2$.

Two birooted graphs B_1 and B_2 are isomorphic when there is an isomorphism $h : B_1 \Rightarrow B_2$. Such a situation is denoted by $B_1 \sim B_2$.

Remark. Thanks to Lemma 3.3, there exists at most one morphism $h : B_1 \Rightarrow B_2$ between any two birooted graphs B_1 and B_2.

Lemma 4.4. (Product Stability w.r.t. Birooted Graphs Morphisms).
Let $f_1 : B_1 \Rightarrow C_1$ and $f_2 : B_2 \Rightarrow C_2$ be two birooted graphs morphisms and let $B_1 \cdot B_2$ and $C_1 \cdot C_2$ be two product instances. Then, there exists a (unique) birooted graphs morphisms $h : B_1 \cdot B_2 \Rightarrow C_1 \cdot C_2$.

This stability property allows us to define the following birooted graph algebras.

Definition 4.5. (Birooted Graph Algebras). Let $HS(A)$ be the set of classes of isomorphic birooted graphs extended with the emptyset equipped with the product defined for every $X, Y \in H(S)$ as follows. In the case there is $B \in X$, $C \in Y$ and a product instance $B \cdot C$, then we take $X \cdot Y = [B]_\sim \cdot [Y]_\sim = [B \cdot Y]_\sim$ and we take $X \cdot Y = \emptyset$ in all other cases.

Notation. In the sequel we shall simply write B (or C) instead of $[B]$ (or $[C]$) and we shall simply write $B \cdot C$ for the product $[B]_\sim \cdot [C]_\sim$ of the corresponding classes of equivalent birooted graphs.

Theorem 4.6. (Semigroup Property). *The algebra $HS(A)$ is a semigroup, that is, the product of birooted graphs is an associative operation.*

Lemma 4.7. (Idempotent Property). *A non-zero birooted graph B of the form $B = \langle in, out \rangle$ is idempotent, that is, $B \cdot B = B$, if and only if $in = out$. Moreover, idempotent birooted graphs commute henceforth form a subsemigroup.*

Theorem 4.8. (Inverse Semigroup Property). *The semigroup $HS(A)$ is an inverse semigroup, that is, for every element B, there is a unique element B^{-1} such that*

$$B \cdot B^{-1} \cdot B = B \text{ and } B^{-1} \cdot B \cdot B^{-1} = B^{-1}$$

The inverse B^{-1} of a non-zero birooted graph $B = \langle in, out \rangle$ is simply given by $B^{-1} = \langle out, in \rangle$.

Inverses allow us to define left and right projections that, following inverse semigroup theory, characterize left and right Green classes.

Definition 4.9. (Left and Right Projection). Let $B \in HS(A)$ be a birooted graph. The left projection B^L of the birooted graph B is defined by $B^L = B^{-1} \cdot B$. The right projection B^R of the birooted graph B is defined by $B^R = B \cdot B^{-1}$.

Lemma 4.10. *Let $B = \langle in, out \rangle$ be a non-zero birooted graph. Then we have $B^L = \langle out, out \rangle$ and $B^R = \langle in, in \rangle$.*

Remark. As a general matter of fact, the relation $B \preceq C$ defined over birooted graphs when there exists a (root preserving) morphism $h : C \Rightarrow B$ is a (partial) order relation. We shall see now that it has an algebraic characterization in inverse semigroup theory: it is the natural order [26].

Definition 4.11 (Natural Order). The natural order $\leq$ is defined over birooted graphs by $B \leq C$ when $B = B^R \cdot C$ (or, equivalently, $B = C \cdot B^L$).

Theorem 4.12 (Natural Order vs Birooted Graph Morphisms). *In the inverse semigroup $HS(A)$, the absorbant element 0 is the least element under the natural order and, for every pair of non zero birooted graphs B and C, $B \leq C$ if, and only if, there is a birooted graph morphism $h : C \Rightarrow B$.*

The inverse semigroup of birooted graphs gives a fairly simple though mathematically robust way to compose birooted graphs one with the other. Now we aim at characterizing a simple set of generators for this semigroup.

Definition 4.13 (Elementary Birooted Graphs). A elementary birooted graph is either zero or any birooted graph among I_m, $P_{m,i,j}$, $T_{m,a}$, $T_{m,\bar{a}}$ F_m or J_m defined below. In the case $m = 3$ these graphs are depicted in Fig. 8.

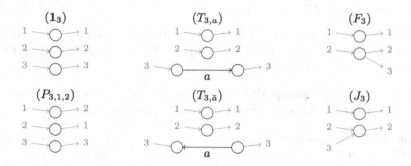

Fig. 8. Elementary birooted graphs.

Formally, the birooted graph $P_{m,i,j} = \langle id_m : I_m \to I_m, out : I_m \to I_m \rangle$ is defined for any $m > 0$ and $1 \leq i, j \leq m$ by $out(i) = j$, $out(j) = i$ and $out(k) = k$ for every other $1 \leq k \leq m$. It is called a *root permutation*. As a particular case, when $i = j$, since $P_{m,i,i} = \langle id_m, id_m \rangle$, the birooted graph $P_{m,i,i}$ is denoted by 1_m instead and called a *root identity*.

The birooted graphs $F_m = \langle id_{m-1} : I_{m-1} \to I_{m-1}, out : I_m \to I_{m-1} \rangle$ and $J_m = \langle in : I_m \to I_{m-1}, id_{m-1} : I_{m-1} \to I_{m-1} \rangle$ are defined for any $m > 1$, by $in(m) = out(m) = m - 1$ and $in(k) = out(k) = k$ for every $1 \leq k \leq m - 1$. They are called a *root fork* and a *root join*.

The birooted graph $T_{m,a} = \langle int : I_m \to G_a, out : I_m \to G_a \rangle$ is defined for any $m > 0$ and $a \in A$, by G_a being the $m + 1$ vertex graph with set of vertices $V = \{1, \cdots, m, m + 1\}$ and sets of edges $E_a = \{(m, m + 1)\}$ and $E_b = \emptyset$ for every $b \neq a$, with $in(m) = m$, $out(m) = m + 1$ and $in(k) = out(k) = k$ for every other $1 \leq k < m$. It is called a *forward edge*. The birooted graph $T_{m,\bar{a}} = T_{m,a}^{-1}$ is called a *backward edge*.

Examples. Some birooted graphs generated by elementary graphs are depicted in Fig. 9.

Theorem 4.14. *Every birooted graphs $\langle in : I_p \to G, out : I_q \to G \rangle$ with n vertices in G is finitely generated from 0 and the elementary birooted graphs 1_k, $P_{k,i,j}$, $T_{k,a}$, $T_{k,\bar{a}}$, F_k and J_k with $1 \leq k \leq max(n, p + 1, q + 1)$.*

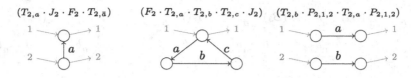

Fig. 9. Some elementary compositions.

Definition 4.15 (Bounded Birooted Graphs Algebras). For any given integer $m > 0$, let $HS_m(A)$ (resp. $HS_{\leq m}(A)$) be the algebraic structure defined as the subsemigroup of $HS(A)$ generated by 1_m, $P_{m,i,j}$, $T_{m,a}$, $T_{m,\bar{a}}$ (resp. 1_k, $P_{k,i,j}$, $T_{k,a}$, $T_{k,\bar{a}}$, F_k and J_k with $1 \leq k \leq m$).

As an corollary of Theorems 4.6 and 4.8, we have:

Theorem 4.16. *For every integer $m > 0$, the algebra $HS_m(A)$ is an inverse monoid with neutral element 1_m.*

Remark. As a particular case, it can be shown that $HS_1(A)$ is the free inverse monoid $FIM(A)$ generated by A. We shall see below that birooted grids of arbitrary size but of type $(2, 2)$ belong to $HS_{\leq 2}(A)$. In other word, in Theorem 4.14, the bound given for k, depending on the number of vertices of G is not optimal.

5   Languages of Birooted Graphs

Now we aim at developing the language theory of higher dimensional strings, that is to say, the study of the definability of subsets of $HS(A)$. For such a purpose, we consider the First Order (FO) logic or the Monadic Second Order (MSO) logic (see [9]) on birooted graphs. We refer the reader to the book [9] for a definition of MSO on graphs.

More precisely, we consider $HS_{\leq m}(A)$ so that the number of input and output roots on graphs is bounded. Then, one can enrich the signature A by $2 * m$ symbols, necessarily interpreted as singletons in order to describes these roots. Clearly, this is easily done within FO or MSO logic and we can thus consider the class of FO-definable or MSO-definable languages of birooted graphs.

Theorem 5.1 (Undecidability). *When $m \geq 2$, the language emptiness problem for FO-definable (hence also MSO-definable) languages of birooted graphs of $HS_{\leq m}(A)$ is undecidable.*

Proof (sketch of). The undecidability of FO follows from the fact that, as soon as $m \geq 2$, as depicted in Fig. 10, grids of arbitrary size can be finitely generated with two edge relations a and b modeling horizontal and vertical directions, hence, together with additional edge relations for encoding arbitrary unary predicates on grid vertices, classical undecidability results apply [9].   □

We first check, following the examples depicted in Fig. 9, that these generators can indeed be defined by means of $P_{k,i,j}$, $T_{k,a}$, $T_{k,\bar{a}}$, F_k and J_k with $1 \leq k \leq 2$. For instance, we have $B_5 = (T_{2,\bar{b}} \cdot J_2)^R \cdot T_{2,a} \cdot T_{2,b} \cdot (T_{2,a} \cdot J_2)^R \cdot P_{2,1,2}$.

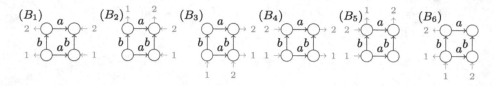

Fig. 10. A finite set of generators B_1, B_2, B_3, B_4, B_5 and B_6.

Then, as depicted in Fig. 11, we can generate birooted grids of arbitrary size by taking the $(2,2)$-birooted graph $B_{m,n}$ defined by $G_{m,n} = (Z_m \cdot Y_m)^n$. Clearly, B_{mn} contains a grid of size m by $2 * n$.

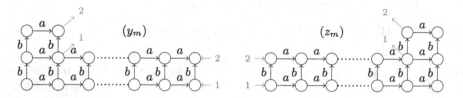

Fig. 11. The $(2,2)$-birooted graphs $Y_m = (B_1)^m \cdot B_2 \cdot B_3$ and $Z_m = (B_4)^m \cdot B_5 \cdot B_6$.

One may ask how generating such graphs of unbounded tree-width can be avoided. It occurs that this can simply be done by restricting the overlaps that are allowed in product instances.

Recently introduced in the context of birooted words [16] or trees [15,17] languages, the definition of the disjoint product, extended to birooted graphs, makes this restriction of overlaps formal.

Definition 5.2 (Disjoint Product). Let $B_1 = \langle in_1, out_1 \rangle$ and $B_2 = \langle in_2, out_2 \rangle$ be two birooted graphs. Let $\langle h_1, h_2 \rangle$ be a pushout of $\langle out_1, in_2 \rangle$ in **UCGrph**(A) and let $B_1 \cdot B_2 \langle in, out \rangle$ with $in = h_1 \circ in_1$ and $out = h_2 \circ out_2$ be the resulting product. Then this product is a *disjoint product* when the pair $\langle h_1, h_2 \rangle$ is also a pushout of in $\langle out_1, in_2 \rangle$ in the category **CGrph**(A). In this case, the disjoint product is denoted by $B_1 \star B_2$.

In other words, a birooted graph product is a disjoint product when the fusion phase in the underlying pushout computation is trivial. Although partially defined, this disjoint product is still associative in the following sense.

Lemma 5.3 (Partial Associativity). *For all birooted graphs B_1, B_2, B_3 the disjoint product $B_1 \star (B_2 \star B_3)$ is defined if and only if the disjoint product $(B_1 \star B_2) \star B_3$ is defined and, in that case, the products are equal.*

Then, the closure under disjoint products and left and right projections are defined as follows.

Definition 5.4 (Disjoint Closure and Decomposition). Let $X \subseteq HS(A)$ be a set of birooted graphs. The *disjoint closure* of the set X is defined to be the least set Y of birooted graphs such that $X \subset Y$ and that Y is closed under disjoint product and left and right projections. This closure is denoted by $\langle X \rangle_{*,L,R}$.

For every birooted graph $B \in \langle X \rangle_{*,L,R}$, a combination of elements of X by disjoint products and let and right projection that equals B is called a *disjoint decomposition* of B over X.

Examples. The subset of $HS_1(A)$ generated by disjoint products of elementary birooted graphs I_1 and $T_{1,a}$ with $a \in A$ is just the free monoid A^*. Adding left and right projections, the disjoint closure of such a set is known in the literature as the free ample monoid $FAM(A)$ whose elements are positive birooted trees (see [12]). Adding backward edges $T_{1,\bar{a}}$ for every $a \in A$, the disjoint closure of the resulting set is the free inverse monoid $FIM(A)$ whose elements are arbitrary birooted trees.

Theorem 5.5 (Decidability and Complexity). *Let $X \subseteq_{fin} HS(A)$ be a finite subset of $HS(A)$. Then, the emptiness problem for MSO-definable subsets of the disjoint closure $\langle X \rangle_{*,R,L}$ is (non-elementary) decidable.*

Moreover, for any MSO-definable language $L \subseteq \langle X \rangle_{*,R,L}$, the membership problem $B \in L$ for any $B \in HS(A)$ is linear in the size of any disjoint decomposition of B over X.

Proof (sketch of). Every disjoint product in $\langle X \rangle_{*,R,L}$ is just a disjoint sum with a bounded glueing of roots. It follows that MSO decomposition techniques (see [30] or [32]) combined with partial algebra techniques [7] are available, as done in [5] for languages of labeled birooted trees, to achieve an algebraic characterization of MSO definable languages in terms of (partial algebra) morphisms into finite structures. Such an approach also proves the complexity claim for the membership problem.   □

Remark. Of course, the membership problem is non elementary in the size of the MSO formula that defines L. This already follows from the case of MSO definable languages of finite words. Also, the problem of finding disjoint decompositions over X for birooted graphs may be delicate and is left for further studies.

As observed above, A^*, $FAM(A)$ and $FIM(A)$ are examples of subsemigroup of $HS(A)$ that are finitely generated by disjoint product, inverses and/or projections [15,17]. By applying Theorem 5.5, this proves (again) that their MSO definable subsets have decidable emptiness problem.

6   The Inverse Monoid of Acyclic Birooted Graphs

Towards application purposes, birooted graphs can be seen as models of computerized system behaviors with vertices viewed as (local) states and edges viewed as (local) transition. In this case, one is tempted to detect and forbid directed cycles which interpretation could be problematic (causally incoherent).

As an illustration of the power of the inverse semigroup framework that is proposed here, we show how these birooted acyclic graphs can simply be defined as the quotient of the inverse semigroup of birooted graphs by the semigroup ideal of cyclic ones. Then, in such a quotient, easily implementable, a product of acyclic birooted graphs is causally coherent if and only if it is non zero.

Lemma 6.1 (Semigroup Ideal). *Let φ be a graph property that is preserved under graph morphisms. Let I_φ be the set $I_\varphi \subseteq HS(A)$ that contains 0 and all birooted graphs whose underlying graph satisfies φ. Then, I_φ is an semigroup ideal of $HS(A)$, that is,*

$$HS(A) \cdot I_\varphi \subseteq HS(A) \text{ and } I_\varphi \cdot HS(A) \subseteq HS(A)$$

and the Rees' quotient $HS(A)/I_\varphi$, that is, the set $HS(A) - I_\varphi + \{0\}$ equipped with the product defined as in $H(A)$ when the result does not belong to I_φ and defined to be 0 otherwise, is still an inverse semigroup.

In other words, much in the same way 0 already appears with products in $HS(A)$ that have no compatible types, when the property φ describes, in some concrete modeling context, a set of faulty models that is preserves under morphism, then the product in $HS(A)/I_\varphi$ equals 0 also when the resulting birooted graph is faulty.

Clearly, the existence of directed cycles is a property preserved by morphism. Then, the algebra of birooted acyclic graphs can simply be modeled as the inverse semigroup $HS(A)/I_C$ where $I_C \subseteq HS(A)$ is the resulting semigroup ideal containing 0 and all (directed) cyclic birooted graphs.

Such a situation is depicted in Fig. 12 where

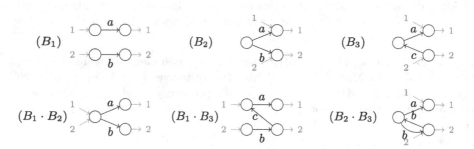

Fig. 12. Causal constraints propagation via products.

examples show how products of birooted graphs may propagate causality constraints eventually leading to non-causal graphs: the product $(B_2 \cdot B_2)$.

In other words, with the proposed approach, one can define a modeling software in such a way that non-causal models raised by combination of causal constraints are easily detected and forbidden, while, at the same time, the underlying algebraic framework still lays in the theory of inverse semigroups.

7 Conclusion

We have shown how a rather simple and intuitive composition operation on graphs, inherited from long standing ideas (see [27]), induces a rich algebraic structure, an inverse semigroup, from which one can define a natural order and other mathematically robust operators such as left and right projections, that capture graph theoretical concepts.

Of course, defining graph products by means of cospans products has already a long history in Theoretical Computer Science (see e.g. [4,6,13]). The originality of our approach consists in restricting to the category of unambiguous graphs and connecting morphisms that allow the resulting semigroup to be an inverse semigroup.

Still, this inverse semigroup is far from being understood in depth. Little is known about its subsemigroups. Thanks to [31], one can easily show that, all A generated E-unitary inverse semigroups (see also [28]) are subsemigroups of the monoid defined by birooted graphs of type (1,1). This suggests that the semigroup $HS(A)$ may satisfy some universality property that is still to be discovered. Also, we have no direct characterizations of the subsemigroups of $HS(A)$ that could be defined by bounding the number of roots on generators.

Following [5], by restricting the product to disjoint product, techniques arising from partial algebras [7] are applicable allowing us to inherit from the existing MSO-language theory of graphs of bounded tree-width [8,9]. It is expected that tile automata, defined in [15,16] over birooted words or trees, can easily be extended to higher dimensional strings and related with MSO-definability. Yet, closure property of MSO-definable languages remains to be detailled. It is by no means clear under which restrictions the product of two definable languages remains definable. Also, defining more suitable subsemigroups of (possible Rees' quotient of) $HS(A)$ that would also have decidable MSO languages is still to be investigated.

With a view towards application, beyond all experiments mentioned in the introduction, the modeling power of birooted graphs also needs to be investigated further in both practical modeling problems and more general modeling theories. For such a purpose, an implementation of the monoid $HS(A)$ with both graphical and programmatic views of its elements is scheduled. As already mentioned, multiple roots gives a flavor of concurrency. It is also expected that higher dimensional strings can be used as (explicitly concurrent) models of partially semi-commutative traces [11,19] henceforth connecting higher dimensional strings with a part of concurrency theory.

Finally, it has been shown recently that (one head) tree and graph walking automata semantics is nicely described in terms of (languages of) birooted graphs with single input and output roots [20]. The generalized birooted graphs presented here may provide nice semantical models of multi-head walking automata: partial runs of these automata clearly define languages of birooted graphs with multiple input and output roots.

Acknowledgements. The idea of developing a notion of higher dimensional strings has been suggested to the author by Mark V. Lawson in 2012. Their presentations have also benefited from numerous and helpful comments from anonymous referees of serval versions of this paper.

References

1. Abramsky, S.: A structural approach to reversible computation. Theor. Comp. Sci. **347**(3), 441–464 (2005)
2. Abrial, J.-R.: Modeling in Event-B - System and Software Engineering. Cambridge University Press, Cambridge (2010)
3. Berthaut, F., Janin, D., Martin, B.: Advanced synchronization of audio or symbolic musical patterns: an algebraic approach. Int. J. Semant. Comput. **6**(4), 409–427 (2012)
4. Blume, C., Bruggink, H.J.S., Friedrich, M., König, B.: Treewidth, pathwidth and cospan decompositions with applications to graph-accepting tree automata. J. Visual Lang. Comput. **24**(3), 192–206 (2013)
5. Blumensath, A., Janin, D.: A syntactic congruence for languages of birooted trees. Semigroup Forum **91**(161), 1–24 (2014)
6. Bruggink, H.J.S., König, B.: On the recognizability of arrow and graph languages. In: Ehrig, H., Heckel, R., Rozenberg, G., Taentzer, G. (eds.) ICGT 2008. LNCS, vol. 5214, pp. 336–350. Springer, Heidelberg (2008)
7. Burmeister, P.: A Model Theoretic Oriented Approach to Partial Algebras. Akademie-Verlag, Berlin (1986)
8. Courcelle, B.: The monadic second-order logic of graphs V: on closing the gap between definability and recognizability. Theor. Comp. Sci. **80**(2), 153–202 (1991)
9. Courcelle, B., Engelfriet, J.: Graph Structure and Monadic Second-order Logic, A Language Theoretic Approach, of Encyclopedia of Mathematics and its Applications, vol. 138. Cambridge University Press, Cambridge (2012)
10. Danos, V., Regnier, L.: Reversible, irreversible and optimal lambda-machines. Theor. Comp. Sci. **227**(1–2), 79–97 (1999)
11. Diekert, V., Lohrey, M., Miller, A.: Partially commutative inverse monoids. Semigroup Forum **77**(2), 196–226 (2008)
12. Fountain, J., Gomes, G., Gould, V.: The free ample monoid. Int. J. Algebra Comput. **19**, 527–554 (2009)
13. Gadducci, F., Heckel, R.: An inductive view of graph transformation. In: Recent Trends in Algebraic Development Techniques, 12th International Workshop, WADT 1997, Selected Papers, pp. 223–237 (1997)
14. Hudak, P., Janin, D.: Tiled polymorphic temporal media. In: Work on Functional Art, Music, Modeling and Design (FARM), pp. 49–60. ACM Press (2014)
15. Janin, D.: Algebras, automata and logic for languages of labeled birooted trees. In: Fomin, F.V., Freivalds, R., Kwiatkowska, M., Peleg, D. (eds.) ICALP 2013, Part II. LNCS, vol. 7966, pp. 312–323. Springer, Heidelberg (2013)
16. Janin, D.: Overlapping tile automata. In: Bulatov, A.A., Shur, A.M. (eds.) CSR 2013. LNCS, vol. 7913, pp. 431–443. Springer, Heidelberg (2013)
17. Janin, D.: On languages of labeled birooted trees: algebras, automata and logic. Inf. Comput. **243**, 222–248 (2014)
18. Janin, D.: Towards a higher-dimensional string theory for the modeling of computerized systems. In: Geffert, V., Preneel, B., Rovan, B., Štuller, J., Tjoa, A.M. (eds.) SOFSEM 2014. LNCS, vol. 8327, pp. 7–20. Springer, Heidelberg (2014)

19. Janin, D.: Free inverse monoids up to rewriting. Research report, LaBRI, Université de Bordeaux (2015)
20. Janin, D.: Walking automata in the free inverse monoid. Research report, LaBRI, Université de Bordeaux (2015)
21. Janin, D., Berthaut, F., Desainte-Catherine, M.: Multi-scale design of interactive music systems: the libTuiles experiment. In: Sound and Music Computing (SMC) (2013)
22. Janin, D., Berthaut, F., DeSainte-Catherine, M., Orlarey, Y., Salvati, S.: The T-calculus: towards a structured programming of (musical) time and space. In: Work on Functional Art, Music, Modeling and Design (FARM), pp. 23–34. ACM Press (2013)
23. Kellendonk, J.: The local structure of tilings and their integer group of coinvariants. Comm. Math. Phys. **187**, 115–157 (1997)
24. Kellendonk, J., Lawson, M.V.: Tiling semigroups. J. Algebra **224**(1), 140–150 (2000)
25. Kellendonk, J., Lawson, M.V.: Universal groups for point-sets and tilings. J. Algebra **276**, 462–492 (2004)
26. Lawson, M.V.: Inverse Semigroups: The Theory of Partial Symmetries. World Scientific, River Edge (1998)
27. Lewis, H.R.: A new decidable problem, with applications (extended abstract). In: IEEE Symposium on Foundations of Computer Science (FOCS), pp. 62–73. IEEE Press (1977)
28. Margolis, S.W., Meakin, J.C.: E-unitary inverse monoids and the Cayley graph of a group presentation. J. Pure Appl. Algebra **58**, 46–76 (1989)
29. Meakin, J.: Groups and semigroups: connections and contrasts. In: Groups St Andrews 2005, London Mathematical Society, Lecture Note Series 340, vol. 2. Cambridge University Press (2007)
30. Shelah, S.: The monadic theory of order. Ann. Math. **102**, 379–419 (1975)
31. Stephen, J.B.: Presentations of inverse monoids. J. Pure Appl. Algebra **63**, 81–112 (1990)
32. Thomas, W.: Ehrenfeucht games, the composition method, and the monadic theory of ordinal words. In: Mycielski, J., Rozenberg, G., Salomaa, A. (eds.) Structures in Logic and Computer Science. LNCS, vol. 1261, pp. 118–143. Springer, Heidelberg (1997)
33. Thomas, W.: Logic for computer science: the engineering challenge. In: Wilhelm, R. (ed.) Informatics. LNCS, vol. 2000, pp. 257–267. Springer, Heidelberg (2001)

A Functorial Bridge Between the Infinitary Affine Lambda-Calculus and Linear Logic

Damiano Mazza$^{(\boxtimes)}$ and Luc Pellissier

CNRS, UMR 7030, LIPN, Université Paris 13, Sorbonne Paris Cité,
Villetaneuse, France
{Damiano.Mazza,Luc.Pellissier}@lipn.univ-paris13.fr

Abstract. It is a well known intuition that the exponential modality of
linear logic may be seen as a form of limit. Recently, Melliès, Tabareau
and Tasson gave a categorical account for this intuition, whereas the first
author provided a topological account, based on an infinitary syntax. We
relate these two different views by giving a categorical version of the topological construction, yielding two benefits: on the one hand, we obtain
canonical models of the infinitary affine lambda-calculus introduced by
the first author; on the other hand, we find an alternative formula for
computing free commutative comonoids in models of linear logic with
respect to the one presented by Melliès et al.

1 Introduction

The Exponential Modality of Linear Logic as a Limit. Following the work of
Girard [5], linearity has become a central notion in computer science and proof
theory: it provides a finer-grained analysis of cut-elimination, which in turn, via
Curry-Howard, gives finer tools for the analysis of the execution of programs. It
is important to observe that the expressiveness of strictly linear or affine calculi is
severely restricted, because programs in these calculi lack the ability to duplicate
their arguments. The power of linear logic (which, in truth, is not linear at
all!) resides in its so-called *exponential modalities*, which allow duplication (and
erasing, if the logic is not already affine).

A possible approach to understand exponentials is to see the non-linear part
of linear logic as a sort of limit of its purely linear part. The following old
result morally says that, in the propositional case, exponential-free linear logic
is "dense" in full linear logic:

Theorem 1 (Approximation [5]). *Define the* bounded exponential

$$!_p A := \overbrace{(A\&1) \otimes \cdots \otimes (A\&1)}^{p \ times},$$

and define $?_p A := (!_p A^\perp)^\perp$. *Note that these formulas are exponential-free (if
A is). Let A be a propositional formula with m occurrences of the $!$ modality
and n occurrences of the $?$ modality. If A is provable in full linear logic, then*

M. Leucker et al. (Eds.): ICTAC 2015, LNCS 9399, pp. 144–161, 2015.
DOI: 10.1007/978-3-319-25150-9_10

for every $p_1, \ldots, p_m \in \mathbf{N}$ *there exist* $q_1, \ldots, q_n \in \mathbf{N}$ *such that* A' *is provable in exponential-free linear logic, where* A' *is obtained from* A *by replacing the i-th occurrence of* ! *with* $!_{p_i}$ *and the j-th occurrence of* ? *with* $?_{q_j}$.

For example, from the canonical proof of $?A^{\perp} \mathbin{\bindnasrepma} (!A \otimes !A)$ (contraction, *i.e.* duplication), we get proofs of $?_{p_1+p_2} A^{\perp} \mathbin{\bindnasrepma} (!_{p_1} A \otimes !_{p_2} A)$ for all $p_1, p_2 \in \mathbf{N}$.

Remember that, if a linear formula A says "A exactly once", then $!A$ stands for "A at will". The formula $A \& 1$ is an affine version of A: it says "A at most once". This is a very specialized use of additive conjunction, in the sequel we prefer to avoid additive connectives and denote the affine version of A by $A^{\bullet}$, which may or may not be defined as $A \& 1$ (for instance, in affine logic, $A^{\bullet} = A$). Therefore, $!_p A = (A^{\bullet})^{\otimes p}$ stands for "A at most p times", hence the name bounded exponential. So the Approximation Theorem supports the idea that $!A$ is somehow equal to $\lim_{p \to \infty} !_p A$.

Categories vs. Topology. This idea was recently formalized in two quite different ways. The first is due to Melliès, Tabareau and Tasson [12], who rephrased the question in categorical terms. It is well known [3] that a $*$-autonomous category admitting the free commutative comonoid A^{∞} on every object A is a model of linear logic (a so-called *Lafont category*). So, given a Lafont category, how does one compute A^{∞}? Using previous work by the first two authors [11], Melliès et al. showed that one may proceed as follows:

- compute the free co-pointed object $A^{\bullet}$ on A (which is $A \& 1$ if the category has binary products);
- compute the symmetric versions of the tensorial powers of $A^{\bullet}$, *i.e.* the following equalizers, where $\mathfrak{S}_n$ is the set of canonical symmetries of $(A^{\bullet})^{\otimes n}$:

$$A^{\leqslant n} \longrightarrow (A^{\bullet})^{\otimes n} \mathbin{\supset} \mathfrak{S}_n$$

- compute the following projective limit, where $A^{\leqslant n} \longleftarrow A^{\leqslant n+1}$ is the canonical arrow "throwing away" one component:

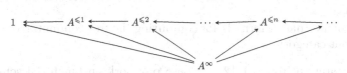

At this point, for A^{∞} to be the commutative comonoid on A it is enough that all relevant limits (the equalizers and the projective limit) commute with the tensor. Although not valid in general, this condition holds in several Lafont categories of very different flavor, such as Conway games and coherence spaces.

The second approach, due to the first author [9], is topological, and is based directly on the syntax. One considers an affine λ-calculus in which variables are treated as bounded exponentials: in a term of this calculus, a variable x may appear any number of times, each occurrence appears indexed by an integer (each instance, noted x_i, is labelled with a distinct $i \in \mathbf{N}$). The argument of applications is not a term but a sequence of terms, and to reduce the redex

$(\lambda x.t)\langle u_0, \ldots, u_{n-1}\rangle$ one replaces each free x_i in t with u_i (a special term $\bot$ is substituted if $i \geq n$). The calculus is therefore affine, in the sense that no duplication is performed, and in fact it strongly normalizes even in absence of types (the size of terms strictly decreases with reduction).

At this point, the set of terms is equipped with the structure of uniform space[1], the Cauchy-completion of which, denoted by $\Lambda_\infty^{\mathrm{aff}}$, contains infinitary terms, *i.e.* allowing infinite sequences $\langle u_1, u_2, u_3, \ldots\rangle$. The original calculus embeds (and is dense) in $\Lambda_\infty^{\mathrm{aff}}$ by considering a finite sequence as an almost-everywhere $\bot$ sequence. Reduction, which is continuous, is defined as above, except that infinitely many substitutions may occur. This yields non-termination, in spite of the calculus still being affine: if $\Delta_n := \lambda x.x_0\langle x_1, \ldots, x_n\rangle$, then $\Delta := \lim_{n\to\infty} \Delta_n = \lambda x.x_0\langle x_1, x_2, x_3, \ldots\rangle$ and $\Omega := \Delta\langle \Delta, \Delta, \Delta, \ldots\rangle \to \Omega$.

Ideally, these infinitary terms should correspond to usual λ-terms. But there is a continuum of them, definitely too many. The solution is to consider a partial equivalence relation $\approx$ such that, in particular, $x_i \approx x_j$ for all i, j and $t\langle u_1, u_2, u_3, \ldots\rangle \approx t'\langle u_1', u_2', u_3', \ldots\rangle$ whenever $t \approx t'$ and, for all $i, i' \in \mathbf{N}$, $u_i \approx u_{i'}'$. After introducing a suitable notion of reduction $\Rightarrow$ on the equivalence classes of $\approx$, one finally obtains the isomorphism for the reduction relations

$$(\Lambda_\infty^{\mathrm{aff}}/\approx, \Rightarrow) \quad \cong \quad (\Lambda, \to_\beta),$$

where $(\Lambda, \to_\beta)$ is the usual pure λ-calculus with β-reduction. Similar infinitary calculi (also with a notion of partial equivalence relation) were considered by Kfoury [6] and Melliès [10], although without a topological perspective. The indices identifying the occurrences of exponential variables are also reminiscent of Abramsky, Jagadeesan and Malacaria's games semantics [1].

Reconciling the Two Approaches. The contribution of this paper is to draw a bridge between the two approaches presented above. Indeed, we develop a categorical version of the topological construction of [9], which turns out to:

1. give a canonical way of building denotational models of the infinitary affine λ-calculus;
2. provide an alternative formula for computing the free commutative comonoid in a Lafont category.

Drawing inspiration from [11,12], we base our work on functorial semantics in the sense of Lawvere, computing free objects as Kan extensions.

Functorial Semantics. The idea of functorial semantics is to describe an algebraic theory as a certain category constituted of the different powers of the domain of the theory as the objects, the operations of the theory as morphisms, and encode the relations between the operations in the composition operation. We will not consider algebraic theories as Lawvere did, but the more general symmetric monoidal theories, or PROPs [7] (*product and permutation categories*).

---

[1] The generalization of a metric space, still allowing one to speak of Cauchy sequences.

Definition 1 (Symmetric Monoidal Theory). *An n-sorted symmetric monoidal theory is defined as a symmetric monoidal category $\mathbb{T}$ whose objects are n-tuples of natural numbers and with a tensorial product defined as the point-wise arithmetical sum.*

A model of $\mathbb{T}$ in a symmetric monoidal category (SMC) $\mathcal{C}$ is a symmetric strong monoidal functor $\mathbb{T} \rightarrow \mathcal{C}$.

A morphism of models of $\mathbb{T}$ in $\mathcal{C}$ is a monoidal natural transformation between models of $\mathbb{T}$ in $\mathcal{C}$. We will denote as $\mathrm{Mod}(\mathbb{T}, \mathcal{C})$ the category with models of $\mathbb{T}$ in $\mathcal{C}$ as objects and morphisms between models as morphisms.

The simplest symmetric monoidal theory, denoted by $\mathbb{B}$, has as objects the natural numbers seen as finite ordinals and as morphisms the bijections between them (the permutations). Alternatively, $\mathbb{B}$ can be seen as the free symmetric monoidal category on one object (the object 1, with monoidal unit 0). As such, a model of $\mathbb{B}$ is nothing but an object A in a symmetric monoidal category $\mathcal{C}$, and the categories $\mathcal{C}$ and $\mathrm{Mod}(\mathbb{B}, \mathcal{C})$ are equivalent.

The key non-trivial example in our context is that of commutative (co)monoids. We remind that a commutative monoid in a SMC $\mathcal{C}$ is a triple $(A, \mu : A \otimes A \rightarrow A, \eta : 1 \rightarrow A)$, with A an object of $\mathcal{C}$, such that the arrows μ and η interact with the associator, unitors and symmetry of $\mathcal{C}$ to give the usual laws of associativity, neutrality and commutativity (see *e.g.* [8]). A morphism of monoids $f : (A, \mu, \eta) \rightarrow (A', \mu', \eta')$ is an arrow $f : A \rightarrow A'$ such that $f \circ \mu = \mu' \circ (f \otimes f)$ and $f \circ \eta = \eta'$. We denote the category of monoids of $\mathcal{C}$ and their morphisms as $\mathrm{Mon}(\mathcal{C})$. The dual notion of comonoid, and the relative category $\mathrm{Comon}(\mathcal{C})$, is obtained by reversing the arrows in the above definition. Now, consider the symmetric monoidal theory $\mathbb{F}$ whose objects are the natural numbers seen as finite ordinals and its morphisms are the functions between them (*i.e.* $\mathbb{F}$ is the skeleton of the category of finite sets). We easily check that $\mathrm{Mod}(\mathbb{F}, \mathcal{C}) \simeq \mathrm{Mon}(\mathcal{C})$ and $\mathrm{Mod}(\mathbb{F}^{\mathrm{op}}, \mathcal{C}) \simeq \mathrm{Comon}(\mathcal{C})$. Indeed, a strict symmetric monoidal functor from $\mathbb{F}$ to $\mathcal{C}$ picks an object of $\mathcal{C}$ and the image of any arrow $m \rightarrow n$ of $\mathbb{F}$ is unambiguously obtained from the images of the unique morphisms $0 \rightarrow 1$ and $2 \rightarrow 1$ in $\mathbb{F}$, which are readily verified to satisfy the monoid laws.

Summing up, finding the free commutative comonoid A^{∞} on an object A of a SMC $\mathcal{C}$ is the same thing as turning a strict symmetric monoidal functor $\mathbb{B} \rightarrow \mathcal{C}$ into a strict symmetric monoidal functor $\mathbb{F}^{\mathrm{op}} \rightarrow \mathcal{C}$ which is universal in a suitable sense. This is where Kan extensions come into the picture.

Free Comonoids as Kan Extensions. Kan extensions allow to extend a functor along another. Let $K : \mathcal{C} \rightarrow \mathcal{D}$ and $F : \mathcal{C} \rightarrow \mathcal{E}$ be two functors. If we think of K as an inclusion functor, it seems natural to try to define a functor $\mathcal{D} \rightarrow \mathcal{E}$ that would in a sense be universal among those that extend F. There are two ways of formulating this statement precisely, yielding left and right Kan extensions. We only describe the latter, because it is the case of interest for us:

Definition 2 (Kan Extension). *Let $\mathcal{C}, \mathcal{D}, \mathcal{E}$ be three categories and $F : \mathcal{C} \rightarrow \mathcal{E}$, $K : \mathcal{C} \rightarrow \mathcal{D}$ two functors. The* right *Kan extension of F along K is a functor*

$\mathrm{Ran}_K F : \mathcal{D} \to \mathcal{E}$ *together with a natural transformation* $\varepsilon : \mathrm{Ran}_K F \circ K \Rightarrow F$ *such that for any other pair* $(G : \mathcal{D} \to \mathcal{E}, \gamma : G \circ K \Rightarrow F)$, γ *factors uniquely through* ε:

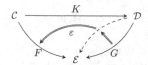

It is easy to check that $\mathbf{Cat}(G, \mathrm{Ran}_K F) \simeq \mathbf{Cat}(G \circ K, F)$, where by $\mathbf{Cat}(f, g)$ (f and g being functors with same domain and codomain) we mean the 2-homset of the 2-category $\mathbf{Cat}$, *i.e.* the set of all natural transformations from f to g. In other words, Ran_K is right adjoint to U_K, the functor precomposing with K (whence the terminology "right"—the left adjoint to U_K is the left Kan extension). This observation is important because it tells us that Kan extensions may be relativized to any 2-category. In particular, we may speak of *symmetric monoidal Kan extensions* by taking the underlying 2-category to be $\mathbf{SymMonCat}$ (symmetric monoidal categories, strict symmetric monoidal functors and monoidal natural transformations).

Now, there is an obvious inclusion functor $i : \mathbb{B} \to \mathbb{F}^{\mathrm{op}}$ (bijections are particular functions), which is strictly symmetric monoidal. So if $\mathcal{E}$ is symmetric monoidal and A is an object of $\mathcal{E}$, we are in the situation described above with $\mathcal{C} = \mathbb{B}$, $\mathcal{D} = \mathbb{F}^{\mathrm{op}}$, $K = i$ and F the strict symmetric monoidal functor corresponding to A, which we abusively denote by A. The fundamental difference is that the diagram lives in $\mathbf{SymMonCat}$ instead of $\mathbf{Cat}$. It is an instructive exercise to verify that the free commutative comonoid on A, if it exists, is $A^\infty = \mathrm{Ran}_i A(1)$, *i.e.* the right symmetric monoidal Kan extension of A along i, computed in 1:

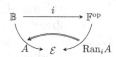

Remember that the free commutative comonoid on A is a commutative comonoid A^∞ with an arrow $\mathrm{d} : A^\infty \to A$ such that, whenever C is a commutative comonoid and $f : C \to A$, there is a unique comonoid morphism $u : C \to A^\infty$ such that $f = \mathrm{d} \circ u$. The arrow d is ε, where $\varepsilon : \mathrm{Ran}_i A \circ i \Rightarrow A$ is the natural transformation coming with the Kan extension.

More generally, if $\mathbb{T}_1$ and $\mathbb{T}_2$ are two symmetric monoidal theories, a symmetric monoidal functor $i : \mathbb{T}_1 \to \mathbb{T}_2$ induces a forgetful functor $U_i : \mathrm{Mod}(\mathbb{T}_2, \mathcal{E}) \to \mathrm{Mod}(\mathbb{T}_1, \mathcal{E})$ such that $M \mapsto M \circ i$. So we may reformulate the problem of finding the "free $\mathbb{T}$-model" on an object A of $\mathcal{E}$ as finding a left monoidal adjoint to U_i with $i : \mathbb{B} \to \mathbb{T}$. That is precisely what we did above, with $\mathbb{T} = \mathbb{F}^{\mathrm{op}}$.

Computing Monoidal Kan Extensions. The above discussion is interesting because it provides a way of explicitly computing A^∞ from A. In fact, there is a well-known formula for computing Kan extensions [8]. When applied to the above special case, it gives

$$A^\infty = \prod_n A^{\otimes n}/\sim,$$

where $A^{\otimes n}/\sim$ is the symmetric tensor product. However, this formula works only for Kan extensions in **Cat** and there are no known formulas in other 2-categories. The main contribution of [11] was to find a sufficient condition under which the formula is correct also in **SymMonCat**. The condition is, roughly speaking, a commutation of the tensor with certain limits depending on the Kan extension at stake. In the above case, it requires the tensor to commute with countable products, which, in models of linear logic, boils down to having countable biproducts. Lafont categories of this kind do exist (*e.g.* the category **Rel** of sets and relations), but they are a little degenerate and not very representative.

The idea of [12] was to decompose the Kan extension in two, so that the commutation condition is weaker and satisfied by more Lafont categories. The intermediate step uses a symmetric monoidal theory denoted by $\mathbb{I}$, whose objects are natural numbers (seen as finite ordinals) and morphisms are the injections. Note that $\mathrm{Mod}(\mathbb{I}^{\mathrm{op}}, \mathcal{C})$ is equivalent to the slice category $\mathcal{C} \downarrow 1$. By definition, this is the category of *copointed objects* of $\mathcal{C}$: pairs $(A, w : A \to 1)$ (with 1 the tensor unit, not necessarily terminal), with morphisms $f : (A, w) \to (A', w')$ arrows $f : A \to A'$ such that $w = w' \circ f^2$.

There are of course strict symmetric monoidal injections $j : \mathbb{B} \to \mathbb{I}^{\mathrm{op}}$ and $j' : \mathbb{I}^{\mathrm{op}} \to \mathbb{F}^{\mathrm{op}}$, such that $j' \circ j = i$. Unsurprisingly, $\mathrm{Ran}_j A(1)$ is the free copointed object on A, which we denoted by $A^\bullet$ above. Since Kan extensions compose (assuming they exist), we have $A^\infty = \mathrm{Ran}_{j'} A^\bullet(1)$:

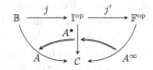

For the second Kan extension to be computed in **SymMonCat** using the **Cat** formula, a milder commutation condition than requiring countable biproducts suffices. It is the commutation condition we mentioned above when we recalled the three-step computation of A^∞ (free copointed object, equalizers, projective limit), which indeed results from specializing the general Kan extension formula.

One More Intermediate Step. The bridge between the categorical and the topological approach will be built upon a further decomposition of the Kan extension: in the second step, we interpose a 2-sorted theory, denoted by $\mathbb{P}$ (this is why we introduced multi-sorted theories, all theories used so far are 1-sorted):

---

[2] The w stands for weakening.

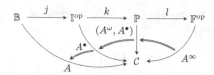

We will call the models of $\mathbb{P}$ *partitionoids*. Intuitively, the free partitionoid on A allows to speak of infinite streams on $A^\bullet$, from which one may extract arbitrary elements and substreams via maps of type $A^\omega \to (A^\bullet)^{\otimes m} \otimes (A^\omega)^{\otimes n}$. Such maps are the key to model the infinitary affine λ-calculus. This intuition is especially evident in **Rel** (the category of sets and relations), where A^ω is the set of all functions $\mathbf{N} \to A^\bullet$ which are almost everywhere $*$ (in **Rel**, $A^\bullet = A \uplus \{*\}$).

2   The Infinitary Affine Lambda-Calculus

We consider three pairwise disjoint, countable sets of *linear*, *affine* and *exponential* variables, ranged over by $k, l, m \ldots$, $a, b, c \ldots$ and $x, y, z \ldots$, respectively. The terms of the infinitary affine λ-calculus belong to the following grammar:

$$
\begin{aligned}
t, u ::= &\ l \mid \lambda l.t \mid tu \mid \mathsf{let}\, k \otimes l = u \,\mathsf{in}\, t \mid t \otimes u && \text{linear} \\
&\mid a \mid \mathsf{let}\, a^\bullet = u \,\mathsf{in}\, t \mid \bullet t && \text{affine} \\
&\mid x_i \mid \mathsf{let}\, x^\omega = u \,\mathsf{in}\, t \mid \langle u_0, u_1, u_2, \ldots \rangle && \text{exponential}
\end{aligned}
$$

The linear part of the calculus comes from [2]. It is the internal language of symmetric monoidal closed categories. As usual, let constructs are binders. The notation $\langle u_0, u_1, u_2, \ldots \rangle$ stands for an infinite sequence of terms. We use $\mathbf{u}$ to range over such sequences and write $\mathbf{u}(i)$ for u_i. Note that each u_i is inductively smaller than $\mathbf{u}$, so terms are infinite but well-founded. The usual linearity/affinity constraints apply to linear/affine variables, with the additional constraint that if x_i, x_j are distinct occurrences of an exponential variable in a term, then $i \neq j$. Furthermore, the free variables of a term of the form $\mathbf{u}$ (resp. $\bullet t$) must all be exponential (resp. exponential or affine).

The reduction rules are as follows:

$$
(\lambda l.t)u \to t[u/l] \qquad\qquad \mathsf{let}\, k \otimes l = u \otimes v \,\mathsf{in}\, t \to t[u/k][v/l]
$$
$$
\mathsf{let}\, a^\bullet = \bullet u \,\mathsf{in}\, t \to t[u/a] \qquad\qquad \mathsf{let}\, x^\omega = \mathbf{u} \,\mathsf{in}\, t \to t[\mathbf{u}(i)/x_i]
$$

In the exponential rule, i ranges over $\mathbf{N}$, so there may be infinitely many substitutions to be performed. There are also the usual commutative conversions involving let binders, which we omit for brevity. The reduction is confluent, as the rules never duplicate any subterm.

The results of [9] are formulated in an infinitary calculus with exponential variables only, whose terms and reduction are defined as follows:

$$
t, u ::= x_i \mid \lambda x.t \mid t\langle u_0, u_1, u_2, \ldots \rangle, \qquad\qquad (\lambda x.t)\mathbf{u} \to t[\mathbf{u}(i)/x_i]
$$

$$\frac{}{\Gamma;\Delta;l:A \vdash l:A}\ \text{lin-ax} \qquad \frac{}{\Gamma;\Delta,a:A;\vdash a:A}\ \text{aff-ax} \qquad \frac{i \in \mathbf{N}}{\Gamma,x:A;\Delta;\vdash x_i:A^\bullet}\ \text{exp-ax}$$

$$\frac{\Gamma;\Delta;\Sigma,l:A \vdash t:B}{\Gamma;\Delta;\Sigma \vdash \lambda l.t:A \multimap B}\ {\multimap}I \qquad \frac{\Gamma;\Delta;\Sigma \vdash t:A \multimap B \quad \Gamma;\Delta';\Sigma' \vdash u:A}{\Gamma,\Delta,\Delta';\Sigma,\Sigma' \vdash tu:B}\ {\multimap}E$$

$$\frac{\Gamma;\Delta;\Sigma \vdash t:A \quad \Gamma;\Delta';\Sigma' \vdash u:B}{\Gamma,\Delta,\Delta';\Sigma,\Sigma' \vdash t \otimes u:B}\ {\otimes}I \qquad \frac{\Gamma;\Delta;\Sigma \vdash u:A \otimes B \quad \Gamma;\Delta';\Sigma',k:A,l:B \vdash t:C}{\Gamma,\Delta,\Delta';\Sigma,\Sigma' \vdash \operatorname{let} k \otimes l = u \operatorname{in} t:C}\ {\otimes}E$$

$$\frac{\Gamma;\Sigma;\vdash t:A}{\Gamma;\Sigma;\vdash \bullet t:A^\bullet}\ {\bullet}I \qquad \frac{\Gamma;\Delta;\Sigma \vdash u:A^\bullet \quad \Gamma;\Delta',a:A;\Sigma' \vdash t:C}{\Gamma,\Delta,\Delta';\Sigma,\Sigma' \vdash \operatorname{let} a^\bullet = u \operatorname{in} t:C}\ {\bullet}E$$

$$\frac{\cdots \quad \Gamma;;\vdash \mathbf{u}(i):A^\bullet \quad \cdots}{\Gamma;;\vdash \mathbf{u}:A^\omega}\ \omega I \qquad \frac{\Gamma;\Delta;\Sigma \vdash u:A^\omega \quad \Gamma,x:A;\Delta';\Sigma' \vdash t:C}{\Gamma,\Delta,\Delta';\Sigma,\Sigma' \vdash \operatorname{let} x^\omega = u \operatorname{in} t:C}\ \omega E$$

Fig. 1. The simply-typed infinitary affine λ-calculus. In every non-unary rule we require that t,u (or, for the ωI rule, $\mathbf{u}(i),\mathbf{u}(j)$ for all $i \neq j \in \mathbf{N}$) contain pairwise disjoint sets of occurrences of the exponential variables in Γ.

(the abstraction binds all occurrences of x). Such a calculus may be embedded in the one introduced above, as follows:

$$x_i^\circ := \operatorname{let} a^\bullet = x_i \operatorname{in} a$$
$$(\lambda x.t)^\circ := \lambda l.\operatorname{let} x^\omega = l \operatorname{in} t^\circ$$
$$(t\langle u_0,u_1,u_2,\ldots\rangle)^\circ := t^\circ \langle \bullet u_0^\circ, \bullet u_1^\circ, \bullet u_2^\circ, \ldots \rangle$$

and we have $t \to t'$ implies $t^\circ \to^* t'^\circ$, so we do not lose generality. However, the categorical viewpoint adopted in the present paper naturally leads us to consider a simply-typed version of the calculus, given in Fig. 1. It is for this calculus that our construction provides denotational models. The types are generated by

$$A,B ::= X \mid A \multimap B \mid A \otimes B \mid A^\bullet \mid A^\omega,$$

where X is an atomic type. Note that the context of typing judgments has three *finite* components: exponential (Γ), affine (Δ) and linear (Σ). Although it may appear additive, the treatment of contexts is multiplicative also in the exponential case, as enforced by the condition in the caption of Fig. 1. The typing system enjoys the subject reduction property, as can be proved by an induction on the depth of the reduced redex.

3   Denotational Semantics

Definition 3 (reduced fpp, Monoidal Theory $\mathbb{P}$). *A finite partial partition (fpp) is a finite (possibly empty) sequence $(S_1,\ldots,S_k)$ of non-empty, pairwise disjoint subsets of $\mathbf{N}$. Fpp's may be composed as follows: let $\beta := (S_1,\ldots,S_k)$, with S_i infinite, and let $\beta' := (S_1',\ldots,S_{k'}')$; we define $\beta' \circ_i \beta := (S_1,\ldots,S_{i-1},T_1,\ldots,T_{k'},S_{i+1},\ldots,S_k)$, where each T_j is obtained as follows: let $n_0 < n_1 < n_2 < \cdots$ be the elements of S_i in increasing order; then, $T_j := \{n_m \mid m \in S_j'\}$. It must be noted that endowed with this composition, fpp's form an operad.*

We will only consider reduced fpp's, *in which each* S_i *is either a singleton or infinite. We will use the notation* $(S_1, \ldots, S_m; T_1, \ldots, T_n)$ *to indicate that the* S_i *are singletons and the* T_j *are infinite, and we will say that such an fpp has size* $m + n$. *Note that the composition of reduced fpp's is reduced. The set of all reduced fpp's will be denoted by* $\mathcal{P}$.

Reduced fpp's induce a 2-sorted monoidal theory $\mathbb{P}$, *as follows: each* $\beta \in \mathcal{P}$ *of size* $m + n$ *induces an arrow* $\beta : (0, 1) \to (m, n)$ *of* $\mathbb{P}$. *There is also an arrow* $w : (1, 0) \to (0, 0)$ *to account for partiality. Composition is defined as above.*

For example, let $\beta := (E, O)$, where E and O are the even and odd integers, and let $\beta' := (\{0\}, \mathbf{N} \setminus \{0\})$ (these are actually total partitions). Then $\beta' \circ_1 \beta = (\{0\}, E \setminus \{0\}, O)$, whereas $\beta \circ_2 \beta' = (\{0\}, O, E \setminus \{0\})$.

Definition 4 (Partitionoid). *A partitionoid in a symmetric monoidal category* $\mathcal{C}$ *is a strict symmetric monoidal functor[3]* $G : \mathbb{P} \to \mathcal{C}$. *Spelled out, it is a tuple* $(G_0, G_1, w, (r_\beta)_{\beta \in \mathcal{P}})$ *with* (G_0, w) *a copointed object and* $r_\beta : G_1 \to G_0^{\otimes m} \otimes G_1^{\otimes n}$ *whenever* β *is of size* $m + n$, *such that the composition of compatible* w *and* r_β *satisfies the equations induced by* $\mathbb{P}$.

A morphism of partitionoids $G \to G'$ *is a pair of arrows* $f_0 : G_0 \to G_0'$, $f_1 : G_1 \to G_1'$ *such that* f_0 *is a morphism of copointed objects and* $r_\beta' \circ f_1 = (f_0^{\otimes m} \otimes f_1^{\otimes n}) \circ r_\beta$ *for all* $\beta \in \mathcal{P}$ *of size* $m + n$.

We say that F *is the* free partitionoid *on* A *if it is endowed with an arrow* $e : F_0 \to A$ *such that, for every partitionoid* G *with an arrow* $f : G_0 \to A$, *there exists a unique morphism of partitionoids* $(u_0, u_1) : G \to F$ *such that* $f = e \circ u$.

For example, for any set X, $(X, X^{\mathbf{N}}, !_X, (r_\beta)_{\beta \in \mathcal{P}})$ is a partitionoid in **Set**, where $!_X$ is the terminal arrow $X \to 1$ and, if $\beta = (\{i_1\}, \ldots, \{i_m\}; \{j_1^1 < j_2^1 < \cdots\}, \ldots, \{j_1^n < j_2^n < \cdots\})$ and $f : \mathbf{N} \to X$, $r_\beta(f) := (f(i_1), \ldots, f(i_m), k \mapsto f(j_k^1), \ldots, k \mapsto f(j_k^n)) \in X^m \times (X^{\mathbf{N}})^n$.

Lemma 1. *If* (F_0, F_1) *is the free partitionoid on* A, *then* $F_0 = A^\bullet$, *the free co-pointed object on* A.

Proof. This follows from observing that $(A^\bullet, F_1)$ is also a partitionoid on A. $\square$

Definition 5 (Infinitary Affine Category). *Let* A *be an object in a symmetric monoidal category. We denote by* $\dagger_A$ *the following diagram:*

$$1 \xleftarrow{\varepsilon_1} A^\bullet \xleftarrow{\varepsilon_2} (A^\bullet)^{\otimes 2} \longleftarrow \cdots \xleftarrow{\varepsilon_n} (A^\bullet)^{\otimes n} \xleftarrow{\varepsilon_{n+1}} (A^\bullet)^{\otimes n+1} \longleftarrow \cdots$$

where $\varepsilon_1 = \varepsilon$ *is the copoint of* $A^\bullet$ *and* $\varepsilon_{n+1} := (\mathrm{id})^{\otimes n} \otimes \varepsilon$, *i.e., the arrow erasing the rightmost component. We set* $A^\omega := \lim \dagger_A$ *(if it exists).*

An infinitary affine category *is a symmetric monoidal closed category such that, for all* A, *the free partitionoid on* A *exists and is* $(A^\bullet, A^\omega)$.

---

[3] An algebra for the fpp operad.

Several well-known categories are examples of affine infinitary categories: sets and relations, coherence spaces and linear maps, Conway games. Finiteness spaces are a non-example. We give the relational example here, which is a bit degenerate but easy to describe and grasp. For the others, we refer to the extended version.

The category **Rel** has sets as objects and relations as morphisms. It is symmetric monoidal closed: the Cartesian product (which, unlike in **Set**, is not a categorical product in **Rel**!) acts both as $\otimes$ (with unit the singleton $\{*\}$) and $\multimap$. Let A be a set and let us assume that $* \notin A$. The free copointed object on A is (up to iso) $A \cup \{*\}$, with copoint the relation $\{(*,*)\}$. The F_1 part of the free partitionoid on A in **Rel** is (up to iso) the set of all functions $\mathbf{N} \to A^\bullet$ which are almost everywhere $*$. Given a reduced fpp $\beta := (\{i_1\}, \ldots, \{i_m\}; \{j_0^1 < j_1^1 < \ldots\}, \ldots, \{j_0^n < j_1^n < \ldots\})$, the corresponding morphism of type $A^\omega \to (A^\bullet)^{\otimes m} \otimes (A^\omega)^{\otimes n}$ is

$$r_\beta := \{(\mathbf{a}, (a_{i_1}, \ldots, a_{i_m}, \langle a_{j_0^1}, a_{j_1^1}, \ldots \rangle, \ldots, \langle a_{j_0^n}, a_{j_1^n}, \ldots \rangle)) \mid \mathbf{a} \in A^\omega\},$$

where we wrote $\langle a_0, a_1, a_2, \ldots \rangle$ for the function $\mathbf{a} : \mathbf{N} \to A^\bullet, i \mapsto a_i$.

Theorem 2. *An infinitary affine category is a denotational model of the infinitary affine λ-calculus.*

Proof. The interpretation of types is parametric in an assignment of an object to the base type X, and it is straightforward (notations are identical). In fact, we will confuse types and the objects interpreting them.

Let now $\Gamma; \Delta; \Sigma \vdash t : A$ be a typing judgment. The type of the corresponding morphism will be of the form $C_1 \otimes \cdots \otimes C_n \longrightarrow A$, where the C_i come from the context and are defined as follows. If it comes from $l : C \in \Sigma$ (resp. $a : C \in \Delta$), then $C_i := C$ (resp. $C_i := C^\bullet$). If it comes from $x : C \in \Gamma$, then $C_i := C^\omega$ if x appears infinitely often in t, otherwise, if it appears k times, $C_i := (C^\bullet)^{\otimes k}$.

The morphism interpreting a type derivation of $\Gamma; \Delta; \Sigma \vdash t : A$ is defined as customary by induction on the last typing rule. The lin-ax rule and all the rules concerning $\otimes$ and $\multimap$ are modeled in the standard way, using the symmetric monoidal closed structure. The only delicate point is modeling the seemingly additive behavior of the exponential context Γ in the binary rules (the same consideration will hold for the elimination rules of $\bullet$ and ω as well). Let us treat for instance the $\otimes I$ rule, and let us assume for simplicity that $\Gamma = x : C, y : D, z : E$, with x (resp. z) appearing infinitely often (resp. m and n times) in t and u, whereas y appears infinitely often in t but only k times in u. Let us also disregard the affine and linear contexts, which are unproblematic. The interpretation of the two derivations gives us two morphisms

$$[t] : C^\omega \otimes D^\omega \otimes (E^\bullet)^{\otimes m} \longrightarrow A, \qquad [u] : C^\omega \otimes (D^\bullet)^{\otimes k} \otimes (E^\bullet)^{\otimes n} \longrightarrow B.$$

Now, we seek a morphism of type $C^\omega \otimes D^\omega \otimes (E^\bullet)^{\otimes(m+n)} \longrightarrow A \otimes B$, because x and y appear infinitely often in $t \otimes u$, whereas z appears $m + n$ times. This is obtained by precomposing $[t] \otimes [u]$ with the morphisms $r_\beta : C^\omega \to C^\omega \otimes C^\omega$ and

$r_{\beta'} : D^\omega \to (D^\bullet)^{\otimes k} \otimes D^\omega$ associated with the fpp's $\beta = (; T_t, T_u)$ such that T_t (resp. T_u) contains all i such that x_i is free in t (resp. in u), and $\beta' = (S'_u; T'_t)$ is defined in a similar way with the variable y.

The weakening on exponential and affine variables in all axiom rules is modeled by the canonical morphisms $A^\bullet \to 1$ and $A^\omega \to 1$. For the rules aff-ax and exp-ax, we use the canonical morphism $A^\bullet \to A$ and the identity on $A^\bullet$, respectively.

The $\bullet I$ rule is modeled by observing that objects of the form $\Gamma^\omega \otimes \Delta^\bullet$ are copointed (from tensoring their copoints), so from an arrow $\Gamma^\omega \otimes \Delta^\bullet \longrightarrow A$ we obtain a unique arrow $\Gamma^\omega \otimes \Delta^\bullet \longrightarrow A^\bullet$ by universality of $A^\bullet$. The $\bullet E$ rule is just composition.

For what concerns the ωI rule, let us assume for simplicity that $\Gamma = x : C$. This defines a sequence of objects $(C_i)_{i \in \mathbf{N}}$ such that C_i is either C^ω or $(C^\bullet)^{\otimes k_i}$ according to whether x appears in $\mathbf{u}(i)$ infinitely often or k_i many times. Let now $S_i := \{j \in \mathbf{N} \mid x_j \text{ is free in } \mathbf{u}(i)\}$, define the fpp $\beta_i = (S_0, \dots, S_i)$ and let

$$\varepsilon'_i := (\mathrm{id})^{\otimes i} \otimes w_i : C_0 \otimes \cdots \otimes C_{i-1} \otimes C_i \longrightarrow C_0 \otimes \cdots \otimes C_{i-1},$$

where $w_i : C_i \to 1$ is equal to $r_\emptyset$ if $C_i = C^\omega$ (with $\emptyset$ the empty fpp) or it is equal to $\varepsilon^{\otimes k_i}$ if $C_i = (C^\bullet)^{\otimes k_i}$. Let $\widehat{\beta}_i$ be the reduced fpp obtained from β_i by "splitting" its finite sets into singletons. If we set $\theta_i := r_{\widehat{\beta}_i}$, we have that for all $i \in \mathbf{N}$, $\varepsilon'_i \circ \theta_{i+1} = \theta_i$. Let now f_i be the interpretations of the derivations of $x : C; ; \vdash \mathbf{u}(i) : A^\bullet$ and consider the diagram.

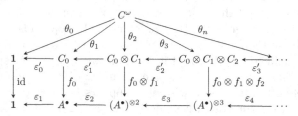

We showed above that all the upper triangles commute. It is easy to check that the bottom squares commute too, making $(C^\omega, ((f_0 \otimes \cdots \otimes f_{i-1}) \circ \theta_i)_{i \in \mathbf{N}})$ a cone for $\dagger_A$. Since $A^\omega = \lim \dagger_A$, this gives us a unique arrow $f : C^\omega \to A^\omega$, which we take as the interpretation of the derivation. The ωE rule is just composition, modulo the interposition of the canonical arrow $A^\omega \to (A^\bullet)^{\otimes k}$ in case x appears k times in t.

It remains to check that the above interpretation is stable under reduction, which may be done via elementary calculations.   □

4   Computing Symmetric Monoidal Kan Extensions

We mentioned that there is a well-known formula for computing regular Kan extensions (*i.e.* in **Cat**). This requires some notions coming from enriched category theory, which we recall next (although here the enrichment will be trivial, *i.e.* on **Set**).

Definition 6 (Cotensor Product of an Object by a set). *Let $\mathcal{C}$ be a (locally small) category. Let A be an object in $\mathcal{C}$ and E a set. The* cotensor product $E \circ A$ *of A by E is defined by:*

$$\forall B \in \mathcal{C}, \mathcal{C}(B, E \circ A) \simeq \mathbf{Set}(E, \mathcal{C}(B, A))$$

Any locally small category with products is cotensored over **Set** *(all of its objects have cotensor products with any set) and the cotensor product is given by:*

$$E \circ A = \prod_E A$$

We will write $\langle f_e \rangle_{e \in E} : B \to E \circ A$ for the infinite pairing of arrows $f_e : B \to A$ and $\pi_e : E \circ A \to A$ the projections.

Definition 7 (end). *Let $\mathcal{C}, \mathcal{E}$ be two categories and $H : \mathcal{C}^{\mathrm{op}} \times \mathcal{C} \to \mathcal{E}$ a functor. The* end *of H, denoted by $\int_{\mathcal{C}} H$, is defined as the universal object endowed with projections $\int_{\mathcal{C}} H \to H(c, c)$ for all $c \in \mathcal{C}$ making the following diagram commute:*

$$
\begin{array}{ccc}
\int_{c \in \mathcal{C}} H(c, c) & \longrightarrow & H(c', c') \\
\downarrow & & \downarrow f^* \\
H(c, c) & \xrightarrow{\;f_*\;} & H(c, c')
\end{array}
$$

for all arrows $f : c \to c'$ in $\mathcal{C}$.

Finally, here is the formula computing Kan extensions:

Theorem 3 ([8], X.4, Theorem1). *With the notations of Definition 2, whenever the objects exist:*

$$\mathrm{Ran}_K F(d) = \int_{c \in \mathcal{C}} \mathcal{D}(d, Kc) \circ Fc.$$

However, as mentioned in the introduction, the formula of Theorem 3 is only valid in **Cat** and we do not have any formula for computing a Kan extension in an arbitrary 2-category, or even in **SymMonCat**, our case of interest. Fortunately, Melliès and Tabareau proved a very general result [11, Theorem 1] giving sufficient conditions under which the Kan extension in **Cat** (something *a priori* worthless for our purposes) is actually the Kan extension in **SymMonCat** (what we want to compute). What follows is a specialized version of their result.

Theorem 4 ([11]). *Let $\mathcal{C}, \mathcal{D}, \mathcal{E}$ be three symmetric monoidal categories and $F : \mathcal{C} \to \mathcal{E}$, $K : \mathcal{C} \to \mathcal{D}$ two monoidal symmetric functors. If (all the objects considered exist and) the canonical morphism*

$$X \otimes \int_{c \in \mathcal{C}} \mathcal{D}(d, Kc) \circ Fc \longrightarrow \int_{c \in \mathcal{C}} X \otimes \mathcal{D}(d, Kc) \circ Fc$$

is an isomorphism for every object X, then the right monoidal Kan extension (in the 2-category **SymMonCat**) *of F along K may be computed as in Theorem 3.*

We may now give the abstract motivation behind Definition 5. The key property therein is that the free partitionoid on A is equal to $(A^\bullet, A^\omega)$. We now instantiate Theorem 4 to give a sufficient condition for that to be the case.

Proposition 1. *Let $\mathcal{C}$ be a symmetric monoidal closed category with all free partitionoids. If, for every objects X and A of $\mathcal{C}$, the canonical morphism*

$$X \otimes \int_{n \in \mathbb{I}^{\mathrm{op}}} \mathbb{P}((0,1),(n,0)) \circ (A^\bullet)^{\otimes n} \longrightarrow \int_{n \in \mathbb{I}^{\mathrm{op}}} X \otimes \left(\mathbb{P}((0,1),(n,0)) \circ (A^\bullet)^{\otimes n} \right)$$

is an isomorphism, then $\mathcal{C}$ is an infinitary affine category.

Proof. In what follows, when denoting the objects of the theory $\mathbb{P}$, we use the abbreviation $n^\bullet := (n,0)$ and $n^\omega := (0,n)$.

Let A be an object of $\mathcal{C}$, seen as a strict monoidal functor $A : \mathbb{B} \to \mathcal{C}$. We let the reader check that, if $(A^\bullet, F_1)$ is the free partitionoid on A, then $F_1 = \mathrm{Ran}_{k'} A(1^\omega)$, where $k' : \mathbb{B} \to \mathbb{P}$ is the strict monoidal functor mapping $n \mapsto n^\bullet$ (indeed, Definition 4 is just this Kan extension spelled out). This functor may be written as $k \circ j$, with $j : \mathbb{B} \to \mathbb{I}^{\mathrm{op}}$ the inclusion functor and $k : \mathbb{I}^{\mathrm{op}} \to \mathbb{P}$ mapping $n \mapsto n^\bullet$, which induces a decomposition of the Kan extension, yielding $F_1 = \mathrm{Ran}_k A^\bullet(1^\omega)$. Now, the hypothesis is exactly the condition allowing us to apply Theorem 4, which gives us

$$F_1 = \int_{n \in \mathbb{I}^{\mathrm{op}}} \mathbb{P}(1^\omega, n^\bullet) \circ (A^\bullet)^{\otimes n},$$

so it is enough to prove that $\lim \dagger_A = \int_{n \in \mathbb{I}^{\mathrm{op}}} \mathbb{P}(1^\omega, n^\bullet) \circ (A^\bullet)^{\otimes n}$.

We start with showing that $\int_{n \in \mathbb{I}^{\mathrm{op}}} \mathbb{P}(1^\omega, n^\bullet) \circ (A^\bullet)^{\otimes n}$ is a cone for $\dagger_A$. Let $\psi_n : (0,1) \to (n,0)$ be the morphism corresponding to the fpp $(\{0\}, \ldots, \{n-1\};)$. By composing the canonical projection with π_{ψ_n} (see Definition 6) we get an arrow

$$p_n : \int_{n \in \mathbb{I}^{\mathrm{op}}} \mathbb{P}(1^\omega, n^\bullet) \circ (A^\bullet)^{\otimes n} \to \mathbb{P}(1^\omega, n^\bullet) \circ (A^\bullet)^{\otimes n} \to (A^\bullet)^{\otimes n}.$$

Observe now that the following diagram commutes:

$$
\begin{array}{ccc}
\mathbb{P}(1^\omega, n^\bullet) \circ (A^\bullet)^{\otimes n} & \xrightarrow{(\varepsilon_{n+1})^*} & \mathbb{P}(1^\omega, (n+1)^\bullet) \circ (A^\bullet)^{\otimes n} \\
& & \downarrow{\pi_{\psi_{n+1}}} \\
& \xrightarrow[\pi_{\psi_n}]{} & (A^\bullet)^{\otimes n}
\end{array}
$$

because $\varepsilon_{n+1} \circ \psi_{n+1} = \psi_n$. Moreover, the diagram

$$
\begin{array}{ccc}
\mathbb{P}(1^\omega, (n+1)^\bullet) \circ (A^\bullet)^{\otimes n+1} & \xrightarrow{\pi_{\psi_{n+1}}} & (A^\bullet)^{\otimes n+1} \\
\downarrow{(\varepsilon_{n+1})_*} & & \downarrow{\varepsilon_{n+1}} \\
\mathbb{P}(1^\omega, (n+1)^\bullet) \circ (A^\bullet)^{\otimes n} & \xrightarrow{\pi_{\psi_{n+1}}} & (A^\bullet)^{\otimes n}
\end{array}
$$

commutes too. So, by pasting them with the defining diagram of $\int_{n \in \mathbb{I}^{\mathrm{op}}} \mathbb{P}(1^\omega, n^\bullet) \circ (A^\bullet)^{\otimes n}$, one gets:

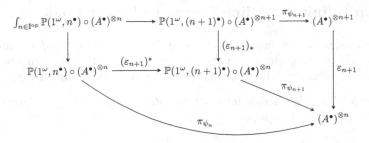

In particular, $(\int_{n\in\mathbb{I}^{\mathrm{op}}} \mathbb{P}(1^\omega, n^\bullet) \circ (A^\bullet)^{\otimes n}, (p_n))$ is a cone for the diagram.

Reciprocally, let $(B, (b_n))$ be any cone for this diagram. (b_n) extends uniquely into a family (β_n) such that:

- $\forall n \in \mathbf{N}, b_n = \pi_{\psi_n} \circ \beta_n$
- (β_n) makes the following diagrams commute:

$$
\begin{array}{ccc}
B & \xrightarrow{\ \beta_m\ } & \mathbb{P}(1^\omega, m^\bullet) \circ (A^\bullet)^{\otimes m} \\
\downarrow{\scriptstyle \beta_n} & & \downarrow{\scriptstyle f_*} \\
\mathbb{P}(1^\omega, n^\bullet) \circ (A^\bullet)^{\otimes n} & \xrightarrow{\ f^*\ } & \mathbb{P}(1^\omega, m^\bullet) \circ (A^\bullet)^{\otimes n}
\end{array}
$$

for all $f : m \to n$ in $\mathbb{P}$.

Indeed, any element s of $\mathbb{P}(1^\omega, n^\bullet)$ is of the form $s = q \circ \psi_m$, where $m \geqslant n$ and $q \in \mathbb{I}^{\mathrm{op}}(m^\bullet, n^\bullet)$. So the family (β_n) is defined by:

$$\forall n \in \mathbf{N}, \beta_n = \langle A^\bullet(q) \circ b_m \rangle_{q \circ \psi_m \in \mathbb{P}(1^\omega, n^\bullet)}$$

is the unique family satisfying

$$\pi_{q \circ \psi_m} \circ \beta_n = q \circ \pi_{\psi_m} \circ \beta_m$$

This definition is sound, as $m > m'$ such that there exists $q, q', \psi_m, \psi_{m'}$ such that $s = q \circ \psi_m = q' \circ \psi_{m'}$, we have

$$q = q' \circ ((\mathrm{id})^{\otimes m'} \otimes (w^\bullet)^{\otimes m - m'})$$

and as such

$$A^\bullet(q) = A^\bullet(q') \circ \varepsilon_{m - m' + 1} \circ \cdots \circ \varepsilon_m$$

and, as (b_n) is a cone for the sequential diagram,

$$A^\bullet(q) \circ b_m = A^\bullet(q') \circ b_{m'}.$$

So B makes the defining diagram of $\int_{n\in\mathbb{I}^{\mathrm{op}}} \mathbb{P}(1^\omega, n^\bullet) \circ (A^\bullet)^{\otimes n}$ commute, as such, (β_n) (and thus (b_n)) factors through it. Since all the cones of $\dagger_A$ factor through $\int_{n\in\mathbb{I}^{\mathrm{op}}} \mathbb{P}(1^\omega, n^\bullet) \circ (A^\bullet)^{\otimes n}$, it is its limit. $\square$

Observe that the condition of Proposition 1 is actually quite easy to grasp: it says that the limit of $\dagger_A$ commutes with the tensor, i.e., if we denote by $X \otimes \dagger_A$ the $\dagger_A$ diagram in which each $(A^\bullet)^{\otimes n}$ and ε_n are replaced by $X \otimes (A^\bullet)^{\otimes n}$ and $\mathrm{id}_X \otimes \varepsilon_n$, respectively, then the condition says $\lim(X \otimes \dagger_A) = X \otimes \lim \dagger_A$.

5   From Infinitary Affine Terms to Linear Logic

In [9], it was shown that usual λ-terms may be recovered as *uniform* infinitary affine terms. The categorical version of this result is that, in certain conditions, a model of the infinitary affine λ-calculus is also a model of linear logic.

Theorem 5. *Let $\mathcal{C}$ be an infinitary affine category. If, for every objects X and A in $\mathcal{C}$, the canonical morphism*

$$X \otimes \int_{(n,m) \in \mathbb{P}} (A^\omega)^{\otimes n} \otimes (A^\bullet)^{\otimes m} \longrightarrow \int_{(n,m) \in \mathbb{P}} X \otimes (A^\omega)^{\otimes n} \otimes (A^\bullet)^{\otimes m}$$

is an isomorphism, then $\mathcal{C}$ is a Lafont category. Moreover, the free commutative comonoid A^∞ on A may be computed as the equalizer of the diagram: where $\delta : A^\omega \to A^\omega \otimes A^\omega$ and $\varepsilon : A^\omega \to \mathbf{1}$ are the morphisms induced by the fpp $(; E, O)$ (even and odd numbers) and the empty fpp, respectively, and swap $: A^\omega \otimes A^\omega \to A^\omega \otimes A^\omega$ is the symmetry of $\mathcal{C}$.

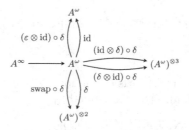

Fig. 2. Recovering the free co-commutative comonoid

Proof. Let $l : \mathbb{P} \to \mathbb{F}^{\mathrm{op}}$ be the strict monoidal functor mapping $(m, n) \mapsto m + n$ and collapsing every arrow $(0, 1) \to (m, n)$ to the unique morphism $1 \to m + n$ in $\mathbb{F}^{\mathrm{op}}$. By composing Kan extensions, we know that $A^\infty = \mathrm{Ran}_l(A^\bullet, A^\omega)(1)$. Remark that $\mathbb{F}^{\mathrm{op}}(1, p)$ is a singleton for all $p \in \mathbf{N}$, so the hypothesis is exactly what allows to apply Theorem 4, giving us

$$A^\infty = \int_{(m,n) \in \mathbb{P}} (A^\omega)^{\otimes n} \otimes (A^\bullet)^{\otimes m} .$$

Now, $\int_{(m,n) \in \mathbb{P}} (A^\omega)^{\otimes n} \otimes (A^\bullet)^{\otimes m}$ is the universal object making

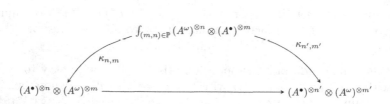

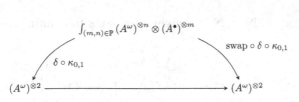

commute. We are going to show that $\int_{(m,n)\in\mathbb{P}} (A^\omega)^{\otimes n} \otimes (A^\bullet)^{\otimes m}$ is a cone for the diagram of Fig. 2. We will only show that commutes. The family $(\iota_n \otimes \iota_m \circ \delta \circ \kappa_{0,1})_{n,m}$ is a cone for $\dagger_A^{\otimes 2}$. Moreover, the $\theta_{n,m} \circ \delta$ are defined in terms of the operations of $\mathbb{P}$, they actually are the canonical maps, and

$$\forall n, m, \iota_n \otimes \iota_m \circ \delta \circ \kappa_{0,1} = \kappa_{0,n+m}$$

The exact same reasoning gives:

$$\forall n, m, \iota_n \otimes \iota_m \circ \mathrm{swap} \circ \delta \circ \kappa_{0,1} = \kappa_{0,n+m}$$

But $(\kappa_{0,n+m})_{n,m}$ factors uniquely through $(A^\omega)^{\otimes 2}$ (the limit of $\dagger_A^{\otimes 2}$) and as such,

$$\forall n, m, \delta \circ \kappa_{0,1} = \mathrm{swap} \circ \delta \circ \kappa_{0,1}$$

which is what we wanted. So $\int_{(m,n)\in\mathbb{P}} (A^\omega)^{\otimes n} \otimes (A^\bullet)^{\otimes m}$ is a cone for the diagram of Fig. 2.

Let us now prove that every cone for the diagram of Fig. 2 is a cone of the diagrams defining $\int_{(m,n)\in\mathbb{P}} (A^\omega)^{\otimes n} \otimes (A^\bullet)^{\otimes m}$.

It is easy to verify that any object B making the diagram defining A^∞ commute is endowed with exactly one map $B \to (A^\omega)^{\otimes n}$ for all $n \in \mathbf{N}$, built from δ and ε which, is moreover, stable under all swaps. In particular, by composing these maps $(B \to (A^\omega)^{\otimes n})_{n\in\mathbf{N}}$ with the arrow $A^\omega \to A^\bullet$, it is clear that there is a unique family of arrows

$$\forall n, m \in \mathbf{N}, B \to (A^\bullet)^{\otimes n} \otimes (A^\omega)^{\otimes m}$$

stable under extractions and weakenings. So any cone for the diagram defining A^ω is a cone for the diagram defining $\int_{(m,n)\in\mathbb{P}} (A^\omega)^{\otimes n} \otimes (A^\bullet)^{\otimes m}$ and as such, factorizes through it. So $\int_{(m,n)\in\mathbb{P}} (A^\omega)^{\otimes n} \otimes (A^\bullet)^{\otimes m}$ is the limit of the diagram of Fig. 2, and thus isomorphic to A^∞. $\square$

Intuitively, this construction amounts to collapsing the family of non-associative and non-commutative "contractions" built with δ, ε and swap.

It should be remarked that the particular δ used is not canonical, other morphisms would yield the same result. Indeed, from [9] we know that recovering usual λ-terms from infinitary affine terms is possible using uniformity which, as recalled in the introduction, amounts to identifying

$$\lambda x.\langle x_0, x_1, x_2, \ldots \rangle \approx \lambda x.\langle x_{\beta(0)}, x_{\beta(1)}, x_{\beta(2)}, \ldots \rangle,$$

for every injection $\beta : \mathbf{N} \to \mathbf{N}$. Theorem 5 amounts to defining a congruence on terms verifying

$$\lambda x. \langle x_0, x_1, x_2, \cdots \rangle \simeq \lambda x. \langle x_0, x_2, x_4, \cdots \rangle$$
$$\lambda x. \langle x_0, x_2, x_4, \cdots \rangle \otimes \langle x_1, x_3, x_5, \cdots \rangle \simeq \lambda x. \langle x_1, x_3, x_5, \cdots \rangle \otimes \langle x_0, x_2, x_4, \cdots \rangle$$

which is sufficient to recover $\approx$.

6 Discussion

We saw how the functorial semantic framework provides a bridge between the categorical and topological approaches to expressing the exponential modality of linear logic as a form of limit. This gives a way to construct, under certain hypotheses, denotational models of the infinitary affine λ-calculus. Moreover, it gives us a formula for computing the free exponential which is alternative to that of Melliès et al. Since both formulas apply only under certain conditions, it is natural to ask whether one of them is more general than the other. Although we do not have a general result, we are able to show that, under a mild condition verified in all models of linear logic we are aware of, our construction is applicable in every situation where Melliès et al.'s is.

Indeed, Melliès et al.'s construction amounts to checking that the Kan extension along m (below, left) is a monoidal Kan extension, whereas the one exposed in this article amounts to checking that the two Kan extensions along k, then l are monoidal (below, right):

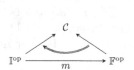
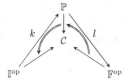

As Kan extensions compose, it suffices to know that the Kan extension along m is monoidal, that $m = k \circ l$, and that there exists two monoidal natural transformations inside the two upper triangles that can be composed to the last one to be sure that the Kan extensions along k and along l are monoidal too. We thus get:

Proposition 2. *Let $\mathcal{C}$ be a symmetric monoidal category with all free partitionoids. Assume that Melliès et al.'s formula works and that A^ω exists. If there exists, for all integers n, m monoidal maps*

$$(A^\infty)^{\otimes n+m} \to (A^\omega)^{\otimes n} \otimes (A^\bullet)^{\otimes m}$$
$$(A^\omega)^{\otimes n} \otimes (A^\bullet)^{\otimes m} \to (A^\bullet)^{\otimes n+m}$$

that composed together are the $n + m$ tensor of the map $A^\infty \to A^{\leqslant 1} \to A^\bullet$ then $\mathcal{C}$ is an infinitary affine category and a Lafont category.

Actually, in all models we are aware of, either both formulas work, or neither does. For instance, our construction fails for finiteness spaces [4], as does the construction given in [12].

Acknowledgments. The authors thank Paul-André Melliès for the inspiration and the lively conversations. This work was partially supported by projects COQUAS ANR-12-JS02-006-01 and ELICA ANR-14-CE25-0005.

An extended version of this work is available on the HAL open archive server.

References

1. Abramsky, S., Jagadeesan, R., Malacaria, P.: Full abstraction for PCF. Inf. Comput. **163**(2), 409–470 (2000)
2. Benton, P.N., Bierman, G.M., de Paiva, V., Hyland, M.: A term calculus for intuitionistic linear logic. In: Proceedings of TLCA, pp. 75–90 (1993)
3. Curien, P.L., Herbelin, H., Krivine, J.L., Melliès, P.A.: Interactive Models of Computation and Program Behavior. Societé Mathématique de France (2009)
4. Ehrhard, T.: Finiteness spaces. Math. Struct. Comput. Sci. **15**(4), 615–646 (2005)
5. Girard, J.Y.: Linear logic. Theor. Comput. Sci. **50**, 1–102 (1987)
6. Kfoury, A.J.: A linearization of the lambda-calculus and consequences. J. Log. Comput. **10**(3), 411–436 (2000)
7. MacLane, S.: Categorical algebra. Bull. Am. Math. Soci. **71**(1), 40–106 (1965)
8. MacLane, S.: Categories for the Working Mathematician, 2nd edn. Springer, Heidelberg (1978)
9. Mazza, D.: An infinitary affine lambda-calculus isomorphic to the full lambda-calculus. In: Proceedings of LICS, pp. 471–480 (2012)
10. Melliès, P.A.: Asynchronous games 1: A group-theoretic formulation of uniformity. Technical Report PPS//04//06//n°31, Preuves, Programmes et Systèmes (2004)
11. Melliès, P.A., Tabareau, N.: Free models of t-algebraic theories computed as Kan extensions (2008). Available on the second author's web page
12. Melliès, P.-A., Tabareau, N., Tasson, C.: An explicit formula for the free exponential modality of linear logic. In: Albers, S., Marchetti-Spaccamela, A., Matias, Y., Nikoletseas, S., Thomas, W. (eds.) ICALP 2009, Part II. LNCS, vol. 5556, pp. 247–260. Springer, Heidelberg (2009)

Automata and Formal Languages

Learning Register Automata
with Fresh Value Generation

Fides Aarts, Paul Fiterău-Broştean, Harco Kuppens, and Frits Vaandrager[✉]

Institute for Computing and Information Sciences, Radboud University Nijmegen,
P.O. Box 9010, 6500 GL Nijmegen, The Netherlands
F.Vaandrager@cs.ru.nl

Abstract. We present a new algorithm for active learning of register automata. Our algorithm uses counterexample-guided abstraction refinement to automatically construct a component which maps (in a history dependent manner) the large set of actions of an implementation into a small set of actions that can be handled by a Mealy machine learner. The class of register automata that is handled by our algorithm extends previous definitions since it allows for the generation of fresh output values. This feature is crucial in many real-world systems (e.g. servers that generate identifiers, passwords or sequence numbers). We have implemented our new algorithm in a tool called Tomte.

1 Introduction

Model checking and automata learning are two core techniques in model-driven engineering. In model checking [13] one explores the state space of a given state transition model, whereas in automata learning [6,17,25] the goal is to obtain such a model through interaction with a system by providing inputs and observing outputs. Both techniques face a combinatorial blow up of the state-space, commonly known as the state explosion problem. In order to find new techniques to combat this problem, it makes sense to follow a cyclic research methodology in which tools are applied to challenging applications, the experience gained during this work is used to generate new theory and algorithms, which in turn are used to further improve the tools. After consistent application of this methodology for 25 years model checking is now applied routinely to industrial problems [16]. Work on the use of automata learning in model-driven engineering started later [22] and has not yet reached the same maturity level, but in recent years there has been spectacular progress.

We have seen, for instance, several convincing applications of automata learning in the area of security and network protocols. Cho et al. [12] successfully used automata learning to infer models of communication protocols used by botnets.

The second author is supported by NWO project 612.001.216: Active Learning of Security Protocols (ALSEP). The remaining authors are supported by STW project 11763: Integrating Testing And Learning of Interface Automata (ITALIA). Some results from this paper appeared previously in the PhD thesis of the first author [1].

M. Leucker et al. (Eds.): ICTAC 2015, LNCS 9399, pp. 165–183, 2015.
DOI: 10.1007/978-3-319-25150-9_11

Automata learning was used for fingerprinting of EMV banking cards [5]. It also revealed a security vulnerability in a smartcard reader for internet banking that was previously discovered by manual analysis, and confirmed the absence of this flaw in an updated version of this device [11]. Fiterau et al. [14] used automata learning to demonstrate that both Linux and Windows implementations violate the TCP protocol standard. Using a similar approach, Tijssen [26] showed that implementations of the Secure Shell (SSH) protocol violate the standard. In [23], automata learning is used to infer properties of a network router, and for testing the security of a web-application (the Mantis bug-tracker). Automata learning has proven to be an extremely effective technique for spotting bugs, complementary to existing methods for software analysis.

A major theoretical challenge is to lift learning algorithms for finite state systems to richer classes of models involving data. A breakthrough has been the definition of a Nerode congruence for a class of register automata [8,9] and the resulting generalization of learning algorithms to this class [18,19]. Register automata are a type of extended finite state machines in which one can test for equality of data parameters, but no operations on data are allowed. Recently, the results on register automata have been generalized to even larger classes of models in which guards may contain arithmetic constraints and inequalities [10].

A different approach for extending learning algorithms to classes of models involving data has been proposed in [4]. Here the idea is to place an intermediate mapper component in between the implementation and the learner. This mapper abstracts (in a history dependent manner) the large set of (parametrized) actions of the implementation into a small set of abstract actions that can be handled by automata learning algorithms for finite state systems. In [2], we described an algorithm that uses counterexample-guided abstraction refinement to automatically construct an appropriate mapper for a subclass of register automata that may only store the first and the last occurrence of a parameter value.

Existing register automaton models [2,8,9] do not allow for the generation of fresh output values. This feature is technically challenging due to the resulting nondeterminism. Fresh outputs, however, are crucial in many real-world systems, e.g. servers that generate fresh identifiers, passwords or sequence numbers. The main contribution of this article is an extension of the learning algorithm of [2] to a setting with fresh outputs. We have implemented the new learning algorithm in our Tomte tool, http://tomte.cs.ru.nl/. As part of the LearnLib tool [21,24], a learning algorithm for register automata without fresh outputs has been implemented. In [3], we compared LearnLib with a previous version of Tomte (V0.3), on a common set of benchmarks (without fresh outputs), a comparison that turned out favorably for Tomte. Tomte, for instance, can learn a model of a FIFO-set buffer with capacity 30, whereas LearnLib can only learn FIFO-set buffers with capacity up to 7. In this paper, we present an experimental evaluation of the new Tomte 0.4. Due to several optimizations, Tomte 0.4 significantly outperforms Tomte 0.3. In addition, Tomte can now learn models for new benchmarks that involve fresh outputs.

2   Register Automata

In this section, we define register automata and their operational semantics in terms of Mealy machines. For reasons of exposition, the notion of *register automaton* that we define here is a simplified version of what we have implemented in our tool: Tomte also supports constants and actions with multiple parameters.

We assume an infinite set $\mathcal{V}$ of *variables*. An *atomic formula* is a boolean expression of the form $x = y$ or $x \neq y$, with $x, y \in \mathcal{V}$. A *formula* φ is a conjunction of atomic formulas. We write $\Phi(X)$ for the set of formulas with variables taken from X. A *valuation* for a set of variables $X \subseteq \mathcal{V}$ is a function $\xi : X \rightarrow \mathbb{Z}$. We write $\mathsf{Val}(X)$ for the set of valuations for X. If φ is a formula with variables from X and ξ is a valuation for X, then we write $\xi \models \varphi$ to denote that ξ satisfies φ.

Definition 1. *A* register automaton *(RA) is a tuple* $\mathcal{R} = \langle I, O, V, L, l_0, \Gamma \rangle$ *with*

- *I and O finite sets of* input symbols *and* output symbols, *respectively,*
- *$V \subseteq \mathcal{V}$ a finite set of* state variables; *we assume two special variables* in *and* out *not contained in V and write $V_{i/o}$ for the set $V \cup \{\mathsf{in}, \mathsf{out}\}$,*
- *L a finite set of* locations *and* $l_0 \in L$ *the* initial location,
- *$\Gamma \subseteq L \times I \times \Phi(V_{i/o}) \times (V \rightarrow V_{i/o}) \times O \times L$ a finite set of* transitions. *For each transition $\langle l, i, g, \varrho, o, l' \rangle \in \Gamma$, we refer to l as the* source, *i as the* input symbol, *g as the* guard, *ϱ as the* update, *o as the* output symbol, *and l' as the* target. *We require that* out *does not occur negatively in the guard, that is, not in a subformula of the form $x \neq y$.*

In the above definition, variables in and out are used to specify the data parameter of input and output actions, respectively. The requirement that out does not occur negatively in guards means that there are two types of transitions: transitions in which there are no constraints on the value of out, and transitions in which the value of out equals the value of one of the other variables in $V \cup \{\mathsf{in}\}$.

Example 1. As a first running example of a register automaton we use a FIFO-set with capacity two, similar to the one presented in [19]. A FIFO-set is a queue in which only different values can be stored, see Fig. 1. Input Push tries to insert a value in the queue, and input Pop tries to retrieve a value from the queue. The output in response to a Push is OK if the input value can be added successfully, or NOK if the input value is already in the queue or if the queue is full. The output in response to a Pop is Return, with as parameter the oldest value from the queue, or NOK if the queue is empty. We omit parameters that do not matter, and for instance write Pop() instead of Pop(in) if parameter in does not occur in the guard and is not touched by the update.

The operational semantics of register automata is defined in terms of (infinite state) Mealy machines.

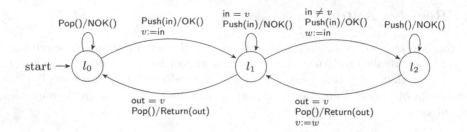

Fig. 1. FIFO-set with a capacity of 2 modeled as a register automaton

Definition 2. *A Mealy machine is a tuple $\mathcal{M} = \langle I, O, Q, q^0, \rightarrow \rangle$, where I, O, and Q are nonempty sets of input actions, output actions, and states, respectively, $q^0 \in Q$ is the initial state, and $\rightarrow \subseteq Q \times I \times O \times Q$ is the transition relation. We write $q \xrightarrow{i/o} q'$ if $(q, i, o, q') \in \rightarrow$, and $q \xrightarrow{i/o}$ if there exists a state q' such that $q \xrightarrow{i/o} q'$. A Mealy machine is input enabled if, for each state q and input i, there exists an output o such that $q \xrightarrow{i/o}$. A Mealy machine is deterministic if for each state q and input action i there is exactly one output action o and exactly one state q' such that $q \xrightarrow{i/o} q'$. A deterministic Mealy machine $\mathcal{M}$ can equivalently be represented as a structure $\langle I, O, Q, q^0, \delta, \lambda \rangle$, where $\delta : Q \times I \rightarrow Q$ and $\lambda : Q \times I \rightarrow O$ are defined by: $q \xrightarrow{i/o} q' \Rightarrow \delta(q, i) = q' \wedge \lambda(q, i) = o$.*

A *partial run* of $\mathcal{M}$ is a finite sequence $\alpha = q_0 \, i_0 \, o_0 \, q_1 \, i_1 \, o_1 \, q_2 \cdots i_{n-1} \, o_{n-1} \, q_n$, beginning and ending with a state, such that for all $j < n$, $q_j \xrightarrow{i_j/o_j} q_{j+1}$. A *run* of $\mathcal{M}$ is a partial run that starts with initial state q^0. A *trace* of $\mathcal{M}$ is a finite sequence $\beta = i_0 \, o_0 \, i_1 \, o_1 \cdots i_{n-1} \, o_{n-1}$ that is obtained by erasing all the states from a run of $\mathcal{M}$. We call a set S of traces *behavior deterministic* if, for all $\beta \in (I \cdot O)^*$, $i \in I$ and $o \in O$, $\beta \, i \, o \in S \wedge \beta \, i \, o' \in S \implies o = o'$. We call $\mathcal{M}$ *behavior deterministic* if its set of traces is so. Let $\mathcal{M}_1$ and $\mathcal{M}_2$ be Mealy machines with the same sets of input actions. Then we say that $\mathcal{M}_1$ *implements* $\mathcal{M}_2$, notation $\mathcal{M}_1 \leq \mathcal{M}_2$, if all traces of $\mathcal{M}_1$ are also traces of $\mathcal{M}_2$.

The operational semantics of a register automaton is a Mealy machine in which the states are pairs of a location l and a valuation ξ of the state variables. A transition may fire for given input and output values if its guard evaluates to true. In this case, a new valuation of the state variables is computed using the update part of the transition. We use 0 as initial value for state variables and do not allow 0 as a parameter value in actions.

Definition 3. *Let $\mathcal{R} = \langle I, O, V, L, l_0, \Gamma \rangle$ be a RA. The operational semantics of $\mathcal{R}$, denoted $[\![\mathcal{R}]\!]$, is the Mealy machine $\langle I \times (\mathbb{Z} \setminus \{0\}), O \times (\mathbb{Z} \setminus \{0\}), L \times \text{Val}(V), (l_0, \xi_0), \rightarrow \rangle$, where $\xi_0(v) = 0$ for all $v \in V$, and relation $\rightarrow$ is given by*

$$\frac{\langle l, i, g, \varrho, o, l' \rangle \in \Gamma \quad \iota = \xi \cup \{(in, d), (out, e)\} \quad \iota \models g \quad \xi' = \iota \circ \varrho}{(l, \xi) \xrightarrow{i(d)/o(e)} (l', \xi')}$$

We call $\mathcal{R}$ input enabled or deterministic if its operational semantics $[\![\mathcal{R}]\!]$ is input enabled or deterministic, respectively. A run or trace of $\mathcal{R}$ is just a run or trace, respectively, of $[\![\mathcal{R}]\!]$. We call $\mathcal{R}$ *input deterministic* if for each state and for each input action at most one transition may fire.

Example 2. The register automaton of Fig. 1 is input deterministic but not deterministic. For instance, as there are no constraints on the value of out for Push-transitions, an input Push(1) may induce both an OK(1) and an OK(2) output (in fact, the output parameter can take any value). Note that for Push-transitions the output value does not actually matter in the sense that out occurs neither in the guard nor in the range of the update function. Hence we can easily make the automaton of Fig. 1 deterministic, for instance by strengthening the guards with out = in for transitions where the output value does not matter.

Example 3. Our second running example is a register automaton, displayed in Fig. 2, that describes a simple login procedure. If a user performs a Register-input then the automaton returns the output symbol OK together with a password. The user may then login by performing a Login-input together with the password that she has just received. After login the user may either change the password or logout. We can easily make the automaton input enabled by adding self loops i/NOK in each location, for each input symbol i that is not enabled. It is not possible to model the login procedure as a deterministic register automaton: the very essence of the protocol is that the system nondeterministically picks a password and gives it to the user.

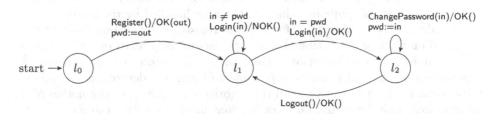

Fig. 2. A simple login procedure modeled as a register automaton

3   Active Automata Learning

Active automata learning algorithms have originally been developed for inferring finite state acceptors for unknown regular languages [6]. Since then these algorithms have become popular with the testing and verification communities for inferring models of black box systems in an automated fashion. While the details change for concrete classes of systems, all of these algorithms follow basically the same pattern. They model the learning process as a game between a learner and a teacher. The learner has to infer an unknown automaton with the help of the teacher. The learner can ask three types of queries to the teacher:

Output Queries ask for the expected output for a concrete sequence of inputs. In practice, output queries can be realized as simple tests.

Reset queries prompt the teacher to return to its initial state and are typically asked after each output query.

Equivalence Queries check whether a conjectured automaton produced by the learner is correct. In case the automaton is not correct, the teacher provides a counterexample, a trace exposing a difference between the conjecture and the expected behavior of the system to be learned. Equivalence queries can be approximated through (model-based) testing in black-box scenarios.

A learning algorithm will use these three kinds of queries and produce a sequence of automata converging towards the correct one. We refer the reader to [20, 25] for introductions to active automata learning.

Figure 3 presents the overall architecture of our learning approach, which we implemented in the Tomte tool. At the right we see the *teacher* or *system under learning (SUL)*, an implementation whose behavior can be described by an (unknown) input enabled and input deterministic register automaton. At the left we see the *learner*, which is a tool for learning finite deterministic Mealy machines. In our current implementation we use LearnLib [21, 24], but there are also other libraries like libalf [7] that implement active learning algorithms. In between the learner and the SUL we place three auxiliary components: the *determinizer*, the *lookahead oracle*, and the *abstractor*. First the determinizer eliminates the nondeterminism of the SUL that is induced by fresh outputs. Then the lookahead oracle annotates events with information about the data values that should be remembered since they play a role in the future behavior of the SUL. Finally, the abstractor maps the large set of concrete values of the SUL to a small set of symbolic values that can be handled by the learner.

The idea to use an abstractor for learning register automata originates from [2] (based on work of [4]). Using abstractors one can only learn restricted types of deterministic register automata. Therefore, [1, 3] introduced the concept of a lookahead oracle, which makes it possible to learn any deterministic register automaton. In this paper we extend the algorithm of [1, 3] with the notion of a determinizer, allowing us to also learn register automata with fresh outputs. In addition, we present some optimizations of the lookahead oracle that considerably improve the performance of our tool.

Fig. 3. Architecture of Tomte

4   A Theory of Mappers

In this section, we recall relevant parts of the theory of mappers from [4]. In order to learn an over-approximation of a "large" Mealy machine $\mathcal{M}$, we may place a transducer in between the teacher and the learner, which translates concrete inputs to abstract inputs, concrete outputs to abstract outputs, and vice versa. This allows us to reduce the task of the learner to inferring a "small" Mealy machine with an abstract alphabet. The determinizer and the abstractor of Fig. 3 are examples of such transducers. The behavior of these transducers is fully specified by a *mapper*, a deterministic Mealy machine in which the concrete symbols are inputs and the abstract symbols are outputs.

Definition 4 (Mapper). *A* mapper *of concrete inputs I, concrete outputs O, abstract inputs X, and abstract outputs Y is a deterministic Mealy machine $\mathcal{A} = \langle I \cup O, X \cup Y, R, r_0, \delta, \lambda \rangle$, where*

- *I and O are disjoint sets of* concrete input and output symbols,
- *X and Y are disjoint sets of* abstract input and output symbols, *and*
- *$\lambda : R \times (I \cup O) \to (X \cup Y)$, referred to as the* abstraction function, *respects inputs and outputs, that is, for all $a \in I \cup O$ and $r \in R$, $a \in I \Leftrightarrow \lambda(r, a) \in X$.*

A mapper allows us to abstract a Mealy machine with concrete symbols in I and O into a Mealy machine with abstract symbols in X and Y. Basically, the *abstraction* of Mealy machine $\mathcal{M}$ via mapper $\mathcal{A}$ is the Cartesian product of the underlying transition systems, in which the abstraction function is used to convert concrete symbols into abstract ones.

Definition 5 (Abstraction). *Let $\mathcal{M} = \langle I, O, Q, q_0, \to \rangle$ be a Mealy machine and let $\mathcal{A} = \langle I \cup O, X \cup Y, R, r_0, \delta, \lambda \rangle$ be a mapper. Then $\alpha_{\mathcal{A}}(\mathcal{M})$, the abstraction of $\mathcal{M}$ via $\mathcal{A}$, is the Mealy machine $\langle X, Y \cup \{\bot\}, Q \times R, (q_0, r_0), \to \rangle$, where $\bot \notin Y$ and $\to$ is given by the rules*

$$\frac{q \xrightarrow{i/o} q', \; r \xrightarrow{i/x} r' \xrightarrow{o/y} r''}{(q,r) \xrightarrow{x/y} (q', r'')} \qquad \frac{\not\exists i \in I : r \xrightarrow{i/x}}{(q,r) \xrightarrow{x/\bot} (q,r)}$$

The first rule says that a state (q, r) of the abstraction has an outgoing x-transition for each transition $q \xrightarrow{i/o} q'$ of $\mathcal{M}$ with $\lambda(r, i) = x$. In this case, there exist unique r', r'' and y such that $r \xrightarrow{i/x} r' \xrightarrow{o/y} r''$ in the mapper. An x-transition in state (q, r) then leads to state (q', r'') and produces output y. The second rule in the definition is required to ensure that the abstraction $\alpha_{\mathcal{A}}(\mathcal{M})$ is input enabled. Given a state (q, r) of the mapper, it may occur that for some abstract input symbol x there exists no corresponding concrete input symbol i with $\lambda(r, i) = x$. In this case, an input x triggers the special "undefined" output symbol $\bot$ and leaves the state unchanged.

A mapper describes the behavior of a transducer component that we can place in between a Learner and a Teacher. Consider a mapper $\mathcal{A} = \langle I \cup O, X \cup Y, R, r_0, \delta, \lambda \rangle$. The transducer component that is induced by $\mathcal{A}$ records the current state, which initially is set to r_0, and behaves as follows:

- Whenever the transducer is in a state r and receives an abstract input $x \in X$ from the learner, it nondeterministically picks a concrete input $i \in I$ such that $\lambda(r, i) = x$, forwards i to the teacher, and jumps to state $\delta(r, i)$. If there exists no concrete input i such that $\lambda(r, i) = x$, then the component returns output $\perp$ to the learner.
- Whenever the transducer is in a state r and receives a concrete answer o from the teacher, it forwards $\lambda(r, o)$ to the learner and jumps to state $\delta(r, o)$.
- Whenever the transducer receives a reset query from the learner, it changes its current state to r_0, and forwards a reset query to the teacher.

From the perspective of a learner, a teacher for $\mathcal{M}$ and a transducer for $\mathcal{A}$ together behave exactly like a teacher for $\alpha_\mathcal{A}(\mathcal{M})$. (We refer to [4] for a formalization of this claim.) In [4], also a *concretization* operator $\gamma_\mathcal{A}(\mathcal{H})$ is defined. This concretization operator is the adjoint of the abstraction operator: for a given mapper $\mathcal{A}$, the corresponding concretization operator turns any abstract Mealy machine $\mathcal{H}$ with symbols in X and Y into a concrete Mealy machine with symbols in I and O. As shown in [4], $\alpha_\mathcal{A}(\mathcal{M}) \leq \mathcal{H}$ implies $\mathcal{M} \leq \gamma_\mathcal{A}(\mathcal{H})$.

5   The Determinizer

The example of Fig. 2 shows that input deterministic register automata may exhibit nondeterministic behavior: in each run the automaton may generate different output values (passwords). This is a useful feature since it allows us to model the actual behavior of real-world systems, but it is also problematic since learning tools such as LearnLib can only handle deterministic systems. Most (but not all) of the nondeterminism of register automata can be eliminated by exploiting symmetries that are present in these automata. These symmetries are captured through the notion of an automorphism.

Definition 6. *A* zero respecting automorphism *is a bijection* $h : \mathbb{Z} \to \mathbb{Z}$ *satisfying* $h(0) = 0$.

Zero respecting automorphisms can be lifted to the valuations, states, actions, runs and traces of a register automaton. They induce an equivalence relation on traces. Below we show that each trace is equivalent to a trace in which all fresh inputs are positive and all fresh outputs are negative. Value 0 plays a special role as the initial value of variables and does not occur in traces.

Definition 7 (Neat Traces). *Consider a trace* β *of register automaton* $\mathcal{R}$:

$$\beta = i_0(d_0)\ o_0(e_0)\ i_1(d_1)\ o_1(e_1)\ \cdots\ i_{n-1}(d_{n-1})\ o_{n-1}(e_{n-1}) \qquad (1)$$

Let S_j *be the set of values that occur in* β *before input* i_j *(together with initial value 0), and let* T_j *be the set of values that occur before output* o_j: $S_0 = \{0\}$, $T_j = S_j \cup \{d_j\}$ *and* $S_{j+1} = T_j \cup \{e_j\}$. *Then* β *has* neat inputs *if each input value is either equal to a previous value, or equal to the largest preceding value plus one, that is, for all* j, $d_j \in S_j \cup \{\max(S_j) + 1\}$. *Similarly,* β *has* neat outputs *if each output value is either equal to a previous value, or equal to the smallest preceding value minus one, that is, for all j, $e_j \in T_j \cup \{\min(T_j) - 1\}$. A trace is neat if it has neat inputs and outputs, and a run is neat if its trace is neat.*

Example 4. The trace $i(1)$ $o(3)$ $i(7)$ $o(7)$ $i(3)$ $o(2)$ is not neat, for instance because the first output value 3 is fresh but not equal to -1, the smallest preceding value (including 0) minus 1. Also, the second input value 7 is fresh but not equal to 4, the largest preceding value plus 1. An example of a neat trace is $i(1)$ $o(-1)$ $i(2)$ $o(2)$ $i(-1)$ $o(-2)$.

The next proposition implies that in order to learn the behavior of a register automaton it suffices to study its neat traces, since any other trace can be obtained from a neat trace via a zero respecting automorphism.

Proposition 1. *For every run α there exists a zero respecting automorphism h such that $h(\alpha)$ is neat.*

In Example 4, for instance, the (non neat) run with trace $i(1)$ $o(3)$ $i(7)$ $o(7)$ $i(3)$ $o(2)$ can be mapped to the neat run with trace $i(1)$ $o(-1)$ $i(2)$ $o(2)$ $i(-1)$ $o(-2)$ by the automorphism h that acts as the identity function except that $h(3) = -1$, $h(7) = 2$, $h(2) = -2$, $h(-1) = 7$ and $h(-2) = 3$.

    Whereas the learner may choose to only provide neat inputs, we usually have no control over the outputs generated by the SUL, so these will typically not be neat. In order to handle this, we place a mapper component, called the *determinizer*, in between the SUL and the learner. The determinizer renames the first fresh output value generated by the SUL to -1, the second to -2, etc. The behavior of the determinizer is fully specified by the mapper $\mathcal{P}$ defined below. As part of its state this mapper maintains a function (one-to-one relation) R describing the current renamings, which grows dynamically during an execution. Whenever the SUL generates an output n that does not occur in $\mathsf{dom}(R)$, the domain of R, this output is mapped to a value m one less than the minimal value in $\mathsf{ran}(R)$, the range of R, and the pair (n, m) is added to R. For any finite one-to-one function R that contains $(0, 0)$, let $\hat{R}$ be a zero respecting automorphism that extends R. Whenever the learner generates an input m, this is concretized by the mapper to value $n = \hat{R}^{-1}(m)$, which is forwarded to the SUL. Again, if n does not occur in the domain of R, then R is extended with the pair (n, m).

Definition 8. *Let $\mathcal{R}$ be an input deterministic register automaton with inputs I and outputs O. The polisher for $\mathcal{R}$ is the mapper $\mathcal{P}$ such that*

– *the sets of concrete and abstract inputs both equal to $I \times \mathbb{Z}$,*
– *the sets of concrete and abstract outputs both equal to $O \times \mathbb{Z}$,*
– *the set of states consists of the finite one-to-one relations contained in $\mathbb{Z} \times \mathbb{Z}$,*
– *the initial state is $\{(0, 0)\}$.*

– *for all mapper states R, $i \in I$, $o \in O$ and $n \in \mathbb{Z}$,*

$$\lambda(R, i(n)) = i(\hat{R}(n))$$

$$\lambda(R, o(n)) = \begin{cases} o(R(n)) & \text{if } n \in dom(R) \\ o(\min(ran(R)) - 1) & \text{otherwise} \end{cases}$$

$$\delta(R, i(n)) = R \cup \{(n, \hat{R}(n))\}$$

$$\delta(R, o(n)) = \begin{cases} R & \text{if } n \in dom(R) \\ R \cup \{(n, \min(ran(R)) - 1)\} & \text{otherwise} \end{cases}$$

Proposition 2. *Any trace of $\alpha_P(\mathcal{R})$ with neat inputs is neat. Moreover, $\alpha_P(\mathcal{R})$ and $\mathcal{R}$ have the same neat traces.*

Example 5. The determinizer does not remove all sources of nondeterminism. The register automation of Fig. 2, for instance, is not behavior deterministic, even when we only consider neat traces, because of neat traces Register(1) OK(1) and Register(1) OK(−1). Figure 4 shows another example, which models a simple slot machine. By pressing a button a user may stop a spinning reel to reveal a symbol. If two consecutive symbols are equal then the user wins, otherwise she loses. The nondeterminism in the automaton of Fig. 2 is harmless since the parameter value of the OK-output does not matter and the behavior after the different outputs is the same. The nondeterminism of Fig. 4, however, is real in the sense that it leads to distinct behaviors with different output symbols.

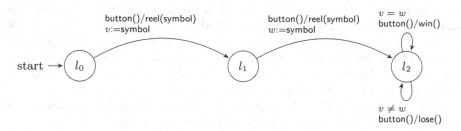

Fig. 4. A simple slot machine modeled as a register automaton

In the scenarios of Example 5 the automata nondeterministically select an output which then 'accidentally' equals a previous value. We call this a *collision*.

Definition 9. *Let β be a trace of $\mathcal{R}$ as in equation (1). Then β ends with a collision if (a) output value e_{n-1} is not fresh ($e_{n-1} \in T_{n-1}$), and (b) the sequence obtained by replacing e_{n-1} by some other value (except 0) is also a trace of $\mathcal{R}$. We say that β has a collision if it has a prefix that ends with a collision.*

Example 6. The trace button() reel(137) button() reel(137) of the register automaton of Fig. 4 has a collision, because the last occurrence of 137 is not fresh and if we replace it by 138 the result is again a trace of the automaton.

In many protocols, fresh output values are selected from a finite but large domain. TCP sequence and acknowledgement numbers, for instance, are 32 bits long. The length of the traces generated during learning is usually not that long and these traces typically only contain a few fresh outputs. As a result, the probability that collisions occur during the learning process is typically very small. For these reasons, we have decided in this paper to consider only observations without collisions. Under the assumption that the SUL will not repeatedly pick the same fresh value, we can detect whether an observation contains a collision by simply repeating the experiment a few times: if, after the renames performed by the determinizer, we still observe nondeterminism then a collision has occurred. By restricting ourselves to collision free traces, it may occur that the automata that we learn incorrectly describe the behavior of the SUL in the case of collisions. We will, for instance, miss the win-transition of Fig. 4. But if collisions are rare then it is extremely difficult to learn those parts of the SUL behavior anyway. In applications with many collisions (for instance when fresh output values are selected randomly from a small domain) it may be better not to use the learning algorithm described in this paper, but rather an algorithm for learning nondeterministic automata such as the one presented in [27].

Our approach for learning register automata with fresh outputs relies on the following proposition.

Proposition 3. *The set of collision free neat traces of an input deterministic register automaton is behavior deterministic.*

This means that our approach works for those register automata in which, when a fresh output is generated, it does not matter for the future behavior whether or not this fresh output equals some value that occurred previously. This is typically the case for real-world systems such as servers that generate fresh identifiers, passwords or sequence numbers.

6   The Lookahead Oracle

The main task of the lookahead oracle is to annotate each output action of the SUL with a set of values that are memorable after occurrence of this action. Intuitively, a parameter value d is memorable if it has an impact on the future behavior of the SUL: either d occurs in a future output, or a future output depends on the equality of d and a future input.

Definition 10. *Let $\mathcal{R}$ be a register automaton, let β be a collision free trace of $\mathcal{R}$, and let $d \in \mathbb{Z}/\{0\}$ be a value that occurs as (input/output) parameter in β. Then d is memorable after β iff $\mathcal{R}$ has a collision free trace of the form $\beta\ \beta'$, such that if we replace all occurrences of d in β' by a fresh value f then the resulting sequence $\beta\ (\beta'[f/d])$ is not a trace of $\mathcal{R}$ anymore.*

Example 7. In our example of a FIFO-set with capacity 2 (Fig. 1), the set of memorable values after trace β = Push(1) OK() Push(2) OK() Push(3) NOK()

is $\{1, 2\}$. Values 1 and 2 are memorable, because of the subsequent trace $\beta' =$ Pop() Return(1) Pop() Return(2). If we rename either the 1 or the 2 in β' into a fresh value, and append the resulting sequence to β, then the result is no longer a trace of the model. In the example of the login procedure (Fig. 2), value 2207 is memorable after Register() OK(2207) because Register() OK(2207) Login(2207) OK() is a trace of the automaton, but Register() OK(2207) Login(1) OK() is not.

When the Lookahead Oracle receives an input action from the Abstractor, the input is forwarded to the Determinizer unchanged. When the Lookahead Oracle receives a concrete output action o from the Determinizer (see Fig. 3), then it forwards a pair consisting of o and a valuation ξ to the Abstractor. The domain of ξ is a set of variables W and the range equals a set of memorable values after the occurrence of o. The set W may grow dynamically during the learning process when the maximal number of memorable values of states in the observation tree increases.

In order to accomplish its task, the Lookahead Oracle stores all the traces of the SUL observed during learning in an *observation tree*. In practice, this observation tree is also useful as a cache for repeated queries on the SUL. Each node N in the tree is characterized by the trace to it from the root and $N.MemV$, the set of values which are memorable after running this trace.

Figure 5 shows two observation trees for our FIFO-set example.

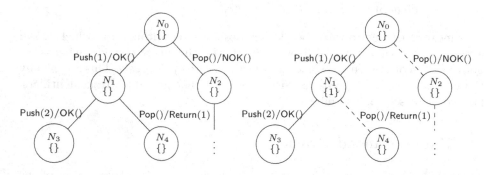

Fig. 5. Observation trees for FIFO-set without and with Pop() lookahead trace

Whenever a new node N is added to the tree, the oracle computes a set of memorable values for it. For this purpose, the oracle maintains a set of *lookahead traces*. All the lookahead traces are run starting at N to explore the future of that node and to discover its memorable values.

Definition 11. *A* lookahead trace *is a sequence of symbolic input actions of the form $i(v)$ with $i \in I$ and $v \in \{p_1, p_2, \ldots\} \cup \{n_1, n_2, \ldots\} \cup \{f_1, f_2, \ldots\}$.*

Intuitively, a lookahead trace is a symbolic trace, where each parameter refers to either a previous value (p_j), or to a new input value (n_j), or to a new, fresh output value (f_j). A lookahead trace can be converted into a concrete lookahead trace on the fly, by replacing each variable by a concrete value. Within lookahead traces, parameter p_1 plays a special role as the parameter that is replaced by a fresh value. Instances of all lookahead traces are run in each new node to compute memorable values. At any point in time, the set $N.MemV$ of known memorable values is a subset of the full set of memorable values of node N. Whenever a memorable value has been added to a node, we require an observation tree to be *lookahead complete*. This means every memorable value found has to have an origin, i.e., it has to stem from either the memorable values of the parent node or the values in the preceding transition:

$$N \xrightarrow{i(d)/o(e)} N' \Rightarrow N'.MemV \subseteq N.MemV \cup \{d, e\}.$$

We employ a similar restriction on any non-fresh output parameters contained in the transition leading up to a node. These too have to originate from either the memorable values of the parent, or the input parameter in the transition. Herein we differentiate from the algorithm in [1] which only enforced this restriction on memorable values at the expense of running additional lookahead traces.

The observation tree at the left of Fig. 5 is not lookahead complete since output value 1 of output Return(1) is neither part of the memorable values of the node N_1 nor is it an input in Pop(). Whenever we detect such an incompleteness, we add a new lookahead trace (in this case Pop()) and restart the entire learning process with the updated set of lookahead traces to retrieve a lookahead-complete observation tree. The observation tree at the right is constructed after adding the lookahead trace Pop(). This trace is executed for every node constructed, as highlighted by the dashed edges. The output values it generates are then tested if they are memorable and if so, stored in the MemV set of the node. When constructing node N_1, the lookahead trace Pop() gathers the output 1. This output is verified to be memorable and then stored to N_1's MemV set. We refer to [1] for more details about algorithms for the lookahead oracle.

7 The Abstractor

7.1 Mapper Definition

The behavior of the abstractor can be formally described by a mapper in the sense of Sect. 4. Let I and O be the sets of input symbols and output symbols, respectively, of the register automaton that we are learning. The lookahead oracle annotates output symbols from O with valuations from a set $W = \{w_1, \ldots, w_n\}$ of variables, thereby telling the abstractor what are the memorable values it needs to store. We define a family of mappers $\mathcal{A}_F$, which are parametrized by a function $F : I \rightarrow 2^W$. Intuitively, $w \in F(i)$ indicates that it is relevant whether the parameter of an input symbol i is equal to w or not. The initial mapper is parametrized by function $F_\emptyset$ given by $F_\emptyset(i) = \emptyset$ for all $i \in I$. Using counterexample-guided abstraction refinement, the sets $F(i)$ are subsequently extended. The abstraction function of the mapper $\mathcal{A}_F$ leaves the input and output symbol unchanged, but modifies the parameter values. The abstraction function replaces the actual value of an input parameter by the name of a variable in $F(i)$ that has the same value, or by $\perp$ in case there is no such variable. Thus the abstract domain of the parameter of i is the finite set $F(i) \cup \{\perp\}$. Likewise, the actual value of an output parameter is not preserved, but only the name of variable in $W \cup \{in\}$ that has the same value, or $\perp$ if there is no such variable. The valuation ξ that has been added as an annotation by the lookahead oracle describes the new state of the mapper after an output action. The abstraction function replaces ξ by an update function ϱ that specifies how ξ can be computed from the old state r and the input and output values received.

Example 8. As a result of interaction with mapper $\mathcal{A}_{F_\emptyset}$, the learner succeeds to construct the abstract hypothesis shown in Fig. 6. This first hypothesis is incorrect since it does not check if the same value is inserted twice. This is because the Abstractor only generates fresh values during the learning phase.

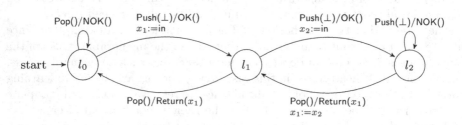

Fig. 6. First hypothesis of the FIFO-set

A flaw in the hypothesis will (hopefully) be detected during the hypothesis verification phase, and the resulting counterexample will then be used for an abstraction refinement. In order to test the correctness of a hypothesis, we need to concretize it. Using the theory of [4] we get a concretization operator for free, but in Tomte we actually use a slightly different concretization, which uses information about the abstract hypothesis to make smart guesses about what to do in situations where the mapper state is not injective. We refer to [1] for a detailed discussion of this issue.

7.2   Counterexample Analysis

During hypothesis verification the mapper selects random values from a small range for every abstract parameter value $\perp$. In this way it will find a concrete counterexample trace, e.g. Pop() NOK() Push(9) OK() Pop() Return(9) Push(3) OK() Push(3) NOK(), that is generated by the SUL but not allowed by the hypothesis, which specifies that the last output should be OK(). In order to simplify the analysis and to improve scalability, Tomte first tries to reduce the length of the counterexample. To this end it uses two reduction strategies, first removing loops, then single transitions. Both of these approaches are described in detail in [15]. Single transition reduction is an optimization applied in this work but not used in [1].

Long sequences of inputs typically lead to loops when they are run in the hypothesis. Tomte eliminates these loops and checks if the result is still a counterexample. Removing cycles from the concrete counterexample results in the reduced counterexample Push(3) OK() Push(3) NOK(). Tomte then eliminates each transition from left to right, preserving the resulting trace if it is still a counterexample. In this case, removing either transition yields the trace Push(3) OK() which is not a counterexample. So Push(3) OK() Push(3) NOK() is preserved. Once shortened, Tomte needs to determine if the counterexample is meant for the Learner to handle or requires abstraction refinement. It does so by converting the reduced concrete counterexample into one in which the input parameters are fresh, Push(1) OK() Push(2) NOK, and running it on the SUL. This fails since the last output returned by the SUL is OK. This means that Tomte needs to refine the input abstraction.

For a detailed discussion of the counterexample analysis algorithm used in Tomte we refer to [1]. Here we just sketch the main ideas. We walk through a counterexample trace and check for each input value if it occurs earlier in the trace. If so, then we check if this relation is already covered by mapper parameter F. If not we have found a potential source for the counterexample. It is possible that the two values are equal by chance, and that their equality is irrelevant for the counterexample. We test this by making the value fresh and recheck if the resulting trace is still a counterexample. If it is still a counterexample, the equality of the two values is not needed for the counterexample, so we leave the fresh value in the trace and continue with the next input. Otherwise, the equality of the two values does matter and we can refine the abstraction by extending the function F. A complicating factor is that a value not related with a value in history can still be important for the counterexample by a relation with an input parameter further on in the trace. So when we toggle a value to a fresh value we also consider all possible ways to toggle along subsequent parameters in the trace with the same value to the same fresh value and verify if the result is still a counterexample. If we find such a possibility then we keep the toggle and continue with next input parameter, else we have really found a meaningful refinement. After analysis of the counterexample, three things may happen: (1) the mapper used by the Abstractor is refined via an extension of function F, (2) the set of lookahead traces is extended since we have found a new memorable value, the added lookahead traces should fetch this value during the new learning iteration, (3) we have discovered that the abstract hypothesis is incorrect, construct an abstract version of the counterexample, and forward it to the Learner. In the current algorithm, as an optimization to [1], the set of lookahead traces is only extended with the shortest lookahead traces required to fetch the memorable value. This leads to comparatively shorter lookahead traces. Do note however, that the length of the longer lookahead trace remains heavily dependent on the reduced counterexample length. Hence, techniques such as loop and single trace reduction are essential.

In the analysis of the counterexample for the hypothesis of Fig. 6, Tomte discovers that it is relevant if the parameter of input Push is equal to variable x_1. Therefore, the set $F(\text{Push})$ is extended to $\{x_1\}$. Consequently, the alphabet of the learner is extended with a new input symbol $\text{Push}(x_1)$ and a corresponding lookahead trace is added to the lookahead oracle. Again, the entire learning process is restarted from scratch. The next hypothesis learned is equivalent to the model in Fig. 1 and the learning algorithm stops.

8   Evaluation and Comparison

We used our tool to learn models of various benchmarks such as SIP, the Alternating Bit Protocol, the Biometric Passport, FIFO-Sets, and Multi-Login Systems. Apart from the last one all these benchmarks have already been tackled in [3] with Tomte 0.3, a previous version of our tool, and with LearnLib. All benchmarks are available via the Tomte website.

Table 1 shows results Tomte 0.4, the release subject of this work, side by side with the results for LearnLib and Tomte 0.3 as reported in [3]. Results for each model are obtained by running the learner 10 times with 10 different seeds. Over these runs we collect the average and standard deviation for the numbers of:

- reset queries run during learning (**learn res**),
- concrete input symbols applied during learning (**learn inp**),
- reset queries run during counterexample analysis (**ana res**), and
- concrete input symbols applied during counterexample analysis (**ana inp**).

We omit running times here, as we consider the number of queries to be a superior metric for measuring efficiency, but the reader may find them at http://automatalearning.cs.ru.nl/. Experiments were done using a random equivalence oracle configured with a maximum test query length of 100. We used 10000 test runs per equivalence query for all models apart from the Multi-Login Systems which required more runs.

Tomte 0.4 shows to be more efficient than LearnLib and Tomte 0.3. The average number input symbols needed to learn decreased between 15 percent to over 90 percent compared to Tomte 0.3 and up to 99 percent compared to LearnLib. LearnLib still performs better for two models but, as noted in [3], it does not scale well for more complex systems. The average number of inputs Tomte 0.4 needs for counterexample analysis is also generally lower. Improvements over Tomte 0.3 can be largely explained by the optimizations in lookahead and counterexample processing that we presented in this article.

The Multi-Login System benchmark can only be handled by Tomte 0.4 (and no other tool to our knowledge) due to the occurrence of fresh outputs. The benchmark generalizes the example of Fig. 2 to multiple users. The difference is an additional input parameter for the user ID, when logging in and registering.

Table 1. Experimental comparison between LearnLib, Tomte 0.3 and Tomte 0.4

	Learnlib				Tomte 0.3				Tomte 0.4			
	learn res	learn inp	ana res	ana inp	learn res	learn inp	ana res	ana inp	learn res	learn inp	ana res	ana inp
Alternating bit protocol sender												
avg	452	2368	40551	405577	465	2459	7	15	65	224	13	30
stddev	453	2781	125904	1258919	0	2	4	11	1	0	1	5
Alternating bit protocol receiver												
avg	6077	102788	72	1420	271	1168	19	56	203	989	4	6
stddev	13184	245291	57	2813	1	0	4	13	0	0	2	2
Biometric passport												
avg	914	8517	365	7768	8769	43371	55	287	729	2884	33	143
stddev	614	12089	112	4334	5	35	7	56	1	3	4	43
Alternating bit protocol channel												
avg	52	252	29	173	67	210	0	0	37	102	0	0
stddev	29	235	12	115	0	0	0	0	0	0	0	0
Palindrome/repnumber checker												
avg	5	5	2050	8032	8366	24713	80	139	413	815	25	23
stddev	0	0	6225	24909	4	9	14	27	1	0	1	2
Session initiation protocol												
avg	92324	1962160	106868	1178964	6195	39754	256	1568	2557	14029	177	925
stddev	137990	4078104	336225	3696587	1103	7857	94	626	108	722	33	192
FIFO-set(2)												
avg	44	136	12	44	99	423	6	17	52	220	9	19
stddev	11	49	9	44	0	2	1	5	1	2	2	13
FIFO-set(7)												
avg	66392	1097470	634	13530	3215	31487	132	1284	1804	19306	143	1123
stddev	195580	3310472	66	2397	7	70	44	616	7	57	52	636
FIFO-set(30)												
avg	unable to learn				591668	20206862	15714	620479	336435	13285345	11443	473839
stddev					72	2112	1427	232984	107	3416	1785	164348
Multi-Login(1)												
avg	unable to learn				unable to learn				3910	21943	323	7002
stddev									1095	14882	629	19774
Multi-Login(2)												
avg	unable to learn				unable to learn				107057	667356	678	3361
stddev									21274	139189	209	1395
Multi-Login(3)												
avg	unable to learn				unable to learn				6495794	55821831	3202	22846
stddev									1237096	12218366	1021	9750

Moreover, a configurable number of users may register, supporting simultaneous login sessions for different registered users. Tomte 0.4 was able to successfully learn instantiations of Multi-Login Systems for 1, 2 and 3 users. The current learning algorithm does not scale well for higher numbers of users. This can be ascribed to the large number of memorable values in combination with the large numbers of abstractions required for this benchmark. The latter is also due to the order in which memorable values are found and thus indexed, which can differ per state.

9   Conclusions and Future Work

We have presented a mapper-based algorithm for active learning of register automata that may generate fresh output values. This class is more general than the one studied in previous work [1–3,8,9]. We have implemented our active learning algorithm in the Tomte tool and have evaluated the performance of Tomte on a large set of benchmarks, measuring the total number of inputs required for learning. For a set of common benchmarks without fresh outputs Tomte outperforms LearnLib (on the numbers reported in [3]), but many further optimizations are possible in both tools. In addition, Tomte is able to learn models of register automata with fresh outputs. Our method for handling fresh outputs is highly efficient and the computational cost of the determinizer is negligible in comparison with the resources needed by the lookahead oracle and the abstractor. Our next step will be an extension of Tomte to a class of models with simple operations on data.

References

1. Aarts, F.: Tomte: bridging the gap between active learning and real-world systems. Ph.D. thesis, Radboud University Nijmegen, October 2014
2. Aarts, F., Heidarian, F., Kuppens, H., Olsen, P., Vaandrager, F.: Automata learning through counterexample guided abstraction refinement. In: Giannakopoulou, D., Méry, D. (eds.) FM 2012. LNCS, vol. 7436, pp. 10–27. Springer, Heidelberg (2012)
3. Aarts, F., Howar, F., Kuppens, H., Vaandrager, F.: Algorithms for inferring register automata. In: Margaria, T., Steffen, B. (eds.) ISoLA 2014, Part I. LNCS, vol. 8802, pp. 202–219. Springer, Heidelberg (2014)
4. Aarts, F., Jonsson, B., Uijen, J., Vaandrager, F.W.: Generating models of infinite-state communication protocols using regular inference with abstraction. FMSD **46**(1), 1–41 (2015)
5. Aarts, F., de Ruiter, J., Poll, E.: Formal models of bank cards for free. In: Software Testing Verification and Validation Workshop, pp. 461–468. IEEE (2013)
6. Angluin, D.: Learning regular sets from queries and counterexamples. Inf. Comput. **75**(2), 87–106 (1987)
7. Bollig, B., Katoen, J.-P., Kern, C., Leucker, M., Neider, D., Piegdon, D.R.: Libalf: the automata learning framework. In: Touili, T., Cook, B., Jackson, P. (eds.) CAV 2010. LNCS, vol. 6174, pp. 360–364. Springer, Heidelberg (2010)
8. Cassel, S., Howar, F., Jonsson, B., Merten, M., Steffen, B.: A succinct canonical register automaton model. In: Bultan, T., Hsiung, P.-A. (eds.) ATVA 2011. LNCS, vol. 6996, pp. 366–380. Springer, Heidelberg (2011)
9. Cassel, S., Howar, F., Jonsson, B., Merten, M., Steffen, B.: A succinct canonical register automaton model. J. Logic Algebraic Methods Program. **84**(1), 54–66 (2015)
10. Cassel, S., Howar, F., Jonsson, B., Steffen, B.: Learning extended finite state machines. In: Giannakopoulou, D., Salaün, G. (eds.) SEFM 2014. LNCS, vol. 8702, pp. 250–264. Springer, Heidelberg (2014)
11. Chalupar, G., Peherstorfer, S., Poll, E., de Ruiter, J.: Automated reverse engineering using Lego. In: WOOT 2014, IEEE Computer Society, August 2014

12. Cho, C., Babic, D., Shin, E., Song, D.: Inference and analysis of formal models of botnet command and control protocols. In: CCS, pp. 426–439. ACM (2010)
13. Clarke, E.M., Grumberg, O., Peled, D.: Model Checking. MIT Press, Cambridge (1999)
14. Fiterău-Broştean, P., Janssen, R., Vaandrager, F.: Learning fragments of the TCP network protocol. In: Lang, F., Flammini, F. (eds.) FMICS 2014. LNCS, vol. 8718, pp. 78–93. Springer, Heidelberg (2014)
15. Koopman, P., Achten, P., Plasmeijer, R.: Model-based shrinking for state-based testing. In: McCarthy, J. (ed.) TFP 2013. LNCS, vol. 8322, pp. 107–124. Springer, Heidelberg (2014)
16. Bryant, R.E.: A view from the engine room: computational support for symbolic model checking. In: Grumberg, O., Veith, H. (eds.) 25 years of Model Checking. LNCS, vol. 5000, pp. 145–149. Springer, Heidelberg (2008)
17. de la Higuera, C.: Grammatical Inference: Learning Automata and Grammars. Cambridge University Press, Cambridge (2010)
18. Howar, F., Steffen, B., Jonsson, B., Cassel, S.: Inferring canonical register automata. In: Kuncak, V., Rybalchenko, A. (eds.) VMCAI 2012. LNCS, vol. 7148, pp. 251–266. Springer, Heidelberg (2012)
19. Howar, F., Isberner, M., Steffen, B., Bauer, O., Jonsson, B.: Inferring semantic interfaces of data structures. In: Margaria, T., Steffen, B. (eds.) ISoLA 2012, Part I. LNCS, vol. 7609, pp. 554–571. Springer, Heidelberg (2012)
20. Isberner, M., Howar, F., Steffen, B.: Learning register automata: from languages to program structures. Mach. Learn. 96(1–2), 65–98 (2014)
21. Merten, M., Steffen, B., Howar, F., Margaria, T.: Next Generation LearnLib. In: Abdulla, P.A., Leino, K.R.M. (eds.) TACAS 2011. LNCS, vol. 6605, pp. 220–223. Springer, Heidelberg (2011)
22. Peled, D., Vardi, M.Y., Yannakakis, M.: Black box checking. In: FORTE, IFIP Conference Proceedings, vol. 156, pp. 225–240. Kluwer (1999)
23. Raffelt, H., Merten, M., Steffen, B., Margaria, T.: Dynamic testing via automata learning. STTT 11(4), 307–324 (2009)
24. Raffelt, H., Steffen, B., Berg, T., Margaria, T.: LearnLib: a framework for extrapolating behavioral models. STTT 11(5), 393–407 (2009)
25. Steffen, B., Howar, F., Merten, M.: Introduction to active automata learning from a practical perspective. In: Bernardo, M., Issarny, V. (eds.) SFM 2011. LNCS, vol. 6659, pp. 256–296. Springer, Heidelberg (2011)
26. Tijssen, M.: Automatic modeling of SSH implementations with state machine learning algorithms. Bachelor thesis, Radboud University, Nijmegen, June 2014
27. Volpato, M., Tretmans, J.: Active learning of nondeterministic systems from an ioco perspective. In: Margaria, T., Steffen, B. (eds.) ISoLA 2014, Part I. LNCS, vol. 8802, pp. 220–235. Springer, Heidelberg (2014)

Modeling Product Lines with Kripke Structures and Modal Logic

Zinovy Diskin[1,2]([✉]), Aliakbar Safilian[1], Tom Maibaum[1],
and Shoham Ben-David[2]

[1] Department of Computing and Software, McMaster University, Hamilton, Canada
{disnkinz,safiliaa,maibaum}@mcmaster.ca

[2] Department of Elecetrical and Computer Engineering,
University of Waterloo, Waterloo, Canada
zdiskin@gsd.uwaterloo.ca, shohambd@gmail.com

Abstract. Product lines are an established framework for software design. They are specified by special diagrams called *feature models*. For formal analysis, the latter are usually encoded by propositional theories with Boolean semantics. We discuss a major deficiency of this semantics, and show that it can be fixed by considering that a product is an instantiation process rather than its final result. We call intermediate states of this process *partial* products, and argue that what a feature model M really defines is a poset of partial products called a *partial product line, PPL(M)*. We argue that such PPLs can be viewed as special *partial product Kripke structures* (ppKS) specifiable by a suitable version of CTL (*partial product CTL or ppCTL*). We show that any feature model M is representable by a ppCTL theory $\Phi(M)$ such that for any ppKs K, $K \models \Phi(M)$ iff $K = PPL(M)$; hence, $\Phi(M)$ is a sound and complete representation of the feature model.

1 Introduction

The *Software Product Line* approach is well-known in the software industry. Products in a product line (pl) share some common *mandatory* features, and differ by having some *optional* features that allow the developer to configure the product the user wants (e.g., MS Office, a Photoshop, or the Linux kernel). Instead of producing a multitude of separate products, the vendor designs a single pl encompassing a variety of products, which results in a significant reduction in development time and cost [23]. Industrial pls may be based on thousands of features inter-related in complex ways. Methods of specifying pls and checking the validity of a pl against a specification is an active research area represented at major software engineering conferences [2,25,31].

The most common method for designing a pl is to build a *feature model (fm)* of the products. A toy example is shown in the inset figure. Model M_1 says that a (root feature called) car *must* have an engine and brakes (black bullets denote *mandatory* subfeatures), and brakes can be *optionally* (note the hollow bullet) equipped with an anti-skidding system (abs). The

© Springer International Publishing Switzerland 2015
M. Leucker et al. (Eds.): ICTAC 2015, LNCS 9399, pp. 184–202, 2015.
DOI: 10.1007/978-3-319-25150-9_12

model specifies a pl consisting of two products: $P = \{\text{car}, \text{eng}, \text{brakes}\}$ and $P' = P \cup \{\text{abs}\}$. As fms of industrial size can be big and complex, they require tools for their management and analysis, and thus should be represented by formal objects processable by tools. A common approach is to consider features as atomic propositions, and view an fm as a theory in the Boolean propositional logic (BL), whose valid valuations are to be exactly the valid products defined by the fm [3]. For example, model M_1 represents the BL theory $\Phi(M_1) = \{\text{eng} \rightarrow \text{car}, \text{brakes} \rightarrow \text{car}, \text{abs} \rightarrow \text{brakes}\} \cup \{\text{car} \rightarrow \text{eng}, \text{car} \rightarrow \text{brakes}\} \cup \{\text{car}\}$: the first three implications encode subfeature dependencies (a feature can appear in a product only if its parent is in the product), and the last two implications encode the mandatory dependencies between features. The root feature must be always included in the product. This approach gave rise to a series of prominent applications for analysis of industrial size pls [11,13,28]. However, the Boolean semantics for fms has an almost evident drawback of misrepresenting fms' hierarchial structure.

The second inset figure shows an fm M_2 that is essentially different from M_1 (and is, in fact, pathological), but has the same set of products, $PL(M_2) = PL(M_1) = \{P, P'\}$ determined by an equivalent Boolean theory $\Phi(M_2) = \{\text{car} \rightarrow \text{eng}, \text{brakes} \rightarrow \text{eng},$ abs$\rightarrow$eng$\} \cup \{\text{eng} \rightarrow \text{car}, \text{eng} \rightarrow \text{brakes}\} \cup \{\text{eng}\}$. The core of the problem is that two semantically different dependencies (the parent feature and a mandatory subfeature) are both encoded by implication.

We are not the first to have noticed this drawback, e.g., it is mentioned in [28], and probably many researchers and practitioners in the field are aware of the situation. Nevertheless, as far as we know, no alternative to the Boolean logic of feature modeling (FM) has been proposed in the literature, which we think is theoretically unsatisfactory. Even more importantly, inadequate logical foundations for FM hinder practical analyses: as important information contained in fms is not captured by their BL-encoding, this information is either missing from analyses, or treated informally, or hacked in an ad hoc way. In a sense, this is yet another instance of the known software engineering problem, when semantics is hidden in the application code rather than explicated in the specification, with all its negative consequences for software testing, debugging, maintenance, and communication between the stakeholders.

The main goal of the paper is to show that Kripke structures and modal logic provide an adequate logical basis for FM. Our main observation is that the key notion of FM—a product built from features—should be considered as an *instantiation process* rather than its final result. We call intermediate states of this process *partial products*, and argue that what an fm M really specifies is a partially ordered set of partial products, which we call a *partial product line (ppl)* generated by model M, $PPL(M)$. The commonly considered products of M (we call them *full*) only form a subset of *PPL(M)*. We then show that any ppl can be viewed as an instance of a special type of Kripke structure, which we axiomatically define and call a *partial product Kripke structure (ppKS)*. The latter are specifiable by a suitable version of modal logic, which we call *partial product CTL* (ppCTL), as it is basically a fragment of CTL enriched with a zero-ary modality that only holds in states representing full products. We show that any fm M can be represented by a ppCTL theory $\Phi_{\mathrm{ML}}(M)$ accurately specifying M's intended semantics: the main result of the paper states that for any ppKS K, $K \models \Phi_{\mathrm{ML}}(M)$ iff $K = PPL(M)$, and hence $\Phi_{\mathrm{ML}}(M)$ is a sound and complete representation of the fm. Then we can replace fms by the respective ppCTL-theories, which are well amenable to formal analysis and automated processing.

In a broader perspective, we want to show that mathematical foundations of FM are mathematically interesting. Especially intriguing are connections of FM to event-based concurrency modeling. In fact, pls can be seen as a special interpretation of configuration structures [32]: features are events, partial products are configurations, and ppls are configuration structures. Then fms can be seen as a far reaching generalization of Winskel's event structures [33] and other formalisms for specifying dependencies between events. On the other hand, we believe that FM can make a non-trivial contribution to concurrency modeling by suggesting a very expressive yet simple and practically usable notation for specifying concurrency (including transaction mechanisms).

Our plan for the paper is as follows. Section 2 motivates our formal framework: we describe the basics of FM, and show how the deficiency of the Boolean semantics can be fixed by introducing *partial* products and transitions between them. In Sect. 3, we formalize fms and their ppl in a way convenient for us to work in the present paper. In Sect. 4, we introduce the notion of a ppKS as immediate abstraction of ppls, and ppCTL as a language to specify ppKS properties. We show how to translate an fm into a ppCTL-theory, and formulate our main results (proofs can be found in the [12]). We discuss some practical applications of the modal logic view of fms in Sect. 5, and connections between FM and concurrency in Sect. 6. Related work in the FM literature is discussed in Sect. 7. Section 8 concludes and lists several open problems. The list of abbreviations used throughout the paper can be found on page 18.

2   Feature Models and Partial Product Lines

This section aims to motivate the formal framework we will develop in the paper. In Sect. 2.1, we discuss the basics of FM, and in Sect. 2.2 introduce partial products and ppls. We will begin with ppls generated by simple fms, which can be readily explained in lattice-theoretic terms. Then (in Sect. 2.3) we show that ppls generated by complex fms are more naturally, and even necessarily, to be considered as transition systems.

2.1   Basics of Feature Modeling

An fm is a graphical structure presenting a hierarchial decomposition of features with some possible *cross-cutting constraints* (*cccs*) between them. Figure 1 gives an example. It is a tree of features, whose root names the product ('car' in this case), and edges relate a feature to its subfeatures. Edges with black bullets denote *mandatory* subfeatures: every car *must* have an eng (engine), a gear, and brakes. The hollow-end edge says that brakes can *optionally* be equipped with abs. Black angles denote so called *OR-groups*: an engine can be either gas (gasoline), or ele (electric), or both. Hollow angles denote *XOR-groups* (eXclusive OR): a gear is either mnl (manual) or atm (automatic) but not both; it must be supplied with oil as dictated by the black-bullet edge. The ×-ended arc says that an electric engine cannot be combined with a manual gear, and the arrow-headed arc says that an automatic gear requires ABS. According to the model, the set of features {car, eng, gas, gear, mnl, oil, brakes} is a valid product, but replacing the gasoline engine by electric, or removal of oil, would make the product invalid. In this way, the model compactly specifies seven valid products amongst the set of 2^9 possible combinations of 9 non-root features (the root is always included), and exhibits dependencies between choices.

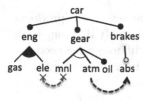

Fig. 1. A sample fm

In the BL of FM, an fm is a representation of a BL theory. For example, the theory encoded by the model in Fig. 1 consists of a set of implications denoting subfeature dependencies and unary mandatory dependencies, as explained in the introduction, plus three implications denoting grouped mandatoriness: {eng→gas ∨ ele, gear→mnl ∨ atm, mnl ∧ atm→⊥} (with ⊥ denoting False), plus two implications encoding cccs: {mnl∧ele→⊥, atm→abs}. However, as we saw above, a BL encoding is deficient.

2.2   Partial Product Lines: Products as Processes

What is lost in the BL-encoding is the *dynamic* nature of the notion of products. An fm defines not just a set of valid products but the very way these products are to be (dis)assembled step by step from constituent features. Correspondingly, a pl appears as a transition system initialized at the root feature (say, car for model M_1 in Fig. 2a) and gradually progressing towards fuller products (say, {car} → {car, eng} → {car, eng, brakes} or {car} → {car, brakes} → {car, brakes, abs} → {car, brakes, abs, eng}); we will call such sequences *instantiation paths*.

The graph in Fig. 2(b1) specifies all possible instantiation paths for M_1 (c, e, b, a stand for car, eng, brakes, abs, resp., to make the figure compact). Nodes in the graph denote *partial* products, i.e., valid products with, perhaps, some mandatory features missing: for example, {c,e} is missing feature b, and {c,b} is missing feature e. In contrast, {e} and {c,a} are invalid as they contain a feature without its parent; such sets do not occur in the graph. As a rule, we will call partial products just *products*. Product {c,e,b} is *full* (complete) as it has all mandatory subfeatures of its member-features; nodes denoting full products are framed. Each edge encodes adding a single feature to the product at the source of the edge; in text, we will often denote such edges by an arrow and write, e.g., {c} —→$_e$ {c, e}, where the subscript denotes the added feature.

We call the instantiation graph described above the *ppl determined by* M_1, and write $PPL(M_1)$. In a similar way, the ppl of M_2, $PPL(M_2)$, is built in

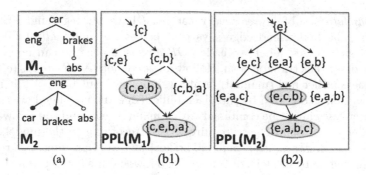

Fig. 2. From fms to ppls: simple cases

Fig. 2(b2). We see that although both fms have the same set of *full* products (i.e., are Boolean semantics equivalent), their ppls are essentially different both structurally and in the content of products. This essential difference between the ppls properly reflects the essential difference between the fms.

2.3   Partial Product Lines: From Lattices to Transition Systems

Generating ppls $PPL(M_{1,2})$ from models $M_{1,2}$ in Fig. 2 can be readily explained in lattice-theoretic terms. Let us first forget about mandatory bullets, and consider all features as optional. Then both models are just trees, and hence are posets, even join semi-lattices (joins go up in feature trees). Valid products of a model M_i are upward-closed sets of features (filters), and form a lattice (consider Fig. 2(b1,b2) as Hasse diagrams), whose join is set union, and meet is intersection. If we freely add meets (go down) to posets $M_{1,2}$ (eng $\wedge$ brakes etc.), and thus freely generate lattices $L(M_i)$, $i = 1, 2$, over the respective posets, then lattices $L(M_i)$ and $PPL(M_i)$ will be dually isomorphic (Birkhoff duality).

The forgotten mandatoriness of some features appears as incompleteness of some products; we call them *partial*. Partial products closed under mandatoriness are *full*. Thus, ppls of simple fms like in Fig. 2(a) are their filter lattices with distinguished subsets of full products. Later, we will discuss whether this lattice-theoretic view works for more complex fms.

Figure 3 (left) shows a fragment of the fm in Fig. 1, in which, for uniformity, we have presented the XOR-group as an OR-group with a new ccc added to the tree (note the ×-ended arc between mnl and atm). To build its ppl, we follow the idea described above, and first consider M_3 as a pure tree-based poset with all the extra-structure (denoted by black bullets and black triangles) removed. Figure 3 (right) describes a part of the filter lattice as a Hasse diagram (ignore the difference between solid and dashed edges for a while); to ease reading, the number of letters in the acronym for a feature corresponds to its level in the tree, e.g., c stands for car, en for eng etc.

Now consider how the additional structure embodied in the fm influences the ppl. Two cccs force us to exclude the bottom central and right products from the ppl; they are shown in brown-red and the respective edges are dashed.

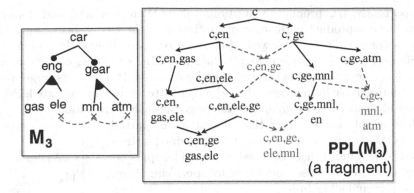

Fig. 3. From fms to ppls: Complex case

To specify this lattice-theoretically, we add to the lattice of features a universal *bottom* element ⊥ (a feature to be a subfeature of any feature), and write two defining equations: ele ∧ mnl = ⊥ and mnl ∧ atm = ⊥. (We owe this idea and much of our lattice-theoretic treatment of fms to Pratt's paper [24].) Then, in the filter lattice, the join of products {c,en,ele,ge} and {c,ge,mnl,en} "blows up" and become equal to the set of all features ("False implies everything"). The same happens with the other pair of conflicting products.

Next we consider the mandatoriness structure of model M_3 (given by black bullets and triangles). This structure determines a set of full products (not shown in Fig. 3) as we discussed above. In addition, mandatoriness affects the set of valid partial products as well. Consider the product $P = \{$c, en, ge$\}$ at the center of the diagram. The left instantiation path leading to this product, $\{$c$\} \longrightarrow_{en}$ $\{$c, en$\} \longrightarrow_{ge} P$ is not good because gear was added to engine *before* the latter is fully assembled (a mandatory choice between being electric or gasoline, or both, has still not been made). Jumping to another branch from inside of the branch being processed may be considered poor design practice that should be prohibited, and the corresponding transition is declared invalid. Similarly, transition $\{$c, ge$\} \longrightarrow_{en} P$ can be also invalid as engine is added before gear instantiation is completed. Hence, product P becomes unreachable, and should be removed from the ppl. (In the diagram, invalid edges are dashed (red with a color display), and the products at the ends of such edges are invalid too[1]).

Thus, a reasonable requirement for the instantiation process is that processing a new branch of the feature tree should only begin after processing of the current branch has reached a full product. We call this requirement *instantiate-to-completion* (I2C) by analogy with the *run-to-completion* transaction mechanism in behavior modeling (indeed, instantiating a mandatory branch of a feature tree can be seen as a transaction).

---

[1] With a more flexible view of product assembly, some possible interleavings could be prohibited and some allowed. (We owe this idea to an anonymous reviewer.) Then we need to add a suitable annotating mechanism to the fm formalism.

Importantly, I2C prohibits transitions rather than products, and it is possible to have a product with some instantiation paths into it being legal (and hence the product is legal as well), but some paths to the product being illegal. Figure 4 shows a simple example. In $PPL(M_4)$, the "diagonal" transition $\{c, ge\} \longrightarrow \{c, en, ge\}$ violates I2C and must be removed. However, its target product is still reachable from product $\{c, en\}$ as the latter is fully instantiated. Hence, the only element excluded by I2C is the diagonal dashed transition.

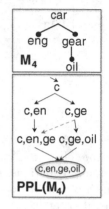

Thus, a ppl can be richer than its lattice of partial products (transition exclusion cannot be explained lattice-theoretically), and hence we need to consider ppls as Kripke structures, and use modal logic for specifying them. Moreover, even if all inclusions are transitions, the Boolean logic is too poor to express important semantic properties embodied in ppls: e.g., we may want to say that every product can be completed to a full product, and every full product is a result of such a completion. Or, if a product P has feature f, then any full product completing P must have feature g, and so on.

Thus, the transition relation is an important and independent component of the general ppl structure. As soon as transitions become first-class citizens, it makes sense to distinguish full products by supplying them, and only

Fig. 4. I2C at work.

them, with identity loops. Such loops do not add any feature to the product, and has a clear semantic meaning: the instantiation process can stay in a full product state indefinitely.

3   Feature Models and Partial Product Lines: Formally

A unified formal approach to fms and their semantics is developed in [27]. Our variant of fms' formalization is designed to support our work in the paper: the structure of our modal theories will follow the structure of fms as defined below.

3.1   Feature Models

Here, mandatory features and XOR-groups are derived constructs. A mandatory feature can be seen as a singleton OR-group. An XOR-group can be expressed by an OR-group with an additional exclusive constraint. Typically, cccs are either *exclusive* (the x-ended arc in Fig. 1), or *inclusive* (the dashed arrow arc in Fig. 1), as done for model M_3 in Fig. 3(a).

Definition 1 (Feature Models). A *feature model* is a tuple $M = (T, \mathcal{OR}, \mathcal{EX}, \mathcal{IN})$ of the following components:

(i) $T = (F, r, \_^\uparrow)$ is a tree whose nodes are *features*: F denotes the set of all features, $r \in F$ is the root, and function $\_^\uparrow$ maps each non-root feature

$f \in F_{-r} \overset{\text{def}}{=} F \setminus \{r\}$ to its parent $f^\uparrow$. The inverse function that assigns to each feature the set of its children (called *subfeatures*) is denoted by $f_\downarrow$. The set of all ancestors and all descendants of a feature f are denoted by $f^{\uparrow\uparrow}$ and $f_{\downarrow\downarrow}$, resp. Features f, g are called *incomparable*, $f \# g$, if neither of them is a descendant of the other. We write $\#2^F$ for the set $\{G \subset F : G \neq \emptyset \text{ and } f \# g \text{ for all } f, g \in G\}$.

(ii) $\mathcal{OR}$ is a function that assigns to each feature $f \in F$ a set $\mathcal{OR}(f) \subset 2^{f_\downarrow}$ (possibly empty) of *disjoint* subsets of f's children called *OR-groups*. If a group $G \in \mathcal{OR}(f)$ is a singleton $\{f'\}$ for some $f' \in f_\downarrow$, we say that f' is a *mandatory* subfeature of f. Elements in set $O(f) \overset{\text{def}}{=} f_\downarrow \setminus \bigcup \mathcal{OR}(f)$ are called *optional* subfeatures of f.

We call the pair $(T, \mathcal{OR})$ a feature tree and denote it by $T_{\mathcal{OR}}$.

(iii) $\mathcal{EX} \subseteq \#2^F$ is a set of *exclusive dependencies* between features.

(iv) $\mathcal{IN} \subset \#2^F \times \#2^F$ is a set of *inclusive dependencies* between features. A member of this set is written as an implication $(f_1 \wedge \ldots \wedge f_m) \rightarrow (g_1 \vee \ldots \vee g_n)$.

Exclusive and inclusive dependencies are also called cccs. The class of all fms over the same feature set F is denoted by $\mathbf{M}(F)$. □

3.2 Full and Partial Products

A common approach for formalizing the pl (of full products) of a given fm is to use BL [3]. Features are considered as atomic propositions, and dependencies between features are specified by logical formulas. So, given an fm M, each of its four components gives rise to a respective propositional theory as shown in the upper four rows of Table 1. Altogether, these theories constitute M's *full product* theory denoted by $\Phi^!_{\mathsf{BL}}(M)$ (note the bang superscript). A set of features P is defined to be a legal *full product* for M iff $P \models \Phi^!_{\mathsf{BL}}(M)$. The set of all full products of M is denoted by $\mathcal{FP}_M$.

Table 1. BL theories extracted from an fm $M = (T, \mathcal{OR}, \mathcal{EX}, \mathcal{IN})$

(1)	$\Phi_{\mathsf{BL}}(T) = \{\top \rightarrow r\} \cup \{f' \rightarrow f : f \in F, f' \in f_\downarrow\}$
(2)	$\Phi_{\mathsf{BL}}(\mathcal{EX}) = \{\bigwedge G \rightarrow \bot : G \in \mathcal{EX}\}$
(3!)	$\Phi^!_{\mathsf{BL}}(\mathcal{OR}) = \{f \rightarrow \bigvee G : f \in F, G \in \mathcal{OR}(f)\}$
(4!)	$\Phi^!_{\mathsf{BL}}(\mathcal{IN}) = \{\bigwedge G \rightarrow \bigvee G' : (G, G') \in \mathcal{IN}\}$
(all!)	$\Phi^!_{\mathsf{BL}}(M) = \Phi_{\mathsf{BL}}(T) \cup \Phi_{\mathsf{BL}}(\mathcal{EX}) \cup \Phi^!_{\mathsf{BL}}(\mathcal{OR}) \cup \Phi^!_{\mathsf{BL}}(\mathcal{IN})$
(3)	$\Phi^{\text{I2C}}_{\mathsf{BL}}(T_{\mathcal{OR}}) = \{f \wedge g \rightarrow (\bigwedge \Phi^!_{\mathsf{BL}}(T^f_{\mathcal{OR}})) \vee$ $(\bigwedge \Phi^!_{\mathsf{BL}}(T^g_{\mathcal{OR}})) : f, g \in F, f^\uparrow = g^\uparrow\}$
(all)	$\Phi_{\mathsf{BL}}(M) = \Phi_{\mathsf{BL}}(T) \cup \Phi_{\mathsf{BL}}(\mathcal{EX}) \cup \Phi^{\text{I2C}}_{\mathsf{BL}}(T_{\mathcal{OR}})$

As discussed in the introduction, the encoding above misrepresents the fm's hierarchical structure. Below we revise the propositional encoding of fms based on our discussion in Sect. 2, and introduce a BL theory for partial products. We call this theory M's *partial product* theory and denote it by $\Phi_{\mathsf{BL}}(M)$ (now without the bang superscript). Theory $\Phi_{\mathsf{BL}}(M)$ consists of three components (see row (all) in the Table): $\Phi_{\mathsf{BL}}(T)$ is the BL-encoding of subfeature dependencies (row (1)), $\Phi_{\mathsf{BL}}(\mathcal{EX})$ is the BL-encoding of exclusive dependencies (row (2)), and $\Phi_{\mathsf{BL}}^{\mathsf{I2C}}(T_{\mathcal{OR}})$ is the Boolean encoding of the I2C-condition, which we describe below.

Consider once again $PPL(M_3)$ in Fig. 3, from which product $\{\mathsf{c}, \mathsf{en}, \mathsf{ge}\}$ is excluded as violating the I2C principle. Note that the conflict between features en and ge is *transient* rather than *permanent*, and its propositional specification is not trivial.[2] To solve this problem, we first introduce the following notion.

Definition 2 (Induced Subfeature Tree and I2C). Let $T_{\mathcal{OR}} = (T, \mathcal{OR})$ be a feature tree over a set of features F, and $f{\in}F$. A feature subtree induced by f is a pair $T_{\mathcal{OR}}^f = (T^f, \mathcal{OR}^f)$ with T^f being the tree under f, i.e., $T^f \overset{\text{def}}{=} (f_{\downarrow\downarrow} \cup \{f\}, f, \_^{\uparrow})$, and mapping $\mathcal{OR}^f$ is inherited from $\mathcal{OR}$, i.e., for any $g \in f_{\downarrow\downarrow}$, $\mathcal{OR}^f(g) = \mathcal{OR}(g)$. □

Now we can specify theory $\Phi_{\mathsf{BL}}^{\mathsf{I2C}}(T_{\mathcal{OR}})$ as shown in row (3) in Table 1. The theory formalizes the idea that if a valid product contains two incomparable features, then at least one of these features must be fully instantiated within the product.

A set of features P is defined to be a legal *partial product* for M iff $P \models \Phi_{\mathsf{BL}}(M)$. The set of all partial products of M is denoted by $\mathcal{FP}_M$. Below the term 'product' will mean 'partial product'. Note that transition exclusion discussed in Sect. 2.3 cannot be explained with BL and needs a modal logic; we will define a suitable logic and show how it works in Sec. 4.

3.3 Ppls as Transition Systems

The problem we address is when a valid product P can be augmented with a feature $f{\notin}P$ so that product $P' = P \uplus \{f\}$ is valid as well. We then write $P \longrightarrow P'$ and call the pair (P, P') *a valid transition*.

Two necessary conditions are obvious: the parent $f^{\uparrow}$ must be in P, and f must not be in conflict with features in P, that is, $P' \models (\Phi_{\mathsf{BL}}(T) \cup \Phi_{\mathsf{BL}}(\mathcal{EX}))$. Compatibility with I2C is more complicated.

Definition 3 (Relative Fullness). Given a product P and a feature $f{\notin}P$, the following theory (continuing the list in Table 1) is defined:

$$(3)_{P,f} \qquad \Phi_{\mathsf{BL}}^{\mathsf{I2C}}(P, f) \overset{\text{def}}{=} \bigcup \{\Phi_{\mathsf{BL}}^!(T_{\mathcal{OR}}^g): g \in P \cap (f^{\uparrow})_{\downarrow}\}$$

We say P is *fully instantiated wrt.* f if $P \models \Phi_{\mathsf{BL}}^{\mathsf{I2C}}(P, f)$. □

Definition 4 (Valid Transitions). Let P be a partial product. Pair (P, P') is a valid transition, we write $P \longrightarrow P'$, iff one of the following two possibilities (a), (b) holds.

---

[2] We *cannot* declare that features en and ge are mutually exclusive and write $\{\mathsf{en} \wedge \mathsf{ge} \to \bot\}$ as down the lattice they are combined in the product $\{\mathsf{c}, \mathsf{en}, \mathsf{ele}, \mathsf{ge}\}$.

(a) $P' = P \uplus \{f\}$ for some feature $f \notin P$ such that the following three conditions hold: (a1) $P' \models \Phi_{\mathsf{BL}}(T)$, (a2) $P' \models \Phi_{\mathsf{BL}}(\mathcal{EX})$, and (a3) $P \models \Phi_{\mathsf{BL}}^{\mathsf{I2C}}(P, f)$.

(b) $P' = P$ and P is a full product, i.e., $P \models \Phi_{\mathsf{BL}}^{\mathsf{I}}(M)$. □

Proposition 1. *If P is a valid product and $P \longrightarrow P'$, then P' is a valid product.*

Definition 5 (Partial Product Line). Let $M = (T, \mathcal{OR}, \mathcal{EX}, \mathcal{IN})$ be an fm. The *partial product line* determined by M is a triple $PPL(M) = (\mathcal{PP}_M, \longrightarrow_M, I_M)$ with the set $\mathcal{PP}_M$ of partial products, transition relations $\longrightarrow_M$ given by Definition 4 (so that full products, and only them, are equipped with self-loops), and the initial product $I_M = \{r\}$ consisting of the root feature.

4 Partial Product Kripke Structures and Their Logic

In this section, we introduce *partial product Kripke structures* (ppKSs), which are immediate abstraction of ppls generated by fms. Then we introduce a modal logic called *partial product CTL* (ppCTL), which is tailored for specifying ppKSs's properties. Finally, we show that any fm is representable by a ppCTL theory such that it is a sound and complete representation of the fm.

4.1 Partial Product Kripke Structures

We deal with a special type of Kripke structures, in which possible worlds are identified with sets of atomic propositions, and hence a labelling function is not needed.

Definition 6 (Partial Product Kripke Structures). Let F be a finite set (of features). A *partial product Kripke structure* (ppKS) over F is a triple $K = (\mathcal{PP}, \longrightarrow, I)$ with $\mathcal{PP} \subset 2^F$ a set (of non-empty partial products), $I \in \mathcal{PP}$ the initial singleton product (i.e., $I = \{r\}$ for some $r \in F$), and $\longrightarrow \subseteq \mathcal{PP} \times \mathcal{PP}$ a binary left-total transition relation. In addition, the following three conditions hold ($\longrightarrow^+$ denotes the transitive closure of $\longrightarrow$):

(Singletonicity) For all $P, P' \in \mathcal{PP}$, if $P \longrightarrow P'$ and $P \neq P'$, then $P' = P \cup \{f\}$ for some $f \notin P$.

(Reachability) For all $P \in \mathcal{PP}$, $I \longrightarrow^+ P$, i.e., P is reachable from I.

(Self-Loops Only) For all $P, P' \in \mathcal{PP}$, if $(P \longrightarrow^+ P' \longrightarrow^+ P)$, then $P = P'$, i.e., every loop is a self-loop.

A product P with $P \longrightarrow P$ is called *full*. The set of full products is denoted by $\mathcal{FP}$. The class of all ppKSs built over F is denoted by $\mathbf{K}(F)$. □

The components of a ppKS K are subscripted with $_K$ if needed, e.g., $\mathcal{PP}_K$. Note that any product in a ppKS eventually evolves into a full product because F is finite, $\longrightarrow$ is left-total, and all loops are self-loops. Obviously, for a given fm M, $PPL(M)$ is a ppKS. We will also need the notion of a sub-ppKS.

Definition 7 (Sub-ppKS). Let K, K' be two ppKSs. We say K is a *sub-ppKS* of K', denoted by $K \sqsubseteq K'$, iff $I_K = I_{K'}$, $\mathcal{PP}_K \subseteq \mathcal{PP}_{K'}$, and $\longrightarrow_K \subseteq \longrightarrow_{K'}$. □

4.2 Partial Product Computation Tree Logic

We define ppCTL: a fragment of CTL enriched with a constant modality ! to capture full products.

Formulas in ppCTL are defined using a finite set of propositional letters F (features), an ordinary signature of propositional connectives: constant $\top$ (truth), unary $\neg$ (negation) and binary $\vee$ (disjunction) connectives, and a modal signature consisting of modal operators: constant modality !, and three unary modalities AX, AF, and AG. The well-formed ppCTL-formulas ϕ are given by the following grammar:

$$\phi := f \mid \top \mid \neg\phi \mid \phi \vee \phi \mid \mathsf{AX}\phi \mid \mathsf{AF}\phi \mid \mathsf{AG}\phi \mid !, \text{ where } f \in F.$$

Other propositional and modal connectives are defined dually via negation as usual: $\bot$, $\wedge$, EX, EF, EG are the duals of $\top$, $\vee$, AX, AG, AF, resp. The set of all ppCTL-formulas over F will be denoted by $\mathsf{ppCTL}(F)$.

The semantics of ppCTL-formulas is given by the class $\mathbf{k}(F)$ of ppKSs built over the same set of features F. Let $k \in \mathbf{k}(F)$ be a ppKS $(\mathcal{PP}, \longrightarrow, I)$. We first define a satisfaction relation $\models$ between a product $P \in \mathcal{PP}$ and a formula $\phi \in \mathsf{ppCTL}(F)$ by structural induction on ϕ as shown in the inset table on the right. Then we define $K \models \phi$ iff $P \models \phi$ for all $P \in \mathcal{PP}_K$ (equivalently, iff $I_K \models \mathsf{AG}\phi$).

$P \models f$	iff	$f \in P$ (for $f \in F$)
$P \models \top$		always holds
$P \models \neg\phi$	iff	$P \not\models \phi$
$P \models \phi \vee \psi$	iff	$(P \models \phi)$ or $(P \models \psi)$
$P \models \mathsf{AX}\phi$	iff	$\forall \langle P \longrightarrow P' \rangle.\ P' \models \phi$
$P \models \mathsf{AF}\phi$	iff	$\forall \langle P{=}P_1 \longrightarrow P_2 \longrightarrow ...\rangle$ $\exists i \geq 1.\ P_i \models \phi$
$P \models \mathsf{AG}\phi$	iff	$\forall \langle P{=}P_1 \longrightarrow P_2 \longrightarrow ...\rangle$ $\forall i \geq 1.\ P_i \models \phi$
$P \models !$	iff	$P \longrightarrow P$

4.3 ppCTL-theory of Feature Models

Given an fm M over a finite set of features F, we build two ppCTL theories from M's data, $\Phi_{\mathrm{ML}_\subseteq}(M)$ and $\Phi_{\mathrm{ML}}(M)$ (ML refers to Modal Logic) specified below such that Theorems 1, 2 and 3 hold.

Theorem 1 (Soundness). $PPL(M) \models \Phi_{\mathrm{ML}}(M)$.

Theorem 2 (Semi-completeness). $K \models \Phi_{\mathrm{ML}_\subseteq}(M)$ implies $K \sqsubseteq PPL(M)$.

Theorem 3 (Completeness). $K \models \Phi_{\mathrm{ML}}(M)$ iff $K = PPL(M)$.

Completeness allows us to replace fms by the respective ppCTL-theories, which are well amenable to formal analysis and automated processing. Semicompleteness is useful as an auxiliary intermediate step to completeness, but also for some practical problems such as *specialization* [29].

Theories $\Phi_{\mathrm{ML}_\subseteq}(M)$ and $\Phi_{\mathrm{ML}}(M)$ are built from small *component* theories:

$$\Phi_{\mathrm{ML}_\subseteq}(M) = \Phi_{\mathsf{BL}}(M) \cup \Phi^!_{\mathrm{ML}_\subseteq}(M) \cup \Phi^{\mathsf{I2C}\rightharpoonup}_{\mathrm{ML}_\subseteq}(\mathcal{T}_{\mathcal{OR}}), \text{ and}$$

$$\Phi_{\mathrm{ML}}(M) = \Phi_{\mathrm{ML}_\subseteq}(M) \cup \Phi^!_{\mathrm{ML}}(T) \cup \Phi^!_{\mathrm{ML}}(M) \cup \Phi^{\leftrightarrow}_{\mathrm{ML}}(\mathcal{T}_{\mathcal{OR}}, \mathcal{EX}),$$

which specify the respective properties of M's ppl in terms of ppCTL:

$$\Phi^!_{\mathrm{ML}_\subseteq}(M) = \{! \to \bigwedge \Phi^!_{\mathrm{BL}}(M)\},$$
$$\Phi^{\mathrm{l2C} \to}_{\mathrm{ML}_\subseteq}(T_{\mathcal{OR}}) = \{f \land \neg \bigwedge \Phi^!_{\mathrm{BL}}(T^f_{\mathcal{OR}}) \to \neg \mathsf{EX} g : f, g \in F, f^\uparrow = g^\uparrow\},$$
$$\Phi^!_{\mathrm{ML}}(T) = \{f \land \neg \bigvee f_\downarrow \to \mathsf{EX} g : f, g \in F, g^\uparrow = f\},$$
$$\Phi^!_{\mathrm{ML}}(M) = \{\bigwedge \Phi^!_{\mathrm{BL}}(M) \to !\},$$
$$\Phi^\leftrightarrow_{\mathrm{ML}}(T_{\mathcal{OR}}, \mathcal{EX}) = \{\bigwedge \Phi^{\mathrm{l2C}}_{\mathrm{BL}}(f) \land \neg f \land \neg \bigvee \Phi^{\mathcal{EX}}_{\mathrm{BL}}(f) \to \mathsf{EX} f : f \in F\}, \text{ where}$$
$$\Phi^{\mathrm{l2C}}_{\mathrm{BL}}(f) = \{g \to \Phi^!_{\mathrm{BL}}(T^g_{\mathcal{OR}}) : g, f \in F, g^\uparrow = f^\uparrow, g \neq f\}$$
$$\Phi^{\mathcal{EX}}_{\mathrm{BL}}(f) = \{\bigwedge(G \setminus \{f\}) : G \in \mathcal{EX}, f \in G\}$$

We have paid a special attention to this fine-grained structure of the theories as it helps to tune the process of reverse engineering fms from ppls as explained above. Details and discussion can be found in the accompanying report [12].

5   Possible Practical Applications

We discuss several practical tasks, in which using modal rather than Boolean logic could be beneficial.

Automated Analysis of Feature Models. Analysis of fms is an important practical issue, and as industrial fms can contain thousands of features, the analysis should be automated [4]. A big group of analysis problems rely on the Boolean semantics of fms. For example, given an fm M, we may be interested in checking whether $PL(M)$ is not empty [30], or whether a given set of features G is a valid full product, i.e., $G \in PL(M)$ [18]. We may also be interested in finding the set of common (core) features among all full products, $\bigcap PL(M)$ [30], or checking whether f is a core feature, i.e., $f \in \bigcap PL(M)$. Specifically, an important problem is to find so called *dead* features, which do not occur in any product [18]. A typical practical approach to these analysis problems is to encode the fm by a Boolean theory, and then use off-the-shelf tools like SAT-solvers [3].

However, there are some other important analysis problems, in which the use of the Boolean semantics can be error-prone. For example, it is often important to know if one fm M_1 is a *refactoring* of another fm M_2, or a *specialization* of M_2, or neither [29]. Standard definitions of refactoring and specialization are based on semantics, which in the Boolean case gives rise to defining refactoring $M_1 \simeq M_2$ as $PL(M_1) = PL(M_2)$ and specialization $M_1 \preceq M_2$ as $PL(M_1) \subseteq PL(M_2)$. However, as we have seen above, the Boolean semantics is too poor and makes the definitions above non-adequate to their goals (see the example in the introduction). Hence, in practice, to investigate refactoring and specialization, engineers should work with pairs $(PL(M), M)$, whose second component represents the feature hierarchical structure not captured by the first component. Working with such pairs brings two issues. First, it leads to obvious maintenance problems: if one of the components changes, the user must remember to propagate the changes to the other component. Second, having a syntactical "non-Boolean" object of analysis does not allow us to use SAT-solvers. However, the ppl semantics allows managing both issues. As our completeness theorem shows, the ppl $PPL(M)$ adequately captures the feature hierarchy, and hence we can analyze a single object, $PPL(M)$ or, equivalently, the modal theory $\Phi^{\mathsf{ML}}(M)$. Moreover, to analyze $PPL(M)$, we can use an off-the-shelf model checker.

Finally, there are analysis problems only addressing the hierarchy, e.g., finding the *Lowest Common Ancestor* (LCA) of a set of features in the feature tree [20]. The ppl semantics allows us to analyze such a problem by using a model checker: given a set of features G and a candidate common ancestor feature c, we need to check whether the Kripke structure $PPL(M)$ satisfies $\bigwedge G \to c$. This way, we could get the set of common ancestors of G. Let us denote it by C. Now, to check whether an element $l \in C$ is the LCA of G, we just need to check if $PPL(M)$ satisfies $l \to \bigwedge C$. Other syntactical analysis problems can be approached in the same way: an fm M is represented by a Kripke structure $PPL(M)$, the problem to be analyzed is encoded by a ppCTL-formula ϕ, and an off-the-shelf model checker is used for checking if $PPL(M) \models \phi$. We plan to implement such an approach for some realistic examples of fms.

Reverse Engineering. Reverse engineering of fms (RE) is an active research area in FM: the goal is to build an fm representing a given pl. Depending on how the pl is specified, the current approaches are grouped into two classes: (a) RE from BL formulas [11], (b) RE from textual descriptions of features [1,21]. She et al. in [28] argue that none of these approaches is satisfactory. The main challenge of RE is to determine an appropriate hierarchical structure of features. The BL approach is incomplete, since, as already discussed, the BL semantics cannot capture the hierarchical structure. The textual approach is also deficient for two reasons: it is an informal approach, and "it suggests only a single hierarchy that is unlikely the desired one" [28]. To relieve the deficiencies of these approaches, the current state-of-the-art method [28] uses a heuristic-based approach, in which both types of information (Boolean formulas and textual descriptions) are given at the input of the RE procedure. In contrast, if the input for the procedure were the ppCTL theory of the pl, which contains everything needed to build the corresponding fm, RE would become simpler and better manageable. Specifically, our careful decomposition of fms' structure and theories into small blocks could also help to tune the RE procedures.

The Vendor's View of FM. For the product user, an fm is just a structure of check-boxes to guide his choices, and hence the modal properties of the ppl are not important. However, they can be important for the vendor, who should plan and provide a reasonable production of all products in the pl. Consider, for example, the following scenario.

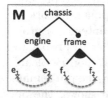

Suppose we want to design a chassis with two mandatory components: an engine and a frame. An engine is either of type e_1 xor e_2, and a frame is of type f_1 xor f_2, as specified in the inset figure. In general, engine e_i better fits in frame f_i, $i = 1, 2$, but the frame supplier can modify the frame for an extra cost. Thus, we have four full products $P_0 \cup P_{ij}$ with $P_0 = \{c, e, f\}$ and $P_{ij} = \{e_i, f_j\}$, $i, j = 1, 2$ (c, e, and f stand for chassis, engine, and frame, resp.).

There are two ways of the chassis assembly. If we first decide on the engine type, then, for engine e_i, we may choose either to order frame f_i, or frame f_j, $j \neq i$, with a suitable modification, depending on what is cheaper (we assume that each frame type has its own supplier). Thus, from each product $P_0 \cup \{e_i\}$, $i = 1, 2$ there are two transitions as shown in the inset figure on the right. However, if we first decide on the frame type, then only the engine of the respective type can be mounted on the frame, and transitions from $P_0 \cup \{f_i\}$ to $P_0 \cup \{f_i, e_j\}$

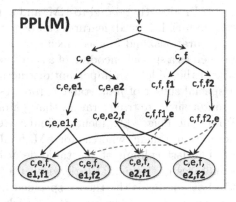

$j \neq i$ are illegal (shown dashed/red in the figure). To exclude the illegal transitions from the ppl, we need to add to the fm the following two modal cccs: $(f_i \wedge e \wedge \neg e_i) \rightarrow \mathsf{AX} \neg e_j$ for $i, j \in \{1, 2\}$ and $i \neq j$. Such constraints cannot be expressed in BL as they do not change the *set* of partial products, and only transition are affected.

6   Feature Vs. Event-based Concurrency Modeling

We will summarize similarities and differences between FM and event-based concurrency modeling (EM) (we will write em for an event model). We will also point to several possibilities of fruitful interactions between the two disciplines.

Following the survey in [32], we distinguish three approaches in EM. The first is based on a topological notion of a configuration structure $(E, \mathcal{C})$ with E a set (possibly infinite) of *events*, and $\mathcal{C} \subset 2^E$ a family of subsets (usually finite) of events, which satisfy some closure conditions (e.g., under intersection and directed union). Sets from $\mathcal{C}$ are called *configurations* and understood as states of the system: $X \in \mathcal{C}$ is a state in which all events from X already occurred.

In the second approach, valid configurations are specified indirectly by some structure D of dependencies between events, which make some configurations invalid. Formally, some notion of *validity* of a set $X \subset E$ wrt. D is specified so that an *event structure* (E, D) determines a configuration structure $\{X \subset E : X \text{ is } valid \text{ wrt.} D\}$. Typical representatives of this approach are Winskel's prime and general event structures, and Pratt's event spaces [24].

The third approach (originated by Gupta and Pratt in [14]) is an ordinary encoding of sets of propositions by Boolean logical formulas. Then an em is just a Boolean theory, i.e., a pair (E, Φ) with Φ a set of propositional formulas over set E of propositions. The left half of Table 2 summarizes this rough mini-survey.

Importantly, transitions between states are typically considered a derived notion: in [14], any set inclusion is a transition, and in [32], special conditions are to hold in order for a set inclusion to be a valid transition. A notable exclusion is *event automata* by Pinna and Poigné [22], i.e., tuples $(E, \mathcal{C}, \rightarrow, I)$ with $\rightarrow$ a *given* transition relation over configurations (states), and $I \in \mathcal{C}$ an initial state.

FM is directly related to EM, and actually can be seen as a special interpretation of EM. Indeed, features can be considered as events, (partial) products as configurations, and fms as special event-structure: fm $M = (T_{\mathcal{OR}}, \mathcal{EX}, \mathcal{IN})$ can be seen as a special encoding of a set of dependencies analogous to $\boldsymbol{D}$ (the middle row of the table). An important distinction of the Boolean FM is the presence of a special subset of *final* states (products), so that FM's topological and logical counterparts are triples rather than pairs (see the Boolean column in the table). Pinna and Poigné [22] mention final states (they call them *quiescent*) but do not actually use them, whereas for FM, final products are a crucial ingredient.

The last column of the table describes FM's basic topological and logical structures in the modal logic view: the upper row is our notion of ppKS, and the bottom one is the theory specified in Sect. 4. Our ppKS is exactly an event automaton with quiescent states, which, additionally, satisfies the conditions of Left-totality of the transition relations and Self-loops only, but Pinna and Poigné do not apply modal logic for specifying automata properties (and do not even mention it); they also do not consider the l2C-principle.

The comparison above shows enough similarities and differences to hope for a fruitful interaction between the two fields. We are currently investigating what FM can usefully bring to EM; and can mention several simple findings. The presence of two separate Boolean theories allows us to formally distinguish between *enabling* and *causality* [14]. Also, we are not aware of propositional specifications of *transient* conflicts (discussed on page 8) such as our Boolean and modal encoding of l2C. These encodings are nothing but a compact formal specification of a transaction mechanism, which is usually considered to be non-trivial. Remarkably, only recently similar generalizations were proposed for EM in the formalism of *dcr-graphs* [15]. The latter also employ two relations between events, *condition* and *response*, that correspond to our *subfeature* and *mandatoriness* relations, and their *markings* roughly correspond to our partial products. Dcr-graphs also use two additional relations *include/exclude*, which allow them to model several important constructs in concurrent distributed workflow, including transient conflicts. These observations show that a simple fm formalism is capable of encoding complex modal theories specifying non-trivial concurrent phenomena. Specifically, a detailed comparative analysis of fms and dcr-graphs should be an interesting and we believe useful research task.

Table 2. Event vs. feature modeling

Approach	Event Model	Feature Model	
		Boolean logic	Modal logic
Topological	$(E, \mathcal{C})$	$(F, \mathcal{PP}, \mathcal{FP})$	$(F, \mathcal{PP}, \rightarrow, I)$
Structural	$(E, \boldsymbol{D})$	(F, M)	
Logical	(E, Φ)	$(F, \Phi_{\mathsf{BL}}, \Phi^{!}_{\mathsf{BL}})$	(F, Φ_{ML})

7 Related Work

Behavioral modeling and transition systems have been numerously used in the FM literature in different contexts, none of which is directly related to our ppKS and ppCTL. Below we briefly outline the approaches, which we classify into three groups, and highlight their distinctions from the behavioral model developed in the present paper (see [12, Sect.VIII] for some details).

Staged Configuration. Czarnecki et al. introduced and developed the concept of *(multi-level) staged configuration* in [8,10]: given an fm M, its full products are instantiated via consecutive specializations (called stages) of M by either discarding an optional feature or accepting it and hence making it mandatory for the stage at hand and all consecutive stages. This process is continued until a fully specialized fm denoting only one configuration is reached. An accurate formal semantics for such multi-level staged configurations was defined by Classen et al. [6]. The idea was further developed by Hubaux et al. [17], who proposed to map fms to tasks and conditions of workflows. Their approach supports parallel execution of stages and choice between them, and iterative configurations.

Although both ppls and configuration stages show how to instantiate full products, they are essentially different. Configuration paths are sequences of *fms* with *decreasing variability*, whereas instantiation paths in ppls are sequences of *products* with *increasing commonality*. Thus, the two frameworks aim at different goals and are somewhat orthogonal (but, of course, ppls cover variability too as full products are included into ppl).

Feature Transition Systems. In a series of papers summarized in [5], Classen *et al.* proposed an elegant and effective solution to checking a given pl of transition systems (TS) in a single run of a model checker rather than checking each of the TSs separately. The entire pl is encoded as a *feature TS* (FTS), in which transitions are labeled by both actions and Boolean expressions over features as Boolean variables. A truth assignment to the feature variables defines the behaviour of a single product, and the FTS as a whole represents the entire pl. They also defined a logic fCTL to allow CTL properties to refer to specific products in the line and extended the model checking procedures to support checking FTSs against fCTL properties. Their tools are capable of reporting, in a single model checking run, all products for which a property holds, as well as those for which it fails to hold. In [7], Cordy et al. extend a common model checking framework known as CEGAR, to support FTSs as well. Thus, FTS and our ppKS are orthogonal ideas: for the former, a product is a TS, while for us a product is a set of features without any functional properties. These two ideas can be combined in a single formalism, but we leave it for future work.

Algebraic Approaches. An algebraic model based on commutative idempotent semirings is developed in [16]. Given an fm M, its pl is encoded as a term in the algebra generated by M's *leaf* features, so that non-leaf features are derived.

In contrast, for us, *all* features are basic, which better conforms to a common FM practice (see [12, Sect. VIII] for a more detailed discussion).

Amongst algebraic models for pls, the closest to ours is the *PL-CCS* calculus [19]. It is a process algebra, which extends the classical CCS by an operator ⊕ to model variability. Each ⊕ occurence in a PL-CCS expression is equipped with a unique index, and runtime occurrences with the same index must make the same choice. This differs ⊕'s behaviour from the classical non-deterministic choice in CCS. In PL-CCS, processes are interpreted as products. The behaviour of a pl is given by a set of process definitions whose semantics is given by multi-valued Kripke structures.

There are interesting similarities and differences between PL-CCS and our ppKS. In PL-CCS, a pl's behaviour is reconstructed from an immediate pl specification. In contrast, we extract the behaviour from the fm, which we have shown can be seen as an indirect pl's specification providing everything needed to reconstruct the behavior. We might say that in PL-CCS, the expressive power of fms is underestimated as they are seen in the Boolean perspective.

Importantly, PL-CCS allows for recursive definitions of processes, which makes it more expressive than our ppCTL. However, allowing recursive product definitions leads us beyond the boundaries of the tree-based fms and our goals in the present paper. Iterative definitions are possible in the so-called *cardinality-based* fms [9], and we built a dynamic semantics for them in [26]. On the other hand, cross-cutting constraints cannot be expressed in PL-CCS, but are readily specified in our approach (we even allow for modal cccs).

8   Conclusion

We have presented a novel view on fms, in which a product is an instantiation process rather than its final result. We called such products *partial*, and showed that the set of partial products together with a set of (carefully defined) valid transitions between them can be considered as a special Kripke structure, whose properties are specifiable by a fragment of CTL enriched with a constant modality. We called this logic ppCTL. Our main result shows that an fm can be considered as a compact representation of a rather complex ppCTL-theory. Thus, the logic of FM is modal rather than Boolean. We have also discussed several practical tasks in FM, which could benefit from the use of the modal logic view of FM. We conclude with a list of interesting open problems.

(i) Find a *complete axiomatic system for* ppCTL, i.e., a set of inference rules complete wrt. the ppKS semantics. Particularly, a complete logic would allow us to use theorem provers for fm analysis problems.

(ii) Axiomatic characterization of the class of ppKSs representable as $PPL(M)$.

(iii) To build complex pls from smaller component pls, we need some *process algebra for ppKS and fm* compositions. As it is clear from Sect. 6, these algebras should be special versions of process algebras.

(iv) Develop a *new modal logic view of event-based models*.

Acknowledgement. We are grateful to Krzysztof Czarnecki for several fruitful discussions of the subject, particularly, of the staged configuration approach. Thanks also go to anonymous reviewers for stimulating criticism and, specifically, for references to DCR-graphs and PL-CCS work. Financial support was provided by Automotive Partnership Canada via the NECSIS project.

Appendix

The list of abbreviations used in the paper:

Abbreviation	Meaning	Abbreviation	Meaning
ccc	crosscutting constraint	BL	Boolean logic
ppKS	partial product Kripke structure	ppCTL	partial product CTL
fm	feature model	FM	feature modeling
pl	product line	I2C	instantiate-to-completion
ppl	partial product line	ML	Modal Logic

References

1. Alves, V., Schwanninger, C., Barbosa, L., Rashid, A., Sawyer, P., Rayson, P., Pohl, C., Rummler, A.: An exploratory study of information retrieval techniques in domain analysis. In: SPLC2008 (2008)
2. Apel, S., Kästner, C., Lengauer, C.: Featurehouse: language-independent, automated software composition. In: ICSE, pp. 221–231 (2009)
3. Batory, D.: Feature models, grammars, and propositional formulas. In: Obbink, H., Pohl, K. (eds.) SPLC 2005. LNCS, vol. 3714, pp. 7–20. Springer, Heidelberg (2005)
4. Benavides, D., Segura, S., Ruiz-Cortés, A.: Automated analysis of feature models 20 years later: a literature review. Inf. Syst. **35**(6), 615–636 (2010)
5. Classen, A., Cordy, M., Heymans, P., Legay, A., Schobbens, P.: Formal semantics, modular specification, and symbolic verification of product-line behaviour. Sci. Comput. Program. **80**, 416–439 (2014)
6. Classen, A., Hubaux, A., Heymans, P.: A formal semantics for multi-level staged configuration. VaMoS **9**, 51–60 (2009)
7. Cordy, M., Heymans, P., Legay, A., Schobbens, P., Dawagne, B., Leucker, M.: Counterexample guided abstraction refinement of product-line behavioural models. In: FSE, pp. 190–201 (2014)
8. Czarnecki, K., Helsen, S.: Staged configuration using feature models. In: Nord, R.L. (ed.) SPLC 2004. LNCS, vol. 3154, pp. 266–283. Springer, Heidelberg (2004)
9. Czarnecki, K., Helsen, S., Eisenecker, U.: Formalizing cardinality-based feature models and their specialization. Softw. Process Improv. Pract. **10**(1), 7–29 (2005)
10. Czarnecki, K., Helsen, S., Eisenecker, U.: Staged configuration through specialization and multilevel configuration of feature models. Softw. Process Improv. Pract. **10**(2), 143–169 (2005)
11. Czarnecki, K., Wasowski, A.: Feature diagrams and logics: there and back again. In: SPLC 2007, pp. 23–34 (2007)
12. Diskin, Z., Safilian, A., Maibaum, T., Ben-David, S.: Modeling product lines with kripke structures and modal logic (GSDLAB–TR 2015–04-01), April 2015

13. Gheyi, R., Massoni, T., Borba, P.: Automatically checking feature model refactorings. J. UCS **17**(5), 684–711 (2011)
14. Gupta, V., Pratt, V.: Gates accept concurrent behavior. In: FOCS, pp. 62–71 (1993)
15. Hildebrandt, T., Mukkamala, R.: Declarative event-based workflow as distributed dynamic condition response graphs (2011, arXiv preprint). arXiv:1110.4161
16. Höfner, P., Khédri, R., Möller, B.: An algebra of product families. Softw. Syst. Model. **10**(2), 161–182 (2011)
17. Hubaux, A., Classen, A., Heymans, P.: Formal modelling of feature configuration workflows. In: SPLC, pp. 221–230 (2009)
18. Kang, K., Cohen, S., Hess, J., Novak, W., Peterson, S.: Feature-oriented domain analysis (foda) feasibility study. Technical report, DTIC Document (1990)
19. Leucker, M., Thoma, D.: A formal approach to software product families. In: Margaria, T., Steffen, B. (eds.) ISoLA 2012, Part I. LNCS, vol. 7609, pp. 131–145. Springer, Heidelberg (2012)
20. Mendonca, M., Wasowski, A., Czarnecki, K., Cowan, D.: Efficient compilation techniques for large scale feature models. In: GPCE, pp. 13–22. ACM (2008)
21. Niu, N., Easterbrook, S.: On-demand cluster analysis for product line functional requirements. In: SPLC (2008)
22. Pinna, M., Poigné, A.: On the nature of events: another perspective in concurrency. Theor. Comput. Sci. **138**(2), 425–454 (1995)
23. Pohl, K., Böckle, G., Van Der Linden, F.: Software Product Line Engineering: Foundations, Principles, and Techniques. Springer, Heidelberg (2005)
24. Pratt, V.R.: Event spaces and their linear logic. In: Nivat, M., Rattray, C., Rus, T., Scollo, G. (eds.) AMAST, pp. 3–25. Springer, Heidelberg (1991)
25. Roos-Frantz, F., Benavides, D., Ruiz-Cortés, A.: Feature model to orthogonal variability model transformation towards interoperability between tools. KISS@. ASE (2009)
26. Safilian, A., Maibaum, T., Diskin, Z.: The semantics of cardinality-based feature models via formal languages. In: Bjørner, N., de Boer, F. (eds.) FM 2015. LNCS, vol. 9109, pp. 453–469. Springer, Heidelberg (2015)
27. Schobbens, P-Y., Heymans, P., Trigaux, J-C.: Feature diagrams: a survey and a formal semantics. In: RE 2006, pp. 136–145. IEEE (2011)
28. She, S., Lotufo, R., Berger, T., Wasowski, A., Czarnecki, K.: Reverse engineering feature models. In: ICSE 2011, pp. 461–470. IEEE (2011)
29. Thum, T., Batory, D., Kastner, C.: Reasoning about edits to feature models. In: ICSE, pp. 254–264 (2009)
30. Trinidad, P., Cortés, A.: Abductive reasoning and automated analysis of feature models: how are they connected? VaMos **9**, 145–153 (2009)
31. Trujillo, V., Batory, D., Díaz, O.: Feature oriented model driven development: a case study for portlets. In: ICSE, pp. 44–53 (2007)
32. van Glabbeek, R., Plotkin, G.: Configuration structures. In: LICS, pp. 199–209 (1995)
33. Winskel, G.: Event structures. In: Rozenberg, G. (ed.) Advances in Petri Nets. LNCS, vol. 255. Springer, Heidelberg (1987)

Deterministic Regular Expressions
with Interleaving

Feifei Peng[1,2], Haiming Chen[1(✉)], and Xiaoying Mou[1,2]

[1] State Key Laboratory of Computer Science, Institute of Software,
Chinese Academy of Sciences, Beijing 100190, China
{pengff,chm}@ios.ac.cn
[2] University of Chinese Academy of Sciences, Beijing, China

Abstract. We study the determinism checking problem for regular expressions extended with interleaving. There are two notions of determinism, i.e., strong and weak determinism. Interleaving allows child elements intermix in any order. Although interleaving does not increase the expressive power of regular expressions, its use makes the sizes of regular expressions be exponentially more succinct. We first show an $\mathcal{O}(|\Sigma||E|)$ time algorithm to check the weak determinism of such expressions, where Σ is the set of distinct symbols in the expression. Next, we derive an $\mathcal{O}(|E|)$ method to transform a regular expression with interleaving to its weakly star normal form which can be used to rewrite an expression that is weakly but not strongly deterministic into an equivalent strongly deterministic expression in linear time. Based on this form, we present an $\mathcal{O}(|\Sigma||E|)$ algorithm to check strong determinism. As far as we know, they are the first $\mathcal{O}(|\Sigma||E|)$ time algorithms proposed for solving the weak and strong determinism problems of regular expressions with interleaving.

Keywords: Regular expressions · Interleaving · Strong determinism · Weak determinism · Algorithms

1 Introduction

DTD and XML Schema are two widely used schema languages recommended by W3C. The Unique Particle Attribution constraint [1] of DTD and XML Schema requires that all regular expressions used are weakly deterministic. The idea is inherited from the SGML Standard [2] to ensure more efficient parsing. Another definition of determinism which is called strong determinism has also been introduced in the context of XML [5]. Roughly speaking, weak determinism means that a symbol in the input word can be matched uniquely without looking ahead [11]. Meanwhile, strong determinism requires that the use of operators also be unique when matching a word. For example, $(a^*)^*$ is weakly deterministic but not strongly deterministic.

Work supported by the National Natural Science Foundation of China under Grant Nos. 61472405, 61070038.

The interleaving operator &, also under the name of unordered concatenation, allows child elements intermix in any order. It existed in SGML but was excluded from the definition of DTDs. RELAX NG resurrects it but in a different way. In SGML, for example, $(ab)\&(cd)$ accepts only sequences "*abcd*" and "*cdab*". For $a\&b^*$, a cannot be present between two b. Usually, the purpose of using & operator is to allow child elements to occur in any order. Hence the above restriction is undesirable. In RELAX NG, $(ab)\&(cd)$ accepts all the sequences in which a occurs before b and c occurs before d, that is, has the interleaving semantics. XML Schema permits a strongly limited interleaving at the top level of a content model but nowhere else [7].

Lots of work (e.g. [4,6,8,9,12]) focused on testing determinism of standard regular expressions and regular expressions with counting. But little progress has been made in the scope of regular expressions with interleaving (called RE(&)). The most difficult parts are that the transitions of the corresponding Glushkov automata can be exponential and RE(&) do not have the property of locality [10] thus the above algorithms are not capable of dealing with RE(&). Some results for expressions with the SGML interleaving operator are provided in [10], which is not the same with the Relax NG interleaving operator considered here. No study has investigated the weak and strong determinism properties of RE(&) [3].

However, investigating the determinism properties of RE(&) has much significance [3]. In practice, it can help to relax the restriction about interleaving used in XML Schema thus will lead to more succinct and flexible schemas. In this paper, we study both the weak and strong determinism of regular expressions extended with interleaving operator. Here we consider the operator interpreting in the same way with RELAX NG in its more general form. The main contribution of this paper are two $\mathcal{O}(|\Sigma||E|)$ time methods to test weak and strong determinism of RE(&). As far as we know, they are the first $\mathcal{O}(|\Sigma||E|)$ time algorithms proposed for solving the above problems.

Our work about weak determinism is related and inspired by the work of unambiguity of extended regular expressions in SGML document grammars by Brüggemann-Klein [10]. Although having different semantics, the treatment of the & operator in [10] is similar to our first algorithm. To obtain a more efficient algorithm, we get inspiration from [9] which combines the *follow* sets for the last symbols of each subexpression together into a single *followlast* set instead of computing *follow* set for each symbol of the expression. We define *followlast* for expression with interleaving and establish the relation between *followlast* and weak determinism for RE(&). As for strong determinism, we extend the notion of weakly star normal form [12] to RE(&) and show that a weakly deterministic RE(&) is strongly deterministic if and only if it is in weakly star normal form. Then we give a $\mathcal{O}(|E|)$ time method to transform an expression to its weakly star normal form. By combining the method and the $\mathcal{O}(|\Sigma||E|)$ time algorithm to check weak determinism, we can check strong determinism in $\mathcal{O}(|\Sigma||E|)$.

The rest of paper is organized as follows. Section 2 contains basic definitions that will be used throughout the paper. Section 3 presents an $\mathcal{O}(|\Sigma||E|^2)$ algorithm for weak determinism based on the *follow*$^-$ relations and an $\mathcal{O}(|\Sigma||E|)$

time algorithm based on *followlast*. In Sect. 4 an $\mathcal{O}(|\Sigma||E|)$ time algorithm checking strong determinism is presented. Details of implementation and experimental results are given in Sect. 5. We conclude in Sect. 6.

2 Preliminaries

2.1 Regular Expressions with Interleaving

Let u and v be two arbitrary strings. By $u\&v$ we denote the set of strings that is obtained by interleaving of u and v in every possible way. That is, $u\&\epsilon = \epsilon\&u = u$, $v\&\epsilon = \epsilon\&v = v$. If both u and v are non-empty, let $u = au', v = bv'$, where a and b are single symbols, then $u\&v = a(u'\&v) \cup b(u\&v')$. The operator $\&$ is then extended to regular languages as a binary operator in the canonical way. It is sufficient enough to say that $\&$ obeys the associative law. That is $E\&(F\&G) = (E\&F)\&G = E\&F\&G$ for any expressions E, F, G in RE($\&$).

For the rest of the paper, Σ always denotes a finite alphabet. The regular expressions with interleaving over Σ are defined as: $\emptyset, \epsilon$ or $a \in \Sigma$ is a regular expression, E_1^*, $E_1 E_2$, $E_1 + E_2$, or $E_1\&E_2$ is a regular expression for regular expressions E_1 and E_2. They are denoted as RE($\&$). The language $L(E)$ described by a regular expression with interleaving E is defined in the following inductive way: $L(\emptyset) = \emptyset$; $L(a) = \{a\}$; $L(E_1^*) = L(E_1)^*$; $L(E_1 E_2) = L(E_1)L(E_2)$; $L(E_1+E_2) = L(E_1)\cup L(E_2)$; $L(E_1\&E_2) = L(E_1)\&L(E_2)$. E? and E^+ are used as abbreviations of $E+\epsilon$ and EE^*, respectively. For example, consider the following expressions and their languages: $L(ab\&cd) = \{acbd, acdb, cabd, cadb, abcd, cdab\}$, $L(a\&(b\&c)) = \{abc, bac, bca, cba, cab, acb\}$.

2.2 Deterministic Regular Expressions with Interleaving

Marked RE($\&$) are those symbols marked with subscripts hence each symbol can only occur once. The expression that removes the subscripts of marked symbols of a marked expression E is denoted by $E^\natural$. We denote $(\cdot)^\natural$ as unmarking operator and $(\cdot)'$ as marking operator. For a language L, let $L^\natural$ denotes $\{w^\natural | w \in L\}$, then obviously $(L(E))^\natural = L(E^\natural)$. The set of symbols that occur in E is denoted by $sym(E)$. The size of a RE($\&$) expression E is denoted by $|E|$. Now a concise definition of weak determinism of expression can be given by its marked form.

Definition 1 (*[4]*). *A marked expression E is weakly deterministic if and only if for all words $uxv \in L(E), uyw \in L(E)$ where $x, y \in sym(E)$ and $u, v, w \in sym(E)^*$, if $x^\natural \neq y^\natural$ then $x \neq y$. An expression E is weakly deterministic if and only if its marked expression E' is weakly deterministic.*

Intuitively, an expression is weakly deterministic if a symbol in the input word can be matched without looking ahead when matching against the expression. For instance, $(a_1?\&b_2)a_3$ is not weakly deterministic since it does not satisfy the condition if $x^\natural \neq y^\natural$ then $x \neq y$ with $u = b_2$, $v = a_3$ and $w = \epsilon$, and with the competing symbols $x = a_1$ and $y = a_3$. The corresponding unmarked expression $(a?\&b)a$ is not weakly deterministic.

A *bracketing of a RE(&) expression* E is a labeling of the iteration nodes of the syntax tree by distinct indices [6]. The bracketing $\widetilde{E}$ of E is obtained by replacing each subexpression $E_1^{*,+}$ of E with a unique index i with $([_iE_1]_i)^{*,+}$. Therefore, a bracketed RE(&) expression is a RE(&) expression over alphabet $\Sigma \cup \Gamma_E$, where $\Gamma_E = \{[_i,]_i \mid 1 \leq i \leq |E|_\Sigma\}$, $|E|_\Sigma$ is the number of symbol occurrences in E. A string w in $\Sigma \cup \Gamma_E$ is correctly bracketed if w has no substring of the form $[_i]_i$.

Definition 2 (*[6]*). *An expression E is* strongly deterministic *if E is weakly deterministic and there do not exist strings u, v, w over $\Sigma \cup \Gamma_E$, strings $\alpha \neq \beta$ over Γ_E, and a symbol $a \in \Sigma$ such that $u\alpha av$ and $u\beta aw$ are both correctly bracketed and in $L(\widetilde{E})$.*

For instance, the expression $(a^*)^*$ is weakly deterministic but not strongly deterministic since $[_2[_1a]_1]_2[_2[_1a]_1]_2, [_2[_1a]_1[_1a]_1]_2 \in L(([_2([_1a]_1)^*]_2)^*)$.

For a RE(&) expression E over Σ and for each $z \in sym(E)$, the following definitions are needed to analyze the determinism of expressions.

$first(E) = \{a|au \in L(E), a \in sym(E), u \in sym(E)^*\}$
$last(E) = \{a|ua \in L(E), a \in sym(E), u \in sym(E)^*\}$
$follow(E, z) = \{a|uzav \in L(E), u, v \in sym(E)^*, a \in sym(E)\}, z \in sym(E)$
$followlast(E) = \{a|uav \in L(E), u \in L(E), u \neq \epsilon, a \in sym(E), v \in sym(E)^*\}$

It is not hard to see that an expression E is not weakly deterministic if and only if there exist two symbols $x, y \in sym(E')$ with $x^\natural = y^\natural$ such that $x, y \in first(E')$ or there is a symbol $z \in sym(E')$ such that $x, y \in follow(E', z)$.

2.3 Computing $follow^-$ Sets

We will need to calculate the $first$ and $follow^-$ sets. The inductive definition of the $first$ set for standard regular expressions can be trivially extended to RE(&). The inductive definition of the $follow^-$ can be found in [10].

Definition 3 (*[10]*). *For a marked expression E, we define $follow^-(E, x)$ for x in $sym(E)$ by induction on E as follows:*

$follow^-(E, \epsilon) = first(E)$

$E = x : follow^-(E, x) = \emptyset$

$E = F + G :$

$$follow^-(E, x) = \begin{cases} follow^-(F, x) & if\ x \in sym(F) \\ follow^-(G, x) & if\ x \in sym(G) \end{cases}$$

$E = FG :$

$$follow(E, x)^- = \begin{cases} follow^-(F, x) & if\ x \in sym(F), x \notin last(F) \\ follow^-(F, x) \cup first(G) & if\ x \in last(F) \\ follow^-(G, x) & if\ x \in sym(G) \end{cases}$$

$E = F \& G$:

$$follow^-(E,x) = \begin{cases} follow^-(F,x) & if\ x \in sym(F), x \notin last(F) \\ & or\ if\ x \in last(F), \epsilon \notin L(G) \\ follow^-(F,x) \cup first(G) & if\ x \in last(F), \epsilon \in L(F) \\ follow^-(G,x) & if\ x \in sym(G), x \notin last(G) \\ & or\ if\ x \in last(G), \epsilon \notin L(F) \\ follow^-(G,x) \cup first(F) & if\ x \in last(G), \epsilon \in L(G) \end{cases}$$

$E = F^*$:

$$follow^-(E,x) = \begin{cases} follow^-(F,x) & if\ x \in sym(F), x \notin last(F) \\ follow^-(F,x) \cup first(F) & if\ x \in last(F) \end{cases}$$

3   Weak Determinism of RE(&)

The weak determinism problem is to decide, given a regular expression with interleaving, i.e. $r \in$ RE(&), whether r is weakly deterministic or not. The classical way [11] is to compute the $first$ and $follow$ sets to check whether there exist symbols x, y such that $x, y \in first(E')$ or $x, y \in follow(E', z)$ for some symbol z. However, when it comes to & operator, there may be symbols x, y, z such that $x, y \in follow(E', z)$ yet E is weakly deterministic. For example, $E = (a \& b)a$, the corresponding marked expression is $E' = (a_1 \& b_2)a_3$, then $follow(E', b_2) = \{a_1, a_3\}$. Thus E might be judged to be not weakly deterministic. Yet $L(E') = \{a_1 b_2 a_3, b_2 a_1 a_3\}$, it is not hard to see E is weakly deterministic in fact.

The best known algorithm to check the weak determinism of standard regular expressions is proposed in [4,11] to check whether the corresponding Glushkov automata is deterministic. First, they proved that every regular expression can be transformed into its star normal form. Next, they showed that the determinism can be tested in $O(|\Sigma||E|)$ based on Glushkov automaton, via transforming an expression E into its star normal form in linear time. However, although we can transform a regular expression with interleaving using similar techniques as introduced in [11], transitions of the corresponding Glushkov can be exponential. Moreover, the method B.Groz [8] proposed to test whether a regular expression is deterministic in linear time by using a new structural decomposition of the parse tree is also not directly applicable to RE(&).

3.1   The First Algorithm

In this section, we consider a subset $follow^-(E, z)$ of $follow(E, z)$ and develop a method based on $follow^-(E, z)$ to check the determinism of RE(&).

Consider the weak determinism between interleavings first. Suppose we have a marked expression $E = E_1 \& ... \& E_n$. If some E_i is not weakly deterministic, then E is not weakly deterministic. Assuming each E_i is weakly deterministic,

E is also not weakly deterministic if there is some symbol a in both $E_i^\natural$ and $E_j^\natural$. Because no matter how other symbols intermix, there would always exist two string u, v such that $ua_i a_j v, ua_j a_i v \in L(E)$. So we need to ensure that $sym(E_i^\natural) \cap sym(E_j^\natural) = \emptyset$ for every two subexpressions E_i and E_j. As with the weak determinism between interleaving and other operators, consider a marked expression $H = EF$. Symbols that belong to E but not in $last(E)$ can not be in the same $follow$ set with symbols in F. A violation can only happen if $x, z \in last(E), y \in first(F), x^\natural = y^\natural$, which will cause $follow(H, z) = \{x, y\}$. If $\epsilon \notin L(E)$, x always occurs before y in any string accepted by $L(H)$ thus will not cause nondeterministic. We, therefore, use $follow^-$ to exclude these symbols in $first(F)$.

First, we have the following property about $follow^-$.

Note that $follow^-$ preserves the semantics of $follow$ when not dealing with & operator. That is, the situation $uzyw \in L(E)$, but $y \notin follow^-(E, z)$ occurs only if z is in subexpression $M\&N$ of E with $z \in sym(M), y \in sym(N)$. The following lemma is straightforward.

Lemma 1. *Let E be a marked expression. There are strings $u, v \in sym(E)^*$ and symbols $x, y, z \in sym(E)$ with $x^\natural = y^\natural$ such that $uzxv, uzyw \in L(E)$. If $x \in follow^-(E, z)$, $y \notin follow^-(E, z)$, then there exists some subexpression $M\&N$ or $N\&M$ of E, such that $x \in sym(M)$, $y \in sym(N)$.*

Proof. We prove it by contradiction. Suppose that x, y belong to the same side of subexpression $M\&N$ or $N\&M$ of E, then z must be in the other side, otherwise we will have $x, y \in follow^-(E, z)$. Assume $x, y \in sym(M)$ and $z \in sym(N)$ without loss of generality. Since $x \in follow^-(E, z)$, by the definition of $follow^-$, $x \in first(M)$ and $\epsilon \in L(M)$. Since $uzxv, uzyw \in L(E)$, we can see $y \in first(M)$ thus $y \in follow^-(E, z)$. This contradicts with the assumption that $y \notin follow^-(E, z)$. □

In fact, we can see from the above analysis that if $uzxv, uzyw \in L(E), x^\natural = y^\natural$ but $x, y \notin follow^-(E, z)$, then there exists some subexpression $M\&N$ or $N\&M$ of E, such that $x, y \in sym(M)$, $z \in sym(N)$. Let substring u' be the longest prefix of z and substrings v', w' be the longest suffix of z in u, v, w amongst $sym(M)$, then $u'xv', u'yw' \in L(M)$. That is, there exists a symbol $s \in sym(M)$ such that $x, y \in follow^-(M, s)$. Since M is a subexpression of E, then $x, y \in follow^-(E, s)$. This is shown in Lemma 2.

Lemma 2. *Let E be a marked expression. If there are strings $u, v \in sym(E)^*$ and symbols $x, y, z \in sym(E)$ with $x^\natural = y^\natural$ such that $uzxv, uzyw \in L(E)$ but $x, y \notin follow^-(E, z)$, then there exist a symbol $s \in sym(E)$ such that $x, y \in follow^-(E, s)$.*

The following theorem is the main result of this section which states the relation between weak determinism and $first, follow^-$ sets.

Theorem 1. *Let E be a marked expression, $z \in sym(E)$. E is not weakly deterministic if and only if there exist $x, y \in sym(E)$ with $x^\natural = y^\natural$ such that:*

(1) $x, y \in first(E)$ *or*
(2) $x, y \in follow^-(E, z)$, *for some symbol* $z \in sym(E)$ *or*
(3) $F \& G$ *or* $G \& F$ *is a subexpression of* E *such that* $x \in sym(F)$, *and* $y \in sym(G)$.

Proof. (=>) Assume E is not weakly deterministic, then $x, y \in first(E)$, $x^\natural = y^\natural$ or there are strings $u, v, w \in sym(E)^*, x, y, z \in sym(E)$ such that $uzxv, uzyw \in L(E), x^\natural = y^\natural$. In the first case, $x, y \in first(E)$, condition (1) holds. For the latter case, there are three conditions:

(A) $x, y \in follow^-(E, z)$ or
(B) only one of x and y is in $follow^-(E, z)$ or
(C) $x, y \notin follow^-(E, z)$

For case (A) we are done. As for case (B), we assume $x \in follow^-(E, z)$ and $y \notin follow^-(E, z)$, then by Lemma 1, condition (3) holds. The other case can be proved similarly. For case (C), by Lemma 2, condition (2) or condition (1) holds.

(<=) It is obvious for condition (1). For condition (2), the proof is the same with that of Theorem 1 in [10]. As for condition (3), if $F \& G$ or $G \& F$ is a subexpression of E and $x \in sym(F)$ and $y \in sym(G)$, then $uzxv \in L(F)$ for some $u, v \in sym(F)^*, wys \in L(G)$ for some $w, s \in sym(G)^*$. Thus, $uwzxvys, uwzysxv \in L(F \& G)$, E is not weakly deterministic. $\square$

The following Corollary indicates the restrictions put on interleaving in XML Schema might be stronger than necessary.

Corollary 1. *Let* $E = E_1 \& E_2 \& ... \& E_n$ *be a marked expression. E is weakly deterministic if and only if* $E_1, E_2, ... E_n$ *are weakly deterministic and* $sym(E_i^\natural) \cap sym(E_j^\natural) = \emptyset$ *when* $j \neq i$.

The process of this approach is formalized in Algorithm 1. The *compete* function checks if there are two elements a, b in the input word such that $a^\natural = b^\natural$. It returns true if such elements exist, or false otherwise. Below we analyze the time used to test weak determinism.

Theorem 2. *Let* E *be an expression over a finite alphabet* Σ. *It can be decided in* $\mathcal{O}(|\Sigma||E|^2)$ *whether* E *is weakly deterministic or not.*

Proof. The *first*, *last* and *follow*$^-$ sets can be implemented bottom up by converting E into a syntax tree, whose internal nodes are labeled with one of the operators $+, \cdot, ^+, ?, ^*$ or $\&$. If sets are maintained as ordered lists, it can be checked whether there exist x, y such that $x^\natural = y^\natural$ that are included in a *first* or *follow*$^-$ set in linear time via merging lists. As soon as this occurs, E is reported to be not weakly deterministic. Hence the maximum length of each *first* or *follow*$^-$ set is $|\Sigma|$. Since E has at most $\mathcal{O}(|E|)$ subexpressions, and each subexpression has at most $|E|$ last symbols. At this point, the total time is $\mathcal{O}(|\Sigma||E|^2)$. Condition (3) can be tested by scanning each symbol x in F to see whether there exists a symbol y in G such that $x^\natural = y^\natural$. Emptiness test of $sym(F^\natural) \cap sym(G^\natural)$ can be done in $\mathcal{O}(|\Sigma|)$ time with a hash table. So each subexpression can be examined in $\mathcal{O}(|\Sigma|)$ time. The upper bound of condition (3) is $\mathcal{O}(|\Sigma||E|)$.

Based on the above discussion, it takes $\mathcal{O}(|\Sigma||E|^2)$ time to check the weak determinism of an expression in RE(&). □

Algorithm 1. *weakDeterm1*

Input: An expression E in RE(&)
Output: true if E is weakly deterministic or false otherwise
 1: construct the corresponding binary tree $T(root)$ of E
 2: **return** $weakDeterm\_helper1(root)$

3.2   The Improved Algorithm

Based on the ideas in the previous section, we can have a simpler method that runs in $\mathcal{O}(|\Sigma||E|)$ time by optimizing the examination of the $follow^-$ relation used in Theorem 1.

Definition 4. *For a marked expression E, we define $followlast(E)$ by induction on E as follows:*

$E = x : followlast(E) = \emptyset$

$E = F^* : followlast(E) = followlast(F) \cup first(F)$

$E = F + G : followlast(E) = followlast(F) \cup followlast(G)$

$E = FG :$

$$followlast(E) = \begin{cases} followlast(G) & if\ \epsilon \notin L(G) \\ followlast(F) \cup first(G) \cup followlast(G) & if\ \epsilon \in L(G) \end{cases}$$

$E = F\&G :$

$$followlast(E) = \begin{cases} followlast(F) \cup followlast(G) & if\ \epsilon \notin L(F), \epsilon \notin L(G) \\ followlast(F) \cup followlast(G) \\ \cup\ first(G) & if\ \epsilon \notin L(F), \epsilon \in L(G) \\ followlast(F) \cup followlast(G) \\ \cup\ first(F) & if\ \epsilon \in L(F), \epsilon \notin L(G) \\ followlast(F) \cup followlast(G) \\ \cup\ first(F) \cup first(G) & if\ \epsilon \in L(F), \epsilon \in L(G) \end{cases}$$

For example, for $E = F\&G$, $last(E) = last(F) \cup last(G)$. If $\epsilon \in L(F)$ and $\epsilon \in L(G)$, for each $z \in last(F)$, we have $follow^-(E, z) = follow^-(F, z) \cup first(G)$. For each $z \in last(G)$, we have $follow^-(E, z) = follow^-(G, z) \cup first(F)$. Together these give that $followlast(E) = followlast(F) \cup first(G) \cup followlast(G) \cup first(F)$.

We can now move to check the weak determinism constraints by computing the $first$ and $followlast$ sets.

Algorithm 2. *weakDeterm_helper1*

Input: the root node F of a binary tree $T(root)$
Output: true if the expression of $T(root)$ is weakly deterministic or false otherwise
    if $F = \epsilon, a$ **then**
        return true
    if $F = F_1 | F_2$ **then**
        if *weakDeterm_helper1*(F_1) and *weakDeterm_helper1*(F_2) **then**
            if $first(F_1) \cap first(F_2) \neq \emptyset$ **then**
                return false
            else return true
        else return false
    if $F = F_1 F_2$ **then**
        if *weakDeterm_helper1*(F_1) and *weakDeterm_helper1*(F_2) **then**
            if $first(F_1) \cap first(F_2) \neq \emptyset$ **then**
                return false
            for each symbol $a \in last(F_1)$ **do**
                if $compete(follow^-(F, a))$ **then**
                    return false
            return true
        else return false
    if $F = F_1 \& F_2$ **then**
        if *weakDeterm_helper1*(F_1) and *weakDeterm_helper1*(F_2) **then**
            if $sym(F_1)^\natural \cap sym(F_2)^\natural \neq \emptyset$ **then**
                return false
            if $first(F_1) \cap first(F_2) \neq \emptyset$ **then**
                return false
            else return true
        else return false
    if $F = F_1^*$ **then**
        if *weakDeterm_helper1*(F_1)==false **then**
            return false
        for each symbol $a \in last(F_1)$ **do**
            if $compete(follow^-(F, a))$ **then**
                return false
        return true
    else return false

Theorem 3. *Let E be a marked expression. E is not weakly deterministic if and only if there exist $x, y \in sym(E)$ with $x^\natural = y^\natural$ and a subexpression F of E such that:*

(A) $x, y \in first(F)$ or
(B) $F = GH$ with $x \in followlast(G)$ and $y \in first(H)$ or
(C) $F = G^*$ or $F = G^+$ with $x \in followlast(G)$ and $y \in first(G)$ or
(D) $F = G\&H$ with $x \in sym(G)$ and $y \in sym(H)$.

Proof. (=>) Assume E is not weakly deterministic, then some of conditions (1)-(3) of Theorem 1 hold. Condition (1) implies condition (A). Condition (2) holds if and only in condition (B) or (C) holds. Condition (3) is equivalent to condition (D).

($\Leftarrow$) Assume some of conditions (A)-(D) hold. If condition (A) holds then condition (1) or (2) of Theorem 1 hold. Condition (B) and condition (C) imply condition (2). Condition (D) is equivalent to condition (3), by Theorem 1, E is not weakly deterministic. $\square$

The process of this approach is formalized in Algorithm 3.

Algorithm 3. *weakDeterm2*

Input: An expression E in RE(&)
Output: true if E is weakly deterministic or false otherwise
  construct the corresponding binary tree $T(root)$ of E
  return *weakDeterm_helper2(root)*

Example 1. The expression $E = (a_1^*\&b_2)^*a_3$ is not weakly deterministic, since $first(E) = \{a_1, a_3\}$. It can also be notified by the fact that $followlast((a_1^*\&b_2)^*) = \{a_1, b_2\}$ and $first(a_3) = \{a_3\}$. The expression $E = a_1b_1\&c_1a_2$ is not weakly deterministic, since $a_2 \in sym(c_1a_2)$ and $a_1 \in sym(a_1b_1)$.

The calculation of $first$ and $followlast$ sets is done at the same time using bottom-up on the syntax tree of E. The algorithm will terminate as soon as at least one of the four conditions is satisfied. Thus the length of $first$ set can be at most $\mathcal{O}(|\Sigma|)$. Since each $followlast$ set contains at most $\mathcal{O}(2|\Sigma|)$ symbols, the computation can be performed in $\mathcal{O}(|\Sigma|)$ time. Emptiness test of $sym(F^\natural) \cap sym(G^\natural)$ can be done in $\mathcal{O}(|\Sigma|)$ time with a hash table. So each subexpression can be examined in $\mathcal{O}(|\Sigma|)$ time. An expression E contains at most $\mathcal{O}(|E|)$ subexpressions. Thus the time complexity of the algorithm is $\mathcal{O}(|\Sigma||E|)$. If the size of the alphabet is fixed, the algorithm has linear running time.

Theorem 4. *Let E be an expression over a finite alphabet Σ. It can be decided in $\mathcal{O}(|\Sigma||E|)$ whether E is weakly deterministic or not.*

4   Strong Determinism of RE(&)

In this section, we derive an algorithm checking strong determinism based on a characterization of strong determinism.

In [12], H. Chen et al. proved that a weakly deterministic regular expression with counting is strongly deterministic if and only if it is in weakly star normal form (wSNF). We will show it also holds for RE(&).

Definition 5 (*[12]*)**.** *An expression E is in weakly star normal form if, for each subexpression H^* of E', $followlast(H) \cap first(H) = \emptyset$, where E' is the marked expression of E.*

Algorithm 4. *weakDeterm_helper2*

Input: the root node F of a binary tree $T(root)$
Output: true if the expression of $T(root)$ is weakly deterministic or false otherwise

1: **if** $F = \epsilon, a$ **then**
2:    **return** true
3: **if** $F = F_1 | F_2$ **then**
4:    **if** $weakDeterm\_helper2(F_1)$ and $weakDeterm\_helper2(F_2)$ **then**
5:       **if** $first(F_1) \cap first(F_2) \neq \emptyset$ **then**
6:          **return** false
7:       **else return** true
8:    **return** false
9: **if** $F = F_1 F_2$ **then**
10:    **if** $weakDeterm\_helper2(F_1)$ and $weakDeterm\_helper2(F_2)$ **then**
11:       **if** $followlast(F_1) \cap first(F_2) \neq \emptyset$ **then**
12:          **return** false
13:       **if** $\epsilon \in L(F_1)$ **then**
14:          **if** $first(F_1) \cap first(F_2) \neq \emptyset$ **then**
15:             **return** false
16:          **else return** true
17:    **else return** false
18: **if** $F = F_1 \& F_2$ **then**
19:    **if** $weakDeterm\_helper2(F_1)$ and $weakDeterm\_helper2(F_2)$ **then**
20:       **if** $sym(F_1)^{\natural} \cap sym(F_2)^{\natural} \neq \emptyset$ **then**
21:          **return** false
22:       **if** $first(F_1) \cap first(F_2) \neq \emptyset$ **then**
23:          **return** false
24:       **else return** true
25:    **else return** false
26: **if** $F = F_1^*$ **then**
27:    **if** $weakDeterm\_helper2(F_1)$==false **then**
28:       **return** false
29:    **if** $followlast(F_1) \cap first(F_1) \neq \emptyset$ **then**
30:       **return** false
31:    **else return** true

We first show that every weakly deterministic expression can be transformed to an equivalent strongly deterministic expression in linear time. The following two definitions are proposed in [4] to transform a standard regular expression to its star normal form. We replace $\epsilon^{\circ} = \emptyset$ with $\epsilon^{\circ} = \epsilon$ in [4] and add the rules for $E = F \& G$ to get a transformation method for weakly star normal form.

Definition 6. $E = \emptyset, x, \epsilon : E^{\circ} = E$
$\quad E = F^* : E^{\circ} = F^{\circ}$
$\quad E = F + G : E^{\circ} = F^{\circ} + G^{\circ}$
$\quad E = FG :$

$$E^\circ = \begin{cases} FG & if\ \epsilon \notin L(F), \epsilon \notin L(G) \\ F^\circ G & if\ \epsilon \notin L(F), \epsilon \in L(G) \\ FG^\circ & if\ \epsilon \in L(F), \epsilon \notin L(G) \\ F^\circ + G^\circ & if\ \epsilon \in L(F), \epsilon \in L(G) \end{cases}$$

$E = F\&G :$

$$E^\circ = \begin{cases} F\&G & if\ \epsilon \notin L(F), \epsilon \notin L(G) \\ F^\circ\&G & if\ \epsilon \notin L(F), \epsilon \in L(G) \\ F\&G^\circ & if\ \epsilon \in L(F), \epsilon \notin L(G) \\ F^\circ + G^\circ & if\ \epsilon \in L(F), \epsilon \in L(G) \end{cases}$$

Definition 7. $E = \emptyset, x, \epsilon : E^\bullet = E$
$E = F + G : E^\bullet = F^\bullet + G^\bullet$
$E = FG : E^\bullet = F^\bullet G^\bullet$
$E = F^* : E^\bullet = F^{\bullet\circ*}$
$E^\bullet = F^\bullet\&G^\bullet$

For example, $E = (a^*\&b^*)^*$, then we have $E^\bullet = (a^*\&b^*)^{*\bullet} = (a^*\&b^*)^{\bullet\circ*} = (a^{*\bullet\circ}\&b^{*\bullet\circ})^* = (a + b)^*$.

Lemma 3. $E^\bullet$ *is the weakly star normal form of E which can be computed from E in linear time and $L(E^\bullet) = L(E)$.*

The proof can make a direct use of the proof of Theorem 3.1 in [4] so we omit it here. The E° of E has the following property.

Lemma 4. *Let E be an RE(&) expression. If $followlast(E') \cap first(E') = \emptyset$, $E = E^\circ$.*

Proof. This can be proved by induction on the structure of E. The cases for $E = \emptyset, a(a \in \Sigma), \epsilon$ are straightforward, where $E = E^\circ$.

$E = F + G$: From the computations of $first$ and $followlast$, we have $followlast(E') = followlast(F') \cup followlast(G')$ and $first(E') = first(F') \cup first(G')$. Since $followlast(E') \cap first(E') = \emptyset$, we have $followlast(F') \cap first(F') = \emptyset$ and $followlast(G') \cap first(G') = \emptyset$. By the inductive hypothesis we have $F = F^\circ$ and $G = G^\circ$, therefore, $E = E^\circ$.

$E = FG$: If $\epsilon \notin L(F), \epsilon \notin L(G)$, by Definition 6 we have $E^\circ = FG$. Therefore $E = E^\circ$. If $\epsilon \in L(F), \epsilon \notin L(G)$, from the computations of $first$ and $followlast$, we have $followlast(E') = followlast(G')$ and $first(E') = first(F') \cup first(G')$. Since $followlast(E') \cap first(E') = \emptyset$, we get $followlast(G') \cap first(G') = \emptyset$. By the inductive hypothesis we have $G = G^\circ$, then $E^\circ = FG^\circ = FG = E$. If $\epsilon \notin L(F), \epsilon \in L(G)$, from the computations of $first$ and $followlast$, we have $followlast(E') = followlast(F') \cup followlast(G') \cup first(G')$ and $first(E') = first(F')$. Since $followlast(E') \cap first(E') = \emptyset$, we can get $followlast(F') \cap first(F') = \emptyset$. By the inductive hypothesis we have $F = F^\circ$, then $E^\circ = F^\circ G = FG = E$. The situation when $\epsilon \in L(F), \epsilon \in L(G)$ can never happens. Otherwise, $followlast(E') = followlast(F') \cup followlast(G') \cup first(G')$ and $first(E') = first(F') \cup first(G')$, $followlast(E') \cap first(E')$ cannot be $\emptyset$, which is a contradiction.

$E = F\&G$: If $\epsilon \notin L(F), \epsilon \notin L(G)$, by Definition 6 we have $E^\circ = FG$. Therefore $E = E^\circ$. The situation when $\epsilon \in L(F), \epsilon \in L(G)$ can never happens. Otherwise, $followlast(E') = followlast(F') \cup followlast(G') \cup first(G') \cup first(F')$ and $first(E') = first(F') \cup first(G')$, $followlast(E') \cap first(E')$ cannot be $\emptyset$, which is a contradiction. The other cases can be proved similarly.

$E = F^*$: This case can never happen. Since $followlast(E') = followlast(F') \cup first(F')$ and $first(E') = first(F')$, $E = F^*$ will contradict to $followlast(E') \cap first(E') = \emptyset$.   □

It is not hard to see that for any expression E, the weakly star normal of E is unique. We prove it by the next lemma.

Lemma 5. *Let E be an RE(&) expression. E is in weakly star normal form iff $E = E^\bullet$.*

Proof. ($\Rightarrow$) We prove it by induction on the structure of E. The cases for $E = \emptyset, a(a \in \Sigma), \epsilon$ are straightforward, where $E = E^\bullet$.

$E = F + G, E = FG$ or $E = F\&G$: Suppose E is in wSNF, then for each subexpression H^* of E', $followlast(H) \cap first(H) = \emptyset$. For each subexpression H_1^* of F' and each subexpression H_2^* of G', since F', G' are subexpressions of E', we have $H_1, H_2 \subseteq H$. Therefore, $followlast(H_1) \cap first(H_1) = \emptyset$ and $followlast(H_2) \cap first(H_2) = \emptyset$ thus F, G are in wSNF. By the inductive hypothesis we have $F = F^\bullet, G = G^\bullet$, then $E = F^\bullet + G^\bullet, E = F^\bullet G^\bullet$ or $E = F^\bullet \& G^\bullet$. Therefore $E = E^\bullet$.

$E = F^*$: Suppose E is in wSNF, then for each subexpression H^* of E', $followlast(H) \cap first(H) = \emptyset$. For any subexpression H_1 of F', we have $H_1 \subseteq H$ and $F' \subseteq H$. Therefore, $followlast(H_1) \cap first(H_1) = \emptyset$ thus F is in wSNF. By the inductive hypothesis we have $F = F^\bullet$. Since $F' \subseteq H$, we have $followlast(F') \cap first(F') = \emptyset$. By Lemma 4, $F = F^\circ$, then $E^\bullet = F^{\bullet \circ *} = F^{\circ *} = F^* = E$.

($\Leftarrow$) By Lemma 3, $E^\bullet$ is in weakly star normal form. Since $E = E^\bullet$, E is in weakly star normal form.   □

Then from Lemma 5, we have

Corollary 2. *If E is not the same with its weakly star normal form $E^\bullet$, E is not in weakly star normal form.*

The following characterization of strong determinism can be found in [12], which can be trivially extended to expressions in RE(&).

Lemma 6. *Let E be an expression in RE(&).*
(1) $E = \epsilon, a \in \Sigma$: E is strongly deterministic.
(2) $E = F+G$: E is strongly deterministic iff F and G are strongly deterministic and $first(F) \cap first(G) = \emptyset$.

(3) E = FG:

(a) If $\epsilon \in L(F)$, then E is strongly deterministic iff F and G are strongly deterministic, $first(F) \cap first(G) = \emptyset$, and $followlast(F) \cap first(G) = \emptyset$.

(b) If $\epsilon \notin L(F)$, then E is strongly deterministic iff F and G are strongly deterministic, and $followlast(F) \cap first(G) = \emptyset$.

(4) $E = F\&G$: E is strongly deterministic iff F and G are strongly deterministic, and $sym(F) \cap sym(G) = \emptyset$.

(5) $E = F^*$: E is strongly deterministic iff F is strongly deterministic and $followlast(F) \cap first(F) = \emptyset$.

Proof. The proof is by induction on the structure of E. We only show the induction step for interleaving. Others can be found in [12].

$E = F\&G$: If E is strongly deterministic, then E is weakly deterministic. By Corollary 1, $sym(F) \cap sym(G) = \emptyset$.

If F, G are strongly deterministic, then F, G are weakly deterministic. Since $sym(F) \cap sym(G) = \emptyset$, E is weakly deterministic from Corollary 1. If E is not strongly deterministic, then there are strings u, v, w over $\Sigma_E \cup \Gamma_E$, strings $\alpha \neq \beta$ over Γ_E, and a symbol $a \in sym(E)$ such that $u\alpha av$ and $u\beta aw$ are both correctly bracketed and in $L(\widetilde{E})$. Assume $a \in sym(F)$ without loss of generality. Let u', v', w' be the substrings of u, v, w amongst $sym(F)$ and α', β' be the substrings of α, β amongst Γ_F, then both of $u'\alpha'av'$ and $u'\beta'aw'$ are bracketed correctly and in $L(\widetilde{F})$, which implies that F is not strongly deterministic. This is a contradiction. So E is strongly deterministic. □

Then we can establish the relation between weak determinism and strong determinism by the following lemma.

Lemma 7. *Let E be a weakly deterministic expression. E is in wSNF iff E is strongly deterministic.*

The proof from right to left is based on Lemma 6 and Definition 5 by contradiction. The details are omitted here. The proof from left to right is by induction of E. For instance, we briefly prove the interesting case $E = F\&G$ in the inductive step. Suppose E is in wSNF. Thus F, G is clearly in wSNF. Since E is weakly deterministic, we have F, G are weakly deterministic and $sym(F) \cap sym(G) = \emptyset$ by Corollary 1. By the inductive hypothesis we have F, G are strongly deterministic. Hence E is strongly deterministic from Lemma 6.

From the above analysis we can get an algorithm to check strong determinism of RE($\&$). First, check weak determinism of E using $weakDeterm2(E)$. If E is weakly deterministic, compute the weakly star normal form $E^\bullet$ of E. If $E^\bullet$ is the same with E, E is strongly deterministic. Otherwise, E is not strongly deterministic. The time complexity of the algorithm is also $\mathcal{O}(|\Sigma||E|)$. The process is formalized in Algorithm 5. For instance, $E = (a^*)^*$. E is weakly deterministic because it contains only one symbol. $E^\bullet = a^{\bullet\circ*\bullet\circ*} = a^{\bullet\circ*\circ*} = a^{*\circ*} = a^*$. Since $E \neq E^\bullet$, E is not strongly deterministic. By Lemma 3, the equivalent strongly deterministic expression of E is $E^\bullet = a^*$.

Algorithm 5. *StrongDeterm*

Input: An expression in RE(&)
Output: true if E is strongly deterministic or false otherwise
 1: **if** *weakDeterm2*(E) is true **then**
 2:     compute the weakly star normal form $E^{\bullet}$ of E
 3:     **if** *equal*$(E, E^{\bullet})$ **then**
 4:         **return** true
 5: **return** false

Theorem 5. *StrongDeterm* (E) *returns true iff E is strongly deterministic.*

Proof. It follows from Lemmas 7 and 5.   □

Theorem 6. *StrongDeterm*(E) *runs in time* $\mathcal{O}(|\Sigma||E|)$.

The proof follows from Theorem 4 and Lemma 3.

5   Implementations and Experiments

In this section we first study the performance of the *followlast* algorithm by comparing it with the *follow*$^{-}$ algorithm. Experiments were performed on a computer with a Intel Core 2 Duo CPU(2.67 GHz) and 4G memory. Next, we discuss implementation of our algorithm for transforming an expression to its weakly star normal form. We have implemented all our algorithms and made them available at http://lcs.ios.ac.cn/~pengff/projects.html.

5.1   Weak Determinism

In this section, we describe experiments for verifying the correctness of algorithms based on *followlast* and *follow*$^{-}$ sets. Complex content models are designed to test the efficiency of the above determinism algorithms.

All algorithms are implemented in Java. First, scan for an input expression and convert it into a syntax tree. Next, make a post order traversal of the syntax tree for computing *first*, *followlast* and *sym* sets in subexpressions for every symbol. These contents are stored in ArrayList objects as attributes of nodes. At the same time, conditions in Theorems 1 or 3 can be checked for each subexpression. Once a subexpression meets the condition, the program is interrupted and shows information for nondeterministic symbols. A general overview of the interfaces of the *followlast* algorithm is presented in Fig. 1.

We design three complex content models for testing the efficiency and scalability of the above algorithms. The design for increasingly large content models is inspired by P. Kilpeläinen [9].

For sequence operator, the content model with interleaving can be defined by the form: $F_1 \& F_2 \& ... \& F_n$, where the repeated subexpression F_i is of the sequence form $a_i b_i ? c_i^* d_i^+$. Figure 2 shows how the algorithms scales up as the

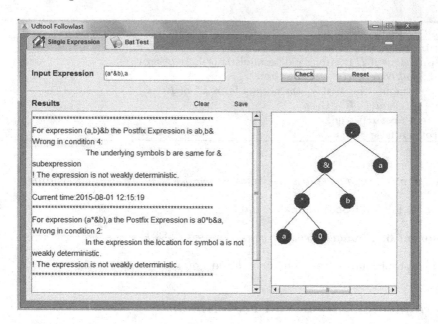

Fig. 1. Checking the weak determinism of $(a^*\&b)a$

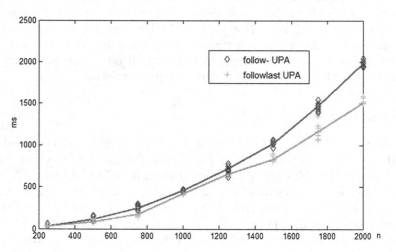

Fig. 2. Scale-up: Number of subexpressions for sequence

number of subexpressions is increased. It can be easily observed that it takes more time for $follow^-$ algorithms in these cases. Nonetheless, the gap is not very large. The reason is that the main difference between the two algorithms lies in sequence operator. We can see it from Algorithms 2 and 4.

We further study the scalability for choice operator. The content model with interleaving can be defined by the form: $G_1\&G_2\&...\&G_n$, where the repeated

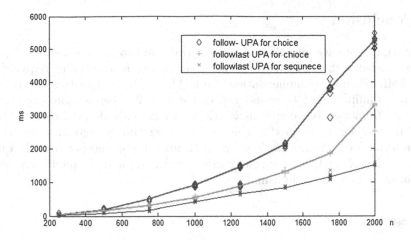

Fig. 3. Scale-up: Number of subexpressions for choice

subexpression G_i is of the sequence form $a_i|b_i?|c_i^*|d_i^+$. The weak determinism checking times are shown in Fig. 3. As for choice form, the time usage of $follow^-$ algorithm increases quadratically, and the time usage of $followlast$ algorithm increases nearly linearly. However, both of them are slower than $followlast$ algorithm for sequence which implies sequence is easier to implement.

5.2   Weakly Star Normal Form and Strong Determinism

$E^\bullet$ is built up from $H^{\bullet\circ}$ for subexpressions H^* of E during a post order traversal through the syntax tree of E. Some tricks must be mentioned. The expression is processed to a postfix expression for keeping the order of operations and removing parentheses, and during this preprocessing, a special symbol 0 is added before a unary operator. When constructing binary trees, the leaf nodes are labeled with 0 or a symbol, and the internal nodes are labeled with operators, i.e. interleaving, choice, sequence or a unary operator. For every subexpression H^* of E, if $H = M^*$, parent node $*$ of node M is deleted. If M is the left node of its parent, we add $parent(node(*)).left = node(*).left$, otherwise we add $parent(node(*)).right = node(*).left$. If $H = FG, \epsilon \in L(F)$ and $\epsilon \in L(G)$, then parent node , of nodes F, G is replaced with $+$. In the end, we can get the postfix expression of $E^\bullet$ by another post order traversal of the syntax tree.

The strong determinism algorithm can be easily implemented by simply combing the above two algorithms. Note that the process of converting a postfix expression into an infix expression may add parentheses to the original expression. For example, from the above algorithm, the postfix expression for wSNF of $E = a, b, c$ is $(ab, c,)$. Thus $E^\bullet = ((a, b), c)$, but actually nothing has been changed in E. Therefore, function $equal(E, E^\bullet)$ is implemented by comparing whether the postfix expressions of E and $E^\bullet$ are equal or not.

6 Conclusion

In this paper, we have investigated the determinism problem for regular expressions extended with interleaving. Weak determinism is a property required by W3C XML Schema Recommendation. An $\mathcal{O}(|\Sigma||E|)$ time algorithm is proposed based on examination of $first$ and $followlast$ sets. We then explored the transformation from weakly deterministic RE(&) to strongly deterministic RE(&). Based on this form, we modify the weakly determinism algorithm to strong determinism. As for future work, we want to investigate whether there is a natural extension of the Glushkov construction for RE(&) and the relation between such automata and determinism.

References

1. World Wide Web Consortium. http://www.w3.org/wiki/UniqueParticle Attribution
2. ISO 8879. Information processingtext and office systems-standard generalized markup language (SGML) (1986)
3. Gelade, W., Martens, W., Neven, F.: Optimizing schema languages for XML: numerical constraints and interleaving. In: Schwentick, T., Suciu, D. (eds.) ICDT 2007. LNCS, vol. 4353, pp. 269–283. Springer, Heidelberg (2006)
4. Brüggemann-Klein, A.: Regular expressions into finite automata. Theoret. Comput. Sci. **120**(2), 197–213 (1993)
5. Koch, C., Scherzinger, S.: Attribute grammars for scalable query processing on XML streams. VLDB J. **16**(3), 317–342 (2007)
6. Gelade, W., Gyssens, M., Martens, W.: Regular expressions with counting: weak versus strong determinism. SIAM J. Comput. **41**(1), 160–190 (2012)
7. Fuchs, M., Brown, A.: Supporting UPA and restriction on an extension of XML Schema. In: Extreme Markup Languages® (2003)
8. Groz, B., Maneth, S., Staworko, S.: Deterministic regular expressions in linear time. In: PODS, pp. 49–60 (2012)
9. Kilpeläinen, P.: Checking determinism of XML Schema content models in optimal time. Inf. Syst. **36**(3), 596–617 (2011)
10. Brüggemann-Klein, A.: Unambiguity of extended regular expressions in SGML document grammars. In: Lengauer, T. (ed.) ESA 1993. LNCS, vol. 726, pp. 73–84. Springer, Heidelberg (1993)
11. Brüggemann-Klein, A., Wood, D.: One-unambiguous regular languages. Inf. Comput. **142**(2), 182–206 (1998)
12. Chen, H., Lu, P.: Checking determinism of regular expressions with counting. Inf. Comput. **241**, 302–320 (2015)

Concurrency

Rigid Families for CCS and the π-calculus

Ioana Domnina Cristescu[1][(✉)], Jean Krivine[1], and Daniele Varacca[2]

[1] Laboratoire PPS, Université Paris Diderot, Paris, France
{ioana.cristescu,jean.krivine}@pps.univ-paris-diderot.fr
[2] LACL, Université Paris Est, Créteil, France
daniele.varacca@u-pec.fr

Abstract. This paper presents a novel causal semantics for concurrency, based on rigid families. Instead of having causality as primitive notion, in our model causality and concurrency are derived from precedence, a partial order local to each run of a process. We show that our causal semantics can interpret CCS and π-calculus terms. We propose some criteria to evaluate the correctness of a causal semantics of process calculi and we argue that none of the previous models for the π-calculus satisfy them all.

1 Introduction

Formal models for concurrency can be divided into *interleaving* models, such as *traces* and *labelled transition systems*, in which concurrency is represented as non deterministic executions, and *non interleaving* ones, like configuration structures [23], event structures [22] or presheaves [8]. Non interleaving semantics have a primitive notion of *concurrency* between computation events. As a consequence one can also derive a *causality* relation, generally defined as the complement of concurrency. These models are therefore sometimes called *causal semantics* or, if causality is represented as a partial order on events, *partial order* semantics. Causal models are also known to be at the foundations of reversible concurrent computations [19].

In this paper we propose to take a notion of *precedence* as the fundamental relation between events. Precedence is a partial order that can be seen as a temporal observation, specific to a given run of a process. In a given run, two events may also not be related by any precedence relation, in which case one can see them as having occurred either simultaneously or in such a way that no common clock can be used to compare them. More traditional causality and concurrency relations are derivable from precedence.

The interpretation of a process is built by induction on the process constructors and defined as operations on *rigid families* [6,15]. We equip rigid families with a *labelling* function on events. In this sense, our semantics resembles the encoding of CCS in configuration structures [22].

I.D. Cristescu – Partially supported by the ANR grant (REVER) ANR-11-INSE-0007.

© Springer International Publishing Switzerland 2015
M. Leucker et al. (Eds.): ICTAC 2015, LNCS 9399, pp. 223–240, 2015.
DOI: 10.1007/978-3-319-25150-9_14

The operations on rigid families that are used to encode CCS can be easily adapted to the π-calculus. Importantly, the restriction operator behaves similarly in the rigid families for both calculi: it removes from the model the executions that are not allowed. In previous models for the π-calculus [2,9] the restriction of a private name introduced new orders between events.

Several causal models have been proposed in the literature for process calculi (see for instance [2–5,7,9,12,13,22]). Each model is shown to be correct, in some sense. But can a model be *more correct* than another one? What are the criteria one uses to make such evaluation? We will show that our model satisfies several correctness criteria, and we will argue that no previous causal model for the π-calculus satifies them all.

The correctness criteria will be introduced as we go along the formal development.

Outline. In Sect. 2 we introduce the category of rigid families and rigid morphisms. In Sects. 3 and 4 we show how to interpret CCS and π-calculus, respectively, such that the models are compositional and sound. In Sect. 5 we present the three remaining correction criteria and conclude with Sect. 6.

2 A Category of Rigid Families

In this section we present *rigid families*, a model for concurrency introduced by Hayman and Winskel [6,15], that is a close relative to *configuration structures* [14]. We first introduce the unlabelled categorical setting. It results in a generic framework to represent concurrent computations as sets of events (configurations) equipped with a partial order that represent temporal precedence between events. Importantly precedence is local to a configuration whereas events can occur in multiple configurations.

When a process P is able to output two messages a and b on two parallel channels, three kinds of observations are possible. The observer sees the output on a before the output on b, or the output on b before the output on a, or she cannot tell in which order they happen, either because they really happen at the same time, or because the observer does not have a global clock on the two events. In the rigid family interpretation of P we would have three corresponding configurations for the parallel emission of a and b.

Definition 1 (Rigid Inclusion of Partial Orders). *Given a partial order x, we write $|x|$ to denote the underlying set and $e \leq_x e'$ whenever $(e, e') \in x$. Rigid inclusion of partial orders $x \preceq y$ is defined iff the following hold:*

$$|x| \subseteq |y| \text{ and } \forall e, e' \in x : e \leq_x e' \iff e \leq_y e'$$
$$\forall e \in y, \forall e' \in x, e \leq_y e' \implies e \in x$$

Definition 2 (Rigid Families). *A rigid family $\mathcal{F} = (E, C)$ is a set of events E and a non empty family of partial orders, called* configurations, *such that $\forall x \in C$, $|x| \in \mathcal{P}(E)$ and C is downward closed w.r.t rigid inclusion: $\forall y \preceq x, y \in C$.*

A morphism on rigid families $\sigma : (E, C) \to (E', C')$ is a partial function on events $\sigma : E \rightharpoonup E'$ that is local injective:

$$\text{For all } x \in C, e, e' \in x, \sigma(e) = \sigma(e') \implies e = e'$$

and that extends to a (total) function on configurations:

$$\sigma(x) = x' \text{ iff } \sigma(|x|) = |x'| \text{ and } \forall e, e' \in \mathrm{dom}(\sigma), \sigma(e) \leq_{x'} \sigma(e') \iff e \leq_x e'$$

We write $\mathcal{F}_1 \cong \mathcal{F}_2$ whenever there is an isomorphism between $\mathcal{F}_1$ and $\mathcal{F}_2$ and we use $\mathbf{0}$ to denote the rigid family with an empty set of events.

Proposition 1. *Rigid families and their morphisms form a category.*

Importantly, the morphisms we employ here differ from the ones introduced by Hayman and Winskel that are defined on configurations and are not required to preserve the order[1].

If precedence is a partial order that is local to a configuration, one may also define a global (partial) order as follows.

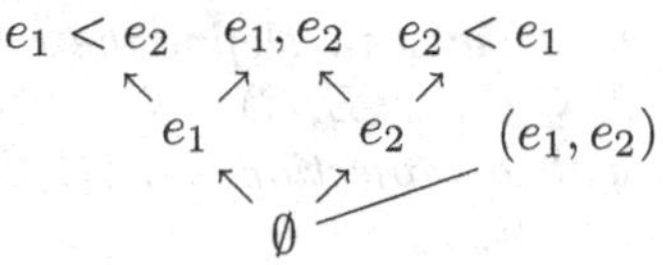

Fig. 1. Example of product

Definition 3 (Causality). *Let $e, e' \in E$ for (E, C) a rigid family. Define $e' < e$ to mean: $\forall x \in C$, if $e, e' \in x$ then $e' <_x e$.*

Rigid families offer a natural notion of *disjoint* causality: i.e. an event e_1 is caused by either e_2 or e_3. This type of dependency is a generalisation of Definition 3:

Definition 4 (Disjoint Causality). *Let (E, C) a rigid family and $e \in E$, $X \subset E$ such that $e \notin X$. Then X is a disjoint causal set for e, denoted $X < e$ iff the following hold:*

1. **disjointness** $\forall e' \in X$, $\exists x \in C$ such that $e' <_x e$ and $\forall e'' \in X \setminus e'$, $e'' \nless_x e$.
2. **completeness** $\forall x \in C$, $e \in x \implies \exists e' \in X$ such that $e' <_x e$;

In particular $e' < e$ whenever $\{e'\} < e$.

Definition 5 (Concurrency). *Let (E, C) a rigid family and $e, e' \in E$. Define $e \Diamond e' \iff \exists x \in C$, $e, e' \in x$ such that $e' \nleq_x e$ and $e \nleq_x e'$.*

Note that concurrency has an existential quantifier: two events are concurrent if there exists a configuration in which they are not comparable. On the other hand, causality is universal: it has to hold for all configurations.

---

[1] We let the reader refer to appendix for details. We also show in appendix how one can compile an event structure from rigid families and *vice versa*. Importantly, the category of Definition 2 and the category of event structures are not equivalent.

Definition 6 (Operations on Rigid Families). *Let* $E^\star = E \cup \{\star\}$.

1. **Product.** *Let* $\star$ *denote* undefined *for a partial function. Define* $(E, C) = (E_1, C_1) \times (E_2, C_2)$ *where* $E = E_1 \times_\star E_2$ *is the product in the category of sets and partial functions with the projections* $\sigma_1 : E \to E_1^\star$, $\sigma_2 : E \to E_2^\star$. *Define the projections* $\pi_1 : (E, C) \to (E_1, C_1)$, $\pi_2 : (E, C) \to (E_2, C_2)$ *and* $x \in C$ *such that the following hold:*
   - *x is a partial order with* $|x| \in \mathcal{P}(E)$;
   - $\pi_1(e) = \sigma_1(e)$ *and* $\pi_2(e) = \sigma_2(e)$;
   - $\pi_1(x) \in C_1$ *and* $\pi_2(x) \in C_2$;
   - $\forall e, e' \in x$, *if* $\pi_1(e) = \pi_1(e') \neq \star$ *and* $\pi_2(e) = \pi_2(e') \neq \star$ *then* $e = e'$.
   - $\forall e, e' \in x$ *such that* $e, e' \in \mathrm{dom}(x)$, $e <_x e' \iff \pi_1(e) <_{\pi_1(x)} \pi_1(e')$ *and* $\pi_2(e) <_{\pi_2(x)} \pi_2(e')$.
   - $\forall y \subseteq x$ *we have that* $\pi_1(y) \in C_1$ *and* $\pi_2(y) \in C_2$.
2. **Restriction.** *Define the restriction of an upward closed set of configurations* $X \subseteq C$ *as* $(E, C) \upharpoonright X = (\cup C', C')$ *with* $C' = C \setminus X$. *We equip the operation with a projection* $\pi : (E, C) \upharpoonright X \to (E, C)$ *such that* π *is the identity on events.*
3. **Prefix.** *Define* $e.(E, C) = (e \cup E, C' \cup \emptyset)$, *for* $e \notin E$ *where*

$$x' \in C' \iff x' = (\{e <_{x'} e' \mid \forall e' \in x\} \cup x) \text{ for some } x \in C.$$

Let $\pi : e.(E, C) \to (E, C)$ *the projection such that* $\pi(e)$ *is undefined and* π *is the identity on the rest of the events.*

Example 1. We obtain the rigid family in Fig. 1 for the product of $(\emptyset \prec \{e_1\})$ and $(\emptyset \prec \{e_2\})$.

Proposition 2. *The following properties hold:*

1. $\mathcal{F}_1 \times \mathcal{F}_2$ *is the cartesian product in the category of rigid families.*
2. $\mathcal{F} \upharpoonright X$ *is a rigid family with the projection* $\pi : \mathcal{F} \upharpoonright X \to \mathcal{F}$ *a morphism.*
3. $e.\mathcal{F}$ *is a rigid family with the projection* $\pi : e.\mathcal{F} \to \mathcal{F}$ *a morphism.*

The following proposition shows that the prefix operation adds event e *before* any other event in the family.

Proposition 3. *Let* $e.(E, C) = (E', C')$. $\forall e' \in E'$, $e \leq e'$.

3   Rigid Families for CCS

Criterion 1 (Compositional Interpretation). *The interpretation of a process should be given by composing the interpretations of its subprocesses.*

We conceived our category of rigid family as model for a large class of concurrent languages, including the π-calculus. In order to illustrate how this should work, we begin by tuning our formalism to finite CCS [17]. We proceed in a similar, yet more technical, manner in Sect. 4 to model the π-calculus.

As it is standard in causal models for concurrency [21], product of rigid families (Definition 6, Eq. (1)) essentially creates all possible pairs of events that respect the rigidity constraint imposed by the morphisms of our category. One then needs to prune out the pairs of events that do not correspond to legitimate synchronisations. We prove in Subsect. 3.2 the correspondence with CCS. For simplicity we do not deal with recursion in this paper and let the reader refer to the appendix for a full treatment of non finitary CCS.

3.1   Definitions

Let N be a set of *names* $N = \{a, b, c, \dots\}$, $\overline{N}$ a set of *co-names* $\overline{N} = \{\overline{a}, \overline{b}, \overline{c}, \dots\}$. The function $\overline{[\cdot]} : N \to \overline{N}$ is a bijection, whose inverse is also denoted by $\overline{[\cdot]}$ so that $\overline{\overline{a}} = a$. Let L be the set of event labels defined by the following grammar:

$$\alpha, \beta ::= a \mid \overline{a} \mid (\alpha, \beta)$$

The pairs of labels are globally denoted by τ. We say that an event is *partial* if it is labelled by a name or a co-name. It represents a possible interaction with the context.

Definition 7 (Labelled Rigid Families).   *A labelled rigid family $\mathcal{F} = (E, C, \ell, \mathsf{P})$ is a rigid family equipped with a distinguished set of names P (the private names of $\mathcal{F}$) and a labelling function $\ell : E \to L$.*

We use labels to determine which events cannot occur in a computation:

Definition 8 (Disallowed Events).   *Let $\mathcal{F} = (E, C, \ell, \mathsf{P})$ and $e \in E$. We say that $\ell(e)$ is disallowed if one of the following properties holds:*

1. [**type mismatch**] $\ell(e) = (\alpha, \beta)$ *with* $\alpha \notin N \cup \overline{N}$ *or* $\overline{\alpha} \neq \beta$;
2. [**private name**] *($\ell(e) = a \in N$ or $\ell(e) = \overline{a} \in \overline{N}$) and $a \in \mathsf{P}$;*

A synchronisation event may only occur between complementary partial events (Eq. (1)) and partial events may not use a private name (Eq. (2)).

Definition 9 (Dynamic Label).   *Define the dynamic label of an event as $\hat{\ell}(e) = \ell(e)$ if $\ell(e)$ is allowed and $\bot$ otherwise.*

We extend now the operations of Definition 6 in order to take labels into account.

Definition 10 (Operations on Labelled Rigid Families)

1. **Restriction of a name.** *Let $a \notin \mathsf{P}$. Then $(E, C, \ell, \mathsf{P}) \restriction a = (E, C, \ell, \mathsf{P} \cup \{a\}) \restriction X$, where $x \in X$ iff $\exists e \in x$ such that $\hat{\ell}(e) = \bot$.*
2. **Prefix.** *Define $\alpha.(E, C, \ell, \mathsf{P}) = (E', C', \ell', \mathsf{P})$ where, for some $e \notin E$, $e.(E, C) = (E', C')$ and $\ell'(e) = \alpha$ and $\ell'(e') = \ell(e')$ for $e' \neq e$.*

3. **Product.** *Let* $(E, C) = (E_1, C_1) \times (E_2, C_2)$ *and* π_1, π_2 *the projections* $\pi_i :$ $(E, C) \to (E_i, C_i)$. *Then*

$$(E_1, C_1, \ell_1, \mathsf{P}_1) \times (E_2, C_2, \ell_2, \mathsf{P}_2) = (E, C, \ell, \mathsf{P}_1 \cup \mathsf{P}_2)$$

where $\ell(e) = \begin{cases} \ell_i(\pi_i(e)) & \text{if } \pi_{3-i}(e) = \star \\ (\ell_1(\pi_1(e)), \ell_2(\pi_2(e_2))) & \text{otherwise} \end{cases}$

4. **Parallel composition**

$$(E_1, C_1, \ell_1, \mathsf{P}_1) \mid (E_2, C_2, \ell_2, \mathsf{P}_2) = (E_1, C_1, \ell_1, \mathsf{P}_1) \times (E_2, C_2, \ell_2, \mathsf{P}_2) \upharpoonright X$$

where $x \in X$ *iff* $\exists e \in x$ *such that* $\hat{\ell}(e) = \perp$.

Definition 11 (Sound Rigid Family). $\mathcal{F}$ *is* sound *iff* $\forall x \in \mathcal{F}, \forall e \in x, \hat{\ell}(e) \neq \perp$.

Proposition 4. *Let* $\mathcal{F}_1 = (E_1, C_1, \ell_1, \mathsf{P}_1)$, $\mathcal{F}_2 = (E_2, C_2, \ell_2, \mathsf{P}_2)$ *sound rigid families such that* $\mathsf{P}_1 \cap \mathsf{P}_2 = \emptyset$ *and let* α *a name such that* $\alpha \notin \mathsf{P}_1$. *Then* $\alpha.\mathcal{F}_1$, $\mathcal{F}_1 \upharpoonright a$ *and* $\mathcal{F}_1 \mid \mathcal{F}_2$ *are sound rigid families.*

3.2   Operational Correspondence with CCS

Criterion 2 (Sound Interpretation). *The interpretation of a process can be equipped with an operational semantics that corresponds to the natural reduction semantics of the process.*

To show the correspondence with the operational semantics of CCS, we need to define a notion of *transition* on rigid families. Intuitively, a computation step consists in triggering a single-event computation $\{e\}$ that belongs to the set of configurations. Once e is consumed, the events in conflict with e are eliminated. The remaining configurations are those that are "above" the configuration $\{e\}$.

Definition 12 (Transitions on Rigid Families). *Let* $(E, C, \ell, \mathsf{P})/e = (E', C', \ell', \mathsf{P})$, *for* $\{e\} \in C$, *be the rigid family obtained after the occurence of event e and defined as follows:*

– $x' \in C' \iff x' = x \setminus \{e\}$, *for* $\{e\} \preceq x \in C$;
– $e' \in E' \iff \exists x \in C, \{e\} \preceq x$ *and* $e' \in x$;

For all rigid family $\mathcal{F} = (E, C, \ell, \mathsf{P})$ with $\{e\} \in C$, note that $\mathcal{F}/e$ is also a labelled rigid family. Now consider (finite) CCS terms defined by the grammar below:

$$P, Q ::= (P|Q) \mid a.P \mid \overline{a}.P \mid P \backslash a \mid 0$$

As usual, occurrences of a and $\overline{a}$ in $P \backslash a$ are bound. For simplicity, we assume that all bound occurrences of names are kept distinct from each other and from free occurrences.

The interpretation of a CCS process as a rigid family is defined by induction on the structure of a term:

$$[\![\alpha.P]\!] = \alpha.[\![P]\!] \quad [\![P|Q]\!] = [\![P]\!]|[\![Q]\!] \quad [\![P \backslash a]\!] = [\![P]\!] \upharpoonright a \quad [\![0]\!] = \mathbf{0}$$

Lemma 1. *Let P a process and $[\![P]\!] = (E, C, \ell, \mathsf{P})$ its interpretation.*

1. *$\forall \alpha$, P' such that $P \xrightarrow{\alpha} P'$, $\exists e \in E$ such that $\ell(e) = \alpha$ and $[\![P]\!]/e \cong [\![P']\!]$;*
2. *$\forall e \in E$, $\{e\} \in C$, $\exists P'$ such that $P \xrightarrow{\ell(e)} P'$ and $[\![P]\!]/e \cong [\![P']\!]$.*

A direct corollary of Lemma 1 is that a process and its encoding can simulate each others reductions.

Theorem 1 (Operational Correspondence with CCS). *Let P a process and $[\![P]\!] = (E, C, \ell, \mathsf{P})$ its encoding.*

1. *$\forall P'$ such that $P \xrightarrow{\tau} P'$, $\exists \{e\} \in C$ closed such that $[\![P]\!]/e \cong [\![P']\!]$;*
2. *$\forall e \in E$, $\{e\} \in C$ closed, $\exists P'$ such that $P \xrightarrow{\tau} P'$ and $[\![P]\!]/e \cong [\![P']\!]$.*

3.3 Causality and Concurrency in CCS

A term in CCS can compose with a *context* and then exhibit more behaviours. We will show below a property that says that precedence in the semantics of a term can be seen as an abstraction of the causality that appears when the term is put into a context.

In the interpretation of a CCS term, if we have a configuration x and two concurrent events in x, we also have the configuration y with the same events and the same order as x except for the concurrent events that become ordered in y. This is stated formally in the following proposition.

Proposition 5. *Let $[\![P]\!] = (E, C, \ell, \mathsf{P})$. $\forall x \in C$ and $\forall e_1, e_2 \in x$ such that $e_1 \Diamond_x e_2$, $\exists y \in C$ such that $|x| = |y|$ and*

1. *y preserves the order in x: $\forall e, e' \in x$, $e \leq_x e' \implies e \leq_y e'$*
2. *x reflects the order in y except for $e_1 <_y e_2$: $\forall e, e' \in y$ such that $\neg(e = e_1 \wedge e' = e_2)$, $e \leq_y e' \implies e \leq_x e'$.*

In CCS we cannot guarantee simultaneity. Two concurrent events (Definition 5) can be observed simultaneously but also in any order. For instance the order $a < b$ induced by the CCS term $a \mid b$ is materialized when the term composes with the context $(\overline{a}.\overline{b} \mid [\cdot]) \backslash ab$. Note that this is a property specific to CCS (and the π-calculus in Subsect. 4.4), but one can encode in rigid families calculi where such a property does not hold.

The causality of Definition 3 is called *structural* in previous models for CCS [3,22]. But we can also express the disjunctive causality of Definition 4 in CCS. Consider the process $a.b \mid \overline{a}$ interpreted in rigid families in Fig. 2, where events are replaced by their corresponding labels. The disjunctive causal set for the event labelled b consists of the events labelled a and τ: $X = \{a, \tau\} < b$. Disjunctive causal sets in CCS always consist of conflicting events. However, it is not the case for the disjunctive causal sets of the π-calculus.

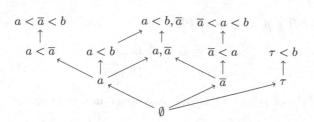

Fig. 2. $a.b \mid \overline{a}$ in rigid families

4   Rigid Families for the π-calculus

We show how definitions in Sect. 3 can be adapted to the π-calculus [18]. The treatment of synchronisation labels is more complicated as noted in [9,16] since names can be substituted during computations. Also, restriction does not necessarily delete events labelled with the private name, due to the phenomenon of *scope extrusion*.

However, in our novel approach, the main difficulty of the encoding resides in the definition of disallowed labels. Given the correct definition, all operations on rigid families for the π-calculus are straightforward extensions from CCS, including the restriction. As in CCS, for simplicity, we do not treat recursion or nondeterministic sum.

4.1   Labels for the π-calculus

We redefine the set L of events labels (see below), in order to handle the labels of the π-calculus. We use α, β to range over L, on which we define the functions subj and obj in the obvious manner:

$$\alpha ::= \overline{b}\langle a \rangle \mid d(c) \mid (\alpha, \beta)$$
$$\mathsf{subj}(\overline{b}\langle a \rangle) = \{b\} \; \mathsf{subj}(d(c)) = \{d\} \; \mathsf{subj}(\alpha, \beta) = \mathsf{subj}(\alpha) \cup \mathsf{subj}(\beta)$$
$$\mathsf{obj}(\overline{b}\langle a \rangle) = \{a\} \; \mathsf{obj}(d(c)) = \{c\} \quad \mathsf{obj}(\alpha, \beta) = \mathsf{obj}(\alpha) \cup \mathsf{obj}(\beta)$$

A labelled rigid family for the π-calculus is defined as in Definition 7, except that the labelling function $\ell : E \to L$ has as codomain the new set L. For a label $\alpha = b(a)$ or $\alpha = \overline{b}\langle a \rangle$, we use the notation $\alpha \in \ell(e)$ if $\ell(e) = (\alpha, \alpha')$, $\ell(e) = (\alpha', \alpha)$ or $\ell(e) = \alpha$. The name b is *binding* in a label α if $\alpha = a(b)$ for some name a. For simplicity we use the *Barendregt convention*: a name b has at most one binding occurrence in all event labels and this occurrence binds all bs in the other event labels.

We call an event e that binds a name b in a configuration an *instantiator* for b. An event e with label $\overline{c}\langle d \rangle$ can thus have two instantiators, one for the subject c and one for the object d. If the event is labelled by a synchronisation $(\ell(e) = (\overline{b}\langle a \rangle, d(c)))$ we can have up to three instantiators (for the names b, a and d). For all e occurring in a configuration x, we write $e' \in \mathsf{inst}_x^s(e)$ and

$e'' = \mathsf{inst}^o_x(e)$ for, respectively, the subject and object instantiator of e. Note that in the interpretation of a π process (respecting the Barendregt convention) as a rigid family in Subsect. 4.3, it can be proved that $e' <_x e$ and $e'' <_x e$.

4.2 Synchronizations

Let Σ be the set of all name substitutions. The function $\sigma_x : x \to \Sigma$ returns a set with all substitutions generated by synchronisation events in x.

Definition 13 (Substitution). *We define σ_x by induction on x:*

$$\sigma_\emptyset = \emptyset$$

$$\sigma_x = \sigma_{x \setminus e} \quad \text{if } \ell(e') \neq (d(a), \bar{b}\langle a' \rangle)$$

$$\sigma_{x \setminus e} \cup \{a'/a\} \quad \text{if } \ell(e') = (d(a), \bar{b}\langle a' \rangle) \text{ and } \{a''/a'\} \notin \sigma_{x \setminus e}$$

$$\sigma_{x \setminus e} \cup \{a''/a\} \quad \text{if } \ell(e') = (d(a), \bar{b}\langle a' \rangle) \text{ and } \{a''/a'\} \in \sigma_{x \setminus e}$$

Define $\ell_x(e) = \ell(e)\sigma_x$ which applies the substitutions to the label of e.

We can prove that for any configuration x in the interpretation of a π process, σ_x is well defined.

The synchronizations of a configuration x are events $e \in x$ such that $\ell(e) = (b(a), \bar{c}\langle d \rangle)$, for some names a, c, d. We use the configuration-indexed predicate $\tilde{\tau}_x : x \to 2$ to denote events of that sort. The *materialized* synchronisations are synchronization events with the $\ell_x(e) = (a(c), \bar{a}\langle d \rangle)$. The predicate $\tau_x : x \to 2$ is the smallest predicate that holds for such events.

Importantly, one cannot simply screen out all $\tilde{\tau}_x$-events in a rigid family that are not also τ_x-events. They might still become fully fledged synchronisations after composing the family with more context. Yet some pairs of events will never satisfy the τ_x predicate, no matter what operations are applied to the rigid family they belong to. Such events can be identified thanks to their *disallowed* label.

The definitions of events that have disallowed labels are quite cumbersome but essentially consist in an exhaustive characterization of events the label of which proves they can no longer appear in a π-calculus computation (Definition 15 and Definition 16). Such events can only appear after applying the product or the restriction, operations needed to represent the parallel composition and name restriction of the π-calculus. They are therefore detected and removed on the fly (see Definition 10). The reader, uninterested in technical details, may now safely skip to Subsect. 4.3, having in mind this informal notion of disallowed events.

A $\tilde{\tau}_x$-event can become a materialized τ_x-event when the names used as subject can be *matched*. This is always a possibility if the names are not private, because input prefix operations can instantiate them to a common value. However, when only one of the names is private then a distinct event, occurring beforehand, has to be in charge of leaking the name to the context. We call such event an *extruder*:

Definition 14 (Extruder). *An event $e \in x$ is an* extruder *of $e' \in x$ if $e <_x e'$ and $\ell_x(e) = \bar{b}\langle a \rangle$ for some b where a is private and $a \in \mathsf{subj}(\ell_x(e'))$.*

Consider a $\tilde{\tau}_x$-event e occurring in a configuration x. If $a, b \in \mathsf{subj}(\ell_x(e))$ we say that a can *eventually match* b iff $a \notin \mathsf{P}$ and either $b \notin \mathsf{P}$ or if $b \in \mathsf{P}$ then there exists $e', e'' \in x$, $e' = \mathsf{inst}_x^s(e)$ and e'' extruder of e such that $e'' <_x e'$. We write $\mathsf{U}(a, b)$ iff $a = b$ or if a can eventually match b or b can eventually match a.

Definition 15 (Disallowed $\tilde{\tau}_x$-Events). *Let* $\mathcal{F} = (E, C, \ell, \mathsf{P})$ *and* $x \in C$, $e \in x$ *with* $\tilde{\tau}_x(e)$. *The label* $\lambda = \ell_x(e)$ *is* disallowed *if one of the following properties holds:*

1. **[type mismatch]** $\lambda = (\alpha, \beta)$ *and it is not the case that* α *is input and* β *output or viceversa;*
2. **[non unifiable labels]** *let* $a, b \in \mathsf{subj}(\lambda)$ *and either:*
   - $\neg\, \mathsf{U}(a, b)$
   - $\exists e' \in x$ *such that* $a, b' \in \mathsf{subj}(\ell_x(e'))$ *for some* b' *and* $\neg\, \mathsf{U}(b, b')$.

Condition 1 is straightforward: an output can only synchronize with an input. Condition 2 says that a $\tilde{\tau}_x$-event cannot materialize if the names used in subject position cannot eventually match.

Private names cannot be used to interact with the context. Therefore we disallows partial events that use a private name as a communication channel (i.e. in the subject of the label). However, if the private name a is sent to the context (by an extruder) then a synchronisation on a becomes possible. This is formally stated by the definition below.

Definition 16 (Disallowed Partial Event). *Let* $\mathcal{F} = (E, C, \ell, \mathsf{P})$ *and* $x \in C$, $e \in x$ *with* $\neg\tilde{\tau}_x(e)$. *The label* $\lambda = \ell_x(e)$ *is* disallowed *if* $a \in \mathsf{subj}(\lambda)$, $a \in \mathsf{P}$ *and* $\nexists e' \in x$ *extruder of* e.

Definition 17 (Dynamic Label). *Define the dynamic label of an event as* $\hat{\ell}_x(e) = \ell_x(e)$ *if* $\ell_x(e)$ *is allowed and* $\bot$ *otherwise.*

We have the same operations as in Definition 10 but applied to the set of labels of the π-calculus and using the dynamic labels of Definition 17.

4.3   Operational Correspondence with the π-calculus

In this section we show the operational correspondence between processes in π-calculus and their encoding in rigid families. We assume the reader is familiar with the π-calculus [18,20]. For the sake of simplicity we use a restricted version of π-calculus defined by the grammar below:

$$P ::= (P|P) \mid \overline{b}\langle a \rangle.P \mid d(c).P \mid P\backslash a \mid 0$$

The restriction of the private name a is denoted as in CCS, to highlight that in their interpretation in rigid families is similar. We use a late LTS for the π-calculus recalled in the appendix. Similarly to Subsect. 3.2 we encode a process into a rigid family as follows:

$$[\![\alpha.P]\!] = \alpha.[\![P]\!] \quad [\![P|Q]\!] = [\![P]\!] | [\![Q]\!] \quad [\![P\backslash a]\!] = [\![P]\!] \uparrow a \quad [\![0]\!] = \mathbf{0}$$

We now revisit Definition 12 since transitions on rigid families with dynamic labels need to apply substitutions on the fly.

Definition 18 (Transitions on Rigid Families). *Let* $(E, C, \ell, \mathsf{P})/e = (E', C', \ell', \mathsf{P})$, *for* $\{e\} \in C$, *be the rigid family obtained after the occurence of event* e *and defined as follows:*

- $x' \in C' \iff x' = x \setminus \{e\}$, *for* $\{e\} \preceq x \in C$;
- $E' = \cup \lfloor x' \rfloor$, *for all* $x' \in C'$;
- *if* $\ell(e) = (\bar{b}\langle a \rangle, c(d))$ *then* $\ell'(e) = \ell(e)\{a/d\}$; *otherwise* $\ell' = \ell$.

In a rigid family we have events that do not have an operational correspondence, but are necessary for compositionality. These are the events for which the predicate $\tilde{\tau}_x(e)$ holds but $\tau_x(e)$ does not. We ignore them when showing the operational correspondence.

Definition 19 (Complete and Closed Configurations). *A configuration* x *in a rigid family* (E, C, ℓ, P) *is* complete *if* $\forall e \in x$, $\tilde{\tau}_x(e) \implies \tau_x(e)$. *We say that* x *is* closed *if* $\forall e \in x$, $\tau_x(e)$ *holds.*

Remark that for minimal events (i.e. e such that $\{e\}$ is a configuration) $\ell_{\{e\}}(e) = \ell(e)$.

Lemma 2. *Let* P *a process and* $\llbracket P \rrbracket = (E, C, \ell, \mathsf{P})$ *its encoding.*

1. $\forall \alpha, P'$ *such that* $P \xrightarrow{\alpha} P'$, $\exists e \in E$ *such that* $\ell(e) = \alpha$ *and* $\llbracket P \rrbracket / e \cong \llbracket P' \rrbracket$;
2. $\forall e \in E$, $\{e\} \in C$ *complete,* $\exists P'$ *such that* $P \xrightarrow{\ell(e)} P'$ *and* $\llbracket P \rrbracket / e \cong \llbracket P' \rrbracket$.

Proof (sketch).

1. We proceed by induction on the derivation of the rule $P \xrightarrow{\alpha} P'$. We consider the following cases:

$$\text{In/Out} \frac{}{\alpha.P \xrightarrow{\alpha} P} \qquad \text{Com} \frac{P \xrightarrow{\bar{b}\langle a \rangle} P' \quad Q \xrightarrow{b(c)} Q'}{P|Q \xrightarrow{\tau} P'|Q'\{a/c\}} \qquad \text{Restr} \frac{P \xrightarrow{\alpha} P'}{P\backslash a \xrightarrow{\alpha} P'\backslash a} a \notin \alpha$$

   - Rule In/Out: let $\llbracket \alpha.P \rrbracket = \alpha.\llbracket P \rrbracket$ hence we have to show that

$$\llbracket P \rrbracket \cong (\alpha.\llbracket P \rrbracket)/e, \text{ where } \ell(e) = \alpha. \tag{1}$$

   - Rule Com: we have $\llbracket P|Q \rrbracket = (\llbracket P \rrbracket \times \llbracket Q \rrbracket) \upharpoonright X$. By induction $\llbracket P' \rrbracket \cong \llbracket P \rrbracket / e_1$ where $\ell(e_1) = \bar{b}\langle a \rangle$ and $\llbracket Q' \rrbracket \cong \llbracket Q \rrbracket / e_2$ where $\ell(e_2) = b(c)$. Then $\{e_1\} \in \llbracket P \rrbracket$ and $\{e_2\} \in \llbracket Q \rrbracket$ it implies that there exists $\{e\} \in \llbracket P \rrbracket \times \llbracket Q \rrbracket$ such that $\pi_1(e) = e_1$ and $\pi_2(e) = e_2$. Hence we have to show that

$$\left((\llbracket P \rrbracket / e_1 \times \llbracket Q \rrbracket / e_2) \upharpoonright X' \right) \{a/d\} \cong \left((\llbracket P \rrbracket \times \llbracket Q \rrbracket) \upharpoonright X \right)/e$$

   where $\mathcal{F}\{a/d\}$ replaces d with a in all labels.
   - Rule Restr: by induction we have that $\llbracket P \rrbracket / e \cong \llbracket P' \rrbracket$, where $\ell(e) = \alpha$. We have then to show that $\{e\} \in \llbracket P\backslash a \rrbracket$ and

$$(\llbracket P \rrbracket / e) \upharpoonright a \cong (\llbracket P \rrbracket \upharpoonright a)/e \tag{2}$$

2. We proceed by structural induction on P. Consider the following cases as example:

   – $P = \alpha.P'$ then $[\![P]\!] = [\![\alpha.P']\!] = \alpha.[\![P']\!]$. There exists only one singleton configuration $\{e\} \in \alpha.[\![P']\!]$, where $\ell(e) = \alpha$. We have that $\alpha.P' \xrightarrow{\alpha} P'$ hence $[\![P']\!] \cong (\alpha.[\![P']\!])/e$, which follows from Equation 1.

   – $P = P'\backslash a$ then $[\![P]\!] = [\![P']\!] \uparrow a$. We have that $\forall \{e\} \in [\![P]\!]$, $a \notin \mathsf{subj}(\ell(e))$ and $\{e\} \in [\![P]\!]$.
   Consider the subcases where either $a \notin \ell(e)$ or $\tilde{\tau}_x(e)$ holds.
   Then $P'\backslash a \xrightarrow{\ell(e)} P''\backslash a$ and $P' \xrightarrow{\ell(e)} P''$. By induction $[\![P'']\!] \cong [\![P']\!]/e$. Hence we have to show that $[\![P'']\!] \uparrow a \cong ([\![P']\!] \uparrow X_a)/e$ that is $([\![P']\!]/e) \uparrow X_a \cong ([\![P']\!] \uparrow X_a)/e$ which follows from Equation 2.

A direct consequence of Lemma 2 is that, as in CCS, a process and its encoding can simulate each others reductions.

Theorem 2 (Operational Correspondance with the π calculus). *Let P a process and $[\![P]\!] = (E, C, \ell, \mathsf{P})$ its encoding.*

1. *$\forall P'$ such that $P \xrightarrow{\tau} P'$, $\exists \{e\} \in C$ closed such that $[\![P]\!]/e \cong [\![P']\!]$;*
2. *$\forall e \in E$, $\{e\} \in C$ closed, $\exists P'$ such that $P \xrightarrow{\tau} P'$ and $[\![P]\!]/e \cong [\![P']\!]$.*

4.4 Causality and Concurrency in the π-calculus

Proposition 5 extends to the π-calculus.

The disjoint causal sets in the π-calculus, capture the causality induced by the prefix operator (as in CCS), but also the causality induced by the restriction of private names. Consider as an example the process $(\overline{b}\langle a \rangle \mid \overline{c}\langle a \rangle \mid a)\backslash a$ with its encoding in rigid families in Fig. 3, where events are replaced by their labels. The disjoint causal set for the event labelled a consists of the events labelled $\overline{b}\langle a \rangle$ and $\overline{c}\langle a \rangle$, that is $\{\overline{b}\langle a \rangle, \overline{c}\langle a \rangle\} < a$. Indeed the event labelled a cannot appear by itself: either $\overline{b}\langle a \rangle$ or $\overline{c}\langle a \rangle$ precedes it. However in rigid families disjunctive causality [9] is not ambiguous: in every configuration the order on the events is fixed.

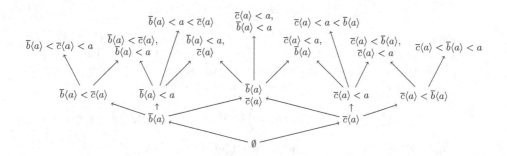

Fig. 3. $(\overline{b}\langle a \rangle \mid \overline{c}\langle a \rangle \mid a)\backslash a$ in rigid families

5   Advanced Criteria

In the previous section we have presented an interpretation that is compositional and sound both for CCS and the π-calculus. We propose in this section a few additional notions of correctness that our semantics enjoys. These criteria are of particular interest in the π-calculus, as previous causal models do not satisfy them. Hence we only formally prove them for the π-calculus, but they can be shown to hold for CCS as well.

5.1   Realisability

Criterion 3 (Realisable Interpretation). *Every labelled run should represent a possible execution.*

A labelled trace of an open system can be closed if the right context composes with the open term. In this case we say that the trace is realisable. This can be seen somehow as the dual to soundness. If soundness means that the model has *enough* labelled events and runs, we also require that it has not *too many* of them. In order to formalise this criterion, we need to use the notion of *context*. A context C in the π-calculus is defined as a process with a hole:

$$C[\cdot] := [\cdot] \ \| \ \alpha.C[\cdot] \ \| \ P|C[\cdot] \ \| \ C[\cdot]\backslash a$$

We do not know how to define a notion of context for rigid families in general, as it is not clear what is a rigid family with a hole. However, if a structure $\mathcal{F}$ has an operational meaning (i.e. $\exists P$ a process such that $\mathcal{F} = [\![P]\!]$) we can use a π-calculus context $C[\cdot]$ to define a new term $[\![C[P]]\!]$.

When analysing the reductions of a process in context, we need to know the contribution the process and the context have in the reduction. To this aim we associate to the context $C[\cdot]$ instantiated by a process P a projection morphism $\pi_{C,P} : [\![C[P]]\!] \to [\![P]\!]$ that can retrieve the parts of a configuration in $[\![C[P]]\!]$ that belong to $[\![P]\!]$.

Definition 20. *Let $C[\cdot]$ a π-calculus context, and P a process. The projection $\pi_{C,P} : [\![C[P]]\!] \to [\![P]\!]$ is inductively defined on the structure of C as follows:*

- *$\pi_{C,P} : [\![\alpha.C'[P]]\!] \to [\![P]\!]$ is defined as $\pi_{C,P}(e) = \pi_{C',P}(e)$;*
- *$\pi_{C,P} : [\![C'[P]|P']\!] \to [\![P]\!]$ is defined as $\pi_{C,P}(e) = \pi_{C',P}(\pi_1(e))$, where $\pi_1 : [\![C'[P]|P']\!] \to [\![C'[P]]\!]$ is the projection morphism defined by the product;*
- *$\pi_{C,P} : [\![C'[P]\backslash a]\!] \to [\![P]\!]$ defined as $\pi_{C,P}(e) = \pi_{C',P}(e)$.*

One can easily verify, by case analysis, that the projection $\pi_{C,P} : [\![C[P]]\!] \to [\![P]\!]$ is a morphism. We naturally extend $\pi_{C,P}$ to configurations.

We can now prove our first criterion: every partial labelled configuration can be "closed".

Theorem 3. $\forall x \in [\![P]\!]$, there exists C a context for $[\![P]\!]$ and $z \in [\![C[P]]\!]$ closed such that $\pi_{C,P}(z) = x$.

Proof (sketch). We proceed in several steps:

1. We show that for $x \in [\![P]\!]$ there exists a context $C_1 = \alpha_1 \cdots \alpha_n.[\cdot]$ and $x_1 \in [\![C_1[P]]\!]$ such that
   - $\pi_{C_1,P}(x_1) = x$ and
   - $\forall e \in x_1$, if $\tilde{\tau}_{x_1}(e)$ then $\forall b \in \mathsf{subj}(\ell_{x_1}(e))$ we have that $b \notin \mathsf{P} \implies \exists e_1 \in x_1, e_1 \in \mathsf{inst}^s_{x_1}(e)$ such that $\ell(e_1) = d'(b)$.
2. Define a *precontext* as a multiset of labels such that $b(a) \in \chi \implies \nexists c(a) \in \chi$. For $x_1 \in [\![P_1]\!]$ we define a precontext and a function $f : \chi \to x_1$ associating a label to any partial event. Intuitively for every open or partial event $e \in x_1$ we associate a label $\alpha \in \chi$ that "closes" the event. Let ς be a set of substitutions generated by x_1 *and* χ.

   We ask that given x_1, the precontext χ and the total and injective function $f : \chi \to x_1$ satisfy the following:
   - $\forall e \in x_1$ with $\ell_{x_1}(e) = (\overline{b}\langle a \rangle, c(d))$ then $\ell_\chi(e) = (\overline{b}\langle a \rangle, b(d))$;
   - $\forall e \in x_1$ with $\ell_{x_1}(e) = \overline{b}\langle a \rangle$ or $\ell_{x_1}(e) = b(d)$ there exists $\alpha \in \chi$, $\alpha = b'(a')$ or $\alpha = \overline{b'}\langle a' \rangle$ respectively, such that $f(\alpha) = e$ and $b'\varsigma = b$;
   - $\forall \alpha_1, \alpha_2 \in \chi$, such that $\alpha_1 = b(a)$, $a = \mathsf{subj}(\alpha_2) \iff f(\alpha_1) <_{x_1} f(\alpha_2)$ and there is $a \in \mathsf{obj}(\ell_{x_1}(f(\alpha_1)))$ private.

   In the above we have that $\ell_\chi(e) = \ell(e)\varsigma(e)$.
3. Let $x_1 \in [\![P_1]\!]$ with χ a precontext and $f : \chi \to x_1$ a total function as defined above. We construct a process P_2 from the precontext χ such that
   - there are no private names in P_2: $(\backslash a) \notin P_2$, for any name a;
   - $\alpha \in P_2 \iff \alpha \in \chi$;
   - $\alpha_1 \cdots \alpha_n.0 \in P_2 \iff \exists a \in \mathsf{obj}(\ell_{x_1}(e_1)), a \in \mathsf{P}$ and $e_n <_{x_1} e_i, a \in \mathsf{subj}(\ell_{x_1}(e_i))$ for $f(\alpha_1) = e_1$ and $f(\alpha_i) = e_i$, with $i \in \{2, n\}$.

   The context required by the theorem is $C = P_2 \mid [\cdot]$. The conditions above guarantee that we construct the context from the precontext and that the sequential operator are only used for an extrusion from P (and the instantiation in P_2) of a private name.
4. We show that $\exists x_2 \in [\![P_2]\!]$ and $g : x_2 \to x_1$ a total, injective function such that

$$\alpha \in P_2 \iff e \in x_2, \ell(e) = \alpha$$
$$g(e_2) = e_1 \iff f(\ell(e_2)) = e_1$$
$$e_2 <_{x_2} e'_2 \iff g(e_2) <_{x_1} g(e_1)$$

Intuitively, x_2 is the configuration that composes with x_1 in order to produce the closed configuration z. We have that x_2 is maximal in $[\![P_2]\!]$, hence it contains all labels in χ. The conditions above ensure that the function g keeps the correspondence between partial events in x_1 and their 'future' synchronisation partners in x_2.

5. Let $x_1 \in [\![P_1]\!]$ and $x_2 \in [\![P_2]\!]$ defined above. We have that $[\![P_1|P_2]\!] = ([\![P_1]\!] \times [\![P_2]\!]) \upharpoonright X$ and π_1, π_2 the projections. Denote $[\![P_1|P_2]\!] = (E,C)$. We show that $\exists z \in [\![P_1|P_2]\!]$ closed with $\pi_1(z) = x_1$ and such that $z \notin X$.

Using Theorem 3 we have the following straightforward corollary, that says that any context has to preserve and reflect the concurrency relation on events, and consequently the precedence between events. It follows from the preservation and reflection of order by the morphisms.

Corollary 1. $\forall x \in [\![P]\!]$ and $\forall C$ context for $[\![P]\!]$ such that $z \in [\![C[P]]\!]$ and $\pi_{C,P}(z) = x$ we have that $e_1 \lozenge_x e_2 \iff \pi_{C,P}^{-1}(e_1) \lozenge_z \pi_{C,P}^{-1}(e_2)$.

5.2   Structural Congruence

Criterion 4 (Denotational Interpretation) *The interpretation should be invariant for structural congruence.*

Corollary 1 says that the interpretation does not contain too much concurrency. We also would like to prove that the interpretation contains *enough* concurrency, or, dually, that it does not have *too much* causality: specifically we require that the restriction operator does not introduce too much causality. Surprisingly this is obtained by simply requiring that the semantics preserves structural congruence. The interesting case is the scope extension rule:

Theorem 4 (Preservation of Structural Congruence)

$$[\![(P)\backslash a \mid Q]\!] \cong [\![(P \mid Q)\backslash a]\!] \text{ if } a \notin \mathrm{fn}(Q)$$

Proof (sketch). We show that

$$\mathcal{F}_1 = ([\![P]\!] \upharpoonright X_1 \times [\![Q]\!]) \upharpoonright X_2 \cong ([\![P]\!] \times [\![Q]\!]) \upharpoonright X_3 = \mathcal{F}_2.$$

where $\mathcal{F}_1 = (E_1, C_1, \ell_1, \mathsf{P}_1)$ and $\mathcal{F}_2 = (E_2, C_2, \ell_2, \mathsf{P}_2)$. Let us denote $\pi_{1,P} : \mathcal{F}_1 \to [\![P]\!]$ and $\pi_{1,Q} : \mathcal{F}_1 \to [\![Q]\!]$ and similarly for $\pi_{2,P}$, $\pi_{2,Q}$. Define a bijection on events as follows $\iota(e_1) = e_2 \iff \pi_{1,P}(e_1) = \pi_{2,P}(e_2)$ and $\pi_{1,Q}(e_1) = \pi_{2,Q}(e_2)$. To show it is an isomorphism we show that $x_1 \in \mathcal{F}_1 \iff x_2 \in \mathcal{F}_2$, where $\iota(x_1) = x_2$.

To better understand the importance of this criterion, consider the causal models presented in [2]. In those models, there is a tight correspondence with the standard transition semantics of the π-calculus. In particular, the first output that extrudes a name has a different label than all subsequent outputs of that name, and moreover it precedes them in the causal relation. If we had made a similar choice here, in the process $P = (\bar{b}\langle a \rangle \mid \bar{c}\langle a \rangle)\backslash a$, we would have only configurations where one of the output would precede the other. By the way parallel composition is defined, this would imply that in $P \mid (b(x) \mid c(y))$ the τ transitions are causally related. However this process is structurally congruent to $(\bar{b}\langle a \rangle \mid \bar{c}\langle a \rangle \mid b(x) \mid c(y))\backslash a$, where the τ transitions are concurrent. Thus Theorem 4 would fail. Though the causal model of [2] is defined on transition systems one can represent the causality relation induced in a rigid family as in Fig. 4. We can see then that the two processes $(\bar{b}\langle a \rangle \mid \bar{c}\langle a \rangle)\backslash a \mid (b(x) \mid c(y))$ and $(\bar{b}\langle a \rangle \mid \bar{c}\langle a \rangle \mid b(x) \mid c(y))\backslash a$ have different interpretations.

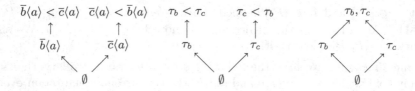

Fig. 4. $(\overline{b}\langle a\rangle \mid \overline{c}\langle a\rangle)\backslash a$, $(\overline{b}\langle a\rangle \mid \overline{c}\langle a\rangle)\backslash a\mid (b(x) \mid c(y))$ and $(\overline{b}\langle a\rangle \mid \overline{c}\langle a\rangle \mid b(x) \mid c(y))\backslash a$ in the model of [2]

5.3 Reversibility and Stability

In the operational setting, reversible variants for CCS [11] and the π-calculus [10] have been studied. We conjecture that rigid families are suitable models for the reversible π- calculus, but leave this as future work. Instead we show, intuitively, why rigid families are a good fit for reversibility.

Reversible calculi allow actions to backtrack in a different order then they appeared in the forward computation. The only constraint is that the causal order between events is respected: we cannot undo the cause before the effect. Configuration structures and rigid families are a suited for reversible calculi as the (rigid) inclusion between sets dictates the allowed forward and backward transitions. Then it is important that in the closed term we can backtrack on any path allowed by the forward computation. We can generalise the realisability criterion (Theorem 3) to the reversible setting: there exists a context that closes a configuration and that preserves all possible paths leading to it.

In previous works [22] this condition is called *stability* and is defined on the domain underlying the rigid families. Intuitively stability means that in any configuration one can deduce a partial order on events. Notably, the semantics proposed in [9] does not satisfy it.

Consider the process $(\overline{b}\langle a\rangle \mid \overline{c}\langle a\rangle \mid a(d))\backslash a$ with the rigid family depicted in Fig. 3. Its representation in [9] is represented in Fig. 5. In this process a is private and it cannot be used as a communication channel. The context has first to receive the name from one of its extruders: on channel b or on channel c. This type of *disjunctive* causality is what the

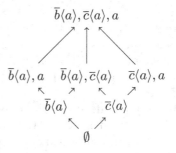

Fig. 5. The configuration structures of $(\overline{b}\langle a\rangle \mid \overline{c}\langle a\rangle \mid a(d))\backslash a$ in [9]

model of [9] depicts. In the top configuration the extruder of a is either $\overline{b}\langle a\rangle$ or $\overline{c}\langle a\rangle$. However in a closed term we can never express this type of disjunctive causality. We can either close the term with a process of the form $b(a').\overline{a'}\langle\rangle$ or $c(a'').\overline{a''}\langle\rangle$. Hence there is no context which can close such a configuration. In our model, instead, in any configuration that has disjunctive causality the order of occurrence between events is fixed.

6   Conclusions

We presented a novel causal semantics for concurrent computations. Importantly, we exhibit correction criteria that differentiate our model from others. We have stated that a correct causal model should be:

1. *Compositional:* we have defined a category of rigid families such that the encoding of a term is given by a categorical operation on the encoding of its subterms.
2. *Realisable:* each configuration in the interpretation of a process is realisable and the precedence in the closed system is the same as in the open one.
3. *Sound:* we showed an operational correspondence with the reduction semantics of CCS and the π-calculus.
4. *Denotational:* the rules of structural congruence are preserved by the causal semantics.

The first two correction criteria can be seen as *internal coherence* results: open traces can compose to form correct closed executions. *External coherence* criteria relate rigid families to other models for concurrency. In this paper we used the model to give interpretations to CCS and π-calculus processes that satisfy the third and fourth correction criteria. As future work, we plan to show a correspondence between reversible π-calculus [10] and its encoding in rigid families. The correction criteria have then to hold in a reversible setting as well. As we showed in Subsect. 5.3 this is particularly interesting for the realisability criterion.

In the π-calculus the input prefix plays the double role of instantiator and of structural predecessor. We can interpret in rigid families a calculus without prefix precedence [24] and we conjecture that the partial orders would then characterise the information flow rather then the temporal precedence.

Equivalence relations defined on previous causal model for CCS have a correspondence in reversible CCS [1]. We plan to show as future work that such equivalence relations can be defined on rigid families and possibly correspond to bisimulations in the reversible π-calculus.

References

1. Aubert, C., Cristescu, I.: Reversible barbed congruence on configuration structures. In: 8th ICE, Satellite Workshop of DisCoTec 2015 (2015)
2. Boreale, M., Sangiorgi, D.: A fully abstract semantics for causality in the π-calculus. Acta. Inf. **35**(5), 353–400 (1998)
3. Boudol, G., Castellani, I.: Flow models of distributed computations: Three equivalent semantics for CCS. Inf. Comput. **114**(2), 247–314 (1994)
4. Bruni, R., Melgratti, H., Montanari, U.: Event structure semantics for nominal calculi. In: Baier, C., Hermanns, H. (eds.) CONCUR 2006. LNCS, vol. 4137, pp. 295–309. Springer, Heidelberg (2006)
5. Busi, N., Gorrieri, R.: A petri net semantics for π-calculus. In: Lee, I., Smolka, S.A. (eds.) CONCUR 1995. LNCS, vol. 962, pp. 145–159. Springer, Heidelberg (1995)

6. Castellan, S., Hayman, J., Lasson, M., Winskel, G.: Strategies as concurrent processes. Electron. Notes. Theor. Comput. Sci. **308**, 87–107 (2014). Proceedings of the 30th Conference on the Mathematical Foundations of Programming Semantics (MFPS XXX)
7. Cattani, G.L., Sewell, P.: Models for name-passing processes: Interleaving and causal. Inf. Comput. **190**(2), 136–178 (2004)
8. Cattani, G.L., Winskel, G.: Presheaf models for CCS-like languages. Theor. Comput. Sci. **300**(1–3), 47–89 (2003)
9. Crafa, S., Varacca, D., Yoshida, N.: Event structure semantics of parallel extrusion in the π-calculus. In: Birkedal, L. (ed.) FOSSACS 2012. LNCS, vol. 7213, pp. 225–239. Springer, Heidelberg (2012)
10. Cristescu, I., Krivine, J., Varacca, D.: A compositional semantics for the reversible π-calculus. In: LICS 2013, pp. 388–397. IEEE Computer Society (2013)
11. Danos, V., Krivine, J.: Reversible communicating systems. In: Gardner, P., Yoshida, N. (eds.) CONCUR 2004. LNCS, vol. 3170, pp. 292–307. Springer, Heidelberg (2004)
12. Degano, P., Priami, C.: Non-interleaving semantics for mobile processes. Theor. Comput. Sci. **216**(1–2), 237–270 (1999)
13. Eberhart, C., Hirschowitz, T., Seiller, T.: An intensionally fully-abstract sheaf model for π. In: CALCO (2015)
14. van Glabbeek, R.J., Plotkin, G.D.: Configuration structures, event structures and petri nets. Theor. Comput. Sci. **410**(41), 4111–4159 (2009)
15. Hayman, J., Winskel, G.: Event structure semantics for security protocols (2013). submitted for publication
16. Krivine, J.: A verification algorithm for declarative concurrent programming. Technical Report 0606095, INRIA-Rocquencourt (2006)
17. Milner, R.: Communication and Concurrency. Prentice-Hall Inc, Upper Saddle River (1989)
18. Milner, R.: Communicating and Mobile Systems: The π-calculus. Cambridge University Press, Cambridge (1999)
19. Phillips, I., Ulidowski, I.: Reversibility and models for concurrency. Electron. Notes Theor. Comput. Sci. **192**(1), 93–108 (2007)
20. Sangiorgi, D., Walker, D.: PI-Calculus: A Theory of Mobile Processes. Cambridge University Press, New York (2001)
21. Sassone, V., Nielsen, M., Winskel, G.: Models for concurrency: towards a classification. Theor. Comput. Sci. **170**(1–2), 297–348 (1996)
22. Winskel, G.: Event structure semantics for CCS and related languages. In: Proceedings of the 9th Colloquium on Automata, Languages and Programming, pp. 561–576 (1982)
23. Winskel, G.: Event structures. In: Brauer, W., Reisig, W., Rozenberg, G. (eds.) Petri Nets: Applications and Relationships to Other Models of Concurrency. LNCS, vol. 255, pp. 325–392. Springer, Heidelberg (1987)
24. Yoshida, N.: Minimality and separation results on asynchronous mobile processes. Theor. Comput. Sci. **274**(1–2), 231–276 (2002). Ninth International Conference on Concurrency Theory 1998

Quotients of Unbounded Parallelism

Nils Erik Flick[(✉)]

Carl von Ossietzky Universität, 26111 Oldenburg, Germany
flick@informatik.uni-oldenburg.de

Abstract. This is a language-theoretic investigation into a situation where a server serves an unbounded number of requests, and handling a request requires a bounded number of (arbitrarily delayed) steps. From a description of the system in interleaving semantics, one endeavours to determine whether some sequence from a given regular language is possible. We model unbounded parallelism using the iterated shuffle operator, investigate quotients of the so-called simple shuffled languages with regular languages, and prove a sufficient condition for obtaining another simple shuffled language by that operation.

1 Introduction

Imagine a restaurant where each of unboundedly many clients orders a finite sequence of menu items, pays up and leaves. All clients are served in parallel (the kitchen is very large), and we would like to determine whether a sequence of events from some regular set might occur. When we model this scenario using formal languages, such questions are about language intersections and quotients.

The unbounded parallelism in our restaurant can be described as an iterated shuffle, as introduced in [14], which yields the possible interleavings of an unbounded number of words from a given language. This operation already produces non-context-free languages when applied to finite sets (see [11]). The sequences of events in the scenario constitute a language of the class $\mathcal{SHUF} = (\cup, \sqcup, \sqcup^{\sqcup})(\mathcal{FIN})$ of *simple shuffled languages* built up from the finite languages using union, shuffle and iterated shuffle. By demonstrating a close relationship between quotients with regular languages and Petri net reachability, we show that $\mathcal{SHUF}$ is not closed under such quotients (which would make answering the question easy

N.E. Flick—This work is supported by the German Research Foundation (DFG), grant GRK 1765 (Research Training Group – System Correctness under Adverse Conditions).

M. Leucker et al. (Eds.): ICTAC 2015, LNCS 9399, pp. 241–257, 2015.
DOI: 10.1007/978-3-319-25150-9_15

in all cases, provided the operation was efficiently computable), but exhibit a nontrivial subset which is closed under all regular quotients.

This paper is structured as follows: Sect. 2 introduces the basic notions used in the rest of the paper, in Sect. 3 we present our results in three subsections, the main result appearing in Subsect. 3.2 while the other two are concerned with relating our problem to Petri nets, Sect. 4 lists related work and Sect. 5 concludes with an outlook.

2   Shuffles and Quotients

In this section, we fix the notation and introduce formal languages with concatenation and shuffle. We then define quotients of languages, and other prerequisites. When they are first introduced and defined, notions appear in *italics*.

The symbol $-$ stands for set difference. A multiset over a finite set A (A-multiset) is a function $A \to \mathbb{N}$. A singleton is a multiset assigning 1 to some $a \in A$ and 0 to all other $a' \neq a \in A$. It may be written a. An empty multiset (of constant value 0) is also written 0. If s and t are multisets, let $s \leq t$ iff $\forall a \in A, s(a) \leq t(a)$. Multisets s, t are added component-wise, $\forall a \in A, (s+t)(a) = s(a) + t(a)$, can be scaled by natural numbers $k \in \mathbb{N}, \forall a \in A, (k \cdot s) \cdot (a) = k \cdot (s \cdot a)$, and t can be subtracted from s provided $s \geq t$, then $\forall a \in A, (s - t)(a) = s(a) - t(a)$. The set of all A-multisets is denoted by $A^{\oplus}$.

2.1   Operators for Constructing Languages

A *language* is a set $L \subseteq \Sigma^*$, where Σ^* is the free monoid generated by a finite alphabet Σ. The binary operation *concatenation* will be denoted by $\odot$ or $\cdot$, the empty word by ϵ. For $w \in \Sigma^*$ let $\|w\|$ denote the *length* of w. If $A \subseteq \Sigma^*$ then $|A|$ denotes the *cardinality* of A which also can be infinite. In particular, $\|\epsilon\| = 0$. For $A \subseteq \Sigma^*$ let $\|A\| = max\{\|w\| \mid w \in A\}$ the *norm* of A.

The *shuffle* operator $\sqcup\!\sqcup : \Sigma^* \times \Sigma^* \to 2^{\Sigma^*}$ has the following recursive definition: for all $a, b \in \Sigma$ and $v, w \in \Sigma^*$, $\epsilon \sqcup\!\sqcup w = w \sqcup\!\sqcup \epsilon = \{w\}$ and $aw \sqcup\!\sqcup bv = \{a\} \odot (w \sqcup\!\sqcup bv) \cup \{b\} \odot (aw \sqcup\!\sqcup v)$. It extends canonically to a binary operation on languages, $\sqcup\!\sqcup : 2^{\Sigma^*} \times 2^{\Sigma^*} \to 2^{\Sigma^*}$ where $A \sqcup\!\sqcup B := \bigcup_{w \in A, v \in B} w \sqcup\!\sqcup v$. As it is associative and commutative, we write $\bigsqcup_{i \in I} L_i$ for the shuffle of several languages indexed by I.

The *iterated shuffle* of a language A is defined in analogy to the Kleene star as $A^{\sqcup\!\sqcup} := \bigcup_{i \in \mathbb{N}} A^{\sqcup\!\sqcup(i)}$, where $A^{\sqcup\!\sqcup(0)} = \{\epsilon\}$ and $A^{\sqcup\!\sqcup(i+1)} = A \sqcup\!\sqcup A^{\sqcup\!\sqcup(i)}$. We also define $A^{\sqcup\!\sqcup(\leq n)}$ as $\bigcup_{i \in \{0 \ldots n\}} (A^{\sqcup\!\sqcup(i)})$. Likewise for concatenation, we define $A^{\odot(0)} = \{\epsilon\}$, $A^{\odot(i+1)} = A \odot A^{\odot(i)}$, $A^{\odot(\leq n)} = \bigcup_{i \in \{0 \ldots n\}} (A^{\odot(i)})$. Then the *Kleene star* $A^* := \bigcup_{i \in \mathbb{N}} A^{\odot(i)}$.

A *family of languages* (see [13]) is a non-empty class of languages that is closed under change of alphabets, excluding $\{\epsilon\}$ and $\emptyset$. The family of finite languages is denoted by $\mathcal{FIN}$. An *operation*, in this context, is a mapping of families to families that is monotonic with respect to inclusion. Many operations on families of languages can be derived as algebraic closures from $k-$ary operations on languages, for example, one can define closure under language concatenation ($\odot$). Generally, taking the closure of the family $\mathcal{F}$ under operations $\mathcal{O}_1 \ldots \mathcal{O}_n$, i.e. the least family containing $\mathcal{F}$ and closed under all of these operations is again an operation, denoted $(\mathcal{O}_1 \ldots \mathcal{O}_n)$. $(\mathcal{O}_1 \ldots \mathcal{O}_n)(\mathcal{F})$ is then the family of all languages obtained by applying any number of operations from $\mathcal{O}_1 \ldots \mathcal{O}_n$, including none, to languages from $\mathcal{F}$. $(\cup, \odot, {}^*)(\mathcal{F})$ is the *rational closure* of $(\mathcal{F})$, and $\mathcal{REG} = (\cup, \odot, {}^*)(\mathcal{FIN})$ is the family of regular languages.

The family $\mathcal{SHUF} := (\cup, \text{\cyrille{w}}, {}^{\text{w}})(\mathcal{FIN})$ does not offer the application of concatenation to infinite sets. Any language in $({}^{\text{w}})(\mathcal{FIN})$ (which is a proper subset of $(\text{\cyrille{w}}, {}^{\text{w}})(\mathcal{FIN})$, in turn a proper subset of $\mathcal{SHUF}$, cf. [16]) consists of n-fold shuffles of words taken from a finite set. Each of these definitions gives rise to a family of terms: a $(\text{\cyrille{w}}, {}^{\text{w}})(\mathcal{FIN})$- (resp. $\mathcal{SHUF}$)-term is either a finite language, or A^{w} or $A \text{\cyrille{w}} B$ (or $A \cup B$), where A and B are $(\text{\cyrille{w}}, {}^{\text{w}})(\mathcal{FIN})$- (resp. $\mathcal{SHUF}$-)terms (we also use the well-known regular expressions for regular languages).

We recall a normal form lemma from [13] or [14]:

Lemma 1 (Normal Form). *Any language $L \in \mathcal{SHUF}$ can be written as $\bigcup_{i \in I} L_i$ with $L_i = B_i \text{\cyrille{w}} A_i^{\text{w}}$, I a finite index set and all A_i and B_i finite; the sets B_i can be chosen to be singletons $\{w_i\}$.*

This result generalises to other commutative operators [18]. In our proofs, the subclass $(\text{\cyrille{w}}, {}^{\text{w}})(\mathcal{FIN})$ will play a role. Its normal form is similar:

Lemma 2 (Normal Form). *Any language $L \in (\text{\cyrille{w}}, {}^{\text{w}})(\mathcal{FIN})$ can be written as $B \text{\cyrille{w}} A^{\text{w}}$, where A and B are finite.*

Proof. If L is finite, set $A = \{\epsilon\}$ and $B = L$. Otherwise, as long as the $(\text{\cyrille{w}}, {}^{\text{w}})(\mathcal{FIN})$-term chosen for L is not already in normal form, there is at least one of the following situations (using associativity and commutativity of $\text{\cyrille{w}}$): a subterm of the form $(X^{\text{w}})^{\text{w}}$ is reducible by idempotence [14]. A subterm of the form $(X^{\text{w}} \text{\cyrille{w}} Y)^{\text{w}}$ or $(X \text{\cyrille{w}} Y^{\text{w}})^{\text{w}}$ can be rewritten into $(X^{\text{w}} \text{\cyrille{w}} Y^{\text{w}})$. Otherwise, a subterm can be transformed using the associativity of binary shuffle into a term with a shuffle of two finite sets, which is another finite set; or there is a shuffle of two sets $B_1^{\text{w}}, B_2^{\text{w}}$ with B_i finite, which can be combined to $(B_1 \cup B_2)^{\text{w}}$; or the whole term is in normal form. Assign to each term the multiset of the heights of subtrees under ${}^{\text{w}}$-nodes. Every reduction step decreases this multiset in the Dershowitz-Manna order [8] induced by $(\mathbb{N}, \leq)$. Hence a normal form is reached. □

Note that these normal forms are not unique, as for example any non-empty set A_i could be replaced by $A_i \cup \{w\}$, where $w \in A_i^{\text{w}} - A_i$.

Despite the rather limited modelling capability, the problem $v \in? w^{\text{w}}$ is NP-complete [23], but the word problem for a fixed language stays within NP even for much larger families built upon both shuffle and concatenation [2].

2.2   Language Quotients

This section introduces the left and right *quotients* of a language by another. The right quotient of $L \subseteq \Sigma^*$ by $M \subseteq \Sigma^*$ is the set

$$L/M := \{u \in \Sigma^* \mid \exists v \in M, uv \in L\},$$

while the left quotient is $L\backslash M := \{v \in \Sigma^* \mid \exists u \in M, uv \in L\}$. It easily follows from the definition that $L \subseteq (L/M) \odot M$, and that the left quotient obeys analogous laws because of the reversal rule $L/M = (L^r\backslash M^r)^r$, subject to reversal closure of the language class.

The operation $\pi : 2^{\Sigma^*} \rightarrow 2^{\Sigma^*}$, $\pi(L) := \{u \in \Sigma^* \mid uv \in L\}$ returns the least prefix-closed language containing L. The right (left) quotient contains only prefixes (suffixes) of the original language, viz. $L/M \subseteq \pi(L)$ and symmetrically.

When taken with respect to a single letter, the quotient is called *derivative*; its application to regular expressions is known as Brzozowski derivative [6]. In that case, we may write L/a instead of $L/\{a\}$. For $L \in \mathcal{SHUF}$, clearly $L^{rev} \in \mathcal{SHUF}$. For $L \in \mathcal{REG}$, L^{rev} may need a larger minimal DFA, but it is still regular. The following fact is well known:

Theorem 1. *If R is regular and Q is any language, then R/Q is regular. A regular language has only finitely many distinct left and right quotients.*

Any quotient R/Q is regular if R is regular, though it might not be effectively computable, depending on Q and its representation. This is due to the fact that the resulting language is still recognised by the same finite monoid, only the accepting state mapping changes (see e.g. Pin and Sakarovitch, [21]).

Forming the quotient with respect to a regular language is also an example of a rational transduction. In the form given by Nivat's theorem (see e.g. Berstel [3]), if $\Sigma' = \{a' \mid a \in \Sigma\}$ $(\Sigma \cap \Sigma' = \emptyset)$ and $L \subseteq \Sigma^*$, then L/R equals $g(h^{-1}(L) \cap \Sigma'^* R)$, with the homomorphisms $h : a \mapsto a, a' \mapsto a$, $g : a' \mapsto a, a \mapsto \epsilon$ for all $a \in \Sigma$. In Latteux [19] for example, more elaborate remarks of this kind are used to show a closure property of a family of languages under certain quotients. But this line of argumentation does not offer any obvious advantage for describing quotients of shuffle languages, as the shuffle operation generally does not mix well with transductions, so a different technique must be used there to describe the effect of a quotient with another language.

2.3 Rules for Quotients

The left and right quotient operations are linear in both arguments: they distribute over $\cup$ (Lemma 3). A product rule is satisfied when deriving the shuffle of two languages (Lemma 8). We recall several rules from [15], where the proofs can be found for the case of the left quotient. The versions adapted to the right quotient, as presented here, follow immediately by reversal. Let Σ be an alphabet, $A, B \subseteq \Sigma^*$, $a \in \Sigma$, $x \in \Sigma^*$. Then the following propositions hold:

Lemma 3. *For $\ominus \in \{/, \backslash\}$,*
$(A \cup B) \ominus C = (A \ominus C) \cup (B \ominus C)$ *and* $A \ominus (B \cup C) = (A \ominus B) \cup (A \ominus C)$.

Lemma 4. $A/(B \odot C) = (A/C)/B$.

Lemma 5. $(A \odot B)/C = A/(C/B) \cup A \odot (B/C)$.

Lemma 6. *For $\ominus \in \{/, \backslash\}$,* $A \ominus B^* = A \ominus \left(\bigcup_{i \in \mathbb{N}} B^{(\odot)i}\right) = \bigcup_{i \in \mathbb{N}} \left(A \ominus B^{(\odot)i}\right)$.

Lemma 7. *For $\ominus \in \{/, \backslash\}$,* $A^\sqcup \ominus x = A^\sqcup \sqcup \left(\bigcup_{i=0}^m (A^{(\sqcup)i}) \ominus x\right)$, $m \le \|x\|$.

Lemma 8. *For $\ominus \in \{/, \backslash\}$,* $(A \sqcup B) \ominus a = (A \ominus a) \sqcup B \cup A \sqcup (B \ominus a)$.

2.4   Orders and Transition Systems

We assume familiarity with the notions of *pre-ordered set* (a set with a reflexive and transitive binary relation on it), *well-quasi-order* (a pre-ordered set with the property that any infinite sequence contains an infinite ascending subsequence), and *transition system* (basically a set of *states* with a binary *transition* relation on it, which may carry supplementary structure such as transition labels or a start state). A *well-structured transition system* (see Finkel and Schnoebelen [10]) is a transition system $(S, \rightarrow, \dots)^1$ with a well-quasi-order $\sqsubseteq \; \subseteq S \times S$ satisfying upward-compatibility: $\forall s_1 \sqsubseteq t_1, \; \forall (s_1, s_2) \in \; \rightarrow, \; \exists t_2, \; t_1 \xrightarrow{*} t_2 \wedge s_2 \sqsubseteq t_2$ ($\xrightarrow{*}$ denoting the reflexive and transitive closure of $\rightarrow$).

2.5   Semilinear Sets

A *semilinear set* of dimension d is a finite union of sets of the form $\boldsymbol{x} + \bigcup_{j \in J} c_j \cdot \boldsymbol{y}_j$, where $\boldsymbol{x} \in \mathbb{Z}^d$, the index set J is finite and $\forall j \in J, c_j \in \mathbb{N} \wedge \boldsymbol{y}_j \in \mathbb{Z}^d$.

2.6   Petri Nets

A *labelled Petri net* (see for example Wimmel [24]) is a tuple $N = (P, T, \Sigma, W^-, W^+, \lambda)$ of two disjoint finite sets P and T of *places* and *transitions*, respectively, a non-empty finite alphabet Σ, pre- and post- *arc weight* functions $W^-, W^+ : T \rightarrow (P \rightarrow \mathbb{N})$ and a *labelling* function $\lambda : T \rightarrow \Sigma$. A *marking* of N is a function $M : P \rightarrow \mathbb{N}$, extended canonically to sequences of transitions. The marking M *fires* to the marking M' via the transition $t \in T$, written $M \xrightarrow{t}_N M'$, iff $M \geq W^-(t)$ and $M' = (M - W^-(t)) + W^+(t)$ *(marking equation)*. Firing is extended to sequences $s \in T^*$ by $M \xrightarrow{\epsilon}_N M'$ and $M \xrightarrow{st}_N M'$ iff $\exists M'', M \xrightarrow{s}_N M''$ and $M'' \xrightarrow{t}_N M$ ($t \in T$). If $M \xrightarrow{s}_N M'$, we also write Ms for M' since it is uniquely determined (otherwise, Ms is undefined). We say that M' is a *successor* of M. A marking M' is said to be reachable from M (implicitly in N) if (M, M') is in the reflexive and transitive closure of the successor relation associated to N. The set of all markings reachable in N from M_0 is denoted by $\mathcal{RS}(N, M_0)$. A transition t is *isolated* if $W^-(t) = W^+(t) = 0$.

3   Observations and Results

The objective of this section is to describe the regular quotients of simple shuffle languages, and to exhibit a subclass $\mathcal{S} \subseteq \mathcal{SHUF}$ that is effectively closed under quotients with regular languages. That is, if $L \in \mathcal{S}$ and $R \in \mathcal{REG}$, L/R (and by symmetry $L \backslash R$) is again in $\mathcal{S}$ and furthermore a $(\cup, \sqcup, ^{\sqcup})(\mathcal{FIN})$ -term for it can be effectively computed.

Note that for any pair (L, R) of languages, $\epsilon \in L/R$ iff $L \cap R \neq \emptyset$. This motivates computing quotients to answer questions about language intersections.

---

1 The dots indicate that the transition system may have further structure, such as initial states or labels. This notation stems from [10].

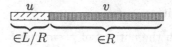

$$\underbrace{\quad u \quad}_{\in L/R} \underbrace{\quad v \quad}_{\in R}$$

Fig. 1. When is there a $\mathcal{SHUF}$-term for L/R?

The first subsection shows the membership problem of L/R to be as hard as Petri Net reachability, on the other hand it can itself be decided using Petri nets, as the third subsection will explain. The second subsection is concerned with a prefix-based representation of the quotients which will be used in the second subsection to derive a closure property for a subclass of $\mathcal{SHUF}$, and in the third subsection to represent quotients as Petri nets.

3.1 Simulating Petri Nets

In this subsection, we show that the membership problem of L/R is at least as hard as Petri Net reachability.

Definition 1 (Parikh Image, Letter Count). *Given a finite alphabet Σ and a word $w \in \Sigma$, $\psi(w)$ is the multiset defined recursively by $\psi(\epsilon)(a) = 0$ for all $a \in \Sigma$, and $\psi(au)(a') = \psi(u)(a') + (\text{if } a = a' \text{ then } 1 \text{ else } 0)$. It induces a mapping from languages to sets of multisets in the canonical way.*

There is an effective construction assigning to each Petri net N, a pair (M_0, M_f) of markings of N and a pair $(L_N, R_N) \in \mathcal{SHUF} \times \mathcal{REG}$ such that $\epsilon \in L_N/R_N \Leftrightarrow M_f \in \mathcal{RS}(N, M_0)$. If M is a Σ-multiset, $\psi^{-1}(M) = \bigsqcup_{a \in M} a^{M(a)}$ is the (finite) set of all words w with letter count $\psi(w) = M$.

Construction 1 (Reduction from Petri Net Reachability). *Given $N = (P, T, \Sigma, W^-, W^+, \lambda)$ and a marking M_0, let the alphabet Σ_N be the disjoint union $P + P' + \{a, b, c\}$, where $P' = \{p' \mid p \in P\}$. Let h be a homomorphism mapping any $p \in P$ to p'. Then define[2]*

$$A^-(t) = b \cdot h(\psi^{-1}(W^-(t))) \cdot a, \ A_N(t) = \psi^{-1}(W^+(t)) \cdot A^-(t),$$
$$A_N = \bigcup_{t \in T} A_N(t), \ B_N(M_0) = \psi^{-1}(M_0) \ \uplus \ c, L_N(M_0) = B_N(M_0) \uplus A_N^{\uplus},$$
$$R'_N = c(b\{pp' \mid p \in P\}a)^* \ and \ R_N(M_f) = \psi^{-1}(M_f) \cdot R'_N.$$

We will show that $M_f \in \mathcal{RS}(N, M_0) \Leftrightarrow \epsilon \in L_N(M_0) \backslash R_N(M_f)$, equivalently $\psi^{-1}(M_f) \subseteq L_N(M_0) \backslash R'_N$:

Lemma 9 (Correctness of Construction 1)

Proof. By definition, $\psi^{-1}(M_f) \subseteq L_N \backslash R'_N \Leftrightarrow \exists w \in L_N, v \in R'_N, \psi^{-1}(M_f)v = w$. By Lemma 4 and $R_N(M_f) = \psi^{-1}(M_f) \cdot R'_N$, we have that $\psi^{-1}(M_f) \subseteq L_N \backslash R'_N \Leftrightarrow \epsilon \subseteq L_N \backslash R_N$ (removing the remaining $\psi^{-1}(M_f)$ prefix).

---

[2] The order of the P' letters in $A^-(t)$ could just as well be fixed, but ψ^{-1} is required so the P letters can appear wherever they are needed.

For any $s = t_1 \cdot \ldots \cdot t_{\|s\|} \in T^*$, define $\mathrm{words}(s) = \psi^{-1}(Ms)cA'(t_{\|s\|}) \cdot \ldots \cdot A'(t_1)$, where $A'(t) = bh'(\psi^{-1}(W^-(t)))a$, h' being a homomorphism mapping each $p' \in P'$ to pp' (notice $\psi(A'(t))(b) = 1$).

Proceed by induction over n to show that if s is a transition sequence of length n, $M_0 s \in \mathcal{RS}(N, M_0) \Leftrightarrow \mathrm{words}(s) \subseteq L_N \Leftrightarrow \epsilon \in L_N \backslash \psi^{-1}(M_0 s)R'_N$.

If this common induction hypothesis holds for all s of length n and $t \in T$, then $M_0 st \in \mathcal{RS}(N, M_0) \Rightarrow \mathrm{words}(st) \subseteq L_N$: if $M_0 st$ is in $\mathcal{RS}(N, M_0)$, then so is $M_0 s$. Using the induction hypothesis, $\mathrm{words}(s) \subseteq L_N$. By definition of words, any word in $\mathrm{words}(st)$ can be written $xcdw$ with $d \in A'(t)$. This can only be achieved in L_N by shuffling $x'd' \in A_N(t)$ into some word $x''cd''w \in L_N$, where $x''d'' \in P^*$. Furthermore, $\psi(d'')$ must equal $W^-(t)$, $\psi(x'') = \psi(x) - W^+(t)$ by the definition of $A_N(t)$. If such an $x''cd''w$ is in L_N, then so is $x''d''cw$ because each letter of d'' is by definition of L_N issued either from a prefix $\psi^{-1}(W^+(t'))$ for some transition t', or from $B_N(M_0 s)$. In both cases, moving the letter to the left of c results again in a word of L_N. Hence $xcdw$ is in $L_N \amalg A_N \subseteq L_N$.

$\mathrm{words}(st) \subseteq L_N \Rightarrow M_0 st \in \mathcal{RS}(N, M_0)$: by induction hypothesis, $\mathrm{words}(s) \subseteq L_N$. For some $w \in \mathrm{words}(st)$ such that $w \in L_N$, proceed as in the proof of the converse, decomposing w as described above. The marking equation then proves this direction as well.

$\mathrm{words}(st) \subseteq L_N \Rightarrow \epsilon \in L_N \backslash (\psi^{-1}(M_0 st)R'_N)$: for any s, all $\mathrm{words}(s)$ are clearly in $\psi^{-1}(M_0 s)R'_N$.

$\epsilon \in L_N \backslash \psi^{-1}(M_0 st)R'_N \Rightarrow \mathrm{words}(st) \subseteq L_N$: If $xcw \in L_N \cap \psi^{-1}(M_0 s)R'_N$, then $\psi(x) = M_0 s$ by construction, and $\forall c', c'', \psi(c') + \psi(c'') = \psi(c) \Rightarrow c'xc''w \in L_N$ because the prefix xc is issued from a word of $B_N(M_0 s)$, and any reordering of that word is also in $B_N(M_0 s)$ by construction. Now $M_0 s \geq W^-(t)$ because t can fire from $M_0 s$. Therefore c'' may be any word from $\psi^{-1}(W^-(t))$, which means that $A(t)$ can be shuffled into the $c'xc''$ prefix to yield $d'xd''$, for any $d'' \in A'(t)$ and d' such that $\psi(d') = W^+(t) + Ms - W^-(t)$, since the letters may occur in arbitrary order by construction. But that is precisely the marking equation. □

An example of a Petri net[3] with a non-semilinear reachability set was developed in Hopcroft and Pansiot [12]. It has transition sequences that iteratively double a component of the marking.

Example 1 (Hopcroft and Pansiot's Example). Let $L_{HP} = plc \amalg \{qbp'a, pmbp'l'a,$ $mpbp'l'a, pmbl'p'a, mpbl'p'a, qllbq'm'a, lqlbq'm'a, llqbq'm'a, qllbm'q'a, lqlbm'q'a,$ $llqbm'q'a, npbq'a, pnbq'a\}^{\amalg}$ and $R_{HP} = c(b(pp' + qq' + ll' + mm' + nn')a)^*$.

Proposition 1 (Counterexample to Closure under Regular Quotients). $L_{HP}/R_{HP} \notin \mathcal{SHUF}$.

Proof. The Parikh image $\psi(L_{HP}/R'_{HP})$ is not semilinear, as it is by Lemma 9 subject to the same constraints as the reachability set of the corresponding net.

---

[3] Equivalently described in [12] as a "Vector Addition System with States" (VASS).

Instead, $\forall w \in L_{HP}/R'_{HP}$, $\psi(w)(p)=1 \Leftrightarrow \psi(w)(l) + \psi(w)(m) \leq 2^{\psi(w)(n)}$ and any such marking with $\psi(w)(p)=1$ is reachable [12], whereas $\mathcal{SHUF}$ languages have semilinear Parikh images, as can be ascertained from their normal form [17]. □

3.2 Quotients Representable as Finite Trees

The discussion in this section is based on the fact that all $\mathcal{SHUF}$ languages derived from L by quotients are built from the prefixes of a set of words that can be read off the $\mathcal{SHUF}$-term. As remarked in Subsect. 2.2, we have $L/M \subseteq \pi(L)$. We will attempt to control the $\mathcal{SHUF}$ languages that arise from the quotients. We have already discussed how they can fail to lie in $\mathcal{SHUF}$ in general and will now propose a sufficient condition. The subsection concludes with an example.

We assume that the language $L \in \mathcal{SHUF}$ is given in normal form $\bigcup_{i \in I} \{w_i\} \shuffle A_i^{\shuffle}$ with i and all A_i finite. For quotients by finite languages, we can use the rules in Subsect. 2.3.

Lemma 10 (Quotient by a Single Letter). *If $L \in \mathcal{SHUF}$, then $L/a \in \mathcal{SHUF}$ can be computed from a normal form of L.*

Proof. Direct consequence of Lemmas 8 and 7.   □

Lemma 11 (Quotient by $A \odot B$). *If $L/A, L/B \in \mathcal{SHUF}$ can be computed from a normal form of L for any $L \in \mathcal{SHUF}$, then so can $L/(A \odot B)$.*

Proof. Direct consequence of Lemma 4.   □

Lemma 12 (Quotient by $A \cup B$). *If $L/A, L/B \in \mathcal{SHUF}$ can be computed from a normal form of L for any $L \in \mathcal{SHUF}$, then so can $L/(A \cup B)$.*

Proof. Let $\bigcup_{i \in I} \{w_i\} \shuffle A_i^{\shuffle}$ be the normal form of $L \in \mathcal{SHUF}$ and $R = \bigcup_{i \in I} L_i$ a finite union of languages. By distributivity of quotient over union (Lemma 3), $L/Q = \bigcup_{i \in I, j \in \{1,\dots,k\}} (\{w_i\} \shuffle A_i^{\shuffle})/L_j$.   □

The quotient L/w by a single word is computed by induction over $\|w\|$ using Lemma 4. The base case $w = \epsilon$ is trivial. The induction step performs a quotient by a single letter and applies Lemma 1 to put the result into normal form for the next step. It follows that $\mathcal{SHUF}$ is closed under quotients with finite languages.

For the remainder of the section, we will mostly consider $(\shuffle,^{\shuffle})(\mathcal{FIN})$ rather than $\mathcal{SHUF}$ languages because the first step in computing any quotient of a $\mathcal{SHUF}$ language will always be an application of distributivity (Lemma 3).

Observation 1 (Quotient Consists of Shuffles of Prefixes). *If $L = B \shuffle A^{\shuffle} \in (\shuffle,^{\shuffle})(\mathcal{FIN})$ in normal form, then for any language $Q \subseteq \Sigma^*$, $L/Q = \bigcup_{k \in K} L_k$ with $L_k = B'_k \shuffle A'^{\shuffle}_k \shuffle A^{\shuffle}$ where $B'_k \subseteq \pi(B) \cup \pi(A)$, $A'_k \subseteq \pi(A)$, for some countable index set K.*

Proof. If the proposition holds word-wise for $L/\{w\}$ for $w \in Q$, then with K' countable the proposition is seen to hold. Induction over $\|w\|$. For the induction basis note that $B \shuffle A^{\shuffle}$ has the required form and quotient by ϵ is the identity.

For $w = ua$, $(B' \shuffle A' \shuffle A^\omega)/w = ((B' \shuffle A' \shuffle A^\omega)/a)/u.(B' \shuffle A' \shuffle A^\omega)/a) = (B' \shuffle (A'/a) \cup (B'/a) \shuffle A') \shuffle A^\omega \cup A' \shuffle B' \shuffle (A^\omega/a)$, applying Lemma 8 twice, which equals $(B' \shuffle (A'/a) \cup (B'/a) \shuffle A') \shuffle A^\omega \cup A' \shuffle B' \shuffle A^\omega \shuffle \bigcup_{z \in A}(\{z\}/a)$. Factoring the A^ω out by Lemma 3 yields a finite union of terms of the form $(B' \shuffle A' \shuffle A^\omega)/a)$ with $B' \subseteq \pi(B) \cup \pi(A)$ and $A' \subseteq \pi(A)$ again. $\qquad\square$

The set K need by no means be finite, for example $L = \{ab\}^\omega/\{b^{2^k} \mid k \in \mathbb{N}\}$ is not in $\mathcal{SHUF}$: the letter count of any word $w \in L$ is $\psi(w) = (n, n - 2^k)$ for some $n, k \in \mathbb{N}$. The assumption $L \in \mathcal{SHUF}$ would entail that within the normal form $L = \bigcup_{i \in I} L_i$ at least one L_i has, for arbitrary $k \in \mathbb{N}$, an infinite intersection $L \cap \{a^n b^{n-2^k} \mid n \in \mathbb{N}\}$. But this means that the count of a's can be increased without bound while leaving the count of b's unchanged, choosing ever increasing powers of 2 for the removed part. This in turn entails that $a^j \in A_i'$ for some $j \in \mathbb{N}$, which forces words with letter count $(n + j, n - 2^{k'})$ to be in L, contradicting $\exists k, \psi(w) = (n, n - 2^k)$ when $j \leq 2^{k'}$. Since this holds for any $k \in \mathbb{N}$, the contradiction is guaranteed to happen and $L \notin \mathcal{SHUF}$.

Observation 1 is too weak for quotients of $\mathcal{SHUF}$ languages by regular languages. These do not, in general, have such a representation for finite K because it is possible that some words that appear as prefixes of words from A can only be iterated together in certain combinations. Take, for example, $\{abc\}^\omega/(cc)^*$. This yields $\{abab, aabb, abc\}^\omega$, a language that does not even possess a normal form involving only prefixes of abc. Instead it becomes necessary to "pre-shuffle" certain prefixes. We shall have to keep this complication in mind.

Definition 2 (Standard Representation). *Given a normal form $\bigcup_{i \in I}\{w_i\} \shuffle A_i^\omega$ of a language $L \in \mathcal{SHUF}$, let $B = \bigcup_{i \in I} B_i$ and $A = \bigcup_{i \in I} A_i$. Let $W = \pi(A) \cup \pi(B) - \{\epsilon\}$. A W standard representation (W-SR) of a $(\shuffle,\omega)(\mathcal{FIN})$ normal form $C \shuffle D^\omega$ where C is a finite shuffle of $\pi(A) \cup \pi(B)$ words and D is a finite shuffle of $\pi(A)$ words is a pair $v = (\boldsymbol{x}_v, \boldsymbol{Y}_v) = (\boldsymbol{x}_v, \{(\boldsymbol{y}_j)_v\}_{1 \leq j \leq n})$ of a W-multiset and a set of n W-multisets such that*

$$C = \bigsqcup_{w \in W} w^{\shuffle(x(w))} \qquad D = \bigsqcup_{j \in \{1,\dots,n\}} \left(\bigsqcup_{w \in W} w^{\shuffle(y_j(w))} \right)^\omega$$

Thus the multiset $\boldsymbol{x}$ stands for the shuffle of the words of C, which are then shuffled with an iterated shuffle of $\boldsymbol{y}_j$-fold shuffles of words of D. Let $L(v)$ denote the language $C \shuffle D^\omega$ defined in this way by $v = (\boldsymbol{x}, \{\boldsymbol{y}_j\}_{1 \leq j \leq n})$.

We write $\boldsymbol{e}_w$ for a singleton ($\boldsymbol{e}_w(w) = 1$ and $\boldsymbol{e}_w(w') = 0$ if $w' \neq w$). For example, when $W = \pi(\{abcd\})$, $L = a \shuffle \{abcd, abc\}^\omega$ would be represented as $(\boldsymbol{e}_a, \{\boldsymbol{e}_{abcd}, \boldsymbol{e}_{abc}\})$ while $L = a \shuffle ab \shuffle ab \shuffle \{abcd, abc, abc \shuffle abc \shuffle ab\}^\omega$ could then be represented as $(\boldsymbol{e}_a, \{\boldsymbol{e}_{abcd}, \boldsymbol{e}_{abc}, 2 \cdot \boldsymbol{e}_{abc} + \boldsymbol{e}_{ab}\})$. The W-SR of a language is in general not unique. In the following, when W is not specified, it is implicitly given as some prefix-closed finite language not containing the empty word.

Definition 3 (SR Comparison). *For any given W, define $\sqsubseteq$ to be the binary relation on W-SRs such that $v \sqsubseteq v'$ holds whenever the $\boldsymbol{x}$ part of v does not exceed* *that of v' in any component, and the semilinear set $\left\{\sum_{j\in J} c_j \cdot \boldsymbol{y}_j \middle| \forall j \in J, c_j \in \mathbb{N}\right\}$ of v is contained in that of v'.*

The relation $\sqsubseteq$ does not imply language inclusion. Instead, $v \sqsubseteq w$ implies that another language with a standard representation can be shuffled with v to obtain w. Let $V(W)$ be the set of all standard representations for fixed W:

Lemma 13 (Meaning of $\sqsubseteq$). $v \sqsubseteq v' \Rightarrow \exists u \in V, L(v) \amalg L(u) = L(v')$.

Proof. $L(v') = \bigsqcup_{w\in W} w^{\amalg(x_{v'}(w))} \amalg U_{v'}$ (constant plus unbounded part) $=$
$\bigsqcup_{w\in W} w^{\amalg(x_{v'}(w)-x_v(w))} \amalg \bigsqcup_{w\in W} w^{\amalg(x_v(w))} \amalg U_{v'} \amalg U_v$ (permissible because the semilinear set inclusion implies that the shuffles that can be obtained from the $\boldsymbol{y}_v$ can also be obtained from the $\boldsymbol{y}_{v'}$, hence $U_{v'} = U_v \amalg U_{v'}$)
$= \left(\bigsqcup_{w\in W} w^{\amalg(x_{v'}(w)-x_v(w))} \amalg U_{v'}\right) \amalg L(v)$.
The term in parentheses is a $(\amalg,^{\amalg})(\mathcal{FIN})$ language normal form built from the same words, of standard representation u. $\qquad\square$

Let the language $L(F)$ of a transition system $F = (X, \dashrightarrow)$ whose nodes are labelled with standard representations be the union of the languages associated with the nodes. A *finitely rooted V-labelled forest* is a transition system $(X, \dashrightarrow, X_0, v)$ with a finite set $X_0 \subseteq X$ of designated *root* nodes without incoming transitions, a labelling function $v : X \to V$, no cycles $x \dashrightarrow^* x$, no multiple incoming transitions for any $x \in X$. The forest is *finite* if X is, and it is a *tree* if it has only one root. *Tip nodes* of F are nodes x such that $\neg\exists y \in X, (x,y) \in\dashrightarrow$.

Given a language $L \in \mathcal{SHUF}$ in normal form $\bigcup_{i\in I} L_i, L_i = \{w_i\}\amalg A_i^{\amalg}$, let F_0 be the finitely rooted forest consisting of one node n_i for each L_i, and no transitions. Suppose for some other language Q, L/Q is in $\mathcal{SHUF}$ and furthermore standard representations $\{q_j\}_{j\in J}$ can be computed such that $L_i/Q = \bigcup_{j\in J} B_j \amalg A_j^{\amalg} = \bigcup_{j\in J} L(q_j)$ in normal form.

We construct a sequence of finite finitely rooted forests as follows: let $F_{k+1}(V)$ be obtained by forming the quotient $L(n)/Q$ of each language of a tip node n and extending F_k by $|J|$ fresh nodes labelled with the standard representations $\{q_j\}_{j\in J}$ of the normal form of $L(n)/Q$, and one transition from n to each q_j. Unions of forests area defined component-wise. Since new nodes in F_{k+1} are always children of nodes in F_k, $\bigcup_{k\in\mathbb{N}} F_k$ is still a union of finitely many trees.

Lemma 14 (The k-th Forest Represents a Quotient). $L(F_k)$ *is equal to* $L/Q^{\odot(\leq k)}$.

Proof. By induction over k that the union of the languages of the tip nodes always represents $L/Q^{\odot(k)}$. The induction step transforms this into $(L/Q^{\odot(k)})/Q = L/Q^{\odot(k+1)}$. By definition, $\bigcup_{0\leq j\leq k} L/Q^{\odot(j)} = L/Q^{\odot(\leq k)}$. $\qquad\square$

Lemma 15 (The Forests Represent a Quotient in the Limit) $\bigcup_{k\in\mathbb{N}} L(F_k) = L/Q^*$

Proof. Follows from Lemma 14, since $L/Q^* = \bigcup_{k\in\mathbb{N}} L/Q^{\odot(\leq k)}$ by definition of Q^* and the fact that a variant of Lemma 3 holds for countable unions. $\qquad\square$

If the sequence $\{F_k\}_{k\in\mathbb{N}}$ reached a fixed point, the resulting language would obviously be in $\mathcal{SHUF}$. However, that does not usually happen: repeated quotients of $\{abc\}^{\sqcup}$ by cc result in a new language every time. We will have to take a closer look at the standard representations along each branch of the forest and exploit the recurring pattern to finish after a finite number of operations.

We define another sequence of finite finitely rooted forests, starting from $G_0 = F_0$. After extending F_k by the nodes labelled with $\{q_j\}_{j\in J}$, we check for each q_j whether there is some node x on the path from a root node to the new node n_j such that $l(x) \sqsubseteq q_j$. In that case the label of n_j is changed to $accel(l(x), q_j)$.

Definition 4 (SR Acceleration). *For $v, v' \in V$ with $v \sqsubseteq v'$, let $accel(v, v')$ be defined as $(\boldsymbol{x}_v, \boldsymbol{Y}_{v'} \cup \{\boldsymbol{y}'\})$ with $\boldsymbol{y}' = \boldsymbol{x}_{v'} - \boldsymbol{x}_v$ (component-wise).*

Next, we introduce a restriction on the language L which leads to quotients with any regular language R being in SHUF again.

Definition 5 (Independent Prefix Property). *A language $L \in (\sqcup,^{\sqcup})(\mathcal{FIN})$ has the independent prefix property if there is a W-standard representation $v = (\boldsymbol{x}, \boldsymbol{Y})$ for some W, $L = L(v)$, such that $\forall w \in W, \exists n_w \in \mathbb{N}, n_w \cdot e_w \in \boldsymbol{Y}$, and v is said to exhibit L's independent prefix property. A language $L \in \mathcal{SHUF}$, in normal form $\bigcup_{i\in I} L_i$, has the independent prefix property if every L_i does.*

This restriction (a multiple of each singleton corresponding to a prefix of an A word being present in Y) may appear somewhat arbitrary or contrived, but in fact it means that our system description counts every prefix of a word of W modulo a fixed number only (which may be chosen arbitrarily large).

Lemma 16 (Well-Quasi-Ordering Nodes Along Paths). *If $L \in (\sqcup,^{\sqcup})$ $(\mathcal{FIN})$ has the independent prefix property and v is a SR exhibiting it, then the labels of nodes along each path in $F_k(v)$ and $G_k(v)$ are well-quasi-ordered by $\sqsubseteq$.*

Proof. Since the semilinear sets along each path already form an ascending chain by Definition 4 in the case of G_k and because they are constant in the case of F_k, it is sufficient to look at the $\boldsymbol{x}$ components of the node labels. The Lemma of Dickson states that every $\mathbb{N}^n$, $n \in \mathbb{N}$ is well-quasi-ordered by the usual product order. $\qquad\qquad\square$

Lemma 17 (Finite Number of Acceleration Steps). *If L has the independent prefix property and $v = (\boldsymbol{x}, \boldsymbol{Y})$ is a normal form SR for L such that $\forall w \in W, \exists n_w \in \mathbb{N}, n_w \cdot e_w \in \boldsymbol{Y}$, then the $\boldsymbol{Y}$ components of the node labels along each path of $G_k(v)$ are well-quasi-ordered by set inclusion.*

Proof. Along each branch of $G_k(v)$, we will keep track of these supplementary pieces of information: the *density* $\delta \in \mathbb{Q}$, the *offset* $\boldsymbol{z} \in W^{\oplus}$ and a finite set $\Xi \subseteq W^{\oplus}$. Partition $\boldsymbol{Y}$ initially into $\boldsymbol{Y}_o + \boldsymbol{Y}_n$, where $\boldsymbol{Y}_n$ contains one multiple $n_w \cdot e_w$ of each e_w, $w \in W$ (defining a W-multiset, $\boldsymbol{n}(w) := n_w$). The initial computation of $\boldsymbol{z}$ and δ at the root nodes is as follows: initialise $\Xi := \emptyset$ and $\delta := 1/\prod_w n_w$ and $\boldsymbol{z}$ as the node label's $\boldsymbol{x}$ component, then iterate over the elements $\boldsymbol{y}_o$ of $\boldsymbol{Y}_o$, distinguishing three cases:

1. $y_o \geq z$ and $y_o = \sum_{y \in Y_n} c_y \cdot y$ or $y_o = z + \sum_{y \in Y_n} c_y \cdot y$ for some numbers $\{c_y\}_{y \in Y_n} \in \mathbb{N}$.
2. $y_o \geq z$, but case 1 does not hold
3. otherwise.

In the first case, do not add y_o to Y since it is superfluous and contributes nothing to the language. In the second or third case, assign $z := z + y_o$ and update $\delta := 2\delta$ and $\Xi := \Xi \cup \sum_w ((y_o - z)(w) \bmod n_w) e_w \cup \sum_w (y_o(w) \bmod n_w) e_w$. For the second case, this guarantees that the updated value Ξ is a proper superset of its previous value, because otherwise the extra conditions of the first case would hold.

At every node of $G_k(v)$, recompute δ, z and Ξ in the same way with the new members of Y as compared to its predecessor in the forest (the Y component of the node label may only accumulate more elements along the path).

This construction works because δ underapproximates the fraction of those multisets inside $z + \mathrm{BOX}(n)$ where $\mathrm{BOX}(n) = \{p \in W^\oplus \mid p \leq n\}$ (n being the multiset $\sum_{w \in W} n_w \cdot e_w$) which can be obtained as sums of x and some positive combination of multisets of Y. The underapproximation of this fraction by δ remains valid when translating $\mathrm{BOX}(n)$ by adding an arbitrary multiset. The number δ can only increase or stagnate, and only increase a number of times bounded by a global constant depending on the numbers n_w because of the way it is updated. Together, δ and z impose a well-quasi-order on the sets Y, compatible with the semantics of Y (equality implying language equality).   □

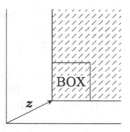

Fig. 2. Illustration of Lemma 17.

Intuitively, the space above a certain multiset z (which itself varies along the path) fills up with ever more grid points, and the filling of the space is in discrete steps since new copies of the Y_n grid are disjointly added whenever the new y_o is not a grid point. For each density of grid points, the well-quasi-ordering property ensures that at some point, the process stops or the density is increased, which can happen only a number of times that is bounded by the numbers n_w.

Lemma 18 (The Sequence $\{G_k\}_{k \in \mathbb{N}}$ Stabilises). *If L has the independent prefix property and $v = (x, Y)$ is a normal form SR for L of the required form, then there is a number $k \in \mathbb{N}$ such that $G_k(v) = G_{k+1}(v) =: G(v)$.*

Proof. Proof by contradiction, assuming that the union of all forests in G contains an infinite tree. That tree, by Kőnig's lemma, contains either an infinite branching or an infinite path. Branching is finite by the assumption on Q, and an infinite path cannot occur because the number of acceleration steps encountered along each path is finite by Lemma 17. □

Theorem 2. *The class $S \subseteq SHUF$ with the independent prefix property is closed under quotients with regular languages.*

Proof. By structural induction over a regular expression for R. The base case, single letters, holds by Lemma 10. The cases of concatenation and union hold by Lemmas 11 and 12, where the independent prefix property is seen to be preserved. The case of Kleene star is handled by constructing G. □

It is of course possible and advisable to keep the same set W for all SRs encountered in the recursive algorithm suggested by our proof. This allows the whole process, as in Example 2, to be represented in a single forest, and prevents the size of W from increasing due to prefix combinations being added.

Note that while closure under regular quotients fails to hold in general, the construction still works for many examples even without L possessing the independent prefix property. However, the computation is not guaranteed to terminate under these circumstances. The following example should be helpful.

Example 2 $(\{abc, a^2c, ab, a^4\}^{\sqcup\!\!\!\!\sqcup}\backslash(ab(c + ab)^*(cc))^*)\ (\in S))$.

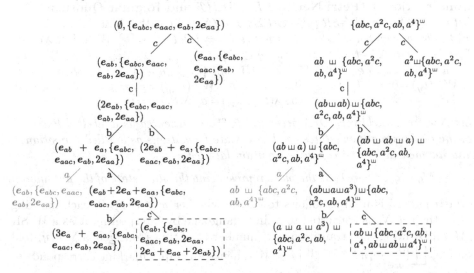

Here, $W = \{abc, aac, ab, aa, a\}$; we have represented node labels on the first tree and $(\sqcup\!\!\!\!\sqcup, {}^{\sqcup\!\!\!\!\sqcup})(FIN)$-expressions for the corresponding languages on the second one. The label of the top node is $v = (\emptyset, \{e_{abc}, e_{aac}, e_{ab}, 2e_{aa}\})$: multiples of e_a could be added to underline the independent prefix property, but since $2e_{aa}$ is obviously equivalent to $4e_a$, we did not include these redundant elements in $\boldsymbol{Y}$. Tip nodes that are not expanded further are repetitions (gray). Some nodes have not yet been fully expanded (the complete example is somewhat larger than the prefix shown, but, according to our result, finite). The node in the dashed box is the result of an acceleration (after two iterations of the inner Kleene star Y was augmented by $ab \shuffle ab \shuffle a^4$, represented as $2e_{ab} + 2e_a + e_{aa}$).

3.3   Any Quotient as a Petri Net

The prefix idea from the previous subsection also allows to answer the emptiness problem of $L \cap R$ by Petri nets.

Construction 2 (Synchronised Product of Petri Nets). *If* $N_i = (P_i, T_i,$ $\Sigma_i, W_i^-, W_i^+, \lambda_i)$ *are Petri nets* ($i \in \{1, 2\}$), *then* $N_1 \otimes N_2$ *is defined as* $(P, T,$ $\Sigma, W^-, W^+, \lambda)$ *with* $P = P_1 + P_2$ *(disjoint union)*, $\Sigma = \Sigma_1 \cup \Sigma_2$ *(in general not disjoint)*, $T = \{(t_1, t_2) \mid t_1 \in T_1, t_2 \in T_2, \lambda_1(t_1) = \lambda_2(t_2)\}$, $W^{\pm}((t_1, t_2)) = W_1^{\pm}(t_1) + W_2^{\pm}(t_2)$ *for* $\pm \in \{+, -\}$, $\lambda((t_1, t_2)) = \lambda_1(t_1) = \lambda_2(t_2)$.

Lemma 19 (Synchronisation). *If* $N_i = (P_i, T_i, \Sigma_i, W_i^-, W_i^+, \lambda_i)$ *are Petri nets* ($i \in \{1, 2\}$), *then for any transition sequence* s *and marking* M *in* $N_1 \otimes N_2$, Ms *exists iff* $M|_i \pi_i(s)$ *do and* $\lambda(\pi_1(s)) = \lambda(\pi_2(s))$, *and* $M|_i \pi_i(s) = Ms|_i$ ($M|_i$ *denoting the restriction of a marking to* P_i *and* $\pi_i(s)$ *being the homomorphism that projects each pair of transitions in* T *to the* i-th *component).*

Proof. As in Winskel [25], except that $\lambda(\pi_1(s)) = \lambda(\pi_2(s)) = \lambda(s)$.   □

Construction 3 (Petri Nets for $L \in \mathcal{SHUF}$ **and Regular Quotient).**
Let $v = (x, Y)$ *be a* W-SR *(alphabet* Σ*) and* $(X = \{x \in W^{\oplus} \mid \exists w \in \{x\} \cup Y, x \le w\})$. *The associated labelled Petri net is* $N_v = (P, T, \Sigma + \{*\}, W^-, W^+, \lambda)$ *with* $P = \{p_x \mid x \in X\}$, *transitions* $\{t_y \mid y \in Y\}$, $\{t_{e_{aw} + x, e_w + x} \mid e_{aw} + x \in W^{\oplus}, a \in \Sigma\}$ *and* $\{t_{e_a + x, x} \mid e_a + x \in W^{\oplus}, a \in \Sigma\}$ *and* $\{t_{e_a, \epsilon} \mid a \in W \cap \Sigma\}$, *where*
  $W^-(t_y) = 0$; $W^+(t_y) = p_y$; $W^-(t_{x,y}) = p_x$; $W^+(t_{x,y}) = p_y$ *if* $y \ne \epsilon$, *otherwise* 0,
  $\lambda(t_y) = *$, $\lambda(t_{e_{aw} + x, e_w + x}) = a$, $\lambda(t_{e_a + x, x}) = a$, $\lambda(t_{e_a, \epsilon}) = a$.

Let N_{Rv} *be an arbitrary DFA accepting* R^{rev}, *represented as a labelled Petri net in the obvious manner (states as places, state transitions as labelled transitions), and augmented by one isolated transition labelled* $*$.

Set $M_0 = q_0$, *where* q_0 *is the place representing the start state of the automaton.*

Construction 3 is used as follows to decide whether $w \in L(v)/R$: each marking M of the net N_v is converted to a finite language by interpreting it as a W-SR $(M, \emptyset)$. By firing each transition t_{y_i} a number n_i of times, one gets a $\sum n_i y_i$-fold shuffle, $\bigsqcup_{y_i \in Y_v} \{w^{\shuffle(n_i y_i(w))} \mid w \in W\}$, hence an empty marking corresponds to the language $\{\epsilon\}$. Note that the number of markings corresponding to each word length is finite. One must show that the transitions of N_v effect right quotients on the shuffle. The composition $N_v \otimes N_{Rv}$ then makes sure that those markings corresponding to right quotients with the words of R are the only ones reachable.

Proposition 2 (Correctness of Construction 3). $w \in L(v)/R \Rightarrow \exists M \in$ $\mathcal{RS}(N_v \otimes N_{Rv})$ *with* $L((M, \emptyset)) \ni w$ *and such an* M *can be computed:*

Let $N_v \otimes N_R = (P, T, \Sigma, W^-, W^+, \lambda)$. Now, $w \in L(v)/R \Rightarrow \exists r \in R, w \in L(v)/r$. Let V be the set of all W-multisets v such that the Parikh image of any word in $L((v, \emptyset))$ equals $\psi(w)$. This is finite and easily determined, since incrementing any component of the first component of the SR $(v, \emptyset)$ clearly increases the Parikh image of all words in the associated language by the same non-zero multiset.

If $r = r_1...r_{\|r\|}$, then by Lemma 4, $L(v)/r = ((L(v)/r_{\|r\|})/...)$. We must show that for all $a \in \Sigma$, at any point $i \in \{1...\|r\|\}$ the successor markings via transitions $t \in N_v$ with $\lambda(t) = a$ correspond to the L_i of a normal form of $L(v)/r_i$. This is easily checked using the definition of the language associated with a SR and the definition of N_v.

Any transition sequence $s(v) \in T^*$ such that $L((M_0 s, \emptyset)) = \{L(v)/r\}$ must end in a marking from V for the places of N_v and a marking of N_{Rv} representing an end state of the automaton. By Lemma 19, this is tantamount to a Petri net reachability query in $N_v \otimes N_{Rv}$ with the start marking $M_0 = q_0$.

The converse is similar: if M such that $w \in \psi(M)$ is reachable in N_v from 0, and an end state is reachable by the same transition sequence in N_{Rv}, then by induction over a transition sequence $\|s\|$ such that $M_0 s = M$, using the definition of N_v w is indeed in the quotient. □

Example 3 $((abc, aac)^{\sqcup}/(cc)^*$ as a Petri net). $W = \{abc, aac, ab, aa, a\}$. The hatched place is the end (and start) state of N_{Rv}. In this example it is not hard to ascertain that ϵ and $aabb$ are in the quotient.

4   Related Work

Languages defined with shuffle and concatenation have been researched for some time by several groups. Notable publications include Jantzen [13], Ésik and Bertol [9], Bloom and Ésik [5]. The operations of concatenation and shuffle have historically received attention because of their connection to concurrent processes and Petri net languages, cf. Jantzen [13]. Free terminal Petri net languages [22] are also closed under left letter, right letter and right word quotient, which reflects the firing dynamics of the net. Process algebras like [1] have parallel operators that also produce interleaved execution sequences and therefore shuffled languages. Recently, the expressive power of the shuffle product has been examined in the context of more restricted families, allowing combinations of a few operations to be applied to a family of elementary languages, for example

Berstel et al. [4], Castiglione and Restivo [7]. Recently, Berglund, Björklund and Högberg have researched the complexity of parsing shuffle-and-concatenation languages [2].

5 Conclusion and Outlook

Our goal was to obtain more insight into the structure of shuffled languages in terms of their quotients. This has arguably been achieved to some extent for the class of simple shuffled languages. We found that $\mathcal{SHUF}$ is not closed under quotients with regular languages (disproving a closure property conjectured in [15] and by Proposition 3 also undecidability of the emptiness of the intersection). On the other hand, there is a nontrivial subclass of $\mathcal{SHUF}$ which is effectively closed under quotients with regular languages, and which contains for instance the prefix closures of the languages $\{a_1 \cdot \ldots \cdot a_n\}^{\sqcup}$, $n \in \mathbb{N}$. We remark that the reduction (Lemma 9) from Petri net reachability to membership in L/R, $L \in \mathcal{SHUF}$, $R \in \mathcal{REG}$ is clearly in polynomial time. It is known that Petri net reachability is EXPSPACE hard [20], so the membership problem in L/R is, too. By Proposition 3, this is the worst case. The use of well-quasi-orderings on transition systems in the proof of Theorem 2 provides yet another example for the application of such techniques.

While the original problem (of deciding membership in the intersection of a simple shuffled and a regular language) may indeed appear contrived, unbounded shuffles do occur in real-world systems and it is desirable to learn more about their properties. We envisage further work on using the introductory setting (observations of a system with an unknown number of parallel processes) to study problems such as learning of shuffled languages from observations. Also, simple shuffled languages constitute a class of somewhat limited modelling capacity. A more general framework as well as classes of languages involving concatenation but still presenting the behaviour of our class $\mathcal{S}$ would also be of interest.

Acknowledgements. We would like to thank the anonymous reviewers for providing valuable comments.

References

1. Baeten, J., Bergstra, J., Klop, J.: An operational semantics for process algebra. Centrum voor Wiskunde en Informatica, Department of Computer Science (1985)
2. Berglund, M., Björklund, H., Högberg, J.: Recognizing shuffled languages. In: Dediu, A.-H., Inenaga, S., Martín-Vide, C. (eds.) LATA 2011. LNCS, vol. 6638, pp. 142–154. Springer, Heidelberg (2011)
3. Berstel, J.: Transductions and Context-Free Languages, vol. 4. Teubner, Stuttgart (1979)
4. Berstel, J., Boasson, L., Carton, O., Pin, J.E., Restivo, A.: The expressive power of the shuffle product. Inf. Comput. **208**(11), 1258–1272 (2010)
5. Bloom, S.L., Ésik, Z.: Axiomatizing shuffle and concatenation in languages. Inf. Comput. **139**(1), 62–91 (1997). http://www.sciencedirect.com/science/article/pii/S0890540197926651

6. Brzozowski, J.A.: Derivatives of regular expressions. J. ACM **11**(4), 481–494 (1964). http://doi.acm.org/10.1145/321239.321249
7. Castiglione, G., Restivo, A.: On the shuffle of star-free languages. Fundam. Inform. **116**(1–4), 35–44 (2012)
8. Dershowitz, N., Manna, Z.: Proving termination with multiset orderings. Comm. ACM **22**(8), 465–476 (1979)
9. Ésik, Z., Bertol, M.: Nonfinite axiomatizability of the equational theory of shuffle. In: Fülöp, Z., Gécseg, F. (eds.) ICALP 1995. LNCS, vol. 944, pp. 27–38. Springer, Heidelberg (1995). http://dx.doi.org/10.1007/3-540-60084-1_60
10. Finkel, A., Schnoebelen, P.: Well-structured transition systems everywhere!. Theor. Comput. Sci. **256**(1), 63–92 (2001)
11. Flick, N.E., Kudlek, M.: On a hierarchy of languages with catenation and shuffle. In: Yen, H.-C., Ibarra, O.H. (eds.) DLT 2012. LNCS, vol. 7410, pp. 452–458. Springer, Heidelberg (2012)
12. Hopcroft, J., Pansiot, J.J.: On the reachability problem for 5-dimensional vector addition systems. Theor. Comput. Sci. **8**(2), 135–159 (1979)
13. Jantzen, M.: The power of synchronizing operations on strings. TCS **14**(2), 127–154 (1981)
14. Jantzen, M.: Extending regular expressions with iterated shuffle. Theor. Comput. Sci. **38**, 223–247 (1985)
15. Kudlek, M., Flick, N.E.: A hierarchy of languages with catenation and shuffle. Fundam. Inform. **128**(1–2), 113–128 (2013)
16. Kudlek, M., Flick, N.E.: Properties of languages with catenation and shuffle. Fundam. Inform. **129**(1–2), 117–132 (2014)
17. Kudlek, M., Martín-Vide, C., Păun, G.: Toward a formal macroset theory. In: Calude, C.S., Pun, G., Rozenberg, G., Salomaa, A. (eds.) Multiset Processing. LNCS, vol. 2235, pp. 123–133. Springer, Heidelberg (2001)
18. Kuich, W., Salomaa, A.: Semirings, Automata, Languages. EATCS monographs on theoretical computer science. Springer, Heidelberg (1986)
19. Latteux, M., Leguy, B., Ratoandromanana, B.: The family of one-counter languages is closed under quotient. Acta Informatica **22**(5), 579–588 (1985)
20. Lipton, R.: The reachability problem requires exponential space. Research Report 62, Department of Computer Science, Yale University, New Haven, Connecticut (1976)
21. Pin, J.É., Sakarovitch, J.: Some operations and transductions that preserve rationality. In: Cremers, A.B., Kriegel, H.-P. (eds.) Theoretical Computer Science. LNCS, vol. 145, pp. 277–288. Springer, Heidelberg (1982)
22. Starke, P.H.: Free petri net languages. In: Winkowski, J. (ed.) Mathematical Foundations of Computer Science 1978. LNCS, vol. 64, pp. 506–515. Springer, Heidelberg (1978)
23. Warmuth, M.K., Haussler, D.: On the complexity of iterated shuffle. J. Comput. Syst. Sci. **28**(3), 345–358 (1984)
24. Wimmel, H.: Entscheidbarkeit bei Petri Netzen Überblick und Kompendium. Springer, Heidelberg (2008)
25. Winskel, G.: Petri nets, algebras, morphisms, and compositionality. Inf. Comput. **72**(3), 197–238 (1987)

Higher-Order Dynamics in Event Structures

David S. Karcher and Uwe Nestmann[✉]

Technische Universität Berlin, Berlin, Germany
{david.s.karcher,uwe.nestmann}@tu-berlin.de

Abstract. Event Structures (ESs) address the representation of direct relationships between individual events, usually capturing the notions of causality and conflict. Recently, Arbach et al. introduced the new Dynamic Causality Event Structure (DCES), in which some event may change the causal dependencies of other events, by adding or dropping causal predecessors. Interestingly, DCES turned out to be incomparable—concerning their expressive power—to van Glabbeek's and Plotkin's Event Structure for Resolvable Conflicts (RCES), up to then considered to be one of the most general ES models.

In this paper, also motivated by process modelling in the health care domain, we present a generalisation of the DCESs, by firstly allowing sets of events for modifying dependencies, and secondly by introducing higher-order dynamics. We show that the newly defined structure is strictly more expressive than the RCESs.

1 Introduction

Concurrency Model. Event Structures (ESs) usually address statically defined relationships that constrain the possible occurrences of events, typically represented as *causality* (for precedence) and *conflict* (for choice). An event is a single occurrence of an action; it cannot be repeated. ESs were first used to give semantics to Petri nets [9], then to process calculi [3,5], and recently to model quantum strategies and games [11]. The semantics of an ES itself is usually provided by the sets of traces compatible with the constraints, or by configuration-based sets of events, possibly in their partially-ordered variant (*posets*).

Dynamics. In [1], we (together with Arbach and Peters) introduced an extension of ESs, the so-called Dynamic Causality ESs (DCESs). In this extended model, the causal dependency between events is dynamised: *adders* are events that dynamically add dependencies to other events; *droppers* are events that remove dependencies for other events (see [2] for full details).

Let us draw an analogy to the notion of types in programming languages: there, one distinguishes between first-order types, consisting of simple so-called base types, and more involved higher-order types, structured by type operators that produce types of higher-order by building upon types of lower-order using a form of λ-abstraction. It is likewise natural to move from first-order dynamics

Supported by the DFG Research Training Group SOAMED.

M. Leucker et al. (Eds.): ICTAC 2015, LNCS 9399, pp. 258–271, 2015.
DOI: 10.1007/978-3-319-25150-9_16

to higher-order dynamics. In the following, we use the convenient terminology of *rules* to discuss this move. First-order rules represent the principle in DCES: the occurrence of one event adds or drops causes of other events. Second-order rules now allow to specify that an event may turn another event into an adder or dropper for a dependency; in other words, with such a rule, the occurrence of an event dynamically activates a new rule. Higher-order rules then consist of chains of activations of respectively lower-order rules.

Motivation. We have two reasons to study higher-order dynamics in ESs. One reason is the incomparability of DCES with the RCES model. The other reason is their usability for process modeling.

The incomparability of DCES and the RCES is essentially due to two aspects: (1) the rather simple difference on whether dynamics is driven by individual events (DCES) or sets of events (RCES); (2) the notion of *order sensitivity*—for a set of events different behaviour is possible depending on the order of occurrence. DCES are order-sensitive, as there could be both an adder and a dropper for the same dependency, so if both take place their order of occurrence determines whether the dependency is afterwards present or absent. In contrast, RCES are *not* order-sensitive, as the underlying transition relation can only see sets of previously occurred events to determine the possible next steps; thus, the order of occurrence of past events cannot be taken into account. From a mathematical point of view, the obvious question arises, whether there is a natural structure that subsumes both DCESs and RCESs, possibly an extension of one of the models. Analysing the incomparability by example, one observes that RCES allows us to model dependencies like "disabling of a disabling after an event", which cannot be represented by DCES, at least not with first-order dynamics. Higher-order dynamics, though, would be capable to express such dependencies, so it seems promising to generalise DCESs towards higher-order dynamics and set-based triggers to be able to simulate the behaviours of RCESs.

In an ongoing case study [6], we study models of a patient treatment process at the German Heart Institute Berlin using DCES. Thereby, the need for set-based adding and at least second-order dynamics was observed. The first was needed to model the following dependency: if disease d is diagnosed and a medicament m is prescribed this could cause trouble, therefore both together change the dependencies. The second was needed to describe a timing condition: if a drug is stopped after a diagnosis then a replacement drug should be given, but only if it happens in this order!

Overview. We generalise the DCESs approach in Sect. 3 first by allowing set-based dependency modifications (instead of only single events) and second by allowing higher-order dynamics. In contrast to the DCES model with four relations (causality, conflict, growing causality and shrinking) and its causal state function we present in this work the higher-order dynamic-causality ES (HDES), which consist of a set of events and a rule set; the latter governs the further behaviour and — as it is dynamically changing—is updated in each transition. In Sect. 4 we show that the newly defined higher-order dynamic-causality Event

Structures (HDESs) are strictly more expressive than the configuration structures in [8], and since the transition graph of an RCES is such a configuration structure, we found a natural structure subsuming RCESs and DCESs. In particular, we found that this subsumption only requires dynamics of third order.

Contributions. We define the HDES model as non-trivial generalisation of the DCES model [1]; the key ingredient here is the algorithm to compute rule updates to define transitions. We then show that any configuration structure (as defined in [8]) can be translated into a HDES, requiring only third-order dynamics.

Related Work. Hildebrandt et al. introduced a declarative process model, the so-called Dynamic Condition Response Graphs (DCR Graphs), cf. [4]. Those DCR Graphs also generalise ESs by splitting the causality/dependency relation into two, a condition and a response relation, further they allow events to occur more than once; one relating aspect is their dynamic inclusion and exclusion of events, but in contrast to our approach this is only first-order dyanmics.

In RCESs [7], there is some kind of dynamic causality, but it is there very implicit, so it can be deduced, but not easily be used for modelling purposes.

2   Technical Preliminaries

In this section we recap the definitions RCESs, and DCESs.

2.1   Event Structures for Resolvable Conflicts

Event Structures for Resolvable Conflicts (RCES) were introduced in [7] to generalise former types of ESs and to give semantics to general Petri Nets. They allow to model the case where a and b cannot occur together until c has taken place, i.e., initially a and b are in a conflict until the occurrence of c resolves this conflict. An RCES consists of a set of events and a witness relation $\vdash$ between sets of events. Here the witness relation also models conflicts between events. The behaviour is defined by a transition relation between sets of events that is derived from the witness relation.

Definition 1. *An* Event Structure for Resolvable Conflicts (RCES) *is a pair* $\rho = (E, \vdash)$ *, where E is a set of* events *and* $\vdash \subseteq (E)^2$ *is the* enabling relation.

In [7] several versions of configurations are defined. Here we consider only reachable and finite configurations. The intuition for the following transition definition is: if $X \subseteq Y$ then there is a transition from X to Y iff there exists a witness (w.r.t $\vdash$) for each subset of Y.

Definition 2. *Let $\rho = (E, \vdash)$ be an RCES and $X, Y \subseteq E$. Then:*

$$X \to_{\mathrm{rc}} Y \text{ if } X \subseteq Y \wedge (\forall Z \subseteq Y . \exists W \subseteq X . W \vdash Z).$$

The set of configurations *of ρ is defined as*

$$C(\rho) = \{X \subseteq E \mid \emptyset \to^*_{rc} X \wedge X \text{ is finite}\}$$

where $\to^*_{rc}$ *is the reflexive and transitive closure of* $\to_{rc}$.

As an example consider the RCES $\rho = (E, \vdash)$, where $E = \{a, b, c\}$, $\{b\} \vdash \{a, c\}$, and $\emptyset \vdash X$ iff $X \subseteq E$ and $X \neq \{a, c\}$. It models the above-described initial conflict between a and c that can be resolved by b. In Fig. 1, the transition graph is shown, i.e. the nodes are all reachable configurations of ρ and the directed edges represent $\to_{rc}$. Note, because of $\{a, c\} \subset \{a, b, c\}$ and $\emptyset \not\vdash \{a, c\}$, there is no transition from $\emptyset$ to $\{a, b, c\}$.

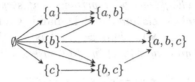

Fig. 1. Transition graphs of RCESs ρ with resolvable conflict.

We consider two RCESs as equivalent if they have the same transition graphs. Note that, since we consider only reachable configurations, the transition equivalence defined below is denoted as reachable transition equivalence in [7].

Definition 3. *Two RCESs* $\rho = (E, \to_{rc})$ *and* $\rho' = (E', \to'_{rc})$ *are transition equivalent, denoted by* $\rho \simeq_t \rho'$, *if* $E = E'$ *and* $\to_{rc} \cap (C(\rho))^2 = \to'_{rc} \cap (C(\rho'))^2$.

We adapt the notion of transition equivalence to arbitrary types of ESs with a transition relation. Let x and y be two arbitrary types of ESs on which a transition relation is defined. We denote the fact that x and y have the same transition graphs by $x \simeq_t y$. Note that for RCESs, transition equivalence is the most discriminating semantics studied in the literature. So we consider two RCESs as behaviourally equivalent if they have the same transition graphs.

2.2 Dynamic Causality

Dynamic Causality ESs were introduced in [1], the main idea is to enhance the Prime ES of Winskel [10] with two new relation $\triangleright$ and $\blacktriangleright$, which change the causal state of events, i.e. if $(c, m, t) \in \triangleright$, notated as $[c \to t] \triangleright m$, the causal predecessor c will be dropped from the target t after the occurrence of the modifier m. Similarly $(c, m, t) \in \blacktriangleright$, notated as $m \blacktriangleright [c \to t]$, the causal predecessor c will be added to the target t after the occurrence of the modifier m.

In Fig. 2 there is a small example for an Dynamic Causality ES with one dropping "[water → plant] ▷ rain" and one adding "pest infestation ▶ [pest control → harvest]". Note the dotted arrow to denote an initially absent dependency. In this

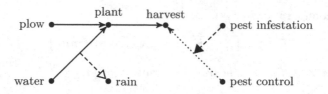

Fig. 2. An example DCES.

example the regular work-flow can be observed as the static dependencies, while the exceptional behaviour is modelled with the dynamic dependencies.

Here we shortly recap the definitions:

Definition 4. *Let $\pi = (E, \#, \rightarrow)$ be a prime event structure (PES) with conflict relation $\#$ and causality relation $\rightarrow$. A* Dynamic Causality Event Structure (DCES) *is a triple $\Delta = (\pi, \triangleright, \blacktriangleright)$, where $\triangleright \subseteq E^3$ is the shrinking causality relation, and $\blacktriangleright \subseteq E^3$ is the growing causality relation such that for all $e, e', e'' \in E$:*

1. $[e \rightarrow e''] \triangleright m \wedge \nexists m' \in E . \, m' \blacktriangleright [e \rightarrow e''] \Longrightarrow e \rightarrow e''$
2. $m \blacktriangleright [e \rightarrow e''] \wedge \nexists m' \in E . \, [e \rightarrow e''] \triangleright m' \Longrightarrow \neg(e \rightarrow e'')$
3. $m \blacktriangleright [e \rightarrow e''] \Longrightarrow \neg([e \rightarrow e''] \triangleright m)$

Sometimes, we use use the expanded form $(E, \#, \rightarrow, \triangleright, \blacktriangleright)$ instead of $(\pi, \triangleright, \blacktriangleright)$.

Condition 1 ensures that, if only a dropping of a causal dependency exists, then the dependency should exist initially. Similarly, Condition 2 ensures the initial absence of a dependency, if there is only an adder. If there are droppers and adders for the same causal dependency we do not specify whether this dependency is contained in $\rightarrow$, because the semantics depends on the order in which the droppers and adders occur. Condition 3 prevents that a modifier adds and drops the same cause for the same target.

The order of occurrence of droppers and adders determines the causes of an event. For example assume $a \blacktriangleright [c \rightarrow t]$ and $[c \rightarrow t] \triangleright d$, then after ad, t does not depend on c, whereas after da, c depends on t. Thus, configurations like $\{a, d\}$ are not expressive enough to represent the state of such a system. Therefore, in a DCES, a state is a pair of a configuration C and a causal state function cs, which computes the causal predecessors of an event, that are still needed.

Definition 5. *Let $\Delta = (E, \#, \rightarrow, \triangleright, \blacktriangleright)$ be a DCES.*

(1) The function mc $: \mathcal{P}(E) \times E \rightarrow \mathcal{P}(E)$ *denotes the* maximal causality *that an event can have after some history $C \subseteq E$, and is defined as*

$$\mathrm{mc}(C, e) = \{\, e' \in E \setminus C \mid e' \rightarrow e \vee \exists a \in C . \, a \blacktriangleright [e' \rightarrow e] \,\}.$$

(2) A state of Δ is a pair (C, cs) where cs $: E \setminus C \rightarrow \mathcal{P}(E \setminus C)$ *such that $C \subseteq E$ and $\mathrm{cs}(e) \subseteq \mathrm{mc}(C, e)$.*

(3) We denote cs *as causality state function, which shows for an event e that has not yet occurred, which events are still missing such that e becomes enabled.*

(4) An initial state of Δ is $S_0 = (\emptyset, \mathrm{cs_i})$, where $\mathrm{cs_i}(e) = \{e' \in E \mid e' \to e\}$.

Note that S_0, is the only state with an empty set of events; for other sets of events there can be multiple states. The behaviour of a DCES is defined by the transition relation on its reachable states with finite configurations.

Definition 6. *Let $\Delta = (E, \#, \to, \rhd, \blacktriangleright)$ be a DCES and $C, C' \subseteq E$.*
Then $(C, \mathrm{cs}) \to_d (C', \mathrm{cs}')$ if :

1. $C \subseteq C'$
2. $\forall e, e' \in C' . \; \neg(e \# e')$
3. $\forall e \in C' \setminus C . \; \mathrm{cs}(e) = \emptyset$
4. $\forall e, e' \in E \setminus C' . \; e' \in \mathrm{cs}(e) \setminus \mathrm{cs}'(e) \implies ([e' \to e] \rhd) \cap (C' \setminus C) \neq \emptyset$
5. $\forall e, e' \in E \setminus C' . \; ([e' \to e] \rhd) \cap (C' \setminus C) \neq \emptyset \implies e' \notin \mathrm{cs}'e$
6. $\forall e \in E \setminus C' . \; e' \in \mathrm{cs}'(e) \setminus \mathrm{cs}(e) \implies (\blacktriangleright [e' \to e]) \cap (C' \setminus C) \neq \emptyset$
7. $\forall e, e' \in E \setminus C' . \; (\blacktriangleright [e' \to e]) \cap (C' \setminus C) \neq \emptyset \implies e' \in \mathrm{cs}'(e)$
8. $\forall e, e' \in E \setminus C . \; ([e' \to e] \rhd) \cap (C' \setminus C) = \emptyset \lor (\blacktriangleright [e' \to e]) \cap (C' \setminus C) = \emptyset$
9. $\forall t, m \in C' \setminus C . \; \forall c \in E . \; m \blacktriangleright [c \to t] \implies (c \in C \lor m \in \{c, t\}).$

Condition 1 insures the accumulation of events. Condition 2 insures conflict freeness. Condition 3 insures that only events which are enabled after C can take place in C'. Condition 4 insures that, if a cause disappears, there has to be a dropper of it. The same is ensured by Condition 6 for appearing causes. Condition 5 insures that if there are adders, the cause has to appear in the new causal state, unless it occurred. Similarly Condition 7 insures, that causes disappear, when there are droppers. To keep the theory simple, Condition 8 avoids race conditions; it forbids the occurrence of an adder and a dropper of the same causal dependency within one transition. Condition 9 ensures that DCESs coincide with Growing Causality ESs of [1], and forbids, in the non trivial casese, the concurrency of an adder and its target.

Based on this state definition, we can now define the proper equivalence for DCESs:

Definition 7. *Let $\Delta = (E, \#, \to, \rhd, \blacktriangleright)$ be a DCES.*

(1) The set of (reachable) states of Δ is defined as

$$S(\Delta) = \{ \, (X, \mathrm{cs}_X) \mid S_0 \to_d^* (X, \mathrm{cs}_X) \land X \text{ is finite} \, \},$$

where $\to_d^*$ is the reflexive and transitive closure of $\to_d$.

(2) Let $\Delta' = (E', \#', \to', \rhd', \blacktriangleright')$ be another DCES.
Then Δ and Δ' are state transition equivalent, denoted by $\Delta \simeq_s \Delta'$, if $E = E'$ and $\to_d \cap (S(\Delta))^2 = \to_d' \cap (S(\Delta'))^2$.

3   Generalisation

We want to model all features of the RCESs, like for example disabling of an
event after the conjunction of two events and a disabling of a disabling after
another event. Therefore, in this paper, we enrich the dynamicity approach with
two concepts. Firstly, we focus on the aspect of set-based dynamics. In contrast
to the DCESs, we now allow sets of events as modifiers for a causal dependency
and the modification only takes place if all elements of the set occurred (note
that for singletons this is exactly the same as in DCESs), in such a manner a
disabling after a and b is easily modelled. Secondly, we study the idea of higher-
order dynamics, i.e. events may alter the capabilities of other events to change
causal dependencies (e.g. so a disabling capability of an event could be dropped
by another).

For two simple example see Fig. 3. The first is an example for set-based
dynamics, in which there are two diagnosis, which demand two different treatments. But if both diagnosis appear at the same time, those two treatments are
not applicable anymore, but a third treatment must be performed. The second
one is an example for 2nd-order dynamics, in which the following process is modeled. After a diagnosis a treatment should follow, but before the treatment a new
doctor, Mr. x, joins the treatment team. Mr. x wants to perform an additional
test, before continuing with the standard treatment. However, if he read a new
paper before joining the team, in which was explained, that the additional test
is not necessary anymore, he does not want to perform the test anymore.

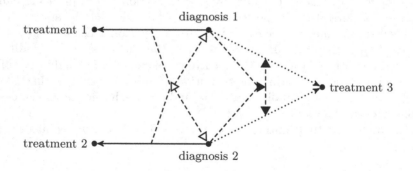

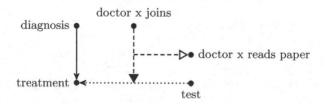

Fig. 3. An example for set-based dynamics and one for 2nd-order dyanmics.

3.1 Higher Order Dynamics

In contrast to the DCESs of [1] we use a rule set, which is updated after each transition, and not a causal state function. For us a state is a configuration with its current set of rules.

Further we will omit the conflict relation from our new structures, because it can be expressed by a mutually adding: $a \# b$ can be modelled by $a \blacktriangleright [b \rightarrow b]$ and $b \blacktriangleright [a \rightarrow a]$. The next definition describes how rule set may look like:

Definition 8. *For a given set of events E a HDES rule is produced by the following grammar, with start symbol R:*

$$R \longrightarrow A \blacktriangleright [R] \mid A \triangleright [R] \mid [c \rightarrow t] \qquad\qquad c, t \in E$$
$$A \longrightarrow F \qquad\qquad\qquad\qquad\qquad\qquad F \subseteq E$$

For a given rule r we call the rank of r, written as $\mathrm{rk}(r)$, the number of $\blacktriangleright$ symbols plus the number of $\triangleright$ symbols occurring in r. For a rule set R we call its projection to the rules of rank zero the causality relation, written as $\rightarrow_R$ (or $\rightarrow$ if the rule set is clear). Let $r = M_1 \operatorname{op}_1 [\ldots M_k \operatorname{op}_k [c \rightarrow t] \ldots]$, with $\operatorname{op}_i \in \{\triangleright, \blacktriangleright\}$ (for $1 \leq i \leq k$) be a rule of rank k, then we denote with r_i the subrule $r_i = M_i \operatorname{op}_i [\ldots M_k \operatorname{op}_k [c \rightarrow t] \ldots]$ (for $1 \leq i \leq k$).

We call a set of HDES rules R a HDES rule set, if for each rule $r = M_1 \operatorname{op}_1 [\ldots M_k \operatorname{op}_k [c \rightarrow t] \ldots] \in R$ with $I, J \subset \{1, \ldots, k\}$, such that $i \in I \Leftrightarrow \operatorname{op}_i = \triangleright$ and $j \in J \Leftrightarrow \operatorname{op}_j = \blacktriangleright$, it follows

1. $\forall i \in I . \exists r' \in R . (r' = r_{i+1}) \vee (r' = M'_1 \blacktriangleright [\ldots [M'_{k'} \blacktriangleright [r_{i+1}]] \ldots])$
2. $\forall j \in J . (\nexists r' \in R . r' = r_{j+1}) \vee (\exists r' \in R . r' = M'_1 \blacktriangleright [\ldots [M'_{k'-1} \blacktriangleright [M'_{k'} \triangleright [r_{i+1}]]] \ldots])$
3. $M \blacktriangleright [r] \implies \neg (M \triangleright [r])$

Those three conditions are just generalisations of those in the definition of the DCESs.

A set based higher order dynamic causality Event Structure (HDES) is a tuple $\Delta = (E, R)$, where E is a set of events and R is an HDES rule set over E.

Like in DCESs configurations are not expressive enough to capture the behavioural state of an HDES.

Definition 9. *Let $\Delta = (E, R)$ be HDES, $C \subseteq E$, S be an HDES rule set over E. Then we call (C, S) a state of Δ. The initial state of Δ is $S_0 = (\emptyset, R)$.*

Like in the case of DCESs the behaviour of an HDES is defined by the transition relation on its states, but before we can define such transitions, we will define for a state a rule update w.r.t. to a set of events. We will only consider rules with modifier sets, which are included in the new set of events, but were not included in the old set of events in the old state. Dropping rules are executed immediately (if they are not dropped themselves), and adding rules are executed in the transition. Further we forbid for an adding rule some concurrent occurrences of events: If the rule adds a causality $M \blacktriangleright [c \rightarrow t]$, then t is only allowed to be in the transition, if it is in M, or if the cause c is in $M \cup T$.

Definition 10. *Let* (C, R_C) *be a state of a HDES* $\Delta = (E, R)$ *and* $C' \subseteq E$ *with* $C \subseteq C'$, *then we call a rule* $M \operatorname{op}[r]$ *active if* $(M \subseteq C') \wedge (M \cap C' \setminus C \neq \emptyset)$, *otherwise we call it passive. We call a rule* r *independent if, there is no active rule* r' *dropping* r.

We call $R_{C'}$ *a rule update for* (C, R_C) *w.r.t.* C', *written as* $(C, R_C) \longrightarrow_{C'} R_{C'}$, *iff* $R_{C'}$ *can be obtained by the Algorithm 1.*

The algorithm consists of two parts, firstly the dropping rules are considered (lines 2 - 8) and step by step the independent rules are executed (if active — line 6) or copied to the new set (if passive — line 8), secondly (lines 11 - 16) each adding rules is executed, if active (line 14), else copied to the new set (line 16), and finally all causality rules (rules of the form $[c \rightarrow t]$*) are copied to the new set (line 19).*

---

    input : *A HDES state* (C, R_C) *of a HDES* $\Delta = (E, R)$ *and a set of events* C', *with* $C \subseteq C' \subseteq E$
    output: *A set of HDES rules* $R_{C'}$

1 $R_{C'} \leftarrow \emptyset$;
2 **while** $\exists r \in R_C, M \subseteq E \,.\, r = M \triangleright [r']$ **do** // while there are dropping rules
3   **let** $r \in R_C \,.\, (r = M \triangleright [r']) \wedge (\nexists r'' \in R_C \,.\, (r'' = M' \triangleright [r]) \wedge (M' \subseteq C') \wedge (M' \cap C' \setminus C \neq \emptyset))$;   // chose a independent rule
4     $R_C \leftarrow R_C \setminus \{r\}$;   // remove the rule from the old set
5   **if** $(M \subseteq C') \wedge (M \cap C' \setminus C \neq \emptyset)$ **then** // active or passive rule?
6 | $R_C \leftarrow R_C \setminus \{r'\}$;   // drop the target from old set
7   **else**
8 | $R_{C'} \leftarrow R_{C'} \cup \{r\}$;   // copy the rule to the new set
9   **end**
10 **end**
11 **foreach** $r = M \blacktriangleright [r'] \in R_C$ **do**   // for each adding rule
12    $R_C \leftarrow R_C \setminus \{r\}$;   // remove the rule from the old set
13   **if** $(M \subseteq C') \wedge (M \cap C' \setminus C \neq \emptyset)$ **then**   // active adding rule?
14 | $R_{C'} \leftarrow R_{C'} \cup \{r'\}$;   // add the target to the new set
15   **else**
16 | $R_{C'} \leftarrow R_{C'} \cup \{r\}$;   // add the rule to the new set
17   **end**
18 **end**
19 $R_{C'} \leftarrow R_{C'} \cup R_C$;   // copy the causal rules to the new set
20 **return** $R_{C'}$;

Algorithm 1. HDRule set update algorithm

Definition 11. *Let* $\Delta = (E, R)$ *be HDES, and* (C, R_C) *and* $(C', R_{C'},)$ *two of its states. Then* $(C, R_C) \rightarrow_{HD} (C', R_{C'})$ *if*

1. $C \subseteq C'$
2. $\forall t \in C' \setminus C \,.\, \forall c \in E \,.\, c \rightarrow_{R_C} t \implies c \in C$
3. $(C, R_C) \longrightarrow_{C'} R_{C'}$

4. $\forall M \subseteq E \,.\, (M \cap (C' \setminus C) \neq \emptyset) \wedge (M \subset C') \wedge ((M \operatorname{op}[r]) \in R_c) \implies (\operatorname{op} = \triangleright)$
$\oplus (\operatorname{op} = \blacktriangleright)$

5. $\forall M, M' \subseteq C', \forall c \in E, t \in C' \setminus C, \forall r \in R_c \,.\, (M \subsetneq C) \implies (r = M \blacktriangleright [c \to t]$
$\implies (c \in C \cup M \vee t \in M))$

Condition 1 ensures an accumulation of events and 2 the left closure under the actual causality relation. Condition 3 ensures that all rules, which modifier sets are included in C' but not in C are executed, and the new rule set is adjusted. Condition 4 ensures that in a transition the same rule can not be added and dropped by the same modifier set. The last condition 5 is a generalisation of the DCES condition, that forbids the concurrency of an adder and its target.

By this definition we may add rules, which never become active: For example if we have an active adding rule like $M_1 \blacktriangleright [M_2 \blacktriangleright [r]]$, and M_2 is contained in the current configuration C union the modifier set M_1. Such a behaviour could be prevented by more a more strict version of condition 5.

4 From Configuration Structures to HDES

In this section we present a translation of an RCES into an HDES, such that both are transition equivalent. We show the result for the even more general class of configuration structures [8]. So here we recap there definition:

Definition 12. *Let E be a set, we call a pair $(E, \mathcal{C})$ with $\mathcal{C} \subseteq \mathcal{P}(E)$ a configuration structure. For x, y in $\mathcal{C}$ we write $x \to_\mathcal{C} y$ if $x \subseteq y$ and*

$$\forall Z(x \subseteq Z \subseteq y \Rightarrow Z \in \mathcal{C}).$$

The relation $\to_\mathcal{C}$ is called the step transition relation.

For the proof, we need some more notation.

Definition 13. *Let $\tau = (E, \mathcal{C})$ be a configuration structure over E, $D \subseteq E$, and $F \subseteq E$. We denote with $\operatorname{En}_\tau(D) \subseteq E \setminus D$ the set of events, which are enabled in D, in a formal way $e \in \operatorname{En}_\tau(D) \iff \exists D' \supseteq D \,.\, D \to_\tau D' \wedge e \in D' \setminus D$. Analogous we denote with $\operatorname{Dis}_\tau(D) \subseteq E \setminus D'$ the set of not enabled (or disabled) events, more formal $\operatorname{Dis}_\tau(D) := D' \setminus (D \cup \operatorname{En}_\tau(D))$. For $F \subseteq E$ we further denote with $\operatorname{Con}_\tau(D, F) \subseteq E \setminus D$ the set of events, which can be concurrent with F in D, in a formal way $e \in \operatorname{Con}_\tau(D, F) \iff \exists D' \supseteq D \,.\, D \to_\tau D' \wedge F \cup \{e\} \subseteq D' \setminus D$.*

Now we can formulate and prove our result. The new HDESs are strictly more expressive than any configuration structure. For any RCES the transition graph is a configuration structure, so the HDESs will be strictly more expressive than the RCESs. It was shown in [1], that there are DCESs, whose behaviour could not simulated by any RCES. Now we show that for each RCES there is a transition equivalent HDES. Both results together (plus the obvious inclusion of DCESs in HDESs), yield the strict inclusion.

Definition 14. *Let $\tau = (E, \mathcal{C})$ be a configuration structure, then we define a HDES namely* $\mathrm{HDES}(\tau) := (E, R)$ *as follows:*

For each $e \in \mathrm{Dis}_\tau(\emptyset)$ we add the rule $e \to e$ to R. Let $F \subseteq \mathrm{En}_\tau(\emptyset)$ with $\emptyset \to_\tau F$, then we add the rule $F \blacktriangleright [e \to e]$ for all $e \in (\mathrm{En}_\tau(\emptyset) \setminus \mathrm{Con}_\tau(\emptyset, F))$ to R.

Let now be $C \subseteq E$ be a non-empty configuration of τ. We add the rule $C \blacktriangleright [e \to e]$ to R, if $e \in \mathrm{Dis}_\tau(C)$. Similarly we add the rule $C \triangleright [e \to e]$, if $e \in \mathrm{En}_\tau(C)$. Further for each $D \subsetneq C$ and for each $e \in \mathrm{Dis}_\tau(D)$ we add the rule $C \triangleright [D \blacktriangleright [e \to e]]$ to R.

For all $F \subseteq \mathrm{En}_\tau(C)$ with $C \to_\tau C'$ and $F \subseteq C'$ and for all $e \in (\mathrm{En}_\tau(C) \setminus \mathrm{Con}_\tau(C, F))$ we add the rule $C \blacktriangleright [F \blacktriangleright [e \to e]]$ to R. Further for all $D \subsetneq C$ and for all $A \subseteq \mathrm{En}_\tau(D)$ with $D \to_\tau D'$ and $A \subseteq D'$ and for all $e \in (\mathrm{En}_\tau(D) \setminus \mathrm{Con}_\tau(D, A))$ we add the rule $C \triangleright [D \blacktriangleright [A \blacktriangleright [e \to e]]]$ to R.

In this translation from configuration structures to HDES we have unique states for each configuration, i.e. for each reachable set of events there exists exactly one rule set and therefore one state.

Definition 15. *Let τ be a configuration structure, $\Delta = (E, R)$ its translation, and $C \neq \emptyset$ a non-empty configuration of τ, we call a rule-set R_C the corresponding rule-set to C, if it can be obtained from R in the following way:*

(1.) For any $D \subsetneq C$ remove any dropping rule $D \triangleright [r]$.
(2.) For any dropping rule $C \triangleright [r]$ remove the rule and its target r.
(3.) For any adding rule $C \blacktriangleright [r]$ remove the rule but add the target r.

Note that by this algorithm and the construction of R in the translation, each rule D op $[r]$ with $D \subseteq C$ is dropped from R_C and each rule E op $[r]$ with $E \not\subseteq C$ is copied to R_C. There are exactly the causality rules $e \to e$ in R_C, for which $e \in \mathrm{Dis}_\tau(C)$ and there are further more the concurrency restricting rules $F \blacktriangleright [e \to e]$, if $F \subseteq \mathrm{En}_\tau(C)$ with $C \to_\tau C'$, $F \subseteq C'$ and $e \in (\mathrm{En}_\tau(C) \setminus \mathrm{Con}_\tau(C, F))$.

We now show that in each reachable state the rule set corresponds to the configuration, and even for any transition starting at a state where the rule set corresponds to the configuration this will hold in the resulting state.

Firstly we show that starting with the initial rule set we reach in one step only states with corresponding rule sets.

Lemma 1. *Let $\tau = (E, \mathcal{C})$ be a configuration structure, $\Delta := \mathrm{HDES}(\tau)$, and $(\emptyset, R) \to_C R_C$ be a rule-update, then R_C corresponds to C.*

Proof. Since $(\emptyset, R) \to_C R_C$ is a rule-update as defined in 10, all active independent dropping rules are considered, this are all rules of the form $D \triangleright [r]$ for $D \subseteq C$ and note that by construction r is no dropping rule. Those dropping rules drop by construction all former added or initial causality rules (e.g. $e \to e$) and concurrency restricting rules (e.g. $F \blacktriangleright [e \to e]$). After the dropping rules we consider the active adding rules there are two types, first causality adding rules (e.g. $C \blacktriangleright [e \to e]$) and concurrency restricting rules (e.g. $C \blacktriangleright [F \blacktriangleright [e \to e]]$). Finally we copy the newly added causality rules to the new rule set. This is exactly the same as in 15, because for each in step (1.) removed rule $D \triangleright [r]$ there is a rule $C \triangleright [r]$, which will be executed in step (2.). Thus R_C corresponds to C.

Secondly we show that starting from a state with a corresponding rule set we reach in one step only states with corresponding rule sets. The proof is almost the same as above.

Lemma 2. *Let* $\tau = (E, \mathcal{C})$ *be a configuration structure,* $\Delta := \mathrm{HDES}(\tau)$, *and* $(C, R_C) \to_{C'} R'_C$ *be a rule-update where* R_C *corresponds to* C, *then* R'_C *corresponds to* C'.

Proof. Since $(C, R_C) \to_{C'} R'_C$ is a rule-update as defined in 10, all active independent dropping rules are considered, this are all rules of the form $D \rhd [r]$ for $C' \subseteq D \subseteq C$ (no smaller modifier sets are possible, because R_C is a corresponding rule set) and note that by construction r is no dropping rule. Those dropping rules drop by construction all former added causality rules (e.g. $e \to e$) and concurrency restricting rules (e.g. $F \blacktriangleright [e \to e]$). After the dropping rules we consider the active adding rules there are two types, first causality adding rules (e.g. $C' \blacktriangleright [e \to e]$) and concurrency restricting rules (e.g. $C' \blacktriangleright [F \blacktriangleright [e \to e]]$). Finally we copy the newly added causality rules to the new rule set. This is exactly the same as in 15, because for each in step (1.) removed rule $D \rhd [r]$ there is a rule $C \rhd [r]$, which will be executed in step (2.). Thus R'_C corresponds to C'. $\qquad \square$

Lemma 3. *Let* $\tau = (E, \mathcal{C})$ *be a configuration structure,* $\Delta := \mathrm{HDES}(\tau)$, *and* (C, S_C) *and* (D, S_D) *two reachable states of* Δ, *then* $C = D \implies S_C = S_D$.

Proof. Because both states are reachable it follows from the previous two Lemmas 1 and 2 that both rule sets corresponds to the configurations, if those configurations are the same then clearly the corresponding rule sets too.

The above lemma justifies to speak about configurations of Δ and and transitions in between them, instead of states, because the rule sets are unique for each reachable configuration (they are the corresponding ones). In order to compare to the original configuration structure τ it is therefore sufficient to show that both are transition equivalent. $\qquad \square$

Lemma 4. *Let* $\tau = (E, \mathcal{C})$ *be a configuration structure and* $\Delta := \mathrm{HDES}(\tau)$, *then* $\emptyset \to_\tau C$ *iff* $(\emptyset, R) \to_\Delta (C, R_C)$.

Proof. Let $\emptyset \to_\tau C$, it follows by Definition 13 $e \in \mathrm{En}_\tau(\emptyset)$ for each $e \in C$ and $C \subseteq \mathrm{Con}_\tau(\emptyset, C)$. Thus by construction in Definition 14 there is no rule $e \to e$ for any $e \in C$ in the rule set R furthermore is there no rule $D \blacktriangleright [e \to e]$ for any $c \in C$ and $D \subseteq C$. We now show all conditions of the transition Definition 11 hold. Condition 1 is clearly satisfied because $\emptyset \subseteq C$, the only initial causality rules are by construction deactivating rules like $e \to e$, which do not occur for events in C therefore condition 2 holds. Next condition 3 constraints only the rule update and is fulfilled by assumption. The structural condition 4 enforces the the same rule is not added and dropped by the same set of modifiers, this is fulfilled by construction 14 of Δ. Regarding the last condition 5, let us first assume there is a rule $M \blacktriangleright [e \to e]$, such that $M \subseteq C$ and $e \in C \backslash M$, then either in the configuration

M the event e is disabled or initially e may not be concurrent with M (by construction of the rule set), but since $\emptyset \to_\tau C$ and therefore $\emptyset \to_\tau M \cup \{e\}$ and $M \to_\tau M \cup \{e\}$ (because τ is a configuration structure). Thus no rule $M \blacktriangleright [e \to e]$ with $M \subseteq C$ and $e \in C \setminus M$ is in R and therefore the property holds. We have shown that all conditions of the transition Definition 11 hold, so we have $(\emptyset, R) \to_\Delta (C, R_C)$.

To show the equivalence we assume $\emptyset \not\to_\tau C$ then there are, by Definitions 13 and 2, $D \subset C$ and $e \in C \setminus D$, such that $e \notin \mathrm{Con}_\tau(\emptyset, D)$. Caused by construction 14 there is a rule $D \blacktriangleright [e \to e]$ in R. Thus by condition 5 of the transition Definition 11 it follows $\emptyset \not\to_\Delta C$.

Lemma 5. *Let $\tau = (E, \mathcal{C})$ be a configuration structure, $\Delta := \mathrm{HDES}(\tau)$, $C \in \mathcal{C}$, and R_C the to C corresponding rule-set, then $C \to_\tau C'$ iff $(C, R_C) \to_\Delta (C', R_{C'})$.*

Proof. Let $\to_\tau C$, it follows by Definition 13 $e \in \mathrm{En}_\tau(C)$ for each $e \in C'$ and $C' \subseteq \mathrm{Con}_\tau(C, C')$. Thus by construction of R in Definition 14 and because of R_C being a to C corresponding rule set by assumption, there is no rule $e \to e$ for any $e \in C'$ in the rule set R_C furthermore is there no rule $D \blacktriangleright [e \to e]$ for any $e \in C'$ and $D \subseteq C'$. We now show all conditions of the transition Definition 11 hold. Condition 1 is clearly satisfied because $C \subseteq C'$, the only causality rules are by construction of R and because R_C is corresponding to C deactivating rules like $e \to e$, which do not occur for events in C' (because they are enabled) therefore condition 2 holds. Next condition 3 constraints only the rule update and is fulfilled by assumption. The structural condition 4 enforces the the same rule is not added and dropped by the same set of modifiers, this is fulfilled by construction 14 of Δ. Regarding the last condition 5, let us first assume there is a rule $M \blacktriangleright [e \to e]$, such that $M \subseteq C'$ and $e \in C' \setminus M$, then either in the configuration M the event e is disabled or in C the event e may not be concurrent with M (by construction of the rule set), but since $\to_\tau C$ and therefore $\to_\tau C M \cup \{e\}$ and $\to_\tau M M \cup \{e\}$ (because τ is a configuration structure). Thus no rule $M \blacktriangleright [e \to e]$ with $M \subseteq C'$ and $e \in C' \setminus M$ is in R_C and therefore the property holds. We have shown that all conditions of the transition Definition 11 hold, so we have $(C, R_C) \to_\Delta (C', R_{C'})$.

To show the equivalence we assume $C \not\to_\tau C'$ then there are, by Definitions 13 and 2, $D \subset C'$ and $e \in C' \setminus D$, such that $e \notin \mathrm{Con}_\tau(C, D)$. Caused by construction 14 there is a rule $D \blacktriangleright [e \to e]$ in R_C. Thus by condition 5 of the transition Definition 11 it follows $(C, R_C) \not\to_\Delta (C', R_{C'})$.

Theorem 1. *Let $\tau = (E, \mathcal{C})$ be a configuration structure then the HDES $\Delta := \mathrm{HDES}(\tau)$ is transition equivalent.*

Proof. By induction with Lemmas 4 and 5.

5   Conclusion

In this paper we present a more general and more elegant dynamic-causality ES than in [1], by allowing dynamicity of higher order and set-based modifications.

This new HDES (higher-order dynamic-causality ES) only consists of an event set and a rule set, which is updated after each transition.

We show that for arbitrary configuration structures with the step transition relation (as defined in [8]), there is a HDES (with at most third order dynamics) with the same transition graph. Since the transition graph of an Event Structure for Resolvable Conflicts (RCESs) [7] is a configuration structure, the newly defined HDESs are strictly more expressive than the RCESs (because of the incomparability of DCESs and RCESs in [1]).

As future work we want to study the HDES more deeply. We want investigate, whether there is a strict hierarchy in the level of dynamicity, i.e. there are some structure with order n dynamics, that can not be formulated with order $n-1$. We also want to relate our approach with the DCR-Graphs [4] and the π-calculus.

References

1. Arbach, Y., Karcher, D., Peters, K., Nestmann, U.: Dynamic causality in event structures. In: Graf, S., Viswanathan, M. (eds.) Formal Techniques for Distributed Objects, Components, and Systems. LNCS, vol. 9039, pp. 83–97. Springer, Heidelberg (2015)
2. Arbach, Y., Karcher, D., Peters, K., Nestmann, U.: Dynamic causality in event structures (Technical report) (2015). Available from http://www.arxiv.org/
3. Boudol, G., Castellani, I.: Flow models of distributed computations: three equivalent semantics for CCS. Inf. Comput. **114**(2), 247–314 (1994)
4. Hildebrandt, T., Mukkamala, R.R., Slaats, T.: Nested dynamic condition response graphs. In: Arbab, F., Sirjani, M. (eds.) FSEN 2011. LNCS, vol. 7141, pp. 343–350. Springer, Heidelberg (2012)
5. Langerak, R.: Transformations and Semantics for LOTOS. Ph.D. thesis, Twente (1992)
6. Trénous, J.: On the Utility and Usability of Event Structures to Model Dynamic Processes in a Clinical Case Study. Bachelor Thesis in Computer Science, to be submitted in August 2015
7. van Glabbeek, R.J., Plotkin, G.: Event structures for resolvable conflict. In: Fiala, J., Koubek, V., Kratochvíl, J. (eds.) MFCS 2004. LNCS, vol. 3153, pp. 550–561. Springer, Heidelberg (2004)
8. van Glabbeek, R.J., Plotkin, G.D.: Configuration structures, event structures and petri nets. Theor. Comput. Sci. **410**(41), 4111–4159 (2009). Festschrift for Mogens Nielsen 60th birthday
9. Winskel, G.: Events in Computation. Ph.D. thesis, Edinburgh (1980)
10. Winskel, G.: An introduction to event structures. In: de Bakker, J.W., de Roever, W.-P., Rozenberg, G. (eds.) Linear Time, Branching Time and Partial Order in Logics and Models for Concurrency. LNCS, vol. 354, pp. 364–397. Springer, Heidelberg (1989)
11. Winskel, G.: Distributed probabilistic and quantum strategies. In: Proceedings of MFPS. ENTCS, vol. 298, pp. 403–425. Elsevier (2013)

Asynchronous Announcements
in a Public Channel

Sophia Knight[1]($\boxtimes$), Bastien Maubert[1], and François Schwarzentruber[2]

[1] LORIA - CNRS / Université de Lorraine, Nancy, France
{sophia.knight,bastien.maubert}@gmail.com
[2] ENS Rennes - IRISA, Rennes, France
francois.schwarzentruber@ens-rennes.fr

Abstract. We propose a variant of public announcement logic for asynchronous systems. We give a syntax where sending and receiving messages are modeled by different modal operators. The natural approach to defining the semantics leads to a circular definition, but we describe two restricted cases in which we solve this problem. The first case requires the Kripke model representing the initial epistemic situation to be a finite tree, and the second one only allows announcements from the existential fragment. Finally, we provide complexity results for the model checking problem.

1 Introduction

Asynchrony has long played a central role in distributed systems, where access to a centralized clock is not always possible, and where communication may not be delivered or received immediately or predictably. Recently, with the proliferation of multi-agent systems (MAS) where independent agents interact, communicate, and make decisions under imperfect information, modelling how knowledge evolves with informative events has also become increasingly important. One of the first and most influential proposals in this direction is public announcement logic (PAL) [11], in which some external omniscient entity publicly makes true announcements to some group of agents. This logic has led to the powerful and much studied dynamic epistemic logic (DEL) [14]. However, both these logics assume synchronicity, even though there has been some discussion on this matter for the latter [4]. In PAL for instance, messages are immediately received by all agents at the same time, as soon as they are sent. As far as we know, little work has been done to address the same problem in asynchronous scenarios.

Our goal in this work is the logical study of scenarios with *asynchronous* announcements. As a first step, similar to PAL, we consider a simple scenario: messages are true at the time of announcement, public (directed to everyone), and we do not model their origin, but rather assume that some external and omniscient entity emits them. Consider the scenario where three autonomous agents, moving through an area, receive messages from a public channel. They do not all read the messages (logical formulas) at the same time, but they do

© Springer International Publishing Switzerland 2015
M. Leucker et al. (Eds.): ICTAC 2015, LNCS 9399, pp. 272–289, 2015.
DOI: 10.1007/978-3-319-25150-9_17

read them in the same order. Figure 1 depicts the architecture of the system: each agent has a private copy of the channel and they read messages in *first in, first out* (FIFO) order, that is, messages are read in the order they are sent.

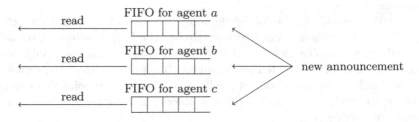

Fig. 1. Agent architecture

In PAL, not only are all messages received at the same time by all agents, but they are also received at the same time they are sent. Therefore, in PAL, the announcement operator combines both sending and receiving. In contrast, in our setting, messages are not received immediately and they may be received at different times by different agents. The syntax reflects this aspect by providing both a sending operator, which adds new messages to the public channel, and a receiving operator for each agent, which allows her to read the first message in the channel that she has not read yet. Thus, in our logic, we provide the following modal operators:

- $K_i\varphi$, where i is an agent and φ a formula. Intuitively, this will mean "agent i knows φ," and as usual, this will be interpreted as "φ is true at every state that agent i considers possible." Below, we discuss the accessibility relation we use to define all the states agent i considers possible at a given state (all states that are *indistinguishable* from the current state for agent i).
- $\langle\psi\rangle\varphi$, which will mean "after the currently true formula ψ is (asynchronously) announced, φ is immediately true."
- $\bigcirc_i\varphi$ which will mean "after agent i *receives* the next announced formula in her queue, φ is immediately true."

Interestingly, the most intuitive semantics for this logic presents a challenging problem of circular definition. We describe this difficulty in more detail below, but the basic issue is that in order to check the truth of some formulas, we must quantify over the set of all indistinguishable states that are consistent. A state is consistent if it is the result of making a true announcement in a state that is itself consistent. So evaluating the truth of a formula requires determining whether a state is consistent, which in turn requires evaluating the truth of formulas. In PAL, a similar problem occurs, as the definition of the update of a model by an announcement and the definition of the truth values are mutually dependent. While in PAL this circularity can be solved simply by resorting to a double induction, things are more complicated here. Indeed, because of asynchrony, an agent does not know what or how many messages other agents have received; therefore, evaluating a knowledge operator in a state requires considering possibly infinitely many indistinguishable states, which makes a double induction impossible. This circularity problem is inherent to the asynchronous setting, and is independent from our choice to consider an external source for the announcements.

We partially tackle this issue by defining two restricted cases in which we manage to avoid circularity. The first one requires the Kripke model representing the initial epistemic situation to be a finite tree; the second one only allows announcements from the existential fragment. In the latter case, the semantics is defined thanks to an application of the Knaster-Tarski fixed point theorem [12].

Finally, we study the model checking problem for our logic and establish the following complexity results:

Restrictions	Complexity of model checking
Propositional announcements	PSPACE-complete
Finite tree initial models	in PSPACE
Announcements from the existential fragment	in EXPTIME, PSPACE-hard

The paper is organized as follows. In Sect. 2, we recall (synchronous) public announcement logic. In Sect. 3 we present the language and the models for asynchronous public announcement logic we propose here. In Sect. 4 we present the circularity problem for defining the semantics of the logic, and we exhibit two cases where it can be solved. We then present some validities in Sect. 5, and we study the model checking problem in Sect. 6. Finally we discuss related work in Sect. 7 and future work in Sect. 8.

2 Background: Public Announcement Logic

In this section, we present background on (synchronous) Public Announcement Logic (PAL) [11]. Let $\mathcal{P}$ be a countable infinite set of *atomic propositions*, and let AGT be a finite set of *agents*.

Definition 1 (Syntax of PAL). *The syntax for PAL is as follows:*

$$\varphi ::= p \mid (\varphi \wedge \varphi) \mid \neg\varphi \mid K_i\varphi \mid \langle\varphi\rangle_{PAL}\varphi$$

where $p \in \mathcal{P}$ and $i \in AGT$.

The intuitive meaning of the last two operators is the following: $K_i\varphi$ means that agent i knows φ, $\langle\psi\rangle_{PAL}\varphi$ means that ψ is true and after ψ has been publicly announced and publicly received by all the agents, φ holds.

The semantics of PAL relies on classic Kripke models and the *possible worlds* semantics, widely used in logics of knowledge [5].

Definition 2. *A Kripke model is a tuple* $\mathcal{M} = (W, \{\rightarrow_i\}_{i \in AGT}, \Pi)$, *where:*

- *W is a non-empty finite set of* worlds,
- *for each* $i \in AGT$, $\rightarrow_i \subseteq W \times W$ *is an* accessibility relation *for agent i,*
- $\Pi : W \rightarrow 2^P$ *is a* valuation *on worlds.*

Note that we do not require the accessibility relations to be equivalence relations as is traditionally done in epistemic logic [14].

Example 1. Let us consider the following Kripke model, where w, u and v are worlds, a and b are agents and p is a proposition. The arrows represent the agents' accessibility relations. At world w, agent a considers u and v possible, and agent b considers only world v possible.

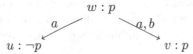

The semantics is given as follows:

- $\mathcal{M}, w \models p$ if $p \in \Pi(w)$;
- $\mathcal{M}, w \models \varphi_1 \wedge \varphi_2$ if $\mathcal{M}, w \models \varphi_1$ and $\mathcal{M}, w \models \varphi_2$;
- $\mathcal{M}, w \models \neg\varphi$ if $\mathcal{M}, w \not\models \varphi$;
- $\mathcal{M}, w \models K_i\varphi$ if for all u such that $w \rightarrow_i u$, $\mathcal{M}, u \models \varphi$;
- $\mathcal{M}, w \models \langle\psi\rangle_{\text{PAL}}\varphi$ if $\mathcal{M}, w \models \psi$ and $\mathcal{M}^\psi, w \models \varphi$ where $\mathcal{M}^\psi$ is the restriction of $\mathcal{M}$ to worlds where ψ holds.

Example 2. Let $\mathcal{M}$ be the model of Example 1. We have $\mathcal{M}, w \models \langle p\rangle_{\text{PAL}} K_a p$. Indeed, we have $\mathcal{M}^p, w \models K_a p$ where $\mathcal{M}^p$ is

$$w : p \xrightarrow{\quad a, b \quad} v : p$$

The model checking in public announcement logic is in P and the satisfiability problem in public announcement logic is PSPACE-complete [9]. A tableau proof system for public announcement logic is provided in [3].

3 Language and Models

3.1 Language

Again, $\mathcal{P}$ is a countable infinite set of atomic propositions, and AGT is a finite set of agents.

Definition 3 (Syntax). *The syntax for the logic is as follows:*

$$\varphi ::= p \mid (\varphi \wedge \varphi) \mid \neg\varphi \mid K_i\varphi \mid \langle\varphi\rangle\varphi \mid \bigcirc_i \varphi,$$

where $p \in \mathcal{P}$ *and* $i \in AGT$.

We use $\mathcal{L}$ to denote the set of all formulas. The intuitive meaning of the last three operators is the following: $K_i\varphi$ means that agent i knows φ, $\langle\psi\rangle\varphi$ means that ψ is true and after ψ has been put on the public channel, φ holds, and $\bigcirc_i\varphi$ means that agent i has a message to read, and after he has read it, φ holds. We classically define $(\varphi \vee \psi)::=\neg(\neg\varphi \wedge \neg\psi)$, $(\varphi \rightarrow \psi)::=(\neg\varphi \vee \psi)$, the dual of the knowledge operator: $\hat{K}_i\varphi::=\neg K_i\neg\varphi$, meaning that agent i considers φ possible, and the dual of the announcement operator: $[\psi]\varphi::=\neg\langle\psi\rangle\neg\varphi$, meaning that if ψ is true, then φ holds after its announcement. $|\varphi|$ is the length of φ.

Note that in (synchronous) public announcement logic (see Definition 1), the emission and the reception of a formula ψ is mixed in the operator $\langle\psi\rangle_{\mathrm{PAL}}$, because in this setting, emission and reception occur simultaneously. In the asynchronous version of announcement logic we propose here (see Definition 3), the emission of ψ is represented by $\langle\psi\rangle$ and the reception of a message by Agent i is represented by the operator $\bigcirc_i$. Note that not only can emission and reception occur at different times, but also different agents may receive the same message at different times.

3.2   Models

The models on which our logic is interpreted represent situations obtained by announcements being made in an initial epistemic model, with agents asynchronously receiving these announcements. We now define initial epistemic models, sequences of announcements, and a third notion that we call *cuts*, representing the announcements each agent has received at the current time.

Initial Kripke Model. An *initial model* is given as a Kripke model $\mathcal{M} = (W, \{\rightarrow_i\}_{i \in AGT}, \Pi)$, as defined in Definition 2. An initial model represents the initial static situation before any announcements are made.

Sequences of Announcements. We consider that, in a given scenario, not every formula may be announced, but rather that there is a certain set of relevant announcements. Furthermore, we allow the number of times an announcement can be made to be bounded.

To represent this, we define an *announcement protocol* to be a multiset of formulas in our language, where the multiplicity of an element ψ is either an integer or ∞.

Example 3. The reader may imagine a card game where it is only possible to announce 'Agent a has a heart card' once and 'Agent a does not know whether Agent b has a heart card or not[1]' twice. We let the proposition $\heartsuit_i$ mean "agent i has a heart card," and define the announcement protocol to be:

$$\{\{\heartsuit_a \, , \quad \hat{K}_a\heartsuit_b \wedge \hat{K}_a\neg\heartsuit_b \, , \quad \hat{K}_a\heartsuit_b \wedge \hat{K}_a\neg\heartsuit_b\}\}.$$

---

[1] that is 'a considers $\heartsuit_b$ possible *and* considers $\neg\heartsuit_b$ possible'.

Given an announcement protocol $\mathcal{A}$, we denote by $\mathsf{Seq}(\mathcal{A})$ the set of finite sequences $\sigma = [\varphi_1, \ldots, \varphi_k]$ such that the multiset $\{\!\{\varphi_1, \ldots, \varphi_k\}\!\}$ is a submultiset of $\mathcal{A}$. We let $|\sigma| = \sum_{i=1}^{k} |\varphi_i|$. For $\sigma, \sigma' \in \mathsf{Seq}(\mathcal{A})$, we write $\sigma \leq \sigma'$ if σ is a prefix of σ'. The sequence $\sigma|_k$ is the prefix of σ of length k.

Given a formula φ and a sequence of formulas σ, $\varphi :: \sigma$ (resp. $\sigma :: \varphi$) is the sequence obtained by adding φ at the beginning (resp. at the end) of σ.

States. We can now define the models of our logic. Let $\mathcal{M}$ be an initial model and $\mathcal{A}$ an announcement protocol. We define the *asynchronous model* $\mathcal{M} \otimes \mathcal{A} = (\mathcal{S}, \{R_i\}_{i \in AGT})$, where each R_i is a *pre-accessibility* relation, which we define in Sect. 3.2, and $\mathcal{S}$ is a set of *states* defined as follows:

$$\mathcal{S} = \big\{(w, \sigma, c) \mid w \in W, \sigma \in \mathsf{Seq}(\mathcal{A}) \text{ and } c : AGT \to \{0, \ldots, |\sigma|\}\big\}.$$

States are also denoted S, S', etc.

The first element of a state represents the world the system is in. The second element is the list of messages that have already been announced. The last element, c, is called a *cut*, and for each $i \in AGT$, $c(i)$ is the number of announcements of σ that Agent i has received so far. Given two cuts c and c', we write $c < c'$ if for all i, $c(i) \leq c'(i)$ and there exists j such that $c(j) < c'(j)$.

Example 4. Consider the state $S = (w, \epsilon, \mathbf{0})$, where ϵ denotes the empty sequence of formulas and $\mathbf{0}$ is the function that assigns 0 to all agents. S represents an initial world w in which no announcement has been made (and therefore no announcement has been received either). It can be represented as follows:

Example 5. Consider the state $S = (w, [\varphi, \psi, \chi], c)$ where $c(a) = 2$ and $c(b) = 1$. S is the state representing that in initial world w, the sequence $[\varphi, \psi, \chi]$ of formulas has been announced and is now in the public channel, Agent a has received φ and ψ, and agent b has only received φ. Only χ remains in the queue of a and has not been read yet, and only ψ and χ remain in the queue of b. We represent S as follows:

and we may also write $S = (w, [\varphi, \psi, \chi], \begin{smallmatrix} a \mapsto 2 \\ b \mapsto 1 \end{smallmatrix})$.

The definition of $\mathcal{S}$ allows for all combinations of worlds, sequences of announcements, and cuts. This definition is an over-approximation of the set of states we want to consider: indeed, some of the states in $\mathcal{S}$ are inconsistent. For example, suppose that w is a world in $\mathcal{M}$ where p does not hold. Then, the state $(w, [p], \mathbf{0})$ is intuitively inconsistent because as the formula p is not true in w, it cannot have been announced. This notion of inconsistency is the source of the circularity problem. Indeed, to define whether a state is consistent requires one to define whether an announcement can be made, and this requires the semantics of our logic to be defined. But to define the semantics of the knowledge operators, we need to define which *consistent* states are related to the current one, which requires us to define which states are consistent, hence the circularity (see Sect. 4).

Pre-accessibility Relation Definition. We now define, for each agent, a *pre-accessibility relation* that does not yet take consistency into account, but is only based on the agents' accessibility relations on the initial model and the current cut. This is the first step toward the final definition of the agents' accessibiliy relations, which is presented below.

Definition 4. *The* pre-accessibility relation *for Agent i, written R_i, is defined as follows: given $S = (w, \sigma, c)$ and $S' = (w', \sigma', c')$, we have SR_iS' if:*

1. $w \to_i w'$, *and*
2. $c(i) = c'(i)$ *and* $\sigma|_{c(i)} = \sigma'|_{c'(i)}$

The first clause is obvious. The second clause says that Agent i is aware of, and only aware of, messages that she has received: therefore she can only consider possible states where she has received exactly the same messages. However, the sequence of messages she has not yet received may be longer or shorter in a related state. Also, as she has no information about what messages the other agents have received, we do not put any constraints on $c'(j)$ if $j \neq i$.

Example 6. Here we give an example of an inconsistent state, to show why further refinement of the above-defined models is necessary. Let us consider the following initial model, where w, u, v and z are worlds, a and b are agents and p is a proposition. The arrows represent the agents' accessibility relations, before any announcements have been made. So at world w, agent a considers u and v possible, and agent b considers world z possible.

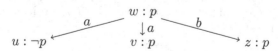

Now assuming that the announcement protocol $\mathcal{A}$ contains p, φ and ψ, a partial depiction of the asynchronous model $\mathcal{M} \otimes \mathcal{A}$ is below. We depict the states w, u, v, and z where no announcements have been made, as well as copies of u where two different sequences of announcements have been made, and received in one state by b and in a different state by a. Of course, the entire model $\mathcal{M} \otimes \mathcal{A}$ is infinite so we do not depict all the states here. The grey state shown in the asynchronous model is not consistent because p has been announced although p is not true in u.

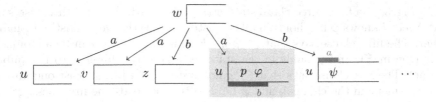

4   Semantics: The Circularity Problem

Figures 2 and 3 show a naive attempt to define the truth conditions, which we explain in Sect. 4.1. Unfortunately, this leads to a circularity problem which we detail in Sect. 4.2. We show how to solve this problem for restricted cases in Sects. 4.3 and 4.4.

$$(w, \epsilon, \mathbf{0}) \models \checkmark$$
$$(w, \sigma, c) \models \checkmark \text{ if there is } c' < c \text{ s.t. } (w, \sigma, c') \models \checkmark, \text{ or}$$
$$\sigma = \sigma'{::}\psi, (w, \sigma', c) \models \checkmark \text{ and} (w, \sigma', c) \models \psi$$

Fig. 2. Truth conditions for consistency

$$(w, \sigma, c) \models p \qquad \text{if } p \in \Pi(w)$$
$$(w, \sigma, c) \models \varphi_1 \wedge \varphi_2 \text{ if } (w, \sigma, c) \models \varphi_1 \text{ and } (w, \sigma, c) \models \varphi_2$$
$$(w, \sigma, c) \models \neg\varphi \qquad \text{if } (w, \sigma, c) \not\models \varphi$$
$$(w, \sigma, c) \models K_i\varphi \qquad \text{if for all } S' \text{ s.t. } (w, \sigma, c)R_iS' \text{ and } S' \models \checkmark, S' \models \varphi$$
$$(w, \sigma, c) \models \langle\psi\rangle\varphi \qquad \text{if } \sigma{::}\psi \in \mathsf{Seq}(\mathcal{A}), \ (w, \sigma, c) \models \psi \text{ and } (w, \sigma{::}\psi, c) \models \varphi$$
$$(w, \sigma, c) \models \bigcirc_i\varphi \qquad \text{if } c(i) < |\sigma| \text{ and } (w, \sigma, c^{+i}) \models \varphi$$
$$\text{where } c^{+i}(j) = \begin{cases} c(j) & \text{if } j \neq i \\ c(j) + 1 & \text{if j=i} \end{cases}$$

Fig. 3. Truth conditions for formulas

4.1   Intuitions

The pre-accessibility relation takes into account states that are not actually consistent, because they contain announcements that were not true at the time they were made. The intuitive meaning of $(w, \sigma, c) \models \checkmark$ is that the state (w, σ, c) is consistent, that is, all announcements were true when they were made. Figure 2 is an attempt to define this concept formally. The first clause is obvious: the initial state where no announcements have been made is consistent. The second clause gives two possibilities for a state to be consistent. Either there was an earlier consistent state, (w, σ, c') and then some agents received some already announced formulas, increasing the cut from c' to c, or a new, true announcement ψ has been made from an earlier consistent state, increasing the history from σ' to $\sigma'.\psi$.

In Fig. 3, the first three clauses are straightforward. The fourth clause says that Agent i knows φ if φ holds in all consistent states that are indistinguishable to her. The fifth clause says that $\langle\psi\rangle\varphi$ holds in a state S if ψ can be announced (it is true in S), and φ holds in the state obtained by adding ψ to the public channel. The last clause says that $\bigcirc_i\varphi$ holds if Agent i has at least one unread announcement in the channel, and φ holds after she reads the first message.

4.2   Circularity

Let us consider the following example, where $AGT = \{a\}$. Let the initial model be $\mathcal{M} = (W, \rightarrow_a, \Pi)$ where $W = \{w\}$, $\rightarrow_a = \{(w, w)\}$ and $\Pi(w) = \emptyset$. Let the announcement protocol be $\mathcal{A} = \{\{K_a p\}\}$. According to Fig. 2, we have: $(w, [K_a p], \mathbf{0}) \models \checkmark$ iff $(w, \epsilon, \mathbf{0}) \models K_a p$. But, as $(w, \epsilon, \mathbf{0}) R_a (w, [K_a p], \mathbf{0})$, the definition of the truth value of $(w, \epsilon, \mathbf{0}) \models K_a p$ depends on the truth value of $(w, [K_a p], \mathbf{0}) \models \checkmark$. To sum up, the definition of $(w, [K_a p], \mathbf{0}) \models \checkmark$ depends on itself.

4.3   When the Initial Model Is a Finite Tree

If we assume the initial model $\mathcal{M} = (W, \{\rightarrow_i\}_{i \in AGT}, \Pi)$ to be such that W is finite and $\bigcup_i \rightarrow_i$ makes a **finite** tree over W, then the circularity problem can be avoided. Indeed, in this case, we can define a well-founded order on tuples of the form (w, σ, c, φ), where φ is either a formula in $\mathcal{L}$ or the symbol $\checkmark$, the idea being that a tuple (w, σ, c, φ) means '$w, \sigma, c \models \varphi$'.

Definition 5. *The order $\prec$ is defined as follows:*

$(w, \sigma, c, \varphi) \prec (w', \sigma', c', \varphi')$ *if either*

① *w is in the subtree of w' in $\mathcal{M}$,*
② *or $w = w'$ and $|\sigma| + |\varphi| < |\sigma'| + |\varphi'|$,*
③ *or $w = w'$, $|\sigma| + |\varphi| = |\sigma'| + |\varphi'|$ and $c < c'$,*

where $|\checkmark| = 1$.

It is easy to see that $\prec$ is a well-founded order, and with this order Figs. 2 and 3 together form a well-founded inductive definition of consistency and semantics of our language.

We detail the non-trivial cases. For the second clause of Fig. 2, observe that by Point ③ of Definition 5, if $c' < c$ then $(w, \sigma, c', \checkmark) \prec (w, \sigma, c, \checkmark)$, and for all w, σ', c and ψ, by Point ② of Definition 5, we have $(w, \sigma', c, \psi) \prec (w, \sigma'::\psi, c, \checkmark)$.

For the fourth clause of Fig. 3, by Point ① of Definition 5 we have that for all $\varphi, \sigma, \sigma', c, c'$, if w' is a child of w then $(w', \sigma', c', \varphi) \prec (w, \sigma, c, K_i\varphi)$.

Finally, for the fifth clause of Fig. 3, by Point ② of Definition 5 we have that $(w, \sigma::\psi, c, \varphi) \prec (w, \sigma, c, \langle\psi\rangle\varphi)$ for all w, σ, c, φ and ψ (note that $|\langle\psi\rangle\varphi| = 1 + |\psi| + |\varphi|$).

The following simple example illustrates how our semantics works, and how it indeed captures the intuitions behind the operators.

Example 7. Suppose that we have only one agent a. Let us consider the following initial model $\mathcal{M}$:

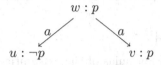

In this model, p holds in the actual world w, but Agent a does not know it. Assume that p can be announced at least once ($p \in \mathcal{A}$). We show that, as expected, after p is announced and Agent a has received this announcement, it holds that Agent a knows that p holds. Formally, we prove that, in $\mathcal{M} \otimes \mathcal{A}$, we have $(w, \epsilon, \mathbf{0}) \models \langle p \rangle \bigcirc_a K_a p$.

To do so we show that $(w, [p], a \mapsto 1) \models K_a p$, from which it follows that $(w, [p], \mathbf{0}) \models \bigcirc_a K_a p$, hence the desired result.

By Definition 4 for pre-accessibility relations, every state S such that $(w, [p], a \mapsto 1) R_a S$ is of the form $S = (w', p{::}\sigma, a \mapsto 1)$, where $w' \in \{u, v\}$ and σ is a sequence of announcements. We just have to show that each such state either is inconsistent or verifies p.

First, for $w' = u$. According to the first clause of Fig. 3, we have that $(u, \epsilon, \mathbf{0}) \not\models p$, and by the second clause of Fig. 2 it follows that $(u, [p], \mathbf{0}) \not\models \checkmark$, from which it follows also that $(u, [p], a \mapsto 1) \not\models \checkmark$ and $(u, p{::}\sigma, a \mapsto 1) \not\models \checkmark$, for any σ.

Now, for $w' = v$, by the first clause of Fig. 3, we have that for all states of the form $S = (v, p{::}\sigma, a \mapsto 1)$, $S \models p$, so that finally every state related to $(w, [p], a \mapsto 1)$ is either inconsistent or verifies p. Note that we could also prove that S is consistent.

In practice, this setting can be used as an approximation scheme: unfolding models and cutting at level l of the obtained trees amounts to assuming that agents cannot reason about deeper nesting of knowledge. This approach is similar to the well known idea of *bounded rationality*, [7], where it is assumed that due to computational limits, agents have only approximate, bounded information about other agents' knowledge, which is represented by allowing only finite-length paths in the Kripke model. We point out, however, that this method of approximation is only appropriate in certain settings. One issue is that it does not allow the accurate representation of transitive accessibility relations, where the leaves of an initial model of any depth l may be reached just by evaluating a formula with one knowledge operator. This setting calls for more work to clarify what this representation really captures, and to develop precise results about which formulas we are able to correctly evaluate using this method of approximation.

4.4   Announcing Existential Formulas

Now, we again allow the initial model to be arbitrary. In particular, we may use an initial model whose underlying frame is $KD45$ (relations are serial, transitive and Euclidean) or $S5$ (relations are equivalence relations) (see [5]). But we restrict the announcement protocol to the existential fragment[2], generated by the following rule:

$$\varphi::=p \mid \neg p \mid \varphi \wedge \varphi \mid \varphi \vee \varphi \mid \hat{K}_i\varphi \mid \bigcirc_i \varphi \mid \langle\varphi\rangle\varphi$$

where $p \in \mathcal{P}$ and $i \in AGT$. Formulas of the existential fragment are called *existential formulas*. If an announcement protocol contains only formulas of the existential fragment, we call it an *existential announcement protocol*. For instance, Example 3 consists of an existential announcement protocol.

Here we can again tackle the circularity problem by defining consistency and truth conditions separately. We first define as a fixed point the semantics of announcements in $\mathcal{A}$, together with consistency. In a second step we define the semantics of the full language with existential announcements as described in Fig. 3, but using the fixed point to evaluate consistency.

Let us fix an initial model $\mathcal{M} = (W, \{\rightarrow_i\}_{i \in AGT}, \Pi)$ and an existential announcement protocol $\mathcal{A}$. Let B be the set of all pairs (S, φ) such that S is a state of $\mathcal{M} \otimes \mathcal{A}$ and φ is either a formula in $\mathcal{A}$ or the symbol $\checkmark$, which is true only at consistent states. Observe that $(\mathcal{P}(B), \subseteq)$ forms a complete lattice. We now consider the function $f : \mathcal{P}(B) \rightarrow \mathcal{P}(B)$ defined in Fig. 4. Function f takes a set Γ of truth pairs (pairs (S, φ) such that $S \models \varphi$), and extends it with the new truth pairs that can be inferred from Γ by applying each of the rules in Figs. 2 and 3 one time. For instance, if $(w, \sigma, c) \models \varphi$ and $(w, \sigma, c) \models \psi$, then $(w, \sigma, c) \models \varphi \wedge \psi$. That is, if (w, σ, c, φ) and (w, σ, c, ψ) are in Γ, then $(w, \sigma, c, \varphi \wedge \psi)$ is in $f(\Gamma)$, which explains line 3 in Fig. 4. Every other line of Fig. 4 similarly follows from one line of the truth conditions.

Now, as we restrict to existential formulas, it is easy to see that f is monotone, that is, if $\Gamma_1 \subseteq \Gamma_2$ then $f(\Gamma_1) \subseteq f(\Gamma_2)$. By the Knaster-Tarski Theorem [12], f has a least fixed point $\Gamma^* := \bigcup_{n \in \mathbb{N}} f^n(\emptyset)$.

We can now define the truth condition for consistency as: $S \models \checkmark$ if $(S, \checkmark) \in \Gamma^*$, and use Fig. 3 to define the semantics of the language with existential announcements.

Remark 1. If announcements of the form $K_i\varphi$ were allowed, then we would have to add the clause

$$\left\{ (w, \sigma, c, K_i\varphi) \mid \begin{array}{l} \text{for all } (w', \sigma', c') \text{ such that } (w, \sigma, c)R_i(w', \sigma', c'), \\ \text{either } (w', \sigma', c', \checkmark) \notin \Gamma \text{ or } (w', \sigma', c', \varphi) \in \Gamma \end{array} \right\}$$

to the definition of f in Fig. 4. But then, if $(w, \sigma, c)R_a(w', \sigma', c')$ we would have:

---

[2] The terminology 'existential fragment' is used in the model checking community [1], because the operators $\hat{K}_i$, $\bigcirc_i$ and $\langle\varphi\rangle$ are existential. For instance, we will require that $(w, \sigma, c) \models \hat{K}_i\varphi$ iff *there exists S'* s.t. $(w, \sigma, c)R_iS'$, $S' \models \checkmark$, and $S' \models \varphi$. When these operators are not in the scope of a negation, only existential quantification needs to be used in the semantic interpretation of formulas.

$$f(\Gamma) = \Gamma \cup \{(w, \sigma, c, p) \mid p \in \Pi(w)\}$$
$$\cup \{(w, \sigma, c, \neg p) \mid p \notin \Pi(w)\}$$
$$\cup \{(w, \sigma, c, \varphi \wedge \psi) \mid (w, \sigma, c, \varphi) \in \Gamma \text{ and } (w, \sigma, c, \psi) \in \Gamma\}$$
$$\cup \{(w, \sigma, c, \varphi \vee \psi) \mid ((w, \sigma, c, \varphi) \in \Gamma \text{ or } (w, \sigma, c, \psi) \in \Gamma\}$$
$$\cup \left\{(w, \sigma, c, \hat{K}_i\varphi) \mid \begin{array}{l} \text{there exists } (w', \sigma', c') \text{ such that } (w, \sigma, c)R_i(w', \sigma', c'), \\ (w', \sigma', c', \checkmark) \in \Gamma \text{ and } (w', \sigma', c', \varphi) \in \Gamma \end{array}\right\}$$
$$\cup \{(w, \epsilon, \mathbf{0}, \checkmark) \mid w \in W\}$$
$$\cup \{(w, \sigma, c, \checkmark) \mid \text{ there is } c' < c \text{ s.t. } (w, \sigma, c', \checkmark) \in \Gamma\}$$
$$\cup \left\{(w, \sigma, c, \checkmark) \mid \begin{array}{l} (w, \sigma', c, \checkmark) \in \Gamma \text{ and } (w, \sigma', c, \psi) \in \Gamma, \\ \text{where } \sigma = \sigma' :: \psi \end{array}\right\}$$
$$\cup \{(w, \sigma, c, \bigcirc_i\varphi) \mid c(i) < |\sigma| \text{ and } (w, \sigma, c^{+i}, \varphi) \in \Gamma\}$$
$$\cup \{(w, \sigma, c, \langle\psi\rangle\varphi) \mid \sigma{::}\psi \in \mathsf{Seq}(\mathcal{A}), (w, \sigma, c, \psi) \in \Gamma \text{ and } (w, \sigma{::}\psi, c, \varphi) \in \Gamma\}$$

Fig. 4. Function applying one step of the truth conditions

- $(w, \sigma, c, K_a p) \in f(\emptyset)$;
- $(w, \sigma, c, K_a p) \notin f(\{(w', \sigma', c', \checkmark), (w', \sigma', c', \neg p)\})$

It would thus no longer hold that $f(\Gamma_1) \subseteq f(\Gamma_2)$ whenever $\Gamma_1 \subseteq \Gamma_2$. As f is clearly not a decreasing function either, we would not be able to apply the Knaster-Tarski theorem.

5   Validities

We say that a formula φ is *valid* if for every initial model $\mathcal{M}$ and every announcement protocol $\mathcal{A}$, such that either $\mathcal{M}$ is a finite tree or $\mathcal{A}$ is an existential announcement protocol, and for every consistent state $S \in \mathcal{M} \otimes \mathcal{A}$,[3] we have $\mathcal{M} \otimes \mathcal{A}, S \models \varphi$. We write $\models \varphi$ to express that φ is valid. As usual, we use $[\varphi]\psi$ as shorthand for $\neg\langle\varphi\rangle\neg\psi$.

Proposition 1. *We have:*

1. $\models \bigcirc_1 \bigcirc_2 \varphi \leftrightarrow \bigcirc_2 \bigcirc_1 \varphi$
2. $\models \bigcirc_1 \top \rightarrow (\bigcirc_1\varphi \leftrightarrow \neg \bigcirc_1 \neg\varphi)$
3. $\models \neg \bigcirc_1 \top \rightarrow [\varphi] \bigcirc_1 K_1\varphi$, *where φ is a propositional formula* [4]

Proof. We prove the first validity and the other two are left to the reader.
Suppose that we have $\mathcal{M} \otimes \mathcal{A}, (w, \sigma, c) \models \bigcirc_1 \bigcirc_2 \varphi$. By Fig. 3, this means that $c(1) < |\sigma|$ and $\mathcal{M} \otimes \mathcal{A}, (w, \sigma, c^{+1}) \models \bigcirc_2\varphi$, and the latter implies that

---

[3] Recall definition of $\mathcal{M} \otimes \mathcal{A}$ from 3.2.
[4] A propositional formula is any formula without modalities, i.e. no occurrences of K_i, $\langle\varphi\rangle$, or $\bigcirc$.

$c^{+1}(2) < |\sigma|$ and $\mathcal{M} \otimes \mathcal{A}, (w, \sigma, (c^{+1})^{+2}) \models \varphi$. Now, because $(c^{+1})^{+2} = (c^{+2})^{+1}$, we get that $\mathcal{M} \otimes \mathcal{A}, (w, \sigma, c^{+2}) \models \bigcirc_1 \varphi$, and therefore $\mathcal{M} \otimes \mathcal{A}, (w, \sigma, c) \models \bigcirc_2 \bigcirc_1 \varphi$. The proof for the other direction is symmetric.

Let us comment on the above validities. Validity 1 says that it is possible to permute the order of agents that receive next messages in their respective queues. Validity 2 says that if an agent has a message to read, then reading the message is a deterministic operation. Validity 3 says that if an agent has no pending messages and some propositional formula[5] is announced, then after reading his next message, the agent will know that formula.

We also establish the following proposition, which essentially says that in the case where all sequences of announcements are possible, if all the $\bigcirc_i$ operators in a formula φ are under the scope of a knowledge operator, then its truth value is left unchanged by the announcement of any other formula ψ. Indeed, the knowledge operator considers all possibilities for the content of the agent's channel, so that the possibility that ψ is in it is considered, no matter whether it was actually the announced formula or not.

In the following, either let $\mathcal{M}$ be a finite tree and $\mathcal{A}_U$ the universal announcement protocol containing every formula with infinite cardinality, or let $\mathcal{M}$ be an arbitrary initial model and $\mathcal{A}_U$ the announcement protocol containing every formula in the existential fragment with cardinality infinity.

Proposition 2. *Let φ be a formula in $\mathcal{L}$, in which every $\bigcirc_i$ is under the scope of some K_j. For every initial model $\mathcal{M}$ and consistent state $S = (w, \sigma, c) \in \mathcal{M} \otimes \mathcal{A}_U$, for every $\psi \in \mathcal{A}_U$, we have $\mathcal{M} \otimes \mathcal{A}_U, S \models \langle \psi \rangle \varphi \leftrightarrow \psi \wedge \varphi$.*

This result follows immediately from the following lemma:

Lemma 1. *Let φ be a formula in $\mathcal{L}$, in which every $\bigcirc_i$ is under the scope of some K_j. For every initial model $\mathcal{M}$ and for every consistent state $S = (w, \sigma, c) \in \mathcal{M} \otimes \mathcal{A}_U$, for every $\psi \in \mathcal{A}_U$ such that $(w, \sigma::\psi, c)$ is consistent, we have $\mathcal{M} \otimes \mathcal{A}_U, (w, \sigma::\psi, c) \models \varphi$ iff $\mathcal{M} \otimes \mathcal{A}_U, (w, \sigma, c) \models \varphi$.*

Proof. By induction on φ. The Boolean cases are omitted.

Case $\varphi = K_a \varphi'$:

It is enough to observe that: $\{S \mid (w, \sigma::\psi, c) R_a S\} = \{S \mid (w, \sigma, c) R_a S\}$.

Case $\varphi = \langle \varphi_1 \rangle \varphi_2$:

By definition of $\mathcal{A}_U$, $\sigma::\varphi_1 \in \mathsf{Seq}(\mathcal{A}_U)$. We therefore have $(w, \sigma::\psi, c) \models \langle \varphi_1 \rangle \varphi_2$ iff $(w, \sigma::\psi, c) \models \varphi_1$ and $(w, \sigma::\psi::\varphi_1, c) \models \varphi_2$. By induction hypothesis, this is equivalent to $(w, \sigma, c) \models \varphi_1$ and $(w, \sigma::\psi, c) \models \varphi_2$. Again by induction hypothesis, the latter is equivalent to $(w, \sigma, c) \models \varphi_2$, which is in turn equivalent to $(w, \sigma::\varphi_1, c) \models \varphi_2$ (observe that $(w, \sigma::\varphi_1, c)$ is consistent as $(w, \sigma, c) \models \varphi_1$). We finally obtain that $(w, \sigma::\psi, c) \models \langle \varphi_1 \rangle \varphi_2$ iff $(w, \sigma, c) \models \varphi_1$ and $(w, \sigma::\varphi_1, c) \models \varphi_2$, that is $(w, \sigma, c) \models \langle \varphi_1 \rangle \varphi_2$.

---

[5] We restrict to propositional formulas in order to avoid Moore's paradox [14].

Finally, the case $\varphi = \bigcirc_i \varphi'$ is not possible as $\bigcirc_i$ is not under the scope of any K_j.

In our framework, the behavior of the public channel is common knowledge. For instance, let us consider a situation with two agents, 1 and 2, where Agent 1 has read all the announced messages. Now, assume that p is announced, thus put in the queue, and Agent 1 reads it. Agent 1 now knows p, but she also knows that if Agent 2 has read all the announced messages (and in particular the last one, which is p), then Agent 2 also knows p. In some sense, it means that initially Agent 1 knows that Agent 2 will receive the same messages as herself. This is reflected in the following validity: $\bigcirc_1 \bot \rightarrow [p!]\, \bigcirc_1 K_1(p \wedge (\bigcirc_2 \bot \rightarrow K_2 p))$.

6   Model Checking

Here we address the model checking problem when $\mathcal{A}$ is a finite multiset, that is, when the support set of $\mathcal{A}$ is finite and the multiplicity of each element is an integer. More precisely, we consider the following decision problem:

- input: an initial pointed model,[6] $(\mathcal{M}, w)$, a *finite* multiset of formulas $\mathcal{A}$ (where multiplicities are written in *unary*), a formula φ_0;
- output: yes if $\mathcal{M} \otimes \mathcal{A}, (w, \epsilon, \mathbf{0}) \models \varphi_0$, no otherwise.

6.1   Propositional Announcements

In this section, we suppose that formulas in $\mathcal{A}$ are propositional (which is a particular case of existential announcements). We consider the model checking problem for asynchronous announcement logic where inputs $\mathcal{M}, w, \mathcal{A}, \varphi_0$ are such that $\mathcal{A}$ only contains propositional formulas. This problem is called the *model checking problem for propositional protocols*.

Theorem 1. *The model checking problem for propositional protocols is in PSPACE.*

Proof. Figure 5 presents an algorithm that takes a pointed model $(\mathcal{M}, w)$, a finite multiset $\mathcal{A}$, a sequence $\sigma \in \mathsf{Seq}(\mathcal{A})$, a cut c on σ and a formula φ as an input. To check the consistency of a state (w, σ, c), we call $\mathsf{checkconsistency}(\mathcal{M}, \mathcal{A}, w, \sigma, c)$ which verifies that every (propositional) formula ψ occurring in σ evaluates to true with the valuation $\Pi(w)$.

It is easily proven by induction that, for all ψ, the following property $P(\psi)$ holds:

$$\mathcal{M}, \mathcal{A}, w, \sigma, c \models \psi \text{ iff } \mathsf{mc}\,(\mathcal{M}, \mathcal{A}, w, \sigma, c, \psi) \text{ returns true.}$$

---

[6] A pointed model is a model with a specified state.

This establishes soundness and completeness of the algorithm. We now analyze its complexity.

First, observe that because $\mathcal{A}$ is finite and each element has finite multiplicity, we have that $\mathsf{Seq}(\mathcal{A})$ only contains sequences of length linear in $|\mathcal{A}|$ (recall that multiplicities are written in unary). It is therefore easy to see that the consistency check $(*\checkmark)$ is done in polynomial time in the size of the input and thus requires a polynomial amount of space. Now, the number of nested calls of mc is bounded by the size of the formula to check, and each call requires a polynomial amount of memory for storing local variables, so that the algorithm runs in polynomial space.

```
function mc (M, A, w, σ, c, φ)
  match φ do
    case p: return p ∈ V(w);
    case ✓: return checkconsistency(M, A, w, σ, c)                    (*✓)
    case ¬ψ: return not mc (M, A, w, σ, c, ψ);
    case (ψ₁ ∧ ψ₂): return mc (M, A, w, σ, c, ψ₁) and mc(M, A, w, σ, c, ψ₂);
    case Kₐψ :
      for u ∈ Rₐ(w), σ' ∈ Seq(A), c' on σ' do
        if c'(i) = c(i) and σ'[1..c(i)] = σ[1..c(i)] and
        mc (M, A, u, σ', c', ✓)  then
          if not mc (M, A, u, σ', c', ψ) then
            return false
      return true
    case ⟨ψ⟩χ :
      if σ::ψ ∈ Seq(A) and mc (M, A, w, σ, c, ψ) then
        return mc (M, A, w, σ.ψ, c, χ);
      else
        return false;
    case ○ᵢψ: return c(i) < |σ| and mc (M, A, w, σ, cᵗⁱ, ψ)
```

Fig. 5. Model checking algorithm

Theorem 2. *The model checking problem for propositional protocols is PSPACE-hard.*

6.2   Finite Tree Initial Model

In this section, we restrict the set of inputs $\mathcal{M}, \mathcal{A}, w, \varphi_0$ of the model checking problem to those where the initial pointed models $(\mathcal{M}, w)$ are finite trees rooted in w.

Theorem 3. *The model checking problem when we restrict initial models to finite trees is in PSPACE.*

Proof. We consider the algorithm of Fig. 5 again but now the consistency checking $(*\checkmark)$ consists of calling the following procedure:

```
function checkconsistency(M, A, w, σ, c)
  if c = 0
    | return true
  else
    for c' < c do
      if mc (M, A, w, σ, c', ✓) then
        | return true
    return mc (M, A, w, σ', c, ✓) and mc (M, A, w, σ', c, ψ)  where σ = σ'.ψ
```

Soundness and completeness are proven by induction on inputs using the order $\prec$ defined in Sect. 4.3.

Concerning the complexity, the argument given in the proof of Theorem 1 no longer holds. In order to bound the number of nested calls of mc, we have to remark that from a call of mc to a sub-call of mc:

(1) either we change the current world w in the initial model for a successor u in the finite tree;
(2) or the quantity $|\sigma| + |\varphi| + \sum_{i \in AGT} c(i)$ is strictly decreasing, where $|\varphi|$ is the length of φ and if $\sigma = [\varphi_1, \ldots, \varphi_k]$ then $|\sigma| = \sum_{i=1}^{k} |\varphi_i|$.

Now, the number of times (1) occurs is bounded by the depth $depth(M, w)$ of the finite tree M, w. As each φ is either a subformula of the input formula φ_0 or a subformula of a formula in A, $|\varphi| \le |\varphi_0| + |A|$ where $|A| := \sum_{\psi \in A} |\psi|$, and where each single formula ψ is counted as many times as it occurs in the multiset A. Furthermore, $|\sigma| \le |A|$ and $c(i) \le |A|$. Thus, the quantity $|\sigma| + |\varphi| + \sum_{i \in AGT} c(i)$ is bounded by $(|AGT| + 2)|A| + |\varphi_0|$. Therefore, the number of nested calls to mc is bounded by $depth(M, w) \times ((|AGT| + 2)|A| + |\varphi_0|)$. So the algorithm requires polynomial amount of memory in the size of the input (recall that the multiplicity of A is encoded in unary).

6.3   Existential Announcements

In this subsection, we design an exponential time algorithm for the model checking problem in the case of existential announcements.

Given an input M, A, w, φ_0, the algorithm first computes the least fixed point Γ^* of the function f defined in Sect. 4.4. Because the number of possible sequences in $\mathsf{Seq}(A)$ is exponential in $|A|$, the set B of pairs (S, φ) where $S \in M \otimes A$ and $\varphi \in A \cup \{\checkmark\}$ is exponential size in the size of the input, and therefore computing the fixed point requires exponential time in the size of the input. This gives us the semantics of consistency for states of $M \otimes A$.

Then, to evaluate φ_0, we use the procedure mc of Fig. 5, in which checking the consistency of a state (w, σ, c), $(*_\checkmark)$, is done by checking whether $(w, \sigma, c, \checkmark) \in \Gamma^*$. The algorithm mc also requires exponential time. To sum up:

Theorem 4. *The model checking problem where the announcements are existential is in EXPTIME.*

7   Related Work

As far as we know, there has not been much work on the relationship between knowledge, announcements and asynchronicity. In [4], asynchronicity in dynamic epistemic logic is studied, but with the idea of synchronicity being that all agents observe a universal clock, whereas our notion of synchronicity is that all agents receive messages at the same time, immediately when they are sent. A logic dealing with knowledge and asynchronicity is also developed in [10], but in this setting, messages do not have logical content: they are atomic propositions, and it is impossible, for example, to make an announcement about knowledge or about the effect of another announcement.

Arbitrary public announcement logic (APAL) [2] has some similarity to our approach. In this logic, one can ask whether some formula holds after any possible announcement; this is not possible in our logic, but the knowledge operator considers any possible future sequence of announcements that follows the protocol, which is a similar idea. Interestingly, the satisfiability problem for APAL is undecidable, but decidability can be achieved by considering a constraint similar to our restriction to existential announcements [6,13].

The Knaster-Tarski theorem is often used to define the denotational semantics of programming languages [15] in the same spirit as our definition of consistency when announcements are existential. We also note that our definition of asynchronous models $\mathcal{M} \otimes \mathcal{A}$, especially the notion of cuts, is in the spirit of [8].

8   Future Work

This work constitutes a first attempt to provide an epistemic logic for reasoning about asynchronous announcements. In the future, first, we would like to overcome the circularity problem, and hence define the semantics, for the most general case (removing the finite tree and existential conditions), and provide model checking algorithms in these cases. One approach to this problem may be using coinduction to define the set of consistent states. Once we have defined the semantics for the general case, we plan to provide a complete axiomatization.

Second, we would like to model more general situations of asynchronous communication. We plan to consider the case where messages are not read in FIFO order, but are received and read in arbitrary order. We also plan to model the origin of the messages, allowing formulas saying that "After Agent a broadcasts that φ holds, ψ holds". In our current setting, when the external broadcaster makes a new announcement, the only effect is to queue it in the channel without affecting anyone's epistemic state. However, in the case where the agents themselves make the announcements, Agent a making an announcement should impact her knowledge: after the announcement she should know, for instance, that the channel is not empty. She should also know that, after another agent checks their channel, that agent will know that ψ has been announced.

Third, we would like to model not only asynchronous broadcast on a public channel but also private asynchronous communications between agents in the system. In essence, this amounts to defining a complete asynchronous version of dynamic epistemic logic [14].

Finally, it would be interesting to add temporal operators to our language, in order to express such things as "After p is announced and Agent 1 receives it, *eventually* she will know that Agent 2 knows p" (assuming that agents are forced to eventually read announcements).

Acknowledgments. We thank Hans van Ditmarsch for invaluable discussion and support of this work. We acknowledge support from ERC project EPS 313360.

References

1. Baier, C., Katoen, J.-P.: Principles of Model Checking. MIT Press, Cambridge (2008)
2. Balbiani, P., Baltag, A., van Ditmarsch, H.P., Herzig, A., Hoshi, T., De Lima, T.: What can we achieve by arbitrary announcements?: A dynamic take on Fitch's knowability. In: Proceedings of the 11th Conference on Theoretical Aspects of Rationality and Knowledge (TARK-2007), Brussels, Belgium, June 25–27, 2007, pp. 42–51 (2007)
3. Balbiani, P., van Ditmarsch, H., Herzig, A., De Lima, T.: Tableaux for public announcement logic. J. Log. Comput. **20**(1), 55–76 (2010)
4. Dégremont, C., Löwe, B., Witzel, A.: The synchronicity of dynamic epistemic logic. In: Proceedings of the 13th Conference on Theoretical Aspects of Rationality and Knowledge (TARK-2011), Groningen, The Netherlands, July 12–14, 2011, pp. 145–152 (2011)
5. Fagin, R., Halpern, J., Moses, Y., Vardi, M.: Reasoning About Knowledge. The MIT Press, Cambridge (2004)
6. French, T., van Ditmarsch, H.P.: Undecidability for arbitrary public announcement logic. In: Advances in Modal Logic 7, Papers From the Seventh Conference on "Advances in Modal Logic," held in Nancy, France, 9–12 September 2008, pp. 23–42 (2008)
7. Jones, B.D.: Bounded rationality. In: Annual Review of Political Science, pp. 2–297 (1999)
8. Lamport, L.: Time, clocks, and the ordering of events in a distributed system. Commun. ACM **21**(7), 558–565 (1978)
9. Lutz, C.: Complexity and succinctness of public announcement logic. In: 5th International Joint Conference on Autonomous Agents and Multiagent Systems (AAMAS 2006), Hakodate, Japan, May 8–12, 2006, pp. 137–143 (2006)
10. Panangaden, P., Taylor, K.: Concurrent common knowledge: Defining agreement for asynchronous systems. Distrib. Comput. **6**(2), 73–93 (1992)
11. Plaza, J.: Logics of public communications. Synthese **158**(2), 165–179 (2007)
12. Tarski, A., et al.: A lattice-theoretical fixpoint theorem and its applications. Pac. J. Math. **5**(2), 285–309 (1955)
13. van Ditmarsch, H., French, T., Hales, J.: Positive announcements (under submission)
14. van Ditmarsch, H., van der Hoek, W., Kooi, B.P.: Dynamic epistemic logic, vol. 337. Springer Science and Business Media (2007)
15. Winskel, G.: The formal semantics of programming languages - an introduction. Foundation of computing series. MIT Press, Amsterdam (1993)

A Totally Distributed Fair Scheduler for Population Protocols by Randomized Handshakes

N. Ouled Abdallah[1,2]([✉]), M. Jmaiel[2,3], M. Mosbah[1], and A. Zemmari[1]

[1] LaBRI, University of Bordeaux - CNRS, 351 Cours de la Libération,
33405 Talence, France
{nouled-a,mosbah,zemmari}@labri.fr
[2] ReDCAD Laboratory, University of Sfax, National School of Engineers of Sfax,
B.P. 1173, 3038 Sfax, Tunisia
mohamed.jmaiel@enis.rnu.tn
[3] Research Center for Computer Science, Multimedia and Digital Data Processing
of Sfax, B.P. 275, Sakiet Ezzit, 3021 Sfax, Tunisia

Abstract. A population protocol is a computational model based on pairwise interactions and designed for networks of passively mobile finite state agents. In the population protocol model, and also in the models that extend it, the interacting pairs are supposed to be chosen by a theoretical fair scheduler. In this paper, we present the *HS Scheduler* which is a totally distributed synchronous randomized handshake procedure. We then prove that this randomized handshake procedure can be a probabilistic consistent scheduler for population protocols that is fair with probability 1. By adopting a protocol aware version of the *HS Scheduler*, we introduce the *iterated population protocols* model where nodes can stop participating in the protocol's computation once they reach a *final state*. We then study the time complexity of the computation of a particular case of this model where a *final state* is reached in only one computation step. We present some upper bounds that are later validated by simulations results.

Keywords: Population protocol · Distributed randomized handshake · Probabilistic fair scheduler · Iterated population protocol

1 Introduction

Sensor networks are composed of small entities with limited resources, memory and computational power. Deployed in a specified area, the sensor nodes have to communicate via a wireless media and cooperate to finally reach a global goal. The sensor nodes can compute a global property of their environment: calculate the global temperature, detect if there is an intrusion in a monitored area, etc. But how to compute in such networks with restricted resources and capacities?

Angluin et al. proposed the population protocols model [3]. A population protocol is a pairwise computational model designed for anonymous passively

© Springer International Publishing Switzerland 2015
M. Leucker et al. (Eds.): ICTAC 2015, LNCS 9399, pp. 290–306, 2015.
DOI: 10.1007/978-3-319-25150-9_18

mobile finite state agents in populations of finite but unbounded size. Initially, the entities have inputs that will be mapped to states according to an input function. Then, communication between pairs of nodes can take place permitting to the nodes to exchange their states and to update them according to the defined transition function of the protocol. Like in any distributed computing system, there are many facts, such as the environment, the mobility of the nodes or their energies, that may interfere in the selection of the entities that will be able to communicate and also in the order in which they will communicate. In literature, these facts are represented by the scheduler.

However, to guarantee the success of the task of the distributed system, a fairness condition should be guaranteed by the scheduler. Some works suppose a local fairness condition which preserves the fact that each node of the system is given a turn to act infinitely often. The local fairness is one variant of weak fairness in distributed computing [4]. In the population protocols model, the imposed fairness is a global one. Angluin et al. give a definition of the global fairness that they assume in their model, and propose a probabilistic scheduler called the Random Scheduler. Spirakis et al. propose a new definition of this fairness and present two new probabilistic fair schedulers [6]: the State Scheduler and the Transition Function Scheduler. All these schedulers are supposed to be theoretical entities able to choose at each iteration of the protocol only one communicating pair of nodes.

Based on the rendezvous algorithm of Métivier et al. [9], and inspired from its use in [11] as a communication synchronizer in population protocols, we present in this paper a distributed randomized handshake algorithm that we called *Handshake Scheduler for population protocols (HS Scheduler)*. This algorithm can in somehow simulate realistic facts influencing the communication in a network: There could be no possible communication at a given time as it could be more than one ordered communicating pair of nodes. We prove that the *HS Scheduler* can be a probabilistic consistent fair scheduler for population protocols with probability 1. We give some analysis of the behavior of this scheduler. We then introduce the model of the *iterated population protocols* where, unlike the basic population protocols model, the nodes can stop participating in the protocol's computation once they reach a specific *final state*. We propose a protocol aware version of the *HS Scheduler* adapted to this new model. We then focus on the case where only one computation step is enough for a node to reach a *final state*. We present some upper bounds of the time complexity of a protocol computation in this model while considering some random possible topologies of the communication graph. As an application, we propose an *iterated mediated population protocol* that computes a Maximal Matching of a communication graph. We implement and simulate this protocol on the ViSiDiA platform [1]. And thanks to the simulations results, the theoretical upper bounds of the time complexity are validated.

The paper is organized as follows. Section 2 recalls the main definition of population protocols. Section 3 presents the already existing schedulers in population protocols. In Sect. 4, the distributed fair scheduler based on randomized handshakes is defined. Then, Sect. 5 introduces the iterated population protocols and some theoretical results about the time complexity upper bounds of a particular variant of this model that are later validated by experimental results in Sect. 6. Finally, the paper is concluded by a recall of the main contributions and ideas about future directions.

2   Population Protocols

Formally, a **population protocol** $\mathcal{A}$ consists of a 6-tuple (X, Y, Q, I, O, δ), where:

- X: a finite input alphabet,
- Y: a finite output alphabet,
- Q: a finite set of states,
- $I: X \to Q$: an input function mapping inputs to states,
- $O: Q \to Y$: an output function mapping states to outputs,
- $\delta: Q \times Q \to Q \times Q$: a transition function defined on pairs of states as a set of transition rules. If $\delta(u, v) = (u', v')$, then $(u, v) \to (u', v')$ is a transition, $\delta_1(u, v) = u'$ and $\delta_2(u, v) = v'$. Note that δ is not symmetric as u here plays the role of an initiator and v plays the role of a responder.

Running the protocol $\mathcal{A}$ on a population $\mathcal{P}$ consisting of a set A of n agents can be described as follows. Initially, the agents are deployed to sense a specific parameter from their environment. The sensed values will be defined as inputs from the alphabet X and thereafter mapped to states from Q according to I. The resulting states will form the initial configuration C_0 of the protocol $\mathcal{A}$. As defined in [3], a population's **configuration** is a snapshot of the agents states of the population. Formally, it is a mapping $C: A \to Q$ that specifies the state of each member of the population. Then, interactions between pairs of agents can take place. Two agents are able to establish a two-way communication once they come sufficiently close to each other. They exchange their states and update them according to the transition function δ. The graph $G = (V, E)$, where V is the set of the vertices representing the set A of the population's agents and E the set of the edges representing all the possible communications links between pairs of nodes, is called the **communication graph** (or also the **interaction graph**). The communication graph can be directed but without self-loops.

Let C and C' be population configurations, and let u, v be distinct agents. We say that C goes to C' via an encounter $e = (u, v) \in E$, denoted $C \xrightarrow{e} C'$ (or $C \to C'$), if $C'(u) = \delta_1(C(u), C(v))$, $C'(v) = \delta_2(C(u), C(v))$ and $C'(w) = C(w)$ for all $w \in A \smallsetminus \{u, v\}$. C' is the configuration resulting from the interaction between the pair of nodes u and v on the configuration C. We say that C goes to C' in one step. C' can also be reachable from C via a sequence of configurations $C_0, C_1, \ldots, C_k$ where $C = C_0$ and $C_k = C'$ and we can write $C \xrightarrow{*} C'$.

An **execution** is a finite or infinite sequence of population configurations C_0, $C_1, C_2, \ldots$ such that for each i, $C_i \to C_{i+1}$. An infinite execution is **fair** if for every possible transition $C \to C'$, if C occurs infinitely often in the execution, then C' also occurs infinitely often. A **computation** is an infinite fair execution [3].

The population protocols do not halt, but they stabilize once the outputs stop changing: it is when the computations lead to a configuration C such as, for every C' reachable from C, $O(C) = O(C')$. The agents can continue interacting but the outputs of their states will not change anymore.

3   Related Works : Existing Schedulers

There are different works that extend the population protocols model. We cite the mediated population protocols [8] where, in addition to assigning states to nodes, states are assigned to the edges of the communication graph. We also cite Paloma [7], where the nodes are equipped with a $O(\log(n))$ memory space, instead of a constant one, with n the total number of nodes in the network. For the basic population protocol model and all those extending it, the authors always suppose the existence of a **fair** scheduler responsible of the choice of the pairs of nodes that will communicate. Angluin et al. characterize the fairness by the following condition: let C, C' two configurations such as $C \rightarrow C'$, if C occurs infinitely often in the execution, then C' will too. For the probabilistic population protocols model, they propose a probabilistic scheduler called the Random Scheduler. In [6], Spirakis et al. give a new definition of the fairness condition (that we will detail later) and they present two new probabilistic schedulers: the State Scheduler and the Transition Function Scheduler.

We give a brief description of these schedulers:

- **The Random Scheduler**: At each step, this scheduler chooses independently, randomly and uniformly only one edge from the interaction graph which is supposed to be complete. This scheduler allows to only one pair of nodes to communicate by exchanging their states and then updating them according to the transition function of the protocol. Given this random, uniform and independent choice, Angluin et al. stated that the Random Scheduler is fair with probability 1. Also, Spirakis et al. proved that this scheduler is fair with probability 1 with respect to their fairness definition. This scheduler acts without any knowledge on the protocol that the population is running, so it is called a protocol oblivious (or agnostic) scheduler.
- **The State Scheduler**: Unlike the Random Scheduler, the State Scheduler is protocol aware which means that it has some knowledge about the protocol run by the population. The scheduler takes this knowledge into account while choosing the pairs of nodes that will communicate. This scheduler first chooses independently and uniformly at random an ordered pair of states (q_1, q_2) from all the interaction candidates of the current configuration. An ordered pair of states (q_1, q_2) is said interaction candidate under a configuration C if $\exists (u, v) \in E$ such that $C(u) = q_1$ and $C(v) = q_2$. Then, the scheduler chooses independently and uniformly at random only one ordered pair of nodes related by an edge in E and which states are (q_1, q_2). This chosen nodes will communicate and update their states according to the transition function. This scheduler was proved to be fair with probability 1.

– **The Transition Function Scheduler**: This scheduler is protocol aware as the choices it makes concerning the communicating pair is based on the transition function rules of the protocol. The scheduler chooses pairs that, when they communicate, will lead to the protocol progress: that is at least the initiator or the responder will change its state or even both of them and it ignores all the transitions where no state changes. To reach a configuration C' from a configuration C, the scheduler proceeds as follows. First, it picks independently and uniformly at random a pair $((q_1, q_2), (q_1', q_2'))$ where (q_1, q_2) is an interaction candidate under C and $\delta(q_1, q_2) = (q_1', q_2')$. Then, it chooses independently and uniformly at random an ordered pair of nodes $(u, v) \in E$, which states correspond to (q_1, q_2) under C. Then, to obtain C', the transition function is applied to (q_1, q_2) to update their states. In case the transition function scheduler can not find any pair able to lead to the protocol progress, it works like the Random Scheduler. This scheduler was proved to be fair with probability 1.

4 The Distributed Probabilistic Fair Handshake Scheduler for Population Protocols

In this section, we introduce the *HS Scheduler* that is a probabilistic fair scheduler based on randomized handshakes. It is implementable in a distributed way and can be used in simulations and experiences of population protocols in real-world sensor networks without any need of identifiers, unlike what is used in [5]. Also, compared to this previous work, the designation of the initiator and the responder is obtained in one computation step.

4.1 The Randomized Rendezvous Algorithm

In anonymous asynchronous networks, based on point-to-point communication via synchronous message passing, both the sender and the receiver need to be ready to exchange messages. It is an agreement to communicate and it is called a **rendezvous**. As Angluin state in [2] that there is no deterministic algorithm to implement synchronous message passing in an anonymous network that passes messages asynchronously, the authors of [9] propose a distributed randomized rendezvous algorithm described by Algorithm 1. Each node v of the network chooses at random a node $c(v)$ from its neighborhood, for which it sends 1, and it sends 0 to the rest of its neighbors. This implies that v would like to synchronize with node $c(v)$. There is a rendezvous between v and $c(v)$ if they mutually choose each other by sending 1 to each other.

Locally, when a node v has more than one neighbor, Algorithm 1 defines how v will choose only one of them to synchronize with. This guarantees that each node in the network will never be able to participate in more than one rendezvous at a given time. As a consequence, the global result of Algorithm 1 is a set of disjoint pairs of nodes that are synchronized.

Algorithm 1. The Randomized Rendezvous Algorithm

1: **loop**
2:     Randomly choose a neighbor $c(v)$;
3:     Send 1 to $c(v)$;
4:     Send 0 to each neighbor $p \neq c(v)$;
5:     Receive messages from all the neighbors;
6:     $r_{c(v)} \leftarrow$ the number received from $c(v)$;
7:     **if** $r_{c(v)} = 0$ **then**
8:         There is no rendezvous
9:     **else**
10:         There is a rendezvous with $c(v)$
11:     **end if**
12: **end loop**

4.2 The Randomized Handshake Scheduler for Population Protocols

The authors of [11] study the broadcast in anonymous mobile asynchronous Wireless Sensor Networks. They propose a randomized algorithm based on the OR population protocol to avoid collision and information duplication problems. They used Algorithm 1 as a synchronizer between pairs of nodes: communications take place between pairs of nodes that succeeded in obtaining rendezvous. These pairs are not ordered, which means that the initiator and the responder are not distinguished. This symmetry does not affect in any way their result, as in the OR protocol, the transition function consists of one symmetric transition rule which is:

$$\delta(q_1, q_0) = \delta(q_0, q_1) = (q_1, q_1).$$

We can conclude from [11] that the randomized rendezvous algorithm can be useful in synchronizing pairs of nodes before starting the protocol exchanges. We can also notice that the global result of this algorithm is in somehow what a scheduler is supposed to do in a distributed system, which is choosing which pairs of nodes will interact and in what order they will interact. However, the pair of nodes picked by a scheduler in population protocols should be ordered so that each of the two nodes is assigned a role of an initiator or a responder. Therefore, to really fit the role of a scheduler for population protocols, we have to make some adjustments to the randomized rendezvous algorithm. We introduce symmetry breaking between the chosen pairs of nodes to obtain ordered pairs. So, instead of sending 1 when inviting the chosen neighbor, a node v will generate an integer $r_v \in \{1, 2, \ldots, N\}$ where N is a constant such that $N \geq 2$. A rendezvous will take place, if two nodes mutually choose each other and generate two different non zero values. The role of the initiator in the application of the transition rules of the population protocol will be attributed to the node that generated the higher value. Also, as we consider a synchronous execution of Algorithm 1, and to avoid energy dissipation and network congestion, a node v will not send the 0 messages anymore to inform its neighbors that its choice does not involve them. If a node v did not receive any message from its chosen neighbor $c(v)$ before the end of the round, it considers that $c(v)$ has chosen another neighbor to communicate with.

The result of these modifications is the distributed algorithm described in Algorithm 2. This algorithm can play the role of a distributed scheduler for population protocols as, based on handshakes, it is able to choose randomly at each step, a set of disjoint ordered pairs of nodes among all those of the interaction graph. We call this algorithm the *Handshake Scheduler for population protocols* (and we denote it later by *HS Scheduler*).

Algorithm 2. The Randomized Handshake Scheduler Algorithm

1: **loop**
2:     Randomly choose a neighbor $c(v)$;
3:     Choose a random number $r_v \in \{1, 2, \ldots, N\}$;
4:     Send r_v to $c(v)$;
5:     $r_{c(v)} \leftarrow$ the message received from $c(v)$;
6:     **if** $r_{c(v)} = NULL$ or $r_{c(v)} = r_v$ **then**
7:         There is no synchronization
8:     **else**
9:         Start the synchronization
10:         **if** $r_v > r_{c(v)}$ **then**
11:             The current node is the initiator
12:         **else**
13:             The current node is the responder
14:         **end if**
15:         Exchange state with $c(v)$;
16:         Apply a transition rule (if applicable) according to the attributed role;
17:         Terminate the synchronization
18:     **end if**
19: **end loop**

Once a synchronization (or a handshake) is established between two nodes, they exchange their states. If there is a transition rule involving these states, they will apply it with respect to their roles.

Probability of at Least a Handshake. For any pair $e = (u, v)$, such that the edge e exists in the communication graph G, there is a **handshake** between nodes u and v (we denote this event $\mathcal{HS}(e)$) if, and only if, $c(u) = v$ and $c(v) = u$ and $r_u \neq r_v$.

We first note that the probability that $c(u) = v$ is

$$Pr(c(u) = v) = \frac{1}{d(u)}, \tag{1}$$

where $d(u)$ is the degree of the node u.

We also note that the probability that u and v generate two different non zero values is:

$$\Pr(r_u \neq r_v) = \sum_{i=1}^{N} \frac{1}{N} \frac{N-1}{N} = 1 - \frac{1}{N}. \tag{2}$$

Then we obtain:

$$\Pr\left(\mathcal{HS}(e)\right) = \frac{1 - \frac{1}{N}}{d(u)d(v)}. \tag{3}$$

In the sequel, we denote $\{e_1, \ldots, e_m\}$ the set of edges in the interaction graph. We also denote by $\mathcal{HS}_G$ the event: *There is at least a handshake in the graph G.* The $\overline{\mathcal{HS}}_G$ and $\overline{\mathcal{HS}}(e)$ are respectively the complement event of $\mathcal{HS}_G$ and $\mathcal{HS}(e)$. Then we have:

$$\Pr\left(\overline{\mathcal{HS}}_G\right) = \Pr\left(\wedge_{i=1}^{m} \overline{\mathcal{HS}}(e_i)\right). \tag{4}$$

This leads us to the case already analyzed in the work of Métivier et al. [9]. So, by reusing this result with the value of $\Pr\left(\mathcal{HS}(e)\right)$ that we already established, we obtain:

$$\Pr\left(\overline{\mathcal{HS}}_G\right) \leq \left(1 - \frac{1 - \frac{1}{N}}{2m}\right)^m \sim e^{-\frac{1 - \frac{1}{N}}{2}}. \tag{5}$$

This yields:

$$\Pr\left(\mathcal{HS}_G\right) \geq 1 - e^{\frac{\frac{1}{N} - 1}{2}}.$$

Lemma 1. *Let $\mathcal{A}$ be the population protocol running on the communication graph $G = (V, E)$. Then, the probability that the scheduler picks at least one ordered pair of nodes at the end of Algorithm 2 is lower bounded by $1 - e^{\frac{\frac{1}{N} - 1}{2}}$.*

The Number of Simultaneous Handshakes. Let $G = (V, E)$ be the communication graph of a population protocol $\mathcal{A}$. Let X be the random variable (r.v) which counts the number of simultaneous handshakes that can take place at the same step. Using the linearity of the expectation, we can easily obtain the expected number of X:

$$\mathbb{E}(X) = \sum_{(u,v) \in E} \left(\frac{1 - \frac{1}{N}}{d(u)d(v)}\right). \tag{6}$$

4.3 The Fairness of the Handshake Scheduler

Given the analysis results presented in the previous section, we can notice that the *HS Scheduler* is able to either select zero, or one, or even more than one edge from the communication graph at a given time. With respect to this description, we give a new definition of one step transition. Let C and C' be configurations. We say that C can go in one step to C', if starting from C we can reach C' in only one iteration of the *HS Scheduler* algorithm via a set of encounters enc. Formally, we denote it $C \to C'$, or equivalently $C \xrightarrow{enc} C'$ such that $enc = \{e_1, e_2, \ldots, e_k\}$ where $\forall i \in \{1, 2, \ldots, k\}$, $e_i \in E$ and $\forall l, l' \in \{1, 2, \ldots, k\}$, if $e_l = (u, v)$ and $e_{l'} = (u', v')$ then $u \neq u'$, $u \neq v'$, $v \neq u'$ and $v \neq v'$. Also, $\forall e_i = (u_i, v_i) \in enc$, we have $C'(u_i) = \delta_1(C(u_i), C(v_i))$ and $C'(v_i) = \delta_2(C(u_i), C(v_i))$, and $C'(w) = C(w)$ for any node w which is not an extremity of an edge from enc.

The *transition graph* $T = (V(T), E(T))$ of a protocol $\mathcal{A}$ running on a graph $G = (V, E)$, denoted $T(\mathcal{A}, G)$ [3], is the directed graph, that may contain self loops, where the nodes are all possible configurations and the edges relating them are all possible one-step transitions (according to our new definition).

Theorem 1. *The HS Scheduler is a probabilistic consistent scheduler.*

Proof. We recall the definition from [6] of a probabilistic scheduler w.r.t a transition graph $T(\mathcal{A}, G)$. A probabilistic scheduler defines for each configuration $C \in V(T)$ an infinite sequence of probability distributions of the form $(d_1^C, d_2^C, \ldots)$, over the set $\Gamma^+(C) = \{C' \mid C \to C'\}$, where $d_t^C \colon \Gamma^+(C) \to [0, 1]$ and such that $\Sigma_{C' \in \Gamma^+(C)} \, d_t^{C'} = 1$ holds for all t and C'. Also from [6], a probabilistic scheduler is consistent with respect to a transition graph $T(\mathcal{A}, G)$, if for all configurations $C \in V(T)$ it holds that $d^C = d_1^C = d_2^C = \ldots$ which means that any time the configuration C is encountered, the scheduler chooses the next configuration with the same probability.

So, let $T(\mathcal{A}, G)$ be any transition graph and C_i be any configuration in $V(T)$. Let C_j be any configuration in $V(T)$ reachable in one step from C_i. We define $Enc_{C_i C_j} = \{enc \mid enc \subset E(G) \text{ and } C_i \xrightarrow{enc} C_j\}$. And, for each element enc from $Enc_{C_i C_j}$, we will define the set E of the edges of the communication graph G, as the union of three disjoint subsets: enc, F_1 and F_2. F_1 will represent the set of edges that are joint to enc. F_2 will represent the set of edges that are disjoint to enc. More formally, for each element enc from $Enc_{C_i C_j}$, we rewrite the set E as following:

$$E = enc \uplus F_1 \uplus F_2 .$$

$F_1 = \{f \in E \mid \text{if } f = (u, v) \text{ then } \exists e \in enc \text{ such that } e = (u, v') \text{ or } e = (u', v)\}$.
$F_2 = \{f \in E \mid \text{if } f = (u, v) \text{ then } \forall e \in enc, \text{ if } e = (u', v') \text{ then } u \neq u', u \neq v', v \neq u' \text{ and } v \neq v'\}$.

Any time C_i is encountered, C_j is selected with the following probability

$$\Pr\!{}_{C_i C_j} = \sum_{enc \subset Enc_{C_i C_j}} \Pr\left(\mathcal{HS}(enc)\right) \Pr\left(\overline{\mathcal{HS}}(F_2)\right) .$$

Thus, we can state that the *HS Scheduler* is probabilistic. Also, the probability of handshakes on any set of edges, depends on the probability of a handshake on each of its edges. And we already proved in the previous section that for any edge $e = (u, v) \in E$, $\Pr\left(\mathcal{HS}(e)\right) = \frac{1 - \frac{1}{N}}{d(u)d(v)}$ which does not depend on time. Therefore, the value of the probability $\Pr\!{}_{C_i C_j}$ will be independent of the number of times C_i has been encountered. This leads us to conclude that the *HS Scheduler* is also consistent.

Theorem 2. *The HS Scheduler is fair with probability 1.*

Proof. We recall the theorem from [6]: Any consistent scheduler, for which it holds that $\mathbb{P}r_{C_iC_j} > 0$, for any protocol $\mathcal{A}$, any communication graph G, and all configurations C_i, $C_j \in V(T(A, G))$ where $C_i \to C_j$ and $C_i \neq C_j$, is fair with probability 1.

Let $T(\mathcal{A}, G)$ be any transition graph and C_i, C_j be any configurations in $V(T)$ such that $C_i \to C_j$ and $C_i \neq C_j$. In Theorem 1, we already proved that the *HS Scheduler* is a probabilistic consistent scheduler. And, based on the definition of $C_i \to C_j$, we have: $\exists \ enc \subset E$ such that $C_i \xrightarrow{enc} C_j$ which means that $\mathbb{P}r_{C_iC_j} > 0$. So, applying the theorem from [6], we conclude that the *HS Scheduler* is fair with probability 1.

5 The Time Complexity of Iterated Population Protocols

Population protocols are defined as being protocols that do not halt, but only stabilize [3]. However, introducing termination to this model would be interesting as it could help the nodes preserve their energies. In [10], the authors suppose that the nodes can have access to some global knowledge via an oracle called an Absence Detector. Based on this knowledge, a node can decide to halt. But, what if a node can halt while having only a local knowledge which is its current state?

5.1 The Iterated Population Protocols

Starting from the basic model of population protocols, we suppose that a node can stop participating in the computation of the protocol when it reaches a specific state that we call *final state*. We define Q_{final} the set of *final states* such that $Q_{final} \subset Q$. Once a node reaches a *final state*, it will never change its state anymore. Also, it will no longer have any impact on the protocol progress: that is if a node in a *final state* is involved in a communication (as an initiator or as a responder), the result will be the identity for both nodes states. We call these protocols *iterated population protocols*.

To stop participating on the protocol's computation, a node should first stop being a candidate for the scheduler's choices. We here propose a scheduler for *iterated population protocols* which is a protocol aware version of the *scheduler HS*. We call it the *Protocol Aware Handshake Scheduler* and will denote it *PA_HS Scheduler*. This scheduler will take into account the states of the participating nodes. When starting the scheduler algorithm on a node, the state of this node is checked. If it is a *final state*, the algorithm will stop running.

In the sequel, we will assume the use of the *PA_HS Scheduler* as a scheduler when talking about *iterated population protocols*. We will also focus, through the following theoretical analysis, on the case where only one computation step suffices to lead a node to reach a *final state*. Initially, the nodes have their initial inputs that they map to states. Then, once a pair of nodes is picked by the scheduler, they will communicate and make a computation step that will lead both of them for sure to *final states*. Also, both of them will stop participating on the protocol computation.

5.2   A General Upper Bound

As a general upper bound for the time complexity of an *iterated population protocol*, we have the following lemma:

Lemma 2. *Let $\mathcal{A}$ be an iterated population protocol. Then, the expected time to compute $\mathcal{A}$ is upper bounded by $O(n)$.*

Proof. Let T be the r.v. that counts the number of rounds before the population agents computing the *iterated protocol* $\mathcal{A}$ halt. For any $t \geq 1$, let X_t denote the number of simultaneous computations in the graph G at round t, and let Y_t be the r.v. $Y_t = n - 2X_t$. It is clear that, for any $t \geq 1$, $T \leq t$ if, and only if, $Y_t \leq 1$.

We define the following (pessimistic) process $(Y_t')_{t \geq 0}$:

$$Y_t' = \begin{cases} n & \text{if } t = 0 \\ Y_{t-1}' - 2 & \text{if } X_t \geq 1 \\ Y_{t-1}' & \text{if } X_t = 0. \end{cases} \tag{7}$$

The process $(Y_t')_{t \geq 0}$ is an irreversible ergodic Markov chain whose states are in the set $\{n, \cdots, 0\}$ and by Lemma 1, the transition probabilities are given by: $\Pr\left(Y_t' = 0 \mid Y_{t-1}' = 0\right) = 1$ and for any $i > 0$, and any $j \in \{n, \cdots, 0\}$,

$$\Pr\left(Y_t' = i \mid Y_{t-1}' = j\right) = \begin{cases} e^{\frac{\frac{1}{N}-1}{2}} & \text{if } i = j \\ 1 - e^{\frac{\frac{1}{N}-1}{2}} & \text{if } i = j - 2 \\ 0 & \text{otherwise.} \end{cases} \tag{8}$$

Let T' denote the time for the process $(Y_t')_{t \geq 0}$, starting at $Y_0' = n$ to reach the absorbing state $Y_t' \leq 1$. Then, an easy computation yields to $\mathbb{E}(T') = \frac{n}{1 - e^{\frac{1}{2}(\frac{1}{N} - 1)}}$.

On the other hand, it is clear that $T < T'$ and hence $\mathbb{E}(T) \leq \mathbb{E}(T')$, which ends the proof. $\square$

5.3   Upper Bound When the Interaction Graph Is with Bounded Degree

Let $G = (V, E)$ be the interaction graph of an *iterated population protocol*. In this section, we consider interaction graphs with degrees bounded by Δ. Then we have the following lemma:

Lemma 3. *Let T the time complexity of an iterated population protocol. The expected value of T satisfies:*

$$\mathbb{E}\,(T) \leq -\frac{\log(\frac{n\Delta}{2})}{\log\left(1 - \left(\frac{1-\frac{1}{N}}{\Delta^2}\right)\right)}.$$

Proof. Let t an iteration of the algorithm and we define the sequence of graphs $(G_t)_{(t\geq 0)}$ as follows: $G_0 = G$ and for all $t \geq 1$, G_{t+1} is the graph obtained by removing, from G_t, the pairs that execute a computation step and their incident edges.

We define the following random variables: for any $t \geq 0$, X_t denotes the number of edges of the graph G_t and Y_t denotes the number of edges removed from the graph G_t at the end of iteration t. Then, we have $X_{t+1} = X_t - Y_t$ and thus:

$$\mathbb{E}\,(X_{t+1} \mid G_t) = \mathbb{E}\,(X_t \mid G_t) - \mathbb{E}\,(Y_t \mid G_t)$$
$$= X_t - \mathbb{E}\,(Y_t \mid G_t). \tag{9}$$

On the other hand, for any pair $e = (u,v)$, if a computation is done by the pair (u,v) at iteration t, then the edge e is removed from the graph G_t. Hence, by (6),

$$\mathbb{E}\,(Y_t \mid G_t) \geq \sum_{(u,v)\in E(G_t)} \frac{1-\frac{1}{N}}{d_t(u)d_t(v)},$$

where $d_t(.)$ stays for the degree in the graph G_t. Then:

$$\mathbb{E}\,(X_{t+1} \mid G_t) \leq X_t - \sum_{(u,v)\in E(G_t)} \frac{1-\frac{1}{N}}{d_t(u)d_t(v)}. \tag{10}$$

Since, for any u, $d_t(u) \leq \Delta$, this becomes:

$$\mathbb{E}\,(X_{t+1} \mid G_t) \leq X_t \left(1 - \frac{1-\frac{1}{N}}{\Delta^2}\right). \tag{11}$$

For $t \geq 0$, we define the r.v. $Z_t = \frac{X_t}{\left(1-(1-\frac{1}{N})\frac{1}{\Delta^2}\right)^t}$. Then, $\mathbb{E}\,(Z_{t+1} \mid G_t) \leq Z_t$. Thus, the r.v. Z_t is a super-martingale, and then:

$$\mathbb{E}\,(Z_{t+1}) = \mathbb{E}\,(\mathbb{E}\,(Z_{t+1} \mid G_t)) \leq \mathbb{E}\,(Z_t). \tag{12}$$

A direct application of a theorem from [12] yields to $\mathbb{E}\,(Z_t) \leq Z_0 = m$. Thus:

$$\mathbb{E}\,(X_t) = \left(1 - \frac{1-\frac{1}{N}}{\Delta^2}\right)^t \mathbb{E}\,(Z_t) \leq m \left(1 - \frac{1-\frac{1}{N}}{\Delta^2}\right)^t. \tag{13}$$

We have $m \leq \frac{n\Delta}{2}$ and the algorithm halts when $X_t < 1$. This implies that t is upper bounded by $-\frac{\log(\frac{n\Delta}{2})}{\log\left(1-(1-\frac{1}{N})\frac{1}{\Delta^2}\right)}$ which ends the proof. □

Corollary 1. *If Δ is a constant, then the expected time to compute the iterated population protocol is $O\,(\log n)$.*

5.4   Upper Bound When the Interaction Graph Is a Random Graph

A random graph is a graph obtained by starting with a set of n vertices and adding edges between them at random. Different random graph models produce different probability distributions on graphs. The most commonly studied model, called $G_{n,p}$, includes each possible edge independently with probability p (and so, the edge is not included with probability $q = 1 - p$). In this section, we consider this model, with $p > 0$. It is straightforward that we have some edges.

Analysis of a Single Round. By reusing the result of Lemma 3 from [13], w.r.t the analysis that we presented in Sect. 4, we obtain the following lemma:

Lemma 4. *Let* $G_{n,p} = (V, E)$ *be a random graph, and let* u *and* v *be two vertices. Then,*

$$\Pr\left(\mathcal{HS}(e)\right) = \frac{\left(1 - q^{n-1}\right)^2}{\left(n - 1\right)^2 p}\left(1 - \frac{1}{N}\right). \tag{14}$$

Thus, if we denote by $X_{n,p}$ the number of simultaneous handshakes in $G_{n,p}$, then

$$\mathbb{E}\left(X_{n,p}\right) = \frac{n}{2\left(n - 1\right)p}\left(1 - q^{n-1}\right)^2\left(1 - \frac{1}{N}\right). \tag{15}$$

On the other hand, the expression (15) can be simplified for some particular values of p:

- If p is a constant, then $\left(1 - q^{n-1}\right)^2 \to 1$ as $n \to \infty$. Hence, $\mathbb{E}\left(X_{n,p}\right) \sim \frac{1}{2p}\left(1 - \frac{1}{N}\right)$ as $n \to \infty$.
- If $np = \lambda + o(1) > 0$, that is the average degree of any vertex v is a constant, then $q^{n-1} \sim e^{-\lambda}$. Hence,

$$\exists \alpha > 0 \text{ such that } \mathbb{E}(X_{n,p}) \sim \alpha n.$$

- If $n^2 p = \lambda + o(1)$, then

$$\exists \beta > 0 \text{ such that } \mathbb{E}(X_{n,p}) \sim \beta.$$

- If $n^\gamma p = \lambda + o(1)$, with $\gamma > 2$ then

$$\mathbb{E}(X_{n,p}) \to 0, \text{ as } n \to \infty.$$

A more interesting case is $p = \frac{\alpha \log n}{n}$ for $\alpha > 1$. Indeed, this value is the connectivity threshold for $G_{n,p}$. In this case, we have the following expression for the expected number of simultaneous computations:

$$\mathbb{E}(X_{n,p}) \sim \frac{n}{2\alpha \log n}(1 - \frac{1}{N}), \text{ as } n \to \infty. \tag{16}$$

The Time Complexity of Iterated Rounds. In this section, we analyze the complexity of an *iterated protocol* in random graphs. We focus our study on the case $p = \frac{\alpha \log n}{n}$ (with $\alpha > 1$), that is on the random graphs which are connected with high probability. We have the following theorem:

Theorem 3. *Let $G_{n,p} = (V, E)$ be a random graph with $p = \frac{\alpha \log n}{n}$ (for $\alpha > 1$). Let T the time complexity of an iterated population protocol. The expected value of T satisfies:*

$$\mathbb{E}(T) \leq -\frac{\log n}{\log\left(1 - \frac{1 - \frac{1}{N}}{\alpha \log n}\right)}.$$

Proof. The proof uses the same arguments as for the proof of Lemma 3. We define the sequence of graphs $(G_t)_{(t \geq 0)}$, and the two r.v. X_t as the size of G_t and Y_t as the number of edges removed at the end of the computation step t. Then we have $X_{t+1} = X_t - 2Y_t$, and

$$\mathbb{E}(X_{t+1} \mid G_t) = X_t - 2\mathbb{E}(Y_t \mid G_t).$$

Then, using (16), this becomes:

$$\mathbb{E}(X_{t+1} \mid G_t) = X_t - \frac{X_t}{\alpha \log X_t}\left(1 - \frac{1}{N}\right).$$

Now, as $X_t \leq n$, $\forall t \geq 0$, we have:

$$\mathbb{E}(X_{t+1} \mid G_t) \leq \left(1 - \frac{1 - \frac{1}{N}}{\alpha \log n}\right) X_t.$$

Then, the theorem holds by the same reasoning as for the proof of (12) and (13). □

Corollary 2. *If the graph is a $G_{n,p}$ with $p = \frac{\alpha \log n}{n}$, then the expected time to compute the iterated population protocol is $O(\log n)$.*

6 Application

The concepts of *iterated protocols*, the fair *HS Scheduler* and the *PA_HS Scheduler*, are also valid and applicable for all the models that extend the population protocols. In this section, as an application, we will consider the *iterated mediated population protocol* that computes a Maximal Matching of a communication graph.

6.1 Maximal Matching with Iterated Mediated Population Protocols

The *mediated population protocols* extend the basic population protocols by adding states to the edges of the interaction graph. So formally, in addition to X, Y, Q, I, O that are already defined in population protocols, a mediated population protocol consists of [8]:

- S: a finite set of edge states,
- $\iota\colon X \to S$: an edge input function mapping inputs to edge states,
- $\omega\colon S \to Y$: an edge output function mapping edge states to outputs,
- r: an output instruction,
- $\delta\colon Q \times Q \times S \to Q \times Q \times S$: a transition function.

We also recall the definition of a *Maximal Matching*. Let $G = (V, E)$ be an undirected graph. Let M be a set of edges from E such that each two edges from M do not share any vertex. M is called a Matching. A Matching M is maximal, if by adding any edge to M, M will not be a Matching any more. Formally, M is a Matching of G if $M = \{e \in E \mid \forall e_1, e_2 \in M, \; if \; e_1 = (u_1, v_1) \; and \; e_2 = (u_2, v_2) \; then \; u_1 \neq u_2, \; u_1 \neq v_2, \; v_1 \neq u_2 \; and \; v_1 \neq v_2\}$. And, M is maximal if, for any Matching M' of G, $M \not\subset M'$.

Let $MaxMatch$ be the mediated population protocol that computes a Maximal Matching in a communication graph $G = (V, E)$ as given in [8]. $MaxMatch$ will be described by:

- $X = \{0\}$, $Y = \{0, 1\}$,
- $Q = \{q_0, q_1\}$, $S = \{0, 1\}$,
- $I(0) = q_0$, $O(q_0) = 0$ and $O(q_1) = 1$,
- $\iota(0) = 0$, $\omega(0) = 0$ and $\omega(1) = 1$,
- $\delta(q_0, q_0, 0) = (q_1, q_1, 1)$.
- r: "Get each edge $e \in E$ such that $\omega(s_e) = 1$ where s_e is the state of the edge e."

By using a *PA_HS Scheduler* aware about the fact that the state q_1 is a *final state* while running the $MaxMatch$ protocol, we obtain the *iterated MaxMatch* protocol.

6.2   The Simulations

To validate the theoretical study about the time complexity upper bounds presented on the previous section, we proceed on simulations of the *iterated MaxMatch* protocol. We opt for ViSiDiA (Visualization and Simulation of Distributed Algorithms) as it represents a suitable platform to simulate population protocols and their variants. It is a Java platform that offers to the user an environment where he can implement his distributed algorithm, generate the graph representing the network on which the algorithm will be executed, launch the simulation, and visualize it [1]. In ViSiDiA, a network is represented by a graph $G = (V, E)$, where V is the set of vertices representing the processors and E is the set of edges representing the communication links existing between the processes. Each vertex v has a label $\lambda(v)$ that describes its state. The edges can also be labeled. Each process is able of making local computations and exchanging messages with its neighbors. This can cause changes on its labels.

Therefore, we implement the *iterated MaxMatch* on ViSiDiA. Then, we begin the simulations on populations where the communication graph is a bounded degree graph. The results are in Table 1 where we can notice that

Table 1. Simulations results of the *iterated MaxMatch* in bounded degree graphs

n	Δ	Experimental average	Theoretical upper bound
100	6	8	202
1000	6	9	284
5000	6	11	341
10000	6	12	366
100	4	6	82
1000	4	7	118
5000	4	9	142
10000	4	11	153

Table 2. Simulations results of the *iterated MaxMatch* in random graphs

n	Experimental average	Theoretical upper bound
100	7	16
1000	10	38
2000	11	46
3000	12	51
4000	14	55
5000	15	59
6000	16	61
7000	17	63
8000	19	65
9000	20	67
10000	22	69

the theoretical upper bound that we calculated is always respected, that is the experimental average of the iterations needed to compute the protocol never exceeded the theoretical upper bound.

We also proceeded on simulations on populations where the communication graph is a $G_{n,p}$ where $p = 2\frac{\log(n)}{n}$. The results in Table 2 validate the upper bound presented in the previous section.

7 Conclusion

In this paper, we proposed a distributed synchronous scheduler for population protocols based on randomized handshakes. We proved that this scheduler is a probabilistic consistent fair scheduler with probability 1. The algorithm that describes the scheduler is synchronous. As a future work, we aim to adapt this scheduler to the asynchronous context of the networks where the population protocols are run and give a theoretical study of its behavior. We also aim to study the time complexity of the stabilisation of some protocols under these two schedulers. Furthermore, we introduced nodes termination in the population protocols. Based only on their local knowledge, nodes can stop participating in the computation of the protocol if they reach states described as final. We only studied the case of iterated population protocol where one computation step suffices to lead a node to a final state, however it could be interesting to study more general cases. Also, characterizing the computational power of the iterated population protocols and the iterated mediated population protocols can be one of the future directions of this work.

References

1. Abdou, W., Ouled Abdallah, N., Mosbah, M.: ViSiDiA: a Java framework for designing, simulating, and visualizing distributed algorithms. In: IEEE/ACM 18th International Symposium on Distributed Simulation and Real Time Applications (DS-RT), pp. 43–46 (2014)
2. Angluin, D.: Local and global properties in networks of processors. In: Proceedings of the Twelfth Annual ACM Symposium on Theory of Computing, pp. 82–93. ACM (1980)
3. Angluin, D., Aspnes, J., Diamadi, Z., Fischer, M.J., Peralta, R.: Computation in networks of passively mobile finite-state sensors. Distrib. Comput. **18**, 235–253 (2006)
4. Angluin, D., Aspnes, J., Fischer, M.J., Jiang, H.: Self-stabilizing population protocols. ACM Trans. Auton. Adapt. Syst. **3**(4), 13 (2008)
5. Becchetti, L., Bergamini, L., Ficarola, F., Salvatore, F., Vitaletti, A.: First experiences with the implementation and evaluation of population protocols on physical devices. IEEE Int. Conf. Green Comput. Commun. (GreenCom) **2012**, 335–342 (2012)
6. Chatzigiannakis, I., Dolev, S., Fekete, S.P., Michail, O., Spirakis, P.G.: Not all fair probabilistic schedulers are equivalent. In: Abdelzaher, T., Raynal, M., Santoro, N. (eds.) OPODIS 2009. LNCS, vol. 5923, pp. 33–47. Springer, Heidelberg (2009)
7. Chatzigiannakis, I., Michail, O., Nikolaou, S., Pavlogiannis, A., Spirakis, P.G.: Passively mobile communicating logarithmic space machines. In: CoRR (2010). abs/1004.3395
8. Chatzigiannakis, I., Michail, O., Spirakis, P.G.: Mediated population protocols. In: Proceedings of the 36th International Colloquium on Automata, Languages and Programming: Part II (2009)
9. Métivier, Y., Saheb, N., Zemmari, A.: Analysis of a randomized rendezvous algorithm. Inf. Comput. **184**(1), 109–128 (2003)
10. Michail, O., Chatzigiannakis, I., Spirakis, P.G.: Terminating population protocols via some minimal global knowledge assumptions. In: Richa, A.W., Scheideler, C. (eds.) SSS 2012. LNCS, vol. 7596, pp. 77–89. Springer, Heidelberg (2012)
11. Ouled Abdallah, N., Hadj Kacem, H., Mosbah, M., Zemmari, A.: Randomized broadcasting in wireless mobile sensor networks. Concurr. Comput. Pract. Exp. **25**(2), 203–217 (2013)
12. Williams, D.: Probability with Martingales. Cambridge University Press, Cambridge (1991)
13. Zemmari, A.: On handshakes in random graphs. Inf. Process. Lett. **108**, 119–123 (2008)

Constraints

Extending the Notion of Preferred Explanations for Quantified Constraint Satisfaction Problems

Deepak Mehta, Barry O'Sullivan, and Luis Quesada$^{(\boxtimes)}$

Insight Centre for Data Analytics, University College Cork, Cork, Ireland
{deepak.mehta,barry.osullivan,luis.quesada}@insight-centre.org

Abstract. The Quantified Constraint Satisfaction Problem (QCSP) is a generalization of classical constraint satisfaction problem in which some variables can be universally quantified. This additional expressiveness can help model problems in which a subset of the variables take value assignments that are outside the control of the decision maker. Typical examples of such domains are game-playing, conformant planning and reasoning under uncertainty. In these domains decision makers need explanations when a QCSP does not admit a winning strategy. We extend our previous approach to defining preferences amongst the requirements of a QCSP by considering more general relaxation schemes. We also present key complexity results on the hardness of finding preferred conflicts of QCSPs under this extension of the notion of preference. This paper unifies work from the fields of constraint satisfaction, explanation generation, and reasoning under preferences and uncertainty.

1 Introduction

Uncertainty is ubiquitous in real-world decision-making. However, quantifying the nature of the uncertainty can be very difficult, if not impossible, in many settings. Domain experts can usually provide qualitative statements of which risks are more important to consider than others, and which outcomes are more likely than others using qualitative statements. In this paper we report on the formal underpinnings of an approach to risk-aware decision-making that is based on an extension of the classic Constraint Satisfaction Problem (CSP) known as the Quantified Constraint Satisfaction Problem (QCSP) [1]. Parameters under the control of the decision maker are modelled as existentially quantified variables since a value (a decision) must be assigned (made) to these variables. All uncertain variables are universally quantified so that decision makers must consider how to preempt every possible assignment to those variables. Of course, such a formulation means that it will be seldom possible for a decision maker to satisfy the constraints of the QCSP since it is likely that some values of the universal (uncertain) variables cannot be preempted. Therefore, we assist the decision maker by abstracting their decision problem so the specific reasons for infeasibility can be focused upon.

© Springer International Publishing Switzerland 2015
M. Leucker et al. (Eds.): ICTAC 2015, LNCS 9399, pp. 309–327, 2015.
DOI: 10.1007/978-3-319-25150-9_19

Example 1 (Weekend Planning). Assume that John wants to prepare a plan for Saturday and Sunday on Friday evening. He is interested in two activities: rowing (*row*) and watching movie (*mov*). Also, assume that there are two weather possibilities: sun (*s*) and rain (*r*). Each activity should be carried out on a different day. If the activity is rowing then the weather should be sunny. Let *Asat* and *Asun* denote the activities performed on Saturday and Sunday, respectively. Let *Wsat* and *Wsun* denote the weather on Saturday and Sunday, respectively. The basic formulation of this problem is as follows:

$$\exists Asat, Asun \in \{row, mov\} : \forall Wsat, Wsun \in \{s, r\} :$$
$$\{(Asat \neq Asun), (Asat = row \Rightarrow Wsat = s), (Asun = row \Rightarrow Wsun = s)\}$$

There is no decision that can be made in this case that properly responds to the risk. This is because for any assignment to *Asat* and *Asun* there is at least one assignment to *Wsat* and *Wsun* that is inconsistent with it. Many relaxations, giving rise to risk responses, of this problem are possible. For example, one relaxation could be to restrict the domain of *Wsat* to $\{s\}$ and another could be to restrict *Wsun* to $\{s\}$. However, if John knows that on Saturday it is less likely to rain, then the former would be preferred over the latter. The QCSP obtained by removing a less likely value *r* from *Wsat* is as follows:

$$\exists Asat, Asun \in \{row, mov\} : \forall Wsat \in \{s\} \; ; \; Wsun \in \{s, r\} :$$
$$\{(Asat \neq Asun), (Asat = row \Rightarrow Wsat = s), (Asun = row \Rightarrow Wsun = s)\}$$

This QCSP is satisfiable, i.e. there is an appropriate risk response in this setting. This is because there exist assignments to the existential variables, $Asat = row$ and $Asun = mov$, such that for any assignment to the uncertain/universal variables, *Wsat* and *Wsun*, the constraints are satisfied. ▲

Our work is motivated by the development of a qualitative approach to reason under uncertainty. Stochastic variables can be modelled as universally quantified variables where preferences capture the likelihood of scenarios. Most preferred explanations correspond to most likely scenarios that we can control by excluding unlikely (least preferred) ones. We present a non-intrusive approach that assumes nothing about the capabilities of the solver but a way of testing the consistency of a QCSP. Such non-intrusive explanation algorithms are the most commonly used in practice, e.g. in ILOG Configurator. The non-intrusive approach is advocated by Junker [2].

In this paper we present a framework for generating preferred explanations in a QCSP setting. An advantage of the framework is that recent developments in QCSP modelling and solving QCSPs can be applied directly to qualitative risk management [3,4]. We present a variety of explanation generation algorithms that take a preference (or likelihood) ordering into account in order to generate the most preferred (most likely) explanation in a given context. We consider both total and partial orders amongst the requirements of a QCSP and present efficient algorithms for each case. We also provide a key complexity result that characterises when generalisations of our framework become intractable. In [5] we restricted our attention to a specific type of relaxation functions. In this paper we broaden the scope of these relaxation functions and elaborate on the complexity challenges that such a generalisation entails.

2   Preliminaries

Definition 1 (Constraint Satisfaction Problem). *A constraint satisfaction problem (CSP) is a 3-tuple $P \hateq \langle \mathcal{X}, \mathcal{D}, \mathcal{C} \rangle$ where $\mathcal{X}$ is a finite set of variables $\mathcal{X} \hateq \{x_1, \ldots, x_n\}$, $\mathcal{D}$ is a set of finite domains $\mathcal{D} \hateq \{D(x_1), \ldots, D(x_n)\}$ where the domain $D(x_i)$ is the finite set of values that variable x_i can take, and a set of constraints $\mathcal{C} \hateq \{c_1, \ldots, c_m\}$. Each constraint c_i is defined by the ordered set $var(c_i)$ of the variables it involves, and a set $sol(c_i)$ of allowed combinations of values. An assignment of values to the variables in $var(c_i)$ satisfies c_i if it belongs to $sol(c_i)$. A solution to a CSP is an assignment to each variable by a value from its domain such that every constraint in $\mathcal{C}$ is satisfied.*

Definition 2 (Quantified CSP). *A QCSP, ϕ, has the form*

$$\mathcal{Q}.\mathcal{C} = Q_1 x_1 \in D(x_1) \cdots Q_n x_n \in D(x_n).\mathcal{C}(x_1, \ldots, x_n)$$

where $\mathcal{C}$ is a set of constraints (see Definition 1) defined over the variables $x_1 \ldots x_n$, and $\mathcal{Q}$ is a sequence of quantifiers over the variables $x_1 \ldots x_n$ where each Q_i $(1 \leq i \leq n)$ is either an existential, $\exists$, or a universal, $\forall$, quantifier.

The expression $\exists x_i.c$ means that "there exists a value $a \in D(x_i)$ such that the assignment (x_i, a) satisfies c". Similarly, the expression $\forall x_i.c$ means that "for every value $a \in D(x_i)$, (x_i, a) satisfies c". Following the work of Gent et al. [4], the semantics of these expressions can be formally defined as follows: if $\mathcal{Q}$ is of the form $\exists x_1 Q_2 x_2 \ldots Q_n x_n$ then $\mathcal{Q}.\mathcal{C}(x_1, \ldots, x_n)$ is true if and only if there exists some value a in $D(x_1)$ such that $Q_2 x_2 \ldots Q_n x_n.\mathcal{C}(x_1, \ldots, x_n)[(x_1, a)]$ is true. If $\mathcal{Q}$ is of the form $\forall x_1 Q_2 x_2 \ldots Q_n x_n$ then $\mathcal{Q}.\mathcal{C}(x_1, \ldots, x_n)$ is true if and only if for each value a in $D(x_1)$, $Q_2 x_2 \ldots Q_n x_n.\mathcal{C}(x_1, \ldots, x_n)[(x_1, a)]$ is true.

When the variable and the domain of the variable is clear from context we often write Q_i rather than $Q_i x_i \in D(x_i)$ in the quantifier sequence. When the position of a universal quantifier, Q_i, in the sequence $\mathcal{Q}$ is j such that $j \neq i$ we write Q_i^j, where $1 \leq j \leq n$, otherwise we simply write Q_i.

3   Relaxations of Requirements

Requirements correspond to either a constraint in the QCSP, or the scope of a universal quantifier, or the position of a universal quantifier. The requirements of an input QCSP are called *original requirements*. When the input QCSP is inconsistent, we seek the closest QCSP by relaxing one or more original requirements. For example, an extensional constraint could be relaxed by removing some of its nogoods, the scope of a universal quantifier could be relaxed by restricting its scope to a subset of the domain of the universally quantified variable, and the position of a universal quantifier could be relaxed by moving it to the left in the sequence of quantifiers. Notice that a universal quantifier could be relaxed by either relaxing its scope or relaxing its position. However, we treat them separately for the purpose of clarity. We frame relaxation of each as instances of *requirement relaxation*, over a partial order defined for that purpose.

Definition 3 (Substitution of a Requirement). *Given a* QCSP ϕ*, the substitution of a requirement r in ϕ results in a new* QCSP $\phi[r]$*.*

- *If the requirement $r \equiv Q_i x_i \in D(x_i)$ of type* scope of universal quantifier *is to be substituted by $Q_i x_i \in D'(x_i)$ then*

$$Q_1 x_1 \ldots Q_i x_i \in D(x_i) \ldots Q_n x_n . C[Q_i x_i \in D'(x_i)]$$

results in the following QCSP:

$$Q_1 x_1 \ldots Q_i x_i \in D'(x_i) \ldots Q_n x_n . C.$$

- *If the requirement $r \equiv Q_i$ of type* position of universal quantifier *is to be substituted by Q_i^k, where $k < i$ then*

$$Q_1 \ldots Q_k \ldots Q_{i-1} Q_i \ldots Q_n . C[Q_i^k]$$

results in positioning Q_i in k and moving the other quantifiers accordingly as shown in the following QCSP:

$$Q_1 \ldots Q_i^k Q_k^{k+1} \ldots Q_{i-1}^i \ldots Q_n . C.$$

- *If the requirement $r \equiv c_j$ of type* constraint *is to be substituted by another constraint c_j' then*

$$Q.(c_1 \ldots c_j \ldots c_m)[c_j']$$

results in the following QCSP:

$$Q.(c_1 \ldots c_j' \ldots c_m).$$

The notion of requirement substitution can be lifted to work on a set of requirements R: $\phi[\emptyset] = \phi$, $\phi[\{r\} \cup R] = (\phi[r])[R]$.

Definition 4 (Ordering over Requirement Relaxations). *Let R be the set of possible relaxations of a requirement r_0 and let $r_1 \in R$ and $r_2 \in R$ be two relaxations of r_0. We say that r_2 is a relaxation[1] of r_1, denoted by $r_1 \sqsubseteq r_2$, if and only if for any* QCSP ϕ *if $\phi[r_1]$ is satisfiable then $\phi[r_2]$ is also satisfiable. We say that r_2 is a strict relaxation of r_1, denoted by $r_1 \sqsubset r_2$, if and only if $r_1 \sqsubseteq r_2$ is true and the converse is not true.*

---

[1] A relaxation of a requirement is also a requirement.

We require that the partial order $\sqsubseteq$ also be a *meet-semilattice*, i.e., *greatest lower bounds* are guaranteed to exist: if $r_1, r_2 \in R$, then $r_1 \sqcap r_2$ is well-defined in which case $r_1 \sqcap r_2 \sqsubseteq r_1$ and $r_1 \sqcap r_2 \sqsubseteq r_2$ hold.

Definition 5 (Relaxations of a Constraint). *Given a constraint $c \; \hat{=} \; \langle var(c), sol(c) \rangle$ we define its relaxations in terms of adding additional allowed tuples to $sol(c)$:*

$$\{sol'(c) : sol(c) \subseteq sol'(c) \subseteq \Pi_{x \in var(c)} D(x)\}.$$

The relaxations of c form the usual lattice using intersection, that is, $\sqcap \; \hat{=} \; \cap$.

Definition 6 (Relaxations of Universal Scopes). *Given a requirement $\forall x \in D(x)$, the set of relaxations of the scope of the universally quantified variable x, is defined as:*

$$\{(\forall x \in D'(x)) : \emptyset \subseteq D'(x) \subseteq D(x)\}$$

Given two requirements $(\forall x \in D(x))$ and $(\forall x \in D'(x))$,

$$(\forall x \in D(x)) \sqcap (\forall x \in D'(x))$$
$$\hat{=} (\forall x \in (D(x) \cup D'(x))).$$

Therefore, the relaxations of $\forall x \in D(x)$ also form a meet-semilattice.

A relaxation of the position of a universal quantifier requirement corresponds to moving a universally quantified variable to the left in the sequence of quantifiers.

Informally, the relaxation of the position of a universal quantifier gives the existential variable the opportunity to take a value based on value taken by the universal variable.

Definition 7 (Relaxations of Universal Positions). *Given a position of the universal quantified variable x_i as a requirement, Q_i^i, its possible relaxations is defined as follows:*

$$\{Q_i^j : 1 \leq j \leq i\} \cup \{Q_i^1 x_i \in \emptyset\}$$

Given two relaxations Q_i^j and Q_i^k of a requirement Q_i^i, $Q_i^j \sqcap Q_i^k = Q_i^{\max(j,k)}$. The elements of the relaxations of Q_i^i also form a lattice using $\sqcap$.

Definition 8 (Relaxation of a QCSP). *Given a requirement r of a* QCSP *ϕ, and a requirement relaxation r' such that $r \sqsubseteq r'$, $\phi[r']$ is a relaxation of ϕ.*

Example 2 (Relaxation of a QCSP). Consider a QCSP defined on the variables x_1 and x_2 such that $D(x_1) = \{3, 5\}$ and $D(x_2) = \{6, 9, 10\}$ as follows: $\exists x_1 \in \{3, 5\} \forall x_2 \in \{6, 9, 10\}. \{x_2 \bmod x_1 = 0\}$. This QCSP is false. This is because for any value for variable x_1 there is at least one value in the domain of x_2 that is inconsistent with it.

If we relax the constraint requirement $(x_2 \bmod x_1 = 0)$ to $(x_2 \bmod x_1 < 2)$ the resulting QCSP $\exists x_1 \in \{3, 5\} \forall x_2 \in \{6, 9, 10\}. \{x_2 \bmod x_1 < 2\}$ becomes true. If we relax the scope of the domain of the universally quantified variable x_2 to $\{6, 9\}$ then the resulting QCSP $\exists x_1 \in \{3, 5\} \forall x_2 \in \{6, 9\}. \{x_2 \bmod x_1 = 0\}$ is true. If we relax the position of the universal quantifier from 2 to 1 the resulting QCSP $\forall x_2 \in \{6, 9, 10\} \exists x_1 \in \{3, 5\}. \{x_2 \bmod x_1 = 0\}$ is true.   ▲

4   Preferred Conflicts as Explanations

Given an unsatisfiable QCSP (a conflict) we compute explanations of this unsat-isfiability by relaxing a subset of its requirements to the point where any fur-ther relaxation would yield a satisfiable QCSP (a minimal conflict). Let ϕ be a QCSP defined over the set of original requirements including those that can be relaxed and those that cannot be relaxed. An original requirement that cannot be relaxed is also called a mandatory requirement. We use Υ to denote a set of original requirements of ϕ that can be relaxed. $\mathcal{R}$ is a relaxation function on Υ that maps each original requirement in Υ to its set of possible requirement relaxations, i.e., $\forall r_i \in \Upsilon$, $\mathcal{R}_i$ is the set of possible requirement relaxations of r_i.

For each $r_i \in \Upsilon$, we use $\dagger_i$ to denote its full relaxation (or *bottom relaxation*). If a requirement r is a constraint c then its bottom relaxation is the Cartesian product of the domains of the variables involved in the constraint c, i.e., $\dagger_r = \Pi_{x \in var(c)} D(x)$. If a requirement r is either a scope of a universal quantifier or a position of a universal quantifier Q_i then $\dagger_r = \forall x_i \in \emptyset$. Throughout the paper we assume that each $r_i \in \Upsilon$ can be fully relaxed, i.e., $\forall r_i \in \Upsilon$, $\dagger_i \in \mathcal{R}_i$.

We say that $\mathcal{I} \in \prod \mathcal{R}_i$ is an instance of $\mathcal{R}$ if and only if $\forall r_i \in \Upsilon$, $\mathcal{I}_i$ is an element of $\mathcal{R}_i$. Let $\mathcal{I}$ and $\mathcal{I}'$ be two instances of $\mathcal{R}$. We say that $\mathcal{I}'$ is a strict relaxation of $\mathcal{I}$, denoted $\mathcal{I} \sqsubset \mathcal{I}'$, if and only if there exists a requirement $r_i \in \Upsilon$ such that $\mathcal{I}_i \sqsubset \mathcal{I}'_i$ and for all the other requirements $r_j \in \Upsilon$, $\mathcal{I}_j \sqsubseteq \mathcal{I}'_j$.

We use $\sqcap(\mathcal{R})$ to denote the *top instance* of $\mathcal{R}$, i.e., if $\mathcal{I} = \sqcap(\mathcal{R})$ then there does not exist any other instance $\mathcal{I}'$ of $\mathcal{R}$ such that $\mathcal{I}' \sqsubset \mathcal{I}$. We use $\sqcup(\mathcal{R})$ to denote the *bottom* (or a most relaxed) instance of $\mathcal{R}$, i.e., if $\mathcal{I} = \sqcup(\mathcal{R})$ then there does not exist any other instance $\mathcal{I}'$ of $\mathcal{R}$ such that $\mathcal{I} \sqsubset \mathcal{I}'$. The former is well-defined when there is a unique minimal relaxation and the latter one is well-defined when there is a unique maximal relaxation for each requirement.

We say that a *conflict* is an instance of $\mathcal{R}$ that makes ϕ inconsistent. When confronted with an inconsistent QCSP a user is generally interested in resolving the conflicts. To allow a user to resolve a conflict by relaxing at most one requirement it is important to ensure the minimality of the conflict. We define the notion of minimal conflict with respect to a (typically incomplete) consistency propagation method Π, such as QAC [6], in a similar way to Junker [2]. In what follows, the con-sistency of a QCSP is defined in terms of Π so consistency means Π-*consistency*. Using an incomplete operator is perfectly reasonable since it only means that the conflict computed is minimal with respect to the consistency operator. Further-more, some interesting QCSP may be easy to solve in practice despite the worst case theoretical complexity, e.g., the QCSPs solved in [7].

Definition 9 (Minimal Conflict). *Given a set of original requirements Υ that can be relaxed, and a consistency propagator Π, a minimal conflict $\mathcal{I}$ of a QCSP ϕ is an instance of $\mathcal{R}$ such that $\phi[\mathcal{I}]$ is inconsistent and there does not exist any $\mathcal{I} \sqsubset \mathcal{I}'$ such that $\phi[\mathcal{I}']$ is inconsistent.*

If $\mathcal{I}$ is a minimal conflict of ϕ under $\mathcal{R}$ then $\phi[\mathcal{I}]$ corresponds to a maximally relaxed explanation of ϕ [8]. There may be some requirements that are more important to the user than others. A user generally prefers to relax less important requirements. Based on the order of importance of requirements, the minimal conflicts of an inconsistent QCSP can be ordered. A minimal conflict is generally resolved by relaxing the least important requirement. We say that a minimal conflict is more important than another if resolving the former involves the relaxation of a more important requirement than resolving the latter.

Example 3 (Minimal Conflict of a QCSP). Consider a QCSP defined on the variables x_1 and x_2 such that $D(x_1) = D(x_2) = \{1, 2, 3\}$ as follows: $\exists x_1 \in \{1, 2, 3\} \forall x_2 \in \{1, 2, 3\}.\{x_1 < x_2\}$. Let $\Upsilon = \{r_1, r_2\}$ be a set of original requirements of ϕ that can be relaxed, where $r_1 \equiv \forall x_2 \in \{1, 2, 3\}$, and $r_2 \equiv x_1 < x_2$. Let $\mathcal{R}$ be the relaxation function on Υ such that $\mathcal{R}_1 = \{\forall x_2 \in \{1, 2, 3\}, \forall x_2 \in \{1, 3\}, \forall x_2 \in \{3\}\}$ and $\mathcal{R}_2 = \{x_1 < x_2, x_1 \leq x_2\}$.

Let $\mathcal{I} = \{\forall x_2 \in \{1, 2, 3\}, x_1 < x_2\}$ and $\mathcal{I}' = \{\forall x_2 \in \{1, 3\}, x_1 < x_2\}$ be two instances of $\mathcal{R}$. Notice that both $\mathcal{I}$ and $\mathcal{I}'$ are conflicts since $\phi[\mathcal{I}] \equiv \exists x_1 \in \{1, 2, 3\} \forall x_2 \in \{1, 2, 3\}.\{x_1 < x_2\}$ and $\phi[\mathcal{I}'] \equiv \exists x_1 \in \{1, 2, 3\} \forall x_2 \in \{1, 3\}.\{x_1 < x_2\}$ are inconsistent. Notice that $\mathcal{I} \sqsubset \mathcal{I}'$, since $\mathcal{I}_1 \sqsubset \mathcal{I}'_1$ and $\mathcal{I}_2 = \mathcal{I}'_2$. Therefore, $\mathcal{I}$ is not a minimal conflict. However, $\mathcal{I}'$ is a minimal conflict. The reason is that if we relax $\mathcal{I}'$ further either by relaxing $\forall x_2 \in \{1, 3\}$ to $\forall x_2 \in \{3\}$ or by relaxing $x_1 < x_2$ to $x_1 \leq x_2$ it will result in a consistent QCSP. ▲

Now we define the notion of preference over conflicts of a quantified CSP building upon the notion of preference over conflicts of a CSP [2]. Given two conflicts I and I' of a quantified CSP, we say that I is more important than I' if resolving I involves relaxing a more important requirement. As the user is supposed to solve all the conflicts, it is better to confront him/her first with those conflicts that involve more critical decisions, i.e., with those conflicts that involve relaxing more important requirements.

Definition 10 (Anti-lex Ordering). *Let $\prec$ be a total order in terms of importance on the set of original requirements Υ. Here, $r_i \prec r_j$ means that r_i is more important than r_j. Let $\mathcal{I}$ and $\mathcal{I}'$ be two instances of a relaxation function $\mathcal{R}$. We say that $\mathcal{I} \prec_{antilex} \mathcal{I}'$ if and only if r_i is the least important original requirement such that $\mathcal{I}_i \neq \dagger_i \wedge \mathcal{I}'_i = \dagger_i$, r_j is the least important original requirement such that $\mathcal{I}'_j \neq \dagger_j \wedge \mathcal{I}_j = \dagger_j$, and $r_i \prec r_j$.*

Many conflicts may exist so we focus on the preferred one. If $\mathcal{I}$ and $\mathcal{I}'$ are two minimal conflicts of $\mathcal{R}$ and $\mathcal{I} \prec_{antilex} \mathcal{I}'$ then it means that $\mathcal{I}$ is more important than $\mathcal{I}'$.

Definition 11 (Preferred Conflict). *Given a total order $\prec$ in terms of importance on set of requirements Υ, a minimal conflict $\mathcal{I}$ of a QCSP ϕ is a preferred conflict if and only if there is no other minimal conflict $\mathcal{I}'$ of ϕ such that $\mathcal{I}' \prec_{antilex} \mathcal{I}$.*

Notice that in Definition 11 the non-bottom relaxations of a given original requirement are incomparable. The notion of preferrered conflict is extended in Sect. 8 by assuming an ordering over the relaxations of a given original requirement.

Example 4 (Antilex Ordering on Instances of $\mathcal{R}$). Consider a QCSP defined on variables x_1, x_2 and x_3 such that $D(x_1) = \{1,2\}$, $D(x_2) = \{1,2,3\}$ and $D(x_3) = \{2,3\}$ as follows: $\exists x_1 \forall x_2 \exists x_3.\{x_1 < x_2, x_2 < x_3\}$. This QCSP is unsatisfiable.

Let $\Upsilon = \{r_1, r_2, r_3\}$ be the set of original requirements that can be relaxed, where $r_1 \equiv \forall x_2 \in \{1,2,3\}$, $r_2 \equiv x_1 < x_2$, and $r_3 \equiv x_2 < x_3$. Let us assume that $r_1 \prec r_2 \prec r_3$ is the order of importance on the requirements. The relaxation function $\mathcal{R}$ is defined as follows: $\mathcal{R}_1 = \{\forall x_2 \in \{1,2,3\}, \forall x_2 \in \emptyset\}$ $\mathcal{R}_2 = \{x_1 < x_2, true\}$, and $\mathcal{R}_3 = \{x_2 < x_3, true\}$. Here $\dagger_1 \equiv \forall x_2 \in \emptyset$, $\dagger_2 \equiv true$, and $\dagger_3 \equiv true$. From the definition of minimal conflict it follows that $\mathcal{I} = \{\forall x_2 \in \{1,2,3\}, x_1 < x_2, true\}$ and $\mathcal{I}' = \{\forall x_2 \in \{1,2,3\}, true, x_2 < x_3\}$ are the only minimal conflicts of $\mathcal{R}$. The least important requirements that need to be relaxed for resolving the conflicts $\mathcal{I}$ and $\mathcal{I}'$ are r_2 and r_3 respectively, and since r_2 is more important than r_3, $\mathcal{I} \prec_{antilex} \mathcal{I}'$. Since there are only two minimal conflicts, $\mathcal{I}$ is also the preferred conflict.   ▲

If a total order on the original requirements is not specified, then one can also use a partial order to compute a preferred conflict. Given a partial order $\preceq$ in terms of importance on Υ, a minimal conflict $\mathcal{I}$ of ϕ is a preferred conflict if and only if there is a total order $\prec$ of $\preceq$ such that $\mathcal{I}$ is a preferred conflict of ϕ with respect to $\prec$.

5   Two-Point Relaxation Functions

We present an algorithm for computing a preferred conflict of ϕ under the two-point relaxation function $\mathcal{R}$, where for every original requirement $r_i \in \Upsilon$, $\mathcal{R}_i = \{\dagger_i, r_i\}$. If $\dagger_i$ is in $\mathcal{R}_i$ then r_i is allowed to relax fully. Notice that any pair of instances, say $\mathcal{I}$ and $\mathcal{I}'$, can only be different if there exists at least one $r_j \in \Upsilon$ such that $\mathcal{I}_j \neq \mathcal{I}'_j$, and that would imply that either $\mathcal{I}_j = \dagger_j$ or $\mathcal{I}'_j = \dagger_j$ holds in a two-point relaxation function. Therefore, any pair of instances of the two-point relaxation function $\mathcal{R}$ are comparable and hence they are totally ordered with respect to $\prec_{antilex}$.

The following proposition shows how to compute a preferred conflict by decomposing a given two-point relaxation function defined on a given set of original requirements, which will form the basis for Algorithm 2.

Proposition 1. *Let $\Upsilon = \{r_1, \ldots, r_m\}$ be an original set of requirements of a QCSP ϕ and let $\mathcal{R} = \{\{\dagger_1, r_1\}, \ldots, \{\dagger_m, r_m\}\}$ be a relaxation function on Υ. Suppose that $\Upsilon^1 = \{r_1, \ldots, r_k\}$ and $\Upsilon^2 = \{r_{k+1}, \ldots, r_m\}$ are disjoint sets of requirements of ϕ and that no requirement of Υ^2 is preferred to a requirement of Υ^1. Let $\mathcal{I}^2$ be the preferred conflict under $\mathcal{R}^2 = \{\{r_1\}, \ldots, \{r_k\}, \{\dagger_{k+1}, r_{k+1}\}, \ldots, \{\dagger_m, r_m\}\}$. Let $\mathcal{I}^1$ be the preferred conflict of ϕ under $\mathcal{R}^1 = \{\{\dagger_1, r_1\}, \ldots, \{\dagger_k, r_k\}, \{\mathcal{I}^2_{k+1}\}, \ldots, \{\mathcal{I}^2_m\}\}$. If $\mathcal{I}^1$ is the preferred conflict of ϕ under $\mathcal{R}^1$ and $\mathcal{I}$ is the preferred conflict of ϕ under $\mathcal{R}$ then $\mathcal{I} = \mathcal{I}^1$.*

Proof. To prove that $\mathcal{I} = \mathcal{I}^1$, i.e., $\mathcal{I}^1$ is the preferred conflict of ϕ under $\mathcal{R}$, we prove that any instance of $\mathcal{R}$ that is not in $\mathcal{R}^1$ cannot be the preferred conflict of $\mathcal{R}$. From the definition of $\mathcal{R}^1$, this is equivalent to proving that the projection of Υ^2 on $\mathcal{I}$ i.e., $\mathcal{I}_{\Downarrow\Upsilon^2}$, is equal to $\mathcal{I}^2_{\Downarrow\Upsilon^2}$. We prove this by contradiction. If we assume that $\mathcal{I}_{\Downarrow\Upsilon^2} \neq \mathcal{I}^2_{\Downarrow\Upsilon^2}$ then either $\mathcal{I}_{\Downarrow\Upsilon^2} \prec_{antilex} \mathcal{I}^2_{\Downarrow\Upsilon^2}$ or $\mathcal{I}_{\Downarrow\Upsilon^2} \succ_{antilex} \mathcal{I}^2_{\Downarrow\Upsilon^2}$. If $\mathcal{I}_{\Downarrow\Upsilon^2} \prec_{antilex} \mathcal{I}^2_{\Downarrow\Upsilon^2}$ then it means that there exists a conflict $\mathcal{I}'$ under $\mathcal{R}^2$ such that $\mathcal{I}'_{\Downarrow\Upsilon^1} = \mathcal{I}^2_{\Downarrow\Upsilon^1}$ and $\mathcal{I}'_{\Downarrow\Upsilon^2} = \mathcal{I}_{\Downarrow\Upsilon^2}$. This would imply that $\mathcal{I}' \prec_{antilex} \mathcal{I}^2$, which contradicts the assumption that $\mathcal{I}^2$ is the preferred conflict of $\mathcal{R}^2$. If $\mathcal{I}_{\Downarrow\Upsilon^2} \succ_{antilex} \mathcal{I}^2_{\Downarrow\Upsilon^2}$ then $\mathcal{I} \succ_{antilex} \mathcal{I}^2$. This would imply $\mathcal{I}$ is not the preferred conflict under $\mathcal{R}$, which also contradicts the assumption.

Let $\Upsilon = \{r_1, \ldots, r_m\}$ be an original set of requirements of ϕ that can be relaxed and let $\mathcal{R} = \{\{\dagger_1, r_1\}, \ldots, \{\dagger_m, r_m\}\}$ be a relaxation function on Υ. The algorithm QUICKQCSPXPLAIN for computing a preferred conflict is depicted in Algorithm 1. If the input QCSP, ϕ, is consistent then there is no conflict in which case the algorithm raises an exception. Otherwise, the algorithm QUICKQCSPXPLAIN' (Algorithm 2) is invoked, which computes the preferred conflict $\mathcal{I}$ of ϕ under $\mathcal{R}$ on the set of requirements Υ.

The invariant of QUICKQCSPXPLAIN' is that ϕ under the top instance of $\mathcal{R}$ is inconsistent. If it is not the case then it means that ϕ is consistent under $\mathcal{R}$. One of the parameters of the algorithm is Δ, which is a Boolean variable. It is **true** if it is unknown that ϕ is inconsistent under the bottom instance $\mathcal{B} = \bigsqcup(\mathcal{R})$. If ϕ is inconsistent under $\mathcal{B}$, then the preferred conflict of ϕ under $\mathcal{R}$ is $\mathcal{B}$ (Line 1-2). If $|\Upsilon| = 1$ then it means that there exists only one requirement with two possible relaxations. As the bottom instance is already known to be consistent from Line 1, the top instance of $\mathcal{R}$ has to be inconsistent and the preferred conflict is $\bigsqcap(\mathcal{R})$.

If the cardinality of the set of the original requirements is greater than one, it is ordered in decreasing order of their importance with respect to $\prec$. To find the preferred conflict the ordered set of original requirements is divided into two sets: $\Upsilon_1 = \{r_1, \ldots, r_k\}$ and $\Upsilon_2 = \{r_{k+1}, \ldots, r_m\}$ such that no requirement of Υ_2 is more important than one of Υ_1. First, a relaxation function $\mathcal{R}^2$ is obtained from $\mathcal{R}$ by enforcing that each requirement in Υ^1 cannot be relaxed (Line 8-9). If $\mathcal{I}^2$ is the preferred conflict of ϕ under relaxation function $\mathcal{R}^2$ then, from Proposition 1, the preferred conflict of $\mathcal{R}$ is the preferred conflict of $\mathcal{R}^1$, obtained from $\mathcal{R}$ by setting each $\mathcal{R}_r$ for each $r \in \Upsilon_2$ to the corresponding one in $\mathcal{I}^2$ (Line 10-12).

Algorithm 1. QUICKQCSPXPLAIN$(\phi, \Upsilon, \mathcal{R}, \prec)$

Require: : A QCSP ϕ; $\forall r_i \in \Upsilon$, $\mathcal{R}_i = \{\dagger_i, r_i\}$.
Ensure: : A preferred conflict of ϕ.
1: **if** $\perp \notin \Pi(\phi)$ **then**
2:     **return** exception "no conflict"
3: $\mathcal{I} \leftarrow$ QUICKQCSPXPLAIN$'(\phi, \textbf{true}, \Upsilon, \mathcal{R}, \prec)$
4: **return** $\mathcal{I}$

Algorithm 2. $\textsc{QuickQcspXplain}'(\phi, \Delta, \Upsilon, \mathcal{R}, \prec)$

1: **if** Δ and $\perp \in \Pi(\phi[\bigsqcup(\mathcal{R})])$ **then**
2:      **return** $\bigsqcup(\mathcal{R})$
3: **if** $|\Upsilon| = 1$ **then**
4:      **return** $\bigsqcap(\mathcal{R})$
5: **let** $r_1, \ldots, r_m$ be an enumeration of Υ that respects $\prec$
6: **let** $k = \lfloor (1 + m)/2 \rfloor$ where $1 \leq k < m$
7: $\Upsilon^1 \leftarrow \{r_1, \ldots, r_k\}$ and $\Upsilon^2 \leftarrow \{r_{k+1}, \ldots, r_m\}$
8: $\mathcal{R}^2 \leftarrow \mathcal{R}$
9: $\forall r \in \Upsilon^1, \mathcal{R}_r^2 \leftarrow \{\bigsqcap(\mathcal{R}_r)\}$
10: $\mathcal{I}^2 \leftarrow \textsc{QuickQcspXplain}'(\phi, \texttt{true}, \Upsilon^2, \mathcal{R}^2, \prec)$
11: $\mathcal{R}^1 \leftarrow \mathcal{R}$
12: $\forall r \in \Upsilon^2, \mathcal{R}_r^1 \leftarrow \{\mathcal{I}_r^2\}$
13: $\Delta_2 \equiv \left(((\mathcal{I}^2)_{\Downarrow \Upsilon^2}) \neq (\bigsqcup(\mathcal{R}^2)_{\Downarrow \Upsilon^2}) \right)$
14: $\mathcal{I}^1 \leftarrow \textsc{QuickQcspXplain}'(\phi, \Delta_2, \Upsilon^1, \mathcal{R}^1, \prec)$
15: **return** $\mathcal{I}^1$

$\textsc{QuickQcspXplain}'$ avoids unnecessary consistency checks in cases where it is known that ϕ is consistent under the bottom instance. Notice that if the execution continues after Line 2, then it is known that ϕ is consistent under the bottom instance. If all requirements in Υ^2 are set to their bottom relaxation in $\mathcal{I}^2$ and all requirements in Υ^1 are set to their bottom relaxation in $\mathcal{I}^1$ then this would imply that ϕ is consistent under the bottom instance of $\mathcal{R}$. Therefore, whenever all the requirements in Υ^2 are set to their bottom relaxation in $\mathcal{I}^2$, Δ_2 is set to `false` (Line 13) to avoid the consistency check in Line 1 when computing $\mathcal{I}^1$.

$\textsc{QuickQcspXplain}$ is a reformulation of $\textsc{QuickXplain}$ [2] in terms of relaxations, and thereby generalising it to QCSP with at most one distinct relaxation available for each of the original requirements, i.e., a requirement is either present or fully relaxed. In the worst-case, $\textsc{QuickQcspXplain}$ will perform $\mathcal{O}(k \log \frac{n}{k})$ number of consistency checks, where n is the number of original requirements and k is the number of original requirements in the preferred conflict that are not fully relaxed. Here consistency checks refers to the number of times consistency of a QCSP is checked using Π.

6   Multi-point Relaxation Functions

We now consider $\textsc{QuickGenQcspXplain}$ (Algorithm 3), a more general algorithm allowing for multi-point relaxation function (requirement relaxation lattices of arbitrary size). The algorithm receives as parameters a QCSP ϕ, a relaxation function $\mathcal{R}$ on Υ and a lexicographic order $\prec$ on Υ. For the multi-point relaxation function, Algorithm 1 becomes inapplicable as it assumes that for each requirement in Υ there are only two possible relaxations: the top relaxation (the original requirement itself) and the bottom relaxation. The task of computing a preferred conflict under the multi-point relaxation function defined over a given

---

Algorithm 3. $\text{QuickGenQcspXplain}(\phi, \Upsilon, \mathcal{R}, \prec)$

Require: : A QCSP ϕ.
Ensure: : A preferred conflict of ϕ.
1: **if** $\perp \notin \Pi(\phi)$ **then**
2:   **return** exception "no conflict"
3: **if** $\perp \in \Pi(\phi[\bigsqcup(\mathcal{R})])$ **then**
4:   **return** $\bigsqcup(\mathcal{R})$
5: $\forall r_i \in \Upsilon, \mathcal{R}'_i \leftarrow \{\dagger_i, \bigsqcap(\mathcal{R}_i)\}$
6: $\mathcal{I}' \leftarrow \text{QuickQcspXplain}(\phi, \Upsilon, \mathcal{R}', \prec)$
7: $\mathcal{R}'' \leftarrow \mathcal{R}$
8: **for all** $r_i \in \Upsilon$ s.t. $\mathcal{I}'_i = \dagger_i$ **do**
9:   $\mathcal{R}''_i \leftarrow \{\dagger_i\}$
10: **return** $\text{QcspRelax}(\phi, \mathcal{R}'', \Upsilon)$

---

set of original requirements can be divided into two subtasks. First, we compute a minimal set of preferred original requirements that cannot be satisfied together. Here minimality means that if any of the preferred original requirements is fully relaxed, then the resulting set of original requirements becomes consistent. Second, we relax the minimal set of inconsistent preferred original requirements as computed in the first step to a point in the relaxation space such that the relaxation is inconsistent but any further relaxation results in a consistent state.

If ϕ is consistent then Algorithm 3 raises an exception. If ϕ is inconsistent under the bottom instance of $\mathcal{R}$ then the bottom instance of $\mathcal{R}$ is the preferred conflict. Otherwise, the algorithm first computes a minimal set of preferred original requirements that leads to an inconsistent QCSP regardless of the way the other requirements are relaxed. To do so, it constructs a two-point relaxation function $\mathcal{R}'$ such that for every original requirement in Υ there are only two possible relaxations: the top relaxation (a.k.a. original requirement) and its bottom relaxation (Line 5). It then invokes QuickQcspXplain (Algorithm 1), which returns the preferred conflict, $\mathcal{I}'$, of ϕ under $\mathcal{R}'$. The elements of $\mathcal{I}'$ that are set to top relaxations are the preferred original requirements that lead to an inconsistent ϕ when the other requirements are set to their bottom relaxations.

Although $\mathcal{I}'$ is the preferred conflict of $\mathcal{R}'$, it may not necessarily be a preferred conflict of $\mathcal{R}$. The reason is that $\mathcal{I}'$ may not be a minimal conflict of $\mathcal{R}$ since there may exist a conflict $\mathcal{I}$ under $\mathcal{R}$ such that $\mathcal{I}' \sqsubset \mathcal{I}$. Therefore, it may be possible to relax $\mathcal{I}'$ further under $\mathcal{R}$. $\mathcal{I}'$ (the instance returned by QuickQcspXplain) can be seen as an approximation of the preferred conflict under $\mathcal{R}$. To compute the exact preferred conflict, first a multi-point relaxation function $\mathcal{R}''$ is obtained from $\mathcal{R}$ where for all requirements i such that $\mathcal{I}'_i = \dagger_i$, $\mathcal{R}''_i$ is set to a singleton set containing only the bottom relaxation of the original requirement r_i. $\mathcal{R}''$ represents the set of most important conflicts, i.e., if there is a conflict $\mathcal{I}'' \in \mathcal{R}$ that is not in $\mathcal{R}''$, and $\mathcal{I}''$ is more important than any of those in $\mathcal{R}''$, then that conflict cannot be minimal. The elements of $\mathcal{R}''$ form a semi lattice whose meet element is $\mathcal{I}'$. Thus, a minimal conflict of $\mathcal{R}''$ is a preferred conflict of $\mathcal{R}$. The relaxation function $\mathcal{R}''$ is passed as a parameter to QcspRelax (Algorithm 4) which computes a minimal conflict under $\mathcal{R}''$. This algorithm is an adaptation of the algorithm QuantifiedXPlain as described in [8].

Algorithm 4. QcspRelax$(\phi, \mathcal{R}, \Upsilon)$

1: **while** $\exists i : |\mathcal{R}_i| > 1$ **do**
2:     $\forall \mathcal{R}_i$ choose r_i from $\{r \in \mathcal{R}_i : \forall r' \in \mathcal{R}_i, r \not\sqsubseteq r'\}$
3:     **if** $\perp \in \Pi(\phi[\{r_1, \ldots, r_k\}])$ **then**
4:         **return** $\{r_1, \ldots, r_k\}$
5:     **while** $\perp \notin \Pi(\phi[\{r_1, \ldots, r_k\}])$ **do**
6:         choose i from Υ s.t. $r_i \neq \bigsqcap(\mathcal{R}_i)$
7:         $r_i' \leftarrow r_i$
8:         $\mathcal{R}_i' \leftarrow \{r : r \in \mathcal{R}_i, r \sqsubseteq r_i\}$
9:         choose r_i from $\{r \in \mathcal{R}_i' : \forall r' \in \mathcal{R}_i', r \not\sqsubseteq r'\}$
10:    $\mathcal{R}_i \leftarrow \{r : r \in \mathcal{R}_i, r_i' \not\sqsubseteq r\}$
11:    $\forall \mathcal{R}_j : \mathcal{R}_j \leftarrow \{r : r \in \mathcal{R}_j, r_j \sqsubseteq r\}$
12: **return** $\{r_1, \ldots, r_k\}$

Algorithm 4 considers a current candidate relaxation comprising one element from each relaxation space, $r_1 \ldots r_k$, and use $\sqsubseteq$ to obtain successive approximations to a minimal conflict. We begin with a maximal relaxation of *each* requirement, and then progressively tighten these one at a time: we select an i such that r_i may be assigned a new relaxation strictly tighter than the old one, tightening by a minimal amount at each step to ensure minimality of the final conflict of ϕ. When an inconsistency is detected, all relaxations tighter than or incomparable to the current approximation from future consideration are eliminated, as unnecessary for a minimal conflict. At the same time, the relaxation space for the last-relaxed requirement, $\mathcal{R}_i$, is restricted to ensure that the requirement may not be as relaxed as the earlier relaxation, r_i', that did not produce an inconsistency, as we have guaranteed that the relaxation cannot take part in the minimal conflict currently under construction. The relaxation space for the last-relaxed requirement, $\mathcal{R}_i$, contains only those relaxations that are weaker than r_i and incomparable to r_i'. Each termination of the inner while loop can call Π at most $\mathcal{O}(d)$ times for each requirement since the QCSP is made tighter in each iteration. Here d is the maximum no. of relaxations of requirements. Whenever an inconsistency is detected in the inner while loop (Line 5) at least one relaxation is removed from one relaxation space of one requirement. This process is repeated with the restricted relaxation spaces, until eventually only one possible relaxation remains for each requirement, or the chosen maximal relaxation (Line 3) is inconsistent, thus fully determining the minimal conflict.

Example 5 (Preferred Conflict wrt. $\prec_{antilex}$). Consider a QCSP defined on variables x_1, x_2 and x_3 such that $D(x_1) = \{1, 2\}$, $D(x_2) = \{1, 2, 3\}$ and $D(x_3) = \{2, 3\}$ as follows: $\exists x_1 \forall x_2 \exists x_3.\{x_1 < x_2, x_2 < x_3\}$. This QCSP is unsatisfiable. Let $\Upsilon = \{r_1, r_2, r_3\}$ be the set of original requirements that can be relaxed, where $r_1 \equiv \forall x_2 \in \{1,2,3\}$, $r_2 \equiv x_1 < x_2$, and $r_3 \equiv x_2 < x_3$. Let us assume that $r_1 \prec r_2 \prec r_3$. The relaxation function $\mathcal{R}$ is defined as follows: $\mathcal{R}_1 = \{\forall x_2 \in \{1,2,3\}, \forall x_2 \in \{1,2\}, \forall x_2 \in \{1,3\}, \forall x_2 \in \emptyset\}$ $\mathcal{R}_2 = \{x_1 < x_2, true\}$, and $\mathcal{R}_3 = \{x_2 < x_3, true\}$.

In order to find a preferred conflict for the multi-point relaxation function $\mathcal{R}$, the first step is to find a most important set of original requirements that are in conflict. This is done by first computing the two-point relaxation function $\mathcal{R}'$ from $\mathcal{R}$, where $\mathcal{R}'_1 = \{\forall x_2 \in \{1,2,3\}, \forall x_2 \in \emptyset\}$, $\mathcal{R}'_2 = \{x_1 < x_2, true\}$, and $\mathcal{R}'_3 = \{x_2 < x_3, true\}$. From Example 4 it follows that $\mathcal{I} = \{\forall x_2 \in \{1,2,3\}, x_1 < x_2, true\}$ is the preferred conflict of $\mathcal{R}'$. A preferred conflict of $\mathcal{R}$ is a minimal conflict $\mathcal{I}'$ of $\mathcal{R}$ such that $\mathcal{I} \sqsubseteq \mathcal{I}'$. It follows that $\mathcal{I}^1 = \{\forall x_2 \in \{1,2\}, x_1 < x_2, true\}$ and $\mathcal{I}^2 = \{\forall x_2 \in \{1,3\}, x_1 < x_2, true\}$ are two preferred conflicts of $\mathcal{R}$. Notice that $\mathcal{I} \sqsubset \mathcal{I}^1$ and $\mathcal{I} \sqsubset \mathcal{I}^2$ and both $\mathcal{I}^1$ and $\mathcal{I}^2$ are minimal conflicts of $\mathcal{R}$.   ▲

Complexity. Only k requirements are relaxed in Algorithm 4, since the set of relaxations for the remaining requirements $n - k$ are made singleton in Line 9 of Algorithm 3. Since there are at most d possible relaxations of each of the k requirements, the upper bound on the number of calls to Π is $\mathcal{O}(k\,d)$. In the worst-case the number of calls to Π performed by QuickGenQcspXplain is the number of calls to Π performed by QuickQcspXplain and QcspRelax, which is $\mathcal{O}(k \log \frac{n}{k}) + \mathcal{O}(k\,d)$. Notice that no assumption on the incrementality of Π is made. If Π is incremental then the complexity can be reduced to $\mathcal{O}(k \max(\log \frac{n}{k}, d))$. It may also be possible to express the complexity of the algorithm in terms of the width the lattice of the relaxations of the requirements.

Correctness. To prove the correctness of Algorithm 3 we need to prove that if $\mathcal{I}'$ is the preferred conflict of $\mathcal{R}'$ then the preferred conflict of $\mathcal{R}$ is the minimal conflict of $\mathcal{R}''$. Notice that $\prec_{antilex}$ partitions the instances of $\mathcal{R}$ into equivalence classes. By construction, $\mathcal{R}'$ contains an instance from every equivalent class of $\mathcal{R}$ under $\prec_{antilex}$. Consequently, an optimal instance of $\mathcal{R}'$ is an optimal instance of $\mathcal{R}$. By construction, the preferred conflict of $\mathcal{R}''$ is either $\mathcal{I}'$ or one of the relaxations of $\mathcal{I}'$. All the relaxations of $\mathcal{I}'$ that are minimal conflict of ϕ are equivalent under $\prec_{antilex}$. Therefore, the preferred conflict of $\mathcal{R}$ is the minimal conflict of $\mathcal{R}''$.

To prove that Algorithm 4 is correct we need to show that it returns a minimal conflict of ϕ under $\mathcal{R}$. The invariant of Algorithm 4 is that there is at least one instance of $\mathcal{R}$ that makes ϕ inconsistent. Notice that the invariant holds when Algorithm 4 is called, otherwise there would not be any conflict. The only place where $\mathcal{R}$ is modified are Lines 10 and 11 after which also the invariant holds since the instance that causes the inconsistency (in Line 5) is not removed. The fact that the returned instance is a conflict follows from two facts: (a) in Line 4 the algorithm returns an instance $\mathcal{I}$ of $\mathcal{R}$ that makes ϕ inconsistent, and (b) in Line 12 it returns the only instance of the current $\mathcal{R}$ after keeping the invariant. What remains to be proved is that the returned conflict is minimal. In Line 2 the algorithm selects a maximal instance of the current $\mathcal{R}$ so minimality is assured. in Line 12, the selected inconsistent maximal instance would not be a minimal conflict if the algorithm removes a relaxation of the selected instance that is inconsistent when there is no a weaker or incomparable relaxation of the removed one that is inconsistent too. But this can not happen since in Line 10 we only remove instances that are consistent, and in Line 11 those that are removed are not weaker than than the current inconsistent one, which is kept.

7   Totally Ordered Requirement Relaxations

We consider a special case of the multi-point relaxation function $\mathcal{R}$ over a set of original requirements Υ such that $\sqsubset$ is a strict total order on $\mathcal{R}_i$ for each requirement r_i. Specifically, for all mutually different $r, r', r'' \in R_i$, if $r \sqsubset r'$ then $r' \not\sqsubset r$, if $r \sqsubset r'$ and $r' \sqsubset r''$ then $r \sqsubset r''$, and either $r \sqsubset r'$ or $r' \sqsubset r$. Notice that $\sqsubset$ is always a total order on relaxations of a position of a universal quantifier. Also, in some cases it may be possible that the user specified relaxations for a constraint requirement or a scope of the universal quantifier requirement are totally ordered with respect to the $\sqsubset$ relation.

In the inner while loop (Line 5) of QCSPRELAX, whenever an inconsistency is detected, the set of relaxations of the last chosen requirement i (Line 3) becomes singleton when $\mathcal{R}_i$ is totally ordered with respect to $\sqsubset$ due to the pruning that takes place afterwards. All the relaxations that are weaker than r_i are removed from $\mathcal{R}_i$ (Line 9), and all the relaxations that are tighter than r_i are removed from $\mathcal{R}_i$ (Line 10), and there are no incomparable relaxations of r_i in $\mathcal{R}_i$ since $\mathcal{R}_i$ is totally ordered with respect to $\sqsubset$. After every failure in the inner while loop at least one $\mathcal{R}_i$ becomes singleton. Since there are at most k requirements that are not singleton initially (when QCSPRELAX is invoked), the algorithm can fail at most k times.

8   Preferences on Requirement Relaxations

The instances of $\mathcal{R}$ so far are ordered based on $\prec_{antilex}$, which in turn is based on the preferences defined on the original set of requirements Υ. It is natural to think that a user may provide an order of importance on relaxations of a given requirement [9]. For example, a tighter relaxation of a given requirement is more important than a weaker relaxation of the same requirement. When a pair of relaxations are incomparable with respect to $\sqsubset$, a user may also provide an ordering between them based on his preference. In general, we assume that there is a total order of importance on the set of relaxations of an original requirement that extends the ordering imposed by $\sqsubset$.

The instances of $\mathcal{R}$ may not always be totally ordered with respect to $\prec_{antilex}$. Any equivalent class resulting from $\prec_{antilex}$ can be further ordered by using a given ordering on the relaxations of original requirements. For example, if both $\mathcal{I}$ and $\mathcal{I}'$ are preferred conflicts (based on Definition 11) and if they are incomparable with respect to $\prec_{antilex}$, then it would be more desirable to choose the one that involves a more important requirement relaxation of a more important original requirement.

Definition 12 (Lex Ordering). *Let $\prec$ be a total order on Υ. Let $\prec_i$ be a total order on $\mathcal{R}_i$ for each $r_i \in \Upsilon$ such that $\forall r, r' \in \mathcal{R}_i, r \sqsubseteq r' \Rightarrow r \prec_i r'$. Let $\mathcal{I}$ and $\mathcal{I}'$ be two instances of $\mathcal{R}$. We say that $\mathcal{I} \prec_{lex} \mathcal{I}'$ iff $\exists k : \mathcal{I}_k \prec_k \mathcal{I}'_k$ and for all $r_j \in \Upsilon$ such that $r_j \prec r_k, \mathcal{I}_j = \mathcal{I}'_j$.*

The idea is to use $\prec_{lex}$ to order the instances of an equivalent class of $\mathcal{R}$ resulting from using $\prec_{antilex}$ in order to find the preferred conflict. The notion of preferred conflict is redefined as follows:

Definition 13 (Preferred Conflict$^+$). *Given a total order $\prec$ in terms of importance on Υ and a total order $\prec_i$ in terms of importance on $\mathcal{R}_i$ for each original requirement $r_i \in \Upsilon$, a minimal conflict $\mathcal{I}$ of ϕ is a preferred conflict$^+$ iff for all other minimal conflict $\mathcal{I}'$ of ϕ either $\mathcal{I} \prec_{antilex} \mathcal{I}'$, or if neither $\mathcal{I} \prec_{antilex} \mathcal{I}'$ nor $\mathcal{I}' \prec_{antilex} \mathcal{I}$ then $\mathcal{I} \prec_{lex} \mathcal{I}'$.*

Example 6 (Breaking Ties with $\prec_{lex}$). Consider again the QCSP as described in Example 5 on variables x_1, x_2 and x_3 such that $D(x_1) = \{1, 2\}$, $D(x_2) = \{1, 2, 3\}$ and $D(x_3) = \{2, 3\}$ as follows: $\exists x_1 \forall x_2 \exists x_3.\{x_1 < x_2, x_2 < x_3\}$. This QCSP is unsatisfiable. Let $\Upsilon = \{r_1, r_2, r_3\}$ be the set of original requirements that can be relaxed, where $r_1 \equiv \forall x_2 \in \{1, 2, 3\}$, $r_2 \equiv x_1 < x_2$, and $r_3 \equiv x_2 < x_3$. Let us assume that $r_1 \prec r_2 \prec r_3$. The relaxation function $\mathcal{R}$ is defined as follows: $\mathcal{R}_1 = \{\forall x_2 \in \{1, 2, 3\}, \forall x_2 \in \{1, 2\}, \forall x_2 \in \{1, 3\}, \forall x_2 \in \emptyset\}$, $\mathcal{R}_2 = \{x_1 < x_2, true\}$, and $\mathcal{R}_3 = \{x_2 < x_3, true\}$. Let $(\forall x_2 \in \{1, 2, 3\}) \prec_1 (\forall x_2 \in \{1, 2\}) \prec_1 (\forall x_2 \in \{1, 3\}) \prec_1 (\forall x_2 \in \emptyset)$ be an ordering of importance on the relaxations of r_1. The ordering of relaxations for the requirements r_2 and r_3 are $(x_1 < x_2) \prec_2 true$, and $(x_2 < x_3) \prec_3 true$ respectively.

From Example 5 it follows that $\mathcal{I}^1 = \{\forall x_2 \in \{1, 2\}, x_1 < x_2, true\}$ and $\mathcal{I}^2 = \{\forall x_2 \in \{1, 3\}, x_1 < x_2, true\}$ are two preferred conflicts of $\mathcal{R}$, which are incomparable with respect to $\prec_{antilex}$. More specifically, in Example 5 $\forall x_2 \in \{1, 2\}$ and $\forall x_2 \in \{1, 3\}$ are incomparable. However, here $\forall x_2 \in \{1, 2\}$ is preferred over $\forall x_2 \in \{1, 3\}$, Therefore, the ties between $\mathcal{I}^1$ and $\mathcal{I}^2$ are broken with respect to $\prec_{lex}$ and $\mathcal{I}^1$ is the only preferred conflict$^+$.   ▲

We show that the number of consistency checks for finding the preferred conflict$^+$ (under Definition 13) is exponential in the worst-case, given $P \neq NP$, when the relaxation lattice of a requirement is arbitrary and the consistency check is polynomial. We remark that finding the preferred conflict$^+$ involves verifying whether there exists a minimal conflict that involves a given relaxation of a given requirement. In the following we define this problem and prove its intractability.

Definition 14 (Restricted Minimal Conflict). *Given a QCSP ϕ, a relaxation function $\mathcal{R}$ on Υ, an original requirement $i \in \Upsilon$ and a relaxation $r \in \mathcal{R}_i$, the restricted minimal conflict problem is to find a minimal conflict $\mathcal{I}$ of ϕ such that $\mathcal{I}_i = r$.*

Although finding a minimal conflict is polynomial, the restricted minimal conflict problem is intractable. We prove this by reduction from 3-SAT [10].

Proposition 2. *Given a polynomial consistency operator and a meet-semilattice relaxation function, finding a restricted minimal conflict of a* QCSP *is NP-complete.*

Proof. To prove this we use a reduction from 3-SAT to the restricted minimal conflict problem. Let $\langle V, C \rangle$ be a 3-SAT instance, where $V = \{x_1, \ldots, x_n\}$ is a set of Boolean variables and C is a set of clauses defined on any 3 variables of V. To construct an instance of the restricted minimal conflict problem that solves the given SAT instance, we define ϕ over the set of original requirements $\{r_0, r_1, \ldots, r_n, r_{n+1}\}$. The original requirement r_0 is equivalent to $u \wedge C$, where u is a Boolean variable that denotes whether C is inconsistent or not. For each $x_i \in V$, $1 \leq i \leq n$, we associate an original requirement r_i that is equivalent to $x_i \wedge \neg x_i$. The requirement r_{n+1} is equivalent to $C \Rightarrow \neg u$, which is a hard requirement. The set of original requirements that can be relaxed is $\Upsilon = \{r_0, \ldots, r_n\}$. The relaxation function $\mathcal{R}$ on Υ is defined as follows: $\mathcal{R}_0 = \{r_0, C\}$ and $\forall x_i \in V$, $1 \leq i \leq n$, $\mathcal{R}_i = \{r_i, \neg x_i, x_i\}$.

By construction $\phi[\mathcal{I}]$ is inconsistent for all $\mathcal{I}$ such that $\mathcal{I}_{\Downarrow V}$ represents an assignment that violates any clause $c \in C$ and $\mathcal{I}_0 = C$. The definition of $\mathcal{R}_0$ guarantees that when C is violated, no minimal conflict can involve the relaxation $u \wedge C$ since it is tighter than C. Additionally, $\phi[\mathcal{I}]$ is inconsistent for all $\mathcal{I}$ such that $\mathcal{I}_{\Downarrow V}$ represents an assignment that makes all the clauses in C consistent and $\mathcal{I}_0 = r_0$. Therefore, a minimal conflict $\mathcal{I}$ where $\mathcal{I}_0 = r_0$ is necessarily one that satisfies all the clauses, i.e., $\mathcal{I}_{\Downarrow V}$ is compatible with every clause in C. Finding a minimal conflict where $\mathcal{I}_0 = r_0$ is equivalent to finding a solution of the 3-SAT instance, which may involve trying all combinations of relaxations of the requirements from r_1 to r_n. Thus finding a restricted minimal conflict is NP-complete.

The bottom relaxation of each $r_i \in \Upsilon$ is omitted in $\mathcal{R}_i$ in the proof of Proposition 2 for the sake of clarity. Nevertheless, the result holds if they are considered. The following corollary follows from Proposition 2.

Corollary 1. *Given a polynomial consistency operator and a meet-semilattice relaxation function, finding a restricted minimal conflict of a* QCSP *requires a non-polynomial number of consistency checks, and hence finding a preferred conflict$^+$ (under Definition 13) also requires a non-polynomial number of consistency checks.*

9   Checking Equivalence for Partial Implementations

In VLSI CAD, checking equivalence for partial implementations is about checking whether a partial implementation can (still) be extended to a complete design, which is equivalent to a given full specification [11]. This can be achieved by combining parts of the implementation that are not yet finished into black boxes. If the implementation differs from the specification for all possible substitutions of the black boxes, a design error is found in the current partial implementation.

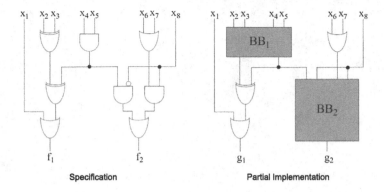

Fig. 1. An example of a partial implementation that can still be extended to a complete design

In the implementation of complex designs, errors may occur due to the distribution of the implementation task to several groups of designers. Each team may locally take implementation decisions that are locally consistent but globally inconsistent. Figure 1 shows an example of a partial implementation that can still be extended to a complete design. Notice there is indeed a possible implementation for each one of the boxes since the boxes can be replaced with the sub-circuits that they cover. Figure 2 shows an example of a partial implementation that cannot be extended to a complete design. Suppose that the coordinator of the implementation decides to split the implementation between two teams in such a way that the first team will be responsible for inputs x_1, x_2, x_3, x_4, x_5 and x_8 and implement the behaviour in the grey box, and the second team is in charge of implementing the remaining part of the circuit using x_6 and x_7. This partition is clearly inconsistent. Black box one should be in charge of the third AND gate if this box is the one that has access to x_8. Once the first black box is implemented, there is no way of completing the implementation if the second black box does not have access to x_8.

Scholl and Becker [11] proposed a necessary condition for checking the consistency of partial implementations. The condition is only a necessary condition since, even if the condition is satisfied, it might be still impossible to complete the implementation. Basically it is a quantified CSP where for each box, the variables associated with its inputs are universally quantified and the variables associated with its output are existentially quantified.

$$\forall I_1 \exists O_1 \dots \forall I_b \exists O_b . \phi$$

In this quantified CSP:

- b is the number of black boxes,
- I_j/O_j is the set of Boolean variables associated with the inputs/outputs of black box j, and
- ϕ is a QCSP that approximates the equivalence between the specification and the partial implementation.

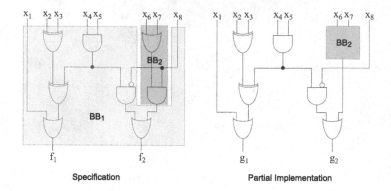

Fig. 2. An example of a partial implementation that cannot be extended to a complete design

In short this means that regardless of the values of the inputs it is possible to compute the expected outputs for the gate.

Approaches have been certainly suggested for addressing this verification problem [11,12]. However, to the best of our knowledge, the problem of computing preferred explanations to errors has not been addressed. Formally, If $\forall I_1 \exists O_1 \ldots \forall I_b \exists O_b. \phi$ is unsatisfiable, one would like to compute the most preferred explanation for the error taking into account that there are boxes that are more critical (e.g., involve more important resources) than others? We propose to compute an explanation that involves the constituents of a most critical box, assuming that the preferences of the individuals that participate in the implementation process are consistent.

10   Conclusions

In this paper we presented a framework for generating most preferred explanations for the inconsistency of a QCSP. The additional expressiveness of the QCSP can help model problems in which a subset of the variables take value assignments that are outside the control of the decision maker, e.g. in game-playing, conformant planning and reasoning under uncertainty. We presented various settings for representing preferences, and in each case we presented an algorithm for computing preferred explanations based on the notion of conflict. Finally, we provided key complexity results on the limits of computing preferred explanations in this setting and motivated the applicability of the framework with an example coming from the industry.

In the future we are planning to empirically prove the effectiveness of our approach by implementing it on top of an already existing QCSPsolver (e.g., [4] or [13]). Even though the focus of our work is on the computation of preferred explanations for an unsatisfiable QCSP, it might be also interesting to find out whether preferred explanations for that type of unsatisfiable problems can be easily expressed and computed in alternative frameworks like those found in the answer set programming community (e.g., [14,15]).

Acknowldedgement. This material is based upon work supported by the Science Foundation Ireland under Grant No.10/CE/I1853 and the FP7 Programme (FP7/2007–2013) under grant agreement No. 318137 (DISCUS). The Insight Centre for Data Analytics is also supported by Science Foundation Ireland under Grant No. SFI/12/RC/2289.

References

1. Chen, H.: The Computational Complexity of Quantified Constraint Satisfaction. Ph.D. thesis, Cornell, August 2004
2. Junker, U.: Quickxplain: preferred explanations and relaxations for over-constrained problems. In: Proceedings of AAAI 2004, pp. 167–172 (2004)
3. Verger, G., Bessière, C.: : A bottom-up approach for solving quantified csps. In: Proceedings of CP, pp. 635–649 (2006)
4. Gent, I.P., Nightingale, P., Stergiou, K.: QCSP-solve: a solver for quantified constraint satisfaction problems. In: Proceedings of IJCAI, pp. 138–143 (2005)
5. Mehta, D., O'Sullivan, B., Quesada, L.: Preferred explanations for quantified constraint satisfaction problems. In: 22nd IEEE International Conference on Tools with Artificial Intelligence, ICTAI 2010, Arras, France, 27–29 October 2010, vol. 1, pp. 275–278. IEEE Computer Society (2010)
6. Bordeaux, L., Monfroy, E.: Beyond NP: arc-consistency for quantified constraints. In: Proceedings of CP 2002, pp. 371–386 (2002)
7. Stynes, D., Brown, K.N.: Realtime online solving of quantified CSPs. In: Gent, I.P. (ed.) CP 2009. LNCS, vol. 5732, pp. 771–786. Springer, Heidelberg (2009)
8. Ferguson, A., O'Sullivan, B.: Quantified constraint satisfaction problems: from relaxations to explanations. In: Proceedings of IJCAI-2007, pp. 74–79 (2007)
9. Brafman, R.I., Domshlak, C.: Introducing variable importance tradeoffs into CP-Nets. In: UAI, pp. 69–76 (2002)
10. Garey, M., Johnson, D.: Computers and Intractability: A Guide to the The Theory of NP-Completeness. W. H Freeman and Company, New York (1979)
11. Scholl, C., Becker, B.: Checking equivalence for partial implementations. In: Proceedings of the 38th Design Automation Conference, DAC 2001, Las Vegas, NV, USA, June 18–22, pp. 238–243. ACM (2001)
12. Miller, C., Kupferschmid, S., Lewis, M., Becker, B.: Encoding Techniques, craig interpolants and bounded model checking for incomplete designs. In: Strichman, O., Szeider, S. (eds.) SAT 2010. LNCS, vol. 6175, pp. 194–208. Springer, Heidelberg (2010)
13. Benedetti, M., Lallouet, A., Vautard, J.: QCSP made practical by virtue of restricted quantification. In: Veloso, M.M. (ed.) IJCAI 2007, Proceedings of the 20th International Joint Conference on Artificial Intelligence, Hyderabad, India, January 6–12 2007, pp. 38–43 (2007)
14. Brewka, G.: Answer sets and qualitative optimization. Log. J IGPL **14**(3), 413–433 (2006)
15. Confalonieri, R., Nieves, J.C., Osorio, M., Vázquez-Salceda, J.: Dealing with explicit preferences and uncertainty in answer set programming. Ann. Math. Artif. Intell. **65**(2–3), 159–198 (2012)

A Graphical Theorem of the Alternative for UTVPI Constraints

K. Subramani and Piotr Wojciechowski[✉]

LDCSEE, West Virginia University, Morgantown, WV, USA
ksmani@csee.wvu.edu, pwojciec@mix.wvu.edu

Abstract. There exist quite a few theorems of the alternative for linear systems in the literature, with Farkas' lemma being the most famous. All these theorems have the following form: We are given two closely related linear systems such that one and exactly one has a solution. Some specialized classes of linear systems can also be represented using graphical structures and the corresponding theorems of the alternative can then be stated in terms of properties of the graphical structure. For instance, it is well-known that a system of difference constraints (DCS) can be represented as a constraint network such that the DCS is feasible if and only if there does not exist a negative cost cycle in the network. In this paper, we provide a new graphical constraint network representation of Unit Two Variable Per Inequality (UTVPI) constraints. This constraint network representation permits us to derive a theorem of the alternative for the feasibility of UTVPI systems. UTVPI constraints find applications in a number of domains, including but not limited to program verification, abstract interpretation, and array bounds checking. Theorems of the Alternative find primary use in the design of certificates in certifying algorithms. It follows that our work is important from this perspective.

1 Introduction

In this paper, we focus on deriving a graphical theorem of the alternative for linear feasibility in UTVPI constraints. UTVPI constraint systems find applications in a number of problem domains, including but not limited to real-time scheduling [GPS95], program verification [CC77] and operations research [LM05]. The graphical characterization of feasibility in UTVPI constraints permits the design of certificates in certifying algorithms.

The field of certifying algorithms is concerned with validating the results of implementations of algorithms. Even algorithms that can be proven correct, suffer the risk of being implemented incorrectly. One of the more famous examples of this phenomenon is the error discovered in the implementation of a planarity testing algorithm in the LEDA software [MN99]. In this case, there was a subtle bug in the implementation of planarity testing. Bug identification is a non-trivial

---

This research was supported in part by the National Science Foundation through Award CCF-1305054.

M. Leucker et al. (Eds.): ICTAC 2015, LNCS 9399, pp. 328–345, 2015.
DOI: 10.1007/978-3-319-25150-9_20

task in and of itself. In the case of LEDA, matters were complicated by the lack
of certificates. Consequently, there is widespread interest in the design and development of certifying algorithms, i.e., algorithms which provide certificates that
validate the answer that is provided. For instance, an algorithm for graph planarity testing could provide a planar embedding when it declares a graph to be
planar, and a subgraph of the input graph that is homeomorphic to $K_{3,3}$ or K_5,
in the event that it declares the graph to be non-planar (Kuratowski's Theorem).
It is understood that the implementations of algorithms for verifying a planar
embedding and checking homeomorphism to $K_{3,3}$ and K_5 are trivial enough to
be checked by a simple, provably correct implementation.

The important contributions of this paper are as follows:

1. We propose a new constraint network structure for UTVPI constraints. This
   network structure is similar to the constraint network structure for difference constraints [CLRS01], but incorporates many features that are unique
   to UTVPI constraint systems.
2. We present a theorem of the alternative for the recognition of linear infeasibility in UTVPI constraints. This theorem is similar in spirit to Farkas'
   lemma for a system of linear constraints, and is crucial from the perspective
   of designing certifying algorithms [Rub90].

The rest of the paper is organized as follows: Sect. 2 discusses the preliminaries of UTVPI constraints and Farkas' lemma. In Sect. 3, we provide a detailed
description of the new constraint network representation of UTVPI constraints.
In Sect. 4, we describe some of the other constraint network representations for
UTVPI constraints. The theorem of the alternative is detailed in Sect. 5. We
conclude in Sect. 6 by summarizing our contributions and identifying avenues
for future research.

2 Preliminaries

In this section, we formally define the linear feasibility problem in UTVPI constraints and also define the various terms that will be used in the rest of the
paper.

Definition 1. *A constraint of the form $a_i \cdot x_i + a_j \cdot x_j \leq c_{ij}$ is said to be a
Unit Two Variable Per Inequality (UTVPI) constraint if $a_i, a_j \in \{-1, 0, +1\}$
and $c_{ij} \in \mathbb{Z}$.*

Definition 2. *A constraint of the form $a_i \cdot x_i + a_j \cdot x_j \leq c_{ij}$ is said to be a
difference constraint if $a_i, a_j \in \{-1, 0, +1\}$, $c_{ij} \in \mathbb{Z}$ and furthermore, $a_i = -a_j$.*

It is easy to see that UTVPI constraints subsume difference constraints.

Definition 3. *A constraint of the form $x_i \leq c_i$ or $-x_i \leq c_i$, where $c_i \in \mathbb{Z}$, is
called an absolute constraint.*

Both difference constraints and UTVPI constraints clearly subsume absolute constraints (see Sect. 3).

Definition 4. *A conjunction of UTVPI constraints is called a UTVPI constraint system and can be represented in matrix form as* $\mathbf{A} \cdot \mathbf{x} \le \mathbf{b}$. *If the constraint system has* m *constraints over* n *variables, then* $\mathbf{A}$ *has dimensions* $m \times n$.

The following lemma is known as Farkas' Lemma and is well-known in the operations research community [Sch87].

Lemma 1. *Let* $\mathbf{A}$ *denote an* $m \times n$ *matrix and let* $\mathbf{b}$ *denote an* m*-vector. Then, either* $\mathbf{I} : \exists \mathbf{x} \in \mathbb{R}^n \ \mathbf{A} \cdot \mathbf{x} \le \mathbf{b}$ *or (mutually exclusively)* $\mathbf{II} : \exists \mathbf{y} \in \mathbb{R}_+^m \ \mathbf{y}^T \cdot \mathbf{A} = \mathbf{0}, \ \mathbf{y}^T \cdot \mathbf{b} < 0.$

First observe that both System $\mathbf{I}$ and System $\mathbf{II}$ cannot be simultaneously true. Since if there exists $\mathbf{x}$ satisfying System $\mathbf{I}$ and $\mathbf{y}$ satisfying System $\mathbf{II}$, then

$$\mathbf{A} \cdot \mathbf{x} \le \mathbf{b}$$
$$\Rightarrow \mathbf{y} \cdot (\mathbf{A} \cdot \mathbf{x}) \le \mathbf{y} \cdot \mathbf{b}, \text{ since } \mathbf{y} \text{ is non} - \text{negative}$$
$$\Rightarrow (\mathbf{y} \cdot \mathbf{A}) \cdot \mathbf{x} \le \mathbf{y} \cdot \mathbf{b}$$
$$\Rightarrow 0 \le r < 0$$

If $\mathbf{I}$ does not hold, then the dual system is either unbounded or infeasible. However, $\mathbf{y} = \mathbf{0}$ is clearly a feasible solution to the dual and hence the function $\mathbf{y} \cdot \mathbf{b}$ is unbounded over the dual. The lemma follows.

It is worth noting that there are several variants of Farkas' lemma in the literature. A formal proof of the above lemma along with a geometric interpretation can be found in [Sch87]. Farkas' lemma can be specialized to difference constraints through a constraint network representation, as described in [CLRS01]. Essentially, the variables become nodes and the constraints become directed edges in this setting. A consequence of Farkas' Lemma is that the difference constraint system is feasible if and only if the corresponding constraint network does not have a negative cost cycle.

We describe a very similar network-constraint correspondence for UTVPI constraints in this paper.

3   Constraint Network Representation

Let $\mathbf{U} : \mathbf{A} \cdot \mathbf{x} \le \mathbf{b}$ denote the UTVPI constraint system and let $\mathbf{X}$ denote the set of all (fractional and integral) solutions to $\mathbf{U}$. Corresponding to this constraint system we construct the constraint network $\mathbf{G} = \langle V, E, \mathbf{c} \rangle$ as follows.

For each variable x_i create a node in V. For ease of reference, both the variable and its corresponding node are referred to as x_i in this paper.

Constraints are represented as edges using the following rules:

(a) A constraint of the form $x_i - x_j \le c_{ij}$ is represented as a directed edge from the node x_j to the node x_i having weight c_{ij}. These edges are called "gray" edges and are represented by $x_j \xrightarrow{c_{ij}} x_i$ or by $x_i \xleftarrow{c_{ij}} x_j$.

(b) A constraint of the form $-x_i - x_j \leq c_{ij}$ is represented by an undirected "black" edge $(x_i \;\overset{c_{ij}}{\blacksquare}\; x_j)$.

(c) A constraint of the form $x_i + x_j \leq c_{ij}$ is represented by an undirected "white" edge $(x_i \;\overset{c_{ij}}{\square}\; x_j)$.

A $(k-1)$-path in our constraint network, is a sequence of k nodes, $x_1, x_2, \ldots x_k$, and $(k-1)$ edges $e_1, e_2, \ldots e_{k-1}$, such that e_i is the edge corresponding to one of the constraints between x_i and x_{i+1} in the UTVPI constraint system.

For a k-path to be considered valid, it must have the following property: For every i from 2 to $k-1$, the coefficients of x_i in the constraints corresponding to the edges e_i and $e_{(i-1)}$ have opposite signs.

Example 1. The path defined by the sequence of nodes x_1, x_2, x_3, x_4 and the sequence of edges $x_1 \;\overset{c_{1,2}}{\blacksquare}\; x_2, x_2 \;\overset{c_{2,3}}{\blacksquare}\; x_3, x_3 \;\overset{c_{3,4}}{\square}\; x_4$ is $x_1 \;\overset{c_{1,2}}{\blacksquare}\; x_2 \;\overset{c_{2,3}}{\blacksquare}\; x_3 \;\overset{c_{3,4}}{\square}\; x_4$. However this path is not valid because the coefficients of x_2 in the constraints corresponding to the edges $x_1 \;\overset{c_{1,2}}{\blacksquare}\; x_2$ and $x_2 \;\overset{c_{2,3}}{\blacksquare}\; x_3$ have the same sign. In fact, both of these constraints are of the form $-x_i - x_j \leq c_{ij}$.

Definition 5. *The weight of a path is the sum of the weights of the edges along that path.*

Example 2. Consider the path $x_1 \;\overset{3}{\blacksquare}\; x_2 \;\overset{1}{\leftarrow}\; x_3 \;\overset{4}{\square}\; x_4$. The weight of this path is 8.

A *closed walk* is simply a valid $(k-1)$-path for which $x_1 = x_k$. In this paper, we refer to closed walks as cycles. Note that a cycle, as defined above can consist of edges and nodes that occur more than once. Thus, the notion of a cycle in this paper differs from the notion of a cycle in a constraint network corresponding to a difference constraint system.

Example 3. Suppose we have the system of constraints:

$$x_1 - x_2 \leq -3 \qquad -x_1 + x_4 \leq 1 \qquad -x_1 - x_4 \leq 1 \qquad x_1 - x_5 \leq 1$$
$$-x_1 + x_5 \leq 0 \qquad x_2 + x_3 \leq 1 \qquad\qquad x_2 - x_3 \leq 1$$

Then, as we can see in Fig. 1, the 8-path

$$x_1 \overset{-3}{\leftarrow} x_2 \overset{1}{\square} x_3 \overset{1}{\rightarrow} x_2 \overset{-3}{\rightarrow} x_1 \overset{0}{\rightarrow} x_5 \overset{1}{\rightarrow} x_1 \overset{1}{\rightarrow} x_4 \;\overset{1}{\blacksquare}\; x_1$$

forms a cycle even though the nodes x_1 and x_2 and the edge $x_2 \overset{-3}{\rightarrow} x_1$ are used multiple times.

At this juncture, it is important to point out that all three types of edges, viz., "white", "black" and "gray" are directionless, i.e., it may be necessary to traverse them in either direction. As shown in Subsect. 3.1,

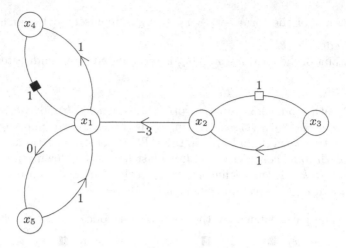

Fig. 1. Example constraint network (without node x_0)

$x_i \overset{c_{ij}}{\leftarrow} x_j \overset{c_{jk}}{\Box} x_k \overset{c_{kl}}{\rightarrow} x_l$ is a valid path from x_i to x_l but requires that gray edges are traversed in both directions.

Finally, we add a node x_0 to the network. Without loss of generality, we assume that node x_0 is assigned the value 0. This gives us a point of reference and allows us to determine values for the remaining variables. For each node x_i in the network we add the four edges $x_0 \overset{(2 \cdot n+1) \cdot C}{\Box} x_i, x_0 \overset{(2 \cdot n+1) \cdot C}{\blacksquare} x_i, x_0 \overset{(2 \cdot n+1) \cdot C}{\rightarrow} x_i$, and $x_i \overset{(2 \cdot n+1) \cdot C}{\rightarrow} x_0$ where C is the largest absolute weight of any edge in the network. These edges allow every node to be reached from x_0 without *introducing* infeasibility into the system. As discussed in Sect. 5, a UTVPI system is infeasible if and only if there exists a specific type of cycle of negative weight in the corresponding constraint network. Observe that any cycle that is introduced by the addition of x_0, must use x_0 and therefore, at least one edge that enters x_0 and at least one edge that leaves x_0. However, these edges have such a large weight $((2 \cdot n+1) \cdot C)$, that the weight of such a cycle cannot be negative, unless a negative weight cycle existed in the network to begin with.

The newly added edges also permit the addition of absolute constraints. An absolute constraint $x_i \leq c$ is converted into a pair of constraints: $x_i + x_0 \leq c$ and $x_i - x_0 \leq c$, which are added to the UTVPI system (after the absolute constraint is deleted from the system). The corresponding edges are added to the constraint network by changing the weight of the appropriate edges from x_0. In the preceding example, this would mean changing the weights of the edges $x_0 \Box x_i$ and $x_0 \rightarrow x_i$ to c.

We will now argue that the above replacement strategy is solution preserving, i.e., if the original UTVPI system is feasible, then it stays feasible after the replacement. Likewise, if the original system is infeasible, then it stays infeasible after the replacement.

Let $\mathbf{P_1} : \mathbf{A} \cdot \mathbf{x} \leq \mathbf{b}$ denote a UTVPI system with $x_1 \leq c$ denoting an absolute constraint in this system. We consider the following cases.

(i) $\mathbf{P_1}$ is non-empty : We can set $x_0 = 0$. Thus, after replacement the constraints $x_1 + x_0 \leq c$ and $x_1 - x_0 \leq c$ both become $x_1 \leq c$ and the system remains feasible with $x_0 = 0$ part of a satisfying assignment.

(ii) $\mathbf{P_1}$ is empty : Observe that if there exists a subsystem of $\mathbf{P_1}$ that is infeasible and which does not include the constraint $x_1 \leq c$, then it stays infeasible after the replacement. Let us therefore consider the case in which the constraint $x_1 \leq c$ is part of the only infeasible subsystem of $\mathbf{P_1}$. In this case, we sum $x_1 + x_0 \leq c$ and $x_1 - x_0 \leq c$, to produce the constraint $2 \cdot x_1 \leq 2 \cdot c$ which is equivalent to the original constraint. Thus, replacing $x_1 \leq c$ does not affect the infeasibility of the system.

In similar fashion, we can show that a constraint of the form: $-x_i \leq c$, can be replaced by the following constraints: $-x_i - x_0 \leq c$ and $x_0 - x_i \leq c$, without affecting the feasibility of the original UTVPI system.

Consider the following constraint system.

$$x_1 + x_3 \leq 0 \qquad x_2 - x_3 \leq -7 \qquad x_4 - x_2 \leq 3 \qquad (1)$$
$$-x_1 - x_4 \leq 5 \qquad x_1 \leq 6$$

The resulting network is shown in Fig. 2. The weight of 63 on some of the edges from x_0 to the other nodes is obtained as $(2 \cdot 4 + 1) \cdot |-7|$. Also note that the edges $x_0 \ \Box \ x_1$ and $x_0 \to x_1$ have weight 6, corresponding to the constraint $x_1 \leq 6$.

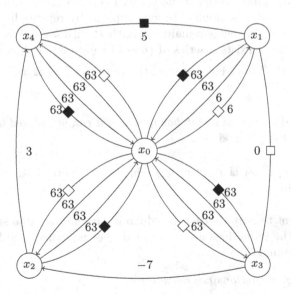

Fig. 2. Example constraint network.

3.1   Edge Reductions

We now introduce the notion of edge reductions.

Definition 6. *An edge reduction is an operation which determines a single edge equivalent to a two-edge path and represents the addition of the two UTVPI constraints which correspond to the edges in question. If this addition results in a UTVPI constraint, the reduction is said to be* valid.

Valid reductions correspond to the following transitive inference rule for UTVPI constraints:

$$\frac{a \cdot x_i + b \cdot x_j \leq c_{ij} \qquad\qquad -b \cdot x_j + b' \cdot x_k \leq c_{jk}}{a \cdot x_i + b' \cdot x_k \leq c_{ij} + c_{jk}}$$

In the case of a valid reduction, since the resultant constraint is a (valid) UTVPI constraint, the path reduces to an edge corresponding to the sum of the two constraints.

Table 1 lists all the valid edge reductions:

Not all edge reductions are valid. For example, the reduction of the path $x_i \overset{c_{ij}}{\square} x_j \overset{c_{jk}}{\square} x_k$, corresponding to the constraints $x_i + x_j \leq c_{ij}$ and $x_j + x_k \leq c_{jk}$, is not valid, since adding the constraints produces the non-UTVPI constraint $x_i + 2x_j + x_k \leq c_{ij} + c_{jk}$. However, the reduction of the path $x_i \overset{c_{ij}}{\square} x_j \overset{c_{jk}}{\blacksquare} x_k$, corresponding to the constraints $x_i + x_j \leq c_{ij}$ and $-x_j - x_k \leq c_{jk}$, is valid, since adding these constraints produces the UTVPI constraint $x_i - x_k \leq c_{ij} + c_{jk}$.

Reductions can also be applied to longer paths by repeatedly applying edge reductions until only one edge remains. A path P with k edges is said to reduce to an edge e, if there exists a series of $(k-1)$ valid edge reductions which can be used to convert P to e. For instance, the path $x_1 \overset{c_1}{\square} x_2 \overset{c_2}{\blacksquare} x_3 \overset{c_3}{\square} x_4$ reduces to the edge $x_1 \overset{c_1+c_2+c_3}{\square} x_4$.

Definition 7. *We say that a path has type t, if it can be reduced to a single edge of type t, where $t \in \{ \square, \blacksquare, \leftarrow, \rightarrow \}$.*

Thus, the path $x_1 \overset{c_1}{\square} x_2 \overset{c_2}{\blacksquare} x_3 \overset{c_3}{\square} x_4$ is a white path. Note that, every valid path must have a type.

It is important to note that when reducing a path down to a single edge, the order in which the reductions are performed does not affect the final weight of the edge (see Lemma 2).

Lemma 2. *Edge reductions are associative.*

Proof. Since each reduction corresponds to the addition of two constraints, the lemma follows from the associativity of addition in inequalities.   □

Example 4. Let p_1 denote the path $x_i \overset{c_{ij}}{\leftarrow} x_j \overset{c_{jk}}{\square} x_k \overset{c_{kl}}{\rightarrow} x_l$. Likewise, let p_2 denote the sub-path $x_i \overset{c_{ij}}{\leftarrow} x_j \overset{c_{jk}}{\square} x_k$ and let p_3 denote the sub-path $x_j \overset{c_{jk}}{\square} x_k \overset{c_{kl}}{\rightarrow} x_l$. It is not hard to see that regardless of the order in which edge reductions are performed on p_1, the edge that it is finally reduced to is $x_i \overset{c_{ij}+c_{jk}+c_{kl}}{\square} x_l$.

Table 1. Valid edge reductions

Constraints	Path	Reduction	Result
$x_j - x_i \le a,\ x_k - x_j \le b$	$x_i \overset{a}{\rightarrow} x_j \overset{b}{\rightarrow} x_k$	$x_i \overset{a+b}{\rightarrow} x_k$	$x_k - x_i \le a+b$
$x_j - x_i \le a,\ -x_k - x_j \le b$	$x_i \overset{a}{\rightarrow} x_j \overset{b}{\blacksquare} x_k$	$x_i \overset{a+b}{\blacksquare} x_k$	$-x_k - x_i \le a+b$
$x_j + x_i \le a,\ x_k - x_j \le b$	$x_i \overset{a}{\square} x_j \overset{b}{\rightarrow} x_k$	$x_i \overset{a+b}{\square} x_k$	$x_k + x_i \le a+b$
$-x_j - x_i \le a,\ x_k + x_j \le b$	$x_i \overset{a}{\blacksquare} x_j \overset{b}{\square} x_k$	$x_i \overset{a+b}{\rightarrow} x_k$	$x_k - x_i \le a+b$
$x_i - x_j \le a,\ x_j - x_k \le b$	$x_i \overset{a}{\leftarrow} x_j \overset{b}{\leftarrow} x_k$	$x_i \overset{a+b}{\leftarrow} x_k$	$x_i - x_k \le a+b$
$-x_i - x_j \le a,\ x_j - x_k \le b$	$x_i \overset{a}{\blacksquare} x_j \overset{b}{\leftarrow} x_k$	$x_i \overset{a+b}{\blacksquare} x_k$	$-x_i - x_k \le a+b$
$x_i - x_j \le a,\ x_j + x_k \le b$	$x_i \overset{a}{\leftarrow} x_j \overset{b}{\square} x_k$	$x_i \overset{a+b}{\square} x_k$	$x_i + x_k \le a+b$
$x_i + x_j \le a,\ -x_j - x_k \le b$	$x_i \overset{a}{\square} x_j \overset{b}{\blacksquare} x_k$	$x_i \overset{a+b}{\leftarrow} x_k$	$x_i - x_k \le a+b$

4   Related Work

We now contrast our constraint network construction with existing representations. First, we will look at the network representation in [Min01]. This representation was also used in [LM05, Min06, BHZ08, SS10].

[Min01] transforms the input UTVPI system into a potential network as follows:

For each variable, two nodes (a positive version and a negative version) are added to the constraint network. For instance, corresponding to the variable x_i, we create the nodes x_i^+ and x_i^-. Each constraint is replaced by a pair of equivalent constraints. For instance, a difference constraint $x_i - x_j \le c$ is replaced by the two constraints $x_i^+ - x_j^+ \le c$ and $x_j^- - x_i^- \le c$. The exception is for absolute constraints, each of which is simply converted to a single equivalent constraint. For instance, $x_i \le c$ yields $x_i^+ - x_i^- \le 2 \cdot c$. Once all the equivalent constraints have been determined, they are represented in a constraint network, as discussed in [CLRS01]. Thus, the network constructed as per [Min01] has $2 \cdot n$ nodes (assuming n variables in the constraint system) and up to $2 \cdot m$ edges (assuming m constraints in the original constraint system). The resultant network is called the potential network. Figure 3 shows the potential network, corresponding to System (1).

It is important to note that even if the constraint system consisted solely of difference constraints, our constraint network differs from the one proposed in [CLRS01] (for instance, the weights on the edges from x_0 to the other nodes are not 0).

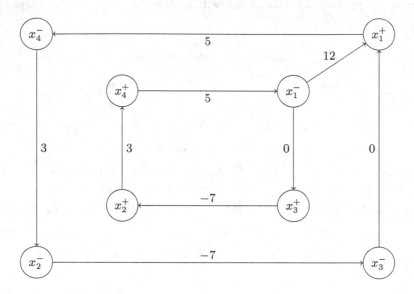

Fig. 3. Example potential network.

An alternate constraint network is used in [Rev09] to find the closure of a system of UTVPI constraints. This network construction avoids doubling the number of nodes and edges by using undirected edges and labeling the endpoints of each edge with the coefficient of the corresponding variable. Thus, the constraint $a \cdot x_i + b \cdot x_j \leq c$ is represented by the edge shown in Fig. 4.

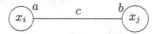

Fig. 4. Example edge.

The construction used by [Rev09] also introduces the vertex x_0 to handle absolute constraints. For the purposes of network construction the absolute constraint $a \cdot x_i \leq c$ is treated as $a \cdot x_i + 0 \cdot x_0 \leq c$. Thus, System (1) results in the constraint network shown in Fig. 5.

Our constraint network differs from the one in [Min01] in several respects:

(a) In our constraint network, the edges are "undirected". This is in marked contrast to the potential network in [Min01], which has directed edges.

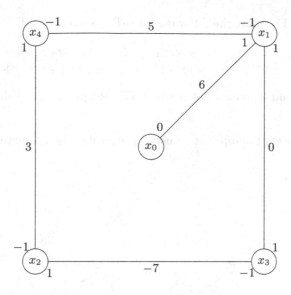

Fig. 5. Example constraint network.

(b) Our constraint network directly reflects the original input UTVPI system. Accordingly, our network retains information about constraint types explicitly.

(c) The network in [Min01] is in essence a difference constraint network representation of UTVPI constraints. Consequently, certificates of infeasibilty cannot be produced directly.

Our constraint network differs from the one in [Rev09] in the following ways:

(a) The constraint network in [Rev09] does not include the large weight edges from x_0 used by our network to ensure reachability.

(b) In the network in [Rev09], each absolute constraint is represented by only one edge. By using two edges, we ensure that no edge is traversed twice in a row.

5   Theorem of the Alternative: Linear Feasibility

In this section, we demonstrate results that exactly characterize linear feasibility in UTVPI constraint systems by using the UTVPI constraint network description in Sect. 3.

Recall that $\mathbf{U} : \mathbf{A} \cdot \mathbf{x} \leq \mathbf{b}$ denotes the UTVPI constraint system, $\mathbf{X}$ denotes the set of all (both fractional and integral) solutions to $\mathbf{U}$, and $\mathbf{G}$ is the constraint network constructed from $\mathbf{U}$ (see Sect. 3).

Example 5. Let **U** denote the following infeasible system of UTVPI constraints:

$$x_1 + x_2 \leq 2 \qquad x_1 + x_4 \leq -1 \qquad x_1 - x_4 \leq -1$$
$$x_3 - x_1 \leq 0 \qquad -x_1 - x_2 \leq 2 \qquad -x_1 - x_3 \leq -3 \qquad (2)$$

The corresponding constraint network **G** (except for the node x_0) is shown in Fig. 6.

We shall be using Example 5 to illustrate several of the lemmata and theorems in this section.

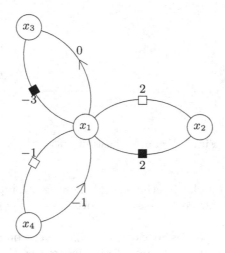

Fig. 6. Example constraint network (without node x_0)

Theorem 1. *Either* **X** *is non-empty or (mutually exclusively) there exists one of the following paths in* **G***:*

(a) *A path from a node x_i to itself that can be reduced to a single gray edge of negative weight. This will be referred to as a path of type (a).*

(b) *A path of negative weight from a node x_i to itself that consists of two sub-paths from x_i to itself, viz., a path which can be reduced to a single white edge and a path which can be reduced to a single black edge. This type of path will be referred to as a path of type (b).*

Example 6. In Fig. 6, the cycle $x_1 \xrightarrow{0} x_3 \;\blacksquare^{-3}\; x_1 \;\square^{-1}\; x_4 \xrightarrow{-1} x_1$ is a path of type (b) because:

1. the cycle has negative weight,
2. the sub-cycle $x_1 \xrightarrow{0} x_3 \;\blacksquare^{-3}\; x_1$ can be reduced to the single black edge $x_1 \;\blacksquare^{-3}\; x_1$, and
3. the sub-cycle $x_1 \;\square^{-1}\; x_4 \xrightarrow{-1} x_1$ can be reduced to the single white edge $x_1 \;\square^{-2}\; x_1$.

To prove Theorem 1, we will first need to prove a number of lemmata, which will build up to the desired result.

Lemma 3. *If there is a path of type* (b) *from* x_i *to itself, then there is a path of type* (a) *from* x_i *to itself.*

Proof. Since edge reductions are associative (Lemma 2), the path of type (b) can be reduced to a single white edge and a single black edge. Let c_1 denote the weight of the white edge and c_2 denote the weight of the black edge. These edges can then be reduced to a single gray edge of weight $(c_1 + c_2)$ (see Table 1). As the original path had negative weight, the reduced edge edge also has negative weight (in fact, it has the same weight). Observe that the reduced edge goes from x_i to itself. It follows that any path of type (b) is also a path of type (a). □

Example 7. Consider the path (cycle) $x_1 \xrightarrow{0} x_3 \; \blacksquare \; x_1 \stackrel{-1}{\square} x_4 \xrightarrow{-1} x_1$ in Fig. 6. As explained in Example 6, this path is a path of type (b), since it can be reduced to the path $x_1 \; \blacksquare \; x_1 \stackrel{-2}{\square} x_1$. However, this path can be reduced to the edge $x_1 \xrightarrow{-5} x_1$. Thus, this is a path of type (a).

Lemma 4. *If* **G** *contains a path of type* (a), *then* **X** *is empty.*

Proof. Since valid edge reductions correspond to additions of UTVPI constraints that produce other UTVPI constraints, a negative gray cycle, i.e., a path of type (a), corresponds to a sequence of UTVPI constraints which can be added together to produce the constraint: $x_i - x_i \leq c_i < 0$. However, this is an obvious contradiction. Thus, if a negative gray cycle exists in **G**, then there is no assignment to the variable x_i, that satisfies this constraint. It follows that **X** is empty.   □

Thus, we have shown that the existence of a path of type (a) or type (b) implies the infeasibility of the UTVPI constraint system.

   We will now show that if **X** is empty, then **G** must contain a path of type (a) or type (b).

   The following lemmata will help us achieve that goal.

Lemma 5. *If* **X** *is empty, then there exists a subset of constraints in* **U**, *which can be added together (possibly with repetitions) to produce a contradiction, i.e., a constraint of the form* $x_i - x_i \leq c < 0$.

Proof. If **X** is empty, then by Farkas' Lemma there exists a rational vector $\mathbf{y} \geq 0$ such that $\mathbf{y}^\mathbf{T} \cdot \mathbf{A} = \mathbf{0}$ and $\mathbf{y}^\mathbf{T} \cdot \mathbf{b} < 0$. We can assume without loss of generality that $\mathbf{y} \in \mathbb{Z}^m$. Let U_j represent the jth constraint of **U**. Consider the set $S = \{U_j : y_j > 0, j = 1 \ldots m\}$, i.e., S is the set of constraints in **U**, for which the corresponding element of $\mathbf{y}$ is non-zero. Summing the constraints of S with the constraint U_j appearing y_j times in the sum, for each $j = 1 \ldots m$, we get the constraint

$$x_i - x_i = 0 = \mathbf{y}^\mathbf{T} \cdot \mathbf{A} \cdot \mathbf{x} \leq \mathbf{y}^\mathbf{T} \cdot \mathbf{b} < 0,$$

where x_i is one of the variables that is involved in a constraint in S.   □

Example 8. In System (2), all of the constraints can be added together, with no repetitions, to produce the constraint $x_1 - x_1 \leq -1$.

Lemma 6. *If* **X** *is empty, then there exists a set* $S \subseteq \mathbf{U}$ *of constraints, such that*

(a) *the constraints in* S *can be added together (possibly with repetitions), to produce the constraint* $x_i - x_i \leq c < 0$, *where* x_i *is a variable defining one or more of the constraints in* S, *and*
(b) *the addition can be carried out in a sequence of steps, such that at each step, the resultant constraint is a UTVPI constraint (permitting constraints of the form:* $x_i + x_i \leq c_{ii}$).

Proof. Observe that part (a) of Lemma 6 follows directly from Lemma 5. In other words, the emptiness of **X** guarantees the existence of a set of constraints $S \subseteq \mathbf{U}$ and and a vector $\mathbf{v} > 0$, such that the weighted sum of the constraints in S with respect to $\mathbf{v}$, produces the contradiction: $x_i - x_i \leq c < 0$, where x_i is a variable defining one or more of the constraints in S. Let C_j denote the jth constraint in S. Note that the weighted sum of the constraints in S refers to the sum taken by multiplying constraint C_j with v_j. Without loss of generality, we assume that **v** is minimal, i.e., there does not exist a vector $\mathbf{v}'$, where $\mathbf{0} \leq \mathbf{v}' \leq \mathbf{v}$, $\mathbf{v}' \neq \mathbf{v}$, for which this property still holds.

It follows that we can construct a sequence of constraints T from S, such that

(a) constraint C_j appears v_j times, and
(b) adding all the constraints in T produces the constraint $x_i - x_i \leq c < 0$.

All that remains to be shown is that the constraints in T can be reordered so that the constraint that results after each addition is a UTVPI constraint.

Observe that the left hand side of the resultant constraint $(x_i - x_i \leq c < 0)$ is just 0. Hence, any variable introduced (positively or negatively) by the inclusion of one constraint in the weighted sum, must be canceled by the addition of some other constraint in which this variable occurs with the opposite sign. Otherwise this variable would remain in the final sum. Utilizing this fact, we impose the following order on the sequence T:

(a) Pick a constraint in which x_i appears positively; this is the first constraint C_j. Let x_b denote the other variable in C_j.
(b) Add C_j to the constraint in T, which eliminates x_b. The resultant constraint could contain a new variable, say x_c.
(c) Repeat step (b), canceling each non-x_i variable as it is introduced. Note that, if at any point $x_c = x_i$, then we need to cancel one occurrence of x_i.

All constraints can be added in this fashion. However, two situations need to be addressed:

(a) At some point, prior to adding the last constraint, the sum yields a constraint of the form $x_i - x_i \leq b$, $b \geq 0$ - In this case, the remaining constraints in T add to $x_i - x_i \leq c - b < 0$, thereby contradicting the minimality of the vector $\mathbf{v}$.

(b) At some point, prior to adding the last constraint, we get a constraint of the form $x_i - x_i \leq b < 0$ - Once again the minimality of the vector $\mathbf{v}$ is contradicted.

Thus, at each step of the addition process, the constraint that results is a UTVPI constraint, with the allowed exception of constraints having the form: $x_i + x_i \leq c_{ii}$. □

Example 9. In System (2) (see Example 5), we can start with x_1 and add the constraints as follows:

1. Start with the constraint $l_1 : x_1 + x_4 \leq -1$.
2. Add the constraint $l_2 : x_1 - x_4 \leq -1$ to l_1 thereby eliminating x_4, and producing the constraint $l_3 : x_1 + x_1 \leq -2$.
3. Add the constraint $l_4 : x_3 - x_1 \leq 0$ to l_3, thereby eliminating x_1, and producing the constraint $l_5 : x_1 + x_3 \leq -2$.
4. Add the constraint $l_6 : -x_1 - x_3 \leq -3$ to l_5, thereby eliminating x_3 and producing the constraint $l_7 : x_1 - x_1 \leq -5$, which is the desired contradiction.

We now obtain a bound on the number of times that a constraint can occur in a contradiction derived from a weighted sum of constraints.

Lemma 7. *If the network G has a path of type (a), then it has a path of type (a) in which each edge is used at most twice.*

Proof. Assume that there is a path of type (a) (say P), in which an edge is used $k \geq 3$ times. Note that a path of type (a) is equivalent to a cycle of negative weight that can be reduced to a single gray edge. This means that one of the nodes defining P is used k times. We will argue that P must contain a sub-path of type (a), in which every node (and hence every edge) is used at most two times.

Observe that the negative gray cycle P can be subdivided into sub-cycles, each of which uses x_i only once. For convenience, we will count the first and last nodes of a cycle as the same occurrence. Each sub-cycle is simply the part of the main cycle P, between, and including, two occurrences of the node x_i.

On account of how cycles are defined, each of these sub-cycles can be reduced to the equivalent of a single white edge, black edge or gray edge from x_i to itself. Those sub-cycles which can be reduced to black edges shall be referred to as black sub-cycles. White and gray sub-cycles are defined similarly. Since the P is gray cycle, it must be the case that the number of white sub-cycles is equal to the number of black sub-cycles. Otherwise, the cycle represented by P, would not reduce to a single gray edge. Thus, as in Lemma 3, each white sub-cycle can be paired with a black sub-cycle to produce a gray sub-cycle that uses x_i twice. If the weight of this sub-cycle is negative it constitutes a path of type (b).

This means that the path P is equivalent to several gray sub-cycles each of which uses x_i at most twice. Since P has a negative weight, at least one of these sub-cycles must have negative weight. Thus, there must exist a gray cycle of negative weight which uses x_i at most twice. It follows that there exists a path of type (a) in G, such that each edge is used at most twice.   □

Example 10. Let **U** represent the following infeasible system of constraints:

$$\begin{array}{lll} x_1 - x_2 \leq -3 & -x_1 + x_2 \leq 4 & -x_1 + x_4 \leq 1 \\ -x_1 - x_4 \leq 1 & x_2 + x_3 \leq 1 & x_2 - x_3 \leq 1 \end{array} \tag{3}$$

The corresponding constraint network, **G**, (except for the node x_0) is shown in Fig. 7.

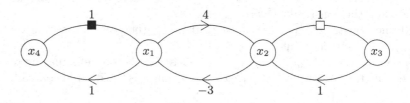

Fig. 7. Example constraint network (without node x_0)

In Fig. 7, the negative weight gray cycle

$$x_1 \overset{-3}{\leftarrow} x_2 \overset{4}{\leftarrow} x_1 \overset{-3}{\leftarrow} x_2 \overset{1}{\square} x_3 \overset{1}{\rightarrow} x_2 \overset{-3}{\rightarrow} x_1 \overset{1}{\rightarrow} x_4 \overset{1}{\blacksquare} x_1$$

uses the edge $x_1 \overset{-3}{\leftarrow} x_2$ three times. However, it can be divided into the gray sub-cycle

$$x_1 \overset{-3}{\leftarrow} x_2 \overset{4}{\leftarrow} x_1,$$

the white sub-cycle

$$x_1 \overset{-3}{\leftarrow} x_2 \overset{1}{\square} x_3 \overset{1}{\rightarrow} x_2 \overset{-3}{\rightarrow} x_1,$$

and the black sub-cycle

$$x_1 \overset{1}{\rightarrow} x_4 \overset{1}{\blacksquare} x_1.$$

Note that, each of these sub-cycles uses x_1 once. We then combine the white and black cycles to form the gray cycle

$$x_1 \overset{-3}{\leftarrow} x_2 \overset{1}{\square} x_3 \overset{1}{\rightarrow} x_2 \overset{-3}{\rightarrow} x_1 \overset{1}{\rightarrow} x_4 \overset{1}{\blacksquare} x_1.$$

This is a negative weight gray cycle that uses edge $x_2 \overset{-3}{\rightarrow} x_1$ twice. Thus, the constraint $x_1 - x_2 \leq -3$ appears twice in the corresponding sum of constraints.

The above lemma leads to the following corollary.

Corollary 1. *Let* $\mathbf{A} \cdot \mathbf{x} \leq \mathbf{b}$ *denote an infeasible UTVPI system. Then there exists a vector* $\mathbf{y} \geq \mathbf{0}$, *such that,*

(a) $\mathbf{y} \cdot \mathbf{A} = \mathbf{0}$,
(b) $\mathbf{y} \cdot \mathbf{b} < 0$,
(c) $y_i \in \{0, 1, 2\}$, $i = 1, 2, \ldots m$.

The proof of Corollary 1 follows from the discussion in the proof of Lemma 7. We note that in case of difference constraints, the dual variables in Corollary 1 can be confined to the set $\{0, 1\}$.

Lemma 8. *If* $\mathbf{X}$ *is empty, then a path of type* (a) *or a path of type* (b) *exists in* $\mathbf{G}$.

Proof. From the previous lemmata, we know that any inconsistency can be expressed as a sequence of constraints that can be added to get a constraint of the form $x_i - x_i \leq c < 0$. Furthermore, this sequence can be reordered so that at each step, the resultant constraint is a UTVPI constraint. Since valid edge reductions in $\mathbf{G}$ correspond precisely to such additions, such a sequence of constraints corresponds to a sequence of edges which can be reduced to a single gray cycle of negative weight, i.e., a path of type (a). If, as mentioned in Lemma 7, the cycle consists of a white sub-cycle and a black sub-cycle, then it is also a path of type (b). ☐

With the preceding lemmata proved, we now return to Theorem 1.

Theorem 1. *Either* $\mathbf{X}$ *is non-empty or (mutually exclusively) there exists one of the following paths in* $\mathbf{G}$:

(a) A path from a node x_i to itself that can be reduced to a single gray edge of negative weight. This will be referred to as a path of type (a).
(b) A path of negative weight from a node x_i to itself that consists of two sub-paths from x_i to itself, viz., a path which can be reduced to a single white edge and a path which can be reduced to a single black edge. This type of path will be referred to as a path of type (b).

Proof. As per Lemmas 3 and 4, if $\mathbf{G}$ contains a path of type (a) or type (b), then $\mathbf{X}$ must be empty. Likewise, as per Lemma 8, if $\mathbf{X}$ is empty, then $\mathbf{G}$ must contain a path of type (a) or type (b). In other words, the theorem of the alternative holds. ☐

6   Conclusion

In this paper, we presented a new constraint network representation for UTVPI constraints. This constraint network has the following property: The given UTVPI constraint system is feasible if and only if the corresponding constraint network does not contain a gray cycle as defined in Sect. 3. The graphical theorem of the alternative presented in this paper finds applications in the design of certifying algorithms. In the event that a UTVPI system is infeasible, an algorithm can supplement the "no" answer with the negative gray cycle.

From our perspective, there are two problems that merit further investigation:

1. Discovering a graphical theorem of the alternative for horn constraints - A constraint of the form $\sum_{i=1}^{n} a_i \cdot x_i \geq b_i$ is called a *Horn constraint*, if the $a_i \in \{0, 1, -1\}$ and at most one of the $a_i = 1$. It is clear that horn constraints generalize difference constraints. A conjunction of Horn constraints constitutes a Horn Constraint System (HCS). HCSs have the following interesting property: Linear feasibility implies integer feasibility [CS13]. It is unknown whether horn constraints can be represented graphically.
2. Finding the gray cycle of shortest length in case of an unsatisfiable UTVPI system - Finding optimal length certificates is an important problem in SMT solvers [DdM06]. From the perspective of a user, short proofs of infeasibility are ideal, since such proofs can be checked by hand. Polynomial time algorithms for optimal length refutations exist for certain constraint classes [Sub09], but the problem is **NP-hard** in general [ABMP98].

References

[ABMP98] Alekhnovich, M., Buss, S., Moran, S., Pitassi, T.: Minimum propositional proof length is NP-hard to linearly approximate. In: Brim, L., Gruska, J., Zlatuška, J. (eds.) MFCS 1998. LNCS, vol. 1450, p. 176. Springer, Heidelberg (1998)

[BHZ08] Bagnara, R., Hill, P.M., Zaffanella, E.: An improved tight closure algorithm for integer octagonal constraints. In: Logozzo, F., Peled, D.A., Zuck, L.D. (eds.) VMCAI 2008. LNCS, vol. 4905, pp. 8–21. Springer, Heidelberg (2008)

[CC77] Cousot, P., Cousot, R.: Abstract interpretation: a unified lattice model for static analysis of programs by construction or approximation of fixpoints. In: POPL, pp. 238–252 (1977)

[CLRS01] Cormen, T.H., Leiserson, C.E., Rivest, R.L., Stein, C.: Introduction to Algorithms. MIT Press, Cambridge (2001)

[CS13] Chandrasekaran, R., Subramani, K.: A combinatorial algorithm for horn programs. Discrete Optimization **10**, 85–101 (2013)

[DdM06] Duterre, B., de Moura, L.: The yices smt solver. Technical report, SRI International (2006)

[GPS95] Gerber, R., Pugh, W., Saksena, M.: Parametric dispatching of hard real-time tasks. IEEE Trans. Comput. **44**(3), 471–479 (1995)

[LM05] Lahiri, S.K., Musuvathi, M.: An efficient decision procedure for UTVPI constraints. In: Gramlich, B. (ed.) FroCos 2005. LNCS (LNAI), vol. 3717, pp. 168–183. Springer, Heidelberg (2005)

[Min01] Miné, A.: The octagon abstract domain. In: Proceedings of the Eighth Working Conference on Reverse Engineering, pp. 310–319 (2001)

[Min06] Miné, A.: The octagon abstract domain. High. Ord. Symbolic Comput. **19**(1), 31–100 (2006)

[MN99] Mehlhorn, K., Näher, S.: The LEDA Platform of Combinatorial and Geometric Computing. Cambridge University Press, Cambridge (1999)

[Rev09] Revesz, P.Z.: Tightened transitive closure of integer addition constraints. In: SARA (2009)

[Rub90] Rubinfield, R.: A Mathematical Theory of Self-checking, Self-testing and self-correcting Programs. Ph.D. thesis, Computer Science Division, University of California, Berkeley (1990)

[Sch87] Schrijver, A.: Theory of Linear and Integer Programming. Wiley, New York (1987)

[SS10] Schutt, A., Stuckey, P.J.: Incremental satisfiability and implication for utvpi constraints. INFORMS J. Comput. **22**(4), 514–527 (2010)

[Sub09] Subramani, K.: Optimal length resolution refutations of difference constraint systems. J. Autom. Reasoning (JAR) **43**(2), 121–137 (2009)

Logic and Semantic

Converging from Branching to Linear Metrics on Markov Chains

Giorgio Bacci[✉], Giovanni Bacci, Kim G. Larsen, and Radu Mardare

Department of Computer Science, Aalborg University, Aalborg, Denmark
{grbacci,giovbacci,kgl,mardare}@cs.aau.dk

Abstract. We study the strong and strutter trace distances on Markov chains (MCs). Our interest in these metrics is motivated by their relation to the probabilistic LTL-model checking problem: we prove that they correspond to the maximal differences in the probability of satisfying the same LTL and LTL$^{\neg x}$ (LTL without next operator) formulas, respectively. The threshold problem for these distances (whether their value exceeds a given threshold) is NP-hard and not known to be decidable. Nevertheless, we provide an approximation schema where each lower and upper-approximant is computable in polynomial time in the size of the MC.

The upper-approximants are Kantorovich-like pseudometrics, i.e. branching-time distances, that converge point-wise to the linear-time metrics. This convergence is interesting in itself, since it reveals a nontrivial relation between branching and linear-time metric-based semantics that does not hold in the case of equivalence-based semantics.

1 Introduction

The growing interest in quantitative systems, e.g. probabilistic and real-time systems, motivated the introduction of new techniques for studying their operational semantics. For the comparison of their behaviour, metrics are preferred to equivalences since the latter are not robust with respect to small variations of the numerical values. *Behavioral metrics* generalize the concept of equivalence by measuring the behavioral dissimilarities of two states.

Several proposals of behavioral distances [8,10,12,13,20] measure the difference according to this general schema: $d(u, v) = \sup_{\phi \in \Phi} |\phi(u) - \phi(v)|$, where Φ is a suitable set of properties of interest and $\phi(u)$ denotes the value of the property ϕ evaluated at state u. A logical characterization as above is desirable in particular when the distances are defined in a different way (e.g., as a fixed-point [8,10,13], a Hausdorff lifting [8] or games [9]) because it relates them in terms of a set Φ of expressible properties. Many logical characterizations in the literature use quantitative logics, whose semantics is given in terms of real-valued functions.

Work supported by the European Union 7th Framework Programme (FP7/2007–2013) under Grants Agreement nr. 318490 (SENSATION), nr. 601148 (CASSTING) and by the Sino-Danish Basic Research Center IDEA4CPS funded by the Danish National Research Foundation and the National Science Foundation China.

© Springer International Publishing Switzerland 2015
M. Leucker et al. (Eds.): ICTAC 2015, LNCS 9399, pp. 349–367, 2015.
DOI: 10.1007/978-3-319-25150-9_21

Such real-valued logics are not supported by quantitative model checking tools (e.g., PRISM [15] and UPPAAL [4]). Therefore, it is desirable to also have logical characterizations relating the distances to the logics adopted by these tools.

In this work we are interested in the relation with the probabilistic model checking problem for LTL [21] against Markov chains (MCs). In particular we provide two logical characterizations. The first relates the trace distance δ_t, which generalizes trace equivalence, to the probabilistic LTL-model checking problem as $\delta_t(u, v) = \sup_{\varphi \in \text{LTL}} |\mathbb{P}(u)(\llbracket \varphi \rrbracket) - \mathbb{P}(v)(\llbracket \varphi \rrbracket)|$, where $\mathbb{P}(u)(\llbracket \varphi \rrbracket)$ is the probability of executing a run from u satisfying the formula φ. The second relates the strutter trace distance δ_{st}, which generalizes stutter trace equivalence, to LTL$^{-\times}$ (LTL without next operator) as $\delta_{st}(u, v) = \sup_{\varphi \in \text{LTL}^{-\times}} |\mathbb{P}(u)(\llbracket \varphi \rrbracket) - \mathbb{P}(v)(\llbracket \varphi \rrbracket)|$. An immediate application is that $\mathbb{P}(u)(\llbracket \varphi \rrbracket)$ (i.e., probabilistically model checking φ at u) can be approximated by $\mathbb{P}(v)(\llbracket \varphi \rrbracket)$ with an error bounded by $\delta_t(u, v)$, for any $\varphi \in \text{LTL}$. This may lead to savings in the overall cost of model checking.

This further motivates the study of efficient methods for computing these distances. Unfortunately, in [6,19] the threshold problem for the trace distance is proven to be NP-hard and, to the best of our knowledge, its decidability is still an open problem. Nevertheless, in [6] it is shown that the problem of approximating this distance with arbitrary precision is decidable. This is done by providing two effective sequences that converge from below and above to the trace distance. In this paper we provide an alternative approximation schema that, differently from [6], is formed by sequences of lower and upper-approximants that are shown to be computable in *polynomial time* in the size of the MC. With respect to [6], our approach is more general with the nice consequence that the same result is obtained for the problem of approximating the stutter trace distance.

Notably, in our construction the upper-approximants are Kantorovich-like pseudometrics, i.e., branching-time distances. These metrics form a net—a concept used in topology that generalizes infinite sequences—that converges pointwise to the linear-time metrics. The result is interesting in itself, since it reveals a nontrivial link (by means of a converging net) between branching and linear-time metric-based semantics that does not hold when a more standard equivalence-based semantics on MCs is used instead. This opens new perspectives in the study of the operational behavior of quantitative systems, and suggests relating behavioral distances by means of converging nets rather than the standard 'greater than or equal to' relation, commonly used in the literature (e.g., in [8]). The technical contributions of the paper can be summarized as follows.

1. We provide a logical characterization of the trace distance terms of LTL. This result, differently from previous proposals (e.g. [8,10]), explicitly relates the trace distance to the probabilistic model checking problem of LTL formulas. We show that a similar characterization holds also for the stutter trace distance on the fragment of LTL without next operator.
2. We construct two nets of bisimilarity-like distances that converge to the strong and stutter trace distance. This construction leverages on a classical duality result that characterizes the total variation distance between two measures as the minimal discrepancy associated with their couplings. To do so we generalize and improve two important results in [5], namely Theorem 8 and Corollary 11.

3. We demonstrate that each element of the proposed converging nets is computable in polynomial time in the size of the MC. Moreover, we provide other two sequences of pseudometrics that, respectively, converges from below to the two linear distances. Also the lower approximants are proven to be polynomially computable. The pairs of converging sequences of upper and lower approximants form the approximation schemata for the problem of computing the strong and stutter trace distances. The approximation schema for the trace distance improves the one proposed in [6].

2 Preliminaries and Notation

The set of functions from X to Y is denoted by Y^X. Any preorder $\sqsubseteq$ on Y is extended to Y^X as $f \sqsubseteq g$ iff $f(x) \sqsubseteq g(x)$, for all $x \in X$. For $f \in Y^X$, let $\equiv_f = \{(x, x') \mid f(x) = f(x')\}$. For $R \subseteq X \times X$ an equivalence relation, $X/_R$ is the quotient set, $[x]_R$ the R-equivalence class of x, and for $A \subseteq X$, $[A]_R = \bigcup_{x \in A} [x]_R$.

Measure Theory. A *field* over a set X is a nonempty family $\Sigma \subseteq 2^X$ closed under complement and finite union. Σ is a σ-algebra if, in addition, it is closed under countable union; in this case (X, Σ) is called a *measurable space* and the elements of Σ *measurable sets*. For $\mathcal{F} \subseteq 2^X$, $\sigma(\Sigma)$ denotes the smallest σ-algebra containing $\mathcal{F}$. For $(X, \Sigma), (Y, \Theta)$ measurable spaces, $f \colon X \to Y$ is *measurable* if for all $E \in \Theta$, $f^{-1}(E) = \{x \mid f(x) \in E\} \in \Sigma$. The *product space*, $(X, \Sigma) \otimes (Y, \Theta)$, is the measurable space $(X \times Y, \Sigma \otimes \Theta)$, where $\Sigma \otimes \Theta$ is the σ-algebra generated by the *rectangles* $E \times F$, for $E \in \Sigma$ and $F \in \Theta$. A *measure* on (X, Σ) is a σ-additive function $\mu \colon \Sigma \to \mathbb{R}_+$, i.e., $\mu(\bigcup_{i \in \mathbb{N}} E_i) = \sum_{i \in \mathbb{N}} \mu(E_i)$ for all of pairwise disjoint $E_i \in \Sigma$; it is a *probability measure* if, in addition, $\mu(X) = 1$. Hereafter $\Delta(X, \Sigma)$ denotes the set of probability measures on (X, Σ). Given a measurable function $f \colon (X, \Sigma) \to (Y, \Theta)$, any measure μ on (X, Σ) defines a measure $\mu[f]$ on (Y, Θ) by $\mu[f](E) = \mu(f^{-1}(E))$, for all $E \in \Theta$; it is called the *push forward* of μ *under* f. A measure ω on $(X, \Sigma) \otimes (Y, \Theta)$ is a *coupling* for (μ, ν) if for all $E \in \Sigma$ and $F \in \Theta$, $\omega(E \times Y) = \mu(E)$ and $\omega(X \times F) = \nu(F)$ (i.e., μ is the *left* and ν the *right marginal* of ω). $\Omega(\mu, \nu)$ denotes the set of couplings for (μ, ν).

Metric Spaces. For a set X, $d \colon X \times X \to \mathbb{R}_+$ is a *pseudometric* on X if for any $x, y, z \in X$, $d(x, x) = 0$, $d(x, y) = d(y, x)$ and $d(x, y) + d(y, z) \geq d(x, z)$; d is a *metric* if, in addition, $d(x, y) = 0$ implies $x = y$. If d is a (pseudo)metric on X, (X, d) is called a *(pseudo)metric space*. We define $\ker(d) = \{(u, v) \mid d(u, v) = 0\}$. For (X, Σ) a measurable space, $\Delta(X, \Sigma)$ can be metrized by the *total variation distance* $\|\mu - \nu\| = \sup_{E \in \Sigma} |\mu(E) - \nu(E)|$. A (pseudo-)metric $d \colon X \times X \to \mathbb{R}_+$ is lifted to $\Delta(X, \Sigma)$ by means of the *Kantorovich (pseudo-)metric*, defined as $\mathcal{K}(d)(\mu, \nu) = \min \left\{ \int d \, d\omega \mid \omega \in \Omega(\mu, \nu) \right\}$.

The Space of Words. Let X^n be the set of words on X of length $n \in \mathbb{N}$, $X^* = \bigcup_{n \in \mathbb{N}} X^n$, $AB = \{ab \in X^* \mid a \in A, b \in B\}$ $(A, B \subseteq X^*)$ and $X^+ = XX^*$.

An infinite word $\pi = x_0 x_1 \ldots$ over X is an element in X^ω. For $i \in \mathbb{N}$, define $\pi[i] = x_i$, $\pi|^i = x_0 \ldots x_{i-1} \in X^i$, and $\pi|_i = x_i x_{i+1} \ldots \in X^\omega$. For $A \subseteq X^n$, the *cylinder set for A* (of rank n) is defined as $\mathfrak{C}(A) = \{\pi \in X^\omega \mid \pi|^n \in A\} \subseteq X^\omega$. For an arbitrary family $\mathcal{F} \subseteq 2^X$, let $\mathfrak{C}^n(\mathcal{F}) = \{\mathfrak{C}(X_1 \cdots X_n) \mid X_i \in \mathcal{F}\}$, for $n \geq 1$, and $\mathfrak{C}(\mathcal{F}) = \bigcup_{n \geq 1} \mathfrak{C}^n(\mathcal{F})$.

If (X, Σ) is a measurable space, $(X, \Sigma)^n$ denotes the product space over X^n, and $(X, \Sigma)^\omega$ the measurable space over X^ω with σ-algebra generated by $\mathfrak{C}(\Sigma)$ (i.e., the smallest s.t., for all $n \in \mathbb{N}$, the prefix $(\cdot)|^n$ and tail $(\cdot)|_n$ functions are measurable). Note that, the stepwise extension $f^\omega \colon X^\omega \to Y^\omega$ of the function $f \colon X \to Y$ is measurable if f is so. Often, X^n and X^ω will also denote $(X, 2^X)^n$ and $(X, 2^X)^\omega$, respectively.

3   Markov Chains and Linear-Time Equivalences

In this section we recall discrete-time Markov chains and the notions of strong and stutter probabilistic trace equivalences on them.

In what follows we fix a finite set $\mathbb{A}$ of atomic propositions.

Definition 1. *A* Markov chain *is a tuple* $\mathcal{M} = (S, \tau, \ell)$ *consisting of a countable set S of* states*, a* transition probability function *$\tau \colon S \to \Delta(S)$ and a* labeling function *$\ell \colon S \to 2^{\mathbb{A}}$.*

Intuitively, if $\mathcal{M}$ is in the state u, it moves to a state $v \in S$ with probability $\tau(u)(v)$. We say that $p \in \mathbb{A}$ holds in u if $p \in \ell(u)$. We will use $\mathcal{M} = (S, \tau, \ell)$ to range over the class of MCs and we will refer to it and its constituents implicitly.

An MC can be thought of as a stochastic process that, from an initial state u, emits execution runs distributed according to the probability $\mathbb{P}(u)$ given below.

Definition 2. *Let $\mathbb{P} \colon S \to \Delta(S^\omega)$ be such that, for all $u \in S$, $\mathbb{P}(u)$ is the unique probability measure[1] on S^ω such that, for all $n \geq 1$ and $U_i \subseteq S$ ($i = 0..n$)*

$$\mathbb{P}(u)(\mathfrak{C}(U_0 \cdots U_n)) = \mathbb{1}_{U_0}(u) \cdot \int \mathbb{P}(\cdot)(\mathfrak{C}(U_1 \cdots U_n)) \, \mathrm{d}\tau(u) \, ,$$

where $\mathbb{1}_A$ denotes the indicator function for a set A.

Intuitively, $\mathbb{P}(u)(E)$ is the probability that, starting from u, the MC executes a run in $E \subseteq S^\omega$. For example, $\mathbb{P}(u)(\mathfrak{C}(u_0..u_n)) = \mathbb{1}_{u_0}(u) \cdot \prod_{i=0}^{n-1} \tau(u_i)(u_{i+1})$.

Remark 3. In Definition 2, since $\mathfrak{C}(U_0) = \mathfrak{C}(U_0 S)$, the case $\mathbb{P}(u)(\mathfrak{C}(U_0))$ is covered implicitly. Indeed, $\mathbb{P}(u)(\mathfrak{C}(U_0 S)) = \mathbb{1}_{U_0}(u) \cdot \int \mathbb{P}(\cdot)(\mathfrak{C}(S)) \, \mathrm{d}\tau(u) = \mathbb{1}_{U_0}(u) \cdot \int 1 \, \mathrm{d}\tau(u) = \mathbb{1}_{U_0}(u)$, since for all $v \in S$, $\mathbb{P}(v)$ is a probability measure.   □

Two states of an MC are considered equivalent if they exhibit the same "observable behaviour". In this work we focus on linear-time properties. In this respect, we recall the most used linear-time equivalences on MCs: *strong* and *stutter probabilistic trace equivalences*.

---

[1] Existence and uniqueness follows by the Hahn-Kolmogorov extension theorem.

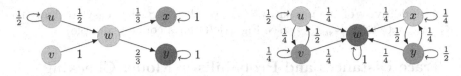

Fig. 1. (Left) u and v are stutter trace equivalent but neither bisimilar nor trace equivalent; (Right) $\delta_t(u, v) = \sqrt{2}/4$ (see [6]) and $\delta_b(u, v) = 1/2$. States are labeled by colors.

Definition 4. *Two states $u, v \in S$ are* probabilistic trace equivalent, *written $u \sim_t v$, if for all $T \in \mathfrak{C}(S/_{\equiv_\ell})$, $\mathbb{P}(u)(T) = \mathbb{P}(v)(T)$.*

Intuitively, $\sim_t$ tests two states w.r.t. all linear-time events, considered up to label equivalence. This is in accordance to the fact that the only things that we observe in a state are the atomic properties (labels). Hereafter, $\mathcal{T}$ denotes $\mathfrak{C}(S/_{\equiv_\ell})$ and its elements are called *trace cylinders*.

The *stutter* (or *weak*) variant of the probabilistic trace equivalence considers a transition step as "visible" only when a change of the current behavior occurs. The guiding idea to define stutter events is to replace the notion of "step" with that of "stutter step". Formally, this corresponds to change the definitions of the tail (i.e., the "next step") and prefix functions over infinite words. Let X be a set and $R \subseteq X \times X$ equivalence. For $n \geq 1$, define the n-th *R-stutter tail* function $\mathsf{tl}_R^n: X^\omega \to X^\omega$, by induction on n, as follows

$$\mathsf{tl}_R^1(\pi) = \begin{cases} \pi|_j & \text{if } \exists j \text{ s.t. } (\pi[0], \pi[j]) \notin R \text{ and } \forall i < j, (\pi[0], \pi[i]) \in R \\ \pi & \text{otherwise (i.e., } \pi \text{ is } R\text{-constant}), \end{cases}$$

$$\mathsf{tl}_R^{n+1}(\pi) = \mathsf{tl}_R^1(\mathsf{tl}_R^n(\pi)).$$

Intuitively, tl_R^1 seeks for the first tail whose head is not R-equivalent to $\pi[0]$ (if it exists!) and $\mathsf{tl}_R^n(\pi)$ is the n-th composition of it. For example, let $\pi = aaabbbc^\omega$, then $\mathsf{tl}_{\equiv}^1(\pi) = bbbc^\omega$ and, for all $n > 1$, $\mathsf{tl}_{\equiv}^n(\pi) = c^\omega$. The n-th *R-stutter prefix* function $\mathsf{pf}_R^n: X^\omega \to X^n$ is defined, by induction on $n \geq 1$, as $\mathsf{pf}_R^1(\pi) = \pi[0]$ and $\mathsf{pf}_R^{n+1}(\pi) = \pi[0]\mathsf{pf}_R^n(\mathsf{tl}_R^1(\pi))$.

Now, the standard definition of cylinder set for $A \subseteq X^n$ can be turned to that of *R-stutter cylinder set for A* (of rank n) as $\mathfrak{C}_R(A) = \{\pi \in X^\omega \mid \mathsf{pf}_R^n(\pi) \in A\}$. For a family $\mathcal{F} \subseteq 2^X$, denote by $\mathfrak{C}_R^n(\mathcal{F}) = \{\mathfrak{C}_R(E_1 \cdots E_n) \mid E_i \in \mathcal{F}\}$ the set of all R-stutter cylinders of rank n over $\mathcal{F}$ and $\mathfrak{C}_R(\mathcal{F}) = \bigcup_{n \geq 1} \mathfrak{C}_R^n(\mathcal{F})$. If (X, Σ) a measurable space, we denote by $(X, \Sigma)_R^\omega$ the measurable space of infinite words over X with σ-algebra generated by $\sigma(\mathfrak{C}_R(\Sigma))$ (i.e., the smallest σ-algebra such that, for all $n \geq 1$, the n-th R-stutter prefix and tail functions are measurable).

Definition 5. *Two states $u, v \in S$ are* probabilistic stutter trace equivalent, *written $u \sim_{st} v$, if for all $T \in \mathfrak{C}_{\equiv_\ell}(S/_{\equiv_\ell})$, $\mathbb{P}(u)(T) = \mathbb{P}(v)(T)$.*

Intuitively, $\sim_{st}$ equates the states that have the same probability on all the $\equiv_\ell$-stutter linear-time events, considered up to label equivalence. Hereafter, $\mathcal{ST}$ denotes $\mathfrak{C}_{\equiv_\ell}(S/_{\equiv_\ell})$ and its elements will be called *stutter trace cylinders*.

By σ-additivity of the measures $\mathbb{P}(u)$, for all $u \in S$, it is easy to show that $\sim_t \subseteq \sim_{st}$. Note that, $\sim_{st} \not\subseteq \sim_t$ (see Fig. 1(left) for a counterexample).

4   Trace Distances and Probabilistic Model Checking

We give the definitions of *strong* and *stutter trace distances* and provide logical characterizations to both of them in terms of suitable fragments of LTL, relating the two behavioral distances to the probabilistic model checking problem.

Linear Distances. The strong and stutter probabilistic trace equivalences on MCs are naturally lifted to pseudometrics $\delta_t, \delta_{st} \colon S \times S \to [0,1]$ as follows

$$\delta_t(u,v) = \sup_{E \in \sigma(\mathcal{T})} |\mathbb{P}(u)(E) - \mathbb{P}(v)(E)|, \qquad \text{(\textsc{strong trace distance})}$$

$$\delta_{st}(u,v) = \sup_{E \in \sigma(\mathcal{ST})} |\mathbb{P}(u)(E) - \mathbb{P}(v)(E)|. \qquad \text{(\textsc{stutter trace distance})}$$

Observe that two states $u, v \in S$ are strong (resp. stutter) trace equivalent iff $\delta_t(u,v) = 0$ (resp. $\delta_{st}(u,v) = 0$). Moreover, by $\sigma(\mathcal{ST}) \subseteq \sigma(\mathcal{T})$, it holds $\delta_{st} \leq \delta_t$.

Note that, the above distances are total variation distances between two measures, namely the restriction of $\mathbb{P}(u)$ and $\mathbb{P}(v)$, on $\sigma(\mathcal{T})$ and $\sigma(\mathcal{ST})$, respectively.

Linear Temporal Logic. (LTL) is a formalism for reasoning about sequences of events [21]. The LTL formulas are generated by the following grammar

$$\varphi ::= p \mid \bot \mid \varphi \to \varphi \mid \mathsf{X}\varphi \mid \varphi \mathsf{U} \varphi, \qquad\qquad \text{where } p \in \mathbb{A}.$$

Let LTL$^{-\mathsf{U}}$ and LTL$^{-\mathsf{X}}$ be the fragments, respectively, built without until (U) and next (X) operators. The semantics of the formulas is given by means of a satisfiability relation defined, for an MC $\mathcal{M}$ and $\pi \in S^\omega$, as follows

$$\mathcal{M}, \pi \models p \qquad\quad \text{if } p \in \ell(\pi[0]),$$
$$\mathcal{M}, \pi \models \bot \qquad\quad \text{never},$$
$$\mathcal{M}, \pi \models \varphi \to \psi \quad \text{if } \mathcal{M}, \pi \models \psi \text{ whenever } \mathcal{M}, \pi \models \varphi,$$
$$\mathcal{M}, \pi \models \mathsf{X}\varphi \qquad \text{if } \mathcal{M}, \pi|_1 \models \varphi,$$
$$\mathcal{M}, \pi \models \varphi \mathsf{U} \psi \quad\; \text{if } \exists i \geq 0 \text{ s.t. } \mathcal{M}, \pi|_i \models \psi, \text{ and } \forall 0 \leq j < i,\, \mathcal{M}, \pi|_j \models \varphi.$$

Define $[\![\varphi]\!] = \{\pi \mid \mathcal{M}, \pi \models \varphi\}$ and $[\![\mathcal{L}]\!] = \{[\![\varphi]\!] \mid \varphi \in \mathcal{L}\}$, for any $\mathcal{L} \subseteq$ LTL. The probabilistic model checking problem for MCs against LTL formulas consists in determining the probability $\mathbb{P}(u)([\![\varphi]\!])$ for an initial state u and $\varphi \in$ LTL. For any $\mathcal{L} \subseteq$ LTL, the pseudometric

$$\delta_{\mathcal{L}}(u,v) = \sup_{\varphi \in \mathcal{L}} |\mathbb{P}(u)([\![\varphi]\!]) - \mathbb{P}(v)([\![\varphi]\!])|$$

measures the maximal difference that can be observed between the states u and v by model checking them over a set $\mathcal{L}$ of linear temporal logic formulas of interest.

In the rest of the section we characterize δ_t and δ_{st} respectively as δ_{LTL} (or $\delta_{\text{LTL}^{-\mathsf{U}}}$) and $\delta_{\text{LTL}^{-\mathsf{X}}}$. We do this by exploiting the following result.

Lemma 6 ([2]). *Let μ and ν be two finite measures on a measurable space (X, Σ). If Σ is generated by a field $\mathcal{F}$, then $\|\mu - \nu\| = \sup_{E \in \mathcal{F}} |\mu(E) - \nu(E)|$.*

By Lemma 6, to provide a logical characterization for δ_t it suffices to show that the σ-algebra $\sigma(\mathcal{T})$ is generated by $[\![\mathrm{LTL}]\!]$ (or $[\![\mathrm{LTL}^{\text{-u}}]\!]$).

Theorem 7. *(i) $\sigma(\mathcal{T}) = \sigma([\![\mathrm{LTL}]\!]) = \sigma([\![\mathrm{LTL}^{\text{-u}}]\!])$, (ii) $\delta_t = \delta_{\mathrm{LTL}} = \delta_{\mathrm{LTL}\text{-u}}$.*

Remark 8. $\delta_t = \delta_{\mathrm{LTL}}$ is not trival. Figure 1(right) shows an MC from [6, Ex. 1][2] where it is proven that $\delta_t(u, x)$ is obtained on a maximizing event in $\sigma(\mathcal{T})$ that is not ω-regular, hence it cannot be expressed by a single LTL formula.   □

In Theorem 7, the proof of $\sigma(\mathcal{T}) \subseteq \sigma([\![\mathrm{LTL}]\!])$ uses the measurability of the n-th tail function $(\cdot)|_n$ w.r.t. $\sigma(\mathcal{T})$. However, $(\cdot)|_n$ is not measurable w.r.t. $\sigma(\mathcal{ST})$, so the logical characterization does not carry over easily to the stutter case.

We solve this problem by giving a coinductive characterization to Lamport's *stutter equivalence* [16] (for a standard definition see e.g. [3, Sect. 7.7.1]). For a relation $R \subseteq S^\omega \times S^\omega$, $\pi \in S^\omega$ is said R-*constant* if, for all $i \in \mathbb{N}$, $\pi \, R \, \pi|_i$.

Definition 9. *A relation $R \subseteq S^\omega \times S^\omega$ is a stutter relation if whenever $\pi \, R \, \rho$*

(i) $\pi[0] \equiv_\ell \rho[0]$;
(ii) π is R-constant iff ρ is R-constant;
(iii) $\pi|_1 \, R \, \rho$ or $\pi \, R \, \rho|_1$ or $\pi|_1 \, R \, \rho|_1$.

Two traces $\pi, \rho \in S^\omega$ are stutter equivalent, written $\pi \simeq \rho$, if they are related by some stutter relation.

Stutter relations are closed under union and reflexive/symmetric/transitive closure, therefore $\simeq$ is an equivalence and a stutter relation.

Proposition 10. *$\pi \simeq \rho$ iff $\forall \varphi \in \mathrm{LTL}^{\text{-x}}. (\mathcal{M}, \pi \models \varphi \Leftrightarrow \mathcal{M}, \rho \models \varphi)$.*

The above states that $\simeq$ characterizes the logical equivalence w.r.t. $\mathrm{LTL}^{\text{-x}}$. Definition 9 and Proposition 10 are essential to prove the next result.

Theorem 11. *(i) $\sigma(\mathcal{ST}) = \sigma([\![\mathrm{LTL}^{\text{-x}}]\!])$, (ii) $\delta_{st} = \delta_{\mathrm{LTL}\text{-x}}$.*

Proof. We prove (i), then (ii) follows by Lemma 6. ($\supseteq$) We prove $[\![\varphi]\!] \in \sigma(\mathcal{ST})$ by induction on φ. We show the case $\varphi = \phi \, \mathsf{U} \, \psi$. Define $q \colon S^\omega \to S^\omega$, as $q(\pi) = \mathsf{pf}^1_{\equiv_\ell}(\pi) q(\mathsf{tl}^1_{\equiv_\ell}(\pi))^3$. The function q is idempotent, moreover, it is $\sigma(\mathcal{ST})$–$\sigma(\mathcal{T})$ measurable, i.e., for all $E \in \sigma(\mathcal{T})$, $q^{-1}(E) \in \sigma(\mathcal{ST})$. It can be shown that $R = \{(\pi, \rho) \mid q(\pi) \equiv_{\ell^\omega} q(\rho)\}$ is a stutter relation. Therefore, by $q(\pi) \equiv_{\ell^\omega} q(q(\pi))$, we get $\pi \, R \, q(\pi)$, hence $\pi \simeq q(\pi)$. Then, the following hold:

$$[\![\phi \, \mathsf{U} \, \psi]\!] = \{\pi \mid \exists i \geq 0. \, q(\pi)|_i \in [\![\psi]\!], \forall 0 \leq j < i. \, q(\pi)|_j \in [\![\phi]\!]\} \quad ([\![\cdot]\!] \,\&\, \text{Prop. 10})$$

$$= \bigcup_{i \geq 0} \bigcap_{0 \leq j < i} (((\cdot)|_i \circ q)^{-1}([\![\psi]\!]) \cap ((\cdot)|_j \circ q)^{-1}([\![\phi]\!])). \quad \text{(preimage)}$$

---

[2] The MC has been adapted to the case of labeled states, instead of labeled transitions.
[3] Note that $q = \lim_{n \geq 1} \mathsf{pf}^n_{\equiv_\ell}$, i.e., it is the unique map s.t., for all $n \geq 1$, $\mathsf{pf}^n_{\equiv_\ell} = (\cdot) n \circ q$.

By inductive hypothesis on ϕ, ψ and $\sigma(\mathcal{ST})$-measurability of $(\cdot)|_k \circ q$, for any $k \in \mathbb{N}$, it follows that $[\![\phi \cup \psi]\!] \in \sigma(\mathcal{ST})$. ($\subseteq$) The σ-algebra $\sigma(\mathcal{ST})$ is alternatively generated by the family $\mathcal{F} = \{\mathfrak{C}_{\equiv_\ell}(C_1 \cdots C_n) \in \mathcal{ST} \mid C_i \neq C_{i+1}\}$. Hence, it suffices to show $\mathcal{F} \subseteq \sigma([\![\text{LTL}^{-\mathsf{x}}]\!])$. Define $B \colon \mathcal{F} \to \text{LTL}^{-\mathsf{x}}$ by induction as follows,

$$B(\mathfrak{C}_{\equiv_\ell}(C_1)) = \bigwedge_{p \in \mathbb{A}} A(p, C_1),$$

$$B(\mathfrak{C}_{\equiv_\ell}(C_1 \cdots C_{n+1})) = \left(B(\mathfrak{C}_{\equiv_\ell}(C_1)) \wedge \neg B(\mathfrak{C}_{\equiv_\ell}(C_2)) \cup B(\mathfrak{C}_{\equiv_\ell}(C_2 \cdots C_{n+1})\right),$$

where $A(p, C) = p$ if there exists $s \in C$ s.t. $p \in \ell(s)$, otherwise $A(p, C) = \neg p$. For $T \in \mathcal{F}$ one can prove that $[\![B(T)]\!] = T$. $\qquad\qquad \square$

5   Convergence from Branching to Linear Distances

We provide two nets of pseudometrics that converge, respectively, to the strong and stutter trace distances. The pseudometrics are shown to be liftings of multi-step extensions of probabilistic bisimilarity and a suitable stutter variant of it.

Our construction is inspired by [5, Cor. 11], where the bisimilarity pseudo-metric δ_b of Desharnais et al. [11] is shown to be an upper bound for the trace distance δ_t. Their result is based on an alternative characterization of δ_b by means of the notion of "coupling structure" [5, Th. 8]. The proof of $\delta_t \leq \delta_b$ uses a classic duality result asserting that *the total variation of two measures coincides to the minimal discrepancy measured among all their couplings* (Lemma 12). Formally, given $\mu, \nu \in \Delta(X, \Sigma)$, the *discrepancy* of $\omega \in \Omega(\mu, \nu)$ is the value $\omega(\not\cong_\Sigma)$, where $\cong_\Sigma = \bigcap \{E \times E \mid E \in \Sigma\}$ is the *inseparability relation* w.r.t. Σ.

Lemma 12 ([18, Th.5.2]). *Let* μ, ν *be probability measures on* (X, Σ). *Then, provided that* $\cong_\Sigma$ *is measurable in* $\Sigma \otimes \Sigma$, $\|\mu - \nu\| = \min \{\omega(\not\cong_\Sigma) \mid \omega \in \Omega(\mu, \nu)\}$.

Along the way to obtain our construction, we nontrivially extend (and improve the proofs of) both Corollary 11 and Theorem 8 in [5]. Moreover, this construction reveals a nontrivial relation between branching and linear-time metric-based semantics (by means of a convergence of the observable behaviors) that does not hold by using the standard equivalence-based semantics.

5.1   The Strong Case

We start by introducing a multi-step generalization of probabilistic bisimulation.

Definition 13. *Let* $k \geq 1$. *An equivalence relation* $R \subseteq S \times S$ *is a* k-*probabilistic bisimulation on* $\mathcal{M}$ *if whenever* $u \, R \, v$, *then, for all* $E_i \in S/{\equiv_\ell}$ *and* $C \in S/R$,

$$\mathbb{P}(u)(\mathfrak{C}(E_0 \cdots E_{k-1}C)) = \mathbb{P}(v)(\mathfrak{C}(E_0 \cdots E_{k-1}C)).$$

Two states $u, v \in S$ *are* k-*probabilistic bisimilar, written* $u \sim_b^k v$, *if they are related by some* k-*probabilistic bisimulation.*

The notion of k-bisimulation weakens that of probabilistic bisimulation of Larsen and Skou [17] by equating states that have the same probability to move to the same k-bisimilarity class after having observed the same labels within k-steps. Note that $\sim_b^1$ coincides with Larsen and Skou bisimilarity. Moreover, for all $k \geq 1$, $\sim_b^k$ is a k-bisimulation and, by σ-additivity of the measures, $\sim_b^1 \subseteq \sim_b^k \subseteq \sim_t$.

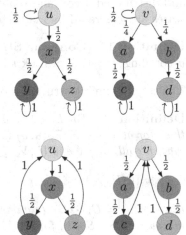

Remark 14. Clearly, $\bigcup_{k \geq 1} \sim_b^k \subseteq \sim_t$. However, the converse inclusion does not hold. A counterexample is shown in the picture aside, where states are labeled by colors.

It is easy to see that u and v are probabilistic trace equivalent, but they are not probabilistic k-bisimilar for any $k \geq 1$. $\quad \Box$

Remark 15. Differently from what one may expect, the k-bisimilarities do not necessarily get weaker by increasing k, i.e., for an arbitrary $k \geq 1$, it does not hold $\sim_b^k \subseteq \sim_b^{k+1}$. An example is shown aside where $u \sim_b^4 v$ but $u \not\sim_b^5 v$, hence $\sim_b^4 \not\subseteq \sim_b^5$. $\quad \Box$

Next we show how to "lift" the above equivalences to behavioral pseudometrics. A pseudometric that lifts bisimilarity is δ_b [11], defined as the least fixed point of the following operator on 1-bounded pseudometrics $d \colon S \times S \to [0,1]$

$$\Theta(d)(u,v) = \begin{cases} 1 & \text{if } u \not\equiv_\ell v \\ \mathcal{K}(d)(\tau(u), \tau(v)) & \text{otherwise}. \end{cases} \quad \text{(Kantorovich Operator)}$$

Intuitively, two states are incomparable if they have different labels, otherwise the difference is given by Kantorovich distance of their transition probabilities.

Analogously, for $k \geq 1$, define the *k-steps transition probability function* $\tau^k \colon S \to \Delta(S^k)$ as the function such that $\tau^k(u)$ is the unique probability measure on S^k that, for all $U_i \subseteq S$ $(i = 1..k)$, $\tau^k(u)(U_1 \cdots U_k) = \mathbb{P}(u)(\mathfrak{C}(uU_1 \cdots U_k))$ (i.e., $\tau^k(u) = \mathbb{P}(u)[(\cdot)k \circ (\cdot)|_1])$. Note that, $\tau = \tau^1$. Then Θ is generalized by

$$\Theta^k(d)(u,v) \begin{cases} 1 & \text{if } u \not\equiv_\ell v \\ \mathcal{K}(\Lambda^k(d))(\tau^k(u), \tau^k(v)) & \text{otherwise}. \end{cases}$$

where $\Lambda^k(d)(u_1..u_k, v_1..v_k) = 1$ if $u_i \not\equiv_\ell v_i$ for some $i = 1..k$, otherwise $d(u_k, v_k)$. We call the above *k-Kantorovich operator*. It is easy to see that Θ^k is monotonic, so that, by Tarski fixed point theorem, it has least fixed point, hereafter denoted by δ_b^k. Note that $\delta_b^1 = \delta_b$, moreover the following hold.

Lemma 16 (k-Bisimilarity Distance). $u \sim_b^k v$ iff $\delta_b^k(u,v) = 0$.

Due to the above result we call δ_b^k the *k-bisimilarity pseudometric*.

Next we characterize δ_b^k by means of the notion of *coupling structure* of rank k. A coupling structure may be thought of as a stochastic process generating of infinite traces of pairs of states starting from a distinguished initial pair (u,v) and distributed according to a coupling in $\Omega(\mathbb{P}(u), \mathbb{P}(v))$. The traces of pairs of states are generated by multi-steps of length k.

Definition 17 (Coupling Structure). *A function* $C: S \times S \to \Delta(S^k \otimes S^k)$ *is a coupling structure of rank* $k \geq 1$ *if for all* $u,v \in S$, $C(u,v) \in \Omega(\tau^k(u), \tau^k(v))$.

The set of coupling structures of rank k is denoted by $\mathbb{C}_k$.

Definition 18. *For* $k \geq 1$ *and* $C \in \mathbb{C}_k$, *let* $\mathbb{P}_C: S \times S \to \Delta(S^\omega \otimes S^\omega)$ *be such that, for all* $u,v \in S$, $\mathbb{P}_C(u,v)$ *is the unique probability measure on* $S^\omega \otimes S^\omega$ *such that, for all,* $n \geq 1$ *and* $U_i, V_i \subseteq S$ $(i = 0..nk)$.

$$\mathbb{P}_C(u,v)(\mathfrak{C}(U_{0,nk}) \times \mathfrak{C}(V_{0,nk})) = \mathbb{1}_{U_0 \times V_0}(u,v) \cdot \int \mathbb{P}_C(\cdot)(\mathfrak{C}(U_{k,nk}) \times \mathfrak{C}(V_{k,nk})) \, d\omega \,,$$

where, $U_{i,j} = U_i \cdots U_j$ *(similarly for* V $)^4$ *and* ω *is the unique (subprobability) measure on* $S \otimes S$ *s.t., for all* $A, B \subseteq S$, $\omega(A \times B) = C(u,v)(U_{1,k-1}A \times V_{1,k-1}B)$.

The following lemma extends [5, Th. 8] to k-bisimilarity pseudometrics and provides the alternative characterization of δ_b^k in terms of coupling structures.

Lemma 19 (Coupling Lemma). $\delta_b^k(u,v) = \inf \{\mathbb{P}_C(u,v)(\neq_{\ell^\omega}) \mid C \in \mathbb{C}_k\}$.

Thanks to Lemma 19 and the next result we can show that the k-bisimilarity pseudometrics δ_b^k form a net that converges point-wise to the trace distance δ_t.

Recall that a poset is *directed* if all its finite subsets have an upper bound. A *net* over a topological space X is a function from a directed poset to X. We denote a net as $(x_i)_{i \in D}$, meaning that $i \in D$ is mapped to x_i. A net $(x_i)_{i \in D}$ over X converges to $x \in X$, written $(x_i)_{i \in D} \to x$, if for every open subset $A \subseteq X$ such that $x \in A$, there exits $h \in D$ such that, for all $j \succeq h$, $x_j \in A$.

Theorem 20. *Let* (X, Σ) *be a measurable space s.t.* $\cong_\Sigma \in \Sigma \otimes \Sigma$, μ, ν *be probability measures on it,* $(D, \preceq)$ *be a directed poset and* $\Omega: D \to 2^{\Omega(\mu,\nu)}$ *be a monotone map such that* $\bigcup_{i \in D} \Omega(i)$ *is dense in* $\Omega(\mu, \nu)$ *w.r.t. the total variation distance. Then, the net* $(u_i)_{i \in D}$ *over* $\mathbb{R}_+$ *defined by* $u_i = \inf \{\omega(\neq_\Sigma) \mid \omega \in \Omega(i)\}$, *converges to* $\|\mu - \nu\|$.

Proof. By Lemma 12, for all $i \in D$, $u_i \geq \|\mu - \nu\|$. Moreover, by monotonicity of Ω, $i \preceq j$ implies $u_i \leq u_j$. Therefore, to prove $(u_i)_{i \in D} \to \|\mu - \nu\|$, it suffices to show $\inf_{i \in D} u_i = \|\mu - \nu\|$. Recall that for $Y \neq \emptyset$ and $f: Y \to \mathbb{R}$ bounded and continuous, if $D \subseteq Y$ is dense then $\inf f(D) = \inf f(Y)$. By hypothesis $\bigcup_{i \in D} \Omega(i) \subseteq \Omega(\mu, \nu)$ is dense; moreover, $\mu \times \nu \in \Omega(\mu, \nu) \neq \emptyset$. We show that

---

^{4} We assume that $U_{i,j} = \{\epsilon\}$ whenever $i > j$.

$ev_{\not\cong} \colon \Omega(\mu, \nu) \to \mathbb{R}$, defined by $ev_{\not\cong}(\omega) = \omega(\not\cong)$ is bounded and continuous. It is bounded since all $\omega \in \Omega(\mu, \nu)$ are probability measures. It is continuous because $\|\omega - \omega'\| \geq |\omega(\not\cong) - \omega'(\not\cong)| = |ev_{\not\cong}(\omega) - ev_{\not\cong}(\omega')|$ (1-Lipschitz continuity). Now, applying Lemma 12, we derive our result. □

Recall that, $\delta_t(u, v)$ is the total variation distance between $\mathbb{P}(u)$ and $\mathbb{P}(v)$ restricted on $\sigma(\mathcal{T})$. Observe that the inseparability relation w.r.t. $\sigma(\mathcal{T})$ is $\equiv_{\ell^\omega}$, which is easily seen to be measurable in $\sigma(\mathcal{T}) \otimes \sigma(\mathcal{T})$. Therefore, by Lemma 12,

$$\delta_t(u, v) = \min \{\omega(\not\equiv_{\ell^\omega}) \mid \omega \in \Omega(\mathbb{P}(u), \mathbb{P}(v))\}.$$

The next lemma shows that (i) a coupling structure $\mathcal{C}$ induces a measure $\mathbb{P}_{\mathcal{C}}(u, v)$ which is a proper coupling for the pair $(\mathbb{P}(u), \mathbb{P}(v))$; (ii) the set of couplings constructed via the coupling structures grows by multiples of the rank k; and (iii) their union is dense in $\Omega(\mathbb{P}(u), \mathbb{P}(v))$.

Lemma 21. Let $u, v \in S$ be a pair of states of an MC $\mathcal{M}$. Then,

(i) for $k \geq 1$ and $\mathcal{C} \in \mathbb{C}_k$, $\mathbb{P}_{\mathcal{C}}(u, v) \in \Omega(\mathbb{P}(u), \mathbb{P}(v))$;
(ii) for $k, h \geq 1$, $\{\mathbb{P}_{\mathcal{C}}(u, v) \mid \mathcal{C} \in \mathbb{C}_k\} \subseteq \{\mathbb{P}_{\mathcal{C}}(u, v) \mid \mathcal{C} \in \mathbb{C}_{hk}\}$;
(iii) $\bigcup_{k \geq 1} \{\mathbb{P}_{\mathcal{C}}(u, v) \mid \mathcal{C} \in \mathbb{C}_k\}$ is dense in $\Omega(\mathbb{P}(u), \mathbb{P}(v))$ w.r.t. the total variation.

Proof. (sketch) (i) It follows directly by definition of $\mathbb{P}_{\mathcal{C}}$ and the definitional conditions of coupling structures. (ii) Let $k, h \geq 1$ and $\mathcal{C} \in \mathbb{C}_k$. Define $\mathcal{D}(u, v)$ as the unique measure on $S^{hk} \otimes S^{hk}$ such that, for all $E, F \subseteq S^{hk}$,

$$\mathcal{D}(u, v)(E \times F) = \mathbb{P}_{\mathcal{C}}(\mathfrak{C}(SE) \times \mathfrak{C}(SF)).$$

Then, $\mathcal{D} \in \mathbb{C}_{hk}$ and $\mathbb{P}_{\mathcal{C}}(u, v) = \mathbb{P}_{\mathcal{D}}(u, v)$. (iii) Let $\Omega = \bigcup_{k \geq 1} \{\mathbb{P}_{\mathcal{C}}(u, v) \mid \mathcal{C} \in \mathbb{C}_k\}$. Note that $\bigcup_{n \in \mathbb{N}} \{\mathfrak{C}(E) \times \mathfrak{C}(F) \mid E, F \subseteq S^n\}$ is a field generating the σ-algebra of $S^\omega \otimes S^\omega$. To prove that Ω is dense w.r.t. the total variation it suffices to show that, for all $\mu \in \Omega(\mathbb{P}(u), \mathbb{P}(v))$, $n \in \mathbb{N}$ and $E, F \subseteq S^n$, there exists $\omega \in \Omega$ s.t. $\omega(\mathfrak{C}(E) \times \mathfrak{C}(F)) = \mu(\mathfrak{C}(E) \times \mathfrak{C}(F))$ (consequence of [2, Lemma 5]). One can check that this equality holds for $\omega = \mathbb{P}_{\mathcal{C}}(u, v)$ and $\mathcal{C} \in \mathbb{C}_n$ s.t. $\mathcal{C}(u, v) = \mu[f]$ is the push forward of μ along $f \colon S^\omega \to S^n$, defined as $f(\pi, \rho) = (\pi|_1 n, \rho|_1 n)$. □

Note that Lemmas 19 and 21(i) imply that, for all $k \geq 1$, $\delta_b^k \geq \delta_t$. This generalizes [5, Cor. 11] to arbitrary k-bisimilarity distances.

Denote by $\mathbb{K}$ the poset over $\mathbb{N} \setminus \{0\}$ with partial order $n \preceq m$ iff there exists $k \in \mathbb{N}$ such that $m = nk$. It is easy to see that $\mathbb{K}$ is directed. According to Theorem 20, Lemmas 19 and 21 suffice to prove the following net-convergence.

Theorem 22 (Convergence). *The net $(\delta_b^k)_{k \in \mathbb{K}}$ converges point-wise to δ_t.*

Remark 23. The use of the preorder $\preceq$ in the definition of the directed poset $\mathbb{K}$ is essential in Theorem 22. Indeed, if $\preceq$ is replaced by the standard total order $\leq$ over natural numbers, the net-convergence does not hold (by Lemma 16, the MC shown in Remark 15 provides a counterexample). □

Remark 24 (Equivalence vs Metric-based semantics). Although $\bigcup_{k \geq 1} \sim_b^k \neq \sim_b$ (see Remark 14), by Theorem 22, we have that $\inf_{k \geq 1} \delta_b^k = \delta_t$. Note that this is not in contradiction with Lemma 16. Actually it shows how much an equivalence and a metric-based semantics may differ. The explanation is topological, and it is due to the fact that equivalences (interpreted as functions) differ from 1-bounded pseudometrics by mapping pairs of states to the two-point space $\{0, 1\}$ (with the discrete topology) which is *disconnected*, whereas $[0, 1]$ is *connected*. □

5.2   The Stutter Case

We show how the construction that led to Theorem 22 can be easily adapted to obtain a net that converges to the strutter trace distance δ_{st}. This proves that the method is general enough to accommodate nontrivial convergence results.

Definition 25. *Let $k \geq 1$. An equivalence relation $R \subseteq S \times S$ is a $\equiv_\ell$-stutter k-probabilistic bisimulation on $\mathcal{M}$ if whenever $u \, R \, v$, then, for all $E_i \in S/_{\equiv_\ell}$ and $C \in S/_R$,*

$$\mathbb{P}(u)(\mathfrak{C}_{\equiv_\ell}(E_0 \cdots E_{k-1} C)) = \mathbb{P}(v)(\mathfrak{C}_{\equiv_\ell}(E_0 \cdots E_{k-1} C)) .$$

Two states $u, v \in S$ are $\equiv_\ell$-stutter k-probabilistic bisimilar, written $u \sim_{sb}^k v$, if they are related by some $\equiv_\ell$-stutter k-probabilistic bisimulation.

The above definition weakens that of k-probabilistic bisimulation by restricting the events to be tested only to those that are $\equiv_\ell$-stutter invariant.

It is easy to show that, for all $k \geq 1$, $\sim_b^k \subseteq \sim_{sb}^k$. Note that, $\sim_{sb}^k \nsubseteq \sim_b^k$ (in Fig. 1(left), $u \sim_{sb}^1 v$ but $u \nsim_b^1 v$). In analogy with the strong case, for all $k \geq 1$, $\sim_{sb}^k$ is a $\equiv_\ell$-stutter k-bisimulation, $\sim_{sb}^1 \subseteq \sim_{sb}^k \subseteq \sim_{st}$.

Now we lift these equivalences to pseudometrics by means of a Kantorovich-like operator. For $k \geq 1$, define the $\equiv_\ell$-*stuttered k-steps transition probability function* $\tau_s^k \colon S \to \Delta(S^k)$ as the function s.t., $\tau_s^k(u)$ is the unique probability measure on S^k that, for all $U_i \subseteq S$, $\tau_s^k(u)(U_1 \cdots U_k) = \mathbb{P}(u)(\mathfrak{C}_{\equiv_\ell}(uU_1 \cdots U_k))$ (i.e., $\tau_s^k(u) = \mathbb{P}(u)[\mathsf{pf}_{\equiv_\ell}^k \circ \mathsf{tl}_{\equiv_\ell}^1])$. Define, for $d \colon S \times S \to [0, 1]$ pseudometric,

$$\Psi^k(d)(u, v) = \begin{cases} 1 & \text{if } u \not\equiv_\ell v \\ \mathcal{K}(\Lambda^k(d))(\tau_s^k(u), \tau_s^k(v)) & \text{otherwise} . \end{cases}$$

The above extends to the stutter case the k-Kantorovich operator. Clearly, Ψ^k is monotonic, so that, by Tarski fixed point theorem, it has a least fixed point, denoted by δ_{sb}^k.

Due to the following result we call δ_{sb}^k the $\equiv_\ell$-*stutter k-bisimilarity distance.*

Lemma 26 (Stutter k-Bisimilarity Distance). $u \sim_{sb}^k v$ iff $\delta_{sb}^k(u, v) = 0$.

Next we provide a characterization of δ_{sb}^k by means of the notion of coupling structure, now modified to accommodate the notion of $\equiv_\ell$-stutter step.

Definition 27. *A function* $C\colon S \times S \to \Delta(S^k \otimes S^k)$ *is a stutter coupling structure of rank* $k \geq 1$ *if, for all* $u, v \in S$, $C(u,v) \in \Omega(\tau_s^k(u), \tau_s^k(v))$.

Hereafter, $\mathbb{C}_k^s$ denotes the set of stutter coupling structures of rank k.

Denote by $st(S^\omega)$ the measurable space over S^ω with σ-algebra $\sigma(\mathfrak{C}_{\equiv_\ell}(2^S))$. The stutter coupling structures are used to define measures in the product space $st(S^\omega) \otimes st(S^\omega)$.

Definition 28. *For* $k \geq 1$ *and* $C \in \mathbb{C}_k^s$, *let* $\mathbb{P}_C\colon S \times S \to \Delta(st(S^\omega) \otimes st(S^\omega))$ *be such that, for all* $u, v \in S$, $\mathbb{P}_C(u,v)$ *is the unique probability measure on* $st(S^\omega) \otimes st(S^\omega)$ *such that, for all,* $n \geq 1$ *and* $U_i, V_i \subseteq S$ *(i = 0..nk)*

$$\mathbb{P}_C(u,v)(\mathfrak{C}_{\equiv_\ell}(U_{0,nk}) \times \mathfrak{C}_{\equiv_\ell}(V_{0,nk})) = \mathbb{1}_{U_0 \times V_0}(u,v) \cdot \int \mathbb{P}_C(\cdot)(\mathfrak{C}_{\equiv_\ell}(U_{k,nk}) \times \mathfrak{C}_{\equiv_\ell}(V_{k,nk})) \, d\omega,$$

where, $U_{i,j} = U_i \cdots U_j$ *(similarly for* V*) and* ω *is the unique (subprobability) measure on* $S \otimes S$ *s.t., for all* $A, B \subseteq S$, $\omega(A \times B) = C(u,v)(U_{1,k-1}A \times V_{1,k-1}B)$.

The following gives a characterization of the k-stutter bisimilarity pseudometric δ_{sb}^k in terms of stutter coupling structures. Note that, by Proposition 10, $\simeq$ is the inseparability relation w.r.t. $\sigma(\mathcal{ST})$ and, since LTL^{-x} is countable, it holds $\simeq \in \sigma(\mathcal{ST}) \otimes \sigma(\mathcal{ST})$.

Lemma 29 (Coupling Lemma). $\delta_{sb}^k(u,v) = \inf \{\mathbb{P}_C(u,v)(\not\simeq) \mid C \in \mathbb{C}_k^s\}$.

According to Theorem 20 what follows suffices to prove the convergence.

Lemma 30. Let $u, v \in S$ be a pair of states of an MC $\mathcal{M}$. Then,

i. for $k \geq 1$ and $C \in \mathbb{C}_k^s$, $\mathbb{P}_C(u,v) \in \Omega(\mathbb{P}(u), \mathbb{P}(v))$;
ii. for $k, h \geq 1$, $\{\mathbb{P}_C(u,v) \mid C \in \mathbb{C}_k^s\} \subseteq \{\mathbb{P}_C(u,v) \mid C \in \mathbb{C}_{hk}^s\}$;
iii. $\bigcup_{k \geq 1} \{\mathbb{P}_C(u,v) \mid C \in \mathbb{C}_k^s\}$ is dense in $\Omega(\mathbb{P}(u), \mathbb{P}(v))$ w.r.t. the total variation,

where $\mathbb{P}(u)$ is assumed to be restricted on the sub-σ-algebra $\sigma(\mathfrak{C}_{\equiv_\ell}(2^S))$.

The next result is a direct consequence of Theorem 20, Lemmas 29, and 30.

Theorem 31 (Convergence). *The net* $(\delta_{sb}^k)_{k \in \mathbb{K}}$ *converges point-wise to* δ_{st}.

6   Approximation Schema for the Linear Distances

In this section we provide each of the two trace distances (strong and stutter) with an approximation schema, that is, a pair of sequences of pseudometrics that converges from below and above to them. We show that each lower- and upper-approximant is computable in polynomial time in the size of the MC.

In the following, we assume that $\mathcal{M}$ has a finite set of states and its transition probabilities are rational (i.e., $\tau(u)(v) \in \mathbb{Q} \cap [0,1]$). The size of $\mathcal{M}$ is determined by the sum of the size of the binary representation of its components. Under this restrictions the pseudometrics proposed in this section have finite domain and image in $\mathbb{Q}$. They are computable if they can be computed on all their domain.

6.1   The Strong Case

Lower-Approximants. The sequence of lower-approximants will be defined by restricting the set of measurable sets over which δ_t evaluates the differences in the probabilities. Formally, for $k \geq 1$, let $\mathcal{E}_k$ be the set of all finite unions of cylinders in $\mathfrak{C}^k(S/_{\equiv_t})$. We define the pseudometrics $l^k : S \times S \to [0,1]$ as follows

$$l^k(u,v) = \max_{E \in \mathcal{E}_k} |\mathbb{P}(u)(E) - \mathbb{P}(v)(E)|$$

The following lemma states that the sequence $(l^k)_{k \geq 1}$ is increasing and that converges point-wise to the trace distance δ_t.

Lemma 32. For all $k \geq 1$, $l^k \leq l^{k+1}$ and $\delta_t = \sup_{k \geq 1} l^k$.

Proof. $l^k \leq l^{k+1}$ follows by $\mathcal{E}_k \subseteq \mathcal{E}_{k+1}$. The equality $\delta_t = \sup_{k \geq 1} l^k$ is a consequence of [2, Theorem 6] and the fact that $\bigcup_{k \geq 1} \mathcal{E}_k$ is a field generating $\sigma(\mathcal{T})$. □

By looking at its definition, it is not clear whether l^k can be computed in polynomial time in the size of $\mathcal{M}$. Indeed, the maximum ranges over a set whose cardinality may be exponential in $|S^k|$ in the worst case. The following characterization shows that to compute l^k we do not need to evaluate the probabilities on all the elements of $\mathcal{E}_k$ but only on the thin cylinders of rank k.

Proposition 33. $l^k(u,v) = \frac{1}{2} \sum_{C \in \mathfrak{C}^k(S)} |\mathbb{P}(u)(C) - \mathbb{P}(v)(C)|$.

Proof. Note that $\mathcal{E}_k$ is finite and closed under complement. Let $\mathcal{F}$ be the family of cylinders $C \in \mathfrak{C}^k(S)$ s.t. $\mathbb{P}(u)(C) \geq \mathbb{P}(v)(C)$. By Hahn decomposition theorem, for $F = \bigcup \mathcal{F}$ we have $\mathbb{P}(u)(F) - \mathbb{P}(v)(F) = \max_{E \in \mathcal{E}_k} |\mathbb{P}(u)(E) - \mathbb{P}(v)(E)|$. Then

$$\begin{aligned}
2 \cdot l^k(u,v) &= 2 \cdot \sum_{F \in \mathcal{F}} \mathbb{P}(u)(F) - \mathbb{P}(v)(F) & \text{(σ-additive)} \\
&= \sum_{F \in \mathcal{F}} (\mathbb{P}(u)(F) - \mathbb{P}(v)(F)) + (\mathbb{P}(v)(F^c) - \mathbb{P}(u)(F^c)) & \text{(compl.)} \\
&= \sum_{C \in \mathfrak{C}^k(S)} |\mathbb{P}(u)(C) - \mathbb{P}(v)(C)|, & (\mathcal{F} \cup \mathcal{F}^c = \mathfrak{C}^k(S))
\end{aligned}$$

where the second equality holds since $\mathbb{P}(v)(F^c) = 1 - \mathbb{P}(v)(F)$. □

Note that the cylinders in $\mathfrak{C}^k(S)$ are all those of the form $\mathfrak{C}(u_1..u_k)$, for some $u_i \in S$ $(i = 1..k)$, and $\mathbb{P}(u)(\mathfrak{C}(u_1..u_k)) = \mathbb{1}_{u_1}(u) \cdot \prod_{i=1}^{k-1} \tau(u_i)(u_{i+1})$. Then, by Proposition 33, to compute $l^k(u,v)$ we need only $2kS^k$ multiplications, S^k subtractions and $S^k - 1$ summations. Hence l^k can be computed in $O(kS^{2+k})$.

Theorem 34. l^k *can be computed in polynomial time in the size of* $\mathcal{M}$.

Upper-Approximants. The decreasing sequence $(u^k)_{k \geq 1}$ of upper-approximants converging to δ_t simply derives from the net of k-bisimilarity pseudometrics presented in Sect. 5. and is defined by $u^k = \delta_b^{2^{k-1}}$ (actually, any infinite subsequence of $(\delta^k)_{k \in \mathbb{K}}$ is fine). The actual contribution of this section is to show that, for all $k \geq 1$, the k-bisimilarity distance δ_b^k can be characterized as the optimal solution of a linear program that can be constructed and solved in polynomial time in the size of the MC.

Our linear program characterization leverages on a *dual* linear program characterization of the Kantorovich distance. For X finite, $d\colon X \times X \to [0,1]$ a pseudometric and $\mu, \nu \in \Delta(X)$, the value of $\mathcal{K}(d)(\mu, \nu)$ coincides with the optimal value of the following linear programs.

PRIMAL		DUAL	
$\min\limits_{\omega} \sum_{x,y \in X} d(x,y) \cdot \omega_{x,y}$		$\max\limits_{\alpha} \sum_{x \in X} (\mu(x) - \nu(x)) \cdot \alpha_x$	
$\sum_y \omega_{x,y} = \mu(x)$	$\forall x \in X$		
$\sum_x \omega_{x,y} = \nu(y)$	$\forall y \in X$	$\alpha_x - \alpha_y \le d(x,y)$	$\forall x, y \in X$
$\omega_{x,y} \ge 0$	$\forall x, y \in X$		

Consider the linear program in Fig. 2, hereafter denoted by D. Note that for an optimal solution of D the value of the unknown $d \in \mathbb{R}^{S \times S}$ is maximized at each component. Therefore, for an optimal solution of D it holds that, if $u \equiv_\ell v$ and $u \not\sim_b^k v$, the maximal value of $d_{u,v}$ is achieved at $\mathcal{K}(\Lambda^k(d))(\tau^k(u), \tau^k(v))$. Otherwise, $d_{u,v} = 1$ when $u \not\equiv_\ell v$, and $d_{u,v} = 0$ when $u \sim_b^k v$. Thus, any optimal solution of D induces a fixed point for Θ^k whose kernel coincides with $\sim_b^k$. In fact, an optimal solution of D characterizes the greatest fixed point of the operator $\Upsilon^k \colon [0,1]^{S \times S} \to [0,1]^{S \times S}$ defined by $\Upsilon^k(d)(u,v) = 0$ if $u \sim_b^k v$, otherwise $\Upsilon^k(d)(u,v) = \Theta^k(d)(u,v)$.

$$
\begin{aligned}
&\operatorname*{argmax}_{d,\alpha} \sum_{u,v \in S} d_{u,v} \\
&\quad d_{u,v} = 0 && \forall u,v \in S.\, u \sim_b^k v \\
&\quad d_{u,v} = 1 && \forall u,v \in S.\, u \not\equiv_\ell v \\
&\quad d_{u,v} = \sum_{x \in S^k} \big(\tau^k(u)(x) - \tau^k(v)(x) \big) \alpha_x^{u,v} && \forall u,v \in S.\, u \equiv_\ell v \text{ and } u \not\sim_b^k v \\
&\quad \alpha_x^{u,v} - \alpha_y^{u,v} \le d_{x_k, y_k} && \forall u,v \in S\, \forall x,y \in S^k.\, \forall i.\, x_i \equiv_\ell y_i \\
&\quad \alpha_x^{u,v} - \alpha_y^{u,v} \le 1 && \forall u,v \in S\, \forall x,y \in S^k.\, \exists i.\, x_i \not\equiv_\ell y_i
\end{aligned}
$$

Fig. 2. Linear program characterization of the k-bisimilarity distance δ_b^k.

Lemma 35. Υ^k has a unique fixed point that coincides with δ_b^k.

This implies that for any optimal solution of D, $d_{u,v} = \delta_b^k(u,v)$, for all $u, v \in S$.

Note that D has a number of constraints bounded by $O(|S|^2 + |S|^{2k+2})$ and a number of unknowns bounded by $O(|S|^2 + |S|^{k+2})$. Moreover, the following lemma ensures that the linear program D can be constructed in polynomial time, provided that k is a constant.

Lemma 36. $\sim_b^k$ can be computed in polynomial time in the size of $\mathcal{M}$.

Theorem 37. δ_b^k can be computed in polynomial time in the size of $\mathcal{M}$.

Proof. (sketch) By Lemma 36, D can be constructed in polynomial time. Since the number of constraints and unknowns in D are bounded by a polynomial in the size of $\mathcal{M}$, D can be solved in polynomial time with the ellipsoid method. □

6.2   The Stutter Case

As one may expect, the sequences $(l_{st}^k)_{k\geq 1}$ and $(u_{st}^k)_{k\geq 1}$ of lower- and upper-approximants for the stutter trace distance δ_{sb} can be defined similarly to those we have shown in the previous section for the strong case. Specifically, for $k \geq 1$

$$l_{st}^k(u,v) = \max_{E \in \mathcal{S}_k} |\mathbb{P}(u)(E) - \mathbb{P}(v)(E)| \qquad \text{and} \qquad u_{st}^k(u,v) = \delta_{st}^{2^{k-1}},$$

where $\mathcal{S}_k$ is the set of all finite unions of stutter trace cylinders in $\mathfrak{C}^k_{\equiv_\ell}(S/_{\equiv_\ell})$.

Convergence and (anti)monotonicity of the sequences follow exactly as before. However, what is not immediate is the proof that, for all $k \geq 1$, l_{st}^k and u_{st}^k can actually be computed in polynomial time. The first difficulty arises, when for computing l_{st}^k, we try to apply the characterization provided by Lemma 32:

$$l^k(u,v) = \tfrac{1}{2} \sum_{C \in \mathfrak{C}^k_{\equiv_\ell}(S)} |\mathbb{P}(u)(C) - \mathbb{P}(v)(C)|.$$

The thin cylinders in $\mathfrak{C}^k_{\equiv_\ell}(S)$ are of the form $\mathfrak{C}(w)$, for some $w \in A_1^* \cdots A_k^*$ and $A_i \in S/_{\equiv_\ell}$ $(i = 1..k)$, hence $\mathfrak{C}^k_{\equiv_\ell}(S)$ is not finite (the word w can be arbitrarily long). Similarly, as for computing u_{st}^k, if we tried to apply directly the LP characterization in Fig. 2 we would have an infinite number of constraints.

To cope with this problem, we propose a reduction from the stutter to the strong case. Formally, we show that, for $k \geq 1$, the problem of computing $\mathbb{P}(u)(\mathfrak{C}_{\equiv_\ell}(u_1..u_k))$ and the k-stutter bisimilarity distance δ_{sb}^k for an MC $\mathcal{M}$ can be reduced to computing $\mathbb{P}(u)(\mathfrak{C}(u_1..u_k))$ and δ_b^k for an MC $\mathcal{N}$ derived from $\mathcal{M}$.

The following states that $\mathcal{N}$ is obtained by replacing the probability transition function τ in $\mathcal{M}$ with the (1-)stutter probability transition function τ_s^1.

Lemma 38. Let $\mathcal{M} = (S, \tau, \ell)$ and $\mathcal{N} = (S, \tau_s^1, \ell)$. Then, for all $k \geq 1$,

(i) $U_i \subseteq S$, $\mathbb{P}_\mathcal{M}(u)(\mathfrak{C}_{\equiv_\ell}(U_1 \cdots U_k)) = \mathbb{P}_\mathcal{N}(u)(\mathfrak{C}(U_1 \cdots U_k))$;
(ii) $\Psi_\mathcal{M}^k = \Theta_\mathcal{N}^k$.

Next we show that $\mathcal{N}$ can be constructed in polynomial time and its size is polynomial in the size of $\mathcal{M}$. Consider the problem of computing $\tau_s^1(u)(v)$.

We consider two possible cases:

Case $u \not\equiv_\ell v$. By definition $\tau_s^1(u)(v) = \mathbb{P}_\mathcal{M}(u)(\mathfrak{C}([u]_{\equiv_\ell}^\pm v))$. This is the probability of reaching the state v starting from u visiting only states in $[u]_{\equiv_\ell}$ prior to reaching v. Using LTL-like notations, this can be written as $\mathbb{P}_\mathcal{M}(u)([u]_{\equiv_\ell} \cup \{v\})$. This is a well studied probabilistic model checking problem that can be solved in polynomial time in the size of $\mathcal{M}$ as the solution of a linear system of equations (see e.g. [3, Sect. 10.1.1 p.762]).

Case $u \equiv_\ell v$. By definition $\tau_s^1(u)(v) = \mathbb{P}_\mathcal{M}(u)(uv[v]_{\equiv_\ell}^\omega)$. This corresponds to the probability of making a transition from u to v and, from v, generating an infinite run that never escapes from the $\equiv_\ell$-equivalence class of v, i.e., $\tau(u)(v) \cdot \mathbb{P}(v)([v]_{\equiv_\ell}^\omega)$. The probability $\mathbb{P}_\mathcal{M}(v)([v]_{\equiv_\ell}^\omega)$ can be conveniently computed as $1 - \sum_{x \not\equiv_\ell u} \tau_s^1(v)(x)$, reusing the probabilities computed in the previous case.

Therefore $\mathcal{N}$ can be constructed in polynomial time in the size of $\mathcal{M}$.

Lemma 39. $\mathcal{N} = (S, \tau_s^1, \ell)$ has size polynomial in the size of $\mathcal{M}$.

Proof. It suffices to show that τ_s^1 is rational of size polynomial in the size of $\mathcal{M}$. Let $u, v \in S$. If $u \not\equiv_\ell v$ then $\tau_s^1(u)(v) = \mathbb{P}_\mathcal{M}(u)([u]_{\equiv_\ell} \cup \{v\})$. Its value is the solution of a system of linear equations where the coefficients are some transition probabilities taken from $\mathcal{M}$ (or a sum of them). Therefore $\tau_s^1(u)(v)$ is an intersection of hyperplanes given by some equalities with rational coefficients whose size is bounded in the size of $\mathcal{M}$. Thus, we conclude that $\tau_s^1(u)(v)$ is rational of size polynomial in size of $\mathcal{M}$. The case $u \not\equiv_\ell v$ follows by the previous one since $\tau_s^1(u)(v) = \tau(u)(v) \cdot (1 - \sum_{x \not\equiv_\ell u} \tau_s^1(v)(x))$. ∎

By Lemmas 38 and 39, and Theorems 34 and 37, the following holds.

Theorem 40. l_{sb}^k and δ_{sb}^k can be computed in polynomial time in the size of $\mathcal{M}$.

Remark 41. Theorems 37 and 40 do not contradict the fact that the problem of approximating the trace distances up to a given precision $\epsilon > 0$ is NP-hard [7]. Indeed, this requires one to compute the lower and upper approximants l_*^k and δ_*^k ($* \in \{b, sb\}$), for increasing values of k, until $\delta_*^k - l_*^k < \epsilon$. Note that the time-complexity of this procedure increases exponentially in the value of k. □

7 Conclusions and Future Work

In this paper we provided the strong and stutter trace distances with a logical characterization in terms of LTL and LTL^{-X} formulas, respectively. These characterizations, differently from other proposals, relate these behavioral distances to the probabilistic model checking problem over MCs.

Then, we proposed a family of behavioral equivalences, namely probabilistic k-bisimilarities, that weaken probabilistic bisimilarity of Larsen and Skou on MCs. This equivalences are in turn generalized to pseudometrics by means of a fixed point definition that uses a generalized Kantorovich operator. These pseudometrics are shown to form a net that converges point-wise to the trace distance. Remarkably, to prove this convergence we extended and improved two important results in [5], namely, Theorem 8 and Corollary 11. The proposed construction is shown to be general enough to accommodate a second nontrivial convergence result between a net of suitable stutter variants of k-bisimilarities pseudometrics and the stutter trace distance. These convergences are interesting because they reveal a nontrivial relation between branching and linear-time metric-based semantics that in Remark 14 is shown not hold when the standard equivalence-based semantics on MCs are used instead.

The above distances are then used to provide the strong and stutter trace distances with an approximation schema, that is, two sequences of pseudometrics that converge from above and below to the two respective linear distances. Each of these lower and under-approximants are shown to be computable in polynomial time in the size of the MC. Notably, for this proof the under-approximants of the trace distance (i.e., the k-bisimilarity pseudometrics) are given a characterization in terms of optimal solutions of a linear program that have size polynomial in the MC. The one we proposed generalizes and improves the linear program characterization given in [5, Eq. 8] for the (undiscounted) bisimilarity pseudometric of Desharnais et al. that, in contrast, has a number of constraints exponential in the size of the MC. Moreover, our approximation schema improves that in [6] both for the generality of its applicability and in terms of computational complexity.

Natural questions are (i) to see if the on-the-fly algorithm for the computation of bisimilarity distance in [1] can be used to compute the k-bisimilarity distances and their stutter variants; (ii) whether this approximation technique carries over to models with non-determinism, such as MDPs (where a recent result by Fu [14] gives new insight on how to obtain minimal information in case the distance is not a bisimilarity metric, and where the PSPACE-complexity results is sharpened to NP ∩ coNP); (iii) whether a similar construction can be applied to stochastic models with continuous time, such as CTMCs.

References

1. Bacci, G., Bacci, G., Larsen, K.G., Mardare, R.: On-the-fly exact computation of bisimilarity distances. In: Piterman, N., Smolka, S.A. (eds.) TACAS 2013 (ETAPS 2013). LNCS, vol. 7795, pp. 1–15. Springer, Heidelberg (2013)
2. Bacci, G., Bacci, G., Larsen, K.G., Mardare, R.: On the total variation distance of semi-markov chains. In: Pitts, A. (ed.) FOSSACS 2015. LNCS, vol. 9034, pp. 185–199. Springer, Heidelberg (2015)
3. Baier, C., Katoen, J.-P.: Principles of Model Checking. MIT Press, Cambridge (2008)
4. Behrmann, G., David, A., Larsen, K.G., Håkansson, J., Pettersson, P., Yi, W., Hendriks, M.: UPPAAL 4.0. In: QEST, pp. 125–126. IEEE Computer Society (2006)
5. Chen, D., van Breugel, F., Worrell, J.: On the complexity of computing probabilistic bisimilarity. In: Birkedal, L. (ed.) FOSSACS 2012. LNCS, vol. 7213, pp. 437–451. Springer, Heidelberg (2012)
6. Chen, T., Kiefer, S.: On the total variation distance of labelled markov chains. In: CSL-LICS 2014, pp. 33:1–33:10. ACM (2014)
7. Cortes, C., Mohri, M., Rastogi, A.: Lp distance and equivalence of probabilistic automata. Int. J. Found. Comput. Sci. 18(04), 761–779 (2007)
8. de Alfaro, L., Faella, M., Stoelinga, M.: Linear and branching metrics for quantitative transition systems. In: Díaz, J., Karhumäki, J., Lepistö, A., Sannella, D. (eds.) ICALP 2004. LNCS, vol. 3142, pp. 97–109. Springer, Heidelberg (2004)

9. de Alfaro, L., Majumdar, R., Raman, V., Stoelinga, M.: Game relations and metrics. In: LICS, pp. 99–108, July 2007
10. Desharnais, J., Gupta, V., Jagadeesan, R., Panangaden, P.: Metrics for labeled markov systems. In: Baeten, J.C.M., Mauw, S. (eds.) CONCUR 1999. LNCS, vol. 1664, pp. 258–273. Springer, Heidelberg (1999)
11. Desharnais, J., Gupta, V., Jagadeesan, R., Panangaden, P.: Metrics for labelled markov processes. Theoret. Comput. Sci. **318**(3), 323–354 (2004)
12. Desharnais, J., Jagadeesan, R., Gupta, V., Panangaden, P.: The metric analogue of weak bisimulation for probabilistic processes. In: LICS, pp. 413–422. IEEE Computer Society (2002)
13. Ferns, N., Precup, D., Knight, S.: Bisimulation for markov decision processes through families of functional expressions. In: van Breugel, F., Kashefi, E., Palamidessi, C., Rutten, J. (eds.) Horizons of the Mind. LNCS, vol. 8464, pp. 319–342. Springer, Heidelberg (2014)
14. Fu, H.: Computing game metrics on markov decision processes. In: Czumaj, A., Mehlhorn, K., Pitts, A., Wattenhofer, R. (eds.) ICALP 2012, Part II. LNCS, vol. 7392, pp. 227–238. Springer, Heidelberg (2012)
15. Kwiatkowska, M., Norman, G., Parker, D.: PRISM 4.0: verification of probabilistic real-time systems. In: Gopalakrishnan, G., Qadeer, S. (eds.) CAV 2011. LNCS, vol. 6806, pp. 585–591. Springer, Heidelberg (2011)
16. Lamport, L.: What good is temporal logic?. In: IFIP, pp. 657–668 (1983)
17. Larsen, K.G., Skou, A.: Bisimulation through probabilistic testing. Inf. Comput. **94**(1), 1–28 (1991)
18. Lindvall, T.: Lectures on the Coupling Method. Wiley Series in Probability and Mathematical Statistics. Wiley, New York (1992)
19. Lyngsø, R.B., Pedersen, C.N.S.: The consensus string problem and the complexity of comparing hidden Markov models. J. Comput. Syst. Sci. **65**(3), 545–569 (2002)
20. Mio, M.: Upper-expectation bisimilarity and Łukasiewicz μ-calculus. In: Muscholl, A. (ed.) FOSSACS 2014 (ETAPS). LNCS, vol. 8412, pp. 335–350. Springer, Heidelberg (2014)
21. Pnueli, A.: The temporal logic of programs. In: SFCS, pp. 46–57. IEEE Computer Society (1977)

MSO Logic and the Partial Order Semantics of Place/Transition-Nets

Mateus de Oliveira Oliveira[✉]

Institute of Mathematics, Academy of Sciences of the Czech Republic,
Prague, Czech Republic
mateus.oliveira@math.cas.cz

Abstract. In this work, we study the interplay between monadic second order logic and the partial order theory of bounded place/transition-nets. First, we show that the causal behavior of any bounded p/t-net can be compared with respect to inclusion with the set of partial orders specified by a given MSO sentence φ. Subsequently, we address the synthesis of Petri nets from MSO specifications. More precisely, we show that given any MSO sentence φ, one can automatically construct a bounded Petri net whose behaviour minimally includes the set of partial orders specified by φ. Combining this synthesis result with the comparability results we study three problems in the realm of automated correction of faulty Petri nets, and show that these problems are decidable.

Keywords: Monadic second order logic · Petri nets · Slice languages

1 Introduction

Petri nets [19], also known as place/transition-nets, are recognized as an elegant mathematical formalism for the specification of concurrent systems. One of the main advantages of modeling concurrent systems via Petri nets, is that there exist very precise, but yet intuitive, ways of formalizing the notion of causality between events in a given run of such a net. One of the most prominent ways of formalizing causality in Petri nets is via the notion of Petri net process [17]. Intuitively, a Petri net process is a DAG whose vertex set is partitioned into *conditions* and *events*. While condition vertices are used to keep track of each token ever created or consumed during a concurrent run, the event vertices are used to keep track of which transitions created or consumed each such token. In such a process π, an event v causally depends on the occurrence of an event v' if there is a path from v' to v in π. The partial order induced on the events of a process is called a *causal-order*. The causal behavior of a Petri net N is the set of all causal orders derived from processes of N.

In this work we study the interplay between the causal behavior of Petri nets, and monadic second order (MSO) logic. Our first result states that given any bounded Petri net N, and any monadic second order logic sentence φ in the vocabulary of partial orders, one can decide whether all causal orders of

© Springer International Publishing Switzerland 2015
M. Leucker et al. (Eds.): ICTAC 2015, LNCS 9399, pp. 368–387, 2015.
DOI: 10.1007/978-3-319-25150-9_22

N satisfy φ. Previously, such a decidability result was known for the class of pure[1] bounded Petri nets [1]. Nevertheless, pure bounded Petri nets form a strict subclass of bounded Petri nets which is not able to model certain important concurrency theoretic primitives, such as waiting loops in communication protocols [2]. Thus, our first main result establishes the decidability of the model checking problem for the partial order behavior of general bounded Petri nets against MSO specifications, and removes the purity restriction imposed on previous works.

In our second result we show that given a MSO formula φ and a bounded Petri net N, one can decide whether all partial orders satisfying φ belong to the causal behavior of N. At an intuitive level our second result states that one can decide whether the causal behavior of a given Petri net over-approximates the set of partial orders defined by a given MSO formula. To the best of our knowledge this result was not known even for pure Petri nets.

Our third result concerns the automated synthesis of bounded Petri nets from partial order specifications. A Petri net is (b, r)-bounded if it is b-bounded and each place appears repeated at most r times. It is an interesting observation that the causal behavior of a Petri net may change with the addition of repeated places. Contrast this fact with other partial order semantics, such as the execution semantics studied in [5,23] for which repeated places are not relevant[2]. We show that given any MSO formula φ, one can determine whether there is a (b, r)-bounded p/t-Net N whose causal behavior includes the set of partial orders specified by φ. In the case the answer is positive, we show how to construct a (b, r)-bounded Petri net whose behavior minimally includes the set of partial orders specified by φ. In particular, if the behavior of some (b, r)-bounded Petri net precisely matches the behavior specified by φ, then this net will be returned by the algorithm.

Combining the three results mentioned above, we establish the decidability of three problems related to the automated design of concurrent systems. First, we introduce a notion of optimally correcting subsystems for bounded Petri nets. We show that these correcting subsystems can be automatically constructed. Subsequently we define a notion of automated repair for bounded Petri nets, and show that this notion is effective. Finally, we consider the synthesis of Petri nets from partial order contracts, where the goal is to construct a net whose behavior comprises all partial orders defined by an MSO formula φ^{yes}, but no partial order defined by a formula φ^{no}.

2   Preliminaries

In this section we will briefly recall the definitions of monadic second order logic, Petri nets, and the partial order semantics of Petri nets.

---

[1] A Petri net is pure if no transition consumes and produces a token at the same place.
[2] In the execution semantics the fact that an event v_1 is smaller than an event v_2 indicates that v_2 occurs after v_1, but v_1 need not be necessarily the cause of v_2. If N is a net and p is a place that is a linear combination of places in N, then N and $N \cup \{p\}$ have the same execution behavior. This is not true for the causal semantics.

2.1   The Monadic Second Order Logic of Partial Orders

In this section we define the monadic second order logic of partial orders, which will be used to describe partial order properties. We represent a partial order ℓ via a relational structure $\ell = (V, <, l)$ where V is a set of vertices, $< \subset V \times V$ is an irreflexive, asymmetric, transitive relation, and $l \subseteq V \times T$ is a function (viewed as a relation) which labels each vertex $v \in V$ with an element of a finite set T. Here, T should be regarded as a finite set of transitions in a concurrent system, or more precisely, a Petri net. First-order variables representing individual vertices are taken from the set $\{x_1, x_2, ...\}$ while second order variables representing sets of vertices are taken from the set $\{X_1, X_2, ...\}$. The set of MSO formulas in the vocabulary of partial orders is the smallest set of formulas containing:

- the atomic formulas $x_i \in X$, $x_i < x_j$, $l(x_i, a)$ for each $i, j \in \mathbb{N}$ with $i \neq j$ and each $a \in T$,
- the formulas $\varphi \wedge \psi$, $\varphi \vee \psi$, $\neg \varphi$, $\exists x_i . \varphi(x_i)$ and $\exists X_i . \varphi(X_i)$, where φ and ψ are MSO formulas.

An MSO sentence is a MSO formula φ without free variables. If φ is a sentence, and $\ell = (V, <, l)$ a partial order, then we denote by $\ell \models \varphi$ the fact that ℓ satisfies φ. Let φ be an MSO formula expressing a property of T-labeled partial orders, and let $c \in \mathbb{N}$. We denote by $\mathcal{P}(c, T, \varphi)$ the set of all T-labeled c-partial-orders satisfying φ.

2.2   Petri Nets

A Petri net is a tuple $N = (P, T, W, \mathfrak{m}_0)$ where P is a set of *places*, T is a set of *transitions* such that $P \cap T = \emptyset$, $W : (P \times T) \cup (T \times P) \to \mathbb{N}$ is a function that associates with each element $(x, y) \in (P \times T) \cup (T \times P)$ a weight $W(x, y)$, and $\mathfrak{m}_0 : P \to \mathbb{N}$ is a function that associates with each place $p \in P$ a non-negative integer $m_0(p)$.

A marking for N is any function of the form $\mathfrak{m} : P \to \mathbb{N}$. Intuitively, a marking $\mathfrak{m}$ assigns a number of tokens to each place of N. The marking $\mathfrak{m}_0$ is called the *initial marking* of N. If $\mathfrak{m}$ is a marking and t is a transition in T, then we say that t is enabled at $\mathfrak{m}$ if $\mathfrak{m}(p) - W(p, t) \geq 0$ for every place $p \in P$. If this is the case, the firing of t yields the marking $\mathfrak{m}'$ which is obtained from $\mathfrak{m}$ by setting $\mathfrak{m}'(p) = \mathfrak{m}(p) - W(p, t) + W(t, p)$ for every place $p \in P$. A firing sequence for N is a mixed sequence of markings and transitions $\mathfrak{m}_0 \xrightarrow{t_1} \mathfrak{m}_1 \xrightarrow{t_2} ... \xrightarrow{t_n} \mathfrak{m}_n$ such that for each $i \in \{1, ..., n\}$, t_i is enabled at $\mathfrak{m}_{i-1}$, and $\mathfrak{m}_i$ is obtained from $\mathfrak{m}_{i-1}$ by the firing of t_i. We say that such a firing sequence is b-bounded if for each $i \in \{0, ..., n\}$ and each $p \in P$, $\mathfrak{m}_i(p) \leq b$. We say that N is b-bounded if each of its firing sequences is b-bounded.

2.3   The Causal Semantics of Petri Nets

In this subsection we introduce the Goltz-Reisig partial order semantics for Petri nets [17]. Within this semantics, partial orders are used to represent the causality between events in concurrent runs of a Petri net. The information about the causality between events is extracted from objects called Petri net *processes*, which encode the production and consumption of tokens along a concurrent run of the Petri net in question. The definition of processes, in turn, is based on the notion of *occurrence net*.

An occurrence net is a DAG $O = (B \mathbin{\dot\cup} V, F)$ where the vertex set $B \mathbin{\dot\cup} V$ is partitioned into a set B, whose elements are called conditions, and a set V, whose elements are called events. The edge set $F \subseteq (B \times V) \cup (V \times B)$ is restricted in such a way that for every condition $b \in B$,

$$|\{(b,v) \mid v \in V\}| \le 1 \quad and \quad |\{(v,b) \mid v \in V\}| \le 1.$$

In other words, conditions in an occurrence net are unbranched. For each condition $b \in B$, we let $InDegree(b)$ denote the number of edges having b as target. A process of a Petri net N is an occurrence net whose conditions are labeled with places of N, and events are labeled with transitions of N. Processes are intuitively used to describe the token game in a concurrent execution of the net.

Definition 2.1 (Process[17]). *A process of a Petri net* $N = (P, T, W, m_0)$ *is a labeled DAG* $\pi = (B \mathbin{\dot\cup} V, F, \rho)$ *where* $(B \mathbin{\dot\cup} V, F)$ *is an occurrence net and* $\rho : (B \cup V) \to (P \cup T)$ *is a labeling function satisfying the following properties.*

1. Places label conditions and transitions label events.

$$\rho(B) \subseteq P \qquad \rho(V) \subseteq T$$

2. For every $p \in P$,

$$|\{b : InDegree(b) = 0, \rho(b) = p\}| = m_0(p).$$

3. For every $v \in V$, *and every* $p \in P$,

$$|\{(b,v) \in F : \rho(b) = p\}| = W(p, \rho(v)) \quad and \quad |\{(v,b) \in F : \rho(b) = p\}| = W(\rho(v), p)$$

Item 1 says that the conditions of a process are labeled with places, while the events are labeled with transitions. Item 2 says that the minimal vertices of the process, are conditions. Intuitively each of these conditions represent a token in the initial marking of N. Thus for each place p of N the process has $m_0(p)$ minimal conditions labeled with the place p. Item 3, determines that the token game of a process corresponds to the token game defined by the firing of transitions in the Petri net N. Thus if a transition t consumes $W(p, t)$ tokens from place p and produces $W(t, p)$ tokens at place p, then each event labeled with t must have $W(p, t)$ in-neighbours that are conditions labeled with p, and $W(t, p)$ out-neighbours that are conditions labeled with p.

Let $R \subseteq X \times X$ be a binary relation over a set X. We denote by $tc(R)$ the transitive closure of R. If $\pi = (B \cup V, F, \rho)$ is a process then the *causal order* of π is the partial order $\ell_\pi = (V, tc(F)|_{V \times V}, \rho|_V)$ which is obtained by taking the

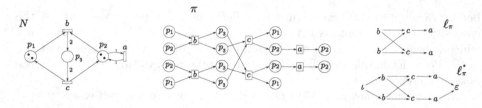

Fig. 1. A 2-bounded Petri net N. A process π of N. The partial order ℓ_π derived from π. The extension ℓ_π^* of ℓ_π.

transitive closure of F and subsequently by restricting $tc(F)$ to pairs of events of V. In other words the causal order of a process π is the partial order induced by π on its events.

If $\ell = (V, <, l)$ is a partial order, then we let $\ell^* = (V', <', l')$ be the *extended version* of ℓ, where $V' = V \cup \{v_\iota, v_\varepsilon\}$, $<' = < \cup (\{v_\iota\} \times V) \cup (V \times \{v_\varepsilon\}) \cup \{(v_\iota, v_\varepsilon)\}$, $l'|_V = l$, $l'(v_\iota) = \iota$ and $l'(v_\varepsilon) = \varepsilon$. In other words, ℓ' is obtained from ℓ by the addition of an element v_ι that is smaller than all other elements, and an element v_ε that is greater than all other elements. The addition of these minimal and maximal elements to a partial order are made to avoid the consideration of special cases in some of our future lemmas.

We denote by $\mathcal{P}_{cau}(N)$ the set of all extended versions of partial orders derived from processes of N.

$$\mathcal{P}_{cau}(N) = \{\ell_\pi^* \,|\, \pi \text{ is a process of } N\}$$

We say that $\mathcal{P}_{cau}(N)$ is the causal language of N. We observe that several processes of N may correspond to the same partial order in $\mathcal{P}_{cau}(N)$.

3   Slice Automata

In this section we define *slices* and *slice automata*. Slice automata will be used to provide static representations of infinite families of partial orders. We notice that slices can be related to several formalisms such as, multi-pointed graphs [16], co-span decompositions [7] and graph transformations [4,6,16,21]. However, some notions such as the notion of unit slice, saturated slice automata, and transitive reduction of slice automata which will be defined below, and which are crucial to the development of our work, are intrinsic to the slice theoretic formalism.

A slice $\mathbf{S} = (V, E, l, s, t, [I, C, O])$ is a DAG where $V = I \dot{\cup} C \dot{\cup} O$ is a set of vertices partitioned into an in-frontier I, a center C and an out-frontier O; E is a set of edges, $s, t : E \to V$ are functions that associate with each edge $e \in E$ a source vertex e^∂ and a target vertex e^t, and $l : V \to T \cup \mathbb{N}$ is a function that labels the center vertices in C with elements of a finite set $T \cup \{\iota, \varepsilon\}$, and the in- and out-frontier vertices with positive integers in such a way that $l(I) = \{1, ..., |I|\}$ and $l(O) = \{1, ..., |O|\}$. We require that each frontier-vertex v in $I \cup O$ is the endpoint of exactly one edge $e \in E$ and that the edges are directed from the in-frontier to the out frontier. More precisely, for each edge $e \in E$, we assume that $e^s \in I \cup C$ and that $e^t \in C \cup O$. For simplicity we may omit the source and target functions s and t when specifying a slice and write simply $\mathbf{S} = (V, E, l)$. We may also speak of a slice $\mathbf{S}$ with frontiers (I, O) to indicate that the in-frontier of $\mathbf{S}$ is I and that the out-frontier of $\mathbf{S}$ is O.

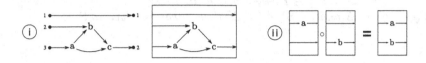

Fig. 2. i) A slice and its pictorial representation. (ii) Composition of slices.

A slice $\mathbf{S}_1 = (V_1, E_1, l_1)$ with frontiers (I_1, O_1) can be glued to a slice $\mathbf{S}_2 = (V_2, E_2, l_2)$ with frontiers (I_2, O_2) provided $|O_1| = |I_2|$. In this case the glueing gives rise to the slice $\mathbf{S}_1 \circ \mathbf{S}_2 = (V_3, E_3, l_3)$ with frontiers (I_1, O_2) which is obtained by taking the disjoint union of $\mathbf{S}_1$ and $\mathbf{S}_2$, and by fusing, for each $i \in \{1, ..., |O_1|\}$, the unique edge $e_1 \in E_1$ for which $l_1(e_1^t) = i$ with the unique edge $e_2 \in E_2$ for which $l_2(e_2^s) = i$. Formally, the fusion of e_1 with e_2 is performed by creating a new edge e_{12} with source $e_{12}^s = e_1^s$ and target $e_{12}^t = e_2^t$, and by deleting e_1 and e_2. Thus in the glueing process the vertices in the glued frontiers disappear.

A *unit slice* is a slice with exactly one vertex in its center. A unit slice is *initial* if it has empty in-frontier and *final* if it has empty out-frontier. We assume that the center vertex of an initial slice is always labeled with the special symbol ι and that the center vertex of a final slice is labeled with the special symbol ε. The width of a slice $\mathbf{S}$ with frontiers (I, O) is defined as $w(\mathbf{S}) = \max\{|I|, |O|\}$. If T is a finite set of symbols, then we let $\vec{\Sigma}(c, T)$ be the set of all unit slices of width at most c whose unique center vertex is labeled with an element of $T \cup \{\iota, \varepsilon\}$. We assume however that the center vertices of non-initial and non-final slices in $\vec{\Sigma}(c, T)$ are labeled with elements from T. The symbols ι and ε are reserved for initial and final slices respectively. Observe that $\vec{\Sigma}(c, T)$ is finite and has asymptotically $|T| \cdot 2^{O(c \log c)}$ slices. A sequence $\mathbf{U} = \mathbf{S}_1\mathbf{S}_2...\mathbf{S}_n$ of unit slices is called a unit decomposition if $\mathbf{S}_i$ can be glued to $\mathbf{S}_{i+1}$ for each $i \in \{1, ..., n-1\}$. In this case, we let $\mathring{\mathbf{U}} = \mathbf{S}_1 \circ \mathbf{S}_2 \circ ... \circ \mathbf{S}_n$ be the DAG derived from $\mathbf{U}$, which is obtained by gluing each two consecutive slices in $\mathbf{U}$.

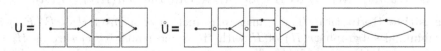

Fig. 3. A unit decomposition $\mathbf{U}$ and the graph $\mathring{\mathbf{U}}$ obtained by gluing each two consecutive slices in $\mathbf{U}$.

Definition 3.1 (Slice Automaton). *Let T be a finite set of symbols and let $c \in \mathbb{N}$. A slice automaton over a slice alphabet $\overrightarrow{\Sigma}(c, T)$ is a finite automaton $\mathcal{A} = (Q, \Delta, q_0, F)$ where Q is a set of states, $q_0 \in Q$ is an initial state, $F \subseteq Q$ is a set of final states, and $\Delta \subseteq Q \times \overrightarrow{\Sigma}(c, T) \times Q$ is a transition relation such that for every $q, q', q'' \in Q$ and every $\mathbf{S} \in \overrightarrow{\Sigma}(c, T)$:*

1. *if $(q_0, \mathbf{S}, q) \in \Delta$ then $\mathbf{S}$ is an initial slice,*
2. *if $(q, \mathbf{S}, q') \in \Delta$ and $q' \in F$, then $\mathbf{S}$ is a final slice,*
3. *if $(q, \mathbf{S}, q') \in \Delta$ and $(q', \mathbf{S}', q'') \in \Delta$, then $\mathbf{S}$ can be glued to $\mathbf{S}'$.*

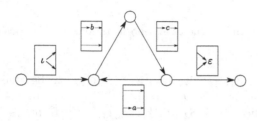

Fig. 4. A slice automaton. Circles are states. A transition $(q, \mathbf{S}, q')$ is represented by an arrow from q to q' with the slice $\mathbf{S}$ depicted next to it.

Languages of a Slice Automaton. A slice automaton $\mathcal{A}$ can be used to represent three types of language. At a syntactic level, we have the slice language $\mathcal{L}(\mathcal{A})$ which consists of the set of all unit decompositions accepted by $\mathcal{A}$.

$$\mathcal{L}(\mathcal{A}) = \{\mathbf{S}_1 \mathbf{S}_2 ... \mathbf{S}_n | \mathbf{S}_1 \mathbf{S}_2 ... \mathbf{S}_n \text{ is accepted by } \mathcal{A}\} \tag{1}$$

At a semantic level, we have the graph language $\mathcal{L}_{\mathcal{G}}(\mathcal{A})$ which consists of all DAGs represented by unit decompositions in $\mathcal{L}(\mathcal{A})$, and the partial order language $\mathcal{L}_{po}(\mathcal{A})$, which consists of all partial orders derived from DAGs in $\mathcal{L}_{\mathcal{G}}(\mathcal{A})$. If H is a DAG, we let $tc(H)$ denote the partial order which is obtained by taking the transitive closure of H. Formally, the graph language and the partial order languages accepted by $\mathcal{A}$ are defined as follows.

$$\mathcal{L}_{\mathcal{G}}(\mathcal{A}) = \{\mathring{\mathbf{U}} \mid \mathbf{U} \in \mathcal{L}(\mathcal{A})\} \quad \mathcal{L}_{po}(\mathcal{A}) = \{tc(\mathring{\mathbf{U}}) \mid \mathring{\mathbf{U}} \in \mathcal{L}_{\mathcal{G}}(\mathcal{A})\}. \tag{2}$$

Saturation. Let H be a DAG whose vertices are labeled with elements from a finite set T. Then we let $ud(H, \overrightarrow{\Sigma}(c, T))$ denote the set of all unit decompositions $\mathbf{U}$ in $\mathcal{L}(\overrightarrow{\Sigma}(c, T))$ for which $\mathring{\mathbf{U}} = H$. We say that a slice automaton $\mathcal{A}$ over $\overrightarrow{\Sigma}(c, T)$ is *saturated* if for every DAG $H \in \mathcal{L}_{\mathcal{G}}(\mathcal{A})$ we have that $ud(H, c) \subseteq \mathcal{L}(\mathcal{A})$.

Transitive Reduction. The transitive reduction of a DAG $H = (V, E, l)$ is the minimal subgraph $tr(H)$ of H with the same transitive closure as H. In other words $tc(tr(H)) = tc(H)$. We say that a DAG H is *transitively reduced* if $H = tr(H)$. Alternatively, we call a transitively reduced DAG a Hasse diagram. We say that a slice automaton $\mathcal{A}$ is transitively reduced if every DAG in

$\mathcal{L}_{\mathcal{G}}(\mathcal{A})$ is transitively reduced. Theorem 3.2 states that any slice automaton $\mathcal{A}$ can be converted into a transitively reduced slice automaton $tr(\mathcal{A})$ representing the same partial order language in such a way that the saturation property is preserved.

Theorem 3.2. *Let $\mathcal{A}$ be a slice automaton over $\overrightarrow{\Sigma}(c,T)$. Then one can construct in time $2^{O(c \log c)} \cdot |\mathcal{A}|$ a slice automaton $tr(\mathcal{A})$ such that $\mathcal{L}_{po}(tr(\mathcal{A})) = \mathcal{L}_{po}(\mathcal{A})$. Additionally, if $\mathcal{A}$ is saturated, then so is $tr(\mathcal{A})$.*

We note that in [13] we proved a weaker version of Theorem 3.2 which only preserves a property called weak-saturation. However for the purposes of the present paper we need the transitive reduction to preserve the stronger version of saturation defined above. Transitively reduced saturated slice automata are important for our setting because they can be used to canonically represent infinite families of partial orders. Additionally, as we will show in the next subsections, transitively reduced saturated slice automata can be used to represent two important classes of partial orders. First, these automata can represent MSO-definable sets of c-partial orders. Second, they can be used to represent the causal behaviour of bounded Petri nets.

4 c-Partial-Orders

In this section we will introduce the notion of c-partial-order. First, we will show that sets of c-partial orders definable via saturated slice automata are closed under union, intersection and an appropriate notion of complementation. This will allow us to operate with infinite sets of c-partial-orders in a similar way in which one operates with infinite sets of strings.

In Sect. 5 we will connect MSO logic with slice languages and show that MSO-definable sets of c-partial-orders can be effectively represented by saturated slice automata. In Sect. 6 we will show that for any $c \in \mathbb{N}$, the set of c-partial-orders of any given bounded Petri net can be represented via a saturated transitively reduced slice automaton. Combined, these two results will imply the decidability of the model checking of the causal behavior of bounded Petri nets.

c-Partial Orders. Let T be a finite set whose elements should be regarded as transitions of a concurrent system. We say that a partial order ℓ is a T-labeled partial-order if each node of ℓ is labeled with some element of T. The Hasse diagram of a partial order ℓ is the directed acyclic graph H with the least number of edges whose transitive closure equals ℓ. In other words, H is the Hasse diagram of ℓ if H is the transitive reduction of ℓ.

We say that a DAG $G = (V, E, l)$ can be covered by k paths if there exist directed paths $\mathfrak{p}_1 = (V_1, E_1), ..., \mathfrak{p}_k = (V_k, E_k)$ in G such that $V = \cup_{i=1}^{k} V_i$ and $E = \cup_{i=1}^{k} E_i$. We note that the paths are not assumed to be edge disjoint nor vertex disjoint. We say that a partial order ℓ is a c-*partial-order* if its Hasse diagram $H = (V, E, l)$ can be covered by c paths. For instance, in Fig. 5, we depict the Hasse diagram of a 3-partial-order.

H

Fig. 5. The Hasse diagram of a 3-partial-order. H can be covered by 3 paths. The two outer horizontal paths plus the path in zig-zag in the center.

We denote by $\mathcal{P}(c, T)$ the set of all c-partial orders whose vertices are labeled with elements from T. Lemma 4.1 below, states that the set $\mathcal{P}(c, T)$ can be represented by a saturated, transitively reduced slice automaton whose size depends only on $|T|$ and c.

Lemma 4.1. *For any finite set T and any $c \in \mathbb{N}$, one can construct a saturated transitively reduced slice automaton $\mathcal{A}(c, T)$ over $\overrightarrow{\Sigma}(c, T)$, of size $|T| \cdot 2^{O(c \log c)}$, such that $\mathcal{L}_{po}(\mathcal{A}(c, T)) = \mathcal{P}(c, T)$.*

Next we show that boolean operations realized with saturated transitively reduced slice automata are reflected into the partial order languages they represent. First, we define a suitable notion of complementation for sets of c-partial-orders.

Definition 4.2 (c-Complementation). *Let $\mathcal{P}$ be a set of T-labeled partial orders. We let $\overline{\mathcal{P}}^c = \mathcal{P}(c, T) \backslash \mathcal{P}$ denote the c-complement of $\mathcal{P}$.*

Let $\mathcal{A} \cup \mathcal{A}'$, $\mathcal{A} \cap \mathcal{A}'$ and $\overline{\mathcal{A}}^c$ be the slice automata whose slice languages are respectively $\mathcal{L}(\mathcal{A}) \cup \mathcal{L}(\mathcal{A}')$, $\mathcal{L}(\mathcal{A}) \cap \mathcal{L}(\mathcal{A}')$ and $\mathcal{L}(\mathcal{A}(c, T)) \backslash \mathcal{L}(\mathcal{A})$. Lemma 4.3 below states that operations performed on saturated, transitively reduced slice automata are reflected on the partial order languages they represent.

Lemma 4.3 (Properties of Saturated Slice Languages). *Let $\mathcal{A}$ and $\mathcal{A}'$ be two transitively-reduced slice automata over $\overrightarrow{\Sigma}(c, T)$. Assume that $\mathcal{A}$ is saturated.*

1. *$\mathcal{L}_{po}(\mathcal{A} \cup \mathcal{A}') = \mathcal{L}_{po}(\mathcal{A}) \cup \mathcal{L}_{po}(\mathcal{A}')$*
2. *$\mathcal{L}_{po}(\mathcal{A} \cap \mathcal{A}') = \mathcal{L}_{po}(\mathcal{A}) \cap \mathcal{L}_{po}(\mathcal{A}')$*
3. *$\mathcal{L}_{po}(\overline{\mathcal{A}}^c) = \overline{\mathcal{L}_{po}(\mathcal{A})}^c$.*
4. *$\mathcal{L}_{po}(\mathcal{A}) \subseteq \mathcal{L}_{po}(\mathcal{A}')$ if and only if $\mathcal{L}(\mathcal{A}) \subseteq \mathcal{L}(\mathcal{A}')$.*
5. *$\mathcal{L}_{po}(\mathcal{A}) \cap \mathcal{L}_{po}(\mathcal{A}') = \emptyset$ if and only if $\mathcal{L}(\mathcal{A}) \cap \mathcal{L}(\mathcal{A}') = \emptyset$.*
6. *If $\mathcal{A}'$ is saturated then $\mathcal{A} \cup \mathcal{A}'$ and $\mathcal{A} \cap \mathcal{A}'$ are also saturated.*

In other words, Lemma 4.3 implies that union, intersection and c-complementation of partial order languages represented by saturated, transitively reduced slice automata are computable, and inclusion and emptiness of intersection of these partial order languages are decidable. We note that Lemma 4.3 would *not* be true if some of the involved automata were not transitively reduced. This lemma would also not hold if none of the automata were saturated. To illustrate the role of saturation, let $\mathbf{U}$ and $\mathbf{U}'$ be two distinct unit decompositions of the same DAG H, and let $\mathcal{L} = \{\mathbf{U}\}$ and $\mathcal{L} = \{\mathbf{U}'\}$. Note that neither $\mathcal{L}$ nor $\mathcal{L}'$ is saturated. We have that $\mathcal{L}_{po} \cap \mathcal{L}'_{po} = \{\mathbf{tc}(H)\}$, but $\mathcal{L} \cap \mathcal{L}' = \emptyset$. Now, to illustrate the role of transitive reduction, let H and H' be two distinct DAGs representing the same partial order ℓ. In other words, $\ell = \boldsymbol{tc}(H) = \boldsymbol{tc}(H')$. Let $\mathcal{L} = \boldsymbol{ud}(H, (\overrightarrow{\Sigma}(c, T))$ be the set of all unit decompositions of H, and $\mathcal{L}' = \boldsymbol{ud}(H', (\overrightarrow{\Sigma}(c, T))$ be the set of all unit decompositions of H'. Note that both $\mathcal{L}$ and $\mathcal{L}'$ are saturated, but at least one of these languages is not transitively reduced, since the transitive reduction of a DAG is unique. Now we have that $\mathcal{L}_{po} \cap \mathcal{L}'_{po} = \{\ell\}$, but $\mathcal{L} \cap \mathcal{L}' = \emptyset$.

5   MSO-Definable Sets of c-Partial-Orders and Slice Languages

In this section we will show that MSO-definable sets of T-labeled c-partial-orders can be represented via saturated, transitively reduced slice automata over the slice alphabet $\overrightarrow{\Sigma}(c, T)$. More precisely, we will show that for each $c \in \mathbb{N}$, and each MSO formula φ, one can construct a slice automaton $\overrightarrow{\Sigma}(c, T)$ representing precisely the set of c-partial orders satisfying φ. This statement is formalized in Theorem 5.1 below, where $\mathcal{P}(c, T, \varphi)$ denotes the set of all T-labeled c-partial-orders satisfying φ.

Theorem 5.1. *Let φ be an MSO formula expressing a partial order property, Then one can construct a saturated transitively reduced slice automaton $\mathcal{A}^*(c, T, \varphi)$ such that $\mathcal{P}(c, T, \varphi) = \mathcal{L}_{po}(\mathcal{A}^*(c, T, \varphi))$.*

To prove Theorem 5.1 we will need to make a small detour. More precisely, we will first define the monadic second order logic of directed acyclic graphs with edge set quantifications, or MSO_2 logic for short. Subsequently, in Lemma 5.2, we will show that given an MSO_2 formula φ one can construct a saturated, transitively reduced slice automaton whose slice language consists of all unit decompositions that yield a graph satisfying φ. To show that these automata are sufficient to represent the set of all c-partial orders satisfying φ, we will prove a simple but crucial proposition stating that all unit decompositions of a DAG that can be covered by c-paths have width at most c. In other words, all unit decompositions of the Hasse diagram of a c-partial-order have width at most c.

We will represent a general DAG G by a relational structure $G = (V, E, s, t, l)$ where V is a set of vertices, E a set of edges, $s, t \subseteq E \times V$ are respectively the source and target relations, $l \subseteq V \times T$ is a vertex labeling relation, where T is a finite set of symbols. If e is an edge in E and v is a vertex in V then $s(e, v)$ is true if v is the source of e and $t(e, v)$ is true if v is the target of e. If $v \in V$ and $a \in T$ then $l(v, a)$ is true if v is labeled with a. First-order variables representing individual vertices will be taken from the set $\{x_1, x_2, ...\}$ and first order variables representing edges, from the set $\{y_1, y_2, ...\}$. Second order variables representing sets of vertices will be taken from the set $\{X_1, X_2, ...\}$ and second order variables representing sets of edges, from the set $\{Y_1, Y_2, ...\}$. The set of MSO_2 formulas is the smallest set of formulas containing:

- the atomic formulas $x_i \in X_j$, $y_i \in Y_j$, $s(y_i, x_j)$, $t(y_i, x_j)$, $l(x_i, a)$ for each $i, j \in \mathbb{N}$ and $a \in T$,

– the formulas $\varphi \wedge \psi$, $\varphi \vee \psi$, $\neg\varphi$, $\exists x_i.\varphi(x_i)$ and $\exists X_i.\varphi(X_i)$, $\exists y_i.\varphi(Y_i)$ and $\exists Y_i.\varphi(Y_i)$, where φ and ψ are MSO_2 formulas.

An MSO_2 sentence is a formula φ without free variables. If φ is a sentence, then we denote by $G \models \varphi$ the fact that G satisfies φ. Lemma 5.2 below states that the set of all unit decompositions $\mathbf{U}$ over a slice alphabet $\overrightarrow{\Sigma}(c,T)$ whose graph $\mathring{\mathbf{U}}$ satisfies a given MSO_2 sentence can be represented by a saturated slice automaton.

Lemma 5.2 (*[14]*). *Given a MSO_2 formula φ, one can effectively construct a slice automaton $\mathcal{A}(c,T,\varphi)$ over $\overrightarrow{\Sigma}(c,T)$ such that*

$$\mathcal{L}(\mathcal{A}(c,T,\varphi)) = \{\mathbf{U} \in \mathcal{L}(\overrightarrow{\Sigma}(c,T)) \mid \mathring{\mathbf{U}} \models \varphi\}.$$

We recall that Courcelle's celebrated model checking theorem states that MSO_2 properties can be model-checked in linear time on graphs of constant treewidth [9]. Since graphs of constant slice-width also have constant treewidth, Lemma 5.2 can be regarded as a special case of Courcelle's theorem transposed to the slice theoretic framework. This transposition is necessary due to a matter of compatibility with our results connecting slice languages to the partial order behaviour of bounded Petri nets established in the subsequent section. Additionally, some statements concerning c-partial orders, such as Proposition 5.3 which is crucial for our results, do not have a direct analog in the context of treewidth. The proof of Lemma 5.2 follows by reducing the model checking problem for graphs of constant slicewidth[3] to the model checking problem for unit decompositions. In other words, we will translate an MSO_2 formula on the vocabulary of graphs to a MSO formula φ' on the vocabulary of unit decompositions in such a way that a graph G satisfies φ if and only if each unit decomposition $\mathbf{U}$ of G satisfies the formula φ. Once this translation has been done, we can construct the automaton $\mathcal{A}(c,T,\varphi)$ using an approach that is similar to the construction of finite automata from MSO formulas over strings which is reminiscent to the proof of Büchi's celebrated result stating that the set of strings satisfying a MSO formula is regular [8,15]. It is worth mentioning that our construction shares similarities with constructions that can be found in [18,22]. Proposition 5.3 below establishes a correspondence between c-coverable DAGs and their sets of unit decompositions.

Proposition 5.3 *Let H be a DAG. If H can be covered by c paths, then any unit decomposition of H has width at most c.*

We let $\gamma(c)$ be the MSO_2 sentence which is true on a DAG H whenever H can be covered by c paths. Then we have that $\mathcal{L}(\mathcal{A}(c,T,\varphi \wedge \gamma(c)))$ is the set of all unit decompositions in $\mathcal{L}(\mathcal{A}(c,T,\varphi))$ whose corresponding DAG can be covered by c-paths.

---

[3] A DAG has slicewidth c if it can be decomposed into a unit decomposition over $\overrightarrow{\Sigma}(c,T)$.

Lemma 5.4 *For any MSO_2 formula φ and any positive integer $c \in \mathbb{N}$, the slice automaton $\mathcal{A}(c, T, \varphi \wedge \gamma(c))$ is saturated.*

Recall that if H is a DAG, then $tr(H)$ denotes the transitive reduction of H.

Proposition 5.5 (Partial Orders vs Hasse Diagrams). *For any MSO formula φ expressing a partial order property, there is an MSO_2 formula φ^{gr} expressing a property of DAGs such that for any partial order $\ell \in \mathcal{P}(c, T)$, $\ell \models \varphi$ if and only if $tr(\ell) \models \varphi^{gr}$.*

Let $c \in \mathbb{N}$, T be a finite set, and φ be a MSO formula. We denote by $\mathcal{P}(c, T, \varphi)$ the set of all c-partial orders satisfying φ whose vertices are labeled with elements from T. We denote by ρ be the MSO_2 formula which is true on a DAG H whenever H is transitively reduced, i.e., whenever $H = tr(H)$.

Proof of Theorem 5.1. Let ρ be the MSO_2 formula which is true in a DAG H whenever H is transitively reduced. Let φ^{gr} be the formula obtained from φ as in Proposition 5.5. Then we have that a DAG H satisfies $\gamma(c) \wedge \rho \wedge \varphi^{gr}$ if and only if H can be covered by c paths, H is transitively reduced and if the partial order $tc(H)$ induced by H satisfies φ. By Lemma 5.4, the slice language of the automaton $\mathcal{A}(c, T, \varphi \wedge \rho \wedge \gamma(c))$ is saturated, regular, and consists precisely of the unit decompositions yielding a graph satisfying $\varphi \wedge \rho \wedge \psi(c)$. Thus we just need to set $\mathcal{A}^*(c, T, \varphi) = \mathcal{A}(c, T, \varphi \wedge \rho \wedge \gamma(c))$. □

6 c-Partial-Orders and Petri Nets

In this section we will connect the notion of c-partial-order with bounded Petri nets. We start by stating Theorem 6.1 which says that the set of all c-partial-orders derived from processes of any bounded Petri net can be effectively represented by a transitively reduced, saturated slice automaton. In other words, slice automata provide us with a suitable way of representing and manipulating sets of partial orders associated with bounded Petri nets. Recall that $\mathcal{P}_{cau}(N)$ denotes the set of all causal orders corresponding to processes of N. We denote by $\mathcal{P}_{cau}(N, c)$ the set of all c-partial orders in $\mathcal{P}_{cau}(N)$.

Theorem 6.1 (Expressibility). *Let $N = (P, T, W, \mathfrak{m}_0)$ be a b-bounded Petri net. Then one can construct in time $2^{O(|P| \cdot c \cdot \log b \cdot c)} \cdot |T|^{|P|}$ a saturated transitively reduced slice automaton $\mathcal{A}(N, c)$ over $\overrightarrow{\Sigma}(c, T)$ such that $\mathcal{L}_{po}(\mathcal{A}(N, c)) = \mathcal{P}_{cau}(N, c)$.*

We note that Theorem 6.1 is a substantial refinement of a theorem proved in [12] stating that the full set of causal-orders of a Petri net can be effectively represented via slice automata. More precisely, Theorem 6.1 can be regarded as a parameterized version of this result which allows us to represent only the partial orders in the causal behaviour of a net up to a fixed width c. The following observation states that the representation of the full causal behaviour of a b-bounded Petri net may be achieved by setting $c = b \cdot |P|$.

Observation 1 *Let $N = (P, T)$ be a b-bounded Petri net and let ℓ be a causal order in $\mathcal{P}(N)$. Then ℓ is a c-partial order for some $c \leq b \cdot |P|$.*

Therefore, if we set $c = b \cdot |P|$ we have that $\mathcal{P}(N, c) = \mathcal{P}(N)$. The proof of Theorem 6.1 has two parts. The first part, is a characterization of Hasse diagrams arising from Petri-net partial orders in terms of interlaced flows, a notion introduced in [12]. The second part is a filtering process that takes a transitively reduced slice automaton $\mathcal{A}$ over $\vec{\Sigma}(c, T)$ and returns another transitively reduced slice automaton $\mathcal{F}(\mathcal{A}, N, c)$ over $\vec{\Sigma}(c, T)$ representing only those Hasse diagrams in $\mathcal{L}_{\mathcal{G}}(\mathcal{A})$ that admit a set of interlaced flows. Since interlaced flows are a certificate to the fact that a Hasse diagram arises from a partial order of the net, we have that $\mathcal{F}(\mathcal{A}, N, c)$ represents precisely those causal orders of N that are in $\mathcal{L}_{po}(\mathcal{A})$.

Below we will consider the notion of interlaced flows. If H is the Hasse diagram of a causal-order, then an interlaced flow is a four-tuple $f = (\boldsymbol{bb}, \boldsymbol{bf}, \boldsymbol{pb}, \boldsymbol{pf})$ of functions of type $E \rightarrow \mathbb{N}$ defined over the edges of H which is used to keep track of the way in which tokens of a Petri net are consumed and produced along a partially ordered run. Intuitively, for each $e \in E$, the value $\boldsymbol{bb}(e)$ counts the total of tokens produced **by** e^s and consumed **by** e^t. The value $\boldsymbol{pb}(e)$ keeps track of some tokens produced in the **p**ast of the node e^s and consumed **by** the node e^t. The value $\boldsymbol{pf}(e)$ keeps track of some of the tokens produced in the **p**ast of e^s and consumed in the **f**uture of e^t, and finally, $\boldsymbol{bf}(e)$ keeps track of some tokens produced **by** e^s and consumed in the **f**uture of e^t.

Definition 6.2 (p-interlaced Flow). *Let $N = (P, T, W, \mathfrak{m}_0)$ be a p/t-net, $H = (V, E, l)$ a Hasse diagram with $l : V \rightarrow T$ and $p \in P$ a place of N. A p-interlaced flow is a four-tuple $f = (\boldsymbol{bb}, \boldsymbol{bf}, \boldsymbol{pb}, \boldsymbol{pf})$ of functions of type $E \rightarrow \mathbb{N}$ which satisfies the following equations around each internal vertex of H:*

1. $\forall v \in V, \sum_{e_1^t = v} \boldsymbol{bf}(e_1) + \boldsymbol{pf}(e_1) = \sum_{e_2^s = v} \boldsymbol{pb}(e_2) + \boldsymbol{pf}(e_2)$.
2. $\forall v \in V \setminus \{v_\varepsilon\}, \; In(v) = \sum_{e^t = v} \boldsymbol{bb}(e) + \boldsymbol{pb}(e) = W(p, l(v))$.
3. $\forall v \in V, \; Out(v) = \sum_{e^s = v} \boldsymbol{bb}(e) + \boldsymbol{bf}(e) = W(l(v), p)$.

Intuitively, Item 6.2.1 states that on a p-interlaced-flow, the total number of tokens produced in the past of a vertex v, that arrives at it without being consumed, will eventually be consumed in the future of v. Item 6.2.2 says that the set of all tokens produced in the past of v and consumed by v is equal to the number of tokens consumed by the transition $l(v)$ on the place p. Similarly, Item 6.2.3 says that the set of all tokens produced by v and consumed at the future of v is equal to the number of tokens produced by the transition $l(v)$ on the place p. The following theorem characterizes causal orders of Petri nets in terms of sets of interlaced flows.

Theorem 6.3. (Interlaced Flow Theorem [12]). *Let $N = (P, T, W, \mathfrak{m}_0)$ be a b-bounded p/t-net and $H = (V, E, l)$ be a Hasse diagram. Then the partial order induced by H is a causal order of N if and only if there exists a set $\{f_p\}_{p \in P}$ of p-interlaced flows on H such that for every edge e of H, the component $\boldsymbol{bb}_p(e)$ of $f_p(e)$, which denotes the direct transmission of tokens, is strictly greater than zero for at least one $p \in P$.*

Intuitively, each flow f_p keeps track of the production and consumption of tokens at place p according to the transitions labeling nodes of H. Each edge e of H indicates a direct causal dependence between the event corresponding to the node e^s and the event corresponding to the node e^t. In other words, there must exist at least one place p such that the component $\boldsymbol{bb}_p(e)$ corresponding to the number of tokens created by the transition $l(e^s)$ and consumed by the transition $l(e^t)$ must be strictly greater than 0.

Lemma 6.4. *Let $\mathcal{A}$ be a transitively reduced slice automaton over $\overrightarrow{\Sigma}(c, T)$, let N be a b-bounded Petri net, and $c \in \mathbb{N}$. Then one can construct a transitively reduced slice automaton $\mathcal{F}(\mathcal{A}, N, c)$ such that*

$$\mathcal{L}_{\mathcal{G}}(\mathcal{F}(\mathcal{A}, N, c)) = \{H \in \mathcal{L}_{\mathcal{G}}(\mathcal{A}) \mid H \text{ admits an } P\text{-interlaced flow.}\}. \qquad (3)$$

Additionally, if $\mathcal{A}$ is saturated, so is $\mathcal{F}(\mathcal{A}, N, c)$.

In particular, if $\mathcal{A}$ is the saturated slice automaton representing the set $\mathcal{P}(c, T)$ of all c-partial orders, then the automaton $\mathcal{F}(\mathcal{A}, N, c)$ is a saturated slice automaton representing all c-partial-orders of N. Therefore, Theorem 6.1 follows from Lemma 6.4.

By combining Theorem 5.1 with Theorem 6.1, we have an algorithm for determining whether all c-partial orders of a bounded Petri net N satisfy a given MSO formula φ. The converse also holds. In other words, we can also determine effectively whether all c-partial-orders satisfying φ belong to the causal behavior of N. These statements are formalized in Theorem 6.5 below.

Theorem 6.5. (Verification). *Let φ be an MSO formula, N be a b-bounded Petri net and $c \in \mathbb{N}$.*

  i) One may effectively determine whether $\mathcal{P}_{cau}(N, c) \subseteq \mathcal{P}(c, T, \varphi)$.
 ii) One may effectively determine whether $\mathcal{P}_{cau}(N, c) \cap \mathcal{P}(c, T, \varphi) = \emptyset$.
iii) One may effectively determine whether $\mathcal{P}(c, T, \varphi) \subseteq \mathcal{P}_{cau}(N, c)$.

In particular, if we set $c = b \cdot |P|$ then by Observation 1 we can use Theorem 6.5.i to model check the whole causal behaviour of any bounded Petri net against MSO specifications. As mentioned in the introduction, previously, such a result had only been known for the restricted case of pure Petri nets. As we will see in Sect. 7, Theorem 6.5.ii and Theorem 6.5.iii will allow us to construct Petri nets whose causal language separates MSO definable sets of partial orders.

6.1 Synthesis

In this subsection we consider the problem of automatically constructing Petri nets whose behavior consists precisely of the set of partial orders specified by a given MSO formula. In other words, we will consider the synthesis of Petri nets from MSO specifications. Indeed we will provide an algorithm for a more robust version of the synthesis problem, in the sense that if there is no p/t-net of a particular type whose causal behavior equals the set of partial orders $\mathcal{P}$ specified by φ, then our algorithm will construct a Petri net whose behaviour minimally over-approximates $\mathcal{P}$. This notion of minimality is formalized below in Definition 6.6. First, we argue that when dealing with the synthesis of bounded Petri-nets with the causal semantics, it makes sense to consider repeated places.

Let $N = (P, T, W, \mathfrak{m}_0)$ be a b-bounded Petri net and let p be a place of N. We let $p^- : T \to \mathbb{N}$ and $p^+ : T \to \mathbb{N}$ be functions that specify respectively how many tokens each transition t takes from p and produces into p respectively. In other words, $p^-(t) = W(p, t)$ and $p^+(t) = W(t, p)$ for each transition $t \in T$. We define the type of the place p is as the triple $(m_0(p), p^-, p^+)$. Two places p_1 and p_2 are said to be identical if they have the same type. Intuitively, two places are identical if and only if they have identical initial marking and identical flow relations. As we can see in Fig. 6, adding repeated places may change the causal behavior of a Petri net. We contrast this observation with the fact that adding repeated places has no effect if we consider other partial order semantics, such as the execution semantics introduced in [23]. Therefore, in our synthesis problems we will consider two parameters for a matter of flexibility, the bound b of a Petri net, and the maximum number r of identical places of each type. We say that a net N is (b, r)-bounded if N is b-bounded and there are at most r places of each given type.

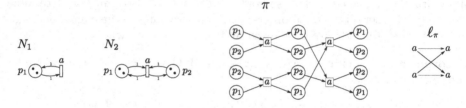

Fig. 6. Two Petri nets N_1 and N_2. The places p_1 and p_2 are repeated. π is a process of N_2. The circles are conditions, the squares are events. ℓ_π is the causal order of π. Note that ℓ_π does not belong to the causal language of N_1. This shows that repeated places do change the causal behavior of a Petri net.

Definition 6.6 (c-causally-minimial). *We say that a (b, r)-bounded Petri net N is c-causally-minimal for a partial order language $\mathcal{P}$, if $\mathcal{P} \subseteq \mathcal{P}_{cau}(N, c)$ and if there is no other (b, r)-bounded Petri net N' with $\mathcal{P} \subseteq \mathcal{P}_{cau}(N', c) \subsetneq \mathcal{P}_{cau}(N, c)$.*

Note that the minimality of N is defined with respect to inclusion of causal languages. Intuitively, N is c-causally-minimal for $\mathcal{P}$ if N is the (b, r)-bounded Petri net whose causal language $\mathcal{P}_{cau}(N, c)$ contains as few elements outside from $\mathcal{P}$ as possible. Our next theorem states that given a MSO formula φ and positive integers c, b, r, one can automatically construct a (b, r)-bounded Petri net that is c-causally minimal for the set of c-partial orders specified by φ.

Theorem 6.7 (Synthesis). *Let φ be a MSO formula on the vocabulary of T-labeled partial orders. Then for any $c, b, r \in \mathbb{N}$ one may automatically determine whether there exists a (b, r)-bounded Petri net N such that $\mathcal{P}(c, T, \varphi) \subseteq \mathcal{P}_{cau}(N, c)$. If such a net exists, one can construct a (b, r)-bounded Petri net that is c-causally-minimal for $\mathcal{P}(c, T, \varphi)$.*

We observe that a minimal net may not be unique. In other words there may exist two Petri nets N_1 and N_2 whose behavior is c-causally-minimal for $\mathcal{P}$, but for which $\mathcal{P}(N_1, c) \neq \mathcal{P}(N_2, c)$. The algorithm solving the synthesis problem is able to list all possible (b, r)-bounded nets that are c-causally-minimal for $\mathcal{P}$.

It is worth comparing Theorem 6.7 with existing literature. In the context of the sequential (or interleaving) semantics of Petri nets the synthesis problem has been considered with respect to many formalisms. In particular, the synthesis from regular sets has been studied extensively in [2,3,10,11] via a set of combinatorial techniques called *theory of regions*. Thus, using the Büchi-Elgot Theorem stating that MSO Logic over strings is as expressive as regular languages [8], the theory of regions can be used to synthesize nets whose interleaving behavior satisfies a given MSO formula over *strings*. The synthesis of Petri nets from infinite sets of partial orders has been considered in [5] but with another partial order semantics, the execution semantics, which is not aimed at representing causality between events, but rather the order of executions of these events (without any causal implication). However with respect to this partial order semantics the synthesis from logical specifications has also not been considered.

7 Behavioural Design and Correction

In this section we apply Theorems 6.1 and 6.7 to establish the decidability of three problems related to the behavioural specification of Petri nets with causal semantics.

7.1 Optimally Correcting Subsystem

Let φ be an MSO formula specifying a set of safe behaviors, and let N be a (b, r)-bounded Petri net whose behaviour $\mathcal{P}_{cau}(N)$ contains some faulty partial order. In the next theorem (Theorem 7.1) we will show that we may be able to fix N by automatically synthesizing the best (b, r)-bounded Petri net N' whose partial order behavior lies in between $\mathcal{P}_{cau}(N, c) \cap \mathcal{P}(c, T, \varphi)$ and $\mathcal{P}_{cau}(N, c)$. In other words, the partial order behavior of N' is a subset of the partial order behavior of N which preserves all safe runs of N. Additionally, the partial order behavior of N' has as few unsafe partial-order runs as possible. We say that N' is an *optimally correcting subsystem* of N with respect to φ. We notice that the net N' does not need to be a sub-net of N, and indeed N' can have even more places than N. Only the behavior of N' is guaranteed to be a subset of the behavior of N.

Theorem 7.1 (Optimally Correcting Subsystem). *Let $c, b, r \in \mathbb{N}$. Given a (b, r)-bounded Petri net $N = (P, T, W, \mathfrak{m}_0)$ and an MSO formula φ, we may automatically synthesize a (b, r)-bounded Petri net N' such that*

i) N' *is c-causally-minimal for* $\mathcal{P}(c, T, \varphi) \cap \mathcal{P}_{cau}(N, c)$,
ii) $\mathcal{P}_{cau}(N', c) \subseteq \mathcal{P}_{cau}(N, c)$.

The notion of *optimally correcting subsystem* is appropriate for three reasons. First, as mentioned above, we have that

$$\mathcal{P}(c, T, \varphi) \cap \mathcal{P}_{cau}(N, c) \subseteq \mathcal{P}(N', c) \subseteq \mathcal{P}(N, c).$$

Second, the minimality condition says that if there is a (b, r)-bounded Petri net N' whose c-partial-order behavior precisely matches $\mathcal{P}_{cau}(N, c) \cap \mathcal{P}(c, T, \varphi)$ then such a Petri net will be returned. In this case, our synthesis algorithm completely corrects the original Petri net. Finally, if all c-partially-ordered runs of N indeed satisfy φ, then our synthesis algorithm returns a net N' satisfying $\mathcal{P}_{cau}(N', c) = \mathcal{P}_{cau}(N, c)$. Thus the set of c-partial order behaviors of the synthesized net does not change if the original net is already correct (although the structure of the net per si may change). In Subsect. 7.2 below we consider a related problem that finds analogies with the field of automatic program repair.

7.2 Behavioral Repair

In this section we consider the notion of automated repair of Petri nets with the partial order semantics. Given MSO formulas φ and ψ and a Petri net N, one is asked to automatically synthesize a Petri net N' whose partial order behavior is lower-bounded by $\mathcal{P}_{cau}(c, T, \varphi) \cap \mathcal{P}_{cau}(N, c)$ and upper bounded by $\mathcal{P}_{cau}(c, T\psi)$. A similar notion of repair has been considered in the context of reactive systems, except for the fact that reactive systems with a sequential semantics are used instead of Petri nets, and LTL formulas are used instead of MSO formulas [24]. In both cases, the intuition is that while φ specifies a set of correct behaviors that should be preserved whenever present in the original system, the formula ψ specifies a set of behaviors that are allowed to be present in the repaired system.

Theorem 7.2 (Behavioral Repair). *Let $c, b, r \in \mathbb{N}$. Given a (b, r)-bounded Petri net $N = (P, T, W, \mathfrak{m}_0)$ and an MSO formula φ, we may automatically determine whether there exists a (b, r)-bounded Petri net N' such that*

(i) N' *is c-causally-minimal for* $\mathcal{P}(c, T, \varphi) \cap \mathcal{P}_{cau}(N, c)$,
(ii) $\mathcal{P}_{cau}(N', c) \subseteq \mathcal{P}_{cau}(c, T, \psi)$.

In case such a net exists, one may automatically construct it.

Note that while Theorems 7.2.i and 7.2.ii imply that $\mathcal{P}(c, T, \varphi) \cap \mathcal{P}_{cau}(N, c) \subseteq \mathcal{P}(N', c) \subseteq \mathcal{P}(c, T, \psi)$, the minimality condition in Theorem 7.2.i implies that if N' is successfully synthesized, then its behavior has as few partial-order runs contradicting φ as possible.

7.3 Synthesis from Partial Order Contracts

Suppose that we are in the early stages of development of a concurrent system. We have arrived to the conclusion that every behavior satisfying a given MSO formula φ^{yes} should be present in the system, but that no behavior in the system should satisfy a formula φ^{no}. Clearly we require that $\mathcal{P}(\varphi^{yes}) \cap \mathcal{P}(\varphi^{no}) = \emptyset$. We say that the pair $(\varphi^{yes}, \varphi^{no})$ is a partial order contract. We can try to develop a first prototype of our system by automatically synthesizing a (b, r)-bounded Petri net N containing all c-partial orders specified by φ^{yes} but no partial order in φ^{no}. The next theorem says that if such a net exists, then it can be automatically constructed.

Theorem 7.3 (Synthesis from Contracts). *Let φ^{yes} and φ^{no} be MSO formulas such that $\mathcal{P}(c, T, \varphi^{yes}) \cap \mathcal{P}(c, T, \varphi^{no}) = \emptyset$. Then one may automatically determine whether there exists a (b, r)-bounded Petri net N such that $\mathcal{P}(c, T, \varphi^{yes})$ $\subseteq \mathcal{P}_{cau}(N, c)$ and $\mathcal{P}(c, T, \varphi^{no}) \cap \mathcal{P}_{cau}(N, c) = \emptyset$. In case such a net exists one may construct it.*

8   Conclusion

In this work we showed that the model checking of the partial order behavior of arbitrary bounded Petri nets against MSO specifications is decidable. Previously, a similar result was only available in the context of pure, bounded Petri nets. It is interesting to note that our model checking result uses completely different techniques than those employed in [1]. We also showed that the reverse direction is decidable. Namely, we showed that one can determine whether the set of c-partial-orders satisfying a given MSO formula φ is included on the c-partial-order behavior of a given Petri net N. Finally, we combined these two results with the synthesis of Petri nets from MSO specification to establish the decidability of interesting problems in the realm of automated design. Namely, we showed how to compute optimally correcting subsystem for Petri nets, we defined a suitable notion of automated repair for Petri nets, and we established the decidability of the synthesis of Petri nets form MSO-definable partial order languages. This last result in particular, can be regarded as a partial order theoretic version of the notion of language separator [20] which has turned to be of fundamental importance in formal language theory.

Acknowledgements. The author gratefully acknowledges financial support from the European Research Council, ERC grant agreement 339691, within the context of the project Feasibility, Logic and Randomness (FEALORA).

References

1. Avellaneda, F., Morin, R.: Checking partial-order properties of vector addition systems with states. In: Proceedings of ACSD 2013, pp. 100–109. IEEE (2013)

2. Badouel, E., Darondeau, P.: On the synthesis of general Petri nets. Technical Report PI-1061, IRISA (1996)
3. Badouel, E., Darondeau, P.: Theory of regions. In: Reisig, W., Rozenberg, G. (eds.) Lectures on Petri Nets I: Basic Models. LNCS, vol. 1491, pp. 529–586. Springer, Heidelberg (1998)
4. Bauderon, M., Courcelle, B.: Graph expressions and graph rewritings. Math. Syst. Theor. **20**(2–3), 83–127 (1987)
5. Bergenthum, R., Desel, J., Lorenz, R., Mauser, S.: Synthesis of Petri nets from infinite partial languages. In: Proceedings of ACSD 2008, pp. 170–179. IEEE (2008)
6. Brandenburg, F.-J., Skodinis, K.: Finite graph automata for linear and boundary graph languages. Theor. Comput. Sci. **332**(1–3), 199–232 (2005)
7. Bruggink, H.S., König, B.: On the recognizability of arrow and graph languages. In: Ehrig, H., Heckel, R., Rozenberg, G., Taentzer, G. (eds.) Graph Transformations. LNCS, vol. 5214, pp. 336–350. Springer, Heidelberg (2008)
8. Büchi, J.R.: Weak second order arithmetic and finite automata. Z. Math. Logik Grundl. Math. **6**, 66–92 (1960)
9. Courcelle, B.: The monadic second-order logic of graphs I. Recognizable sets of finite graphs. Inf. Comput. **85**(1), 12–75 (1990)
10. Darondeau, P.: Deriving unbounded Petri nets from formal languages. In: Sangiorgi, D., de Simone, R. (eds.) CONCUR'98 Concurrency Theory. LNCS, vol. 1466, pp. 533–548. Springer, Heidelberg (1998)
11. Darondeau, P.: Region based synthesis of P/T-nets and its potential applications. In: Nielsen, M., Simpson, D. (eds.) ICATPN 2000. LNCS, vol. 1825, p. 16. Springer, Heidelberg (2000)
12. de Oliveira Oliveira, M.: Hasse diagram generators and Petri nets. Fundam. Inf. **105**(3), 263–289 (2010)
13. de Oliveira Oliveira, M.: Canonizable partial order generators. In: Dediu, A.-H., Martín-Vide, C. (eds.) LATA 2012. LNCS, vol. 7183, pp. 445–457. Springer, Heidelberg (2012)
14. de Oliveira Oliveira, M.: Subgraphs satisfying MSO properties on z-topologically orderable digraphs. In: Gutin, G., Szeider, S. (eds.) IPEC 2013. LNCS, vol. 8246, pp. 123–136. Springer, Heidelberg (2013)
15. Elgot, C.C.: Decision problems of finite automata and related arithmetics. Trans. Am. Math. Soc. **98**, 21–52 (1961)
16. Engelfriet, J., et al.: Context-free graph grammars and concatenation of graphs. Acta Inf. **34**, 773–803 (1997)
17. Goltz, U., Reisig, W.: Processes of place/transition-nets. In: Diaz, J. (ed.) Automata, Languages and Programming. LNCS, vol. 154, pp. 264–277. Springer, Heidelberg (1983)
18. Madhusudan, P., Thiagarajan, P.S.: Distributed controller synthesis for local specifications. In: Orejas, F., Spirakis, P.G., van Leeuwen, J. (eds.) ICALP 2001. LNCS, vol. 2076, p. 396. Springer, Heidelberg (2001)
19. Petri, C.A.: Fundamentals of a theory of asynchronous information flow. Proc. of IFIP Congr. **62**, 166–168 (1962). Munchen
20. Place, T., Zeitoun, M.: Separating regular languages with first-order logic. In: Proceedings of CSL/LICS, p. 75. ACM (2014)
21. Thomas, W.: Finite-state recognizability of graph properties. Theor. des Automates et Appl. **172**, 147159 (1992)

22. Thomas, W.: Languages, automata, and logic. Handb. Formal Lang. Theor. **3**, 389–455 (1997)
23. Vogler, Walter (ed.): Modular Construction and Partial Order Semantics of Petri Nets. LNCS, vol. 625. Springer, Heidelberg (1992)
24. von Essen, C., Jobstmann, B.: Program repair without regret. In: Sharygina, N., Veith, H. (eds.) CAV 2013. LNCS, vol. 8044, pp. 896–911. Springer, Heidelberg (2013)

A Resource Aware Computational Interpretation for Herbelin's Syntax

Delia Kesner[1] and Daniel Ventura[2]([⊠])

[1] University Paris-Diderot, SPC, PPS, CNRS, Paris, France
kesner@pps.univ-paris-diderot.fr
[2] University Federal de Goiás, INF, Goiânia, Brazil
daniel@inf.ufg.br

Abstract. We investigate a new computational interpretation for an intuitionistic focused sequent calculus which is compatible with a resource aware semantics. For that, we associate to Herbelin's syntax a type system based on non-idempotent intersection types, together with a set of reduction rules –inspired from the *substitution at a distance* paradigm– that preserves (and decreases the size of) typing derivations. The non-idempotent approach allows us to use very simple combinatorial arguments, only based on this measure decreasingness, to characterize strongly normalizing terms by means of typability. For the sake of completeness, we also study typability (and the corresponding strong normalization characterization) in the reduction calculus obtained from the former one by projecting the explicit substitutions.

1 Introduction

Intuitionistic logic can be expressed in different formal systems such as natural deduction and sequent calculi. Equivalence between these two formal styles has been widely studied [19,21,41,43,49], *i.e.* every derivation in one system can be encoded into a derivation of the other one. However, this correspondence is not one-to-one, in particular several *cut-free* proofs in intuitionistic sequent calculus correspond to the same *normal* natural deduction derivation. This gives rise to a restriction of sequent calculi, the so-called *focused* sequent calculi [3], which preserves its structure and expressive power, while establishing a better relationship with natural deduction. Indeed, a one-to-one correspondence is achieved between the cut-free proofs in focused sequent calculi and the normal derivations in natural deduction.

In 1994 Herbelin [25] introduced the $\overline{\lambda}$-calculus, obtained by a computational interpretation of the focused sequent calculus for the minimal intuitionistic logic *LJT*. In contrast to the usual λ-calculus notation for natural deduction, $\overline{\lambda}$ notation brings head variables to the surface, treats sequences of arguments as lists, and encodes cuts with *explicit substitutions*. Its operational semantics is specified by means of a complete set of cut-elimination rules. The calculus is permutation-free and can be used to describe proof-search in pure Prolog and some of its extensions [36]. The reduction system of the $\overline{\lambda}$-calculus was then extended [17]

© Springer International Publishing Switzerland 2015
M. Leucker et al. (Eds.): ICTAC 2015, LNCS 9399, pp. 388–403, 2015.
DOI: 10.1007/978-3-319-25150-9_23

with permutation rules, thus showing how to model beta-reduction, *i.e.* thus giving a natural basis for implementation of functional languages.

Unfortunately, Herbelin's calculus is not compatible with a resource aware semantics, mainly because propagation of explicit cuts w.r.t. the *structure* of $\overline{\lambda}$-terms induces useless duplications of empty resources (*cf.* technical discussion in Sect. 3 after Corollary 1). This is substantiated when trying to interpret the λ-calculus by means of proof-nets [23] or non-idempotent intersection types (pioneered by [9, 28, 30]).

The first contribution of the paper is to propose a new computational interpretation for the focused intuitionistic sequent calculus *LJT*, called E-*calculus*, which is compatible with a resource aware semantics. The calculus keeps Herbelin's syntax but changes the operational semantics of $\overline{\lambda}$ to a resource-controlled interpretation, inspired from the structural lambda-calculus [2], and the linear substitution calculus [1, 37]. The terms of the E-calculus can be seen as λ-terms with explicit cuts of the form $t[x/u]$, where $[x/u]$ is propagated according to the number of free occurrence of x in t (and not w.r.t. the structure of terms). For the sake of completeness, we also study in the second part of the paper the I-*calculus*, a formalism using full –in contrast to partial– substitution, in which normal forms are exactly the same as those of the E-calculus, and whose reduction sequences are obtained by projecting E-reduction sequences into terms without explicit cuts. In other words, E-reduction implements the meta-level operators of the I-calculus by using a resource aware semantics specified by means of explicit reduction rules. Thus, the paper gives a self-contained study of calculi based on sequent calculus, completely independent from their isomorphic natural deduction counter-part.

The second contribution of the paper is to provide type systems based on non-idempotent intersection types for both E and I calculi. Intersection types were introduced to give characterizations of strong β-normalizing terms in the λ-calculus [11, 33, 42]; since then they have been used to characterize termination properties in a broader sense [13], as well as to construct models of the λ-calculus itself [6] Commonly, intersection types are *idempotent*, *i.e.* $\sigma \wedge \sigma = \sigma$, but we use here *non-idempotent* types [9, 28, 30], suitable to obtain *quantitative* information about reduction sequences. The non-idempotent type systems are used in this paper to characterize strongly normalizing terms, *i.e.* an I-term (resp. E-term) is typable if and only if it is strongly I-normalizing (resp. E-normalizing). Thanks to the non-idempotent approach, the characterization proofs use simple combinatorial arguments, and do not need any reducibility technique as required in the idempotent case. More precisely, in the case of the E-calculus, the characterization proof is based on the postponement of the erasing steps of the calculus and a combinatorial argument based on a *weighted* subject reduction property. In the case of the I-calculus, which does not admit postponement of erasing steps, a characterization of strongly normalizing I-terms is obtained from the characterization of strongly normalizing E-terms by a projection lemma.

Some Related Work: In the last years, there has been a growing interest in non-idempotent intersection types. The relation between the size of a non-idempotent intersection typing derivation and the head/weak-normalization execution time of lambda-terms by means of abstract machines was established by D. de Carvalho [16]. Non-idempotence is used to reason about the longest reduction sequence of strongly normalizing terms in both the lambda-calculus [7,15] and in different lambda-calculi with explicit substitutions [8,27]. Non-idempotent types also appear in linearization of the lambda-calculus [28], type inference [30,38], different characterizations of solvability [40] and verification of higher-order programs [39]. While the inhabitation problem for intersection types is known to be undecidable in the idempotent case [46], decidability was recently proved [10] through a sound and complete algorithm in the non-idempotent case. Concerning the use of *idempotent* intersection types for focused intuitionistic sequent calculi, two different papers [18,22] provide characterizations of strongly normalizing terms by means of typability, but none of them give quantitative information about reduction, as done in this paper. Moreover, in contrast to [22], which is based on explicit control operators for weakening and contraction, we keep the simple, original syntax of Herbelin.

The work presented in this paper originates from a first computational interpretation of LJT appearing in an unpublished technical report [26]. The approach in [26] gives an elegant formulation of the typing rules by introducing witness derivations *everywhere*, so that it is too costly and resource demanding (see the discussion at the end of Sect. 8). The type systems in this paper *only* require *witness* derivations for potentially erasable arguments of functions and substitutions. As a consequence, the upper bound for the longest reduction sequence of a strongly normalizing term obtained in this paper, represented just by the size of a typing derivation, is tighter than the one in [26].

Structure of the Paper: The explicit E-calculus is introduced in Sect. 2 and its associated typing system in Sect. 3. The characterization of strongly E-normalizing terms is developed in Sects. 4 and 5 presents the syntax and the operational semantics of the I-calculus, while Sect. 6 presents a non-idempotent typing system for I together with its properties. In Sect. 7 an inductive definition of strongly I-normalizing terms is used to complete the characterization result for the I-calculus. Finally, we conclude in Sect. 8.

2   The E-Calculus

This section introduces the syntax and the operational semantics of the E-calculus. The term language follows from [25], while the reduction rules aim to give a resource aware semantics based on the *substitution at a distance* paradigm [1,37].

Given a countable infinite set of symbols $x, y, z, \ldots$, three syntactic categories are defined as **E-objects**, **E-terms** and **E-lists**, respectively:

$$o, p := t \mid l \qquad t, u := xl \mid tl \mid \lambda x.t \mid t[x/t] \qquad l, m := \texttt{nil} \mid t; l$$

The construction $[x/u]$ is said to be an **explicit cut**. Remark that the symbol x alone is not an object of the syntax (term variables in natural deduction style would be encoded by x nil), and explicit cuts do not apply to lists, but only to terms, *i.e.* $l[x/u]$ is not in the grammar. We write $tl_1 \ldots l_n$ for $(\ldots(tl_1)\ldots l_n)$ and $x_{\mathtt{nil}}$ for x nil. The **size** of the object o is denoted by $|o|$.

The notions of **free** and **bound** variables are defined as usual, in particular,

$$
\begin{aligned}
\mathtt{fv}(xl) &:= \{x\} \cup \mathtt{fv}(l) & \mathtt{fv}(t[x/u]) &:= (\mathtt{fv}(t) \setminus \{x\}) \cup \mathtt{fv}(u) \\
\mathtt{fv}(tl) &:= \mathtt{fv}(t) \cup \mathtt{fv}(l) & \mathtt{fv}(\mathtt{nil}) &:= \emptyset \\
\mathtt{fv}(\lambda x.t) &:= \mathtt{fv}(t) \setminus \{x\} & \mathtt{fv}(t;l) &:= \mathtt{fv}(t) \cup \mathtt{fv}(l)
\end{aligned}
$$

The **number of free occurrences of** x in o is written $|o|_x$. We work with the standard notions of α-conversion (*i.e.* renaming of bound variables for abstractions and substitutions), and Barendregt's convention [5].

We also consider two categories of E-contexts:

$$
\mathtt{L} ::= \Box \mid \mathtt{L}[x/t] \qquad\qquad
\begin{aligned}
\mathtt{O},\mathtt{P} &::= \mathtt{C} \mid \mathtt{V} \\
\mathtt{C},\mathtt{D} &::= \Box \mid x\mathtt{V} \mid \mathtt{C}l \mid \lambda y.\mathtt{C} \mid \mathtt{C}[y/u] \mid t[y/\mathtt{C}] \mid t\mathtt{V} \\
\mathtt{V},\mathtt{U} &::= \mathtt{C}; l \mid t; \mathtt{V}
\end{aligned}
$$

When the replacement of the hole of $\mathtt{O}$ by the object o is well defined (*i.e.* gives an object), then we denote it by $\mathtt{O}[o]$. Similarly, $\mathtt{L}[t]$ denotes the term obtained by replacing the hole of $\mathtt{L}$ by the term t. We write $\mathtt{C}^x$ for a context $\mathtt{C}$ which does not capture the free variable x, i.e. there are no abstractions or explicit substitutions in the context that binds the variable x. For instance, $C = \lambda y.\Box$ can be specified as $\mathtt{C}^x$ while $C = \lambda x.\Box$ cannot. In order to emphasize this particular property we write $\mathtt{C}^x[\![t]\!]$ instead of $\mathtt{C}^x[t]$, and we may omit x when it is clear from the context.

The **reduction relation** $\to_E$ is defined as the closure by contexts $\mathtt{O}$ of the following rewriting rules:

$$
\begin{aligned}
\mathtt{L}[\lambda x.t]\mathtt{nil} &\mapsto_{\mathrm{dB_{nil}}} \mathtt{L}[\lambda x.t] \\
\mathtt{L}[\lambda x.t](u;l) &\mapsto_{\mathrm{dB_{cons}}} \mathtt{L}[t[x/u]l] \\
\mathtt{C}^x[\![x\,l]\!][x/u] &\mapsto_{\mathrm{c}} \mathtt{C}^x[\![u\,l]\!][x/u] && \text{if } |\mathtt{C}^x[\![x\,l]\!]|_x > 1 \\
\mathtt{C}^x[\![x\,l]\!][x/u] &\mapsto_{\mathrm{d}} \mathtt{C}^x[\![u\,l]\!] && \text{if } |\mathtt{C}^x[\![x\,l]\!]|_x = 1 \\
t[x/u] &\mapsto_{\mathrm{w}} t && \text{if } |t|_x = 0 \\
\mathtt{L}[xl]m &\mapsto_{@_{\mathrm{var}}} \mathtt{L}[x(l@m)] \\
\mathtt{L}[tl]m &\mapsto_{@_{\mathrm{app}}} \mathtt{L}[t(l@m)]
\end{aligned}
$$

where the operation $\_@\_$ is defined by the following equations:

$$
\mathtt{nil}@l := l \qquad\qquad (u;l)@m := u;(l@m)
$$

An example of reduction sequence is

$$
\begin{aligned}
(\lambda x.x(x_{\mathtt{nil}};\mathtt{nil}))(u;\mathtt{nil}) &\to_{\mathrm{dB_{cons}}} x(x_{\mathtt{nil}};\mathtt{nil})[x/u]\mathtt{nil} \to_{@_{\mathrm{var}}} \\
x(x_{\mathtt{nil}};\mathtt{nil})[x/u] &\to_{\mathrm{c}} x(u\,\mathtt{nil};\mathtt{nil})[x/u] \to_{\mathrm{d}} u(u\,\mathtt{nil};\mathtt{nil})
\end{aligned}
$$

There are many differences with the reduction rules in [25]. First of all, the use of the meta-operation @ for concatenating lists in the rules $\mapsto_{@_{\mathrm{var}}}$ and $\mapsto_{@_{\mathrm{app}}}$ replaces the explicit concatenation rules in [25]. This is particularly convenient since we only reduce objects that are terms (even if these terms occur inside lists), so that the proofs are simpler/shorter because there are less rules and only of one kind. A major difference with [25] is the use of rules *at a distance*, specified by means of (term and list) contexts, where the propagation of substitutions is not performed by structural induction on terms, since they are consumed according to the multiplicity of their corresponding variables. As a consequence, the behaviour of substitution is specified by a resource aware semantics, thus preventing the useless duplication of empty resources, which happens in [25] when using reduction steps of the form $(tl)[x/u] \to t[x/u]l[x/u]$, where $|tl|_x = 0$. This is particularly unsuitable when considering non-idempotent types (*cf.* the discussion at the end of Sect. 3). In contrast to other calculi at a distance which only contains w and c-rules, such as for example the linear substitution calculus [1,37], we also consider here a dereliction rule d. This is appropriate to obtain a *weighted* subject reduction property relative to our typing system (*cf.* Sect. 3), which would fail for the alternative rewriting rule $C[\![x\,l]\!][x/u] \mapsto_c C[\![u\,l]\!][x/u]$ when $|C[\![x\,l]\!]|_x = 1$.

The reduction relation $\to_E$ can also be refined. We write $\to_X$ for the closure by contexts O of the rewriting rule $\mapsto_X$ for every X. We define $B@ := \{dB_{nil} \cup dB_{cons} \cup @_{var} \cup @_{app}\}$ and $\to_{B@} := \bigcup_{X \in B@} \to_X$. The **non-erasing** reduction relation $\to_{E\backslash w}$ is given by $\to_{B@\cup\{d,c\}}$, *i.e.* $\to_{E\backslash w} = \to_E \backslash \to_w$, and plays a key role in the characterization of strongly E-normalizing terms (*cf.* Sect. 4).

Let $\mathcal{R}$ be any reduction system. We denote by $\to_{\mathcal{R}}^*$ (resp. $\to_{\mathcal{R}}^+$) the **reflexive-transitive** (resp. **transitive**) closure of a given reduction relation $\to_{\mathcal{R}}$. The reduction relation $\mathcal{R}$ is **confluent** if and only if for all objects o_1, o_2, o_3 such that $o_1 \to_{\mathcal{R}}^* o_2$ and $o_1 \to_{\mathcal{R}}^* o_3$, there is o_4 being able to close the diagram, *i.e.* $o_2 \to_{\mathcal{R}}^* o_4$ and $o_3 \to_{\mathcal{R}}^* o_4$. An object o is **strongly** $\mathcal{R}$-normalizing, written $o \in \mathcal{SN}(\mathcal{R})$, if there is no infinite $\mathcal{R}$-reduction sequence starting at o, and o is $\mathcal{R}$-finitely branching if the set $\{o' \mid o \to_{\mathcal{R}} o'\}$ is finite. If an object o is $\mathcal{R}$-strongly normalizing and $\mathcal{R}$-finitely branching then the **depth of** o, written $\eta_{\mathcal{R}}(o)$, is the maximal length of $\mathcal{R}$-reduction sequences starting at o.

3   A Non-idempotent Typing System for E-Terms

This section introduces the typing system $\mathcal{Q}_E$ for the E-calculus. Given a countable infinite set of *base types* $\alpha, \beta, \gamma, \ldots$ we consider **types** and **multiset types** defined as follows:

(types)	$\tau, \sigma, \rho ::= \alpha \mid \mathcal{M} \to \tau$
(multiset types)	$\mathcal{M} ::= [\tau_i]_{i \in I}$ where I is a finite set

Our types are *strict* [12,48], *i.e.* the type on the right hand side of a functional type is never a multiset. They also make use of usual notations for multisets, as in [16], so that [] denotes the empty multiset, and $[\sigma, \sigma, \tau]$ must be understood as $\sigma \wedge \sigma \wedge \tau$, where the symbol $\wedge$ enjoys commutativity and associativity but not idempotence, *i.e.* $\sigma \wedge \sigma$ is not equal to σ.

Type assignments, written Γ, Δ, are functions from variables to multiset types, assigning the empty multiset to all but a finite set of variables. The domain of Γ is given by $\text{dom}(\Gamma) := \{x \mid \Gamma(x) \neq [\,]\}$. The **intersection of type assignments**, written $\Gamma + \Delta$, is defined by $(\Gamma + \Delta)(x) := \Gamma(x) + \Delta(x)$, where the symbol $+$ denotes multiset union. Hence, $\text{dom}(\Gamma + \Delta) = \text{dom}(\Gamma) \cup \text{dom}(\Delta)$. When $\text{dom}(\Gamma)$ and $\text{dom}(\Delta)$ are disjoint we write $\Gamma ; \Delta$ instead of $\Gamma + \Delta$. We write $\Gamma \setminus\!\!\setminus x$ for the assignment $(\Gamma \setminus\!\!\setminus x)(x) = [\,]$ and $(\Gamma \setminus\!\!\setminus x)(y) = \Gamma(y)$ if $y \neq x$.

The symbol $_{-}$ is called the **empty stoup**. A **stoup** Σ is either a type σ or the empty stoup. **Type environments** are pairs of the form $\Gamma \mid \Sigma$, where Γ is a type assignment and Σ is a stoup. **Type judgments** are triples of the form $\Gamma \mid \Sigma \vdash o{:}\tau$, where o is an object, $\Gamma \mid \Sigma$ a type environment and τ a type. The $\mathcal{Q}_{\mathsf{E}}$ type system for the E-calculus is given in Fig. 1 ; it derives type judgments of the form $\Gamma \mid _{-} \vdash t{:}\tau$ and $\Gamma \mid \sigma \vdash l{:}\tau$, where t is a term and l is a list. We write $\Gamma \mid \Sigma \vdash_{\mathcal{Q}_{\mathsf{E}}} o{:}\tau$ or $\Phi \triangleright_{\mathcal{Q}_{\mathsf{E}}} \Gamma \mid \Sigma \vdash o{:}\tau$ to denote derivability in system $\mathcal{Q}_{\mathsf{E}}$. The **hlist-size** of the type derivation Φ is a positive natural number written $\text{sz2}(\Phi)$ which denotes the size of Φ where every node `hlist` counts 2^1. Example 1 further justifies the use of this `hlist`-size function.

$$\frac{}{\emptyset \mid \tau \vdash \texttt{nil}{:}\tau} \; (\text{ax}) \qquad \frac{\Gamma \mid _{-} \vdash t{:}\tau}{\Gamma \setminus\!\!\setminus x \mid _{-} \vdash \lambda x.t{:}\Gamma(x) \to \tau} \; (\to \text{r}) \qquad \frac{\Gamma \mid \sigma \vdash l{:}\tau}{\Gamma + \{x{:}[\sigma]\} \mid _{-} \vdash xl{:}\tau} \; (\texttt{hlist})$$

$$\frac{\Gamma \mid _{-} \vdash t{:}\sigma \quad \Delta \mid \sigma \vdash l{:}\tau}{\Gamma + \Delta \mid _{-} \vdash tl{:}\tau} \; (\text{app}) \qquad \frac{\Gamma \mid _{-} \vdash t{:}\rho \quad \Delta \mid \tau \vdash l{:}\sigma}{\Delta + \Gamma \mid [\,] \to \tau \vdash t; l{:}\sigma} \; (\to 1_{\notin})$$

$$\frac{(\Gamma_j \mid _{-} \vdash t{:}\tau_j)_{j \in J} \quad J \neq \emptyset \quad \Delta \mid \tau \vdash l{:}\sigma}{\Delta +_{j \in J} \Gamma_j \mid [\tau_j]_{j \in J} \to \tau \vdash t; l{:}\sigma} \; (\to 1_{\in})$$

$$\frac{\Delta \mid _{-} \vdash u{:}\sigma \quad \Gamma \mid _{-} \vdash t{:}\tau \quad x \notin \text{dom}(\Gamma)}{\Gamma + \Delta \mid _{-} \vdash t[x/u]{:}\tau} \; (\text{es}_{\notin})$$

$$\frac{(\Delta_j \mid _{-} \vdash u{:}\sigma_j)_{j \in J} \quad J \neq \emptyset \quad x{:}[\sigma_j]_{j \in J}; \Gamma \mid _{-} \vdash t{:}\tau}{\Gamma +_{j \in J} \Delta_j \mid _{-} \vdash t[x/u]{:}\tau} \; (\text{es}_{\in})$$

Fig. 1. The type system $\mathcal{Q}_{\mathsf{E}}$ for the E-Calculus

Notice that the system $\mathcal{Q}_{\mathsf{E}}$ is syntax oriented, *i.e.* for each type judgment of the form $\Gamma \mid \Sigma \vdash o{:}\tau$ there is a *unique* typing rule whose conclusion matches the type judgment. Indeed, there are two different rules to type a list $t; l$, but the type in the stoup Σ discriminates between them. There is a similar distinction for terms of the form $t[x/u]$ depending on whether x is in the free variables of t or not. A consequence of this property is that statements usually proved in generation Lemmas (*cf.* [6,35]) are straightforward in system $\mathcal{Q}_{\mathsf{E}}$.

---

[1] The node `hlist` counts 2 since it corresponds, in the standard sequent calculus, to an application of an axiom rule followed by a contraction.

The (app) typing rule is the *head-cut* rule in the underlying logical system; similarly, $(es_{\notin})$ and $(es_{\in})$ give an interpretation of the so-called *mid-cut* [25]. The type derivation for t (resp. for u) in rule $(\to 1_{\notin})$ (resp. $(es_{\notin})$) is called a **witness** derivation, and turns out to be essential to guarantee strong-normalization of the whole typed term $t; l$ (resp. $t[x/u]$). Indeed, if witness derivations are not required, then non-terminating terms like $t(\Omega; l)$ or $t[x/\Omega]$ would be typable in the system, for $\Omega = (\lambda x.xx_{\texttt{nil}})\lambda x.xx_{\texttt{nil}}$. The rules $(\to 1_{\notin})$ and $(\to 1_{\in})$ (resp. $(es_{\notin})$ and $(es_{\in})$) can be specified by means of a unique typing rule $(\to 1)$ (resp. (es)), usually used in the proofs in order to save some space. They have the form:

$$\frac{(\Gamma_j \mid {}_- \vdash t{:}\tau_j)_{j \in J} \qquad \Delta \mid \tau \vdash l{:}\sigma}{\Delta +_{j \in J} \Gamma_j \mid [\tau_i]_{i \in I} \to \tau \vdash t; l{:}\sigma} \ (\to 1)$$

$$\frac{(\Delta_j \mid {}_- \vdash u{:}\sigma_j)_{j \in J} \qquad x{:}[\sigma_i]_{i \in I}; \Gamma \mid {}_- \vdash t{:}\tau}{\Gamma +_{j \in J} \Delta_j \mid {}_- \vdash t[x/u]{:}\tau} \ (es)$$

where $(I = \emptyset \Rightarrow |J| = 1)$ and $(I \neq \emptyset \Rightarrow I = J)$.

The system $\mathcal{Q}_E$ is *relevant* [14], *i.e.* typing environments only contain the consumed premises. Moreover, in contrast to [6,8], no subtyping relation is needed for abstractions and/or applications.

Lemma 1. *If* $\Gamma \mid \Sigma \vdash_{\mathcal{Q}_E} o{:}\tau$*, then* $\text{dom}(\Gamma) = \text{fv}(o)$*.*

We now introduce a technical tool which will be used in Sect. 4 in order to give a characterization of strongly E-normalizing terms. Indeed, we do not want to distinguish terms having explicit cuts at different *head positions*, mainly because they do have exactly the same maximal reduction lengths. More precisely, the **head graphical equivalence** $\sim$ on E-terms, inspired from the σ-equivalence on λ-terms [44] and the σ-equivalence on λ-terms with explicit substitutions [2], is given by the contextual, transitive, symmetric and reflexive closure of the following axiom

$$(tl)[x/u] \approx t[x/u]l, \text{ where } |l|_x = 0$$

Notice that $(xl)[x/u]$ cannot be $\sim$-converted into $x[x/u]l$ when $x \notin \text{fv}(l)$, since x alone is not a term of the calculus.

The main properties of system $\mathcal{Q}_E$ for E-terms follow.

Lemma 2 (Invariance for $\sim$). *Let* o, o' *be E-objects s.t.* $o \sim o'$*. Then,*

1. $o \to_{E\backslash w} o_0$ *iff* $o' \to_{E\backslash w} o'_0$*, where* $o_0 \sim o'_0$*. In particular,* $\eta_{E\backslash w}(o) = \eta_{E\backslash w}(o')$*.*

2. $\Phi \triangleright \Gamma \vdash_{\mathcal{Q}_E} o{:}\tau$ *iff* $\Phi' \triangleright \Gamma \vdash_{\mathcal{Q}_E} o'{:}\tau$*. Moreover,* $\text{sz2}(\Phi) = \text{sz2}(\Phi')$*.*

Lemma 3 (Weighted Subject Reduction). *Let* $\Phi \triangleright \Gamma \mid \Sigma \vdash_{\mathcal{Q}_E} o{:}\tau$*. If* $o \to_{E\backslash w}$ o'*, then* $\Phi' \triangleright \Gamma \mid \Sigma \vdash_{\mathcal{Q}_E} o'{:}\tau$ *and* $\text{sz2}(\Phi) > \text{sz2}(\Phi')$*.*

The sz2 function plays a central role in obtaining a strictly decreasing measure in the subject reduction lemma above. More precisely, if we consider the standard measure on typing derivations, written sz, which counts 1 for every node of the derivation tree, then the weighted subject reduction property does not hold. Here is an example.

Example 1. Let t be the term $x(x_{\mathtt{nil}};\mathtt{nil})$, and consider the following type derivation Φ_t such that $\mathbf{sz}(\Phi_t) = 5$.

$$\Phi_t := \dfrac{\dfrac{\dfrac{\emptyset \mid \sigma \vdash \mathtt{nil}{:}\sigma}{x{:}[\sigma] \mid \_ \vdash x_{\mathtt{nil}}{:}\sigma} \qquad \overline{\emptyset \mid \tau \vdash \mathtt{nil}{:}\tau}}{x{:}[\sigma] \mid [\sigma]{\to}\tau \vdash x_{\mathtt{nil}};\mathtt{nil}{:}\tau}}{x{:}[\sigma,[\sigma]{\to}\tau] \mid \_ \vdash x(x_{\mathtt{nil}};\mathtt{nil}){:}\tau}$$

Let u be a term s.t. $\Phi_u^1 \rhd \Delta_1 \mid \_ \vdash u{:}\sigma$ and $\Phi_u^2 \rhd \Delta_2 \mid \_ \vdash u{:}[\sigma]{\to}\tau$. Therefore,

$$\Phi := \dfrac{\Phi_u^1 \rhd \Delta_1 \mid \_ \vdash u{:}\sigma \qquad \Phi_u^2 \rhd \Delta_2 \mid \_ \vdash u{:}[\sigma]{\to}\tau \qquad \Phi_t \rhd x{:}[\sigma,[\sigma]{\to}\tau] \mid \_ \vdash t{:}\tau}{\Delta_1 + \Delta_2 \mid \_ \vdash t[x/u]{:}\tau}$$

where $\mathbf{sz}(\Phi) = 6 +_{i=1,2} \mathbf{sz}(\Phi_u^i)$.

Given the reduction step $t[x/u] \to_{\mathtt{c}} x(u\,\mathtt{nil};\mathtt{nil})[x/u] = t'$, there is a derivation Φ' typing t' such that $\mathbf{sz}(\Phi') = \mathbf{sz}(\Phi)$. Indeed,

$$\Phi' := \dfrac{\Phi_u^2 \rhd \Delta_2 \mid \_ \vdash u{:}[\sigma]{\to}\tau \qquad \dfrac{\dfrac{\dfrac{\Phi_u^1 \rhd \Delta_1 \mid \_ \vdash u{:}\sigma \qquad \overline{\emptyset \mid \sigma \vdash \mathtt{nil}{:}\sigma}}{\Delta_1 \mid \_ \vdash u\,\mathtt{nil}{:}\sigma} \qquad \overline{\emptyset \mid \tau \vdash \mathtt{nil}{:}\tau}}{\Delta_1 \mid [\sigma]{\to}\tau \vdash u\,\mathtt{nil};\mathtt{nil}{:}\tau}}{x{:}[[\sigma]{\to}\tau] + \Delta_1 \mid \_ \vdash x(u\,\mathtt{nil};\mathtt{nil}){:}\tau}}{\Delta_1 + \Delta_2 \mid \_ \vdash t'{:}\tau}$$

Corollary 1. *If o is $\mathcal{Q}_{\mathrm{E}}$-typable then $o \in \mathcal{SN}(\mathtt{E}\backslash\mathtt{w})$.*

As we mentioned in the introduction, the $\overline{\lambda}$-calculus [25] is not compatible with a resource aware semantics, as illustrated by the following example. Consider a $\overline{\lambda}$-reduction of the form $o = (tl)[x/u] \to t[x/u]l[x/u] = o'$, and suppose $|tl|_x = 0$. Let Φ be a typing derivation for the object o, thus having the following form:

$$\Phi := \dfrac{\dfrac{\Phi_t \rhd \Gamma_t \mid \_ \vdash t{:}\sigma_t \qquad \Phi_l \rhd \Gamma_l \mid \sigma_t \vdash l{:}\tau}{\Gamma_t + \Gamma_l \mid \_ \vdash tl{:}\tau} \qquad \Phi_u \rhd \Delta \mid \_ \vdash u{:}\sigma}{(\Gamma_t + \Gamma_l) + \Delta \mid \_ \vdash (tl)[x/u]{:}\tau}$$

The typing derivation for the object o', let say Φ', must use *twice* the typing tree Φ_u, and thus $\mathbf{sz2}(\Phi) > \mathbf{sz2}(\Phi')$ cannot hold. In other words, propagation of substitution w.r.t. the structure of terms induces useless duplications of empty resources, turning out to be inappropriate in the framework of a resource aware semantics.

The last key property relates typing with expansion:

Lemma 4 (Subject Expansion). *Let* $\Gamma \mid \Sigma \vdash_{\mathcal{Q}_E} o':\tau$. *If* $o \rightarrow_{E\backslash w} o'$, *then* $\Gamma \mid \Sigma \vdash_{\mathcal{Q}_E} o:\tau$.

4   Characterizing Strongly E-Normalizing Terms

This section is devoted to the characterization of E-strong normalization. We use the technical tools developed in Sect. 3 to characterize strongly normalizing E-terms by means of $\mathcal{Q}_E$-typability, namely, the subject reduction and expansion properties.

We start by giving an alternative definition of $\mathcal{SN}(E\backslash w)$, where $\rightarrow_{E\backslash w}$ is defined as $\rightarrow_E \backslash \rightarrow_w$. Indeed, the **inductive set of** $E\backslash w$-strongly-normalizing objects, written $\mathcal{ISN}(E\backslash w)$, is the smallest subset of objects satisfying the following properties:

(EL) $\texttt{nil} \in \mathcal{ISN}(E\backslash w)$.

(NEL) If $t, l \in \mathcal{ISN}(E\backslash w)$, then $t; l \in \mathcal{ISN}(E\backslash w)$,

$\quad(L)$ If $t \in \mathcal{ISN}(E\backslash w)$, then $\lambda x.t \in \mathcal{ISN}(E\backslash w)$.

$\quad(HL)$ If $l \in \mathcal{ISN}(E\backslash w)$, then $xl \in \mathcal{ISN}(E\backslash w)$.

$\quad(W)$ If $t, s \in \mathcal{ISN}(E\backslash w)$ and $|t|_x = 0$, then $t[x/s] \in \mathcal{ISN}(E\backslash w)$.

$(\text{dB}_{\texttt{nil}})$ If $(\lambda x.t)l_1 \ldots l_n$ $(n \geq 0) \in \mathcal{ISN}(E\backslash w)$, then $(\lambda x.t)\texttt{nil} l_1 \ldots l_n \in \mathcal{ISN}(E\backslash w)$.

$(\text{dB}_{\text{cons}})$ If $t[x/u]ml_1 \ldots l_n$ $(n \geq 0) \in \mathcal{ISN}(E\backslash w)$, then $(\lambda x.t)(u;m)l_1 \ldots l_n \in \mathcal{ISN}(E\backslash w)$.

$(@_{\text{var}})$ If $x(m_1@m_2)l_1 \ldots l_n$ $(n \geq 0) \in \mathcal{ISN}(E\backslash w)$, then $(xm_1)m_2l_1 \ldots l_n \in \mathcal{ISN}(E\backslash w)$.

$(@_{\text{app}})$ If $t(m_1@m_2)l_1 \ldots l_n$ $(n \geq 0) \in \mathcal{ISN}(E\backslash w)$, then $(tm_1)m_2l_1 \ldots l_n \in \mathcal{ISN}(E\backslash w)$.

$\quad(C)$ If $\texttt{C}[\![u\,l]\!][x/u] \in \mathcal{ISN}(E\backslash w)$ and $|\texttt{C}[\![x\,l]\!]|_x > 1$, then $\texttt{C}[\![x\,l]\!][x/u] \in \mathcal{ISN}(E\backslash w)$.

$\quad(D)$ If $\texttt{C}[\![u\,l]\!] \in \mathcal{ISN}(E\backslash w)$ and $|\texttt{C}[\![x\,l]\!]|_x = 1$, then $\texttt{C}[\![x\,l]\!][x/u] \in \mathcal{ISN}(E\backslash w)$.

$\quad(E)$ If $(tl)[x/s] \in \mathcal{ISN}(E\backslash w)$ and $|l|_x = 0$, then $t[x/s]l \in \mathcal{ISN}(E\backslash w)$.

Remark in particular that case (E) guarantees the closure of $\mathcal{ISN}(E\backslash w)$ for the head graphical equivalence.

It is not difficult to show that the sets $\mathcal{SN}(E\backslash w)$ and $\mathcal{ISN}(E\backslash w)$ coincide (the proof follows the same scheme used for example in [26]) so that we can show the following result:

Lemma 5. *Let o be an* E-*object. If* $o \in \mathcal{SN}(E\backslash w)$ *then o is* $\mathcal{Q}_E$-*typable.*

Proof. By induction on the structure of $o \in \mathcal{ISN}(E\backslash w) = \mathcal{SN}(E\backslash w)$.

– If $o = \texttt{nil}$, $o = t; l$, $o = \lambda x.t$, $o = xl$, or $o = u[x/v]$ with $|u|_x = 0$, then the proof is straightforward by using the *i.h.*
– If $o \in \mathcal{ISN}(E\backslash w)$ comes from one of the rules $(\text{dB}_{\texttt{nil}})$, $(\text{dB}_{\text{cons}})$, (C), (D), $(@_{\text{var}})$ or $(@_{\text{app}})$, then the property holds by the *i.h.* and the Subject Expansion Lemma 4.

- If $t[x/s]l \in \mathcal{ISN}(E\backslash w)$ comes from the rule (E), then $(tl)[x/s] \in \mathcal{ISN}(E\backslash w)$, so that $(tl)[x/s]$ is $\mathcal{Q}_E$-typable by the *i.h.* and the property holds by Lemma 2.

Strong E-normalization can be now obtained from strong $E\backslash w$-normalization as follows:

Lemma 6 (From $E\backslash w$ to E). *Let o be an E-object. If $o \in \mathcal{SN}(E\backslash w)$, then $o \in \mathcal{SN}(E)$.*

Proof. One first shows a postponement property for w-reduction steps given by: if $o \rightarrow_w^+ \rightarrow_{E\backslash w} o'$, then $o \rightarrow_{E\backslash w} \rightarrow_w^+ o'$. Then the property is proved by contradiction using the postponement property.

We can now conclude with the main result of this section.

Theorem 1. *Let o be an E-object. Then o is $\mathcal{Q}_E$-typable iff $o \in \mathcal{SN}(E)$.*

Proof. Let o be $\mathcal{Q}_E$-typable. Then $o \in \mathcal{SN}(E\backslash w)$ by Corollary 1 and $o \in \mathcal{SN}(E)$ by Lemma 6. For the converse, $o \in \mathcal{SN}(E) \subseteq \mathcal{SN}(E\backslash w)$ because $\rightarrow_{E\backslash w} \subseteq \rightarrow_E$. We conclude by Lemma 5.

5 The I-Calculus

We now introduce the syntax and the operational semantics of the I-calculus, slightly differently defined in [20]. The I-calculus can be obtained from E by an appropriate projection function (*cf.* Lemma 8).

Given a countable infinite set of symbols $x, y, z, \ldots$, three syntactic categories as **I-objects**, **I-terms** and **I-lists** are respectively defined by the following grammars:

$$o: := t \mid l \qquad t, u, v: := \lambda x.t \mid xl \mid (\lambda x.t)l \qquad l, m: := \texttt{nil} \mid t; l$$

Remark that general terms of the form tl are not I-terms.

As before, we work with Barendregt's convention and the standard notion of α-conversion. The notions of **I-term** and **I-list contexts** are defined as expected according to the grammars above. The **reduction relation** $\rightarrow_I$ is given by the context closure of the following rules:

$$(\lambda x.t)\texttt{nil} \mapsto_{\beta_{\text{nil}}} \lambda x.t \qquad\qquad (\lambda x.t)(u; l) \mapsto_{\beta_{\text{cons}}} t\{x/u\} \circ l$$

where the operations $\_\circ\_$ and $\{\_/\_\}$ are defined as follows:

$$
\begin{aligned}
(xl) \circ m &:= x\,(l@m) & \texttt{nil}\{x/v\} &:= \texttt{nil} \\
((\lambda y.t)l) \circ m &:= (\lambda y.t)(l@m) & (u; l)\{x/v\} &:= u\{x/v\}; l\{x/v\} \\
(\lambda x.t) \circ m &:= (\lambda x.t)m & (y\,l)\{x/v\} &:= yl\{x/v\} \\
& & (xl)\{x/v\} &:= v \circ l\{x/v\} \\
& & ((\lambda y.t)l)\{x/v\} &:= (\lambda y.t\{x/v\})l\{x/v\} \\
& & (\lambda y.t)\{x/v\} &:= \lambda y.t\{x/v\}
\end{aligned}
$$

The substitution operator $\{\_/\_\}$ is defined on α-equivalence classes of terms in order to avoid the capture of free variables. Notice that substitution distributes with respect to @ and $\circ$, i.e. one can show that $(t@l)\{x/u\} = t\{x/u\}@l\{x/u\}$ and $(t\circ l)\{x/u\} = t\{x/u\}\circ l\{x/u\}$. As expected, the I-calculus enjoys confluence.

An **erasing step** is the closure by contexts of the reduction rule $(\lambda x.t)(u; l) \mapsto t\{x/u\} \circ l$, where $x \notin \mathtt{fv}(t)$, i.e. an erasing step discards the argument u since $x \notin \mathtt{fv}(o)$ implies $o\{x/u\} = o$. Notice that erasing steps cannot be postponed, so that we cannot apply to the I-calculus the same (simple) proof technique used in Sect. 4 to characterize E-strong normalization. An example of non-erasing step is

$$(\lambda y.y_{\mathtt{nil}})(x_{\mathtt{nil}}; x_{\mathtt{nil}}; \mathtt{nil}) \to x(x_{\mathtt{nil}}; \mathtt{nil})$$

while $(\lambda y.z_{\mathtt{nil}})(x_{\mathtt{nil}}; x_{\mathtt{nil}}; \mathtt{nil}) \to z(x_{\mathtt{nil}}; \mathtt{nil})$ is an erasing step. Non-erasing steps play a key role in Lemma 11.

The I-calculus can be simulated in the E-calculus in terms of more atomic steps. Indeed,

Lemma 7. *Let o be an I-term. If $o \to_\mathtt{I} o'$, then $o \to_\mathtt{E}^+ o'$.*

Proof. One first shows that for all I-objects t, u, l, $t[x/u] \to_\mathtt{E}^* t\{x/u\}$ and $tl \to_\mathtt{E}^* t \circ l$. The proof of the statement of the lemma then proceeds by induction on I-reduction. The interesting case is when $o = (\lambda x.t)(u; l) \to_\mathtt{I} t\{x/u\} \circ l = o'$, for which we conclude by $o = (\lambda x.t)(u; l) \to_\mathtt{E} t[x/u]l \to_\mathtt{E}^* t\{x/u\}l \to_\mathtt{E}^* t\{x/u\} \circ l = o'$ using the properties mentioned above.

Reciprocally, the E-calculus can be projected into the I-calculus. For that, we first remark that the system $\mathtt{sub} = \{\mathtt{w}, \mathtt{d}, \mathtt{c}, @_{\mathtt{var}}, @_{\mathtt{app}}\}$ is locally confluent and terminating. Hence $\mathtt{sub}$-normal forms of objects are unique; we thus write $\mathtt{sub}(o)$ for the $\mathtt{sub}$-normal forms of the object o.

Lemma 8 (Projection). *Let o be an E-term. If $o \to_\mathtt{E} o'$, then $\mathtt{sub}(o) \to_\mathtt{I}^* \mathtt{sub}(o')$.*

Proof. One first shows that for all E-terms t, u, one has $\mathtt{sub}(t[x/u]) = \mathtt{sub}(t)\{x/\mathtt{sub}(u)\}$. The proof of the statement of the lemma is then by induction on E-reduction using the previous remark.

Using confluence of I, together with Lemmas 7 and 8 we obtain the following property.

Corollary 2. *The reduction relation $\to_\mathtt{E}$ is confluent.*

6   A Non-idempotent Typing System for I-Terms

In this section we restrict to I-objects the system $\mathcal{Q}_\mathtt{E}$ introduced in Sect. 3. The resulting system $\mathcal{Q}_\mathtt{I}$ is given in Fig. 2. As before, relevance holds for I-objects.

Lemma 9. *If $\Gamma \mid \Sigma \vdash_{\mathcal{Q}_\mathtt{I}} o{:}\tau$, then $\mathtt{dom}(\Gamma) = \mathtt{fv}(o)$.*

Typing Rules $\{(\mathtt{ax}), (\to \mathtt{r}), (\mathtt{hlist}), (\to \mathtt{1}_{\notin}), (\to \mathtt{1}_{\in})\}$ plus

$$\frac{\Gamma \mid {}_{-} \vdash \lambda x.t{:}\sigma \qquad \Delta \mid \sigma \vdash l{:}\tau}{\Gamma + \Delta \mid {}_{-} \vdash (\lambda x.t)l{:}\tau} \ (\mathtt{app})$$

Fig. 2. The type system $\mathcal{Q}_{\mathtt{I}}$ for the I-Calculus

In order to characterize the set $\mathcal{SN}(\mathtt{I})$ of strongly I-normalizing term by means of $\mathcal{Q}_{\mathtt{I}}$-typability, we first need to show that every $\mathcal{Q}_{\mathtt{I}}$-typable term is strongly I-normalizing. This is obtained as follows:

Lemma 10. *Let o be an I-term. Then, $o \in \mathcal{SN}(\mathtt{E})$ implies $o \in \mathcal{SN}(\mathtt{I})$.*

Proof. A direct consequence of Lemma 7.

Corollary 3. *If $\Phi \triangleright_{\mathcal{Q}_{\mathtt{I}}} \Gamma \mid \Sigma \vdash o{:}\tau$, then $o \in \mathcal{SN}(\mathtt{I})$.*

Proof. If o is $\mathcal{Q}_{\mathtt{I}}$-typable, then o is also trivially $\mathcal{Q}_{\mathtt{E}}$-typable. Theorem 1 gives $o \in \mathcal{SN}(\mathtt{E})$ and Lemma 10 gives $o \in \mathcal{SN}(\mathtt{I})$.

The converse property, *i.e.* the fact that every strongly I-normalizing term is $\mathcal{Q}_{\mathtt{I}}$-typable, will be proved in Sect. 7. For that, the following key property will be used.

Lemma 11 (Subject Expansion for Non-erasing Reductions). *If $\Phi \triangleright_{\mathcal{Q}_{\mathtt{I}}}$ $\Gamma \mid \Sigma \vdash o'{:}\tau$ and $o \to_{\mathtt{I}} o'$ is a non-erasing step, then there exists Φ' such that $\Phi' \triangleright_{\mathcal{Q}_{\mathtt{I}}} \Gamma \mid \Sigma \vdash o{:}\tau$.*

Proof. By induction on the non-erasing reduction relation $\to_{\mathtt{I}}$.

7 Characterizing Strongly I-Normalizing Terms

This section completes the characterization result for the I-calculus, namely, we show that every strongly I-normalizing term is $\mathcal{Q}_{\mathtt{I}}$-typable, so that, together with Corollary 3, we obtain a full characterization of strongly I-normalizing terms by means of $\mathcal{Q}_{\mathtt{I}}$-typability.

In order to achieve the main result of this section, we define an inductive set of objects $\mathcal{ISN}(\mathtt{I})$ containing the set of strongly I-normalizing objects and contained in the set of those that are $\mathcal{Q}_{\mathtt{I}}$-typable. The set $\mathcal{ISN}(\mathtt{I})$ is inspired by the idempotent intersection typing system proposed by Valentini [47], then revisited by Kikuchi [31].

We start by defining the set $\mathcal{ISN}(\mathtt{I})$ as the smallest subset of I-objects satisfying the following properties:

(ax) $\mathtt{nil} \in \mathcal{ISN}(\mathtt{I})$.
$(\to \mathtt{r})$ If $t \in \mathcal{ISN}(\mathtt{I})$, then $\lambda x.t \in \mathcal{ISN}(\mathtt{I})$.

(**hlist**) If $l \in \mathcal{ISN}(\mathtt{I})$, then $xl \in \mathcal{ISN}(\mathtt{I})$.
($\to$ **l**) If $t \in \mathcal{ISN}(\mathtt{I})$ and $l \in \mathcal{ISN}(\mathtt{I})$, then $t; l \in \mathcal{ISN}(\mathtt{I})$.
(**app**$_{nil}$) If $\lambda x.t \in \mathcal{ISN}(\mathtt{I})$, then $(\lambda x.t)\mathtt{nil} \in \mathcal{ISN}(\mathtt{I})$
(**app**$_\in$) If $t\{x/u\} \circ l \in \mathcal{ISN}(\mathtt{I})$ and $x \in \mathtt{fv}(t)$, then $(\lambda x.t)(u; l) \in \mathcal{ISN}(\mathtt{I})$.
(**app**$_\notin$) If $t \circ l \in \mathcal{ISN}(\mathtt{I})$ and $u \in \mathcal{ISN}(\mathtt{I})$ and $x \notin \mathtt{fv}(t)$, then $(\lambda x.t)(u; l) \in \mathcal{ISN}(\mathtt{I})$.

Every strongly I-normalizing object o turns out to be in $\mathcal{ISN}(\mathtt{I})$.

Theorem 2. *If $o \in \mathcal{SN}(\mathtt{I})$, then $o \in \mathcal{ISN}(\mathtt{I})$.*

Proof. By induction on $\langle \eta_{\mathtt{I}}(o), |o| \rangle$.

If $o = \mathtt{nil}$, then the statement is trivial. If $o = u; l$, then the *i.h.* gives u and l in $\mathcal{ISN}(\mathtt{I})$ so that $u; l \in \mathcal{ISN}(\mathtt{I})$ using rule ($\to$ **l**). The same reasoning can be applied if $o = \lambda x.t$ or $o = xl$. If $o = (\lambda x.t)l$, we consider two cases.

- $l = \mathtt{nil}$. The *i.h.* gives $\lambda x.t \in \mathcal{ISN}(\mathtt{I})$, then $(\lambda x.t)\mathtt{nil} \in \mathcal{ISN}(\mathtt{I})$ using rule (**app**$_{nil}$).
- $l = u; l'$. We consider again two cases.
  - $x \in \mathtt{fv}(t)$. Since $\eta_{\mathtt{I}}(t\{x/u\} \circ l) < \eta_{\mathtt{I}}(o)$, then by the *i.h.* we have $t\{x/u\} \circ l \in \mathcal{ISN}(\mathtt{I})$, then we obtain $o \in \mathcal{ISN}(\mathtt{I})$ using rule (**app**$_\in$).
  - $x \notin \mathtt{fv}(t)$. Since $\eta_{\mathtt{I}}(t\{x/u\} \circ l) = \eta_{\mathtt{I}}(t \circ l) < \eta_{\mathtt{I}}(o)$, then by the *i.h.* we have $t \circ l \in \mathcal{ISN}(\mathtt{I})$. Moreover, $\eta_{\mathtt{I}}(u) < \eta_{\mathtt{I}}(o)$, so that also by the *i.h.* we have $u \in \mathcal{ISN}(\mathtt{I})$. Then $o \in \mathcal{ISN}(\mathtt{I})$ using rule (**app**$_\notin$).

Moreover, every object in $\mathcal{ISN}(\mathtt{I})$ is $\mathcal{Q}_{\mathtt{I}}$-typable.

Theorem 3. *If $o \in \mathcal{ISN}(\mathtt{I})$, then there exists Φ' such that $\Phi' \rhd_{\mathcal{Q}_{\mathtt{I}}} \Gamma \mid \Sigma \vdash o{:}\tau$.*

Proof. By induction on the definition of the predicate $\mathcal{ISN}(\mathtt{I})$. The cases (**ax**), ($\to$ **r**), (**hlist**) and ($\to$ **l**) are straightforward. Let consider $(\lambda x.t)\mathtt{nil} \in \mathcal{ISN}(\mathtt{I})$ coming from $\lambda x.t \in \mathcal{ISN}(\mathtt{I})$ by rule (**app**$_{nil}$). By the *i.h.* we have $\Gamma \mid {}_- \vdash_{\mathcal{Q}_{\mathtt{I}}} \lambda x.t{:}\tau$, thus we conclude by:

$$\Phi' := \cfrac{\Gamma \mid {}_- \vdash \lambda x.t{:}\tau \qquad \cfrac{}{\emptyset \mid \tau \vdash \mathtt{nil}{:}\tau} \text{ (ax)}}{\Gamma \mid {}_- \vdash (\lambda x.t)\mathtt{nil}{:}\tau} \text{ (app)}$$

Consider $(\lambda x.t)(u; l) \in \mathcal{ISN}(\mathtt{I})$ coming from $t\{x/u\} \circ l \in \mathcal{ISN}(\mathtt{I})$ and $x \in \mathtt{fv}(t)$ by rule (**app**$_\in$). By the *i.h.* $\Gamma \mid {}_- \vdash_{\mathcal{Q}_{\mathtt{I}}} t\{x/u\} \circ l{:}\tau$, thus we get $\Gamma \mid {}_- \vdash_{\mathcal{Q}_{\mathtt{I}}} (\lambda x.t)(u; l){:}\tau$ by Lemma 11.

Consider $(\lambda x.t)(u; l) \in \mathcal{ISN}(\mathtt{I})$ coming from $t \circ l \in \mathcal{ISN}(\mathtt{I})$, $u \in \mathcal{ISN}(\mathtt{I})$ and $x \notin \mathtt{fv}(t)$ by rule (**app**$_\notin$). By the *i.h.* $\Gamma_1 \mid {}_- \vdash_{\mathcal{Q}_{\mathtt{I}}} t \circ l{:}\tau$ and $\Gamma_2 \mid {}_- \vdash_{\mathcal{Q}_{\mathtt{I}}} u{:}\sigma$. It can be proved, by induction that there are $\Gamma_1' \mid {}_- \vdash_{\mathcal{Q}_{\mathtt{I}}} t{:}\tau'$ and $\Gamma_1'' \mid \tau' \vdash_{\mathcal{Q}_{\mathtt{I}}} l{:}\tau$, where $\Gamma_1 = \Gamma_1' + \Gamma_1''$. Thus we get the following $\mathcal{Q}_{\mathtt{I}}$-derivation:

$$\Phi' := \cfrac{\cfrac{\Gamma_1' \mid {}_- \vdash t{:}\tau'}{\Gamma_1' \mid {}_- \vdash \lambda x.t{:}[\,]{\to}\tau'} (\to \mathbf{r}) \qquad \cfrac{\Gamma_2 \mid {}_- \vdash u{:}\sigma \qquad \Gamma_1'' \mid \tau' \vdash l{:}\tau}{\Gamma_2 + \Gamma_1'' \mid [\,]{\to}\tau' \vdash u; l{:}\tau} (\to \mathbf{l}_\notin)}{\Gamma_1' + \Gamma_1'' + \Gamma_2 \mid {}_- \vdash (\lambda x.t)(u; l){:}\tau} \text{ (app)}$$

We can thus conclude this section with the equivalence between strongly I-normalizing terms and typable terms in system $\mathcal{Q}_{\mathrm{I}}$.

Corollary 4. $\Gamma \mid \Sigma \vdash_{\mathcal{Q}_{\mathrm{I}}} o{:}\tau$ *if and only if* $o \in \mathcal{SN}(\mathtt{I})$.

8 Conclusion

This paper proposes a resource aware computational semantics for Herbelin's syntax. The resulting E-calculus can be seen as a refinement of the non resource aware I-calculus, whose meta-level operations are implemented by more atomic reduction rules in E. In contrast to more complex resource-controlled interpretations (*cf.* [22]) realized by means of explicit control operators for weakening and contraction, our implementation is achieved by rewriting rules inspired from the *substitution at a distance* paradigm, recently used in successful investigations in computer science.

We define typing systems for both calculi I and E, based on relevant, strict, non-idempotent intersection types. Typing rules of the systems are syntax directed. In both cases, typability is used to completely characterize strongly normalizing terms. Our results are presented in a self-contained form, without resorting to their isomorphic natural deduction counter-parts. The proofs only use combinatorial arguments, neither reducibility candidates nor memory operators have been necessary.

Balance between typing and reduction systems in a resource aware framework is sensitive; this is illustrated by the approach in [26] and the one taken here. Indeed, adding a dereliction rule to the reduction system, as done in this paper, allows to restrict the witness type derivations to erasable subterms only, while dereliction can be simply omitted, as done in [26], when using a resource consuming approach based on witnesses everywhere.

References

1. Accattoli, B., Bonelli, E., Kesner, D., Lombardi, C.: A nonstandard standardization theorem. In: POPL, pp. 659–670. ACM (2014)
2. Accattoli, B., Kesner, D.: The structural λ-calculus. In: Dawar, A., Veith, H. (eds.) CSL 2010. LNCS, vol. 6247, pp. 381–395. Springer, Heidelberg (2010)
3. Andreoli, J.-M.: Logic programming with focusing proofs in linear logic. J. Logic Comput. **2**(3), 297–347 (1992)
4. Baader, F., Nipkow, T.: Term Rewriting and All That. Cambridge University Press, Cambridge (1998)
5. Barendregt, H.: The Lambda Calculus: Its Syntax and Semantics (revised Edition), Volume 103 of Studies in Logic and the Foundations of Mathematics. Elsevier Science, New York (1984)
6. Barendregt, H., Coppo, M., Dezani-Ciancaglini, M.: A filter lambda model and the completeness of type assignment. Bull. Symbolic Logic **48**, 931–940 (1983)
7. Bernadet, A., Lengrand, S.: Complexity of strongly normalising λ-terms via non-idempotent intersection types. In: Hofmann, M. (ed.) FOSSACS 2011. LNCS, vol. 6604, pp. 88–107. Springer, Heidelberg (2011)

8. Bernadet, A., Lengrand, S.: Non-idempotent intersection types and strong normalisation. Log. Methods Comput. Sci. **9**(4), 1–46 (2013)
9. Boudol, G., Curien, P.-L., Lavatelli, C.: A semantics for lambda calculi with resources. Math. Struct. Comput. Sci. **9**(4), 437–482 (1999)
10. Bucciarelli, A., Kesner, D., Ronchi Della Rocca, S.: The inhabitation problem for non-idempotent intersection types. In: Diaz, J., Lanese, I., Sangiorgi, D. (eds.) TCS 2014. LNCS, vol. 8705, pp. 341–354. Springer, Heidelberg (2014)
11. Coppo, M., Dezani-Ciancaglini, M.: A new type-assignment for lambda terms. Archiv für Mathematische Logik und Grundlagenforschung **19**, 139–156 (1978)
12. Coppo, M., Dezani-Ciancaglini, M.: An extension of the basic functionality theory for the λ-calculus. Notre Dame, J. Formal Logic **21**, 685–693 (1980)
13. Coppo, M., Dezani-Ciancaglini, M., Venneri, B.: Functional characters of solvable terms. Math. Logic Q. **27**(2–6), 45–58 (1981)
14. Damiani, F., Giannini, P.: A decidable intersection type system based on relevance. In: Hagiya, M., Mitchell, J.C. (eds.) TACS 1994. LNCS, vol. 789, pp. 707–725. Springer, Heidelberg (1994)
15. De Benedetti, E., Ronchi Della Rocca, S.: Bounding normalization time through intersection types. In: ITRS, EPTCS, pp. 48–57. Cornell University Library (2013)
16. de Carvalho, D.: Sémantiques de la logique linéaire et temps de calcul. Université Aix-Marseille II, These de doctorat (2007)
17. Dyckhoff, R., Urban, C.: Strong normalization of herbelin's explicit substitution calculus with substitution propagation. J. Logic Comput. **13**(5), 689–706 (2003)
18. Espírito Santo, J., Ivetic, J., Likavec, S.: Characterising strongly normalising intuitionistic terms. Fundamenta Informaticae **121**(1–4), 83–120 (2012)
19. Espírito Santo, J.: Santo. The lambda-calculus and the unity of structural proof theory. Theory Comput. Syst. **45**(4), 963–994 (2009)
20. Espírito Santo, J.: Revisiting the correspondence between cut elimination and normalisation. In: Welzl, E., Montanari, U., Rolim, J.D.P. (eds.) ICALP 2000. LNCS, vol. 1853, pp. 600–611. Springer, Heidelberg (2000)
21. Gentzen, G.: The collected papers of Gerhard Gentzen (1969)
22. Ghilezan, S., Ivetić, J., Lescanne, P., Likavec, S.: Intersection types for the resource control lambda calculi. In: Cerone, A., Pihlajasaari, P. (eds.) ICTAC 2011. LNCS, vol. 6916, pp. 116–134. Springer, Heidelberg (2011)
23. Girard, J.-Y.: Proof-nets: The parallel syntax for proof-theory. In: Logic and Algebra, pp. 97–124. Marcel Dekker (1996)
24. Girard, J.-Y., Lafont, Y., Taylor, P.: Proofs and Types, Volume 7 of Cambridge Tracts in Theoretical Computer Science. Cambridge University Press, Cambridge (1989)
25. Herbelin, H.: A lambda-calculus structure isomorphic to Gentzen-style sequent calculus structure. In: Pacholski, L., Tiuryn, J. (eds.) CSL1994. LNCS, vol. 933, pp. 61–75. Springer, Heidelberg (1995)
26. Kesner, D., Ventura, D.: Quantitative types for intuitionistic calculi. Technical Report hal-00980868, Paris Cité Sorbonne (2014)
27. Kesner, D., Ventura, D.: Quantitative types for the linear substitution calculus. In: Diaz, J., Lanese, I., Sangiorgi, D. (eds.) TCS 2014. LNCS, vol. 8705, pp. 296–310. Springer, Heidelberg (2014)
28. Kfoury, A.: A linearization of the lambda-calculus and consequences. Technical report, Boston University (1996)
29. Kfoury, A.: A linearization of the lambda-calculus and consequences. J. Logic Comput. **10**(3), 411–436 (2000)

30. Kfoury, A., Wells, J.B.: Principality and type inference for intersection types using expansion variables. Theoret. Comput. Sci. **311**(1–3), 1–70 (2004)
31. Kikuchi, K.: Uniform proofs of normalisation and approximation for intersection types. In: ITRS, Vienna, Austria (2014)
32. Klop, J.W.: Combinatory reduction systems. Ph.D. thesis, University of Utrecht (1980)
33. Krivine, J.-L.: Lambda-calculus, Types and Models. Ellis Horwood (1993)
34. Lengrand, S.: Deriving strong normalisation. In: HOR, pp. 84–88 (2004). http://www-i2.informatik.rwth-aachen.de/HOR04/
35. Lengrand, S., Lescanne, P., Dougherty, D., Dezani-Ciancaglini, M., van Bakel, S.: Intersection types for explicit substitutions. Inf. Comput. **189**, 17–42 (2004)
36. Miller, D., Nadathur, G., Pfenning, F., Scedrov, A.: Uniform proofs as a foundation for logic programming. Ann. Pure Appl. Logic **51**, 125–157 (1991)
37. Milner, R.: Local bigraphs and confluence: Two conjectures: (extended abstract). ENTCS **175**(3), 65–73 (2007)
38. Neergaard, P.M., Mairson, H.G.: Types, potency, and idempotency: why nonlinearity and amnesia make a type system work. In: ICFP, pp. 138–149. ACM (2004)
39. Ong, L., Ramsay.: Verifying higher-order functional programs with pattern matching algebraic data types. In: POPL, pp. 587–598. ACM (2011)
40. Pagani, M., della Rocca, S.R.: Solvability in resource lambda-calculus. In: Ong, L. (ed.) FOSSACS 2010. LNCS, vol. 6014, pp. 358–373. Springer, Heidelberg (2010)
41. Pottinger, G.: Normalization as a homomorphic image of cut-elimination. Ann. Math. Logic **12**, 323–357 (1977)
42. Pottinger, G.: A type assignment for the strongly normalizable λ-terms. In: Seldin, J.P., Hindley, J.R. (eds.) To H.B. Curry: Essays on Combinatory Logic, Lambda Calculus and Formalism, pp. 561–578. Academic Press, London (1980)
43. Prawitz, D.: Natural deduction: a proof-theoretical study. Ph.D. thesis, Stockholm University (1965)
44. Regnier, L.: Une équivalence sur les lambda-termes. Theoret. Comput. Sci. **2**(126), 281–292 (1994)
45. Tait, W.: Intensional interpretations of functionals of finite type I. J. Symbolic Logic **32**(2), 198–212 (1967)
46. Urzyczyn, P.: The emptiness problem for intersection types. J. Symbolic Logic **64**(3), 1195–1215 (1999)
47. Valentini, S.: An elementary proof of strong normalization for intersection types. Arch. Math. Logic **40**(7), 475–488 (2001)
48. van Bakel, S.: Complete restrictions of the intersection type discipline. Theoret. Comput. Sci. **102**(1), 135–163 (1992)
49. Zucker, J.: The correspondence between cut-elimination and normalization. Ann. Math. Logic **7**, 1–112 (1974)

Undecidability Results
for Multi-Lane Spatial Logic

Heinrich Ody[(✉)]

Department of Computing Science, University of Oldenburg, Oldenburg, Germany
heinrich.ody@uni-oldenburg.de

Abstract. We consider (un)decidability of Multi-Lane Spatial Logic (MLSL), a multi-dimensional modal logic introduced for reasoning about traffic manoeuvres. MLSL with length measurement has been shown to be undecidable. However, the proof relies on exact values. This raises the question whether the logic remains undecidable when we consider robust satisfiability, i.e. when values are known only approximately. Our main result is that robust satisfiability of MLSL is undecidable. Furthermore, we prove that even MLSL without length measurement is undecidable. In both cases we reduce the intersection emptiness of two context-free languages to the respective satisfiability problem.

Keywords: Robustness · Interval logic · Spatial logic · Undecidability

1 Introduction

To reason about configurations of cars on a motorway in a formal manner we need an abstract representation of a motorway and the cars on it. In [13] the authors model a motorway as a set of lanes, where the extension of a lane is represented by the real numbers. The space occupied by a car then is a subinterval of the real numbers, together with the lane the car is located on. Additionally, the authors define Multi-Lane Spatial Logic (MLSL) to express topological properties of the cars on the motorway from the local perspective of a car. Possible properties are, e.g. 'there is a car on an adjacent lane' or 'there is some free space in front of us' or 'there is a collision', i.e. an overlap of occupied spaces. MLSL then has been extended to support length measurement and dynamic modalities [12,16]. With the dynamic modalities specifications of, e.g. a distance controller can be expressed in the logic. With length measurement properties such as 'there are five units of free space in front of me' can be expressed. MLSL with length measurement has been proven to be undecidable [16].

However, in all variants of MLSL considered it was assumed that data are known exactly, e.g. the position of every car is known exactly. Considering that the logic is used to express spatial properties of traffic on a motorway and that

---

Work of the author is supported by the Deutsche Forschungsgemeinschaft (DFG) within the Research Training Group DFG GRK 1765 SCARE.

M. Leucker et al. (Eds.): ICTAC 2015, LNCS 9399, pp. 404–421, 2015.
DOI: 10.1007/978-3-319-25150-9_24

data are obtained by imperfect sensors this assumption is too idealistic. This raises the question, whether the logic becomes decidable when values are known only approximately. Consider Metric Temporal Logic (MTL) [15] as an example where weakening the assumption of working with exact values leads to decidability. MTL is undecidable [10] and the proof heavily relies on formulas of the form $\Box(P \implies \Diamond_{=1}Q)$, which specifies that always exactly 1 time unit after P holds Q will hold. However, the fragment Metric Interval Temporal Logic, where exact timing properties are not expressible, is decidable [3].

As an intermediate result we show that even without length measurement MLSL is undecidable. This is surprising, because in this fragment we can only express topological relations between cars. Even though this fragment does not contain length measurement it still assumes exact positions. To formalize that values are known only approximately we consider a *robust* satisfaction relation, where a model robustly satisfies a formula iff all similar models also satisfy the formula. Our main result is that robust satisfiability of MLSL without length measurement is undecidable. To prove this we reduce the intersection emptiness of two context-free languages to the robust satisfiability problem. Our proof also works for a variety of restrictions on MLSL, e.g. when the extension of the lanes is represented by natural numbers instead of real numbers or when the minimal or maximal size of cars is bounded or any combination of these.

Robustness of spatial logics received little attention so far, because most spatial logics consider qualitative properties. However, as MLSL is inspired by temporal logics, robustness of timed systems is related. Our definition of robust satisfaction is similar to other approaches to introduce robustness into a formalism. A robust timed automaton accepts a trajectory iff it also accepts all slightly perturbed trajectories [8]. They were defined with the hope that by making exact timing properties inexpressible, the universality problem (does an automaton accept all trajectories) would become decidable, similarly to the satisfiability problem for MITL. However, the universality problem remains undecidable for robust timed automata [11]. In [7] a robust interpretation for a fragment of Duration Calculus has been defined. Whether this fragment is decidable with this robust interpretation is not known. For a survey on robustness in timed systems we suggest [1].

MLSL is inspired by DC [6], Interval Temporal Logic [17] and Shape Calculus [19]. Other logics similar to MLSL are CDT [20] and Halpern-Shoam-logic (HS) [9]. CDT and HS are one dimensional modal interval logics used to express temporal properties. In both of these logics topological properties of intervals can be expressed. However, while in MLSL there is a constant number of atomic propositions, in CDT and HS there is an arbitrary but finite number of atomic propositions. Further, both of these logic leave the temporal structure quite unconstrained, i.e. time may be branching or linear, dense or discrete.

2 Multi-Lane Spatial Logic

With Multi-Lane Spatial Logic (MLSL) we reason about static motorway traffic configurations. To this end we model the configuration of a motorway at a specific point in time as a *traffic snapshot*. In such a traffic snapshot the motorway is represented in a two dimensional manner, i.e. one vertical and one horizontal dimension. The vertical dimension represents the lane a car is located on and the horizontal dimension represents the position along the lane. A car then *reserves* the part of a lane where it is currently driving. When a car is changing lanes it has multiple reservations on adjacent lanes.

In the original definition of MLSL the model contains information about whether a car intents to change lanes (there called claims), the perspective from which formulas are evaluated (there called ego) and how a traffic snapshot changes as time progresses. We do not need this expressiveness to show undecidability. Hence, we consider a simplified logic without this information. As original MLSL is a conservative extension of our simplified logic our results also hold for the original logic.

With small adaptions our definitions in this section are taken from [13,16].

2.1 The Model

To formally define a traffic snapshot, we assume a countably infinite set of globally unique *car identifiers* $\mathbb{I}$ and an arbitrary but fixed set of lanes $\mathbb{L} = \{0, \ldots, k\}$, for some $k \in \mathbb{N}_{\geq 1}$.

Definition 1 (Traffic Snapshot). *A traffic snapshot $\mathcal{TS}$ is a structure $\mathcal{TS} = (res, pos, br\text{-}dis)$, where $res, pos, br\text{-}dis$ are functions*

- $res : \mathbb{I} \to \mathcal{P}(\mathbb{L})$ *such that $res(C)$ is the set of lanes the car C reserves,*
- $pos : \mathbb{I} \to \mathbb{R}$ *such that $pos(C)$ is the position of the car C along the lanes,*
- $br\text{-}dis : \mathbb{I} \to \mathbb{R}_{>0}$ *such that $br\text{-}dis(C)$ comprises the braking distance and the physical size of C,*

where $\mathcal{P}(\mathbb{L})$ denotes the powerset of $\mathbb{L}$.

In the original definition the authors imposed sanity conditions on a traffic snapshot, such as $1 \leq |res(C)| \leq 2$ holds for all cars C [13]. Our results hold regardless of whether these conditions are assumed. The extension of a lane occupied by a reservation is given by the *safety envelope* of a car, which is defined as

$$se(C) = [pos(C), pos(C) + br\text{-}dis(C)].$$

We reason only about a finite part of a traffic snapshot. In the original definition of MLSL this is called a *view* and we keep this name here.

Definition 2 (View). *For a given traffic snapshot $\mathcal{TS}$ with a set of lanes $\mathbb{L}$, a view V is defined as a structure $V = (L, X)$, where*

- $L \subseteq \mathbb{L}$ *is an interval of lanes that are visible in V,*
- $X \subseteq \mathbb{R}$ *is an interval representing the extension of the lanes visible in V.*

*A subview of V is obtained by restricting the lanes and extension we observe.
Let L', X' be subintervals of L and X, then we define*

$$V^{L'} = (L', X) \qquad and \qquad V_{X'} = (L, X').$$

*If the interval of lanes is empty or the extension is a point interval we say that
the view is* empty.

To give the semantics of MLSL in a uniform way we define chopping of views
into subviews.

Definition 3 (Chopping Views). *Let $V_i = (L_i, X_i)$ be views with $i \in \{0, 1, 2\}$
and $X_i = [r_i, t_i]$. Then we define* vertical chopping *(denoted by $\ominus$) and* horizontal
chopping *(denoted by $\oplus$) of V_0 into V_1 and V_2 as*

$$V_0 = V_1 \ominus V_2 \text{ iff } L_0 = L_1 \cup L_2 \text{ and } L_1 \cap L_2 = \emptyset \text{ and } X_0 = X_1 = X_2 \text{ and}$$
$$(L_1 = \emptyset \text{ or } L_2 = \emptyset \text{ or } \max(L_1) + 1 = \min(L_2)),$$
$$V_0 = V_1 \oplus V_2 \text{ iff } t_1 = r_2 \text{ and } r_0 = r_1 \text{ and } t_0 = t_2 \text{ and } L_0 = L_1 = L_2.$$

The intuition is that after vertical chopping the lane intervals of the subviews
are adjacent and non-overlapping because each lane should belong to exactly one
subview. After horizontal chopping the new subviews are adjacent and share a
common point, which is the endpoint of the left subview and the startpoint of
the right subview. Note that a nonempty view can be chopped into an empty and
a nonempty subview and that an empty view can be chopped into two empty
subviews.

Let $CVar$ be the set of variables ranging over $\mathbb{I}$. A valuation ν is a function
$\nu : CVar \to \mathbb{I}$ that maps car variables to car identifiers. Additionally we define
valuation updates with the override notation $\oplus$ from Z [21] as $\nu \oplus \{c \mapsto C\}(c') = C$ if $c = c'$ and $\nu(c')$ otherwise.

Definition 4 (Model). *Let TS be a traffic snapshot, V a view and ν a valuation, then we call $\mathcal{M} = (TS, V, \nu)$ a model of MLSL.*

2.2 The Logic

The atoms of MLSL are used to express that some part of a lane is filled by a
reservation or completely free of reservations. The chop operators in the logic
are defined using the chop operators on views. The horizontal chop formula
$\phi_0 \frown \phi_1$ expresses that on the left subview ϕ_0 and on the right subview ϕ_1 holds.
With the vertical chop formula $\genfrac{}{}{0pt}{}{\phi_1}{\phi_0}$ we require that the lower subview satisfies ϕ_0
and that the upper subview satisfies ϕ_1. Note that for both chop formulas the
satisfying subviews might be empty. Additionally, the logic is closed under first
order operators.

Definition 5 (Syntax). *Given variables $c, c' \in CVar$ the syntax of MLSL is given by*

$$\phi ::= c = c' \mid \textit{free} \mid re(c) \mid \neg\phi \mid \phi \wedge \phi \mid \exists c. \phi \mid \phi \frown \phi \mid \frac{\phi}{\phi}.$$

Definition 6 (Semantics). *Let $c, c' \in CVar$. Then, given a traffic snapshot $\mathcal{TS}$, a view $V = (L, X)$ with $X = [r, t]$, and a valuation ν we define the satisfaction of a formula by a model $\mathcal{M} = (\mathcal{TS}, V, \nu)$ as*

$$\begin{aligned}
\mathcal{M} &\models c = c' & \text{iff} \quad & \nu(c) = \nu(c'), \\
\mathcal{M} &\models \textit{free} & \text{iff} \quad & |L| = 1 \text{ and } r < t \text{ and} \\
& & & (\forall C \in \mathbb{I}. \, L \not\subseteq res(C) \text{ or } se(C) \cap (r, t) = \emptyset), \\
\mathcal{M} &\models re(c) & \text{iff} \quad & |L| = 1 \text{ and } r < t \text{ and} \\
& & & L \subseteq res(\nu(c)) \text{ and } X \subseteq se(\nu(c)), \\
\mathcal{M} &\models \neg\phi & \text{iff} \quad & \mathcal{M} \not\models \phi, \\
\mathcal{M} &\models \phi_0 \wedge \phi_1 & \text{iff} \quad & \mathcal{M} \models \phi_0 \text{ and } \mathcal{M} \models \phi_1, \\
\mathcal{M} &\models \exists c. \phi & \text{iff} \quad & \exists C \in \mathbb{I}. \, \mathcal{TS}, V, \nu \oplus \{c \mapsto C\} \models \phi, \\
\mathcal{M} &\models \phi_0 \frown \phi_1 & \text{iff} \quad & \exists V_0, V_1. \, V = V_0 \oslash V_1 \text{ and} \\
& & & \mathcal{TS}, V_0, \nu \models \phi_0 \text{ and } \mathcal{TS}, V_1, \nu \models \phi_1, \\
\mathcal{M} &\models \frac{\phi_1}{\phi_0} & \text{iff} \quad & \exists V_0, V_1. \, V = V_0 \ominus V_1 \text{ and} \\
& & & \mathcal{TS}, V_0, \nu \models \phi_0 \text{ and } \mathcal{TS}, V_1, \nu \models \phi_1.
\end{aligned}$$

In addition we make use of the standard abbreviations such as $true, \vee, \forall$. Additionally, we use a derived modality to express that *somewhere* on the motorway ϕ holds. It is defined by using both chop operators as

$$\langle \phi \rangle \overset{\text{def}}{=} true \frown \begin{pmatrix} true \\ \phi \\ true \end{pmatrix} \frown true.$$

3   Undecidability of MLSL

Undecidability of Duration Calculus was proven by a reduction to the halting problem of a two counter machine [5]. This construction has been adapted to prove undecidability of MLSL with length measurement, where length measurement allows to compare the size of the current view to a constant [16]. The authors defined that the representation of a configuration of a two counter machine is of length $5k$, where k is a constant. The value m of a counter was represented in an interval of length k and consisted of m reservations and the remaining space of a configuration was used for markers to separate the counters. To increase the value of a counter they required that for all reservations part of the representation of the counter there is a reservation $5k$ space units later (in the next representation of the counter's value) and additionally, there is exactly one reservation in the later representation for which there is no reservation $5k$ space units earlier. For this construction it is important to be able to specify the distance between reservations.

The construction from [16] does not work in our setting, as we consider MLSL without length measurement. Instead, we reduce the emptiness problem of the intersection of two context-free languages, which is undecidable [14], to the satisfiability problem of MLSL. For the reduction we create a formula such that satisfying models represent derivations of two context-free grammars for the same word. A letter is represented as a fixed number of successive reservations. Different letters are represented by a different number of reservations in their representation. Letters are separated by free space, and rewrite steps are represented by different lanes.

Definition 7 (Context-Free Grammar). *A context-free grammar (CFG) is a tuple $G = (\mathcal{N}, \mathcal{T}, \mathcal{R}, S)$, where $\mathcal{N}$ is the set of nonterminals, $\mathcal{T}$ with $\mathcal{T} \cap \mathcal{N} = \emptyset$ is the set of terminals, $S \in \mathcal{N}$ is the starting nonterminal and $\mathcal{R} \subseteq \mathcal{N} \times (\mathcal{T} \cup \mathcal{N})^*$ is the set of rewrite rules. A rewrite rule $(N, w) \in \mathcal{R}$ is usually written as $N \rightarrow w$. We extend $\rightarrow$ as follows. Let $\mathcal{R}^N = \{w \mid (N, w) \in \mathcal{R}\}$, i.e. the set of words that may replace N. Let $u, v, w \in (\mathcal{T} \cup \mathcal{N})^*, N \in \mathcal{N}$ then $uNv \rightarrow uwv$ with $(N, w) \in \mathcal{R}$ is a rewrite step. A sequence of rewrite steps is a derivation. The language $\mathcal{L}(G)$ of a CFG G is the set of terminal words reachable with a derivation that starts with S. We refer to $\sigma \in \mathcal{N} \cup \mathcal{T}$ as letter.*

All grammars we consider are in Chomsky normal form. With the definition we use this implies that the empty word is not in the language of a grammar in Chomsky normal form. Still the language intersection problem of these grammars remains undecidable.

Definition 8 (Chomsky Normal Form). *A CFG $G = (\mathcal{N}, \mathcal{T}, \mathcal{R}, S)$ is in Chomsky normal form (CNF) iff all rewrite rules have the form $N_0 \rightarrow \tau$ or $N_0 \rightarrow N_1 N_2$, where $N_0, N_1, N_2 \in \mathcal{N}, \tau \in \mathcal{T}$.*

Definition 9 (Derivation Tree). *For a grammar $G = (\mathcal{N}, \mathcal{T}, \mathcal{R}, S)$ in CNF a derivation tree is a $\mathcal{N} \cup \mathcal{T}$-labelled tree, where the following conditions hold. The label of the root is S. All leaves and no other nodes are labelled by terminals. Further, for any node labelled by a nonterminal N let $w = \sigma_0 \sigma_1 \ldots \sigma_{k-1}$ be the word formed by its k children, then $(N, w) \in \mathcal{R}$.*

3.1 Construction

First we give an intuitive explanation of how a model satisfying the formula we construct in this section looks like. We encode two CFGs G_D and G_U in one formula such that a satisfying model represents two derivation trees, one from each grammar. The representation of a derivation tree for G_D has its root on the top lane and grows *downward*, whereas the tree for G_D has its root on the bottom lane and grows *upward*. In a satisfying model every letter is represented as a number of successive reservations without space in between.

Representations of adjacent letters are separated by free space and different letters are represented by a different number of reservations. If a nonterminal from the downwards growing grammar is replaced by a word, the word is represented on the lane below the nonterminal such that the horizontal space used to represent the word is strictly contained in the space used for the nonterminal. Equality of the derived words is represented as horizontal alignment of terminals that differ only in their subscripts (e.g. a_U and a_D in Fig. 1). In Fig. 1 we depict an MLSL model representing the derivation trees of the derivations from Example 1. Note that we never show the view in our visualisations of models.

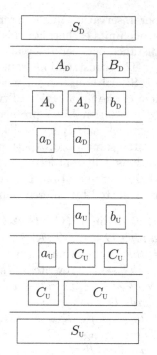

Example 1. Let $G_D = (\mathcal{N}_D, \mathcal{T}, \mathcal{R}_D, S_D)$ and $G_U = (\mathcal{N}_U, \mathcal{T}, \mathcal{R}_U, S_U)$ be two grammars in CNF, where $\mathcal{N}_D = \{A_D, B_D, S_D\}$, $\mathcal{T} = \{a, b\}$, $\mathcal{N}_U = \{C_U, S_U\}$ and $\mathcal{R}_D, \mathcal{R}_U$ are given in BNF-like notation:

$$S_D \to A_D B_D \qquad S_U \to C_U C_U$$
$$A_D \to A_D A_D \mid a \qquad C_U \to C_U C_U \mid a \mid b$$
$$B_D \to B_D B_D \mid b$$

Fig. 1. Visualisation of an MLSL model representing two derivation trees of derivations taken from Example 1. The boxes correspond to letters and reservations are not shown

Two derivations of G_D and G_U are $S_D \to A_D B_D \to A_D A_D B_D \to A_D A_D b \to a A_D b \to aab$ and $S_U \to C_U C_U \to a C_U \to a C_U C_U \to aa C_U \to aab$.

Given two CFGs $G_D = (\mathcal{N}_D, \mathcal{T}, \mathcal{R}_D, S_D)$ and $G_U = (\mathcal{N}_U, \mathcal{T}, \mathcal{R}_U, S_U)$ such that $\mathcal{N}_D \cap \mathcal{N}_U = \emptyset$, we assume two sets $\mathcal{T}_D, \mathcal{T}_U$ and bijective functions $\pi_D : \mathcal{T} \to \mathcal{T}_D$ and $\pi_U : \mathcal{T} \to \mathcal{T}_U$ such that $\mathcal{T}_D, \mathcal{N}_D, \mathcal{T}_U$ and $\mathcal{N}_U$ are all disjoint. The idea behind the two functions is that we want to differentiate between the MLSL encoding of terminals from G_U and from G_D.

Letter. Let $\mu : \mathcal{T}_D \cup \mathcal{N}_D \cup \mathcal{T}_U \cup \mathcal{N}_U \to \mathbb{N}_{>0}$ be an injective function. In MLSL we represent every letter $\sigma \in \mathcal{T}_D \cup \mathcal{N}_D \cup \mathcal{T}_U \cup \mathcal{N}_U$ as $\mu(\sigma)$ successive reservations from different cars, without free space in between. We formalize this as

$$letter(\sigma, c) \stackrel{\text{def}}{=} re(c) \frown \exists c_1, \ldots, c_{\mu(\sigma)-1}.$$

$$re(c_1) \frown \ldots \frown re(c_{\mu(\sigma)-1}) \wedge \bigwedge_{\substack{i, j \in \{1, \ldots \mu(\sigma)-1\}}}^{i \neq j} c_i \neq c_j,$$

where $c \in CVar$ is a car variable. We use c as an identifier to uniquely differentiate letters within a formula.

Assume that the letter a_D is represented by one reservation, and that the letter b_D is represented by two reservations. We have to distinguish between two occurrences of a_D and one occurrence of b_D. To be able to recognize letters we demand that before and after every letter there is some free space. For this we define

$$letter_{\text{free}}(\sigma, c) \stackrel{\text{def}}{=} free \frown letter(\sigma, c) \frown free\,.$$

The formulas *letter* and *letter*$_{\text{free}}$ are used in other formulas, which will always bind the car variable, here c.

Start. To ensure that there is a starting letter we express that the topmost lane contains the starting nonterminal S_D as

$$start_D \stackrel{\text{def}}{=} \exists c.\ \frac{letter_{\text{free}}(S_D, c)}{true}\,.$$

Step. Now we encode the rewrite relation as a formula. Recall that we consider grammars in Chomsky normal form, so any right-hand side of a rewrite rule has one or two letters. For a word w we define

$$word(w) \stackrel{\text{def}}{=} \begin{cases} l\exists c_0.\ letter_{\text{free}}(\sigma_0, c_0) & \text{if } |w| = 1\,, \\ \exists c_0, c_1.\ c_0 \neq c_1 \wedge letter_{\text{free}}(\sigma_0, c_0) \frown letter_{\text{free}}(\sigma_1, c_1) & \text{if } |w| = 2\,, \end{cases}$$

where σ_j is the j-th letter of w.

To define that a nonterminal N is replaced by a word w, according to the rewrite rule $\mathcal{R}_D$, we define

$$step_D(N, c) \stackrel{\text{def}}{=} \frac{free}{free} \frown \left(\frac{letter(N, c)}{\bigvee_{w \in \mathcal{R}_D^N} word(w)} \right) \frown \frac{free}{free}\,.$$

This means that we replace a nonterminal on the lane below it with any of the words from the rewrite relation. As we use *letter*$_{\text{free}}$ in the definition of *word*, we ensure that the replaced letter is horizontally larger than its replacement. Note that we subscripted the formula with D, because we only use it to encode derivation trees growing downward.

As all nonterminals should be replaced on the next lane we define

$$step\ all_D \stackrel{\text{def}}{=} \forall c.\ \bigwedge_{N \in \mathcal{N}_D} (\langle letter_{\text{free}}(N, c) \rangle \implies \langle step_D(N, c) \rangle)\,.$$

In the premise we test, whether somewhere the car variable c is used as identifier for an occurrence of the nonterminal N. Intuitively, we bind the variable c to the occurrence matched in the premise. For this to work as intended we have to assume that c is used as identifier for only this one occurrence. We formalize this later. In the conclusion we use this c, bounded to one specific occurrence of a nonterminal, to state that below this occurrence there should be a word as defined by the rewrite relation.

Side Conditions. As already mentioned we do not consider cars with two reservations or overlapping reservations. To exclude these we define

$$mutex \stackrel{\text{def}}{=} \neg \exists c, c'. c \neq c' \wedge \langle re(c) \wedge \rangle, \qquad no\ 2\ res \stackrel{\text{def}}{=} \neg \exists c. \frac{\langle re(c) \rangle}{\langle re(c) \rangle}.$$

We define $asm \stackrel{\text{def}}{=} mutex \wedge no\ 2\ res$.

As we want to encode derivations, all reservations should be part of the representation of the derivation. Thus, all reservations should belong to the representation of some letter, as ensured by

$$all\ res\ in\ letter \stackrel{\text{def}}{=} \forall c. \langle re(c) \rangle \implies$$
$$\exists c'. \bigvee_{\substack{\sigma \in \mathcal{N}_D \cup \mathcal{N}_U \cup \\ \mathcal{T}_D \cup \mathcal{T}_U}} \langle letter_{\text{free}}(\sigma, c') \wedge (true \frown re(c) \frown true) \rangle,$$

under the assumption that asm holds. We point out that $re(c)$ in the premise and $re(c)$ in the conclusion refer to the same reservation, as every car has only one reservation. Further, $letter_{\text{free}}(\sigma, c')$ is evaluated on the same view as $(true \frown re(c) \frown true)$. Thus, this formula ensures that every reservation is inside the representation of a letter.

We want to ensure that all representations of letters are part of a derivation, i.e. we want to forbid orphaned letters. For this we demand that for all letters not on the topmost lane, there is a nonterminal above them. The formula

$$letter\ next\ to\ letter_D \stackrel{\text{def}}{=} \forall c. \bigwedge_{\sigma \in \mathcal{N}_D \cup \mathcal{T}_D} \left(\left\langle \begin{matrix} \langle \exists c'.\ free \vee re(c') \rangle \\ letter_{\text{free}}(\sigma, c) \end{matrix} \right\rangle \implies \right.$$
$$\left. \exists c''. \bigvee_{N \in \mathcal{N}_D} \left(\langle letter_{\text{free}}(N, c'') \rangle \wedge \left\langle \begin{matrix} letter(N, c'') \\ letter_{\text{free}}(\sigma, c) \end{matrix} \right\rangle \right) \right)$$

ensures this, where we use $\langle \exists c'.\ free \vee re(c') \rangle$ to match at least one lane, without regard for what is on the lane. Similar as in $step\ all_D$ we bind a car variable c to the occurrence of a letter σ in the premise of the implication. However, in the premise we additionally require that the letter is not located on the topmost lane. This is necessary, because we do not want to demand that there is another nonterminal above the starting nonterminal. In the conclusion we bind a new variable c'' to the occurrence of a nonterminal N and require that the representation of N is above and strictly larger than the representation of σ. Intuitively, $letter\ next\ to\ letter_D$ ensures that from any representation of a letter there is a sequence of vertically adjacent representations leading to S_D on the top lane. As $step\ all_D$ requires that all of these sequences obey the rewrite rules, now we can extract derivations from satisfying models.

Second Grammar. The formulas so far can be used to ensure that the MLSL representation of a derivation from the grammar G_D starts at the top lane and grows downwards. Now we add formulas that demand that a derivation from G_U starts at the bottom and grows upwards.

For all formulas ϕ_D defined so far we create a formula ϕ_U by replacing indices D with U in ϕ and swapping lower and higher formulas in vertical chop operators. For example we define

$$start_U \stackrel{\text{def}}{=} \exists c. \frac{true}{letter_{\text{free}}(S_U, c)},$$

$$step_U(N, c) \stackrel{\text{def}}{=} \frac{free}{free} \frown \left(\frac{\bigvee_{w \in \mathcal{R}_U^N} word(w)}{letter(N, c)} \right) \frown \frac{free}{free},$$

and the other formulas are similarly defined.

The formula

$$free\ lane \stackrel{\text{def}}{=} \frac{true}{\underset{true}{free}}$$

requires that there is at least one lane without any reservations. This, together with *letter next to letter*$_D$, ensures that below this free lane there are no representations of letters from G_D. If there was such a letter, then from that letter the sequence of vertically adjacent representations would be interrupted by a free lane. Symmetrically, there is no letter from G_U above this free lane.

We say for two words w, w' that w is a *subsequence* of w' iff we can create w from w' by only removing letters from w'. We now define that the derived word of one grammar is a subsequence of the derived word of the other grammar. For $i \in \{D, U\}$ and $\tau \in \mathcal{T}$ we use τ_i as abbreviation for $\pi_i(\tau)$. Let c, c' be car variables, then we define

$$\phi(\tau, c, c') \stackrel{\text{def}}{=} \left\langle \begin{matrix} free \\ true \\ free \end{matrix} \frown \left(\begin{matrix} letter(\tau_D, c) \\ true \\ letter(\tau_U, c') \end{matrix} \right) \frown \begin{matrix} free \\ true \\ free \end{matrix} \right\rangle,$$

which requires that the representations of the terminal τ using the variables c, c' are horizontally aligned. The horizontal alignment is enforced by ensuring that the formulas $letter(\pi_D(\tau), c_D)$ and $letter(\pi_U(\tau), c_U)$ are evaluated on the same extension. This is done by evaluating the horizontal chops before the vertical chops. Further, we define

$$subseq_D \stackrel{\text{def}}{=} \bigwedge_{\tau \in \mathcal{T}} \forall c. (\langle letter_{\text{free}}(\tau_D, c) \rangle \implies \exists c'. (\langle letter_{\text{free}}(\tau_U, c') \rangle \wedge \phi(\tau, c, c'))),$$

$$subseq_U \stackrel{\text{def}}{=} \bigwedge_{\tau \in \mathcal{T}} \forall c'. (\langle letter_{\text{free}}(\tau_U, c') \rangle \implies \exists c. (\langle letter_{\text{free}}(\tau_D, c) \rangle \wedge \phi(\tau, c, c'))).$$

Note that in $subseq_D$ and $subseq_U$ the subformula $\phi(\tau, c, c')$ is the same. However, the car variable names and the subscripts D and U outside $\phi(\tau, c, c')$ are swapped. The formula $subseq_D$ ensures that for every terminal τ, when the downward derivation contains the downward encoding $\pi_D(\tau)$ of τ, then the upwards derivation contains the upward encoding $\pi_U(\tau)$ of τ, horizontally aligned. In other words $subseq_D$ requires that each terminal from the downwards derivation has an horizontally aligned corresponding terminal from the upwards derivation. The horizontal alignment prevents that two downwards terminals share the same corresponding upwards terminal. The explanation for $subseq_U$ is symmetric. Because one word is a subsequence of the other word and vice versa, the two derived words are equal.

For the final formula we conjoin the downward and the upward formulas:

$$F(G_D, G_U) \overset{\text{def}}{=} \bigwedge_{i \in \{U, D\}} step\ all_i \wedge start_i \wedge letter\ next\ to\ letter_i \wedge subseq_i \wedge$$
$$free\ lane \wedge all\ res\ in\ letter \wedge asm.$$

Now we can state our first lemma. We can create an MLSL formula that is satisfiable iff there is a word derivable in two CFGs.

Lemma 1. *Given CFGs $G_D = (\mathcal{N}_D, \mathcal{T}, \mathcal{R}_D, S_D)$ and $G_U = (\mathcal{N}_U, \mathcal{T}, \mathcal{R}_U, S_U)$ we can create a formula $F(G_D, G_U)$ such that*

$$\exists w \in \mathcal{L}(G_D) \cap \mathcal{L}(G_U)\ \text{iff}\ \exists \mathcal{M}.\ \mathcal{M} \models F(G_D, G_U).$$

Proof. In the full version of this paper [18].

As the intersection emptiness of two context-free languages is undecidable [14], we obtain our first main theorem.

Theorem 1. *The satisfiability problem of MLSL is undecidable.*

4   Undecidability of Robust Satisfiability

The undecidability result for MLSL by Hilscher and Linker relies on length measurement [16] and in their construction the authors rely on exact values. Even though in our construction we do not use length measurement, we still rely on correct positional information. For example, we represented letters as non-overlapping, successive reservations without free space in between, i.e. when the position of a single car is shifted a small amount the resulting model does not satisfy the formula.

Here we show that MLSL remains undecidable, even when the position of cars along the lanes are perturbed. We chose positional perturbations as it seems realistic, considering that the position could be obtained via GPS and the lane via more accurate means. We say that a model $\mathcal{M}$ *robustly* satisfies a formula iff there is some $\epsilon > 0$ such that all models differing by at most ϵ (w.r.t. to a *metric*) from $\mathcal{M}$ also satisfy the formula. A formula is robustly satisfiable iff there is a model that robustly satisfies the formula. In the following we adapt our construction from Sect. 3 to this definition of robust satisfiability.

Our definition of robust satisfiability is similar to the definition of tube acceptance from [8], where a robust timed automaton accepts a trajectory iff the automaton also accepts all similar trajectories.

4.1 Robust Satisfiability of MLSL

We define a metric on models of MLSL and robust satisfiability of MLSL formulas w.r.t. to this metric. A metric can be understood to assign to every two models a distance.

Definition 10 (Metric on Models). *Let $\mathcal{M} = (\mathcal{TS}, V, \nu)$, $\mathcal{M}' = (\mathcal{TS}', V', \nu')$ with $\mathcal{TS} = (res, pos, br\text{-}dis)$, $\mathcal{TS}' = (res', pos', br\text{-}dis')$, $V = (L, [r, t])$ and $V = (L', [r', t'])$. Then we define $d(\mathcal{M}, \mathcal{M}') = \infty$ if $L \neq L'$ or $res \neq res'$ or $br\text{-}dis \neq br\text{-}dis'$ and*

$$d(\mathcal{M}, \mathcal{M}') = \max_{C \in \mathbb{I}}\{|r - r'|, |t - t'|, |pos(C) - pos'(C)|,$$
$$|(pos(C) + br\text{-}dis(C)) - (pos'(C) + br\text{-}dis'(C))|\}$$

otherwise.

Since we only use absolute values, the distance between two models always is positive. Further, it is not difficult to show that the triangle inequality is satisfied. Thus, d is indeed a metric. This means that the distance of two models is infinite, if they disagree on discrete values. Otherwise the distance is the greatest difference of any dense value.

With this we can define robust satisfaction of a formula.

Definition 11 (Robust Satisfaction). *Let $\mathcal{M}$ be a model and let ϕ be an MLSL formula. Then we say that $\mathcal{M}$ robustly satisfies ϕ, denoted by $\mathcal{M} \models^R \phi$, if and only if*

$$\exists \epsilon \in \mathbb{R}_{>0}. \forall \mathcal{M}'. d(\mathcal{M}, \mathcal{M}') \leq \epsilon \text{ implies } \mathcal{M}' \models \phi.$$

A formula is robustly satisfiable *iff there is a model that robustly satisfies it.*

Example 2. Consider the following formulas:

$$\phi_0 \stackrel{\text{def}}{=} c_0 \neq c_1 \wedge \langle re(c_0) \frown re(c_1) \rangle,$$

$$\phi_1 \stackrel{\text{def}}{=} \phi_0 \wedge \neg \exists c_2, c_3. c_2 \neq c_3 \wedge \langle re(c_2) \wedge re(c_3) \rangle.$$

The formula ϕ_0 requires that there are two successive reservations from different cars without free space in between. The model $\mathcal{M}_0$, depicted in Fig. 2(a), robustly satisfies ϕ_0 because the positions of the cars can be perturbed by a small amount without affecting satisfaction by the model. Hence, ϕ_0 is robustly satisfiable. The model, $\mathcal{M}_1$ (Fig. 2(b)) does not robustly satisfy ϕ_0 because if the position of c_1 is increased (moved to the right) by an arbitrary small amount there is free space between the reservations, which violates ϕ_0.

The formula ϕ_1 additionally requires that there are no overlapping reservations. Thus, $\mathcal{M}_0$ does not satisfy the formula. While $\mathcal{M}_1$ satisfies ϕ_1, moving a single car in any direction either creates overlapping reservations or free space, both of which violate the formula. In general ϕ_1 requires that the reservation of c_0 ends exactly where the reservation of c_1 starts. Naturally, this is not robustly satisfiable.

(a)   (b)

Fig. 2. (a) The model $\mathcal{M}_0$, which contains two overlapping reservations and (b) The model $\mathcal{M}_1$, which contains two reservations. The second reservation starts exactly where the first ends

4.2   Construction

We need to replace those parts of our construction, that are affected by small perturbations of positions. For this we adapt our representation of letters, because they are represented as successive reservations starting exactly where another reservation ends. Additionally, we have to adapt the *subseq*-formulas, because they ensured perfect alignment of letters, which is not possible in a setting with perturbations.

Letter. To represent letters we remove the assumption that there are no overlapping reservations. However, we still do not consider cars with multiple reservations. For a finite set $\mathcal{C} \subseteq CVar$ of car variables we define

$$only(\mathcal{C}) \stackrel{\text{def}}{=} \bigwedge_{c \in \mathcal{C}} re(c) \wedge \left(\forall c'. \left(true \frown re(c') \frown true \right) \implies \bigvee_{c'' \in \mathcal{C}} c'' = c' \right),$$

$$only_{\text{free}}(\mathcal{C}) \stackrel{\text{def}}{=} free \frown only(\mathcal{C}) \frown free ,$$

to ensure that the current view is filled by reservations from all car variables in $\mathcal{C}$, but does not contain reservations from any other cars. See Fig. 3 for a visualisation. Now we can define our representation of letters as

$$letter(\sigma, c) \stackrel{\text{def}}{=} startmarker(c) \frown \exists c_0, \ldots, c_{\mu(\sigma)-1}. \bigwedge_{\substack{i \neq j \\ i,j \in \{0, \ldots \mu(\sigma)-1\}}} c_i \neq c_j \wedge$$

$$only_{\text{free}}(\{c_0\}) \frown \ldots \frown only_{\text{free}}(c_{\mu(\sigma)-1}) \frown \exists e. \, endmarker(e) ,$$

$$startmarker(c) \stackrel{\text{def}}{=} only(\{c\}) \frown \exists c'. c \neq c' \wedge only(\{c, c'\}) \frown only(\{c'\}) ,$$

$$endmarker(e) \stackrel{\text{def}}{=} \exists c'. e \neq c' \wedge only(\{c'\}) \frown only(\{e, c'\}) \frown only(\{e\}) ,$$

where σ is either a terminal or a nonterminal and c, e are car variables. Our representation of letters begins and ends with two different markers. In between these markers the representation contains $\mu(\sigma)$ reservations. This representation of letters does not depend on exact positions of cars. In Fig. 3 we depict two adjacent letters with free space in between them.

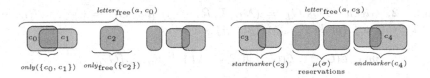

Fig. 3. Visualization of a model satisfying $letter_{\mathrm{free}}(a, c_0) \frown letter_{\mathrm{free}}(a, c_3)$ with $\mu(a) = 2$, where $letter_{\mathrm{free}}$ uses the new $letter$-formula. Reservations are shown as rectangles with rounded corners; different heights are used for better visualisation

Subsequence. In Sect. 3 we ensured that terminals at the same position in the derived words are horizontally aligned. With imprecise positions we cannot ensure such an alignment. We define the new subsequence formulas similar to their definition in Sect. 3. For $i \in \{D, U\}$ and $\tau \in \mathcal{T}$ we use τ_i as abbreviation for $\pi_i(\tau)$. Let c, c' be car variables, then we define

$$\psi(\tau, c, c') \stackrel{\text{def}}{=} \left\langle \begin{matrix} free \\ true \\ free \end{matrix} \frown \left(\begin{matrix} letter_{\mathrm{free}}(\tau_D, c) \\ true \\ letter(\tau_U, c') \end{matrix} \quad \vee \quad \begin{matrix} letter(\tau_D, c) \\ true \\ letter_{\mathrm{free}}(\tau_U, c') \end{matrix} \right) \frown \begin{matrix} free \\ true \\ free \end{matrix} \right\rangle,$$

which requires that one representation of τ is horizontally strictly contained within the other representation. Horizontal containment is ensured by aligning $letter$ with $letter_{\mathrm{free}}$. The disjunction represents that it does not matter which representation is the larger one. Further, we define

$$subseq_D \stackrel{\text{def}}{=} \bigwedge_{\tau \in \mathcal{T}} \forall c. (\langle letter_{\mathrm{free}}(\tau_D, c) \rangle \implies \exists c'. (\langle letter_{\mathrm{free}}(\tau_U, c') \rangle \wedge \psi(\tau, c, c'))),$$

$$subseq_U \stackrel{\text{def}}{=} \bigwedge_{\tau \in \mathcal{T}} \forall c'. (\langle letter_{\mathrm{free}}(\tau_U, c') \rangle \implies \exists c. (\langle letter_{\mathrm{free}}(\tau_D, c) \rangle \wedge \psi(\tau, c, c'))).$$

As before, the subformula $\psi(\tau, c, c')$ is the same in $subseq_D$ and $subseq_U$ and we swapped the car variable names and the subscripts D and U outside $\psi(\tau, c, c')$.

All other formulas remain as they are, only that they use the new $letter$ formula. The final formula looks almost as before, with the exception that we do not forbid overlapping reservations. This does not pose a problem, because still all reservations are required to be inside a letter and there we exactly specified which reservations we allow. As already mentioned the final formula does not contain $mutex$ anymore. We define

$$F_{\text{robust}}(G_D, G_U) \overset{\text{def}}{=} \bigwedge_{i \in \{U,D\}} \textit{step all}_i \wedge \textit{start}_i \wedge \textit{letter next to letter}_i \wedge \textit{subseq}_i \wedge$$

$$\textit{free lane} \wedge \textit{all res in letter} \wedge \textit{no 2 res}.$$

The following lemma reduces the intersection problem of two CFGs to the robust satisfiability problem.

Lemma 2. *Given CFGs* $G_D = (\mathcal{N}_D, \mathcal{T}, \mathcal{R}_D, S_D)$ *and* $G_U = (\mathcal{N}_U, \mathcal{T}, \mathcal{R}_U, S_U)$ *we can create a formula* $F_{\text{robust}}(G_D, G_U)$ *such that*

$$\exists w \in \mathcal{L}(G_D) \cap \mathcal{L}(G_U) \text{ iff } \exists \mathcal{M}. \mathcal{M} \models^R F_{\text{robust}}(G_D, G_U).$$

Proof. In the full version of this paper [18].

Thus, we get our second theorem.

Theorem 2. *The robust satisfiability problem of MLSL is undecidable.*

5   Discussion

For our constructions we do not require some of the features of original MLSL, i.e. claims and the ego constant. Hence, we consider a simplified logic without these features. As original MLSL is a conservative extension of our simplified logic, our results also apply to original MLSL. Further, our constructions do not require that a car has multiple reservations. Still we allowed for multiple reservations to not deviate from the original logic too far.

In the following we discuss some possible restrictions on the model side, to perhaps arrive at a decidable logic. Further, we discuss whether the resulting restricted logic remains undecidable.

As both of our constructions rely only on topological arguments, the constructions still can be used to prove undecidability after various restrictions on the model are imposed. However, both of our constructions require that all terminals of a derivation are horizontally contained in the representation of the starting nonterminal of that derivation without overlapping with another letter. As a derivation contains an unbounded amount of terminals, for our constructions to work the maximum size of the representation of the starting nonterminal needs to be unbounded, or the minimum size of terminals needs to be unbounded.

Our construction from Sect. 3 requires that there is no free space in between reservations of the same letter. This implies that the minimum and maximum size of letters are bounded by the minimum and maximum size of reservations. Hence, when we impose bounds on both, by the above argument the construction does not work anymore, i.e. there are context-free grammars for which $F(G_D, G_U)$ is unsatisfiable even though the intersection of the languages is not empty.

Our construction from Sect. 4 does not restrict the size of reservations or the free space in between reservations in any way. Thus, the maximum size of a letter is not bounded by the maximum size of a reservation. To restrict the maximum size of a letter we could bound the maximum size of a view. Then, by the above argument we can not use our construction to answer whether the robust satisfiability problem of the resulting restricted logic is undecidable. Note that undecidability of robust satisfiability implies undecidability of classical satisfiability.

Below we name the mentioned restrictions, and we introduce discreteness of the horizontal dimension as new restriction:

P_0: The maximum size of reservations (given by $br\text{-}dis$) is bounded.
P_1: The maximum size of the view is bounded.
Q_0: The minimum size of reservations is bounded.
Q_1: The horizontal domain is discrete.

Note that in effect P_1 implies P_0, and that Q_1 implies Q_0. In Table 1 we summarize our arguments from above using the names introduced for our restrictions.

Table 1. The table shows under which combination of restrictions on the model our constructions still serve to prove undecidability of the (robust) satisfiability problem ($\checkmark$). With X we mark combinations where our constructions do not work. In the table let $R \in \{P_0, P_1, Q_0, Q_1\}$ and $i \in \{0, 1\}$

	R	P_0 and Q_i	P_1 and Q_i
Construction from Sect. 3	$\checkmark$	X	X
Construction from Sect. 4	$\checkmark$	$\checkmark$	X

At last we point out that our constructions strongly depend on the unboundedness of the number of lanes and cars.

6   Conclusion

As our first result we proved undecidability of MLSL without length measurement via a reduction from the emptiness of the intersection of two context free grammars. As our second result we proved that the logic remains undecidable even when the position along the lane is known only approximately. This proof also works with restrictions of MLSL, e.g. when the extension of the lanes is discrete instead of dense or when the minimal or maximal size of reservations are bounded or any combination of these.

In future work it is worthwhile to create a connection between MLSL and other well studied logics. This may lead to a better understanding of MLSL and increase interest in this logic. Consider for example the modal logic of intervals Halpern-Shoam-logic (HS) [9]. There we have a set of atomic propositions and intervals over a domain, e.g. the real numbers. Every interval satisfies an atomic proposition or its negation. Additionally, we can use for each of Allen's interval relations (before, after, begins etc.) [2] a modal operator that captures the semantics of the relation. In a reduction from HS to MLSL, car reservations might correspond to intervals, (negated) propositions to lanes and the horizontal chop operator to modal operators in HS.

Additionally, we are interested in decidable fragments of MLSL. Possible fragments could be obtained by restricting the nesting of vertical and horizontal chops or imposing an upper bound on the possible number of lanes. Further, it might be interesting to consider simpler modal operators, such as the unary left and right neighbourhood modalities from [4,9] instead of the chop operators.

Acknowledgements. I thank Manuel Gieseking, Martin Hilscher, Sven Linker, Ernst-Rüdiger Olderog and Maike Schwammberger for helpful discussions and proofreading. Additionally, I would like to thank the anonymous reviewers of this paper and of a previous version of this paper for valuable feedback.

References

1. Akshay, S., Bérard, B., Bouyer, P., Haar, S., Haddad, S., Jard, C., Lime, D., Markey, N., Reynier, P.A., Sankur, O., Thierry-Mieg, Y.: Overview of robustness in timed systems. Citeseer (2012)
2. Allen, J.F.: Maintaining knowledge about temporal intervals. Commun. ACM **26**(11), 832–843 (1983)
3. Alur, R., Feder, T., Henzinger, T.A.: The benefits of relaxing punctuality. J. ACM **43**(1), 116–146 (1996)
4. Zhou, C., Hansen, M.R.: An adequate first order interval logic. In: de Roever, W.-P., Langmaack, H., Pnueli, A. (eds.) COMPOS 1997. LNCS, vol. 1536, p. 584. Springer, Heidelberg (1998)
5. Chaochen, Z., Hansen, M.R., Sestoft, P.: Decidability and undecidability results for duration calculus. In: Enjalbert, P., Wagner, K.W., Finkel, A. (eds.) STACS 1993. LNCS, vol. 665. Springer, Heidelberg (1993)
6. Chaochen, Z., Hoare, C.A.R., Ravn, A.P.: A calculus of durations. Inf. Process. Lett. **40**(5), 269–276 (1991)
7. Fränzle, M., Hansen, M.R.: A robust interpretation of duration calculus. In: Van Hung, D., Wirsing, M. (eds.) ICTAC 2005. LNCS, vol. 3722, pp. 257–271. Springer, Heidelberg (2005)
8. Gupta, V., Henzinger, T., Jagadeesan, R.: Robust timed automata. In: Maler, O. (ed.) HART 1997. LNCS, vol. 1201. Springer, Heidelberg (1997)
9. Halpern, J.Y., Shoham, Y.: A propositional modal logic of time intervals. J. ACM **38**(4), 935–962 (1991)
10. Henzinger, T.: The Temporal Specification and Verification of Real-time Systems. Ph.D. thesis, Stanford University (1991)
11. Henzinger, T.A., Raskin, J.-F.: Robust undecidability of timed and hybrid systems. In: Lynch, N.A., Krogh, B.H. (eds.) HSCC 2000. LNCS, vol. 1790, pp. 145–159. Springer, Heidelberg (2000)
12. Hilscher, M., Linker, S., Olderog, E.-R.: Proving safety of traffic manoeuvres on country roads. In: Liu, Z., Woodcock, J., Zhu, H. (eds.) Theories of Programming and Formal Methods. LNCS, vol. 8051, pp. 196–212. Springer, Heidelberg (2013)
13. Hilscher, M., Linker, S., Olderog, E.-R., Ravn, A.P.: An abstract model for proving safety of multi-lane traffic manoeuvres. In: Qin, S., Qiu, Z. (eds.) ICFEM 2011. LNCS, vol. 6991, pp. 404–419. Springer, Heidelberg (2011)

14. Hopcroft, J., Ullman, J.: Introduction to Automata Theory, Languages, and Computation. Addison Wesley, New York (1979)
15. Koymans, R.: Specifying real-time properties with metric temporal logic. Real-Time Syst. **2**(4), 255–299 (1990)
16. Linker, S., Hilscher, M.: Proof theory of a multi-Lane spatial logic. In: Liu, Z., Woodcock, J., Zhu, H. (eds.) ICTAC 2013. LNCS, vol. 8049, pp. 231–248. Springer, Heidelberg (2013)
17. Moszkowski, B.: A temporal logic for multi-level reasoning about hardware. IEEE Comput. **18**(2), 10–19 (1985)
18. Ody, H.: Undecidability results for multi-Lane-spatial-logic. Reports of SFB/TR 14 AVACS 112, SFB/TR 14 AVACS (2015). http://www.avacs.org
19. Schäfer, A.: A calculus for shapes in time and space. In: Liu, Z., Araki, K. (eds.) ICTAC 2004. LNCS, vol. 3407, pp. 463–477. Springer, Heidelberg (2005)
20. Venema, Y.: A modal logic for chopping intervals. J. Log. Comput. **1**(4), 453–476 (1991)
21. Woodcock, J., Davies, J.: Using Z – Specification, Refinement, and Proof. Prentice Hall, New York (1996)

Software Architecture
and Component-Based Design

Aspect-Oriented Development of Trustworthy Component-based Systems

José Dihego$^{(\boxtimes)}$ and Augusto Sampaio

Centro de Informática, Universidade Federal de Pernambuco, Recife, PE, Brazil
{jdso,acas}@cin.ufpe.br

Abstract. In this paper we integrate the aspect-oriented paradigm into a process algebra based component model ($\mathcal{BRIC}$), where correctness is guaranteed by construction. We contribute with an approach to capture, specify and use aspects to safely evolve component-based systems. We establish that components extended by aspects preserve a convergence relation that guarantees service conformance. We illustrate our results by presenting a case study of an autonomous healthcare system.

Keywords: Component-based aspect-oriented design · Correctness by construction · Behavioural convergence · CSP

1 Introduction

Aspect-oriented programming (AOP) [10] has been presented as a theory to capture, define and modularise crosscutting concerns, which spread through applications. Aspects are hard to modularise by using conventional programming units, such as functions or classes, because they tend not to be units of the system's functional decomposition, but rather to be properties that affect the semantics of regular components [10]. Some aspects are so recurring that have become classical examples of when and how applying AOP: context-sensitive behaviour, performance optimizations, monitoring, logging, persistence, distribution, security and transactional management, just to name a few. Successful AOP technologies are used both in industry and academy, as AspectJ [9] for Java and AspectC [2] for the C language.

Aspect-oriented programming is part of a broader development strategy, Aspect-Oriented Software Development (AOSD) [6], which brings aspect-oriented analyses to early phases of software development. The incorporation of aspect-oriented design (AOD) in the modelling phase creates a more accurate and natural correspondence between specification and implementation, helping to improve development activities that tend to be very costly, such as testing and maintenance.

A promising scenario that we investigate here is the integration of AOSD and component-based model driven development (CB-MDD) [15], a well recognised approach to develop complex systems, which are built from simpler ones, called components, with well-defined interface and behaviour. This integration provides

M. Leucker et al. (Eds.): ICTAC 2015, LNCS 9399, pp. 425–444, 2015.
DOI: 10.1007/978-3-319-25150-9_25

a way to capture and modularise, in early specification phases, some system aspects that otherwise would be scattered through regular components. As far as we are aware, there is no work that explores the integration between AOD and CB-MDD formal approaches, which is the main goal of this work.

We contribute by formalising how to capture, model and weave aspects with components in the $\mathcal{BRIC}$ component model [12,13], which formalises the core CB-MDD concepts and, moreover, supports compositions, where behavioural properties, such as deadlock freedom, are ensured by construction. Our aspect-oriented design (AOD) approach supports safely extension of $\mathcal{BRIC}$ components by preserving a notion of service conformance given in terms of a convergence relation [4] that is suitable to capture model evolution.

This work is organised as follows: Sect. 2 presents the $\mathcal{BRIC}$ component model [13]. Section 3 presents the first contribution of this work, where we stablish how to define and weave aspects into component-based specifications. In Sect. 4 we prove that aspects guarantee service conformance by establishing how it relates with behavioural convergence [4]. Our results are illustrated by a case study of an autonomous healthcare system in Sect. 5. We conclude with our contributions and discuss related work in Sect. 6.

2   The BRIC Component Model

The $\mathcal{BRIC}$ component model specifies components, connectors, their behaviour (given in the Communicating Sequential Processes (CSP) language [14]) and the rules by which they are assembled. Global behavioural properties (e.g., deadlock freedom) of compositions using the $\mathcal{BRIC}$ rules are guaranteed by construction, which implies they can be proved by local analyses. This is a direct consequence of the conditions imposed by the rules as demonstrated in [13].

2.1   CSP

A process algebra like CSP can be used to describe systems composed of interacting components, which are independent self-contained processes with interfaces used to interact with the environment. Such formalisms provide a way to explicitly specify and reason about interactions between different components. Furthermore, phenomena that are exclusive to the concurrent world, that arise from the combination of components and not individual components, like deadlock and livelock, can be more easily understood and controlled using such formalisms. Tool support is another reason for the success of CSP in industrial applications, and consequently, for our choice to use it as the formal notation. For instance, FDR3 [7] provides an automatic analysis of model refinement and of properties like deadlock and divergence.

The two basic CSP processes are STOP (deadlock) and SKIP (successful termination). The prefixing c -> P is initially able to perform only the event c; afterwards it behaves like process P. The prefixing choice c?x -> P inputs a value through channel c and assigns it to the variable x, and then behaves like

P, which has the variable x in scope. Multiple inputs and outputs are also possible. For instance, c?x?y!z inputs two values that are assigned to x and y and outputs the value z. A Boolean guard may be associated with a process: given a predicate g, if the condition g holds, the process g&P behaves like P; it deadlocks otherwise. It can also be defined as if g then P else STOP.

The sequential composition P1;P2 behaves like P1 and, provided it terminates successfully, P2 takes over. The external choice P1 [] P2 initially offers events of both processes. The occurrence of the first event resolves the choice in favor of the process that performs it. The environment has no control over the internal choice P1 |~| P2. The parallel composition P1 [| cs |] P2 synchronizes P1 and P2 on the channels in the set cs; events that are not in cs occur independently. Processes composed in interleaving P1 ||| P2 run independently. The event hiding operator P \ cs encapsulates the events that are in the channel set cs, which become no longer visible to the environment. These operators can also appear indexed by non empty sets of events. For example, the process [] x:A @ P(x) behaves as the external choice between x -> P(x) for each x ∈ A (A is a finite and non-empty set of events).

In this work we use two denotational models of CSP: traces ($\mathcal{T}$) and stable failures or just failures ($\mathcal{F}$). Let Σ be the alphabet of all possible events and Σ^* the set of all possible finite sequences of events in Σ, then: a trace of a process P is a sequence (a member of Σ^*) of events that it can perform and $\mathcal{T}(P)$ denotes the set of all its finite traces. For example $\mathcal{T}(e1 \rightarrow e2 \rightarrow STOP) = \{\langle\rangle, \langle e1\rangle, \langle e1, e2\rangle\}$. The function application $\mathcal{F}(P)$ consists of all stable failures (s, X), where s is a trace of P $(s \in \mathcal{T}(P))$ and X is a set of events P can refuse in some stable state after s. A stable state is one that can only be changed by a visible event (registered in the process trace). The unique invisible event CSP has is τ; it can happen, for example, in the internal choice P |~| Q, which will be implemented as a process which can take an invisible τ event to decide the choice of P or Q. We also use α P(α P $\subseteq \Sigma$) to denote the (alphabet) set of events P can communicate and |c| to stand for the set of events that can communicate through a channel c.

As an example consider the CSP process SRV. It communicates the event srv.in.v.1, where srv is a channel of type I_DTSER and SUB_INT is a finite subset of $\mathbb{N}$; the tags in and out distinguish input from output events and v indicates the communication of a SUB_INT value. The process SRV offers the environment the choice between srv.in.v.1 and srv.in.v.2; if it synchronises on the first then SRV communicates, non deterministically, srv.out.v.1 or srv.out.v.2 and recurses. The same reasoning applies in the case the environment synchronises on srv.in.v.2.

```
datatype I_DTSER = in.v.SUB_INT | out.v.SUB_INT
channel srv : I_DTSER
SRV =
  srv.in.v.1 -> (srv.out.v.1 -> SRV |~| srv.out.v.2 -> SRV)
  []
  srv.in.v.2 -> (srv.out.v.3 -> SRV |~| srv.out.v.4 -> SRV)
```

2.2 $\mathcal{BRIC}$

A component is defined as a contract (Definition 1) that specifies its behaviour, communication points (channels) and their types.

Definition 1 (Component Contract). *A component contract Ctr : ⟨B, R, J, C⟩ comprises an observational behaviour B specified as a CSP process, a set of communication channels C, a set of interfaces J (data types), and a total function R : C ↦ J mapping channels to interfaces, such that B is an I/O process.*

We require the CSP process $\mathcal{B}$ to be an I/O process, which is a non-divergent process with infinite traces, but a finite state space. Moreover, it offers to the environment the choice over its inputs (external choice) but reserves the right to choose between its outputs (internal choice). It represents a widely range of specifications, including the server-client protocol, where the client sends requests (inputs) to the server, which decides the outputs to be returned to the client. The CSP process SRV we met before is actually an I/O process by Definition 2 and, by Definition 1, it can be encapsulated as the $\mathcal{BRIC}$ component $Ctr_{\text{SRV}} = \langle \text{SRV}, \{\text{srv} \mapsto \text{I\_DTSER}\}, \{\text{I\_DTSER}\}, \{\text{srv}\}\rangle$.

Definition 2 (I/O Process). *We say that a CSP process P is an I/O process if it satisfies the following five conditions, which are formally presented in [13]:*

*(1) **I/O channels**: Every channel in P has its events partitioned between inputs and outputs.*
*(2) **infinite traces**: P has an infinite set of traces (but finite state-space).*
*(3) **divergence-freedom**: P is divergence-free.*
*(4) **input determinism**: If a set of input events in P are offered to the environment, none of them are refused.*
*(5) **strong output decisive**: All choices (if any) among output events on a given channel in P are internal; the process, however, must offer at least one output on that channel.*

Contracts can be composed using any of the four rules available in the model: interleaving, communication, feedback, or reflexive. Each of these rules impose different side conditions, which must be satisfied by the contracts and channels involved in the composition in order to guarantee deadlock freedom by construction.

The rules provide asynchronous pairwise compositions, mediated by infinite buffers, and focus on the preservation of deadlock freedom in the resulting component. Using the rules, developers may synchronise two channels of two components, or even of the same component. The four rules are illustrated in Fig. 1. The result of an application of a composition rule is a new component.

The interleave composition rule is the simplest form of composition. It aggregates two independent entities such that, after composition, these entities still do not communicate between themselves. They directly communicate with the environment as before, with no interference from each other. The communication composition states the most common way for linking complementary channels of

Fig. 1. Composition rules

two different entities. The next two compositions allow the link of two complementary channels of a same entity. First, the feedback composition provides the possibility of creating safe cycles for systems with a tree topology. In practice, however, there are more complex systems that indeed present cycles of dependencies in the topology of the system structure. The last composition rule, *reflexive composition*, is more general than the feedback one and allows cyclic topologies. However, in general it requires a global analysis to ensure deadlock freedom.

3   Aspect Oriented Modelling for BRIC

In the aspect oriented paradigm, a join point is a well-defined point in a program execution, established in terms of its control flow and/or predicates over its state space. A pointcut stands for a collection of join points grouped by a common predicate. We can consider a join point as the result of a behavioural pattern matching, a dynamic entity that serves the purpose of identifying the points where advices (behaviour modifiers) can be incorporated. An advice is an action (behaviour) that is designed to be offered on a set of related join points, grouped into a pointcut. The process of causing the relevant advice to be offered on join points is called *weaving*, which is by definition a dynamic process [16].

In our AOD approach for $\mathcal{BRIC}$, a join point represents an I/O process stable state, one without internal transitions leading out from it. A pointcut groups a set of related join points. An advice is an I/O process that is designed to be weaved with $\mathcal{BRIC}$ components at a set of join points highlighted (marked) in their behaviour. An aspect is a pair formed of an advice and a pointcut, which expresses the intention to weave such an advice into the pointcut join points. The fact that we have a common language for describing component behaviour, pointcuts and advices has a tremendous positive impact on dealing with the complexity of our aspect-component weaver (ACW) [10]. This task is reduced to combine CSP processes, which benefit from an extensive set of operators [14].

A pointcut raises a set of join points that need to be marked, so the ACW (presented later in this section) is able to identify where advices must be plugged. We use as marks the events communicated through a special kind of channel, a *check mark* channel. A check mark channel c communicates only two events, $c.bf$ and $c.af$, which are offered sequentially ($c.bf$ then $c.af$) acting as join points marks (Definition 3). The event $c.bf$ must be an input; so, according to the input determinism property (Definition 2), a decision to mark a join point is always predictable (deterministic).

Recall the component Ctr_{SRV} we presented in Sect. 2 and suppose we want to mark, for logging purposes, every time it communicates srv.in.v.2 followed by

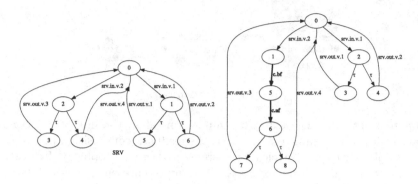

Fig. 2. Marking join points

`srv.out.v.3` or `srv.out.v.4`. Figure 2 shows the SRV's LTS (for visualisation purposes) and the resulting process after marking the described join point with the events communicated through the `channel c : {bf,af}`.

Definition 3 (Check Mark Channel). *Consider an I/O channel c, we say that c is a check mark channel if and only if:* $\{|c|\} = \{c.bf, c.af\} \wedge c.bf \in inputs$.

The pattern and the mark events used to define and signalise join points are encapsulated in a *pointcut designator* (or just a *pointcut*, Definition 4). It is defined by a CSP process P, which marks join points by offering the events of a check mark channel c. For example, consider the process `P1` defined as follows:

```
P1 = srv.in.v.2 -> c.bf-> c.af ->
     (srv.out.v.3 -> P1 [] srv.out.v.4 -> P1)
αP1 = union({srv.in.v.2,srv.out.v.3, srv.out.v.4}, {|c|})
```

Furthermore, consider the LTS depicted on the right-hand side of Fig. 2. It is the result of the pointcut $p : \langle P1, c \rangle$ application over SRV. After engaging in `srv.in.v.2`, `P1` does `c.bf` and `c.af` sequentially, then offers the choice between `srv.out.v.3` and `srv.out.v.4`, returning, in either case, to its initial state. Note that `P1` is not an I/O process (it does not satisfy the strong output decisiveness property of Definition 2 by offering outputs in external choice), although its composition with SRV (SRV `[|diff(αP1, |c|)|]` P1) is, as we can see by analysing the LTS of the composition on the right-hand side of Fig. 2.

Definition 4 (Pointcut Designator). *Consider a check mark channel c and a CSP process P. We say that pcd : $\langle P, c \rangle$ is a pointcut designator if and only if:*

(i) P is deterministic and deadlock-free;

(ii) $\forall t \in traces(P) \bullet t \downarrow c.af \leq t \downarrow c.bf \leq t \downarrow c.af + 1$. P never communicates more c.af events than c.bf events, and neither do they fall more than one behind;

(iii) $\exists t \,\hat{}\, \langle c.bf \rangle \in traces(P) \implies (t \,\hat{}\, \langle c.bf \rangle, \Sigma \setminus \{c.af\}) \in failures(P)$. If P communicates c.bf then it rejects everything but c.af;

(iv) $\exists t \,^\frown\langle c.bf \rangle \in traces(P) \implies (t, \Sigma \setminus \{c.bf\}) \in failures(P)$. *If P can commu-nicate c.bf then it rejects everything but c.bf.*

Given a component behaviour $\mathcal{B}$ (an I/O process), P performs $c.bf$ and $c.af$, sequentially, every time $\mathcal{B}$ synchronises on its behavioural pattern. The behavioural properties of I/O processes (Definition 2), especially strong output decisiveness and input determinism, associated with the pointcut well-formed conditions (Definition 4) allow us to state that:

(i) if P offers, initially, a set of input events *iset*, then $\mathcal{B}$ matches P, if it offers, initially, at least one element of *iset*;
(ii) if P offers, initially, a set of output events *oset*, then $\mathcal{B}$ matches P, if it offers, initially, only a non-empty subset of *oset*;
(iii) inductively, the above holds for any subsequent state of P and $\mathcal{B}$, provided $P \setminus \{|c|\}$ and $\mathcal{B}$ have performed the same trace.

The check mark channel c should only be used to mark join points, so $\{|c|\}$ (the set of events communicated through c), *iset* and *oset* are disjoint. Since P is deterministic [14]: (i) is a consequence of input determinism; (ii) an implication of the strong output decisiveness property; (iii) emerges by the fact that the events $\{|c|\}$ are business meaningless, so $\{|c|\} \cap \alpha\mathcal{B} = \emptyset$. Therefore, for $\mathcal{B}$, it does not matter if P engages in any event via channel c, so $\mathcal{B}$ is concerned only with the process $P \setminus \{|c|\}$, which behaves as P, where all events from $\{|c|\}$ are hidden from the environment.

We have established the means to define, locate and mark join points. Next, we discuss how to define the behaviour we want to weave at them. An *advice* (Definition 5) comprises an I/O process A (obeying some well-formed conditions), which is designed to be weaved at join points marked by a check mark channel c and to interact with its environment through a set of I/O channels C.

Definition 5 (Advice). *Let A, C and c stand, respectively, for a CSP process, a set of I/O channels and a check mark channel. We say that adv : $\langle A, C, c \rangle$ is an advice if and only if:*

(i) $A \setminus \{|c|\}$ *is an I/O process.*
(ii) $c \notin C$ *and* $\alpha(A \setminus \{|c|\}) \subseteq \{|C|\}$*. The events performed by A are signalled in its interface, except those communicated through c;*
(iii) $\forall t \in traces(A) \bullet t \downarrow c.af \leq t \downarrow c.bf \leq t \downarrow c.af + 1$*. A never communicates more c.af's than c.bf's, and neither do they fall more than one behind;*
(iv) $\exists t \,^\frown\langle c.bf \rangle \in traces(P) \implies (t, \Sigma \setminus \{c.bf\}) \in failures(P)$*. If P can communicate c.bf then it rejects everything but c.bf;*
(v) $\exists t \,^\frown\langle c.af \rangle \in traces(P) \implies (t, \Sigma \setminus \{c.af\}) \in failures(P)$*. If P can communicate c.af then it rejects everything but c.af.*

Definition 5 requires the advice visible behaviour $(A \setminus \{|c|\})$ to be an I/O process, which is not enforced on the pointcuts behaviour (Definition 4). This is because advices are designed to change how a component behaves by being weaved into

its behaviour, which must be, by Definition 1, an I/O process; on the other
hand, pointcuts are designed to just mark join points into components behaviour,
leaving unchanged their external behaviour (the behaviour exhibited though
their interface), therefore they do not need to behave as I/O processes.

As an example, consider the join points marked over the component Ctr_{SRV}
by the pointcut p : $\langle P1, c \rangle$ (on the right-hand side of Fig. 2). The advice
a : $\langle A, \{db\}, c \rangle$ comprises the logging I/O process A, which is designed to be
weaved at each join point marked by c and to interact with its environment by
the channel db. By communicating c.bf and c.af it knows where A must be
weaved. The I/O process A communicates the event db.out.save.v.2 to log
in a database the fact that its target server Ctr_{SRV} has performed srv.in.v.2,
followed by srv.out.v.3 or srv.out.v.4. Figure 3 shows the result component
behaviour after weaving the advice a (on the right-hand side) at the join point
marked by the pointcut p (on the left-hand side).

```
subtype I_DB = in.save.v.SUB_INT | out.save.v.SUB_INT
channel db : I_DB
A = c.bf-> db.out.save.v.2 -> c.af -> A
```

Advices can present different levels of encapsulation, depending on how their
functional behaviour $(A \upharpoonright C)$ are positioned in relation to $c.bf$ and $c.af$ marking
events. We say that *adv* is *self-contained* if A, when projected over C, is confined
between $c.bf$ and $c.af$ (Definition 6). This is the case of a : $\langle A, \{db\}, c \rangle$.

Finally, we define an *aspect* as a tuple comprising a pointcut (Definition 4)
and an advice (Definition 5). In Definition 7: (i) ensures *pcd* and *adv* agree on
the mark events used to signalise join points and (ii) requires the *pcd.P* and
adv.A alphabets, except for $\{\!| c |\!\}$, to be disjoint; it avoids possible interferences
between *adv.A* and *pcd.P*.

Definition 6 (Self-contained Advice). *We say that an advice adv* : $\langle A, C, c \rangle$
is self-contained if and only if:

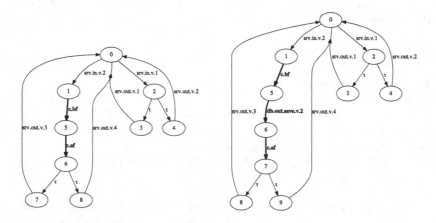

Fig. 3. Weaving advices into join points

(i) $(\langle\rangle, \Sigma \setminus \{c.bf\}) \in failures(A)$. *Initially, A rejects everything but c.bf;*
(ii) $\exists t \,^\frown \langle c.af \rangle \in traces(A) \implies (t \,^\frown \langle c.af \rangle, \Sigma \setminus \{c.bf\}) \in failures(A)$. *If A performs c.af, then it rejects everything but c.bf.*

Definition 7 (Aspect). *An aspect asp* : $\langle pcd, adv \rangle$ *is a tuple consisting of a pointcut pcd and an advice adv such that:*

(i) $adv.c = pcd.c$ *(ii)* $(\alpha\ pcd.P \cap \alpha\ adv.A) \subseteq \{adv.c\}$

Given an aspect asp : $\langle pcd, adv \rangle$ and a component T : $\langle \mathcal{B}, \mathcal{R}, \mathcal{I}, \mathcal{C} \rangle$, the task of our ACW is to search over $\mathcal{B}$ for all join points defined by pcd, mark them, weave the adv's behaviour at these join points and, finally, extend $\mathcal{C}$ and $\mathcal{I}$ by the adv's channels and their types, respectively. Definition 8 presents the ACW partial (not all aspects and components are weave-able) operator.

Definition 8 (Aspect-Component Weaver). *Consider a $\mathcal{BRIC}$ component T : $\langle \mathcal{B}, \mathcal{R}, \mathcal{I}, \mathcal{C} \rangle$ and an aspect asp* : $\langle pcd, adv \rangle$*, such that adv* : $\langle A, C, c \rangle$ *and pcd* : $\langle P, c \rangle$ *. We define the ACW (infix) partial operator* $\_ \overset{\times}{\times} \_$: *Aspect* $\times$ $\mathcal{BRIC} \rightarrow \mathcal{BRIC}$

$$asp \overset{\times}{\times} T = \langle (\mathcal{B}\ [|ss|]\ P)\ [|c|]\ A, \mathcal{R} \cup R', \mathcal{I} \cup I', \mathcal{C} \cup C \rangle$$

where

$ss = \alpha P \setminus \{c\} \;\wedge\; I' = \{type(ch) \mid ch \in C\} \;\wedge\; R' = \{ch \mapsto type(ch) \mid ch \in C\}$
provided $\mathcal{C} \cap C = \emptyset \;\wedge\; (\mathcal{B}\ [|ss|]\ P) \setminus \{c\} \equiv_F \mathcal{B}$

The **provided** clause above restricts the possible compositions between aspects and components: the act of discovering and marking join points does not change the overall behaviour of the target components ($\mathcal{B}\ [|ss|]\ P \setminus \{c\} \equiv_F \mathcal{B}$, where $\equiv_F$ means failure equivalence [14]). Therefore, the mark events $c.bf$ and $c.af$ do not change, from the environment perspective, the behaviour of T; pcd marks at least one join point on $\mathcal{B}$, so $\mathcal{B}\ [|ss|]\ P$ deadlocks if both processes cannot agree on the behavioural pattern described by P at least once (the previous failure equivalence implies $\mathcal{B}\ [|ss|]\ P$ is deadlock free); ($\mathcal{C} \cap C = \emptyset$) the interaction points of adv and T are disjoint, so when A is weaved into T there is no clash between their interfaces.

The behaviour of $asp \overset{\times}{\times} T$ is given by $(\mathcal{B}\ [|ss|]\ P)\ [|c|]\ A$, where $ss = \alpha P \setminus \{c\}$.

The ACW operator $\overset{\times}{\times}$ acts in two phases over $\mathcal{B}$: by synchronising it with P on $\alpha P \setminus \{c\}$, it marks join points, and then by putting this result in parallel with A on $\{c\}$, it inlays A at each join point (between $c.bf$ and $c.af$) marked on $\mathcal{B}$. Concerning the result component structure, the interaction points of $asp \overset{\times}{\times} T$ are $\mathcal{C} \cup C$, the types of C (I') and $\mathcal{I}$ from its interface, and interaction points and interfaces are related by $\mathcal{R} \cup R'$, where R' relates C and I'. Note that in Definition 8 we have assumed the behaviour of $asp \overset{\times}{\times} T$ is an I/O process, which is proved by Lemma 1.

Lemma 1 ($\mathcal{BRIC}$ is Closed Under ACW). *Consider the $\mathcal{BRIC}$ component $T : \langle \mathcal{B}, \mathcal{R}, \mathcal{I}, \mathcal{C} \rangle$ and the aspect $asp : \langle pcd, adv \rangle$, such that $pcd : \langle P, c \rangle$ and $adv : \langle A, C, c \rangle$. If $asp \overset{\times}{\times} T$ is defined, then it is a $\mathcal{BRIC}$ component.*

Proof.

Part i ($asp \overset{\times}{\times} T$ structure)

$[I' = \{type(ch) \mid ch \in C\}, \quad R' = \{ch \mapsto type(ch) \mid ch \in C\}, \quad Definition\ 8]$

$\mathcal{R}_{asp \overset{\times}{\times} T} = \mathcal{R} \cup R' \ \wedge \ \mathcal{I}_{asp \overset{\times}{\times} T} = \mathcal{I} \cup I' \ \wedge \ \mathcal{C}_{asp \overset{\times}{\times} T} = \mathcal{C} \cup C$

$\implies \mathcal{R}$ *is a total function between* $\mathcal{C}$ *and* $\mathcal{I} \wedge \mathcal{R}'$ *is a total function between* C *and* I'

$\implies [\mathcal{C} \cap C = \emptyset] \ \mathcal{R}_{asp \overset{\times}{\times} T}$ *is a total function between* $\mathcal{C}_{asp \overset{\times}{\times} T}$ *and* $\mathcal{I}_{asp \overset{\times}{\times} T}$ □

Part ii ($asp \overset{\times}{\times} T$ behaviour)

$[Definition\ 8, \quad ss = \alpha P \setminus \{\!\{c\}\!\}] \quad \mathcal{B}_{asp \overset{\times}{\times} T} = (\mathcal{B} \, [\!|ss|\!] \, P) \, [\!|c|\!] \, A \ \wedge \ (\mathcal{B} \, [\!|ss|\!] \, P) \setminus \{\!\{c\}\!\} \equiv_F \mathcal{B}$

$\implies [Definition\ 2, \ \equiv_F \ semantics\ [14]] \quad (\mathcal{B} \, [\!|ss|\!] \, P) \setminus \{\!\{c\}\!\}$ *is an I/O process*

$\implies [c.bf \in inputs, hiding\ semantics\ [14], (i - iii - iv)\ of\ Definition\ 4]$

$\mathcal{B} \, [\!|ss|\!] \, P$ *is an I/O process*

$\implies [\mathcal{C} \cap C = \emptyset, \ \alpha P \cap \{\!\{C\}\!\} = \emptyset, \ parallelism\ semantics\ [14], \ (iii - iv - v)\ of\ Definition\ 5]$

$(\mathcal{B} \, [\!|ss|\!] \, P) \, [\!|c|\!] \, A$ *is an I/O process*   □

By Definition 7 the aspect $LOG : \langle p, a \rangle$ modularises the logging feature we have developed alongside our aspect theory and, by Definition 8, the result of weaving LOG with Ctr_{SRV} is the component $LOG \overset{\times}{\times} Ctr_{\text{SRV}}$ below. We emphasise that it is possible to reuse both p and a with other pointcuts and advices to capture new aspects, given the loose coupling between them.

$$\left\langle \begin{array}{l} \text{(SRV [|diff(αP1, \{|c|\})|] P1) [|\{|c|\}|] A,} \\ \{\text{srv} \mapsto \text{I\_DTSER, db} \mapsto \text{I\_DB}\}, \{\text{I\_DTSER, I\_DB}\}, \{\text{srv, db}\} \end{array} \right\rangle$$

4 Safely Evolving BRIC Components Using Aspects

In this section we discuss how AOD can be used to safely evolve $\mathcal{BRIC}$ specifications. In [4], we develop a set of $\mathcal{BRIC}$ extension relations based on the concept of *behavioural convergence*, which fulfils the substitutability principle [17]: a component extension should be usable wherever the original component was expected, without any other component, acting as a client, being able to tell the difference. In this section we prove that a convergent component can be concisely and elegantly generated from an original component by using aspects in the style defined in the previous section; so the original and the (aspect oriented) evolved component satisfy the $\mathcal{BRIC}$ substitutability principle.

Convergence is a behavioural relation between I/O processes. An I/O process $\mathcal{B}'$ is convergent to $\mathcal{B}$ ($\mathcal{B}'$ io_ecvg $\mathcal{B}$, Definition 9) if in each converging point of their execution it can offer more or equal inputs $(Y \cap inputs \supseteq X \cap inputs)$ but is restricted to offer less or equal outputs $(Y \cap outputs \subseteq X \cap outputs)$. It can also offer any other event $(\Sigma\backslash Y \subseteq X)$ after a new input and before converging to its original behaviour, adding more implementation details. A convergent point $(t'\ \texttt{ecvg}\ t)$ represents a state reachable by both the original and the convergent process when doing two convergent sequences of events; these sequences differ only because the convergent process is allowed to do extra new-in-context-inputs (inputs not allowed by the original process at that point) in converging points followed by any event not allowed by the original process.

In Definition 9, $\mathcal{T}(\mathcal{B})$ and $\mathcal{F}(\mathcal{B})$ stand for the traces and failures of a process $\mathcal{B}$, respectively. Recall that Σ stands for the alphabet of all possible events, Σ^* is the set of possible sequences of events from Σ, the input events are contained in Σ $(inputs \subseteq \Sigma)$ and $in(\mathcal{B}, t)$ is a function that yields the set of input events that can be communicated by the I/O process $\mathcal{B}$ after some $t \in \mathcal{T}(\mathcal{B})$. Therefore $in : I/OProcess \times \Sigma^* \rightarrow inputs$; similarly, $out(\mathcal{B}, t)$ yields the set of output events of $\mathcal{B}$ after t. Additionally, if $t_1 \leq t_2$, it means that t_1 is a subtrace of t_2.

Definition 9 (Behavioural Convergence). *Consider two I/O processes $\mathcal{B}$ and $\mathcal{B}'$. We say that $\mathcal{B}'$ is an I/O convergent behaviour of $\mathcal{B}$ ($\mathcal{B}'$ io_ecvg $\mathcal{B}$), if and only if:*

$$\forall (t',X) \in \mathcal{F}(\mathcal{B}'), \exists (t,Y) \in \mathcal{F}(\mathcal{B}) \bullet$$
$$\left(\left(\begin{array}{c} t'\ \texttt{ecvg}\ t \wedge \\ \left(\left(\begin{array}{c} Y \cap inputs \supseteq X \cap inputs \wedge \\ Y \cap outputs \subseteq X \cap outputs \end{array} \right) \right) \\ \vee (\Sigma\backslash Y \subseteq X) \end{array} \right) \right)$$

$$\text{where, } t'\ \texttt{ecvg}\ t \iff \left(\begin{array}{c} (\#t' > \#t) \ \wedge \ \exists t_1, t_2, t_3 : \Sigma^*, \exists ne \in \Sigma \ | \\ t' = t_1 \,\widehat{}\,\langle ne \rangle \,\widehat{}\, t_2 \,\widehat{}\, t_3 \wedge t_1 \leq t \wedge \\ ne \in inputs \wedge ne \notin in(T, t_1) \wedge \\ set(t_2) \cap (in(T, t_1) \cup out(T, t_1)) = \emptyset \wedge \\ t_1 \,\widehat{}\, t_3\ \texttt{ecvg}\ t \end{array} \right) \vee (t' = t)$$

Consider the process SRV (on the left-hand side of Fig. 2) and SRV' (on the right-hand side of Fig. 3). Based on Definition 9, we have that SRV' io_cvg SRV. To explain why this is the case, let (t', X) and (t, Y) be failures of SRV' and SRV, respectively. Then by a non-exhaustive analysis:

- If $(t', X) = (\langle srv.in.v.2 \rangle, \Sigma \setminus \{c.bf\})$, then we know that t' ecvg t, for $t = \langle srv.in.v.2 \rangle$. The failure after t is $(t, Y) = (\langle srv.in.v.2 \rangle, \Sigma \setminus \{srv.out.v.3\})$, and $Y \cap inputs \supset X \cap inputs$ and $Y \cap outputs \subset X \cap outputs$;
- If $(t', X) = (\langle srv.in.v.2, c.bf \rangle, \Sigma \setminus \{db.out.save.2\})$, then we know t' ecvg t, for $t = \langle srv.in.v.2 \rangle$ because $c.bf \in inputs$ and $c.bf \in in(\text{SRV}, \texttt{t})$. The failure after t is $(t, Y) = (\langle srv.in.v.2 \rangle, \Sigma \setminus \{srv.out.v.3\})$, and $\Sigma\backslash Y \subset X$;

Behavioural convergence is a relation between I/O processes. Based on that in [4] we define an inheritance relation for $\mathcal{BRIC}$ (Definition 10), which considers components structure and behaviour. Structurally, it guarantees that the inherited component preserves at least its parent's channels and their types. Regarding behaviour, they are related by convergence. Additionally, it guarantees, for the purpose of substitutability, that the inherited component only refines the behaviour exhibited by common channels (default channel congruence [4]) or that additional inputs over common channels are not exercised by any possible client of its parent (input channel congruence [4]).

Definition 10 (Component Inheritance). *Consider T and T' two $\mathcal{BRIC}$ components. We say that T' inherits from T ($T \leftarrow_{ecvg} T'$) if and only if:*

$$\mathcal{R}_T \subseteq \mathcal{R}_{T'} \land \mathcal{B}_{T'} \text{ io\_ecvg } \mathcal{B}_T$$

provided *their corresponding channels are default or input channel congruent*

We know SRV' io_ecvg SRV and $LOG \overset{\times}{\times} Ctr_{\text{SRV}}$ channels extends those of Ctr_{SRV}. Additionally, SRV' is also default channel congruent w.r.t SRV [4]. Therefore, by Definition 10 we have $Ctr_{\text{SRV}} \leftarrow_{ecvg} LOG \overset{\times}{\times} Ctr_{\text{SRV}}$. In fact, this relation between aspects and inheritance always holds as proved by Theorem 1.

Theorem 1 (Aspects and Inheritance). *Let T and asp stand for a component and an aspect, respectively. If $asp \overset{\times}{\times} T$ is defined, then $T \leftarrow_{ecvg} asp \overset{\times}{\times} T$, provided $asp.adv$ is self-contained (see Definition 6).*

Proof.

T'

$$= \left[\begin{array}{l} \textit{Definition 8,} \quad T : \langle \mathcal{B}, \mathcal{R}, \mathcal{I}, \mathcal{C} \rangle, \quad asp : \langle \langle P, c \rangle, \langle A, C, c \rangle \rangle, \\ T' = asp \overset{\times}{\times} T, \quad I' = \{type(ch) \mid ch \in C\}, \quad R' = \{ch \mapsto type(ch) \mid ch \in C\} \end{array} \right]$$

$\langle (\mathcal{B} \, [\![\alpha P \setminus \{\!|c|\!\}]\!] \, P) \, [\![c|]\!] \, A, \mathcal{R} \cup R', \mathcal{I} \cup I', \mathcal{C} \cup C \rangle$

$\implies$ [*Definition 8,* $\mathcal{C} \cap C = \emptyset$, *set theory*]

(i) $\mathcal{R}_T \subseteq \mathcal{R}_{T'}$

$\implies \left[\begin{array}{l} \textit{Definition 7, Definition 8, Definition 10,} \quad \textit{alphabetised parallelism semantics,} \\ (\mathcal{B} \, [\![\alpha P \setminus \{\!|c|\!\}]\!] \, P) \setminus \{\!|c|\!\} \equiv_F \mathcal{B}, \quad \alpha P \setminus \{\!|c|\!\} \subseteq \alpha \mathcal{B}, \quad \alpha A \cap \alpha \mathcal{B} = \emptyset, \quad \alpha A \cap \alpha P = \{\!|c|\!\} \end{array} \right]$

(ii) $\forall c : \mathcal{C}_T \bullet (\mathcal{B}_{T'}$ and $\mathcal{B}_T$ are default channel congruent on $c)$

$\implies \left[\begin{array}{l} \textit{(iii - iv) of Definition 4,} \quad \textit{Definition 9,} \quad \textit{Definition 3 (c.bf} \in inputs), parallelism \\ semantics, \quad (\mathcal{B} \, [\![\alpha P \setminus \{\!|c|\!\}]\!] \, P) \setminus \{\!|c|\!\} \equiv_F \mathcal{B}, \quad \alpha P \setminus \{\!|c|\!\} \subseteq \alpha \mathcal{B}, \quad \{\!|c|\!\} \cap \alpha \mathcal{B} = \emptyset \end{array} \right]$

$\exists t ^\frown \langle c.bf \rangle \in \mathcal{T}(P) \implies (t ^\frown \langle c.bf \rangle, \Sigma \setminus \{c.af\}) \in \mathcal{F}(P) \land (t, \Sigma \setminus \{c.bf\}) \in \mathcal{F}(P) \land$

$\forall (t', X) \in \mathcal{F}(\mathcal{B} \, [\![\alpha P \setminus \{\!|c|\!\}]\!] \, P), \exists (t, Y) \in \mathcal{F}(\mathcal{B}) \bullet t'$ ecvg $t \land$

$$\left(\left(\begin{array}{c} Y \cap inputs \supseteq X \cap inputs \land \\ Y \cap outputs \subseteq X \cap outputs \end{array} \right) \right)$$
$$\lor (\Sigma \backslash Y \subseteq X)$$

$\implies \mathcal{B} \, [\![\alpha P \setminus \{\!|c|\!\}]\!] \, P$ io_ecvg $\mathcal{B}$

$\implies$ [*Definition 6*]

$\exists t ^\frown \langle c.af \rangle \in \mathcal{T}(A) \implies (t ^\frown \langle c.af \rangle, \Sigma \setminus \{c.bf\}) \in \mathcal{F}(A) \land (\langle \rangle, \Sigma \setminus \{c.bf\}) \in \mathcal{F}(A)$

$\implies [\alpha A \cap \alpha \mathcal{B} \, [\![\alpha P \setminus \{\!|c|\!\}]\!] \, P = \emptyset,$ *(i - iv - v) Definition 5*]

$(\mathcal{B} \, [\![\alpha P \setminus \{\!|c|\!\}]\!] \, P) \, [\![c|]\!] \, A$ io_ecvg $\mathcal{B} \, [\![\alpha P \setminus \{\!|c|\!\}]\!] \, P$

$\implies [(\mathcal{B} \, [\![\alpha P \setminus \{\!|c|\!\}]\!] \, P) \setminus \{\!|c|\!\} \equiv_F \mathcal{B},$ *Definition 8*]

(iii) $(\mathcal{B} \, [\![\alpha P \setminus \{\!|c|\!\}]\!] \, P) \, [\![c|]\!] \, A$ io_ecvg $\mathcal{B}$

$\implies$ [(i), (ii), (iii), *Definition 10*] $\mathbf{T} \leftarrow_{ecvg} \mathbf{asp} \overset{\times}{\times} \mathbf{T}$ $\square$

Considering the substitutability principle for $\mathcal{BRIC}$ (with focus on deadlock freedom, but not limited to it) as stated in Theorem 2 [4] and Theorem 1 we enunciate the relevant Corollary 1.

Theorem 2 (Substitutability). *Let T, T' to be two components such that $T \leftarrow_{ecvg} T'$. Consider $S[T]$ a deadlock free component contract, where T is a basic component contract that was composed within S using one of the $\mathcal{BRIC}$ composition rules, then $S[T']$ is deadlock free.*

Corollary 1 (Aspects Preserve Substitutability). *Let T and asp stand for a component and an aspect, respectively. Consider $S[T]$ a deadlock free component contract, where T is a basic component contract that was composed within S using one of the $\mathcal{BRIC}$ composition rules. If $asp \underset{\times}{\times} T$ is defined, then $S[asp \underset{\times}{\times} T]$ is deadlock free.*

The above corollary follows direct from Theorems 1 and 2. As far as the authors are aware, this is the first time a relation between component inheritance and aspects-oriented design is established for a formal approach to CB-MDD. It offers two ways to evolve component specifications that, although distinct in concept, are related by their implementation mechanisms. It is important to make clear that aspects are designed to cope with crosscutting concerns that scatter through families of components by enlarging them, where inheritance is exclusively designed to create such families. It is fair to say that aspects operate horizontally and inheritance vertically over component family trees.

5 Case Study

We model an autonomous healthcare robot that monitors and medicates patients, being able to contact the relevant individuals or systems in case of emergency. It receives data from a number of sensors and actuates by injecting intravenous drugs and/or by calling the emergency medical services and the patient's relatives or neighbours. We use the following data types: BI (breath intensity), BT (body temperature), DD (drug dose), BGL (blood glucose level), CL (call list, the relevant individuals to be called in the case of emergency), DRUG (the drugs in the robot's actuators), QUEST (the robot's question list, to ask the patient when its voice recognition module is used).

```
nametype BI = {1..5},   BT  = {34..41}, DD = {0..5}
datatype BGL = low | normal | threshold | high
datatype CL = c911 | cFamily | cNeighbor | ack
datatype DRUG = insulin | painkiller | antipyretic
datatype QUEST = chest | head | vision | 1st
```

These data types are composed into more elaborated ones that will be communicated through the channels used to connect sensors, actuators and phones to the robot: BS, the body attached sensors; IS, the vision recognition devices; VS, the noise recognition devices; TK, the voice interaction devices; PH, the phone interface and INV, the intravenous injection actuator.

```
datatype BS = breath.BI | bodyTemp.BT | bloodGlucose.BGL
datatype IS = numbnessFace.Bool |  fainting.Bool
datatype VS = cough.Bool | troubleSpeaking.Bool
```

```
datatype TK = visionTrouble.Bool  |  chestDiscomfort.Bool |
              headache.Bool | ask.QUEST
datatype PH = call.CL
datatype IVN = administer.DRUG.DD
```

Each type above can be an input (tag in) or an output (tag out), depending on whether it is produced or consumed by a specific component.

```
channel bodySen: in.BS | out.BS
channel imageRec : in.IS | out.IS
channel voiceRec : in.VS | out.VS
channel talk : in.TK | out.TK
channel phone : in.PH | out.PH
channel intravenousNeedle : in.IVN | out.IVN
```

The behaviour of our healthcare robot is defined in terms of the I/O process HC_BOT. It waits for the breath level indicator; if this level is critical (< 3), then it behaves as MOD_CALL_P1 (module phone call priority one). MOD_CALL_P1 contacts a patient's neighbour, the registered emergency service and relatives, in that order, then it waits for at least two of them to acknowledge before coming back to its initial state. Otherwise, the patient is breathing normally, and the robot reads the noise sensor to check whether he is coughing (voiceRec.in.cough?b). If so, it reads the body temperature (bodySen.in.bodyTemp?t) and blood glucose sensors (bodySen.in.bloodGlucose?g). If the body temperature exceeds 38 degrees Celsius, then it administers a dose of antipyretic (intravenousNeedle. out.administer.antipyretic.d_ ap). If the blood glucose level is in the threshold or high, it administers insulin (intravenousNeedle.out.administer.insulin.d_in), otherwise it just comes back to its initial state. After administrating any drug, and before coming back to its initial state, the robot must contact the patient's neighbour and relatives by behaving as MOD_CALL_P2 (module phone call priority two), where at least one of them must acknowledge. The phone call modules are very simple and we omit them here for space limitations.

If the patient is breathing normally but in silence, the robot asks the image recognition module to inform about: any unusual sign in his face (imageRec.in. numbnessFace?nf) or if he fainted (imageRec.in.fainting?f). If at least one condition holds, the robot administers a painkiller (intravenousNeedle.out. administer.painkill er.d_pk), calls the relevant individuals by behaving as MOD_CALL_P1. In any case, it goes to its initial state.

```
HC_BOT = bodySen.in.breath?x ->
if (x < 3) then bodySen.out.breath.x ->  MOD_CALL_P1; HC_BOT
else voiceRec.in.cough?b ->
 if b then bodySen.in.bodyTemp?t-> bodySen.in.bloodGlucose?g->
  if(t > 38)
  then |~| d_ap : DD @
   intravenousNeedle.out.administer.antipyretic.d_ap ->
   MOD_CALL_P2 ; HC_BOT
  else
   if (g == high or g ==threshold)
```

```
then |~| d_in : DD @
  intravenousNeedle.out.administer.insulin.d_in ->
  MOD_CALL_P2 ; HC_BOT
else HC_BOT
else
  imageRec.in.numbnessFace?nf -> imageRec.in.fainting?f ->
  if (nf or f) then |~| d_pk : DD @
    intravenousNeedle.out.administer.painkiller.d_pk ->
    MOD_CALL_P1; HC_BOT
  else HC_BOT
```

In $\mathcal{BRIC}$, the healthcare robot is defined in terms of the Ctr_{HC_BOT} contract, where I_BS stands for in.BS $\cup$ out.BS as well as the other channel types.

$$Ctr_{HC\_BOT} \hat{=}$$
$$\left\langle HC\_BOT, \left\{ \begin{array}{l} \text{bodySen} \mapsto \text{I\_BS, imageRec} \mapsto \text{I\_IS,} \\ \text{voiceRec} \mapsto \text{I\_VS, phone} \mapsto \text{I\_PH,} \\ \text{intravenousNeedle} \mapsto \text{I\_IVN} \end{array} \right\}, \left\{ \begin{array}{l} \text{I\_BS, I\_IS,} \\ \text{I\_VS, I\_PH,} \\ \text{I\_IVN} \end{array} \right\}, \right.$$
$$\left. \{\text{bodySen, imageRec, voiceRec, phone, intravenousNeedle}\} \right\rangle$$

The robot Ctr_{HC_BOT} is able to diagnose and select the appropriate drug to be administered. Nevertheless, there is no criteria to define an appropriate drug dose given the seriousness of the patient condition. The I/O process HC_BOT_TK addresses this question by using two criteria: each degree above 38 degrees Celsius will correspond to an unite of antipyretic (intravenousNeedle.out.administer.antipyretic.t%37) and, the insulin dose will be one or two unities if the blood glucose level is on the threshold or greater than two, otherwise.

In addition to the refinement of the HC_BOT drug dose selection mechanism, the I/O process HC_BOT_TK is also able to interact with patients by the voice simulation/recognition device through the new channel talk. It offers, initially, the possibility of behaving as MOD_TALK: it receives a chat request, then collects information about chest discomfort (talk.in.chestDiscomfort?cd), headache (talk.in.headache?hd) and vision problems (talk.in.visionTrouble?vt). If the patient reports chest discomfort associated with headache or vision problems, the robot understands that a serious situation is under way and calls all the relevant individuals (MOD_CALL_P1). In any case, it goes to its initial state.

```
HC_BOT_TK =
bodySen.in.breath?x ->
 if (x < 3)
 then bodySen.out.breath.x ->  MOD_CALL_P1; HC_BOT_TK
 else voiceRec.in.cough?b ->
  if (b) then bodySen.in.bodyTemp?t ->
  bodySen.in.bloodGlucose?g ->
  if(t > 38)
  then intravenousNeedle.out.administer.antipyretic.t
   MOD_CALL_P2 ; HC_BOT_TK
   else
    if (g == high) then |~| d_in_h : {3,4,5} @
```

```
intravenousNeedle.out.administer.insulin.d_in_h ->
MOD_CALL_P2 ; HC_BOT_TK
else
if (g == threshold) then |~| d_in_t : {1,2} @
  intravenousNeedle.out.administer.insulin.d_in_t ->
  MOD_CALL_P2 ; HC_BOT_TK
  else HC_BOT_TK
else
imageRec.in.numbnessFace?nf -> imageRec.in.fainting?f ->
if (nf or f) then |~| d_pk : DD @
intravenousNeedle.out.administer.painkiller.d_pk ->
MOD_CALL_P1; HC_BOT_TK
else HC_BOT_TK
[] MOD_TALK ; HC_BOT_TK

MOD_TALK = talk.in.ask.1st ->
  talk.out.ask.chest -> talk.in.chestDiscomfort?cd ->
  talk.out.ask.head -> talk.in.headache?hd ->
  talk.out.ask.vision -> talk.in.visionTrouble?vt ->
  if (cd and (hd or vt)) then MOD_CALL_P1 else SKIP
```

In $\mathcal{BRIC}$, this extension is defined by the contract $Ctr_{HC\_BOT\_TK}$. According to Definition 10, we have $Ctr_{HC\_BOT\_ACC} \leftarrow_{ecvg} Ctr_{HC\_BOT\_TK}$.

$$Ctr_{HC\_BOT\_TK} \widehat{=}$$
$$\left\langle HC\_BOT\_TK, \left\{ \begin{array}{l} \text{bodySen} \mapsto \text{I\_BS, imageRec} \mapsto \text{I\_IS,} \\ \text{voiceRec} \mapsto \text{I\_VS, phone} \mapsto \text{I\_PH,} \\ \text{intravenousNeedle} \mapsto \text{I\_IVN,} \\ \text{talk} \mapsto \text{I\_TK} \end{array} \right\}, \left\{ \begin{array}{l} \text{I\_BS, I\_IS,} \\ \text{I\_VS, I\_PH,} \\ \text{I\_IVN, I\_TK} \end{array} \right\}, \right\rangle$$
$$\{\text{bodySen, imageRec, voiceRec, phone, intravenousNeedle, talk}\}$$

We can have many kinds of robots attached to a patient, sharing an extremely critical resource, his/her intravenous access, a common point where drugs can be injected without doctors supervision. In life threatening scenarios, doctors can be compelled to intervene, but they will need to know which robot injected which drug and in which dose, so they can proceed with the appropriate intervention.

This is a classic problem, which has a well-known solution: logging. Therefore, we must, originally, redefine each component in our specification to add the logging feature. As new communication channels must be used to implement the logging functionality, we need to update component interfaces to refer to these new channels as well as their types. If we have many types of robots, we must face an exhaustive and error prone task in updating each of them and, worst, it must be done each time we need to change the logging and/or robots specifications. Thanks to our AOD approach to CB-MDD, we can solve this problem in a more modular and maintainable way by defining a logging aspect.

We identify each robot with a unique identifier, a member of BOT_ID, whose elements range from one to TOP (a natural number). First we define a check mark channel and an pointcut to capture the condition we are interested on: a robot has administered a drug. We define the tags bf and af, which are used by the check mark channel drug_mark. It communicates {bf, af} alongside the injected drug DRUG and its respective dose DD. The process DET_DRUD synchronises on any access to the intravenous needle (intravenousNeedle.out.administer.dg.dose) and marks it as a join point by performing the events drug_mark.bf.dg.dose and drug_mark.af.dg.dose sequentially.

```
channel drug_mark :   {bf,af}.DRUG.DD
DET_DRUD =   [] dg: DRUG , dose: DD @
   intravenousNeedle.out.administer.dg.dose ->
   drug_mark.bf.dg.dose -> drug_mark.af.dg.dose -> DET_DRUD
αDET_DRUD = union({|intravenousNeedle.out|}, {|drug_mark|})
```

We want to log triples: which robot has administered, which drug, in which quantity. The channel drug_log will be used by our advice to log the fact that some robot (BOT_ID) has administered a dose of a given drug (DRUG.DD) to its target patient. The advice behaviour is given by the I/O process LOG_DRUG(x); it synchronises on the check mark event drug_mark.bf.dg.dose, then logs the fact that the robot x has injected the amount dose of the drug dg (drug_log.out.log.x.dg.dose). After, it agrees on drug_mark.af.dg.dose (the join point end mark) and goes to its initial state.

```
channel drug_log: in.log.BOT_ID.DRUG.DD|out.log.BOT_ID.DRUG.DD
LOG_DRUG(x)=   [] dg: DRUG, dose: DD @ drug_mark.bf.dg.dose ->
               drug_log.out.log.x.dg.dose ->
               drug_mark.af.dg.dose -> LOG_DRUG(x)
```

We define our logging drug aspect as $LOG\_D : \langle PD, AD \rangle$, where PD : $\langle$DET_DRUD, drug_mark$\rangle$ and AD : $\langle$LOG_DRUG(x), {drug_log}, drug_mark$\rangle$. Then, LOG_D can be weaved (Definition 8) into Ctr_{HC_BOT} and $Ctr_{HC\_BOT\_TK}$, resulting, respectively, in the following components:

```
HC_BOT [|diff(α DET_DRUD, {|drug_mark|})|] DET_DRUD
        [|{|drug_mark|}|] LOG_DRUG (1),
```
$$\left\langle \left\{ \begin{array}{l} \text{bodySen} \mapsto \text{I\_BS, imageRec} \mapsto \text{I\_IS, phone} \mapsto \text{I\_PH,} \\ \text{voiceRec} \mapsto \text{I\_VS, intravenousNeedle} \mapsto \text{I\_IVN,} \\ \text{drug\_log} \mapsto \text{I\_IDL} \end{array} \right\}, \left\{ \begin{array}{l} \text{I\_BS, I\_IS,} \\ \text{I\_VS, I\_PH,} \\ \text{I\_IVN, I\_IDL} \end{array} \right\}, \right\rangle$$
```
{bodySen, imageRec, voiceRec, phone, intravenousNeedle, drug_log}
```

```
HC_BOT_TK [|diff(α DET_DRUD, {|drug_mark|})|] DET_DRUD
          [|{|drug_mark|}|] LOG_DRUG (1),
```
$$\left\langle \left\{ \begin{array}{l} \text{bodySen} \mapsto \text{I\_BS, imageRec} \mapsto \text{I\_IS, phone} \mapsto \text{I\_PH,} \\ \text{voiceRec} \mapsto \text{I\_VS, intravenousNeedle} \mapsto \text{I\_IVN,} \\ \text{talk} \mapsto \text{I\_TK, drug\_log} \mapsto \text{I\_IDL} \end{array} \right\}, \left\{ \begin{array}{l} \text{I\_BS, I\_IS,} \\ \text{I\_VS, I\_PH,} \\ \text{I\_IVN, I\_TK,} \\ \text{I\_IDL} \end{array} \right\}, \right\rangle$$
```
{bodySen, imageRec, voiceRec, phone, intravenousNeedle, talk, drug_log}
```

We know that $Ctr_{HC\_BOT} \twoheadleftarrow_{ecvg} Ctr_{HC\_BOT\_TK}$ and, by Definition 8 and Theorem 1, we have that:

$$Ctr_{HC\_BOT} \twoheadleftarrow_{ecvg} LOG\_D \overset{\times}{\times} Ctr_{HC\_BOT}$$

$$Ctr_{HC\_BOT\_TK} \twoheadleftarrow_{ecvg} LOG\_D \overset{\times}{\times} Ctr_{HC\_BOT\_TK}$$

Two important results conclude this case study. They come from the transitivity of inheritance and the ACW identity over inheritance [3]. If T, T' and T'' are components and asp an aspect, transitivity ensures that if $T \twoheadleftarrow_{ecvg} T'$ and $T' \twoheadleftarrow_{ecvg} T''$ then $T \twoheadleftarrow_{ecvg} T''$. The ACW identity over inheritance guarantees that if $T \twoheadleftarrow_{ecvg} T'$ and $asp \overset{\times}{\times} T$ is defined, then $asp \overset{\times}{\times} T \twoheadleftarrow_{ecvg} asp \overset{\times}{\times} T'$.

$$Ctr_{HC\_BOT} \twoheadleftarrow_{ecvg} LOG\_D \overset{\times}{\times} Ctr_{HC\_BOT\_TK}$$

$$LOG\_D \overset{\times}{\times} Ctr_{HC\_BOT} \twoheadleftarrow_{ecvg} LOG\_D \overset{\times}{\times} Ctr_{HC\_BOT\_TK}$$

6   Conclusions and Related Work

This work defines a formal approach to introduce aspect-oriented modelling into component-based specifications, where preservation of some properties is guaranteed by construction. We developed for the $\mathcal{BRIC}$ component model a proper way to characterise pointcuts, advices and aspects. We proved aspects preserves a conformance notion, given in terms of the substitutability principle [17]. We also established a connection between component inheritance and aspects, presenting them as interchangeable ways to safely extend component specifications.

We brought the recognised aspects benefits (crosscutting concerns modularisation, reuse, maintainability [10]) to a trustworthy component-base model, where behaviour properties can be compositionally verified [13]. We have not addressed the entire aspects theory, but created a formal basis to its adoption in CB-MDD.

To illustrate how the design of component-based specifications can benefit from aspect theory, we developed a case study of an autonomous healthcare system, which evolve by the addition of new functionalities via inheritance and by the modularisation of its crosscutting concerns in a reusable and maintainable manner with aspects. We show how these concepts can be put together benefiting design, clarity and maintainability.

In [1], the authors define an aspect-weaver algorithm for a process algebra based on CSP with process equivalence given in the traces model. Aspects are processes and the weaver works by resolving synchronism, in our strategy, we consider components and aspects in different categories given their distinct nature. We define equivalence in the failures model, which considers refusals in addition to traces. Differently from [1] our approach is compositional, so global properties can be checked locally.

The Protocol Modelling framework [11] also uses a CSP-like parallelism to weave aspects and supports local reasoning. Nevertheless, it requires specifications to be deterministic, a limitation we have relaxed. It address, as [1], only behavioural constructions. We differentiate by considering both structure and behaviour at a component level, where extensibility and conformance are properly addressed. The work reported in [5] identifies categories of aspects that preserve corresponding classes of properties; our focus here is to use aspects that obey a particular protocol to evolve system design models by preserving some conformance notions.

This work benefits from the formalisation, given by a denotational semantics, of the aspects theory for a programming language that embodies the key features of join points, pointcuts, and advices [16].

As future work, we plan to mechanise our AOM approach in CSP-Prover [8], an interactive theorem prover for CSP based on Isabelle/HOL. We expect to develop a tool support for assisted development of component-based specifications with aspects and inheritance support. It receives a specification in a high level language (as SysML), translates it to $\mathcal{BRIC}$ and performs a formal verification in CSP-Prover.

References

1. Andrews, J.H.: Process-algebraic foundations of aspect-oriented programming. In: Matsuoka, S. (ed.) Reflection 2001. LNCS, vol. 2192, pp. 187–209. Springer, Heidelberg (2001)
2. Coady, Y., Kiczales, G., Feeley, M., Smolyn, G.: Using aspectC to improve the modularity of path-specific customization in operating system code. SIGSOFT Softw. Eng. Notes **26**(5), 88–98 (2001)
3. Dihego, J., Sampaio, A.: Aspect-oriented development of trustworthy component-based systems - Extended version, Technical report (2015). http://www.cin.ufpe.br/jdso/technicalReports/TR062.pdf
4. Dihego, J., Sampaio, A., Oliveira, M.: Constructive extensibility of trustworthy component-based systems. In: Proceedings of the 30th Annual ACM Symposium on Applied Computing, SAC 2015. ACM (2015)
5. Djoko, S.D., Douence, R., Fradet, P.: Aspects preserving properties. Sci. Comput. Program. **77**(3), 393–422 (2012)
6. Filman, R., Elrad, T., Clarke, S., Aksit, M.: Aspect-oriented Software Development, 1st edn. Addison-Wesley Professional, Reading (2004)
7. Gibson-Robinson, T., Armstrong, P., Boulgakov, A., Roscoe, A.W.: FDR3 — A modern refinement checker for CSP. In: Ábrahám, E., Havelund, K. (eds.) TACAS 2014 (ETAPS). LNCS, vol. 8413, pp. 187–201. Springer, Heidelberg (2014)
8. Isobe, Y., Roggenbach, M.: CSP-prover: a proof tool for the verification of scalable concurrent systems. Comput. Softw. **25**(4), 85–92 (2008)
9. Kiczales, G., Hilsdale, E., Hugunin, J., Kersten, M., Palm, J., Griswold, W.G.: An overview of aspectJ. In: Lindskov Knudsen, J. (ed.) ECOOP 2001. LNCS, vol. 2072, pp. 327–354. Springer, Heidelberg (2001)
10. Kiczales, G., Lamping, J., Mendhekar, A., Maeda, C., Lopes, C., Loingtier, J.-M., Irwin, J.: Aspect-oriented programming. In: Akşit, M., Matsuoka, S. (eds.) ECOOP 1997. LNCS, vol. 1241, pp. 220–242. Springer, Heidelberg (1997)
11. McNeile, A., Roubtsova, E.: CSP parallel composition of aspect models. In: Proceedings of the 2008 AOSD Workshop on Aspect-oriented Modeling, AOM 2008, pp. 13–18. ACM, New York (2008)

12. Ramos, R.: Systematic Development of Trustworthy Component-based Systems. Ph.D. thesis, Centro de Informática - Universidade Federal de Pernambuco, Brazil (2011)
13. Ramos, R., Sampaio, A., Mota, A.: Systematic development of trustworthy component systems. In: Cavalcanti, A., Dams, D.R. (eds.) FM 2009. LNCS, vol. 5850, pp. 140–156. Springer, Heidelberg (2009)
14. Roscoe, A.W.: Theory and practice of concurrency. Prentice-Hall Series in Computer Science. Prentice-Hall, Englewood Cliffs (1998)
15. Szyperski, C.: Component Software: Beyond Object-oriented Programming. ACM Press/Addison-Wesley Publishing Co., New York (1998)
16. Wand, M., Kiczales, G., Dutchyn, C.: A semantics for advice and dynamic join points in aspect-oriented programming. ACM Trans. Program. Lang. Syst. 26(5), 890–910 (2004)
17. Wegner, P., Zdonik, S.B.: Inheritance as an incremental modification mechanism or what like is and isn't like. In: Gjessing, S., Nygaard, K. (eds.) ECOOP 1988. LNCS, vol. 322, pp. 55–77. Springer, Heidelberg (1988)

A Game of Attribute Decomposition
for Software Architecture Design

Jiamou Liu(✉) and Ziheng Wei

School of Computer and Mathematical Sciences,
Auckland University of Technology, Auckland, New Zealand
{jiamou.liu,bys7090}@aut.ac.nz

Abstract. Attribute-driven software architecture design aims to provide decision support by taking into account the quality attributes of softwares. A central question in this process is: *What architecture design best fulfills the desirable software requirements?* To answer this question, a system designer needs to make tradeoffs among several potentially conflicting quality attributes. Such decisions are normally ad-hoc and rely heavily on experiences. We propose a mathematical approach to tackle this problem. Game theory naturally provides the basic language: Players represent requirements, and strategies involve setting up coalitions among the players. In this way we propose a novel model, called *decomposition game (DG)*, for attribute-driven design. We present its solution concept based on the notion of cohesion and expansion-freedom and prove that a solution always exists. We then investigate the computational complexity of obtaining a solution. The game model and the algorithms may serve as a general framework for providing useful guidance for software architecture design. We present our results through running examples and a case study on a real-life software project.

Keywords: Software architecture · Coalition game · Decomposition game

1 Introduction

The architecture of a software lays out the basic system composition; it is crucial to the software, as it heavily influences important quality attributes such as performance, reliability, usability and security [3]. A focusing question of software design is the following: *What architecture best fulfills the desirable software requirements?* In most cases, however, a "perfect" architecture that fulfills every requirement is unlikely to exist. For example, performance and security are both key non-functional requirements, which may demand fast response time to the users, and the application of a sophisticated encryption algorithm, respectively. These two requirements are in intrinsic conflict, as a strong focus of one will negatively impact the fulfilment of the other. A main task of the software architect, therefore, is to balance such "interactions" among requirements, and decide on appropriate tradeoffs among such conflicting requirements.

© Springer International Publishing Switzerland 2015
M. Leucker et al. (Eds.): ICTAC 2015, LNCS 9399, pp. 445–463, 2015.
DOI: 10.1007/978-3-319-25150-9_26

While it is a common practice to design software architecture through the architects' experiences and intuition, formal approaches for architecture design are desirable as they facilitate standardisation and automation of this process, providing rigorous guidelines, allowing automatic analysis and verifications [8]. Notable formal methods in software architecture include a large number of formal architecture description languages (ADL), which are useful tools in communicating and modeling architectures. However, as argued by [16], industry adoptions of ADL are rare due to limitations in usability and formality.

In this paper, we investigate the problem of software architecture design from a game theory perspective. Computational game theory studies the algorithmic nature of conflicting entities and establishes *equilibria*: A state of balance that minimises the negative effects among players. The field has attracted much attention in the recent 10–15 years due to applications in multi-agent systems, electronic markets and social networks [11–13]. Our goal here is to interpret the interactions among software requirements as a game whose equilibrium give rise to a desirable software architecture. Our motivation comes from the following two lines of research:

(1). Attribute driven design (ADD) : ADD is a systematic method for software architecture design. The method was invented by Bass, Klein and Bachmann in [5] and subsequently updated and improved through a sequence of works [4,15]. The goal is to assist designers to analyse quality attribute tradeoffs and provide design suggestions and guidance. Inputs to ADD are functional and non-functional requirements, as well as design constraints; outputs to ADD are conceptual architectures which outline coarse-grained system compositions. The method involves a sequence of well-defined steps that recursively decompose a system to components, subcomponents, and so on. These steps are not algorithmic: They are meant to be followed by system designers based on their experience and understanding of design principles. As mentioned by the authors in [5], an ongoing effort is to investigate rigorous approaches in producing conceptual architectures from requirements, hence enabling automated design recommendation under the ADD framework. To this end, we initiate a game-theoretic study to formulate the interactions among software requirements so that a conceptual architecture can be obtained in an algorithmic way.

(2). Coalition game theory : A coalition game is one where players exercise collaborative strategies, and competition takes place among coalitions of players rather than individuals. In ADD, we can imagine each requirement is "handled" by a player, whose goal is to set up a coalition with others to maximise the collective payoff. The set of coalitions then defines components in a system decomposition which entails a software architecture. This fits into the language of coalition games. However, the usual axioms in coalition games (with transferrable utility) specify super-additivity and monotonicity, that is, the combination of two coalitions is always more beneficial than each separate coalition, and the payoff increases as a coalition grows in size. Such assumptions are not suitable in this context as combination of two conflicting requirements may result in a lower

payoff. Hence a new game model is necessary to reflect the conflicting nature of requirements. In this respect, we propose that our model also enriches the theory of coalition games.

Our Contribution. We provide a formal framework which, following the ADD paradigm [5], recursively decomposes a system into sub-systems; the final decomposition reveals design elements in a software architecture. The basis of the framework is an algorithmic realisation of ADD. A crucial task in this algorithmic realisation is *system decomposition*, which derives a rational decomposition of an attribute primitive. We model system decomposition using a game, which we call *decomposition game*. The game takes into account *interactions* between requirements, which express the positive (enhancement) or negative (canceling) effects they act on each other. A *solution concept* (equilibrium) defines a rational decomposition, which is based on the notions of *cohesion* and *expansion-freedom*.

We demonstrate that any such game has a solution, and a solution may not be unique. We also investigate algorithms that compute solutions for the decomposition game. Finding cohesive coalitions with maximal payoff turns out to be NP-hard (Theorem 13). Hence we propose a relaxed notion of *k-cohesion* for $k \geq 1$, and present a polynomial time algorithm for finding a k-cohesive solution of the game (Theorem 16). To demonstrate the practical significance our the framework, we implement the framework and perform a case study on a real-world Cafeteria Ordering System.

Related Works. The work [9] starts a systematic study of tradeoff among quality attributes using empirical analysis. This investigation is then extended in [17]. The work [1] follows the ADD framework to define architecture by computing tradeoff between non-functional requirements based on relationships between non-functional and functional requirements. Another related work is [10] which uses hierarchical clustering to group requirements into software components based on their interactions. Here the authors label each component with a set of attributes and identify similarities between components based on their common attributes. Hence this work does not put emphasis on the enhancement and conflicts between requirements.

Paper Organisation. Section 2 introduces the formal ADD framework. Section 3 discusses decomposition game and its solution concept. Section 4 presents algorithms for solving decomposition games. Section 5 presents the case study and finally Sect. 6 concludes with future works.

2   Algorithmic Attribute Driven Design (ADD) Process

ADD is a general framework for transforming software requirements into a *conceptual software architecture*. Pioneers of this approach introduced it through several well-formed, but informally-defined concepts and steps [5,15]. A natural question arises whether it can be made more algorithmic, which provides unbiased, mathematically-grounded outputs. To answer this question, one would first need to translate the original informal descriptions to a mathematical language.

2.1 Software Requirements and Constraints

Functional Requirements. Functional requirements are specifications of what tasks the system perform (e.g. "the system must notify the user once a new email arrives"). A functional requirement does not stand alone; often, it acts with other functional requirements to express certain combined functionality (e.g. "the user should log in before making a booking"). Thus, a functionality may depend on other functionalities. We use a partial ordering $(F, \prec)$ to denote the functional requirements where each $r \in F$ is a functional requirement, and $r_1 \prec r_2$ denotes that r_1 depends on r_2. Note that $\prec$ is a transitive relation.

Non-functional Requirements. Non-functional requirements specify the desired quality attributes; ADD uses *general scenarios* and *scenarios* as their standard representations. A general scenario is a high-level description on what it means to achieve a non-functional requirement [5]. For example, the general scenario "*A failure occurs and the system notifies the user; the system continues to perform in a degraded manner*" refers to the availability attribute. There has been an effort to document all common general scenarios; a rather full list is given in [4]. Note that a general scenario is vaguely-phrased and is meant to serve as a template for more concrete "instantiations" of quality attributes. Such "instantiations" are called scenarios. More abstractly, we use a pair $(S, \approx)$ to denote the non-functional requirements where S is a set of scenarios and $\approx$ is an equivalence relation on S, denoting the *general scenario relation*: $q_1 \approx q_2$ means that q_1 and q_2 instantiates the same general scenario.

Design Constraints. Design constraints are factors that must be taken into account and enforce certain design outcomes. A design constraint may affect both functional and non-functional requirements. More abstractly, we use a collection of sets $C \subseteq 2^{F \cup S}$ to denote the set of design constraints, where each set $c \in C$ is a design constraint. Intuitively, if two requirements r_1, r_2 belong to the same $c \in C$, then they are constrained by the same design constraint c.

Derived Functionalities. The enforcement of certain quality attributes may lead to additional functionalities. For example, to ensure availability, it may be necessary to add extra functionalities to detect failure and automatically bypass failed modules. Hence we introduce a *derivation relation* $\hookrightarrow \subseteq S \times F$ such that $r \hookrightarrow s$ means the functional requirement s is derived from the scenario r.

2.2 Attribute Primitives

The intentional outcome of ADD describes the *design elements*, i.e., subsystems, components or connectors. It is important to note that the goal of ADD is not the complete automation of the design process, but rather, to provide useful guidance. Thus, the conceptual view reveals only the organisational structure but not the concrete design.

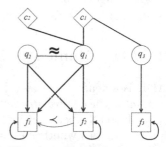

Fig. 1. Example 1: the requirements, constraints and their relations.

An attribute primitive is a set of design elements that collaboratively perform certain functionalities and meet one or more quality requirements; it is also the minimal combination with respect to these goals [5]. Examples of attribute primitives include data router, firewall, virtual machine, interpreter and so on. ADD prescribes a list of attribute primitives together with descriptions of their properties and side effects (such as in [4]). Hence, ADD essentially can be viewed as assigning the right attribute primitives to the right requirement combinations. Note also that an attribute primitive may be broken down further.

Definition 1 (Attribute Primitive). *An attribute primitive is a tuple*

$$\mathcal{A} = (\mathsf{F}, \mathsf{S}, \mathsf{C}, \prec, \approx, \hookrightarrow)$$

where F *is a set of functional requirements,* S *is a set of scenarios,* $\mathsf{C} \subseteq 2^{\mathsf{F} \cup \mathsf{S}}$ *is a set of design constraints,* $\prec$ *is the dependency relation on* F, $\approx$ *is the general scenario relation of* S, *and* $\hookrightarrow \subseteq \mathsf{S} \times \mathsf{F}$ *is a derivation relation.*

Let $\mathcal{A} = (\mathsf{F}, \mathsf{S}, \mathsf{C}, \prec, \approx, \hookrightarrow)$ be an attribute primitive. We also need the following:

- A *requirement* of $\mathcal{A}$ is an element in the set $\mathsf{R} := \mathsf{F} \cup \mathsf{S}$.
- For $r \in \mathsf{F}$, the *dependency set* of r is the set $f(r) := \{r' \in \mathsf{F} \mid r \preceq r'\}$.
- For $r \in \mathsf{S}$, the *general scenario* of r is the set $g(r) := \{r' \in \mathsf{S} \mid r \approx r'\}$, i.e., the $\approx$-equivalence class of r.
- For $r \in \mathsf{R}$, the *constraints* of r is the set $c(r) := \{t \in \mathsf{C} \mid r \in t\}$.
- For $r \in \mathsf{S}$, the *derived set* of r is $d(r) := \{s \in \mathsf{F} \mid r \hookrightarrow s\}$, and for $s \in \mathsf{F}$, let $d^{-1}(s) := \{r \in \mathsf{S} \mid r \hookrightarrow s\}$

Definition 2 (Design Element). *A design element of* $\mathcal{A}$ *is a subset* $D \subseteq \mathsf{R}$. *An decomposition of* $\mathcal{A}$ *is a sequence of design elements* $\boldsymbol{D} = (D_1, D_2, \ldots D_k)$ *where* $k \geq 1$, $\bigcup_{1 \leq i \leq k} D_k = \mathsf{R}$, *and each* $D_i \cap D_j = \varnothing$ *for any* $i \neq j$.

Example 1. Figure 1 shows an attribute primitive $\mathcal{A} = (\mathsf{F}, \mathsf{S}, \mathsf{C}, \prec, \approx, \hookrightarrow)$

- $\mathsf{F} = \{f_1, f_2, f_3\}$ and $\mathsf{S} = \{q_1, q_2, q_3\}$ are the requirements
- $\mathsf{C} = \{c_1, c_2\}$ where $c_1 = \{q_1, q_3\}, c_2 = \{q_1\}$
- $f_1 \prec f_2, q_1 \approx q_2, q_1 \hookrightarrow f_1, q_1 \hookrightarrow f_2, q_2 \hookrightarrow f_1, q_2 \hookrightarrow f_2, q_3 \hookrightarrow f_3$.

2.3   The ADD Procedure

Essentially ADD provides a means for system decomposition: The entire system is treated as an attribute primitive, which is the input. At each step, the procedure decomposes an attribute primitive $\mathcal{A}$ by identifying a decomposition $(D_1, D_2, \ldots, D_k)$. The process then maps each resulting design element D_i to an attribute primitive $\mathcal{A}_i = (\mathsf{F}_i, \mathsf{S}_i, \mathsf{C}_i, \prec_i, \approx_i, \hookrightarrow_i)$, which contains all elements in D_i and may require some further requirements and constraints. Hence we require that $D_i \subseteq \mathsf{F}_i \cup \mathsf{S}_i$ and $\prec_i, \approx_i, \mathsf{C}_i, \hookrightarrow_i$ are consistent with $\prec, \approx, \mathsf{C}$ and $\hookrightarrow$ on D_i, resp.; in this case we say that $\mathcal{A}_i$ is *consistent* with D_i. Thus the attribute primitive $\mathcal{A}$ is decomposed into k attribute primitives $\mathcal{A}_1, \mathcal{A}_2, \ldots, \mathcal{A}_k$. On each

$\mathcal{A}_i$ where $1 \le i \le k$, the designer may choose to either terminate the process, or start a new step recursively to further decompose $\mathcal{A}_i$. See Procedure 1.

We point out that the ADD procedure, as presented by its original proponents, involves numerous additional stages other than the ones described above [15]. The reason we choose this over-simplified description is that we believe these are the steps that could be rigorously presented, and they abstractly capture in a way most of the steps mentioned in the original informal description.

Procedure 1. ADD($\mathcal{A}$) (General Plan)

1: $(D_1, D_2, \ldots, D_k) \leftarrow$ Decompose($\mathcal{A}$) // compute a rational decomposition of $\mathcal{A}$
2: **for** $1 \le i \le k$ **do**
3:     $\mathcal{A}_i \leftarrow$ an primitive attribute consistent with D_i
4:     **if** $\mathcal{A}_i$ needs further decomposition **then**
5:         ADD($\mathcal{A}_i$)

The Decompose($\mathcal{A}$) operation produces a rational decomposition $(D_1, \ldots, D_k)$ of the input attribute primitive $\mathcal{A}$ that satisfies the requirements of $\mathcal{A}$. We also note that Decompose($\mathcal{A}$) amounts to a crucial step in the ADD process, as the decomposition determines to a large extend how well the quality attributes are met. This step is also a challenging one as interactions among quality attributes create potential conflicts. Thus, in the next section, we define a game model which allows us to automate the Decompose($\mathcal{A}$) operation.

3   Decomposition Games

3.1   Requirement Relevance

The Decompose($\mathcal{A}$) procedure looks for a rational decomposition that meets the requirements in $\mathcal{A}$ as much as possible. Let $\mathcal{A} = (\mathsf{F}, \mathsf{S}, \mathsf{C}, \prec, \approx, \hookrightarrow)$ be an attribute primitive. Relevance between requirements are determined by relations $\prec, \approx, \hookrightarrow$ and the constraint set C. In the following, the *Jaccard index* $J(S_1, S_2)$ between two sets S_1, S_2 — a common statistical measure for similarity of sets — is defined as $J(S_1, S_2) = \frac{|S_1 \cap S_2|}{|S_1 \cup S_2|}$; Intuitively, the relevance of a requirement r to other requirements is influenced by the "links" between r and the functional, the non-functional requirements, as well as design constraints.

Definition 3 (Relevance). *Two requirements $r_1, r_2 \in \mathsf{R}$ are relevant if*

- $r_1, r_2 \in \mathsf{F}$, and either $d^{-1}(r_1) \cap d^{-1}(r_2) \ne \varnothing$ (derived from some common scenario), or $f(r_1) \cap f(r_2) \ne \varnothing$ (relevant through dependency), or $c(r_1) \cap c(r_2) \ne \varnothing$ (share some common design constraints).
- $r_1, r_2 \in \mathsf{S}$, and either $r_1 \approx r_2$ (instantiate the same general scenario), or $d(r_1) \cap d(r_2) \ne \varnothing$ (jointly derives some functionality) or $c(r_1) \cap c(r_2) \ne \varnothing$.
- $r_1 \in \mathsf{F}$, $r_2 \in \mathsf{S}$, and either $f(r_1) \cap d(r_2) \ne \varnothing$ (r_1 depends on a requirement that is derived from r_2), or $c(r_1) \cap c(r_2) \ne \varnothing$.

If two requirements are relevant, the "amount of relevance" depends on overlaps between their derived sets, dependency sets and constraints. If two requirements are not relevant, then we regard them as having a negative relevance $\lambda < 0$, which represents a "penalty" one pays when two irrelevant requirements get in the same design element.

Definition 4. *We define the* relevance index $\sigma(r_1, r_2)$ *of* $r_1 \neq r_2 \in R$ *as follows:*

1. if two functional requirements $r_1, r_2 \in F$ *are relevant, then*

$$\sigma(r_1, r_2) = \alpha J(d^{-1}(r_1), d^{-1}(r_2)) + \beta J(f(r_1), f(r_2)) + \gamma J(c(r_1), c(r_2));$$

2. if two scenarios $r_1, r_2 \in S$ *are relevant, then*

$$\sigma(r_1, r_2) = \beta J(d(r_1), d(r_2)) + \gamma J(c(r_1), c(r_2));$$

3. If $r_1 \in F$ *and* $r_2 \in S$ *are relevant, then*

$$\sigma(r_1, r_2) = \sigma(r_2, r_1) = \beta J(f(r_1), d(r_2)) + \gamma J(c(r_1), c(r_2));$$

4. otherwise, $\sigma(r_1, r_2) = \lambda$

The constants α, β, γ *are positive real numbers, that represent weights on the overlaps in* d_1, d_2*'s generated sets, dependency sets and constraints, respectively. We require* $\alpha + \beta + \gamma = 1$.

For simplicity, we do not include these constants in expressing the function σ, and all subsequent notions that depend on σ.

Example 2. Continue from $\mathcal{A}$ in Example 1. To emphasise the non-functional requirements we give a larger weight to α, setting $\alpha = 0.5$, $\beta = 0.4$, $\gamma = 0.1$. We also set $\lambda = -0.5$. Then $\sigma(r_1, r_2) = 0.4 \times \frac{2}{2} = 0.4$ for any $(r_1, r_2) \in \{(q_1, q_2), (q_3, f_3)\} \cup (\{q_1, q_2\} \times \{f_1, f_2\})$; $\sigma(q_1, q_3) = 0.1 \times \frac{1}{2} = 0.05$; $\sigma(f_1, f_2) = 0.5 \times \frac{2}{2} + 0.4 \times \frac{2}{2} = 0.9$; and relevance between any other pairs is -0.5. Figure 2 (a) illustrates the (positive) relevance in a weighted graph.

3.2   Decomposition Games

We employ notions from coalition games to define what constitutes a *rational* decomposition. In a coalition game, players cooperate to form coalitions which achieve certain collective payoffs [6].

Definition 5 (Coalition Game). *A* coalition game *is a pair* (N, ν) *where* N *is a finite set of players, and each subset* $D \subseteq N$ *is a* coalition; $\nu : 2^N \to \mathbb{R}$ *is a* payoff function *associating every* $D \subseteq N$ *a real value* $\nu(D)$ *satisfying* $\nu(\varnothing) = 0$.

This provides the set up for decompositions: Imagine a coalition game consisting of $|R|$ agents as players, where each agent is in charge of a different requirement. The players form coalitions which correspond to sets of requirements, i.e., design elements. The payoff function would associate with every coalition a numerical value, which is the payoff gained by each member of the coalition. Therefore, an equilibrium of the game amounts to a decomposition with the right balance among all requirements – this would be regarded as a rational decomposition.

It remains to define the payoff function. Naturally, the payoff of a coalition is determined by the *interactions* among its members. Take $r_1, r_2 \in D$. If one of r_1, r_2 is a functional requirement, then their interaction is defined by their relevance index $\sigma(r_1, r_2)$, as higher relevance means a higher level of interaction. Suppose now both r_1, r_2 are scenarios (non-functional). Then the interaction becomes more complicated, as a quality attribute may enhance or defect another quality attribute. In [14, Chapter 14], the authors identified effects acting from one quality attribute to another, which is expressed by a *tradeoff matrix T*:

- T has dimension $m \times m$ where m is the number of general scenarios
- For $i \neq j \in \{1, \ldots, m\}$, the (i, j)-entry $T_{i,j} \in \{-1, 0, 1\}$.

Let $g_1, g_2, \ldots, g_m$ be general scenarios. $T_{i,j} = 1$ (resp. $T_{i,j} = -1$) means g_1 has a positive (resp. negative) effect on g_2, $T_{i,j} = 0$ means no effect. E.g., the tradeoff matrix defined on six common quality attributes is:

	Performance	Modifiability	Security	Availability	Testability	Usability
Performance	0	−1	0	0	0	−1
Modifiability	−1	0	0	1	1	0
Security	−1	0	0	1	−1	−1
Availability	0	0	0	0	0	0
Testability	0	1	1	1	0	1
Usability	−1	0	0	0	−1	0

Note that the matrix is not necessarily symmetric: The effect from g_1 to g_2 may be different from the effect from g_2 to g_1. For example, an improvement in system performance may not affect security, but increasing security will almost always adversely impact performance. We assume that the matrix T is given prior to ADD; this assumption is reasonable as there is an effective map from any general scenario to the main quality attribute it tries to capture. We use this tradeoff matrix to define the interaction between two scenarios in S.

Definition 6 (Coalitional Relevance). *For a coalition $D \subseteq R$ and $r \in D$, the* coalitional relevance *of r in D is the total relevance from r to all other requirements in D, i.e., $\rho(r, D) = \sum_{s \in D, s \neq r} \sigma(r, s)$.*

Definition 7 (Effect Factor). *For scenarios r_1, r_2 in the same coalition D, the* effect factor *from r_1 to r_2 expresses the effect of r_1 towards r_2, i.e.,*

$$\varepsilon(r_1, r_2, D) = \begin{cases} -|\rho(r_1, D)| & if \ T(g(r_1), g(r_2)) = -1 \\ 0 & if \ T(g(r_1), g(r_2)) = 0 \\ \rho(r_1, D) & if \ T(g(r_1), g(r_2)) = 1 \end{cases}$$

We are now ready to define the interaction between two scenarios $r_1, r_2 \in R$.

Definition 8 (Interaction). *Let $r_1 \neq r_2 \in R$ be requirements. The interaction between r_1, r_2 is simply the relevance $\sigma(r_1, r_2)$ if one of r_1, r_2 is functional; otherwise (both r_1, r_2 are non-functional), it is the sum of their effect factors, i.e.,*

the interaction $\nu(r_1, r_2, D) := \begin{cases} \sigma(r_1, r_2) & \text{if } \{r_1, r_2\} \cap \mathcal{F} \neq \varnothing \\ \varepsilon(r_1, r_2, D) + \varepsilon(r_2, r_1, D) & \text{otherwise} \end{cases}$

The coalition utility $\nu(D)$ of any coalition $D \subseteq R$ is defined as the sum of interactions among all pairs of requirements in the coalition, i.e.,

$$\nu(D) = \sum_{r_1 \neq r_2 \in D} \nu(r_1, r_2, D)$$

Definition 9 (Decomposition Games (DG)). *Let $\mathcal{A} = (F, S, C, \prec, \approx, \hookrightarrow)$ be an attribute primitive. The DG $G_{\mathcal{A}}$ is the coalition game $(F \cup S, \nu)$ where $\nu : 2^{F \cup S} \to \mathbb{R}$ is the coalition utility function.*

Example 3 (Coalition Utility). Continue the setting in Example 2. Let the general scenarios be $g_1 = \{q_1, q_2\}$ and $g_2 = \{q_3\}$. We assume matrix T specifies $T(g_1, g_2) = 1$ and $T(g_2, g_1) = -1$. Consider the coalition $C = \{q_1, q_3, f_3\}$. We have: $\rho(q_1, C) = 0.05 - 0.5 = -0.45$; and $\rho(q_3, C) = 0.4 + 0.05 = 0.45$. So $\varepsilon(q_1, q_3, C) = -0.45 \times 1 = -0.45$ and $\varepsilon(q_3, q_1, C) = 0.45 \times (-1) = -0.45$.

Thus $\nu(q_1, q_3, C) = -0.45 - 0.45 = -0.9$. Therefore, $\nu(C) = \sigma(q_1, f_3) + \sigma(q_3, f_3) + (-0.9) = (-0.5) + 0.4 + (-0.9) = -1$ but $\nu(C \setminus \{q_1\}) = \nu(\{q_3, f_3\}) = \sigma(q_3, f_3) = 0.4$; See Fig. 2(b).

As it has turned out, despite the fact that matrix T indicates q_1 will act positively to q_3, and that q_1, q_3 have a positive (0.05) relevance, adding q_1 into the coalition of $\{q_3, f_3\}$ drastically decreases the coalition utility.

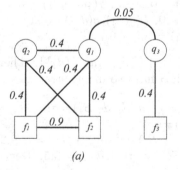
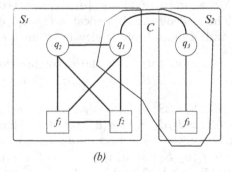

(a)   (b)

Fig. 2. (a) Weights on the edges are relevance (function σ) between requirements in Example 2; the diagram omits the negative weighted pairs. (b) The decomposition $\{S_1, S_2\}$ is a solution with $\nu(S_1) = 2.5$, $\nu(S_2) = 0.4$. The coalition C has $\nu(C) = -1$

3.3   Solution Concept

The decomposition game as defined above is a form of *hedonic game*, which consists of a set of players, and each player's payoff depends only on the member of her coalition [2,7]. This setting is different from the typical coalition formation games with transferrable utilities (TU): Firstly, in TU games, one normally assumes the axioms of superadditivity ($\nu(D_1 \cup D_2) \geq \nu(D_1) + \nu(D_2)$) and monotonicity ($D_1 \subseteq D_2 \Rightarrow \nu(D_1) \leq \nu(D_2)$) which would obviously not hold for decomposition as players may counteract with each other, reducing their combined utility. Secondly, the typical solution concepts in coalition games (such as Pareto optimality, and Shapely value) focus on distribution of payoffs to each individual player assuming a grand coalition consisting of all players. In decomposition such a grand coalition is normally not desirable and the focus is on the overall payoff of each coalition D, rather than the individual requirements. The above differences motivate us to consider a different solution concept of DG G_A. At any instance of the game, the players form a decomposition $(D_1, D_2, \ldots, D_k)$. We assume that the players may perform two collaborative strategies:

1. *Merge strategy*: Two coalitions may choose to merge if they would obtain a higher combined payoff.
2. *Bind strategy*: Players within the same coalition may form a sub-coalition if they would obtain a higher payoff.

Example 4 (A Dilemma). We present an example demonstrating the dynamics of a DG G_A. This example shows a real-world dilemma: As a coalition pursues higher utility through expansion (merging with others), it may be better to choose a "less-aggressive" expansion strategy over the "more-aggressive" counterpart, even though the latter clearly brings a higher payoff. Assume the following situation (which is clearly plausible in an attribute primitive):

- $R = \{d_1, d_2, d_3, d_4\}$ where $S = \{d_1, d_4\}$ and $d_1 \not\approx d_4$.
- We set $\sigma(\{d_1, d_2\}) = \sigma(\{d_1, d_3\}) = \sigma(\{d_2, d_3\}) = 0.1$, and $\sigma(\{d_2, d_4\}) = 0.5$.
- The tradeoff matrix indicates $T(g(d_1), g(d_4)) = 0$, $T(g(d_4), g(d_1)) = -1$.
- And, d_1 and d_4 are irrelevant, namely $\sigma(d_1, d_4) = \lambda = -0.7$.

Suppose we start with the decomposition $\{S = \{d_1, d_2\}, \{d_3\}, \{d_4\}\}$. Then $\nu(S) = \rho(d_1, d_2, S) = \nu(d_1, d_2, S) = 0.1$. Coalition S has two merge strategies:

(1) For $S_1 = S \cup \{d_3\}$: $\nu(d_1, d_2, S_1) = \sigma(d_1, d_2) = 0.1$, $\nu(d_1, d_3, S_1) = \sigma(d_1, d_3) = 0.1$, $\nu(d_2, d_3, S_1) = \sigma(d_2, d_3) = 0.1$. Thus $\nu(S_1) = 0.3$.
(2) For $S_2 = S \cup \{d_4\}$: $\nu(d_1, d_4, S_2) = \varepsilon(d_4, d_1, S_2) = -0.7 + 0.5 = -0.2$, $\nu(d_1, d_2, S_2) = \sigma(d_1, d_2) = 0.1$, $\nu(d_2, d_4, S_2) = \sigma(d_2, d_4) = 0.5$. Hence $\nu(S_2) = 0.1 - 0.2 + 0.5 = 0.4$

Merging with $\{d_4\}$ clearly results in a higher payoff for the combined coalition. However, if this merge happens, as $\nu(\{d_2, d_4\}) = 0.5 > \nu(S_2) = 0.4$, d_2 and d_4 would choose to bind together, hence leaving S_2. This would be undesirable if d_1 is a critical non-functional requirement for d_2.

Example 4 shows that a solution concept would be a decomposition where no "expansion" nor "crumbling" occur to any coalition. Formally, we define the following solution concepts:

Definition 10 (Solution). Let $D = (D_1, \ldots, D_k)$ be a decomposition of $\mathcal{A}$.

1. A coalition $D \subseteq \mathsf{R}$ is cohesive if for all $C \subseteq D$, $\nu(C) < \nu(D)$; D is cohesive if so is every D_i.
2. A coalition D_i is expansion-free with respect to D if $\max\{\nu(D_i), \nu(D_j)\} > \nu(D_i \cup D_j)$; D is expansion-free if so is every D_i.

A solution of a DG is a decomposition that is both cohesive and expansion-free.

Example 5 (Solution). Continue from Example 3, the utilities for

$$S_1 = \{q_1, q_2, f_1, f_2\} \qquad \text{and} \qquad S_2 = \{q_3, f_3\} \qquad \text{are:}$$

- S_1: $\nu(q_1, q_2, S_1) = 0$, $\nu(q_1, f_1, S_1) = \nu(q_1, f_2, S_1) = \nu(q_2, f_1, S_1) = 0.4$,
  $\nu(q_1, f_2, S_1) = 0.4$, $\nu(f_1, f_2, S_1) = 0.9$. Thus $\nu(S_1) = 0.4 \times 4 + 0.9 = 2.5$
- S_2: $w(q_3, f_3, S_2) = 0.4$. Thus $\nu(S_2) = 0.4$

Both S_1 and S_2 are cohesive. Furthermore, we have $\nu(q_1, q_3, \mathsf{R}) = 0.75 - 1.05 = -0.3$ and $\nu(q_2, q_3, \mathsf{R}) = 0.2 - 1.05 = -0.85$. Thus $\nu(\mathsf{R}) = 2.9 - 0.5 \times 6 - 0.85 - 0.3 = -1.45$. Consequently, $\{S_1, S_2\}$ is also expansion-free, and is thus a solution of the game.

A solution of a DG $G_{\mathcal{A}}$ corresponds to a rational decomposition of the attribute primitive $\mathcal{A}$. As shown by Theorem 11, any attribute primitive admits a solution, and rather expectedly, a solution may not be unique.

Theorem 11 (Solution Existence). There exists a solution in any DG $G_{\mathcal{A}}$.

Proof. We show existence of a solution by construction. Let $(D_1, D_2, \ldots, D_k)$ be a longest sequence such that for any $i = 1, \ldots, k$, D_i is a minimal coalition with maximal utility in $2^{\mathsf{R}} \setminus \{D_1, \ldots, D_{i-1}\}$ (i.e., $\forall D \subseteq 2^{\mathsf{R}} \setminus \{D_1, \ldots, D_{i-1}\} : \nu(D_1) \geq \nu(D)$ and $\forall D \subseteq D_1 : \nu(D_1) > \nu(D)$).

We claim that $D = (D_1, \ldots, D_k)$ is a solution in $G_{\mathcal{A}}$. Indeed, for any $1 \leq i \leq k$, any proper subset of D_i would have payoff strictly smaller than $\nu(D_i)$ by minimality of D_i. Thus D is cohesive. Moreover, if $\nu(D_i \cup D_j) > \min\{\nu(D_i), \nu(D_j)\}$ for some $i \neq j$, then $D_{\min\{i,j\}}$ does not have maximal utility in $\mathsf{R} \setminus \{D_1, \ldots, D_{\min\{i,j\}-1}\}$. Hence D is expansion-free. $\square$

Proposition 12. The solution of a DG may not be unique.

Proof. Let $\mathcal{A} = (\mathsf{F}, \mathsf{S}, \mathsf{C}, \prec, \approx, \hookrightarrow)$ be an attribute primitive where $\mathsf{S} = \varnothing$ and $\mathsf{F} = \{d_1, d_2, \ldots, d_6\}$. We may define $\mathsf{C}, \prec, \approx, \hookrightarrow$ in such a way that

- For all $\{i, j\} \subseteq \{1, 2, 3, 4\}$ and $\{i, j\} \subseteq \{4, 5, 6\}$, $i \neq j \Rightarrow \nu(\{d_i, d_j\}) = 0.1$
- For all $i \in \{1, 2, 3\}$, $j \in \{5, 6\}$, $\nu(\{d_i, d_j\}) = -0.1$

Consider $C = \{C_1 = \{d_1, d_2, d_3\}, C_2 = \{d_4, d_5, d_6\}\}$ and $D = \{D_1 = \{d_1, d_2, d_3, d_4\}, D_2 = \{d_5, d_6\}\}$. Note that $\nu(C_1) = 0.3$ and $\nu(C_2) = 0.3$; C is cohesive and C is expansion-free as $\nu(\mathsf{F}) = 0.3 = \nu(C_1)$. Note also that $\nu(D_1) = 0.6$ and $\nu(D_2) = 0.1$; D is cohesive and D is expansion-free as $\nu(D_1) > \nu(\mathsf{F})$ $\square$

4   Solving Decomposition Games

Based on our game model, the operation Decompose($\mathcal{A}$) in Procedure 1 is reduced to the following DG problem:

INPUT: An attribute primitive $\mathcal{A} = (\mathsf{F}, \mathsf{S}, \mathsf{C}, \prec, \approx, \hookrightarrow)$
OUTPUT: A solution $\boldsymbol{D} = (D_1, D_2, \ldots, D_k)$ of the game $G_{\mathcal{A}}$

Here, we measure computational complexity with respect to the number of requirements in $\mathsf{F} \cup \mathsf{S}$. The proof of Theorem 11 already implies an algorithm for solving the DG problem: check all subsets of R to identify a minimal set with maximal utility; remove it from R and repeat. However, it is clear that this algorithm takes exponential time. We will demonstrate below that a polynomial-time algorithm for this problem is, unfortunately, unlikely to exist.

We consider the decision problem DG_D: *Given $\mathcal{A}$ and a number $w > 0$, is there a solution $\boldsymbol{D}$ of $G_{\mathcal{A}}$ in which the highest utility of a coalition reaches w?* Recall that the payoff function ν of $G_{\mathcal{A}}$ is defined assuming constants $\alpha, \beta, \gamma > 0$ and $\lambda < 0$. The theorem below holds assuming $\lambda < -\gamma$.

Theorem 13. *The DG_D problem is NP-hard.*

Proof. The proof is via a reduction from the maximal clique problem, which is a well-known NP-hard problem. Given an undirected graph $H = (V, E)$, we construct an attribute primitive $\mathcal{A}$ such that any cohesive coalition in $G_{\mathcal{A}}$ reveals a clique in H. Suppose $V = \{1, 2, \ldots, n\}$. The requirements of $\mathcal{A}$ consist of n^2 scenarios: $\mathsf{R} = \mathsf{S} := \{a_{i,i'} \mid 1 \leq i \leq n, 1 \leq i' \leq n\}$. In particular, all requirements are non-functional. We define an edge relation E' on S such that

1. $(i, j) \in E$ iff $(a_{i,i'}, a_{j,j'}) \in E'$ for some $1 \leq i' \leq n$ and $1 \leq j' \leq n$
2. If $(a_{i,i'}, a_{j,j'}) \in E'$ then $(a_{i,i''}, a_{j,j''}) \notin E'$ for any $(i'', j'') \neq (i', j')$.
3. Any $a_{i,i'}$ is attached to at most one edge in E'.

Note that such a relation E' exists as any node $i \in V$ is only connected with at most $n - 1$ other nodes in H. Intuitively, a set of requirements $A_i = \{a_{i,1}, \ldots, a_{i,n}\}$ serves as a "meta-node" and corresponds to the node i in H. In constructing $\mathcal{A}$, we may define the general scenarios in such a way that

- $T(g(a_{i,j_1}), g(a_{i,j_2})) = 0$ for any $1 \leq i \leq n$ and $j_1 \neq j_2$.
- $T(g(a_{i_1,j_1}), g(a_{i_2,j_2})) = -1$ for any $(i_1, i_2) \notin E$.
- $T(g(a_{i_1,j_1}), g(a_{i_2,j_2})) = 1$ for any $(a_{i_1,j_1}, a_{i_2,j_2}) \in E'$
- $T(g(a_{i_1,j_1}), g(a_{i_2,j_2})) = 0$ for any $(i_1, i_2) \in E$ but $(a_{i_1,j_1}, a_{i_2,j_2}) \notin E'$

For every $1 \leq i \leq n$ and $1 \leq j < j' \leq n$, put in a constraint $c_i(j, j') = \{a_{i,j}, a_{i,j'}\}$. Thus the relevance between $a_{i,j}$ and $a_{i,j'}$ is

$$\sigma(a_{i,j}, a_{i,j'}) = \frac{|c(a_{i,j}) \cap c(a_{i,j'})|}{|c(a_{i,j}) \cup c(a_{i,j'})|} = \frac{\gamma}{2(n-1)}$$

Furthermore if $i \neq i'$, then for any j, j' we set $\sigma(a_{i,j}, a_{i,j'}) = \lambda$. Suppose $U = \{i_1, \ldots, i_\ell\}$ induces a complete subgraph of H. We define the *meta-clique coalition* of U as

$$D_U = \bigcup_{1 \leq j \leq \ell} A_{i_j}$$

By the above definition, for any $1 \leq s < t \leq \ell$, take j, j' such that $(a_{i_s,j}, a_{i_t,j'}) \in E'$.

$$
\begin{aligned}
w(i_s, i_t, D_U) &= \varepsilon(a_{i_s,j}, D_U) + \varepsilon(a_{i_t,j'}, D_U) \\
&= \rho(a_{i_s,j}, D_U) + \rho(a_{i_t,j'}, D_U) \\
&= (n-1) \times \frac{\gamma}{2(n-1)} + (n-1) \times \frac{\gamma}{2(n-1)} = \gamma
\end{aligned}
$$

Thus $\nu(D_U) = \frac{n(n-1)\gamma}{2}$. Taking out any element from D_U results in a strict decrease in utility, and hence D_U is cohesive.

Now take any coalition $D \subseteq R$ that contains two requirements $a_{i,i'}, a_{j,j'}$ such that $(i,j) \notin E$. Let $s = |A_i \cap D|$ and $t = |A_j \cap D|$. Note also that $\sigma(a_{j,j'}, a_{i,i''}) = \lambda$ for any $a_{i,i''} \in A_i \cap D$. Therefore we have

$$
\nu(D) - \nu(D \setminus \{a_{j,j'}\}) \leq \gamma + 2w(a_{j,j'}, a_{i,i'}, D) \times s \leq \gamma + 2\lambda + \gamma = 2(\lambda + \gamma) < 0
$$

The last inequality above is by assumption that $\lambda < -\gamma$. Thus D is not cohesive.

By the above argument, a coalition $D \subseteq R$ is cohesive in $G_{\mathcal{A}}$ iff D is the meta-clique coalition D_U for some clique U in H. Furthermore, a decomposition $\boldsymbol{D} = (D_1, D_2, \ldots, D_k)$ is a solution in $G_{\mathcal{A}}$ iff V can be partitioned into sets $U_1, \ldots, U_k$ where each U_i is a clique, and $D_i = D_{U_i}$ for all $1 \leq i \leq k$. In particular, H has a clique with ℓ nodes if and only if $G_{\mathcal{A}}$ has a solution that contains a coalition whose utility reaches $\frac{\ell(\ell-1)\gamma}{2}$. This finishes the reduction. $\square$

Theorem 13 shows that, in a sense, identifying a "best" solutions in a DG $G_{\mathcal{A}}$ is hard. The main difficulty comes from the fact that one would examine all subsets of players to find an optimal cohesive coalition. This calls for a relaxed notion of a solution that is computationally feasible. To this end we introduce the notion of k-cohesive coalitions. Fix $k \in \mathbb{N}$ and enforce this rule: Binding can only take place on k or less players. That is, a coalition C is k-cohesive whenever $\nu(C)$ is greater than the utility of any subsets with at most k players.

Definition 14. *Fix $k \in \mathbb{N}$. In a DG $G_{\mathcal{A}} = (F \cup S, \nu)$, we say a coalition $D \subset F \cup S$ is k-cohesive if $\nu(D') < \nu(D)$ for all $D' \subset D$ with $|D'| \leq k$. An decomposition $\boldsymbol{D}$ is k-cohesive if every coalition in $\boldsymbol{D}$ is k-cohesive; if $\boldsymbol{D}$ is also expansion-free, then it is a k-cohesive solution of the game $G_{\mathcal{A}}$.*

Remark. In a sense, the value k in the above definition indicates a level of *expected cohesion* in the decomposition process. A higher value of k implies less restricted binding within any coalition, which results in higher "sensitivity" of design elements to conflicts. In a software tool which performs ADD based on DG, the level k may be used as an additional parameter.

Let R be a set of requirements. A coalition D is called *maximally k-cohesive* in R if $|D| \leq k$, D is k-cohesive and $\nu(D) \geq \nu(D')$ for any $D' \subseteq R$. Suppose the operation $\mathsf{max}(R, k)$ computes a maximally k-cohesive set in R. The algorithm $\mathsf{DGame}(\mathcal{A}, k)$ (Procedure 2), which uses $\mathsf{Cohesive}(\mathcal{A}, k)$ (Procedure 3) as a subroutine, computes a k-cohesive solution of $G_{\mathcal{A}}$. Note that the $\mathsf{Cohesive}(\mathcal{A}, k)$

Procedure 2. DGame($\mathcal{A}, k$)

INPUT: Attribute primitive $\mathcal{A}$, $k > 0$
OUTPUT: Attribute Decomposition D
1: $D \leftarrow$ Cohesives($\mathcal{A}, k$)
2: Combine $\leftarrow$ true
3: **while** Combine **do**
4:      Combine $\leftarrow$ false
5:      **for** $(D, D') \in D^2$, $D \neq D'$ **do**
6:          **if** $\nu(D' \cup D) > \nu(D)$ and $\nu(D' \cup D) > \nu(D')$ **then**
7:              $D \leftarrow D' \cup D$ and remove D' from D
8:              Combine $\leftarrow$ true
9: **return** D

Procedure 3. Cohesive($\mathcal{A}, k$)

INPUT: Attribute primitive $\mathcal{A}$, $k > 0$
OUTPUT: Attribute Decomposition D
1: $D \leftarrow [\ ]$, $R \leftarrow \mathsf{F} \cup \mathsf{S}$
2: **while** $|R| > 0$ **do**
3:      $S \leftarrow \max(R, k)$ // compute a maximally k-cohesive coalition
4:      $R \leftarrow R \setminus S$
5:      $D \leftarrow [D, S]$
6: **return** D

---

operation maintains a list D, which when returned, denotes a decomposition. Note also that the returned $D = (D_1, \ldots, D_m)$ satisfies the following condition:

$$\forall 1 \leq i \leq m : D_i \text{ is maximally } k\text{-cohesive in } D_i \cup \cdots \cup D_m$$

We call this D a *maximally k-cohesive decomposition*.

Lemma 15. *Suppose D is a maximally k-cohesive decomposition. Take any $1 \leq i < j \leq n$. If $\nu(D_i \cup D_j) > \max\{\nu(D_i), \nu(D_j)\}$ then $D_i \cup D_j$ is k-cohesive.*

Proof. Let $S_i = \bigcup_{i \leq j \leq m} D_j$ for any $i = 1, \ldots, m$. Suppose $\nu(D_i \cup D_j) > \max\{\nu(D_i), \nu(D_j)\}$ for $1 \leq i < j \leq m$. By assumption D_i is maximally k-cohesive in S_i. For any finite set $U \subseteq D_i \cup D_j \subseteq S_i$ such that $|U| \leq k$, we have $\nu(U) \leq \nu(D_i) < \nu(D_i \cup D_j)$. Hence $D_i \cup D_j$ is also k-cohesive. $\square$

Theorem 16. *Given an attribute primitive $\mathcal{A}$, the DGame($\mathcal{A}, k$) algorithm computes a k-cohesive solution of the decomposition game $G_{\mathcal{A}}$ in time $O(n^k)$, where n is the number of requirements in $\mathcal{A}$.*

Proof. The DGame($\mathcal{A}, k$) algorithm calls Cohesive($\mathcal{A}, k$) to produce a maximally k-cohesive decomposition D, and then performs several iterations to "combine" the coalitions in D. By Lemma 15, the decomposition D after each iteration is k-cohesive. There is a point when for all $D, D' \in D$ we have $\nu(D \cup D') \leq \max\{\nu(D), \nu(D')\}$. At this moment, the **while**-loop will terminate and D is expansion-free. The time complexity is justified as there are $O(n^k)$ subsets of F ∪ S with size $\leq k$. Thus computing a maximally k-cohesive decomposition takes time $O(n^k)$.   □

5   Case Study: Cafeteria Ordering System

To demonstrate applicability of our game model in real-world, we build a DG for a cafeteria ordering system (COS). A COS permits employees of a company to order meals from the company cafeteria online and is a module of a larger cafeteria management system. The requirements of the project have been produced through a systematic requirement engineering process and is well-documented (See full details from [14, Appendix C]). Since COS is a subsystem within a larger system, the requirements also incorporate interfaces with other subsystems of the overall system. The initial attribute primitive has 60 requirements with $|S| = 11$, $|F| = 49$ and 7 design constraints. Non-functional requirements conflict with each other, e.g., the

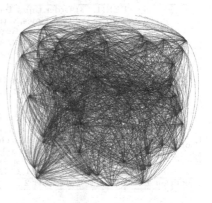

Fig. 3. Interactions between requirements in the COS [14]. Blue edges indicate positive interactions and red edges indicate negative interactions (Color figure online).

general scenario USE conflicts with the general scenario PER. Also the requirements exhibit some complex relationships, e.g. SEC1 ↪ Order.Pay.Deduct.

We demonstrate the complicated interactions among requirements using a complete graph where nodes are all requirements in R = S ∪ F; see Fig. 3. The edges are in two colours: (r_1, r_2) gets blue if $w(r_1, r_2, R) > 0$ and red if $w(r_1, r_2, R) < 0$.

We run the DGame$(\mathcal{A}, k)$ algorithm to identify a k-cohesive solution for different levels k of expected cohesion. In order to clearly identify sub-components, we give a higher penalty λ between conflicting requirements: $\alpha = 0.4$, $\beta = 0.3$, $\gamma = 0.3$, $\lambda = -1.3$. We choose $k \in \{1, \ldots, 7\}$. As argued above, setting a higher value of k should in principle improve the quality of the output decomposition, although this also means a longer computation time. We implement our algorithm using Java on a laptop with Intel Core i7-3630QM CPU 2.4 GHz 8.0 GB RAM. The running time for different values of k is: 503 milliseconds for $k = 3$ and approximately 1140 s for $k = 6$ (Table 1).

Cohesion Level $k = 3$. The 3-cohesive solution consists of 5 coalitions. An examination at the requirements in each coalition reveals: *Coalition 0* relates to usability and ensures availability of user interactions; it apparently corresponds to a user interface module. *Coalition 1* is performance-oriented and is separated from the usability requirements; it thus corresponds to a back-end module that handles all the internal operations. *Coalition 2* deals with the payroll system

Table 1. Resulting 3- and 6-cohesive solutions, ordered by payoff values.

	3-Cohesive Solution	6-Cohesive Solution
Coalition 0	AVL1 ROB1 SAF1 SEC(1,2,4) USE(1,2) Order.Confirm Order.Menu.Data Order.Deliver.(Select,Location) Order.Pay Order.Place Order.Retrieve Order.Units.Multiple UI2 UI3	AVL1 ROB1 SAF1 SEC(1,2,4) PER(1,2,3) USE(1,2) Order.Confirm Order.Deliver Order.Deliver.(Select,Location) Order.Menu.Date Order.Pay Order.Retrieve Order.Place Order.Units Order.Units.Multiple UI2 UI3
Coalition 1	PER(1,2,3) Order.Units.TooMany Order.Deliver.(Times,Notimes) Order.Place.(Cutoff,Data,Register,No) Order.Pay.(OK,NG) Order.Done.Failure Order.Confirm.(Prompt,Response,More)	Order.pay.(Deliver,Pickup,Deduct) Order.Done.Patron SI2.2 SI2.3
Coalition 2	Order.pay.(Deliver,Pickup,Deduct) Order.Done.Patron SI2.2 SI2.3	Order.Units.TooMany Order.Deliver.(Times,Notimes) Order.Place.(Cutoff,Data,Register,No) Order.Pay.(OK,NG) Order.Done.Failure Order.Confirm.(Prompt,Response,More)
Coalition 3	Order.Menu Order.Unit Order.Done Order.Done.(Menu,Times,Cafeteria) Order.Done.(Store,Inventory) Order.Deliver Order.Menu.Available Order.Confirm.Display Order.Pay.Method SI1.3 SI2.5 CI2	Order.Done Order.Done.(Menu,Times,Cafeteria) Order.Done.(Store,Inventory) SI1.3 SI2.1 SI2.4 SI2.5 CI1 CI2
Coalition 4	SI1.1 SI1.2	Order.Menu.Available SI1.1 SI1.2

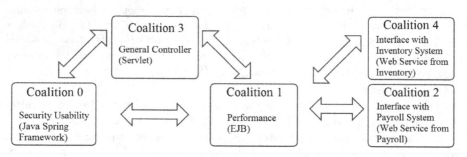

Fig. 4. The 3-cohesive solution. *Coalition 0*: Java Spring framework uses server page as user interface and provides a powerful encryption infrastructure (Spring Crypto Module). Server page is suitable for implementing interactive user interface. *Coalition 1*: Enterprise Java Bean (EJB) is a middleware (residing in the application server) used to communicate between different components. It provides rich features for processing HTTP requests. *Coalition 2*: The COS uses a package solution from corresponding payroll system. *Coalition 3*: A servlet is a controller in Java application server which separates business logic from control. *Coalition 4*: A web service interface outside COS.

outside COS and defines a controlling interface from COS to payroll. *Coalition 3* consists of several functional requirements that control life cycle of the COS. *Coalition 4* is an interface to access the inventory system outside COS.

It is clear that this solution separates the control, user inputs and computation modules, and fits the MVC (Model-View-Controller) architectural pattern. In addition, there is a design constraint that requires the use of Java and Oracle database engine. So, we instantiate the design elements as in Fig. 4.

Cohesion Level $k = 6$. The 6-cohesive solution also contains five coalitions, with a similar structure as the 3-cohesive counterpart. There are, nevertheless, several important differences: Firstly, the performance (PER) scenarios now belong to coalitions 0. This means that some performance-related computation is moved to the front-end. This is reasonable as this lightens the computation load of the back-end and thus improving performance and availability. Secondly, the functional requirement Order.Menu.Available is moved to coalition 4, which is the interface between COS and the inventory system. This requirement specifies that the menu should only display those food items that are available in inventory.

Instead of server page, we use scripting to reduce the server's computation load. This can be achieved by changing the front-end to a JavaScript oriented designs. The main difficulty lies in that we need to put extra effort when using JavaScript to communicate with web server (such as AJAX) in order to ensure usability, performance and security. We instantiate design elements as in Fig. 5.

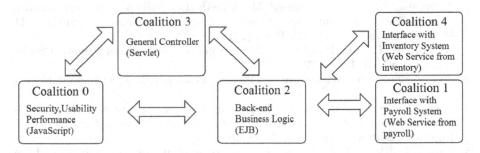

Fig. 5. The 6-cohesive solution. *Coalition 0* uses JavasScript as a front end for user interface. It also takes some computation for sever in order to achieve better performance. *Coalition 1* is an interface for accessing the payroll system. *Coalition 2* ensures the business logic in COS. *Coalition 3* coordinates input from front end (coalition 0) to back end (coalition 1). *Coalition 4* is an interface for accessing the inventory system.

6   Conclusion and Future Work

Analysing tradeoff among requirements in software architecture design has been an important research topic. The use of computational games in software architecture design is a novel technique aimed to contribute to this line of research. We argue that equilibrium concepts in games provide an appropriate formal set-up of arguing rational designs. The proposed game-based approach not only builds on established software architecture methodology (ADD), but is also shown — through a case study — to provide reasonable design guidelines to a real world application. We suggest that this framework would be useful in the following:

– Designing a software system that involves a large number of functionalities and quality attributes, which will result in a complicated architecture design

– Designing a software system that hinges on the satisfaction of certain core quality attributes
– Evaluating and analysing the rationale of an architecture design in a formal way; identifying potential risks with a design.

It is noted that the framework described here assumes the completion of requirement analysis. In real life requirements are usually identified as the software is implemented (e.g. the agile software development methodology). It would thus be interesting to develop a dynamic version of the game model, which supports architectural design using incremental refinements. Another future work is to develop a mechanism which maps coalitions generated by the algorithm to appropriate attribute primitives. This would then lead to a full automation of the ADD process linking requirements to conceptual architecture designs.

References

1. Alebrahim, A., Hatebur, D., Heisel, M.: A method to derive software architectures from quality requirements. In: Software Engineering Conference (APSEC), 2011 18th Asia Pacific, pp. 322–330. IEEE (2011)
2. Banerjee, S., Konishi, H., Sönmez, T.: Core in a simple coalition formation game. Soc. Choice Welf. **18**(1), 135–153 (2001)
3. Bass, L.: Software Architecture in Practice. Pearson Education India, Gurgaon (2007)
4. Bass, L., Klein, M., Moreno, G.: Applicability of general scenarios to the architecture tradeoff analysis method (No. CMU/SEI-2001-TR-014). Carnegie-Melon University, Software Engineering Institute (2001)
5. Bass, L.J., Klein, M., Bachmann, F.: Quality attribute design primitives and the attribute driven design method. In: van der Linden, F.J. (ed.) PFE 2002. LNCS, vol. 2290, p. 169. Springer, Heidelberg (2002)
6. Branzei, R., Dimitrov, D., Tijs, S.: Models in Cooperative Game Theory, vol. 556. Springer Science & Business Media, Heidelberg (2015)
7. Bogomolnaia, A., Jackson, M.: The stability of hedonic coalition structures. Games Econ. Behav. **38**, 201–230 (2002)
8. Garlan, D.: Formal modeling and analysis of software architecture: components, connectors, and events. In: Bernardo, M., Inverardi, P. (eds.) SFM 2003. LNCS, vol. 2804, pp. 1–24. Springer, Heidelberg (2003)
9. Kazman, R., Klein, M., Barbacci, M., Longstaff, T., Lipson, H., Carriere, J.: The architecture tradeoff analysis method. In: ICECCS1998 Proceedings of Fourth IEEE International Conference on Engineering of Complex Computer Systems, pp. 68–78. IEEE, August 1998
10. Lung, C.H., Xu, X., Zaman, M.: Software architecture decomposition using attributes. Int. J. Softw. Eng. Knowl. Eng. **17**(05), 599–613 (2007)
11. Papadimitriou, C.: Algorithms, games, and the internet. In: Proceedings of STOC 2001, pp. 749–753. ACM, July 2001
12. Roughgarden, T., Tardos, E., Vazirani, V.V.: Algorithmic Game Theory, vol. 1. Cambridge University Press, Cambridge (2007)
13. Shoham, Y., Leyton-Brown, K.: Multiagent systems: Algorithmic, Game-Theoretic, and Logical Foundations. Cambridge University Press, Cambridge (2008)

14. Wiegers, K., Beatty, J.: Software Requirements. Pearson Education, Noida (2013)
15. Wojcik, R., Bachmann, F., Bass, L., Clements, P., Merson, P., Nord, R., Wood, B.: Attribute-Driven Design (ADD), Version 2.0 (No. CMU/SEI-2006-TR-023). Carnegie-Melon University, Software Engineering Institute (2006)
16. Woods, E., Hilliard, R.: Architecture Description Languages in Practice Session Report. 5th Working IEEE/IFIP Conference on Software Architecture (WICSA 2005), p. 243 (2005)
17. Zhu, L., Aurum, A., Gorton, I., Jeffery, R.: Tradeoff and sensitivity analysis in software architecture evaluation using analytic hierarchy process. Softw. Q. J. **13**(4), 357–375 (2005)

Multi-rate System Design Through Integrating Synchronous Components

Ke Sun [(✉)]

University of Rennes 1, Campus de Beaulieu, 35042 Rennes Cedex, France
ke.sun.x@gmail.com

Abstract. This paper presents a component-based scheme for the development of multi-rate critical embedded systems. A multi-rate system is formally specified as a modular assembly of several locally mono-clocked components into a globally multi-clocked system. Mono-clocked components are modeled in particular using the synchronous programming language Quartz. Each synchronous component is first transformed into an intermediate model of clocked guarded actions. Based on the abstraction of component behaviors, consistent communication networks can be established, in which the production and consumption of inter-component dataflow are related by affine relations. Furthermore, symbolic component schedules and corresponding minimal buffering requirements are computed.

Keywords: Synchronous programming · Component-based design · Affine relation · Symbolic schedule

1 Introduction

With the continuous growth of critical embedded system scale, one of the prevalent trends in system engineering is towards *component-based design*. In this kind of design approach, systems are composed of various components, which may provide different functions and run at different rates. The integration stage takes in charge of integrating components to build multi-rate systems, while respecting all the functional requirements, especially the ones of inter-component communications. In practice, systems generally comprise a large number of components, therefore developing low-level integration code, i.e. manually building inter-component communications, is tedious and error-prone.

In this paper, we apply synchronous programming [3] in the component-based design of multi-rate critical embedded systems. Synchronous programming languages such as Esterel [4], Lustre [10] or Quartz [15] are all based on the *synchronous hypothesis*. Under this assumption, behaviors are projected onto a discrete sequence of logical instants. As the sequence is discrete, nothing occurs between two consecutive instants. Such temporal abstraction makes synchronous programming lend itself to modeling predictable component behaviors. Synchronous components are referred to as *mono-clocked* components, since each of them

© Springer International Publishing Switzerland 2015
M. Leucker et al. (Eds.): ICTAC 2015, LNCS 9399, pp. 464–482, 2015.
DOI: 10.1007/978-3-319-25150-9_27

holds a *master clock* driving its own execution. Furthermore, when integrating synchronous components, deterministic concurrency is a system-level key feature that refers closely to component behaviors, inter-component communication strategies and component schedules. In the past few decades, several integration approaches based on synchronous programming have been proposed.

Esterel components are restricted by the mono-clocked feature that connected components must be driven by the same clock. To go beyond this restriction, an extended version of the Esterel language, named *multi-clocked Esterel*, is proposed [5]. Although components are still mono-clocked in the extended version, they can now communicate with each other even if driven by different clocks. To transfer data among different clock scales, two kinds of feasible communication devices are introduced, which are sampler and reclocker. The integrated system thereby comprises a number of components running at different rates. Nonetheless, the frequency relations between clocks and the potential dependency cycles between simultaneous components have not been well studied in multi-clocked Esterel programming.

A software architecture language Prelude [9] is introduced for integrating synchronous components into multi-rate critical embedded systems. In detail, it provides a high layer of abstraction upon the Lustre language. Such high-level abstraction implements component-based design by assembling locally synchronous components (named tasks in Prelude) into globally multi-rate systems. Furthermore, real-time constraints (e.g. periods, deadlines, release dates, etc.) are well analyzed in Prelude programs to synthesize appropriate scheduling policies and tailor-made buffering communication protocols. Nevertheless, each task has only one computation mode therefore provides a single function. To build a real size system, thousands of Prelude tasks are required, whereas most RTOS accept at most one hundred tasks [14]. In addition, inter-task communications are all user-specified, the combined network can thus be prone to inconsistency.

In this paper, we propose a new component-based design approach for multi-rate critical embedded systems. Mono-clocked components in our proposal are modeled in the Quartz language. Different from Prelude tasks, each Quartz component can have multiple computation modes and provide various functions. That is, a component can switch its own mode along with its execution to perform different computations. In this way, a Quartz component can play the roles of multiple Prelude tasks such that the amount of components is controllable when building real size systems. In the proposed approach, a synthesis technique is first introduced to abstract component behaviors. The obtained interface behaviors are based on computation modes and motivate the synthesis of communication networks, in which affine relations are used to define communication patterns that are the production-consumption relations of inter-component dataflow. To guarantee the network consistency, communication patterns are partially user-defined then the remaining ones are determined in an automatic way. In addition, a valid schedule can be computed, which respects data dependencies and describes partially ordered firing relations between components. Finally, minimal capacities of communication buffers are determined. Based on all these essential factors, the integration of synchronous components gets achieved.

A simplified aircraft turning control system is used as a running example to illustrate our approach throughout the paper. The system is designed for controlling aircraft turning, according to the pilot commands. To regulate the inflight direction, corresponding aircraft actuators (i.e. ailerons and rudder) have to turn in a concerted way. As a consequence, the aircraft is rolled and a banked turn is realized. To achieve smooth aircraft turning, the control system periodically computes and issues the moving angles for ailerons and rudder in response to the pilot commands and to the aircraft movement. As depicted in Fig. 1, the system comprises four distributed controllers [1]: main controller (MC), left wing controller (LWC), right wing controller (RWC) and rudder controller (RC). According to the pilot-commanded roll angle c_M and the current angles of actuators, MC computes the moving angle of each actuator until the aircraft movement accords with c_M. The remaining controllers are respectively in charge of turning a particular actuator gradually towards the goal moving angle emitted from MC, then send the current angles of actuators back to MC.

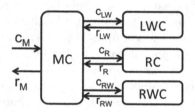

Fig. 1. A simplified aircraft turning control system

The rest of this paper is organized as follows. Section 2 introduces a tagged model [12] to formally define clocks, then briefly reviews the intermediate model clocked guarded action system, which serves as the starting point. Section 3 presents how to abstract interface behaviors from the intermediate model. In Sect. 4, the synthesis of communication networks is presented, including defining communication patterns, computing valid schedules and minimal buffering requirements. Section 5 ends the paper with conclusions and future work.

2   Foundations

2.1   Tagged Model

We start with the following sets: $\mathcal{V}$ is a non-empty set of data values; $\mathcal{B} = \{ff, tt\} \subset \mathcal{V}$ is a set of Boolean values where ff and tt respectively denote *false* and *true*; $\mathbb{T}$ is a dense set equipped with a partial order relation, denoted by $\leq$. The elements in $\mathbb{T}$ are called *tags*. We now introduce the notion of *time domain*.

Definition 1 (Time Domain). *A time domain is a partially ordered set $(\mathcal{T}, \leq)$ where $\mathcal{T} \subset \mathbb{T}$ that satisfies: $\mathcal{T}$ is countable; $\mathcal{T}$ has a lower bound $0_\mathcal{T}$ for $\leq$, i.e., $\forall t \in \mathcal{T}, 0_\mathcal{T} \leq t$; $\leq$ over $\mathcal{T}$ is well-founded; the width of $(\mathcal{T}, \leq)$ is finite.*

$(\mathbb{T}, \leq)$ provides a continuous time dimension. $(\mathcal{T}, \leq)$ defines a discrete time dimension that corresponds to the logical instants [11], at which the presence of data can be observed during the system execution. Thus, the mapping of $\mathcal{T}$ on $\mathbb{T}$ allows one to move from "abstract" to "concrete".

A *chain* $(C, \leq) \subseteq (\mathcal{T}, \leq)$ is a totally ordered set of tags admitting a lower bound 0_C. We denote the set of all chains in $\mathcal{T}$ by $\mathcal{C}_\mathcal{T}$.

Definition 2 (Event). *An event on a given time domain $\mathcal{T}$ is a pair $(t, v) \in \mathcal{T} \times \mathcal{V}$, which associates a tag t with a data value v.*

All the events whose tags belong to the same chain, can constitute a dataflow. Formally,

Definition 3 (Signal). *A signal $s : C \to \mathcal{V}$ is a function from a chain of tags to a set of values, where $C \in \mathcal{C}_\mathcal{T}$. The domain of s is denoted by $tags(s)$.*

$\mathcal{S} = \cup_{C \in \mathcal{C}_\mathcal{T}}(s : C \to \mathcal{V})$ is the set of signals over the time domain $(\mathcal{T}, \leq)$. The presence status of signal $s : C \to \mathcal{V}$ is denoted by its associated clock, which is a special signal $\hat{s} : C \to \{tt\}$. A signal is present only when its clock ticks.

The constraints on clocks are referred to as *clock relations*, such as synchronization relation $s_1 \hat{=} s_2$, i.e., s_1 and s_2 are synchronous $(tags(s_1) = tags(s_2))$, and inclusion relation $s_1 \hat{\leq} s_2$ $(tags(s_1) \subseteq tags(s_2))$.

2.2   Clocked Guarded Actions and Quartz

Clocked Guarded Actions. To process synchronous programs, it is quite natural to first compile them into intermediate models. In this way, the whole processing can be modularly divided into several steps and the models can be reused for different purposes, such as validation, comparison, model transformation and code generation. Furthermore, the processing on intermediate models is independent of particular synchronous languages.

Clocked guarded actions (CGAs) are designed in the spirit of traditional guarded commands, which are well-established intermediate code for the description of concurrent systems. CGAs have become a common representation for various synchronous languages [7,18].

Definition 4 (CGA System). *A CGA system is a set of CGAs of the form $\langle \gamma \Rightarrow A \rangle$ composed by using the parallel operator $\|$. Guard γ is a Boolean condition. Action A is either an assignment or an assumption. They are defined over a set of variables $\mathcal{X}$. Each variable $x \in \mathcal{X}$ owns an associated clock $\hat{x} \in \mathcal{S}$. CGAs can be of the following forms:*

$$\gamma \Rightarrow x = \tau \quad \text{(immediate)}$$
$$\gamma \Rightarrow next(x) = \tau \quad \text{(delayed)}$$
$$\gamma \Rightarrow assume(\sigma) \quad \text{(assumption)}$$

The execution of a CGA system is to iteratively evaluate the guards to trigger the actions. Once a guard is evaluated to *true*, the corresponding action instantaneously starts. Both kinds of assignments evaluate the (Boolean or arithmetic) expression τ at the current instant. An immediate assignment $x = \tau$ instantaneously transfers the value of τ to variable x. Furthermore, it implicitly imposes the constraint $\gamma \to \hat{x}$: the clock of x ticks whenever γ holds. On the other hand, the effect of a delayed assignment $next(x) = \tau$ takes place at next instant $\hat{x}$ ticks. An assumption $assume(\sigma)$ declares a constraint σ that has to be satisfied when γ holds. Clock relations can be declared as assumptions.

When a given variable x is present but no assignment to it can be fired, its value is determined by the *absence reaction*. The reaction determines the value, according to the storage type of the assigned variable: a *non-memorized* variable is reset to the default value that depends on its data type; a *memorized* variable keeps its previous value, or takes its default value at the initial instant.

Guarded actions describe behaviors via two parts: the *dataflow* part computes internals and outputs; the *controlflow* part computes *labels*. CGAs in the controlflow part are particularly in the form of $\gamma \Rightarrow next(l) = true$, where label l is a non-memorized Boolean internal denoting a pause location of controlflow. If l holds at the current instant, it means that the controlflow reached the pause location of l at the end of the previous instant, then it resumes from this location at the beginning of the current instant. Such instant where l holds is named an instant of l. Note that more than one label can hold at the same instant. This enables the description of the parallelism feature [15] of synchronous programs.

Quartz Programs into CGA Systems. The latest version of the Quartz language (aka. clocked Quartz) introduces clocked variables. In a Quartz program, classic variables, including labels, are synchronous and driven by the master clock clk, whereas each clocked variable x has its own clock $\hat{x}$ such that x can be absent, i.e. $x \hat{\leq} clk$.

Quartz programs follow an imperative style. The modeling of MC in Quartz is sketched in Fig. 2. In the interface part, inputs (prefixed by ?) and outputs (prefixed by !) are declared as clocked variables of real data type. In the body part, internals such as *cmd* and *langle* are first declared as classic variables of real data type, then a do while loop is embedded in an infinite loop. Furthermore, the assignment $cmd = c_M$ takes place at the instants of label $L0$, $langle = r_{LW}$ occurs at the instants of $L1$ and the assignment to r_M is fired when $L2$ holds. These assignments consequently impose the clock constraints $L0 \to \hat{c}_M$, $L1 \to \hat{r}_{LW}$ and $L2 \to \hat{r}_M$. As illustrated, I/O are in general modeled as clocked variables to avoid unnecessary communications. I/O are assumed to be clocked variables in the following.

The translation of classic Quartz programs (i.e., with no clocked variable) into CGA systems has been proposed and implemented in the Averest framework[1] [7]. In the translated CGA system, a master clock clk is introduced and

---

[1] http://www.averest.org/.

```
module MainController
(clocked real ?cM , ?rLW , !rM , ...)
{
    real cmd, langle , ...;
    ...
    loop {
        L0:  pause;
        cmd = cM ;
        ...
        do {
            L1:  pause;
            langle = rLW ;
            ...
        } while (...)
        L2:  pause;
        rM = ...;
        ...
    }
}
```

Fig. 2. Main controller

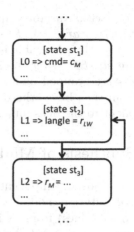

Fig. 3. Finite automaton

synchronization relation between each classic variable x and clk is declared as a guarded assumption $true \Rightarrow assume(x \hat{=} clk)$.

In the translation of (clocked) Quartz programs, the clock relations of clocked variables are not so straightforward. Different from synchronization relations between classic variables and master clock, the clock relations of clocked variables are closely related to controlflow. As illustrated in the modeling of MC, the presence status of clocked variables depend on the labels. However, both controlflow and dataflow are represented symbolically in CGA systems. To clarify clock relations, controlflow has to be further processed to get explicit.

To this end, finite automaton has been introduced as a variant of CGA system, which is translated from original CGA system via an abstract simulation [2]. In a finite automaton, the controlflow is explicitly represented: each state st is a function $st : \mathcal{L} \to \mathcal{B}$, where $\mathcal{L} \subset \mathcal{X}$ is the set of labels and $\forall l \in \mathcal{L}$, $st(l)$ denotes the value of label l in st; edges between states are labeled with conditions that must be fulfilled to enable the transitions.

Definition 5 (Finite Automaton). *A finite automaton is a tuple $\{\mathbb{S}, st_0, \mathbb{T}, \mathbb{D}\}$, where $\mathbb{S}$ is a set of states, $st_0 \in \mathbb{S}$ is the initial state, $\mathbb{T} \subseteq (\mathbb{S} \times G \times \mathbb{S})$ is a finite set of state transitions where G is the set of transition conditions, G_{st} denotes the set of transition conditions belonging to $st \in \mathbb{S}$, i.e. $\forall (st, g, st') \in \mathbb{T}, g \in G_{st}$. $\mathbb{D}$ is a mapping $\mathbb{S} \to 2^{\Lambda}$, which assigns each state $st \in \mathbb{S}$ a set of dataflow guarded actions $\mathbb{D}(st) \subseteq \Lambda$ that can be simultaneously executed in state st.*

A finite automaton can have at most one active state st at the same instant. Only a subset $\mathbb{D}(st)$ of guarded actions needs to be considered at each instant. Following this way, the executable code generated from finite automata can be optimized to generally have a smaller size and a better runtime performance [2].

In finite automata, the clock relations of clocked variables can be synthesized from guarded actions. Given a guarded action $\langle \gamma \Rightarrow x = \tau \rangle \in \mathbb{D}(st)$, all the clocked variables appearing in γ and τ are present in state st: $\forall y \in Vars_C(\gamma) \cup Vars_C(\tau), st \rightarrow \hat{y}$, where $Vars_C(\gamma)$ denotes the set of clocked variables in γ. Moreover, if x is a clocked variable, then it is conditionally present in state st: $st \land \gamma \rightarrow \hat{x}$. As illustrated in Fig. 3, guarded actions to cmd, $langle$ and r_M are grouped under different states, the clock relations of clocked variables can then be synthesized: $L0 \rightarrow \hat{c}_M$, $L1 \rightarrow \hat{r}_{LW}$ and $L2 \rightarrow \hat{r}_M$, which are equivalent to the ones imposed from the source MainController module.

3   Synthesis of Mode-Based Interface Behaviors

In our proposal, component behaviors are abstracted into computation modes. In a computation mode, a component consumes a predefined number of tokens (aka. data values) from its input channels, then produces a predefined number of tokens on its output channels. Formally,

Definition 6 (Computation Mode). *A computation mode m is a function $m : \Phi \rightarrow \mathbb{N}$, where $\Phi \subset \mathcal{X}$ is the set of I/O variables and $\forall x \in \Phi$, $m(x)$ denotes the token number of x within mode m.*

The precedence relations between input tokens and output tokens are embodied rather in mode instances, which are implementations of computation modes and describe concrete component behaviors. This notion will be elaborated in Sect. 3.2. Note that computation modes do not impose any other constraints on component behaviors, such as the number of consumed instants, consumption orders among input tokens or production orders among output tokens.

A component is named multi-mode component if its behaviors can be abstracted into multiple computation modes. Component behaviors can be first clustered into computation modes, then the analyses of transition relations among modes conduce to a transition system, called a mode switching in this paper. Mode-based interface behaviors are thus obtained.

In this section, we introduce a synthesis technique to abstract mode-based interface behaviors from CGA systems. To provide explicit controlflow representation for the synthesis, the starting point is in fact the variant finite automata. Recall that CGA systems are intermediate models, the proposed synthesis technique is thereby independent of particular synchronous languages.

3.1   SMT-based Synthesis of Controlflow-Driven Productions

As the number of produced tokens of each output should be predefined in any mode, the output production of a given multi-mode component has to be purely controlflow-driven. That is, in the viewpoint of finite automata, state transitions determine which output productions take place. In the sequel, a mapping between state transitions and presence status of outputs can be built, when controlflow-driven productions are synthesized.

The synthesis is based on *Satisfiability Modulo Theories* (SMT) [13]. The concept of SMT is to check the satisfiability of arbitrary logic formulas of linear real and integer arithmetic, scalar types, and other user-defined data structures. In detail, given a formula over first-order predicates, a SMT solver can answer whether there exist assignments to the free variables such that the whole formula is evaluated to true. When it returns the result *invalid*, it means that no assignment to the free variables can make the formula hold. When the result is *valid*, it means that the formula is always true no matter what assignments to the free variables. When the result is *satisfiable*, it means that the formula is evaluated to true when there are appropriate assignments to the free variables.

SMT-based Local Synthesis. The first stage is to perform the synthesis within each state separately. Guarded actions within a state can be encoded into a single equation system. As each guarded action $\mathcal{A} = \langle \gamma \Rightarrow x = \tau \rangle$ can be seen as a conditional equation, it can be basically encoded into a SMT implication constraint of the form $I(\mathcal{A}) = \langle \gamma \rightarrow x = \tau \rangle$. Then, it is straightforward to collect the conjunction of all the implication constraints within the same state st. We call this conjunction the *assertion system* in st: $Z_{st} = \bigwedge_{\mathcal{A} \in D(st)} I(\mathcal{A})$.

Recall that if variable x is present in state st but its value cannot be determined by any action, then absence reaction works. Hence, the constraint encoded from absence reaction should be added to Z_{st}. Assume that i is a Boolean input, $st(l) = true$ and $\langle l \wedge i \Rightarrow x = true \rangle \in \mathbb{D}(st)$ is the only guarded action that can assign the non-memorized classic variable x in state st. The absence reaction to x in st is thus $\langle l \wedge \neg i \Rightarrow x = false \rangle$ (the default value of Boolean type is *false*). The corresponding SMT constraint $\langle l \wedge \neg i \rightarrow x = false \rangle$ is added to Z_{st}.

Besides guarded actions, each transition condition $g \in G_{st}$ is encoded into a SMT Boolean constraint of the form $E(g) = \langle g = true \rangle$. The assertion system of a transition from state st to st' is $Z_{st \rightarrow st'} = Z_{st} \wedge (\bigvee_{(st,g,st') \in \mathbb{T}} E(g))$. Using SMT solver, we check its satisfiability. If it proves that $Z_{st \rightarrow st'}$ is invalid, the transition $st \rightarrow st'$ cannot occur in any case.

To synthesize controlflow-driven productions, we check the satisfiability of the guards of guarded actions under the assertion system $Z_{st \rightarrow st'}$. Given an output x, the guards of guarded actions assigning it in state st are $\{\gamma_1, \dots, \gamma_n\}$. We check the satisfiability of their disjunction, i.e. $\bigvee_{i=1}^{n} \gamma_i$. If the SMT solver proves that the disjunction is valid, it means that the production of x is executed every time st transforms to st'. That is, the production of x is driven by $st \rightarrow st'$. If the disjunction is proved to be invalid, it means that the production of x is provably disabled during $st \rightarrow st'$. Besides valid and invalid results, SMT solver may prove that the disjunction is satisfiable, i.e., the satisfiability of the guards depends on the assignments to the free variables. In that case, the constraints from predecessor states have to be taken into account.

SMT-based Constraint Propagation. There are two possibilities how constraints are propagated across states: (1) absence reaction assigns the previous value to a memorized variable; (2) a delayed action $next(x) = \tau$ evaluates τ in a previous state, then assigns x in the current state. In both cases, the current value of a given variable is determined by the computation in a previous state.

Constant propagation is a static program analysis technique that is used by compilers to iteratively identify constant expressions. In [2], the propagation technique was extended by using SMT solver to propagate constraints, such as values of memorized variables and values assigned by delayed actions, across states. This SMT-based constraint propagation technique is reused in our synthesis. With the addition of propagated constraints, the guards that were proved satisfiable in local synthesis can now be proved valid or invalid. Additional controlflow-driven productions are thereby synthesized.

Furthermore, propagated constraints imply the dependencies between consecutive transitions. For instance, st_{pre}, st, st', st'' are four states, and the transitions between them contain $(st_{pre}, true, st)$, (st, x, st'), $(st, \neg x, st'')$. Given $\langle l \Rightarrow next(x) = true \rangle \in \mathbb{D}(st_{pre})$ (where $st_{pre}(l) = true$), the constraint $\langle x = true \rangle$ is then propagated into $Z_{st \rightarrow st'}$ and $Z_{st \rightarrow st''}$. In the sequel, $Z_{st \rightarrow st'}$ is proved valid whereas $Z_{st \rightarrow st''}$ is invalid. This means that only $st \rightarrow st'$ can be enabled after $st_{pre} \rightarrow st$. Such dependency relation is encoded into a mapping relation $\Gamma : \mathbb{T} \rightarrow 2^{\mathbb{T}}$, such that $\langle st \rightarrow st' \rangle \in \Gamma(st_{pre} \rightarrow st)$.

The whole procedure is embedded in an iteration to repeatedly propagate the constraints, recheck the satisfiability, then update assertion systems and transition dependencies. The complete algorithm is sketched in Fig. 4.

```
method  ConstraintPropagation (st_pre, st)  {
    if  Γ(st_pre → st) = null
        Γ(st_pre → st) := new List < Transition > ()
    else
        return
    Z_st_pre→st := AssertionSystem (st_pre → st)
    Z_st := AssertionSystem (st)
    foreach (st, g, st') ∈ T
        Z_st→st' := Z_st ∧ (V_(st,g,st')∈T E(g)) ∧ propagatedConstraintsfrom(Z_st_pre→st)
        if SMTCheck(Z_st→st') = satisfiable | valid
            Γ(st_pre → st).add(st → st')
    foreach (st, g, st') ∈ T
        ConstraintPropagation (st, st')
}
```

Fig. 4. Algorithm of SMT-based constraint propagation

3.2   Synthesis of Modes and Mode Switching

After analyzing controlflow-driven productions, each output production should be provably controlflow-driven. During each state transition, presence status of I/O can be decided. The set of present I/O during a transition $st \rightarrow st'$, denoted as $IO(st \rightarrow st')$, is made up of two parts: required I/O within st and computed outputs along with the transition.

Computed outputs along with $st \rightarrow st'$ are simply the ones whose productions are driven by $st \rightarrow st'$. Required I/O within st are synthesized from the guarded actions. Given $\langle \gamma \Rightarrow x = \tau \rangle \in \mathbb{D}(st)$ (or $\langle \gamma \Rightarrow next(x) = \tau \rangle \in \mathbb{D}(st)$), it requires the set $(Vars(\gamma) \cup Vars(\tau)) \cap \Phi$ of I/O for evaluating γ and τ, where $Vars(\gamma)$

is the set of variables in γ and Φ denotes the set of I/O variables. The required I/O within st are naturally the union of required I/O of each guarded action belonging to $\mathbb{D}(st)$.

Moreover, $IO(st \rightarrow st')$ can be divided into disjoint sets $I(st \rightarrow st')$ and $O(st \rightarrow st')$, where $I(st \rightarrow st')$ (resp. $O(st \rightarrow st')$) is the set of present inputs (resp. present outputs) during $st \rightarrow st'$. Note that the computations of present inputs actually depend on current state rather than state transitions, $I(st \rightarrow st')$ is equivalent to the set of present inputs in st, denoted as $I(st)$.

Given a finite automaton, once presence status of I/O during each transition are known, the transitions can be clustered to build mode instances by following the aforementioned feature that the consumption of input tokens precedes the production of output tokens in each computation mode.

Definition 7 (Mode Instance). *In finite automata, a mode instance is a sequence of consecutive transitions $st_1 \rightarrow \ldots \rightarrow st_{n-1} \rightarrow st_n$, in which $I(st_n) \neq \emptyset$ and $\exists i \in [1, n[$ s.t. $O(st_i \rightarrow st_{i+1}) \neq \emptyset$, $\forall a \in [1, i-1], O(st_a \rightarrow st_{a+1}) = \emptyset$, $\forall b \in [i+1, n[, I(st_b) = \emptyset$.*

Note that all the intervals in this paper are integer intervals.

Given a finite automaton $\{\mathbb{S}, st_0, \mathbb{T}, \mathbb{D}\}$, all potential transition tracks (i.e. sequences of consecutive transitions starting from the initial state st_0) can be deduced, in which each pair of consecutive transitions $(st_{pre} \rightarrow st, st \rightarrow st')$ satisfies the dependency relation $\langle st \rightarrow st' \rangle \in \Gamma(st_{pre} \rightarrow st)$. Then, according to Definition 7, transition tracks can be divided into segments, i.e. mode instances.

Computation modes and instances are characterized by controlflow barriers and I/O token numbers. Controlflow barriers divide transitions into different mode instances. A barrier is the ending of a mode instance and the beginning of the next. Given a mode instance $st_1 \rightarrow \ldots \rightarrow st_n$, its beginning and ending are respectively st_1 and st_n. The synthesis of mode instances is in fact to distinguish controlflow barriers along with transition tracks.

To compute the I/O token numbers, a function $\mathcal{P}_{st \rightarrow st'} : \Phi \rightarrow \{0, 1\}$ is introduced. $\mathcal{P}_{st \rightarrow st'}(x)$ returns 1 if the given I/O x is present in $st \rightarrow st'$ (i.e. $x \in IO(st \rightarrow st')$), or returns 0 if it is absent. The token number of x within a mode instance $st_1 \rightarrow \ldots \rightarrow st_n$ is thereby the accumulation of its presence during each transition, i.e. $\sum_{i=1}^{n-1} \mathcal{P}_{st_i \rightarrow st_{i+1}}(x)$. By this means, I/O token numbers are computed.

According to the characteristics, mode instances can be categorized into different computation modes. The given mode instances are the implementations of the same mode if they have identical I/O token numbers, and their beginnings are the same state or the endings of the same previous mode. Once the set of instances that implement the same mode m is decided, the beginning set B^m and ending set E^m of mode m are obtained, which are respectively the set of beginnings and the set of endings of the involved instances.

Furthermore, the transition relations between modes can be synthesized. In our design, these transition relations are conveniently expressed using Property

Specification Language (PSL), in particular using Sequential Extended Regular Expressions (SEREs) [8]. Given two modes m_1 and m_2, their transition relations can be well deduced by comparing their beginning sets and ending sets:

- *alternative relation* $m_1|m_2$, meaning that when $m_1|m_2$ is executed, either m_1 or m_2 is executed. $m_1|m_2$ holds if $B^{m_1} = B^{m_2}$.
- *sequential relation* $m_1; m_2$, meaning that m_1 and m_2 are executed in successive order. $m_1; m_2$ holds if $E^{m_1} = B^{m_2}$.
- *repeatable relation* $m_1[*]$, which means that m_1 can be repeatedly executed. $m_1[*]$ holds if $B^{m_1} = E^{m_1}$.

Based on these basic relations, a mode switching can be iteratively formed, which represents complete transition relations between modes.

4   Affine Communications for Multi-mode Components

The synthesis of mode-based interface behaviors has been implemented as a .NET program, which is based on the Averest framework and the SMT solver Z3 that provides .NET managed API. After abstracting multi-mode component behaviors, the synthesized interface behaviors promote the establishment of inter-component communications. In this section, the proposed communication strategy is devoted to the multi-mode components, whose mode switchings follow the BNF form

$$\langle switching \rangle ::= [\langle prefix \rangle ``;"] ``\{" \langle repeated \rangle ``\}" ``[*]"$$
$$\langle prefix \rangle ::= \langle mode \rangle ``;" \langle prefix \rangle | \langle mode \rangle$$
$$\langle repeated \rangle ::= \langle modes \rangle ``;" \langle repeated \rangle | \langle modes \rangle$$
$$\langle modes \rangle ::= \langle mode \rangle | \langle mode \rangle ``[*]"$$

where $\langle mode \rangle$ is a computation mode, $\langle mode \rangle ``[*]"$ denotes a repeatable mode, $\langle prefix \rangle$ is a mode sequence serving as an optional prefix, $``\{" \langle repeated \rangle ``\}" ``[*]"$ is a mode sequence repeated infinitely. In the following, each mode switching M is by default composed of an optional prefix sequence M^ρ followed by an infinitely repeated sequence M^r, i.e., $M := M^\rho; M^r[*]$.

Based on mode-based interface behaviors, we obtain mode-based dataflow. Given a multi-mode component p, its mode switching is synthesized as

$$M := m_1; \ldots; m_\kappa; \{m_{\kappa+1}; \ldots; m_\iota[*]; \ldots; m_n\}[*],$$

then dataflow x produced or consumed by p correspondingly composes an amplitude sequence

$$M|_x := m_1(x); \ldots; m_\kappa(x); \{m_{\kappa+1}(x); \ldots; m_\iota(x)[*]; \ldots; m_n(x)\}[*],$$

where $\forall i \in [1, n]$, $m_i(x)$ is the token number of x in mode m_i.

As $M := M^\rho; M^r[*]$, $M|_x$ can be divided into prefix $M^\rho|_x := m_1(x); \ldots; m_\kappa(x)$ and period $M^r|_x := m_{\kappa+1}(x); \ldots; m_\iota(x)[*]; \ldots; m_n(x)$.

The sum of tokens in an amplitude sequence $M|_x$ is denoted as $||M|_x||$. Accordingly, $||M^p|_x||$ is a constant $\sum\limits_{i=1}^{\kappa} m_i(x)$ whereas $||M^r|_x||$ is in general a linear polynomial, in which each variate denotes the times of repetition of a repeatable mode in a period. Due to the existence of repeatable modes, mode-based dataflow is more general than ultimately periodic dataflow in affine dataflow graphs [6], in which $||M^r|_x||$ is always a constant (since no repeatable mode exists).

In the aircraft turning control system, the synthesized mode switchings M_{MC} and M_{LWC} are respectively equal to $\{m_1; m_2[*]; m_3\}[*]$ and $\{m'\}[*]$, where

- mode m_1: according to the pilot-commanded roll angle of aircraft, calculate the goal moving angles of ailerons and rudder.
- repeatable mode m_2: according to the current angles of ailerons and rudder, calculate the current roll angle of aircraft; if the current roll angle is not equal to the desired one, continue to adjust the moving angles of actuators.
- mode m_3: according to the current angles of ailerons and rudder, calculate the current roll angle of aircraft; if the current roll angle satisfies the pilot command, terminate the adjustment of actuators.
- mode m': according to the moving angle c_{LW} from MC, control the corresponding actuator and return its angles before and after the adjustment.

Then the corresponding mode-based dataflow contain

$$M_{MC}|_{c_{LW}} := M^r_{MC}|_{c_{LW}}[*] := \{1; 1[*]; 0\}[*] \qquad ||M^r_{MC}|_{c_{LW}}|| = 1 + 1 * \alpha$$
$$M_{MC}|_{r_{LW}} := M^r_{MC}|_{r_{LW}}[*] := \{0; 1[*]; 1\}[*] \qquad ||M^r_{MC}|_{r_{LW}}|| = 1 + 1 * \alpha$$
$$M_{LWC}|_{c_{LW}} := M^r_{LWC}|_{c_{LW}}[*] := \{1\}[*] \qquad ||M^r_{LWC}|_{c_{LW}}|| = 1$$
$$M_{LWC}|_{r_{LW}} := M^r_{LWC}|_{r_{LW}}[*] := \{2\}[*] \qquad ||M^r_{LWC}|_{r_{LW}}|| = 2$$

where variate $\alpha \in \mathbb{N}$ denotes the times of repetition of m_2 in a period. RWC and RC have the similar mode-based interface behaviors as LWC. The corresponding dataflow c_{RW}, c_R, r_{RW} and r_R are thereby respectively similar to c_{LW} and r_{LW}.

4.1   Affine Communication Patterns

To integrate multi-mode components, the communication patterns between them have to be determined, which are periodic production-consumption relations of inter-component dataflow. Indeed, flow-preserving is the basic communication pattern, in which every produced token is consumed once. In compound patterns, every produced token may be consumed several times or rejected by the consumer. In our proposal, compound communication patterns are expressed using affine relations [16], which have been employed to denote periodic firing relations of actors in affine dataflow graphs [6].

Definition 8 (Affine Relation). *An affine relation applies its parameters (n, φ, d) to a clock $\hat{s}_1$ in order to produce another clock $\hat{s}_2$ by inserting $n - 1$ tags between each pair of successive tags in $tags(\hat{s}_1)$, then counting on this created timeline each dth tag, starting from the φth tag.*

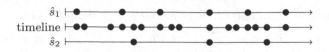

Fig. 5. $(3, 4, 5)$-affine relation

We say that the signals s_1 and s_2 are (n, φ, d)-affine or equivalently s_2 and s_1 are $(d, -\varphi, n)$-affine, where n and d are strictly positive integers while φ can be negative. When φ is negative, starting from the φth tag means that $|\varphi|$ (the absolute value of φ) tags are concatenated with the head of the created timeline as the beginning of counting. Figure 5 presents a $(3, 4, 5)$-affine relation.

As one can notice, periodic order relations exist between ticks of affine clocks and they can be computed according to a timeline. Assume that a timeline $\hat{s}$: $C \to \{tt\}$ is created, where $C = \{t_{-|\varphi|}, t_{1-|\varphi|}, \dots, t_0, t_1, \dots\}$ such that $\forall i_1, i_2 \in [-|\varphi|, +\infty[, i_1 \le i_2 \to t_{i_1} \le t_{i_2}$. Then, (n, φ, d)-affine clocks $\hat{s}_1$ and $\hat{s}_2$ can be determined, $\hat{s}_1 : C_1 \to \{tt\}$, where $C_1 = \{t_{i*n}|i \in \mathbb{N}\}$, $\hat{s}_2 : C_2 \to \{tt\}$, where $C_2 = \{t_{\varphi+i*d}|i \in \mathbb{N}\}$. The periodicity of order relations between ticks of $\hat{s}_1$ and $\hat{s}_2$ is straightforward: each period consists of $n * d$ tags; within each period, $\hat{s}_1$ ticks d times while $\hat{s}_2$ ticks n times; the order relations between ticks of $\hat{s}_1$ and $\hat{s}_2$ are the same within each period. Formally, $\forall t_a \in C_1, (t_b \in C_2) \wedge (t_a \le t_b) \to (t_{a+n*d} \le t_{b+n*d}) \wedge (t_{a+n*d} \in C_1) \wedge (t_{b+n*d} \in C_2)$. Such periodic feature starts from the tag $t_\phi \in C_1$, where ϕ is equal to 0 if $\varphi < 0$, to n if $0 \le \varphi < n$ or to $\lfloor \frac{\varphi}{n} \rfloor * n$ if $n \le \varphi$. The sign $\lfloor \frac{\varphi}{n} \rfloor$ refers to the largest integer not bigger than the fraction $\frac{\varphi}{n}$. Since C_1 and C_2 are subsets of C, their ticks are totally ordered. Furthermore, for each tick $t_b \in C_2$, if $0 \le b$, there exists the immediate predecessor tick t_b^{pre} in C_1 such that $t_b^{pre} \le t_b$ and $\forall t_a \in C_1, t_b^{pre} \le t_a \to t_b \le t_a$.

In our proposal, a (n, φ, d)-affine communication pattern means a pair of (n, φ, d)-affine clocks $\hat{s}_1$ and $\hat{s}_2$ to respectively denote the rates of the production and consumption of a dataflow. That is, d tokens are produced and consumed n times in each period. Moreover, each token consumed in a tick $t_b \in tags(\hat{s}_2)$ is produced in the immediate predecessor tick $t_b^{pre} \in tags(\hat{s}_1)$. For the consumptions that precede the start of production, since no preceding production exists, initial value is consumed instead. Such production-consumption relation can be formally summarized: the αth produced token is consumed as the f_0 first input tokens, then the $(\alpha + i)^{th}$ produced token is successively consumed f_i times, where $i \in [1, +\infty[, \alpha = \frac{\phi}{n}$ and

$$f_0 = \begin{cases} \lceil \frac{\phi - \varphi}{d} \rceil & \text{if } \varphi < n \\ 0 & \text{if } n \le \varphi \end{cases}$$

$$\beta = \begin{cases} \phi - \varphi - (f_0 - 1) * d & \text{if } f_0 > 0 \\ \phi - \varphi & \text{if } f_0 = 0 \end{cases}$$

$$f_i = \lceil \frac{\beta + i * n}{d} \rceil - \lceil \frac{\beta + (i - 1) * n}{d} \rceil$$

Note that 0th produced token is also known as the initial value and the sign $\lceil \frac{\phi - \varphi}{d} \rceil$ refers to the smallest integer not less than $\frac{\phi - \varphi}{d}$. According to these formulas, such periodic production-consumption relation can be computed. Furthermore, a mapping from production indices to corresponding consumption indices in a period can be built: $\Omega_{(n,\varphi,d)} : [1,d] \to 2^{[1,n]}$. For instance, given $(3,4,5)$-affine communication, $\alpha = 1$, $\Omega_{(3,4,5)}(1) = \{1\}$, $\Omega_{(3,4,5)}(2) = \Omega_{(3,4,5)}(5) = \emptyset$, $\Omega_{(3,4,5)}(3) = \{2\}$ and $\Omega_{(3,4,5)}(4) = \{3\}$, i.e., the first period starts from the 2nd production and in each period, among the five produced tokens, the 1st, 3rd and 4th tokens are respectively consumed once.

Flow-preserving communication pattern can be seen as a special affine communication pattern. (n, φ, d)-affine pattern is indeed a flow-preserving pattern if and only if $n = d$ and $\varphi \in [0, n[$. By default, flow-preserving communication pattern is identified as $(2, 1, 2)$-affine pattern.

4.2   Consistent Communication Networks

In practice, incomplete affine relations are specified for inter-component communication patterns. Incomplete affine relation means that the parameter φ is undetermined. Indeed, the value of φ depends not only on the parameters n and d, but also on the interface behaviors of connected components.

Given two components p_1 and p_2, their mode switchings are respectively M_{p_1} and M_{p_2}. A dataflow x is produced by p_1 and consumed by p_2. By default, its tokens produced in the prefix of p_1 (i.e. $M_{p_1}^\rho$) are consumed by the prefix of p_2 (i.e. $M_{p_2}^\rho$). According to this, the parameter φ in the affine pattern of x is determined as:

$$\varphi = n * ||M_{p_1}^\rho|x|| - d * ||M_{p_2}^\rho|x||.$$

Furthermore, once a communication pattern is determined, the hyperperiod of connected components (i.e. the rate between their periods) is obtained.

Proposition 1. *When a dataflow x connects components p_1 and p_2 by following (n, φ, d)-affine communication pattern, then the hyperperiod of p_1 and p_2 is $\mu : \nu$, s.t.,*

$$\mu = \frac{d * ||M_{p_2}^r|x||}{\delta}, \nu = \frac{n * ||M_{p_1}^r|x||}{\delta}$$

where δ is the greatest common divisor (GCD) of $n * ||M_{p_1}^r|x||$ and $d * ||M_{p_2}^r|x||$.

Proof. Recall that (n, φ, d)-affine communication implies periodic production-consumption relation: d tokens are produced and consumed n times in each period of communication. In the viewpoint of connected components, d tokens are produced in $\frac{d}{||M_{p_1}^r|x||}$ period of producer and n consumptions occur in $\frac{n}{||M_{p_2}^r|x||}$ period of consumer. The rate between $\frac{d}{||M_{p_1}^r|x||}$ and $\frac{n}{||M_{p_2}^r|x||}$ is $\frac{d*||M_{p_2}^r|x||}{n*||M_{p_1}^r|x||}$. This can be equivalently transformed to reduce into lowest terms:

$$\frac{d * ||M_{p_2}^r|x|||}{n * ||M_{p_1}^r|x|||} = \frac{\mu * \delta}{\nu * \delta} = \frac{\mu}{\nu},$$

where δ is the GCD of $d*||M_{p_2}^r|x|||$ and $n*||M_{p_1}^r|x|||$. Hence, $\mu \in \mathbb{N}^*$ (resp. $\nu \in \mathbb{N}^*$) denotes the amount of periods of producer (resp. of consumer) in a hyperperiod.

The consistency of communication network requires the hyperperiod of each pair of (directly or indirectly) connected components to be consistent, i.e., their execution must follow a stable rate. To guarantee the consistency feature in the design of communication network, communication patterns are partially user-specified so as to obtain the hyperperiods of connected components. The remaining communication patterns are then automatically determined according to the hyperperiods.

In the aircraft turning control system, the communication pattern of dataflow c_{LW} is user-specified as flow-preserving, i.e. $(2, 1, 2)$-affine pattern. Recall that $||M_{MC}^r|c_{LW}||| = 1 + \alpha$ and $||M_{LWC}^r|c_{LW}||| = 1$, where α denotes the times of repetition of repeatable mode m_2 in a period of MC. According to Proposition 1, the hyperperiod of MC and LWC is consequently $1 : (1+\alpha)$. Then, the pattern of dataflow r_{LW} can be deduced: in one hyperperiod, $2*(1+\alpha)$ (i.e. $||M_{LWC}^r|r_{LW}||| * (1 + \alpha)$) tokens of r_{LW} are produced by LWC whereas the consumption only occurs $(1 + \alpha) * 1$ (i.e. $||M_{MC}^r|r_{LW}||| * 1$) times. To guarantee the consistency of communication network, the pattern of r_{LW} can be $(1, 0, 2)$-affine.

4.3   Mode-Based Schedules and Buffering Requirements

After building consistent communication networks, determinate affine communication patterns and hyperperiods motivate the synthesis of periodic mode-based schedules, which consist of partially ordered firing relations between mode instances of connected components.

To streamline the presentation of the synthesis procedure, repeatable modes are provisionally left out of account in the periods of components. Given a pair of connected components p_1 and p_2, their periods are respectively $M_{p_1}^r := m_1; \ldots; m_j$ and $M_{p_2}^r := m_1'; \ldots; m_l'$. When their hyperperiod is $\mu : \nu$, the mode instances involved in a hyperperiod compose two totally ordered sets $M_{p_1}^r[= \mu] = \{m_b^a | a \in [1, \mu], b \in [1, j]\}$ and $M_{p_2}^r[= \nu] = \{m_b'^a | a \in [1, \nu], b \in [1, l]\}$, where m_b^a is the ath mode instance of m_b in a hyperperiod. The periodic mode-based schedules are consequently the precedence relations (denoted by $\preceq$) between the mode instances in $M_{p_1}^r[= \mu] \cup M_{p_2}^r[= \nu]$.

Given a dataflow x that is produced by p_1 and consumed by p_2, the production (resp. the consumption) number of x in a hyperperiod is thus $\mu * ||M_{p_1}^r|x|||$ (resp. $\nu * ||M_{p_2}^r|x|||$). Furthermore, for each index $\epsilon \in [1, \mu * ||M_{p_1}^r|x|||]$, the ϵth token of x in a hyperperiod is produced in m_b^a, in which $a = \lceil \frac{\epsilon}{||M_{p_1}^r|x|||} \rceil$ and b satisfies that

$$\sum_{i=1}^{b-1} m_i^a < \epsilon - (a - 1) * ||M_{p_1}^r|x||| \leq \sum_{i=1}^{b} m_i^a.$$

In the sequel, a mapping from production indices of dataflow x to corresponding mode instances of producer p_1 is built:

$$\Theta^x_{p_1} : [1, \mu * ||M^r_{p_1}|x|||] \rightarrow M^r_{p_1}[= \mu].$$

In like manner, for each index $\eta \in [1, \nu * ||M^r_{p_2}|x|||]$, the ηth consumption of x takes place in which mode instance can be determined as well. Then, a mapping from consumption indices of dataflow x to corresponding mode instances of consumer p_2 is built:

$$\Delta^x_{p_2} : [1, \nu * ||M^r_{p_2}|x|||] \rightarrow M^r_{p_2}[= \nu].$$

Recall that δ is the GCD of $d * ||M^r_{p_2}|x||$ and $n * ||M^r_{p_1}|x||$, let q denote $\frac{||M^r_{p_1}|x|| * ||M^r_{p_2}|x||}{\delta}$. According to Proposition 1, $d * q$ tokens of x are produced and $n * q$ consumptions occur in a hyperperiod. Therefore, a hyperperiod contains q periods of (n, φ, d)-affine communication pattern. Recall that $\Omega_{(n,\varphi,d)} :$ $[1, d] \rightarrow 2^{[1,n]}$ is a mapping from production indices to corresponding consumption indices in a period of (n, φ, d)-affine pattern. Through accumulating operation, an extended mapping $\Omega^q_{(n,\varphi,d)} : [1, d * q] \rightarrow 2^{[1,n*q]}$ is obtained. As a result, precedence relations between mode instances can be deduced: in a hyperperiod, the mode instance producing the ϵth token has to precede all the mode instances that consume it. Formally,

$$\Omega^q_{(n,\varphi,d)}(\epsilon) \neq \emptyset \rightarrow \forall i \in \Omega^q_{(n,\varphi,d)}(\epsilon), \Theta^x_{p_1}(\epsilon) \preceq \Delta^x_{p_2}(i).$$

In like manner, precedence relations between mode instances of prefixes can be deduced from the mapping relations. More generally, when repeatable modes exist in p_1 or p_2, their hyperperiod $\mu : \nu$ turns into a fraction with variates, in which each variate denotes the times of repetition of a repeatable mode in a period. The instances of repeatable modes should therefore be taken into account in the schedule synthesis. The synthesis method above also applies to this more general case.

Recall that the hyperperiod of MC and LWC is $1 : (1 + \alpha)$. $M^r_{MC}[= 1] = \{m^1_1, m^1_{2_1}, \ldots m^1_{2_\alpha}, m^1_3\}$ and $M^r_{LWC}[= 1 + \alpha] = \{m'^1, \ldots, m'^{1+\alpha}\}$. In a hyperperiod, $1 + \alpha$ tokens of c_{LW} are produced by MC and consumed $1 + \alpha$ times by LWC. In detail, m^1_1 produces the 1st token, $m^1_{2_\beta}$ produces the $(1 + \beta)^{th}$ token ($\beta \in [1, \alpha]$), and m'^ς performs the ςth consumption ($\varsigma \in [1, 1 + \alpha]$). In addition, $(2, 1, 2)$-affine communication pattern implies that each produced token of c_{LW} corresponds to one consumption. Based on all these mapping relations, the synthesized schedule contains $m^1_1 \preceq m'^1$ and $m^1_{2_\beta} \preceq m'^{1+\beta}$, where $\beta \in [1, \alpha]$.

Meanwhile, $2 * (1 + \alpha)$ tokens of r_{LW} are produced by LWC and consumed $1 + \alpha$ times by MC in a hyperperiod. In detail, m'^ς produces the $(2 * \varsigma - 1)^{th}$ and the $(2 * \varsigma)^{th}$ token ($\varsigma \in [1, 1 + \alpha]$), $m^1_{2_\beta}$ performs the βth consumption ($\beta \in [1, \alpha]$), and m^1_3 performs the last one. $(1, 0, 2)$-affine communication pattern implies that the $(2 * \varsigma - 1)^{th}$ token corresponds to one consumption. Based on these mapping relations, the synthesized schedule contains $m'^\beta \preceq m^1_{2_\beta}$ and $m'^{1+\alpha} \preceq m^1_3$, where $\beta \in [1, \alpha]$.

The synthesized schedule ensures that the producer writes always before the consumer reads. Nonetheless, it does not ensure the requirement that the current mode instance of consumer reads before the next mode instance of producer overwrites the token produced by its previous mode instance. This requirement closely refers to the buffering requirement. A buffer is allocated for an inter-component dataflow to cache and transfer tokens. Its size has to be big enough to avoid untimely overwriting.

Before calculating minimal buffering requirements, we introduce the notion of *lifespan*. The lifespan of a given token is an interval of mode instances. Its endpoints are m_b^a and m_j^i, such that m_b^a produces the token and m_j^i is the last instance consuming it.

Definition 9. (Lifespan). *Given a dataflow x in (n, φ, d)-affine communication pattern, the lifespan of its ϵth produced token in a hyperperiod, is*

$$sp_x(\epsilon) = \begin{cases} [\Theta^x(\epsilon), \Delta^x(max(\Omega_{(n,\varphi,d)}^q(\epsilon)))] & if \ \Omega_{(n,\varphi,d)}^q(\epsilon) \neq \emptyset \\ \emptyset & if \ \Omega_{(n,\varphi,d)}^q(\epsilon) = \emptyset \end{cases}$$

Given two tokens of the same dataflow, if there may exist an overlap between their lifespans, at least two buffer cells are required. The intersection operation $\cap$ is thus defined to check the overlap between lifespans: given $sp_x(\epsilon) = [m_a^e, m_b^h]$ and $sp_x(\epsilon') = [m_k^i, m_l^j]$, where $\epsilon < \epsilon'$, s.t. $m_a^e \preceq m_k^i$, then

$$(m_a^e = m_k^i) \vee (m_a^e \prec m_k^i \wedge m_k^i \preceq m_b^h) \rightarrow sp_x(\epsilon) \cap sp_x(\epsilon') \neq \emptyset$$
$$m_a^e \prec m_k^i \wedge m_b^h \preceq m_k^i \rightarrow sp_x(\epsilon) \cap sp_x(\epsilon') = \emptyset$$

Note that to minimize the buffering requirement, if no precedence exists between m_b^h and m_k^i, $m_b^h \preceq m_k^i$ is added as an additional constraint in schedule.

The indices of produced tokens whose lifespans overlap $sp_x(\epsilon)$ compose a set

$$\bigcap_x(\epsilon) = \{i | i \in [1, \mu * ||M_{p_1}^r|x||], sp_x(\epsilon) \cap sp_x(i) \neq \emptyset\}.$$

Let $|\bigcap_x(\epsilon)|$ denote the size of $\bigcap_x(\epsilon)$. Hence, the minimal buffering requirement for dataflow x in the hyperperiod phase is

$$\max_{1 \leq i \leq \mu * ||M_{p_1}^r|x||} (|\bigcap_x(i)|).$$

In aircraft turning control system, for the produced tokens of c_{LW} in a hyperperiod, their lifespans are defined as

$$sp_{c_{LW}}(\epsilon) = \begin{cases} [m_1^1, m'^1] & if \ \epsilon = 1, \\ [m_{2\epsilon-1}^1, m'^\epsilon] & if \ \epsilon \in [2, 1 + \alpha] \end{cases}$$

Then, the intersection of any pair of successive tokens is $sp_{c_{LW}}(\epsilon) \cap sp_{c_{LW}}(\epsilon+1)$, where $\epsilon \in [1, \alpha]$. Based on the synthesized precedence relations, we deduce that $\max_{1 \leq i \leq 1+\alpha} (|\bigcap_{c_{LW}}(i)|) = 1$, i.e. the minimal buffering requirement for c_{LW} is 1. Similarly, we can deduce that no overlap exists between the lifespans of any pair of tokens for r_{LW}, the corresponding buffering requirement is equal to 1 as well.

5 Conclusion

This paper presents a component-based multi-rate system design methodology. Each multi-mode component is first modeled as a synchronous component, in particular as a Quartz module in our proposal. After abstracting component behaviors, the generated mode-based interface behaviors motivate the synthesis of affine communication networks. Furthermore, valid schedules and corresponding buffering requirements are computed for building reliable networks.

One perspective for future work is to develop reusable component adapters and communicators to apply affine communication patterns, symbolic schedules and calculated buffering requirements in assembling multi-rate systems. This work would also utilize the optimization technique [17] to reduce the communication quantity.

References

1. Bae, K., Krisiloff, J., Meseguer, J., Iveczky, P.C.: Designing and verifying distributed cyber-physical systems using multi-rate PALS: an airplane turning control system case study. Sci. Comput. Programm. **103**, 13–50 (2015)
2. Bai, Y., Brandt, J., Schneider, K.: SMT-based Optimization for Synchronous Programs. In: Proceedings of the 14th International Workshop on Software and Compilers for Embedded Systems. pp. 11–20. SCOPES 2011, ACM, New York, NY, USA (2011)
3. Benveniste, A., Caspi, P., Edwards, S., Halbwachs, N., Le Guernic, P., de Simone, R.: The synchronous languages 12 years later. Proc. IEEE **91**(1), 64–83 (2003)
4. Berry, G., Gonthier, G.: The Esterel synchronous programming language: design, semantics. Implement. Sci. Comput. Program. **19**(2), 87–152 (1992)
5. Berry, G., Sentovich, E.: Multiclock Esterel. In: Margaria, T., Melham, T.F. (eds.) CHARME 2001. LNCS, vol. 2144, pp. 110–125. Springer, Heidelberg (2001)
6. Bouakaz, A., Talpin, J., Vitek, J.: Affine dataflow graphs for the synthesis of hard real-time applications. In: 2012 12th International Conference on Application of Concurrency to System Design (ACSD), pp. 183–192 (2012)
7. Brandt, J., Gemünde, M., Schneider, K., Shukla, S., Talpin, J.P.: Integrating system descriptions by clocked guarded actions. In: Morawiec, A., Hinderscheit, J., Ghenassia, O. (eds.) Forum on Specification and Design Languages (FDL), pp. 1–8. IEEE Computer Society, Oldenburg, Germany (2011)
8. Design Automation Standards Committee and IEEE Standards Association Corporate Advisory Group, New York, NY 10016–5997, USA: IEEE Standard for Property Specification Language (PSL), IEEE Std 1850–2005 edn. (2005)
9. Forget, J., Boniol, F., Lesens, D., Pagetti, C.: A Real-Time Architecture Design Language for Multi-Rate Embedded Control Systems. In: Proceedings of the 2010 ACM Symposium on Applied Computing. pp. 527–534. SAC 2010, ACM, New York, NY, USA (2010)
10. Halbwachs, N., Caspi, P., Raymond, P., Pilaud, D.: The synchronous dataflow programming language Lustre. Proc. IEEE **79**(9), 1305–1320 (1991)
11. Le Guernic, P., Talpin, J.P., Le Lann, J.C.: Polychrony for system design. J. Circuits, Syst. Comput. **12**(3), 261–304 (2003)

12. Lee, E., Sangiovanni-Vincentelli, A.: A framework for comparing models of computation. IEEE Trans. Comput. Aided Des. Integr. Circuits Syst. **17**(12), 1217–1229 (1998)
13. de Moura, L., Dutertre, B., Shankar, N.: A tutorial on satisfiability modulo theories. In: Damm, W., Hermanns, H. (eds.) CAV 2007. LNCS, vol. 4590, pp. 20–36. Springer, Heidelberg (2007)
14. Pagetti, C., Forget, J., Boniol, F., Cordovilla, M., Lesens, D.: Multi-task implementation of multi-periodic synchronous programs. Discrete Event Dyn. Syst. **21**(3), 307–338 (2011)
15. Schneider, K.: The synchronous programming language Quartz. Internal Report 375, Department of Computer Science, University of Kaiserslautern, Kaiserslautern, Germany (2009)
16. Smarandache, I.M., Le Guernic, P.: Affine transformations in Signal and their application in the specification and validation of real-time systems. In: Bertran, M., Rus, T. (eds.) Transformation-Based Reactive Systems Development. LNCS, vol. 1231, pp. 233–247. Springer, Berlin Heidelberg (1997)
17. Sun, K., Besnard, L., Gautier, T.: Optimized Distribution of Synchronous Programs via a Polychronous Model. In: 2014 Twelfth ACM/IEEE International Conference on Formal Methods and Models for Codesign (MEMOCODE), pp. 42–51. IEEE Computer Society, Lausanne, Switzerland (2014)
18. Yang, Z., Bodeveix, J.P., Filali, M., Hu, K., Ma, D.: A Verified Transformation: From Polychronous Programs to a Variant of Clocked Guarded Actions. In: Proceedings of the 17th International Workshop on Software and Compilers for Embedded Systems SCOPES 2014, pp. 128–137. ACM, New York, NY, USA (2014)

Verification

Verifying Android's Permission Model

Gustavo Betarte[1], Juan Diego Campo[1],
Carlos Luna[1(✉)], and Agustín Romano[2]

[1] InCo, Facultad de Ingeniería, Universidad de la República, Montevideo, Uruguay
{gustun,cluna,jdcampo}@fing.edu.uy
[2] FCEIA, Universidad Nacional de Rosario, Rosario, Argentina
agustinr88@gmail.com

Abstract. In the Android platform application security is built primarily upon a system of permissions which specify restrictions on the operations a particular process can perform. Several analyses have recently been carried out concerning the security of the Android system. Few of them, however, pay attention to the formal aspects of the permission enforcing framework. In this work we present a comprehensive formal specification of an idealized formulation of Android's permission model and discuss several security properties that have been verified using the proof assistant Coq.

Keywords: Android · Security properties · Formal verification · Coq

1 Introduction

Android [22] is an open platform for mobile devices developed by the Open Handset Alliance led by Google, Inc. Concerning security, Android embodies mechanisms at both OS and application level. As a Linux system, Android behaves as a multi-process system and therefore the security model resembles that of a multi-user server. Access control at application level is implemented by an Inter-Component Communication reference monitor that enforces MAC policies regulating access among applications and components.

Application security is built primarily upon a system of permissions, which specify restrictions on the operations a particular process can perform. Permissions are basically tags that developers declare in their applications, more precisely in the so-called application *manifest*, to gain access to sensitive resources. At installation time the user of the device is requested to grant the permissions required by the application or otherwise the installation of the application is canceled. After a successful installation, an application will be able to access system and application resources depending on the permissions granted by the user.

Several analyses have recently been carried out concerning the security of the Android system. Some of them [15,21] point out the rigidity of the permission system regarding the installation of new applications in the device. Other studies [13,18] have shown that many aspects of Android security, like *privilege escalation*, depend on the correct construction of applications by their developers.

© Springer International Publishing Switzerland 2015
M. Leucker et al. (Eds.): ICTAC 2015, LNCS 9399, pp. 485–504, 2015.
DOI: 10.1007/978-3-319-25150-9_28

Additionally, it has been pointed out [18,19] that the mechanism of permission delegation offered by the system has characteristics that require further analysis in order to ensure that no new vulnerabilities are added when a permission is delegated. Few works, however, pay attention to the formal aspects of the permission enforcing framework.

Reasoning about implementations provides the ultimate guarantee that deployed mechanisms behave as expected. However, formally proving non-trivial properties of code might be an overwhelming task in terms of the effort required, especially if one is interested in proving security properties rather than functional correctness. In addition to that, many implementation details are orthogonal to the security properties to be established, and may complicate reasoning without improving the understanding of the essential features for guaranteeing important properties. Complementary approaches are needed where verification is performed on idealized models that abstract away the specifics of any particular implementation, and yet provide a realistic setting in which to explore the security issues that pertain to the realm of those (critical) mechanisms.

Security models play an important role in the design and evaluation of high assurance security systems. *State machines*, in turn, are a powerful tool that can be used for modeling many aspects of computing systems. In particular, they can be employed as the building block of a security model. The basic features of a state machine model are the concepts of state and state change. A *state* is a representation of the system under study at a given time, which should capture those aspects of the system that are relevant to the analyzed problem. State changes are modeled by a state transition function that defines the next state based on the current state and input. If one wants to analyze a specific safety property of a system using a state machine model, one must first specify what it means for a state to satisfy the property, and then check if all state transitions *preserve* it. Thus, state machines can be used to model the enforcement of a security policy on a system.

The main contribution of the work presented in this paper is the development of a comprehensive formal specification of the Android security model and the machine-assisted verification of several security properties. Most of those properties have been discussed in previous works where they have been presented and analyzed using a variety of formal settings and approaches. In this work we provide a complete and uniform formulation of multiple properties using the higher order logic of the Calculus of Inductive Constructions [23], and the formal verification is carried out using the Coq proof assistant [12,26]. Furthermore, we present and discuss proofs of properties that have not been previously given a formal treatment. The idealized security model formalizes behaviour of the security mechanisms of *Kitkat* [1] (as of June 2015 the single most widely used Android version) according to the official documentation and available implementations. We claim that our results also apply to *Lollipop*, the latest version of Android. The formal security model and the proofs of the security properties presented in this work may be obtained from [20].

2   Background

Architecture of Android. The architecture of Android takes the form of a software stack which comprises an operating system, a run-time environment, middleware, services and libraries, and applications.

At the bottom of the software stack, providing a level of abstraction between the hardware and the upper layers of the software stack, is positioned the Linux Kernel. The multitasking execution environment provided by Linux allows multiple processes to execute concurrently. In fact, each application running on an Android device does so within its own instance of the Dalvik virtual machine (DVM). The applications running on a DVM are sandboxed, that is, they can not interfere with the operating system or other applications nor can they directly access the device hardware.

The Application Framework is a set of services that collectively form the environment in which Android applications run and are managed. This framework implements the concept that Android applications are constructed from reusable, interchangeable and replaceable components. This concept is taken a step further in that an application is also able to publish its capabilities along with any corresponding data so that they can be found and reused by other applications. The Android framework includes several key services, or components, like the Activity Manager, which controls all aspects of the application lifecycle and activity stack and the Content Providers, which allows applications to publish and share data with other applications.

Located at the top of the Android software stack are the applications. These comprise both the native applications provided with the particular Android implementation (for example web browser and email applications) and the third party applications installed by the user after purchasing the device.

Application Components. An Android application is built up from *components*. A component is a basic unit that provides a particular functionality and that can be run by any other application with the right permissions. There exist four types of component: Activities, Services, Content Providers and Broadcast Receivers [2]. An **activity** is essentialy a user interface of the application. Typically, each application has a principal activity which is the first screen the user sees when the application is started. Even if applications usually have a principal activity, any activity can be started if the initiator has the right permissions. In a same session multiple instances of the same activity can be running concurrently. A **service** is a component that executes in background without providing an interface to the user. Any component with the right permissions can start a service or interact with it [2]. If a component starts a service that is already running no new instance is created, the component just interacts with the running instance of the service [4,9]. A **content provider** is a component intended to share information among applications. A component of this type provides an interface through which applications can manage persisted data [25]. The information may reside in a SQLite data base, the web or in any other available persistent storage [2], and it can be presented by a content provider in the form of a file or a table. Finally, a

broadcast receiver is a component whose objective is to receive messages, sent either by the system or an application, and trigger the corresponding actions. Those messages, called *broadcasts*, are trasmitted all along the system and the broadcast receivers are the components in charge of dispatching those messages to the targeted applications.

Three out of the four preceding types of components, activities, services and broadcast receivers, are activated by a special kind of message called *intent*. An intent makes it possible for different components, belonging to the same application or not, to interact at runtime [2]. Typically, an intent is used as a broadcast or as a message to interact with activities and services.

Android's Security Model. Android implements a least privilege model by ensuring that each application executes on a sandbox, enforcing then that each application only has unrestricted access to the resources it owns. For an application to access other components of the system it must require, and be granted, the corresponding access permission. The sandbox mechanism is implemented at kernel level and relies on the correct application of a Mandatory Access Control policy which is enforced by a reference monitor using a user identifier (UID) [17] assigned to each installed application. Interaction among applications is achieved through *Inter Process Communication*) (IPC) mechanisms [10]. Even if the kernel provides traditional UNIX-like IPC (like sockets and signals) application developers are recommended to make use to higher level IPC mechanisms provided by Android. One such mechanism are intents, that allow to specify security policies that regulate communication between process/applications [8].

Every Android application must be digitally signed and be accompanied by the certificate that authenticates its origin. Those certificates, however, are not required to be signed by a Certification Authority, current practice indicates that certificates are signed by the developers. The Android platform uses the certificates to establish that different applications have been developed by the same author. This information is relevant both to assign permissions of the type *signature* (see below) or to authorize applications to share the same UID, and therefore be allowed to also share their resources or even be executed within the same process [5].

Permissions. Applications usually need to use system resources to execute properly. This entails that it's (almost) necessary the existence of a decision procedure (a reference monitor) that guarantees the authorized access to those resources. Decisions are taken following security policies which make use of a quite simple notion of permission. The permission system of Android embodies the following procedures: (i) an application declares the set of permissions needed to acquire further capacities from those that are by default assigned to it, (ii) at installation time the required permissions are granted or refused, depending of the time of permissions and the certificate attached to the application, or, as it's more frequently the case, by direct authorization of the owner of the device. There are also permissions that are automatically granted by the system, (iii) if a requested permission is refused the application should not be installed on the device, and typically that is the case, but there exist ways to install an application with non granted permissions [6].

In the general case, if an application is installed then it may exercise all the permissions it requests. Note that it's not possible to dynamically assign permissions in Android. Every permission is identified by a name/text and has assigned a protection level. There are two principal classes of permissions: the ones defined by the applications, by the sake of self-protection, and those predefined by Android, which are intended to protect access to resources and services of the system. Depending on the protection level of the permission, the system defines the corresponding decision procedure [7]. There exist four classes of permission level: (i) *Normal*, assigned to low risk permissions that grant access to isolated characteristics, (ii) *Dangerous*, permissions of this level are those that provide access to private data or control over the device, (iii) *Signature*, a permission of this level can be granted only if the application that requires and the application that have defined it are both signed with the same certificate, and (iv) *Signature/System*, this level is assigned to permissions that regulate the access to critical system resources or services.

On the other side, an application can also declare the permissions that are needed to access it. The granularity of the permissions system makes it possible to specify the privileges required to access to a component of the application and different set of permissions can be defined for different components.

It is also possible for a developer to force the system to execute a verification in runtime. For doing that, Android provides methods that can verify the permissions of an application in runtime. This mechanism might be used by a developer, for instance, to force the system to check that an application has specific privileges once a certain internal counter has reached a given value.

Since version *Honeycomb*, a component can access any other component of the same application without being required to have explicitly granted access to that component.

Permission Delegation. Android provides two mechanisms by which an application can delegate its own permissions to another one. These mechanisms are: *pending intents* and *URI permissions*. An intent may be defined by a developer to perform a particular action, for instance to start an activity. A `PendingIntent` is an object which is associated to the action, a reference that might be used by another application to execute that action. The object might be used by authorized applications even if the application that created it, which is the only one that can cancel the reference, is no longer active. The *URI permissions* mechanism can be used by an application that has read/write access on a *content provider* to (partially) delegate those permissions to another application. An application may attach to the result returned to an activity owned by another application an intent with URIs of resources of a content provider it owns together with an operation identifier. This grants to the receiving application the privileges to perform the operation on the indicated resources independently of the permissions the application has. The Android specification establishes that only activities may receive an *URI permission* by means of intents.

These kinds of permissions may also be explicitly granted using the method `grantUriPermission()` and revoked using the method `revokeUriPermission()`. In any case, for this delegation mechanism to work an explicit declaration must be done in the application owner of the content provider authorizing the access to the resources in question.

The Android Manifest. Every Android application must include in its root directory a XML file called `AndroidManifest`. All the components included in the application as well as some static attributes of them are declared in that file. Additionally, both the permissions requested at installation time and the ones required by the application to be accessed are also included. The authorization to use the mechanism of *URI permissions* explained above is also specified in the manifest file of an application. One of the most important elements of a manifest is <**application**>: it describes attributes of the application and also the elements that describe the components embodied by the application. Each component is declared using one of the following elements: <**activity**>, <**service**>, <**provider**>, and <**receiver**>. Additionally, the body of the manifest includes: (i) <**uses-permission**>, that specifies those permissions, defined by the system or an application, which shall be required at installation time; (ii) <**permission**>, that defines statically an application level permission and its protection level. There must be one declaration for each defined permission; and (iii) <**permission-tree**>, which is used to reserve a name space that can be used to define application level permissions on runtime. It defines prefixes to attach to any permission defined dynamically using the method `addPermission()`. Several declarations of this kind of element shall define as many prefixes. Additionally, the element <**application**> has the attribute **android:permission** which is used to specify, if any, the permission required to access any component of the application [3]. As to the elements declared by the components included in the application, there are two common attributes: (i) **android:permission**, similar to the one defined for the application, but this one has precedence over it, and (ii) **android:exported**, if this attribute is set to `true` the component shall be available to be accessed from an external application.

3   A Formally Verified Security Model of Android

In this section we outline the formalization of the idealized security model of Android. We first provide a brief description of the specification setting and the proof-assistant Coq, then we describe the state of the model and provide an axiomatic semantics of successful operations in the Android system. The operations are specified as state transition functions.

3.1   The Proof Setting

The Coq proof assistant is a free open source software that provides a (dependently typed) functional programming language and a reasoning framework based on higher order logic to perform proofs of programs. Coq allows developing mathematical facts. This includes defining objects (sets, lists, functions, programs); making statements (using basic predicates, logical connectives and quantifiers); and finally writing proofs. The Coq environment supports advanced notations, proof search and automation, and modular developments. It also provides program extraction towards languages like Ocaml and Haskell for execution of (certified) algorithms. These features are very useful to formalize and reason about complex specifications and programs.

We developed our specification in the Calculus of Inductive Constructions (CIC) using Coq. The CIC is a type theory, in brief, a higher order logic in which the individuals are classified into a hierarchy of types. The types work very much as in strongly typed functional programming languages which means that there are basic elementary types, types defined by induction, like sequences and trees, and function types. An inductive type is defined by its constructors and its elements are obtained as finite combinations of these constructors. Data types are called *Sets* in the CIC (in Coq). On top of this, a higher-order logic is available which serves to predicate on the various data types. The interpretation of the propositions is constructive, i.e. a proposition is defined by specifying what it means for an object to be a proof of the proposition. A proposition is true if and only if a proof can be constructed.

3.2   Model States

Applications. An application, as depicted in Fig. 1, is defined by its identifier, the certificate of its public key, the `AndroidManifest`, and the resources that will be used at run-time. Although Android applications do not statically declare the resources they are going to use, we decided to include this declaration in the current version of our model for the sake of simplicity.

ContProv	::=	CompId × Uri → Res	Content provider
Comp	::=	Activity | Service | BroadReceiv | ContProv	Application component
Comps	::=	{Comp}	Set of components
PermLvl	::=	*dangerous* | *normal* | *signature* | *signature/system*	Permission level
Perm	::=	PermId × PermLvl	Permission
Perms	::=	{Perm}	Set of permissions
OptionPerm	::=	*Some(p)* | *none*	A possible empty permission
CompPerms	::=	Comp → Perms	Components permissions
CPPerms	::=	ContProv → Perms	Content providers permissions
ExtPerms	::=	OptionPerm × CompPerms × CPPerms × CPPerms	External permissions
UriPerms	::=	ContProv → Perms	URI permissions
DelPerms	::=	ContProvs × UriPerms	Delegated permissions
Manifest	::=	Comps × Perms × Perms × Comps × ExtPerms × DelPerms	Manifest
AppRes	::=	{Res}	Application resources
App	::=	AppId × Cert × Manifest × AppRes	Application

Fig. 1. Formal definition of applications

Manifest. The type Manifest is an abstraction of the `AndroidManifest` file. Manifests are modelled as 6-tuples that respectively declare application components, the set of permissions it needs, the permissions that will be required by the application at runtime and those that are delegated.

An application component (Comp) is either an activity, a service, a broadcast receiver or a content provider. All of them are denoted by a component identifier of type CompId. A content provider (ContProv), in addition, encompasses a mapping to the managed resources (of type Res) from the URIs (of type Uri) assigned to them for external access. We omit the definition types Uri and Res, which are formally defined in the Coq specification. While the components constitute the static building blocks of an application, all runtime operations are initiated by component instances, which are represented in our model as members of the abstract type iComp.

The first component of a manifest (of type Comps) stores the set of application components included in the application. The second component (of type Perms) stores the set of permissions the application needs to be executed properly. A permission (Perm) is defined as a tuple comprised of a permission identifier (PermId) and the permission level (PermLvl) that indicates the security level, which can be either dangerous, normal, signature, or signature/system. The third and fourth component store the set of permissions that are defined in the application and the application components that are exported, respectively. The fifth component (of type ExtPerms) stores the information that is required to access the application, namely the permission required (if any) to access any component of the application, the permission required to access a particular component and the permissions required for performing a read or write operation on a content provider. Finally, the sixth component (of type DelPerms) stores the information concerning the delegation of permissions for accessing content providers and resources of content providers of the application as the result of using the URI permissions mechanism.

States. The states of the platform are modelled as 8-tuples that respectively store data about the set of installed applications and their permissions, running components, a registry of temporary and permanent delegated permissions and information about the applications installed in the system image of the platform; the formal definition appears in Fig. 2.

The first and second component of a state record the set of installed applications and the permissions granted to them by the system or the user, respectively. The third component stores the permissions defined by each installed application and the fourth component the set of running component instances. The fifth and sixth components keep track of the permanent and temporary permissions delegations, respectively. A permanent delegated permission (of type DelPP) represents that an application has delegated permission to perform either a read, write or read/write operation (of type OpTy) on the resource identified by an URI of the indicated content provider. A temporary delegated permission, in turn, refers to permission that have been delegated to a component instance. The seventh component stores the values of resources of applications. The final

InstApps	::= {App}	Installed applications
AppPS	::= {AppId × Perms}	Permissions granted at install time
AppDefPS	::= {AppId × Perms}	Permissions defined by each application
CompInstance	::= CompId × iComp	Component instance
CompInsRun	::= {CompInstance}	Running component instances
OpTy	::= $read \mid write \mid rw$	Access type
DelPP	::= AppId × ContProv × Uri × OpTy	Delegated permanent permission
DelPPS	::= {DelPP}	Delegated permanent URI permissions
DelTP	::= iComp × ContProv × Uri × OpTy	Delegated permanent permission
DelTPS	::= {DelTP}	Delegated temporary URI permissions
ARV	::= AppId × Res × Val	Value of application resource
ARVS	::= {ARV}	Values of applications resources
ImgApps	::= {App}	Applications in system image
AndroidST	::= InstApps × AppPS × AppDefPS × CompInsRun	Android platform state
	× DelPPS × DelTPS × ARVS × ImgApps	

Fig. 2. The state

component stores the applications installed in the Android system image, information that is relevant when granting permissions of level *signature/system*.

We use some functions and predicates to manipulate and observe the components of the state. Some of these operations, used in this paper, are presented and described in Table 1.

Valid State. The model formalizes a notion of valid state that captures several well-formedness conditions. It is formally defined as a predicate *validState* on the elements of type AndroidST. This predicate holds on a state s if the following conditions are met:

- the applications installed in s and their corresponding components have unique identifiers;
- every component belongs to only one application;
- every user-defined permission is declared in an installed application;
- all the parts involved in active permission delegations are installed in the system;
- if there is a temporary permission delegation taking place, the recipient is running;
- If a component is running, it can not be a content provider;
- all the running instances belong to a unique component, which is part of an installed application; and
- all the resources in the system have a unique value and are owned by an installed application.

All these safety properties have a straightforward interpretation in our model[1] [20]. Valid states are invariant under execution, as will be shown later.

3.3 Platform Semantics

Our formalization considers a representative set of actions to install and uninstall applications, start and stop the execution of component instances, to read and

---

[1] We omit the formal definition of *validState* due to space constraints.

Table 1. Helper functions and predicates

$compInstalled(c, s)$	holds if component c belongs to an installed application in state s
$isCProvider(c)$	holds if component c is a content provider
$running(ic, c, s)$	is satisfied if ic is an instance of component c running in state s
$canStart(c', c, s)$	holds if the application containing component c' (installed in state s) has the required permissions to create a new running instance of component c
$insNotInState(ic, s)$	requires ic to be a new instance in the state s
$runComp(ic, c, s)$	returns the running component instances of state s with the addition of the new instance ic of the component c
$inApp(c, ap)$	holds if component c belongs to application ap
$inManifest(c, ap)$	holds if component c belongs to the application components that are exported by the application ap in its manifest file
$existsRes(u, cp, s)$	holds if there exists a resource, pointed to by the URI u, in the content provider cp
$canOp(c, cp, pt, s)$	is satisfied if the application containing component c has the appropriate permissions to perform the operation pt (of type OpTy) on the content provider cp in the state s
$delPerms(c, cp, u, pt, s)$	establishes that the component c has been delegated permissions to perform the operation pt on the resource identified by u of content provider cp in the state s
$canGrant(u, cp, s)$	is satisfied if possible to delegate permissions on the content provider cp for resource identified by u in the state s
$delPPerms(ap, cp, u, pt, s)$	holds if application ap has permanent delegated permissions to perform the operation pt on the resource identified by u of the content provider cp in the state s
$delTPerms(ic, cp, u, pt, s)$	is satisfied if the running instance ic has temporary delegated permissions to perform the operation pt on the resource identified by u of the content provider cp in the state s
$compCanCall(c, sac, s)$	is satisfied if component c can perform the system call sac in the state s
$grantTPerm(ic, cp, u, pt, s)$	returns the temporary permissions delegations of state s with the incorporation of the new temporary delegated permission corresponding to the running instance ic

Table 2. Actions

`install` *ap*	Installs application *ap* in the system
`uninstall` *ap*	Uninstalls application *ap* from the system
`start` *ic c*	The running component *ic* starts the execution of component *c*
`stop` *ic*	The running component *ic* finishes its execution
`read` *ic cp u*	The running component *ic* reads the resource corresponding to URI *u* from content provider *cp*
`write` *ic cp u val*	The running component *ic* writes value *val* on the resource corresponding to URI *u* from content provider *cp*
`grantT` *ic cp act u pt*	The running component *ic* delegates temporary permissions to activity *act*. This delegation enables *act* to perform operation *pt* (of type OpTy) on the resource assigned to URI *u* from content provider *cp*
`grantP` *ic cp ap u pt*	The running component *ic* delegates permanent permissions to application *ap*. This delegation enables *ap* to perform operation *pt* on the resource assigned to URI *u* from content provider *cp*
`revoke` *ic cp u pt*	The running component *ic* revokes delegated permissions on URI *u* from content provider *cp* to perform operation *pt*
`call` *ic sac*	The running component *ic* makes the API call *sac* (of type *SACall*)

write resources from content providers, to delegate temporary/permanent permissions and revoke them and to perform system application calls; see Table 2. The behavior of an action *a* (of type Action) is formally described by giving a precondition and a postcondition, which represent the requirements enforced on a system state to enable the execution of *a* and the effect produced after this execution takes place. We represent the execution of an action with the relation $\hookrightarrow$ (one-step execution):

$$\frac{Pre(s, a) \qquad Post(s, a, s')}{s \overset{a}{\hookrightarrow} s'}$$

Intuitively, this relation models a system state transition fired by a particular action *a*. This transition takes place between a state *s* which fulfills the precondition of the action, and a state *s'* in which the postcondition holds.

Figure 3 presents the semantics of the following actions: `start` (start the execution of a component instance), `read` (a running component reads resources of a content provider), and `grantT` (a running component delegates temporary permissions to an activity). Notice that what is specified is the effect the execution of an action has on the state of the system.

Action start *ic c*

The running component *ic* starts the execution of component *c*.

Rule

$$
\frac{\begin{array}{c} compInstalled(c, s) \land \neg isCProvider(c) \land s = (aps, ps, psD, iCs, delPP, delTP, v, img) \land \\ \exists(c' : Comp), running(ic, c', s) \land compInstalled(c', s) \land \neg isCProvider(c') \land canStart(c', c, s) \\ \exists(ic' : iComp), insNotInState(ic', s) \land runComp(ic', c, s) = iCs' \land \\ s' = (aps, ps, psD, iCs', delPP, delTP, v, img) \end{array}}{s \xrightarrow{start\ ic\ c} s'}
$$

Precondition The component *c* belongs to an installed application in state *s* and is not a content provider. Additionally, *ic* is a running instance of a component *c'* and the application containing this latter component (installed in *s*) has the required permissions to create a new running instance of component *c*.

Postcondition The instance *ic'*, which was not running in state *s*, is a new running instance of the component *c* in the resulting state and that is the only difference between both states.

Action read *ic cp u*

The running instance *ic* reads resource *u* from content provider *cp*.

Rule

$$
\frac{\begin{array}{c} compInstalled(cp, s) \land existsRes(u, cp, s) \land \exists(c : Comp), compInstalled(c, s) \land \\ running(ic, c, s) \land \neg isCProvider(c) \land (canOp(c, cp, read, s) \lor delPerms(c, cp, u, read, s)) \end{array}}{s \xrightarrow{read\ ic\ cp\ u} s}
$$

Precondition The content provider *cp* is installed in state *s* and it contains a resource that is pointed to by the *URI* *u*. The component *ic* is a running instance of the installed component *c*, which is not a content provider. Additionally, the application containing component *c* either has the appropriate permissions to read *cp* in the state *s* or it has been delegated the permissions to perform the operation *read* on the resource identified by *u*. Notice that any component of an application is implicitly granted the permissions that were delegated to a running instance of that application.

Postcondition After the execution of this action, the system state remains unchanged.

Action grantT *ic cp act u pt*

The running component *ic* delegates temporary permissions to activity *act*. This delegation enables *act* to perform operation *pt* on the resource assigned to URI *u* from content provider *cp*.

Rule

$$
\frac{\begin{array}{c} compInstalled(cp, s) \land canGrant(u, cp, s) \land existsRes(u, cp, s) \land compInstalled(act, s) \land \\ \exists(c : Comp), compInstalled(c, s) \land running(ic, c, s) \land canStart(c, c, s) \land \\ (canOp(c, cp, pt, s) \lor delPerms(c, cp, u, pt, s)) \land s = (aps, ps, psD, iCs, delPP, delTP, v, img) \\ \exists (ic' : iComp), insNotInState(ic', s) \land runComp(ic', act, s) = iCs' \land \\ grantTPerm(ic', cp, u, pt, s) = delTP' \land s' = (aps, ps, psD, iCs', delPP, delTP', v, img) \end{array}}{s \xrightarrow{grantT\ ic\ cp\ act\ u\ pt} s'}
$$

Precondition The content provider *cp* is installed in state *s*. Permissions can be delegated on *cp* for resource identified by the URI *u*. The component *act* is an activity installed in *s*. The component *ic* is a running instance of a component (*c*) that belongs to an installed application in *s*. The application containing component *act* has the required permissions to create a new running instance of component *c*. Additionally, the application containing component *c* either has the appropriate permissions to perform the operation *pt* on *cp* or it has has been delegated permissions to perform *pt* on the resource identified by *u*.

Postcondition A new running component instance of activity *act* is incorporated into the system state. Moreover, a new temporary delegated permission corresponding to the running instance generated is added into the temporary permissions delegations of state *s*. This delegation enables *act* to perform *pt* on the resource assigned to *u* from *cp*. Apart from that, both states are equal.

Fig. 3. Formal semantics of actions start, read, and grantT

One-step execution preserves valid states, i.e. the state resulting from the execution of an action on a valid state is also valid.

Lemma 1. $\forall (a : \mathsf{Action})(s \; s' : \mathsf{AndroidST})$,
$$validState \; s \Rightarrow s \overset{a}{\hookrightarrow} s' \Rightarrow validState \; s'$$

System state invariants, such as state validity, are useful to analyze other relevant properties of the model. In particular, the results presented in the following section are obtained from valid states of the system.

4   Security Properties

In this section we present and discuss some relevant properties that can be established concerning the Android security framework. Many of these properties have already been analyzed elsewhere. Some of them, however, have not been studied in previous works. All of the properties were successfully stated and proved using our specification, which represents, up to our knowledge, the first comprehensive analysis under the same formal model of multiple safety and security properties of the Android system. The corresponding Coq development can be found at [20]. To simplify the presentation that follows we will assume all variables of type AndroidST to be valid states, and variables of type App to be installed applications in a given state, when there is no possibility of confusion. Components will also be assumed to be installed.

4.1   Privileges

One of the most important properties claimed about the Android security model is that it meets the so-called *principle of least privilege*, i.e. that "each application, by default, has access only to the components that it requires to do its work and no more" [2]. Using our specification we have proved several lemmas which were aimed at showing the compliance with this principle when a running instance creates another component instance, reads/writes a content provider or delegates/revokes a permission. In this setting, least privilege means that a running instance will need to have the appropriate permissions to execute the desired action in each of these scenarios. In particular, the following specific properties were proved:

- if components c and c' belong to the same application, then c can start c';
- if a component c' is not exported and the component c belongs to another application, then c cannot start c';
- if components c and c' belong to two different applications ap and ap', and c' requires a permission that ap does not have, then c cannot start c';
- if components c and c' belong to two different applications ap and ap', c' requires no permission, but ap' requires a permission that ap does not have, then c cannot start c';

- if ic can read/write the resource pointed by the URI u in cp, then its associated component belongs to an application that has permission to do so, either from its installation or through a delegation of permissions[2];
- if a content provider cp and a component c belong to the same application, then all running instances of c can read or write cp;
- if ic delegated permissions, temporary or permanent, to read or write a resource pointed by the URI u in cp, then ic can perform this operation;
- if ic revoked permissions to read or write the resource pointed by the URI u in cp, then ic can perform this operation.

All properties have a straightforward representation in our model.

While the fulfillment of the *principle of least privilege* when creating a new instance is widely studied in the literature [19, 24], the analysis of this principle when accessing a content provider or delegating/revoking a permission has not been covered in other publications. Since our model includes these two scenarios, we are able to formally state and prove lemmas like the following:

Lemma 2. $\forall(s : \mathsf{AndroidST})(ap\ ap' : \mathsf{App})(c : \mathsf{Comp})(ic : \mathsf{iComp})$
$(cp : \mathsf{ContProv})(u : \mathsf{Uri}), ap \neq ap' \wedge inApp(c, ap) \wedge running(ic, c, s) \wedge$
$inApp(cp, ap') \wedge \neg inManifest(cp, ap') \wedge existsRes(u, cp, s) \Rightarrow$
$Pre(s, \mathbf{read}\ ic\ cp\ u) \iff delPerms(c, cp, u, read, s))$

If cp is not exported and c belongs to a different application than a, then cp can be read by c if and only if the application corresponding to the latter has delegated permissions to do so[3].

Lemma 3. $\forall(s : \mathsf{AndroidST})(ap\ ap' : \mathsf{App})(c : \mathsf{Comp})(ic : \mathsf{iComp})$
$(cp : \mathsf{ContProv})(act : \mathsf{Activity})(pt : \mathsf{OpTy})(u : \mathsf{Uri}),$
$inApp(c, ap) \wedge running(ic, c, s) \wedge inApp(cp, ap) \wedge$
$existsRes(u, cp, s) \wedge canGrant(u, cp, s) \Rightarrow$
$Pre(s, \mathbf{grantT}\ ic\ cp\ act\ u\ pt) \wedge Pre(s, \mathbf{grantP}\ ic\ cp\ ap'\ u\ pt)$

If c and cp belong to the same application and cp authorizes the delegation on u, then ic can delegate both temporary and permanent permissions on u.

The above lemmas establish that even if a component of an application is not exported, it can still be accessed from a different application. In particular, Lemmas 2 and 3 show that it is possible for an external application to obtain delegated permissions to access a non-exported content provider. This contradicts the description of exported components given in the official developer's guide [3].

The interested reader is referred to [20] where he can find the Coq files with the complete proofs of the lemmas we have just discussed.

---

[2] In particular, ic can read/write the resource pointed by u in cp if ic has permission due to a delegation via intents.

[3] In [20] we prove a similar result for action `write`.

Revocation. One of the peculiarities of the Android security model is that the explicit revocation of delegated permissions is relatively coarse-grained, in the sense that it is impossible to only revoke permissions to a particular application.

Although this property was studied in [19], no formal statement or proof is provided in that work. In our formal setting we are able to state and prove the following lemma:

Lemma 4. $\forall(s\ s' : \mathsf{AndroidST})(ic : \mathsf{iComp})(cp : \mathsf{ContProv})(u : \mathsf{Uri})(pt : \mathsf{OpTy})$, $s \xrightarrow{\text{revoke } ic\ cp\ u\ pt} s' \Rightarrow (\forall(ap : \mathsf{App}), \neg delPPerms(ap, cp, u, pt, s')) \wedge$ $(\forall(ic' : \mathsf{iComp})(c : \mathsf{Comp}), running(ic', c, s') \Rightarrow \neg delTPerms(ic', cp, u, pt, s')$

If ic revokes the permission to perform operation pt over the resource pointed by u in cp, this revocation will be applied to all the applications in the system.

A direct consequence of this property is that a running component can revoke permissions that were not delegated by itself, which may result in confusing and problematic scenarios [19]. For instance, suppose applications A and B both have the same delegated permission p. In the case that an application C revokes p with the intention that B does not longer use it, A shall also lose that permission without further notice. Application A will just find out when attempting a task that requires p, provoking then a runtime exception.

Privilege Escalation. According to [18], in the Android system a *privilege escalation* attack occurs when "an application with a permission performs a privileged task on behalf of an application without that permission". This privileged task can be, for instance, invoking a system service, or accessing an application. We have proved that a *privilege escalation* scenario involving either task is possible in our model. The proof was divided in two separate lemmas, one for each kind of privileged operation.

Lemma 5. $\forall(s : \mathsf{AndroidST})(ic : \mathsf{iComp})(c : \mathsf{Comp})(sac : \mathsf{SACall})$, $\neg Pre(s, \mathtt{call}\ ic\ sac) \wedge \neg Pre(s', \mathtt{call}\ ic\ sac) \wedge compCanCall(c, sac, s') \wedge$ $s \xrightarrow{\text{start } ic\ c} s' \Rightarrow \exists(ic' : \mathsf{iComp}), running(ic', c, s') \wedge Pre(s', \mathtt{call}\ ic'\ sac)$

If ic cannot perform the API call sac but it starts the execution of a component c which is able to do it, then it will be possible to invoke sac through an instance of c.

Lemma 6. $\forall(s\ s' : \mathsf{AndroidST})(c\ c'\ c'' : \mathsf{Comp})(ic : \mathsf{iComp}), \neg isCProvider(c'') \wedge$ $\neg canStart(c, c'', s) \wedge \neg canStart(c, c'', s') \wedge canStart(c', c'', s') \wedge s \xrightarrow{\text{start } ic\ c'} s' \Rightarrow$ $\exists(ic' : \mathsf{iComp}), running(ic', c', s') \wedge Pre(s', \mathtt{start}\ ic'\ c'')$

If ic cannot access a component c' but it starts the execution of a component c' which is able to do it, it will be possible to start c" through an instance of c'.

The proof of Lemma 6 is fairly straightforward: we need to give a running instance of a component which is always able to access component c'' in state s'. We claim that this witness is the resulting instance of executing the **start** operation in state s (hypothesis $s \xrightarrow{\text{start } ic\ c'} s'$). Calling this instance ic', we have to prove that both $running(ic', c', s')$ and $Pre(s', \text{start } ic' \ c'')$ are verified. The first predicate is trivially satisfied by the definition of ic'. Next, the precondition of operation start, as described in Sect. 3, requires ic' to be able to start component c'', which must be installed in state s' and be different from a content provider. While the first and third requests are explicitly assumed in the hypotheses of the lemma, we prove that component c'' is installed in state s' beginning by the fact that, by hypothesis, c'' is installed in state s and, since the start operation does not change the installed applications, c'' must be in s' as well. The proof of Lemma 5 is analogous to the one just described.

More informally, in the above lemmas ic represents the unprivileged running component and c' the component that has the permissions to make a API call (predicate $compCanCall$ in Lemma 5) or access another component, respectively. If ic access c' (creating a running instance of c'), then the privileged operation will be available to be executed (by, at least, the running instance just created).

In models that avoid privilege escalation it is not enough to call an instance with the required permissions to perform a privileged operation. In such models, extra controls are implemented in order to prevent the called instance from being used as a deputy of an unprivileged component [13]. The issues just discussed were originally presented in [13,16,18] but referred to earlier versions of the Android platform and used different approaches to perform their analysis. Since our formalism fully captures both the interaction between components and the execution of API calls in the Android system we are convinced that the latest versions of the platform are still vulnerable to privilege escalation attacks.

4.2   Permission Redelegation

The last property we want to discuss makes explicit that it is possible to redelegate a permission an unlimited number of times. This particular aspect of the Android security model was also studied by Fragkaki et al. [19] and has been successfully represented in our formalism.

Lemma 7. $\forall(s : \mathsf{AndroidST})(ap \ ap' : \mathsf{App})(c : \mathsf{Comp})(ic : \mathsf{iComp})(act : \mathsf{Activity})$
$(cp : \mathsf{ContProv})(u : \mathsf{Uri})(pt : \mathsf{OpTy}), inApp(c, ap) \wedge running(ic, c, s) \wedge$
$existsRes(u, cp, s) \wedge (delTPerms(ic, cp, u, pt, s) \vee delPPerms(ap, cp, u, pt, s)) \Rightarrow$
$Pre(s, \text{grantT } ic \ cp \ act \ u \ pt) \wedge Pre(s, \text{grantP } ic \ cp \ ap' \ u \ pt)$

If ic or an application ap have a delegated permission, they can redelegate it in a temporary or permanent way.

As a corollary of this property, if a running component receives a temporal permission delegation, then any instance of that component can redelegate the given permission to the application itself in a permanent way. Consequently, a permission that was originally temporarily delegated, ends up being permanently delegated [19]. This behavior means that in practice, the two delegation mechanisms are not substantially different. For example, a running component can receive a permission delegation through an *intent* because the sender wants this permission to get revoked when the recipient finishes execution.

However, the receiver could redelegate the given permission to its own application in a permanent way so that it can only be revoked via the method: revokeUriPermission(); which would contradict the original purpose of the sender.

5   Related Work

Several works have analyzed the limitations and weaknesses of the security model of Android. The study of most of the properties that we have formally verified and presented in this paper is scattered in several publications from the literature. The results presented in those publications are formulated using different formal settings in accordance to the type of study and properties in which they are interested.

Felt et al. [18] study, although not formally, Android applications to determine whether Android developers follow least privilege policies when declaring applications permission requests. The authors develop in particular an OS mechanism for defending against permission re-delegation (when an application with permissions performs a privileged task for an application without permissions). Our work initiates the development of an exhaustive formal specification of the Android security model from which it is possible to formally reason, for instance, about the property of least privilege for any application.

In Chaudhuri's work [14] it is defined a typed language to model the communication between different components of the Android's platform. Given an expression defined in this language, if a type can be inferred for it, the operation being modelled is in compliance with some desirable security properties concerning the integrity and confidentiality of the information being exchanged. Analogously, Armando et al. [10] and Bugliesi et al. [13] present a type and effect system in which they model basic Android's components and the semantics of some operations. Although these three publications follow similar approaches, they all define different new languages, which are focused on the features being analyzed in each work. Additionally, no formal guarantee is provided of the correctness of the results obtained.

In the case of the work by Fragkaki et al. [19], instead of defining a typed language, the authors generalize the Android permission scheme and propose an abstract formal framework to represent the systems that meet these general characteristics. This model is used to enunciate security properties that, according to the authors, any instantiation should obey. In this way, the Android platform is represented as a particular instance of the proposed abstract model and its formal analysis consists of checking if the enunciated security properties actually hold on it. As pointed out in Sect. 4, many of the properties studied in [19] were selected to be proved in our own model. The success in doing so shows that our formal framework is expressive enough to enunciate and prove the properties in question, offering the support of a widely used tool, as it is Coq, in all the stages of the proof development process.

Finally, Shin et al. [24] adapt the approach followed by Zanella et al. [11] to build a formal framework that represents the Android's permission system, which is based on the Calculus of Inductive Constructions and it is developed in Coq, as we do. However, that formalization does not consider several aspects of the platform covered in our model, namely, the different types of components, the interaction between a running instance and the system, the reading/writing operation on a content provider and the semantics of the permission delegation mechanism. These last two aspects of our model allow us to formulate and prove security properties which cannot be formally studied in Shin's model, such as Lemmas 4, 6, and 7, addressed in Sect. 4. Furthermore, there are important differences between the two models regarding the way of representing, for example, the applications and its components, the state of the platform, the AndroidManifest file and the operations execution. We claim that the results we have obtained constitute a quite complete, expressive and extensible model of Android's permission system. Moreover, we understand our contribution as an alternative adaptation of the work presented in [11] rather than an extension of the one proposed by Shin.

6   Conclusion and Future Work

This work initiates the development of an exhaustive formal specification of the Android security model that includes elements and properties that have been partially analyzed in previous works. The formal model considers the latest version of the security mechanisms of the platform. Furthermore, we present the proof of security properties concerning the Android permission mechanisms that have not previously been formally verified. Thus, this specification represents, up to our knowledge, the first comprehensive analysis under the same formal model of several security properties of the Android system. The formal development is about 5.5kLOC of Coq (see Fig. 4) , including proofs, and constitutes a suitable basis for reasoning about Android security model.

Formal model	1k
Valid state invariance	3k
Security properties	1.5k
Total	5.5k

Fig. 4. LOC of Coq development

There are several directions for future work. We are already working in enriching the model with behaviour not considered so far, like the actions of sending and receiving broadcasts and implicit intents, application update, and management of signatures and certificates. We are also interested in producing an alternative definition of the model which would be better suited for the verification of a toy implementation of the control access decision procedure. From this result, and using the program extraction mechanisms provided by Coq, we expect to derive a certified Haskell prototype of the reference monitor.

Acknowledgments. Work partially funded by project ANII-Clemente Estable FCE_1_2014_1_103803: Mecanismos autónomos de seguridad certificados para sistemas computacionales móviles.

References

1. Android Developers. Android KitKat. https://developer.android.com/about/versions/kitkat.html. Accessed on August 2015
2. Android Developers. Application Fundamentals. http://developer.android.com/guide/components/fundamentals.html. Accessed on August 2015
3. Android Developers. Application Manifest. http://developer.android.com/guide/topics/manifest/manifest-intro.html. Accessed on August 2015
4. Android Developers. Context. http://developer.android.com/reference/android/content/Context.html. Accessed on August 2015
5. Android Developers. manifest. http://developer.android.com/guide/topics/manifest/manifest-element.html#uid. Accessed on August 2015
6. Android Developers. Permissions. http://developer.android.com/guide/topics/security/permissions.html. Accessed on August 2015
7. Android Developers. R.styleable. http://developer.android.com/reference/android/R.styleable.html. Accessed on August 2015
8. Android Developers. Security Tips. http://developer.android.com/training/articles/security-tips.html. Accessed on August 2015
9. Android Developers. Services. http://developer.android.com/guide/components/services.html. Accessed on August 2015
10. Armando, A., Costa, G., Merlo, A.: Formal modeling and reasoning about the android security framework. In: 7th International Symposium on Trustworthy Global Computing (2012)
11. Zanella Béguelin, S., Betarte, G., Luna, C.: A formal specification of the MIDP 2.0 security model. In: Dimitrakos, T., Martinelli, F., Ryan, P.Y.A., Schneider, S. (eds.) FAST 2006. LNCS, vol. 4691, pp. 220–234. Springer, Heidelberg (2007)
12. Bertot, Y., Castran, P.: Interactive Theorem Proving and Program Development. Coq'Art: The Calculus of Inductive Constructions. Texts in theoretical computer science. Springer, Berlin (2004)
13. Bugliesi, M., Calzavara, S., Spanò, A.: Lintent: towards security type-checking of android applications. In: Beyer, D., Boreale, M. (eds.) FORTE 2013 and FMOODS 2013. LNCS, vol. 7892, pp. 289–304. Springer, Heidelberg (2013)
14. Chaudhuri, A.: Language-based security on android. In: Proceedings of the ACM SIGPLAN Fourth Workshop on Programming Languages and Analysis for Security, PLAS 2009, pp. 1–7. ACM, New York, NY, USA (2009)
15. Conti, M., Nguyen, V.T.N., Crispo, B.: CRePE: context-related policy enforcement for android. In: Burmester, M., Tsudik, G., Magliveras, S., Ilić, I. (eds.) ISC 2010. LNCS, vol. 6531, pp. 331–345. Springer, Heidelberg (2011)
16. Davi, L., Dmitrienko, A., Sadeghi, A.-R., Winandy, M.: Privilege escalation attacks on android. In: Burmester, M., Tsudik, G., Magliveras, S., Ilić, I. (eds.) ISC 2010. LNCS, vol. 6531, pp. 346–360. Springer, Heidelberg (2011)
17. Felt, A.P., Chin, E., Hanna, S., Dawn Song, and David Wagner. Android permissions demystified. In: Proceedings of the 18th ACM conference on Computer and communications security, CCS 2011, pages 627–638. ACM, New York, NY, USA (2011)

18. Felt, A.P., Wang, H.J., Moshchuk, A., Hanna, S., Chin, E.: Permission redelegation: Attacks and defenses. In: USENIX Security Symposium. USENIX Association (2011)
19. Fragkaki, E., Bauer, L., Jia, L., Swasey, D.: Modeling and enhancing android's permission system. In: Foresti, S., Yung, M., Martinelli, F. (eds.) ESORICS 2012. LNCS, vol. 7459, pp. 1–18. Springer, Heidelberg (2012)
20. GSI. Formal verification of the security model of Android: Coq code. https://www. fing.edu.uy/inco/grupos/gsi. Accessed on August 2015
21. Nauman, M., Khan, S., Zhang, X.: Apex: extending android permission model and enforcement with user-defined runtime constraints. In: Proceedings of the 5th ACM Symposium on Information, Computer and Communications Security, ASIACCS 2010, pp. 328–332. ACM, New York, NY, USA (2010)
22. Open Handset Alliance. Android project. http://source.android.com/. Accessed on August 2015
23. Paulin-Mohring, C.: Inductive definitions in the system Coq rules and properties. In: Bezem, M., Groote, J.F. (eds.) TLCA 1993. LNCS, vol. 664, pp. 328–345. Springer, Heidelberg (1993)
24. Shin, W., Kiyomoto, S., Fukushima, K., Tanaka,T.: A formal model to analyze the permission authorization and enforcement in the android framework. In: Proceedings of the 2010 IEEE Second International Conference on Social Computing, pp. 944–951, Washington, DC, USA, 2010. IEEE Computer Society
25. Six, J.: Application Security for the Android Platform. O'Reilly Media, San Francisco (2011)
26. Team, The Coq Development: The Coq Proof Assistant Reference Manual - Version **V8**, 4 (2012)

CSP and Kripke Structures

Ana Cavalcanti[1]([⊠]), Wen-ling Huang[2], Jan Peleska[2], and Jim Woodcock[1]

[1] University of York, York, UK
Ana.Cavalcanti@york.ac.uk
[2] University of Bremen, Bremen, Germany

Abstract. A runtime verification technique has been developed for CSP via translation of CSP models to Kripke structures. With this technique, we can check that a system under test satisfies properties of traces and refusals of its CSP model. This complements analysis facilities available for CSP and for all languages with a CSP-based semantics: Safety-Critical Java, Simulink, SysML, and so on. Soundness of the verification depends on the soundness of the translation and on the traceability of the Kripke structure analysis back to the CSP models and to the property specifications. Here, we present a formalisation of soundness by unifying the semantics of the languages involved: normalised graphs used in CSP model checking, action systems, and Kripke structures. Our contributions are the unified semantic framework and the formal argument itself.

Keywords: Semantic models · UTP · Formal testing · Runtime verification

1 Introduction

CSP [19] is a well established process algebra with consistent denotational, operational and axiomatic semantics that have been thoroughly studied. A commercial model checker, FDR3 [9] and its predecessors, has been in widespread use for years and has encouraged industrial take up. For finite processes, FDR3 provides a semantics in terms of normalised graphs: deterministic finite automata with edges labelled by events and nodes by sets of maximal refusals.

Recently, this semantics has been used to develop a runtime verification technique for CSP [17]. It checks the behaviour of programs or simulations during their execution against given specifications; this is typically applied in situations where model checking would be infeasible due to the size of the state space.

In the approach of [17], a specification of traces and refusals of a CSP process is translated to a safety LTL formula. Runtime verification of the resulting property is then carried out using a technique that assumes that the system under test (SUT) behaves like an unknown Kripke structure. (Although the technique does not require the construction of the Kripke structure, its soundness is established in terms of an unknown Kripke structure that models the SUT.) Soundness of the CSP technique is argued via translation of the FDR3 normalised graphs to nondeterministic programs, and then to Kripke structures.

© Springer International Publishing Switzerland 2015
M. Leucker et al. (Eds.): ICTAC 2015, LNCS 9399, pp. 505–523, 2015.
DOI: 10.1007/978-3-319-25150-9_29

Based on the Kripke structures, we can apply an existing runtime verification technique that defines practical health monitors (error-detection mechanisms) [13]. They do not provide false positives or negatives and can be activated at any time during the execution of an SUT. Using the technique in [17], health monitors can be created based on specifications of CSP processes and, therefore, based on any language for which a CSP-based semantics exists. Some very practical examples are Safety-Critical Java [8], Simulink [5], and SysML [15]. For Safety-Critical Java, this technique can complement assertion-based analysis techniques that use JML [3] and SafeJML [10], which support reasoning about data models and execution time, with facilities to reason about reactivity.

Soundness is rigorously argued in [17] based on the following premises:

1. the semantics of finite CSP processes as a normalised graph, as originally described in [18, Chapter 21] and then implemented in FDR3, is consistent with the CSP semantics;
2. a mapping of the normalised graphs into nondeterministic programs defined in [17] preserves the semantics of CSP;
3. a semantics in terms of Kripke structures for these nondeterministic programs, defined in [17] in terms of their operational semantics, preserves the semantics of the programs; and
4. a mapping of a safety LTL formula of a particular form to a trace and refusal specification defined in [17] captures the semantics of the safety formula in the failures model.

With these results, we can then conclude that the notion of satisfaction in the failures model corresponds to the notion of satisfaction in Kripke structures.

In this paper, we still take (1) as a premise: it is widely accepted and validated both in the standard semantic theories of CSP [19] and in the extensive use of FDR3 (and its predecessors). We, however, go further and formalise the notions of semantics preservation in (2) and (3). We carry out this work using Hoare and He's Unifying Theories of Programming [12], a relational semantic framework that allows us to capture and relate theories for a variety of programming paradigms. A UTP theory for CSP is already available, as are many others (for object-orientation [21], time [20], and so on). Finally, as pointed out in [17], (4) is trivial because the mapping from the safety LTL formula subset under consideration to trace and refusal specifications is very simple.

In formalising (2) and (3), we define UTP theories for normalised graphs and Kripke structures. The nondeterministic programs are action systems and are encoded in the UTP theory for reactive processes. Galois connections between these theories establish semantic preservation. Unification is achieved via an extra UTP theory that captures a kind of stable-failures model, where traces are associated with maximal refusals. Galois connections with this extra theory identify the traces and maximal refusals of a normalised graph, an action system, and a Kripke structure. Figure 1 gives an overview of our results.

In the unified context of the theory of traces and maximal refusals, we define satisfaction for CSP normalised graphs and for Kripke structures. The properties that we consider are the conditions, that is, predicates on a single state, of

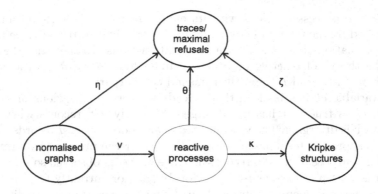

Fig. 1. New UTP theories and their relation to reactive processes

that theory of traces and maximal refusals. The Galois connections are used to establish the relationship between satisfaction in CSP and in Kripke structures.

Besides contributing to the UTP agenda of unification of programming theories, we open the possibility of using the runtime verification technique of Kripke structures for other languages with a UTP semantics, such as, *Circus* [16], rCOS [14], Handel-C [4], and SystemC [22].

The approach is restricted to the, still significant, class of divergence-free programs. Divergence freedom is a standard assumption in testing techniques, where observation of divergence is perceived as deadlock.

Next, we give an overview of the UTP and the existing theory of reactive processes. Our theories are presented afterwards: normalised graphs in Sect. 3, Kripke structures in Sect. 4, and traces and maximal refusals in Sect. 5. Section 3 also gives the Galois connection between graphs and reactive processes, and Sect. 4 between reactive processes and Kripke structures. Finally, Sect. 5 gives the Galois connections between graphs, reactive processes and Kripke structures and traces and maximal refusals. In Sect. 5, we also define satisfaction and present our main result: soundness of the CSP runtime verification technique. We conclude and present related and future work in Sect. 6.

2   A UTP Theory of Reactive Processes

In the UTP, relations are defined by predicates over an alphabet (set) of observational variables that record information about the behaviour of a program. In the simplest theory of general relations, these are the programming variables v, and their dashed counterparts v', with v used to refer to an initial observation of the value of v, and v' to a later observation.

Theories are characterised by an alphabet and by healthiness conditions defined by monotonic idempotent functions from predicates to predicates. The predicates of a theory with alphabet a are the predicates on a that are fixed points of the healthiness conditions. As an example, we consider the existing theory of reactive processes used in our work to model action systems.

A reactive process interacts with its environment: its behaviour cannot be characterised by the relation between its initial and final states only; we need to record information about the intermediate interactions. To that end, the alphabet of the theory of reactive processes includes four extra observational variables: ok, $wait$, tr, and ref and their dashed counterparts.

The variable ok is a boolean that records whether the previous process has diverged: ok is true if it has not diverged. Similarly, ok' records whether the process itself is diverging. The variable $wait$ is also boolean; $wait$ records whether the previous process terminated, and $wait'$ whether the process has terminated or not. The purpose of tr is to record the trace of events observed so far. Finally, ref records a set of events refused, previously (ref) or currently (ref').

The monotonic idempotents used to define the healthiness conditions for reactive processes are in Table 1. The first healthiness condition R1 is characterised by the function $\mathsf{R1}(P) \mathrel{\widehat{=}} P \wedge tr \le tr'$. Its fixed points are all predicates that ensure that the trace of events tr' extends the previously observed trace tr. R2 requires that P is unaffected by the events recorded in tr, since they are events of the previous process. Specifically, R2 requires that P is not changed if we substitute the empty sequence $\langle\rangle$ for tr and the new events in tr', that is, the subsequence $tr' - tr$, for tr'. Finally, the definition of R3 uses a conditional. It requires that, if the previous process has not terminated ($wait$), then a healthy process does not affect the state: it behaves like the identity relation II.

Table 1. Healthiness conditions of the theory of reactive processes

$$\mathsf{R1}(P) \mathrel{\widehat{=}} P \wedge tr \le tr'$$
$$\mathsf{R2}(P) \mathrel{\widehat{=}} P[\langle\rangle/tr, (tr' - tr)/tr']$$
$$\mathsf{R3}(P) \mathrel{\widehat{=}} (\, \mathit{II} \mathrel{\vartriangleleft} wait \mathrel{\vartriangleright} P)$$

The theory of reactive processes is characterised by the healthiness condition $\mathsf{R} \mathrel{\widehat{=}} \mathsf{R1} \circ \mathsf{R2} \circ \mathsf{R3}$. The reactive processes that can be described using CSP can be expressed by applying R to a design: a pre and postcondition pair over ok, $wait$, tr and ref, and their dashed counterparts. In such a process $\mathsf{R}(pre \vdash post)$, the precondition pre defines the states in which the process does not diverge, and $post$ the behaviour when the previous process has not diverged and pre holds.

Typically, a theory defines a number of programming operators of interest. Common operators like assignment, sequence, and conditional, are defined for general relations. Sequence is relational composition.

$$P;\ Q \mathrel{\widehat{=}} \exists\, w_0 \bullet P[w_0/w'] \wedge Q[w_0/w], \textbf{ where } out\alpha(Q) = in\alpha(Q)' = w'$$

The relation $P;\ Q$ is defined by a quantification that relates the intermediate values of the variables. It is required that the set of dashed variables $out\alpha(P)$ of P, named w', matches the undashed variables $in\alpha(Q)$ of Q. The sets w, w', and w_0 are used as lists that enumerate the variables of w and the corresponding decorated variables in the same order.

A central concern of the UTP is refinement. A program P is refined by a program Q, which is written $P \sqsubseteq Q$, if, and only if, $P \Leftarrow Q$, for all possible values of the variables of the alphabet. We write $[P \Leftarrow Q]$ to represent the universal quantification over all variables in the alphabet. The set of alphabetised predicates in the theory of relations form a complete lattice with this ordering.

As well as characterising a set of healthy predicates via their fixed points, healthiness conditions can be viewed as functions from arbitrary relations to predicates of the theory that they define. Since they are monotonic idempotents, their images, that is, the theory that they characterise, are also complete lattices under refinement. In these theories, recursion is modelled by weakest fixed points $\mu X \bullet F(X)$, where F is a monotonic function from predicates to predicates.

In presenting our theories in the next sections, we define their alphabet and healthiness conditions, and prove that the healthiness conditions are monotonic and idempotent. Finally, we establish Galois connections between them.

3   A UTP Theory for Normalised Graphs

A normalised graph $(N, n_0, t : N \times \Sigma \twoheadrightarrow N, r : N \to \mathbb{F}(\mathbb{F}\Sigma))$ is a quadruple, where N is a set of nodes, n_0 is the initial node, t defines the transitions between nodes from N labelled with events from a set Σ, and r defines labels for states as sets of (maximal) refusal sets, that is, finite sets of finite sets of events in Σ.

Alphabet. We take N and Σ as global constants, and define the alphabet to contain the variables $n, n' : N$ to represent the source and target nodes, $e : \Sigma^\varepsilon$ to represent labels of the transitions, and $r' : \mathbb{F}(\mathbb{F}\Sigma^\varepsilon)$ to represent labels of the target nodes. The predicates define a graph by identifying the source and target nodes n and n' of the transitions, their associated events e, and the labelling r' of the target nodes. The initial node is always ι, a constant of type N. In ι, only the special event ε is available. The set $\Sigma^\varepsilon = \Sigma \cup \{\varepsilon\}$.

Example 1. We consider the graph for $a \to c \to STOP \;\square\; b \to c \to STOP$ shown in Fig. 2 . It is defined in our UTP theory by the following relation.

$$
\begin{aligned}
EG \;\widehat{=}\; & n = \iota \wedge e = \varepsilon \wedge n' = n_1 \wedge r' = \{\{c, \varepsilon\}\} \;\vee \\
& n = n_1 \wedge e \in \{a, b\} \wedge n' = n_2 \wedge r' = \{\{a, b, \varepsilon\}\} \;\vee \\
& n = n_2 \wedge e = c \wedge n' = n_3 \wedge r' = \{\{a, b, c, \varepsilon\}\}
\end{aligned}
$$

We observe that n_1, n_2 and n_3 are arbitrary node identifiers: values in N. $\square$

By including just r' in the alphabet, instead of r and r', we avoid the need to specify labelling for a node repeatedly whenever it is used as a source or a target of a transition (and to include a healthiness condition to ensure that the duplicated information is consistent). Since the initial node is always ι, for which labelling is irrelevant, it is enough to define the labels of the target nodes to get information for all (reachable) nodes. Normalised graphs are connected.

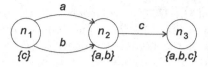

Fig. 2. Normalised graph for $a \to c \to STOP \;\square\; b \to c \to STOP$

Healthiness Conditions. Table 2 defines the healthiness conditions of our theory. HG1 requires all nodes to have a non-empty label: every label contains at least one set X, and, as specified by HG2, each X contains ε. HG2 is concerned with ι and ε; from ι, the only possible event is ε, which is then always refused, and, besides, no transition leads back to ι. HG3 requires that, for any node n and event e, there is at most one transition: the graph is deterministic. Similarly, HG4 establishes that all transitions that target a node n' define the same label: labelling is unique. HG5 requires that, if there is no transition from a node n' for an event e_1, then e_1 is in all refusals X of the label r' of n'. We write $G(w, x, y, z)$ to denote the predicate $G[w, x, y, z / n, e, n', r']$. Finally, HG6 rules out the empty graph *false*.

Table 2. Healthiness conditions of the normalised-graph theory

HG1$(G) \;\hat{=}\; G \wedge r' \neq \emptyset$
HG2$(G) \;\hat{=}\; G \wedge (n = \iota \Rightarrow e = \varepsilon) \wedge \forall X : r' \bullet \varepsilon \in X \wedge n' \neq \iota$
HG3$(G) \;\hat{=}\; G \wedge DetEdges(G) \quad DetEdges(G) \;\hat{=}\; \forall n, e \bullet \#\{n', r' \mid G \bullet n'\} \leq 1$
HG4$(G) \;\hat{=}\; G \wedge DetRefs(G) \quad DetRefs(G) \;\hat{=}\; \forall n' \bullet \#\{n, e, r' \mid G \bullet r'\} \leq 1$
HG5$(G) \;\hat{=}\; G \wedge AccEvents(G)$
$\qquad AccEvents(G) \;\hat{=}\; \forall e_1 \bullet (\forall n'_1, r'_1 \bullet \neg\, G(n', e_1, n'_1, r'_1)) \Rightarrow \forall X : r' \bullet e_1 \in X$
HG6$(G) \;\hat{=}\; G \wedge \exists n, e, n', r' \bullet G$

All of HG1 to HG6 are conjunctive (that is, of the form $\mathsf{HC}(P) \,\hat{=}\, P \wedge F(P)$, for a function $F(P)$ that is monotonic or does not depend on P). So, they are all monotonic, idempotent, and commute [11]. Commutativity establishes independence of the healthiness conditions. We can then define the healthiness condition HG of our theory as the composition of HG1 to HG6. Commutativity implies that HG is an idempotent, just like each of HG1 to HG6.

Connection to Reactive Processes. In [17], graphs are transformed to nondeterministic programs of a particular form. They are action systems [2]: initialised nondeterministic loops, with part of the state at the beginning of each iteration visible. These are, therefore, reactive processes, that communicate to the environment the value of the relevant state components at the start of a loop.

For a graph G, the corresponding action system $AS(G)$ in [17] is as follows.

Definition 1.

$$AS(G) \mathrel{\widehat{=}} \mathbf{var}\ n, tr, ref \bullet n, tr, ref := \iota, \langle\rangle, \Sigma;\ \mu\,Y \bullet vis!(tr, ref) \rightarrow$$
$$Skip \triangleleft ref = \Sigma^{\varepsilon} \triangleright \sqcap e : \Sigma^{\varepsilon} \setminus ref \bullet$$
$$tr := tr \frown \langle e \rangle;$$
$$n, ref : [true, \exists\, r' \bullet G \wedge ref' \in r'];\ Y$$

The program uses a local variable n as a pointer to the current node as it iterates over G. The initial value of n is ι. As the loop progresses, the program accumulates the traces of events tr and records a refusal ref in r. Their values are initialised with the empty sequence $\langle\rangle$ and the whole set of events Σ (but not ε). The values of tr and ref are communicated in each step of the iteration via a channel vis. It is the values that can be communicated that capture the traces and maximal refusals semantics of G.

The loop is defined by a tail recursion ($\mu\,Y \bullet \ldots;\ Y$). Its termination condition is $ref = \Sigma^{\varepsilon}$, that is, it terminates when there is a deadlock. Otherwise, it chooses nondeterministically ($\sqcap$) an event e that can be offered, that is, an event from $\Sigma^{\varepsilon} \setminus ref$, updates tr to record that event, and then updates n and ref as defined by G using a design $n, ref : [true, \exists\, r' \bullet G \wedge ref' \in r']$. The postcondition $\exists\, r' \bullet G \wedge ref' \in r'$ defines the new values of n and ref; the value of ref is also chosen nondeterministically from the events in r' as defined by G.

Example 2. For the process in Example 1, the corresponding reactive process obtained from the graph in Fig. 2, is equivalent to that shown below, where we unfold the recursion and eliminate nondeterministic choices over one element relying on the property ($\sqcap e_1 : \{e_2\} \bullet P(e_1)) = P(e_2)$.

$$\mathbf{var}\ n, tr, ref \bullet n, tr, ref := \iota, \langle\rangle, \{a, b, c\};$$
$$vis!(tr, ref) \rightarrow tr := tr \frown \langle\varepsilon\rangle;\ n, ref := n_1, \{c, \varepsilon\};$$
$$\sqcap e : \{a, b\} \bullet vis!(tr, ref) \rightarrow tr := tr \frown \langle e \rangle;\ n, ref := n_2, \{a, b, \varepsilon\};$$
$$vis!(tr, ref) \rightarrow tr := tr \frown \langle c \rangle;\ n, ref := n_3, \{a, b, c, \varepsilon\};$$
$$vis!(tr, ref) \rightarrow Skip \qquad \qquad \square$$

Besides the healthiness conditions of reactive processes, as defined in Sect. 2, the processes of interest here satisfy the healthiness condition below.

$$\mathsf{R4}(P) \mathrel{\widehat{=}} P \wedge \operatorname{ran} tr' \subseteq \{\!\!\{vis\}\!\!\}$$

It ensures that all events observed in the trace are communications over the channel vis. Together with R1, R4 guarantees that this holds for tr and tr'. We use ran s to denote the set of elements in a sequence s.

The function $\nu(G)$ defined below maps a graph G to a reactive process. It provides an abstract specification for $AS(G)$ using the observational variables of the reactive process theory, rather than programming constructs.

Definition 2. $\nu(G) \,\hat{=}\, R(true \vdash \nu_P(G))$ *where*

$$\nu_P(G) \,\hat{=}\, tr < tr' \Rightarrow$$
$$\left(\begin{array}{l} \exists\, tr_M, ref_M \bullet (tr_M, ref_M) = \mathrm{msg} \circ \mathrm{last}\,(tr') \wedge \\ \left(\begin{array}{l} (\exists\, e, n', r' \bullet G[node(G)(\langle\rangle)/n]) \\ \lhd tr_M = \langle\rangle \rhd \\ \left(\begin{array}{l} \exists\, r' \bullet ref_M \in r' \wedge \\ G[node(G)(\mathrm{front}\,tr_M), \mathrm{last}\,tr_M, node(G)(tr_M)/n, e, n'] \end{array} \right) \end{array} \right) \end{array} \right)$$

We use a node-labelling partial function $node(G)$ that maps traces to nodes of G. It is well defined, because an essential property of a normalised graph is that, for every trace, there is a unique node to which it leads [19, p.161]. We define $\nu(G)$ as a design that specifies that it never diverges: the precondition is *true*. The postcondition $\nu_P(G)$ defines that if an event has occurred ($tr < tr'$), then the behaviour is given by the failure (tr_M, ref_M) communicated in the last event recorded in tr'. We use the function $msg(vis.(tr_M, ref_M)) \,\hat{=}\, (tr_M, ref_M)$. With tr_M and ref_M, $\nu_P(G)$ specifies that what happens next depends on whether the failure emitted contains the empty trace ($tr_M = \langle\rangle$). If it does, then G has to have a node reachable via $\langle\rangle$. Otherwise, the last two elements of the trace must describe a transition in G. The target of this transition has a set of refusal sets r'; the refusal set in the failure must be an element of r'.

The function $\nu(G)$ is the left (upper) adjoint of a Galois connection between the theories of normalised graphs and reactive processes. To establish this result, and others in the sequel, we use the relationship between an R2-healthy assertion ψ used as a postcondition and the process $Proc(\psi)$ that implements ψ. We define $Proc(\psi) \,\hat{=}\, R(true \vdash \psi)$, as a reactive design that requires that the process does not diverge and establishes ψ. Moreover, for a reactive design P, we define a simple way to extract its postcondition $Post(P) \,\hat{=}\, P[true, true, false/ok, ok', wait]$.

Theorem 1. *The pair* $(Proc, Post)$ *is a Galois connection.*

Proof. A design $\phi \vdash \psi$ is defined by $ok \wedge \phi \Rightarrow ok' \wedge \psi$. So, in $R(\phi \vdash \psi)$, the values of ok and ok' in ψ are defined by the design to be *true*. Moreover, the value of *wait* is defined by R3 to be *false*. Therefore, below we consider, without loss of generality, that ψ does not have free occurrences of ok, ok', or *wait*.

$Post \circ Proc(\psi)$

$= (R(true \vdash \psi))[true, true, false/ok, ok', wait]$ [definitions of $Post$ and $Proc$]

$= R1 \circ R2((true \vdash \psi)[true, true, false/ok, ok', wait])$ [definition of R]

$= R1 \circ R2(\psi[true, true, false/ok, ok', wait])$ [substitution in a design]

$= R1 \circ R2(\psi)$ [ok, ok', and *wait* are not free in ψ]

$= R1(\psi)$ [ψ is R2]

$\Rightarrow \psi$ [definition of R1 and predicate calculus]

Next, we prove that $Proc \circ Post(P) = \Pi$, so we have a co-retract. We use P_a^b to stand for the substitution $P[a, b/wait, ok']$, and use t and f for $true$ and $false$.

$Proc \circ Post(P)$

$= \mathsf{R}(true \vdash P[true, true, false/ok, ok', wait])$ [definitions of $Proc$ and $Post$]

$= \mathsf{R}(true \vdash (\mathsf{R}(\neg P_f^f \vdash P_f^t)[true, true, false/ok, ok', wait]))$

[reactive-design theorem: $P = \mathsf{R}(\neg P_f^f \vdash P_f^t)$]

$= \mathsf{R}(true \vdash \mathsf{R}(\neg P_f^f \Rightarrow P_f^t))$ [substitution]

$= \mathsf{R}(true \vdash \mathsf{R}2(\neg P_f^f \Rightarrow P_f^t))$

[$\mathsf{R} = \mathsf{R}1 \circ \mathsf{R}2 \circ \mathsf{R}3$ and $\mathsf{R}1 \circ \mathsf{R}3(P \vdash \mathsf{R}1 \circ \mathsf{R}3(P)) = \mathsf{R}1 \circ \mathsf{R}3(P \vdash Q)$]

$= \mathsf{R}(true \vdash \neg P_f^f \Rightarrow P_f^t)$ [assumption: P is R2]

$= \mathsf{R}(\neg P_f^f \vdash P_f^t)$ [property of a design]

$= P$ [reactive-design theorem]

□

For graphs and reactive processes we have the following result.

Theorem 2. $\nu(G)$ defines a Galois connection.

Proof. From the definition of $\nu(G)$, we have that $\nu(G) = Proc \circ \nu_P(G)$. Since $\nu_P(G)$ is monotonic and universally disjunctive, it defines a Galois connection between normalised graphs and (R2-healthy) assertions. By Theorem 1, $Proc$ defines a Galois connection between assertions and reactive processes. The composition of Galois connections is a Galois connection itself.   □

With the above theorem, we formalise the point (2) described in Sect. 1.

4   A UTP Theory for Kripke Structures

A Kripke structure $(S, s_0, R : \mathbb{P}(S \times S), L : S \to \mathbb{P}AP, AP)$ is a quintuple, where S is the set of states, s_0 is the initial state, R is a transition relation between states, and L is a labelling function for states. The labels are sets of atomic propositions from AP that are satisfied by the states. R is required to be total, so that there are no stuck states in a Kripke structure.

In our theory, states are identified with the valuations of variables v, which define the properties satisfied by the states, and so define L and AP. Moreover, we include $pc, pc' : 0..2$ to record a program counter. The value of pc in a state defines whether it is initial, $pc = 0$, intermediate, $pc = 1$, or final, $pc = 2$. Satisfaction of properties is checked in the intermediate states.

In Kripke structures for reactive processes, the other variables of interest are t_k : seq Σ^ε, whose value is the trace performed so far, and ref_k : $\mathbb{P}\Sigma^\varepsilon$, whose value is the current refusal, and their dashed counterparts t_k' and ref_k'. We present, however, a theory that is not specific to these variables.

Example 3. Figure 3 gives the Kripke structure for the process in Example 1 and corresponding program in Example 2. In Fig. 3, we give the values of the variables t_k and ref_k in each state as a pair. For the states in which $pc = 0$ or $pc = 2$, however, the values of these variables is arbitrary and not given. The states for which $pc = 2$ have self-transitions to avoid stuck states. □

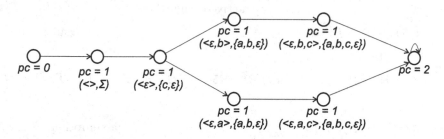

Fig. 3. Kripke structure for $a \rightarrow c \rightarrow STOP \ \square \ b \rightarrow c \rightarrow STOP$

Example 4. The relation EK for the Kripke structure in Fig. 3 is as follows.

$$pc = 0 \wedge pc' = 1 \wedge t'_k = \langle \rangle \wedge ref'_k = \Sigma \vee$$
$$pc = pc' = 1 \wedge t_k = \langle \rangle \wedge ref_k = \Sigma \wedge t'_k \in \{\langle a \rangle, \langle b \rangle\} \wedge ref'_k = \{a, b, \varepsilon\} \vee$$
$$pc = pc' = 1 \wedge t_k = \langle a \rangle \wedge ref_k = \{a, b, \varepsilon\} \wedge t'_k = \langle a, c \rangle \wedge ref'_k = \{a, b, c, \varepsilon\} \vee$$
$$pc = pc' = 1 \wedge t_k = \langle b \rangle \wedge ref_k = \{a, b, \varepsilon\} \wedge t'_k = \langle b, c \rangle \wedge ref'_k = \{a, b, c, \varepsilon\} \vee$$
$$pc = 1 \wedge t_k \in \{\langle a, c \rangle, \langle b, c \rangle\} \wedge pc' = 2 \vee$$
$$pc = 2 \wedge pc' = 2$$
□

Healthiness Conditions. Table 3 presents the healthiness conditions. From the initial state, we move to an intermediate state, and there is no transition back to the initial state or out of the final state. All this is ensured by HK1. With HK2 we establish that the value of v when $pc = 0$ or $pc = 2$ is arbitrary. Similarly, with HK3 we establish that the value of v' when $pc' = 2$ is arbitrary. *SelfT* specifies transitions that keep the value of pc, but that preserve v only in intermediate states. These are a kind of self transitions. We use $v = v'$ to refer to the conjunction $v_1 = v'_1 \wedge \ldots \wedge v_n = v'_n$ including an equality for each variable in v and v'. Finally, HK4 requires that either the Kripke structure is empty, or there is a transition from the initial state.

HK1 is conjunctive and so idempotent, and monotonic since *ValT* does not depend on K. For HK2 and HK3, monotonicity follows from monotonicity of sequence. Idempotence follows from the result below [12, p.90].

Lemma 1. *SelfT; SelfT = SelfT where $out\alpha(SelfT) = in\alpha(SelfT) = \{v', pc'\}$.*

The proof of this results and others omitted below can be found in [6].

Table 3. Healthiness conditions of the Kripke-structure theory

$\mathsf{HK1}(K) \mathrel{\widehat{=}} K \wedge ValT$
$\qquad ValT \mathrel{\widehat{=}} pc = 0 \wedge pc' = 1 \vee pc = 1 \wedge pc' \neq 0 \vee pc = 2 \wedge pc' = 2$
$\mathsf{HK2}(K) \mathrel{\widehat{=}} SelfT;\; K$
$\qquad SelfT \mathrel{\widehat{=}} pc = pc' \wedge (pc = 1 \Rightarrow v = v')$
$\mathsf{HK3}(K) \mathrel{\widehat{=}} K;\; SelfT$
$\mathsf{HK4}(K) \mathrel{\widehat{=}} K \wedge \exists\, v, v' \bullet K[0/pc]$

HK4 is conjunctive and so idempotent, and monotonic since $\exists\, v, v' \bullet K[0/pc]$ is monotonic on K. Commutativity of HK1 with HK2 and HK3 is proved in [6]. Commutativity of HK1 and HK4 is simple because they are both conjunctive. Commutativity of HK2 and HK3 is established in [12, p.90]. Finally, commutativity of HK2 and HK4, and of HK3 and HK4 are established in [6].

We cannot introduce a healthiness condition $\mathsf{HK}(K) = (true;\; K) \Rightarrow K$ that requires that a Kripke structure is not empty; (like H4 in the case of designs) it is not monotonic. So, we keep *false* in the lattice; it represents miracle, as usual.

Connection from Reactive Processes. As mentioned above, for modelling processes, the additional variables are t_k and ref_k, and their dashed counterparts t'_k and ref'_k. In this more specific setting, we have the extra healthiness condition below.

$\mathsf{HK5}(K) \mathrel{\widehat{=}} K \wedge ValRT$
$ValRT \mathrel{\widehat{=}} t'_k = \langle\rangle \wedge pc = 0 \wedge pc' = 1 \vee$
$\qquad\quad t'_k \neq \langle\rangle \wedge pc = pc' = 1 \wedge t_k = \text{front } t'_k \vee$
$\qquad\quad pc' = 2$

The property $ValRT$ defines valid reactive transitions. From the initial state, we reach just the empty trace, and each transition between intermediate states capture the occurrence of a single event: the last event in t'_k.

As discussed previously, we can represent a CSP process G by a reactive process P that outputs in a channel *vis* the failures of G with maximal refusals. In other words, the events of P define the failures of G. Below, we define how, given a reactive process P whose events are all communications on *vis*, we construct a corresponding Kripke structure $\kappa(P)$ whose states record the failures of G. To model the state of the action system before it produces any traces or maximal refusals, we let go of HK1 and allow transitions from states for which $pc = 2$ back to an intermediate state with $pc' = 1$.

Definition 3. $\kappa(P) \cong \kappa_I \circ Post(P)$ *where*

$$\kappa_I(P) \cong \exists\, wait', tr, tr', ref, ref' \bullet P \wedge I_\kappa$$

$$I_\kappa \cong \begin{pmatrix} tr = tr' \wedge pc = 2 \wedge pc' = 1 \vee \\ \#(tr' - tr) = 1 \wedge \\ \quad pc = 0 \wedge pc' = 1 \wedge (t'_k, ref'_k) = \mathrm{msg} \circ \mathrm{last}\,(tr') \vee \\ \#(tr' - tr) > 1 \wedge \\ \quad pc = 1 \wedge pc' = 1 \wedge \\ \quad (t_k, ref_k) = \mathrm{msg} \circ \mathrm{last} \circ \mathrm{front}\,(tr') \wedge (t'_k, ref'_k) = \mathrm{msg} \circ \mathrm{last}\,(tr') \vee \\ \neg\, wait' \wedge \#(tr' - tr) > 1 \wedge \\ \quad pc = 1 \wedge pc' = 2 \wedge (t_k, ref_k) = \mathrm{msg} \circ \mathrm{last} \circ \mathrm{front}\,(tr') \vee \\ \neg\, wait' \wedge pc = 2 \wedge pc' = 2 \end{pmatrix}$$

In defining $\kappa(P)$, of interest is the behaviour of P when the previous process did not diverge (ok' is *true*) and terminated ($wait$ is *false*) and P has not diverged (ok' is *true*). This is the postcondition of P, as defined by $Post$. The postcondition has in its alphabet the variables $wait'$, tr, tr', ref, and ref'. In defining κ_I, these variables are quantified, and used in I_κ to define the values of pc, t_k, ref_k, pc', t'_k, and ref'_k from the theory of Kripke structures.

If only one output has occurred ($\#(tr' - tr) = 1$), then the event $vis.(t'_k, ref'_k)$ observed defines the state that can be reached from the initial state of the Kripke structure. When more events have occurred, we define a transition between the intermediate states characterised by the last two events. Prefix closure of P ensures that we get a transition for every pair of events. When P terminates ($\neg\; wait'$), we get two transitions, one from the last event to the final state ($pc' = 2$), and the loop transition for the final state.

To establish that $\kappa(P)$ defines a Galois connection, we use a result in [12] proved in [6]. It considers functions L and R between lattices A and B (ordered by $\sqsubseteq$) with alphabets a and c, when L and R are defined in terms of a predicate I over the alphabet defined by the union of a and c. We can see these functions as establishing a data refinement between A and B with coupling invariant I.

Theorem 3. *L and R defined below are a Galois connection between A and B.*

$$L(P_C) \cong \exists\, c \bullet P_C \wedge I \quad \text{and} \quad R(P_A) \cong \forall\, a \bullet I \Rightarrow P_A$$

This result can be used to prove the following theorem.

Theorem 4. *$\kappa(P)$ defines a Galois connection.*

Proof. From Theorem 1, we know that $Post$ defines a Galois connection. Theorem 3 establishes that κ_I defines a Galois connection as well. Their composition, which defines κ, therefore, also defines a Galois connection. $\square$

The above theorem formalises the point (3) mentioned in Sect. 1.

5   A UTP Theory for Traces and Maximal Refusals

This is a theory of conditions (predicates on a single state) with alphabet ok_M, tr_M and ref_M. These variables are similar to those of the theory of reactive processes, but ref_M records only maximal refusals. We use the notion of refinement in this theory to define satisfaction for relations in all our theories.

As can be expected, there is a rather direct Galois connection between reactive processes and definitions of traces and maximal refusals in this theory.

Definition 4. $\theta(P) \mathrel{\widehat{=}} \theta_P \circ Post(P)$ where

$$\theta_P(P) \mathrel{\widehat{=}} \exists\, wait', tr, tr', ref, ref' \bullet P \wedge I_\theta$$
$$I_\theta \mathrel{\widehat{=}} ok_M = (tr' > tr) \wedge (ok_M \Rightarrow (tr_M, ref_M) = \mathrm{msg} \circ \mathrm{last}\,(tr'))$$

In defining the failures of $\theta(P)$, we need the postcondition of P. From that, we obtain failures once P has started communicating, so ok_M is characterised by $(tr' > tr)$. If we do have a failure, it is that last communicated via vis in tr'.

Theorem 5. $\theta(P)$ *defines a Galois connection.*

Proof. Similar to that of Theorem 4. $\qquad\qquad\qquad\qquad\qquad\qquad\square$

The healthiness conditions of a theory of traces and refusals are well known [19]. We record, however, via the healthiness condition HM below the role of ok_M, as a flag that indicates whether observations are valid.

$$\mathsf{HM}(M) \mathrel{\widehat{=}} ok_M \Rightarrow M$$

(This is just the healthiness condition H1 of the UTP theory of designs, which first introduces the use of ok). The predicates of our theory of traces and maximal refusals are used as conditions in our satisfaction relations presented next.

5.1   Satisfaction for Normalised Graphs

The function $\eta(G)$ defines a Galois connection between the theory of normalised graphs and the theory of traces and maximal refusals.

Definition 5 $\eta(G) \mathrel{\widehat{=}} \exists\, n, e, n', r' \bullet ok_M \Rightarrow G \wedge I_\eta$ where

$$I_\eta \mathrel{\widehat{=}} \left((n = node(G)(\langle\rangle)) \triangleleft tr_M = \langle\rangle \triangleright \begin{pmatrix} n = node(G)(\mathrm{front}\ tr_M)\ \wedge \\ e = \mathrm{last}\ tr_M\ \wedge \\ n' = node(G)(tr_M)\ \wedge \\ ref_M \in r' \end{pmatrix} \right)$$

As required, we define $\eta(G)$ by characterising traces tr_M and refusals ref_M using the variables n, e, n' and r' from the theory of graphs. If ok_M is *true*, then tr_M is empty if the current node n can be reached with the empty trace (that is, it is the initial node). Otherwise, the trace is that used to reach n concatenated with $\langle e \rangle$. Moreover, ref_M is a refusal in the label r' of the target node.

To establish that $\eta(G)$ is the left adjoint of a Galois connection between the theories of normalised graphs and of maximal refusals, we use the following general result, similar to that in Theorem 3.

Theorem 6. *L and R defined below are a Galois connection between A and B.*

$$L(P_C) \cong \exists\, c \bullet b \Rightarrow P_C \wedge I \quad \text{and} \quad R(P_A) \cong \forall\, a \bullet I \Rightarrow P_A$$

where b is a boolean variable in the alphabet a of A, and $\mathsf{HC}(P_A) = b \Rightarrow P_A$ *is a healthiness condition of the lattice B.*

The proof is similar to that of Theorem 3 and can be found in [6].

Theorem 7. $\eta(G)$ *defines a Galois connection.*

Proof. Direct application of Theorem 6.

Using $\eta(G)$, we can use refinement in the theory of traces and maximal refusals to define satisfaction as shown below.

Definition 6. *For a property ϕ and a graph G, we define $G\,\mathrm{sat}\,\phi \cong \phi \sqsubseteq \eta(G)$.*

Normalised graphs G have the same traces and maximal refusals as the reactive program $\nu(G)$ that it characterises.

Theorem 8. $\eta(G) = \theta \circ \nu(G)$

Proof.

$$\theta \circ \nu(G)$$
$$= \exists\, tr, tr' \bullet \nu_P(G) \wedge \hspace{3cm} \text{[definitions of } \theta \text{ and } \nu\text{]}$$
$$\quad ok_M = (tr' > tr) \wedge (ok_M \Rightarrow (tr_M, ref_M) = \mathrm{msg} \circ \mathrm{last}\,(tr'))$$

$$= \exists\, tr, tr' \bullet \hspace{5cm} \text{[definition of } \nu_P\text{]}$$
$$\left(\begin{array}{l} tr < tr' \Rightarrow \\ \quad \left(\begin{array}{l} \exists\, tr_M, ref_M \bullet (tr_M, ref_M) = \mathrm{msg} \circ \mathrm{last}\,(tr') \wedge \\ \quad \left(\begin{array}{l} (\exists\, e, n', r' \bullet G[node(G)(\langle\rangle)/n]) \\ \quad \vartriangleleft tr_M = \langle\rangle \vartriangleright \\ \quad \left(\begin{array}{l} \exists\, r' \bullet ref_M \in r' \wedge \\ \quad G[node(G)(\mathrm{front}\, tr_M), \mathrm{last}\, tr_M, node(G)(tr_M) \\ \quad /n, e, n'] \end{array}\right) \end{array}\right) \end{array}\right) \\ \wedge \\ ok_M = (tr' > tr) \wedge (ok_M \Rightarrow (tr_M, ref_M) = \mathrm{msg} \circ \mathrm{last}\,(tr')) \end{array}\right)$$

$$= \neg\, ok_M \vee \hspace{5cm} \text{[case analysis on } ok_M\text{]}$$
$$\exists\, tr, tr' \bullet$$
$$\left(\begin{array}{l} \left(\begin{array}{l} \exists\, tr_M, ref_M \bullet (tr_M, ref_M) = \mathrm{msg} \circ \mathrm{last}\,(tr') \wedge \\ \quad \left(\begin{array}{l} (\exists\, e, n', r' \bullet G[node(G)(\langle\rangle)/n]) \\ \quad \vartriangleleft tr_M = \langle\rangle \vartriangleright \\ \quad \left(\begin{array}{l} \exists\, r' \bullet ref_M \in r' \wedge \\ \quad G[node(G)(\mathrm{front}\, tr_M), \mathrm{last}\, tr_M, node(G)(tr_M)/n, e, n'] \end{array}\right) \end{array}\right) \end{array}\right) \\ \wedge \\ (tr' > tr) \wedge (tr_M, ref_M) = \mathrm{msg} \circ \mathrm{last}\,(tr') \end{array}\right)$$

$$= \neg\, ok_M \lor \qquad\qquad\qquad\qquad\qquad\text{[predicate calculus]}$$
$$\exists\, tr, tr' \bullet$$
$$\left(\begin{array}{l}(\exists\, e, n', r' \bullet G[node(G)(\langle\rangle)/n])\\ \quad \lhd tr_M = \langle\rangle \rhd\\ \quad\left(\begin{array}{l}\exists\, r' \bullet ref_M \in r'\, \wedge\\ \quad G[node(G)(\text{front } tr_M), \text{last } tr_M, node(G)(tr_M)/n, e, n']\end{array}\right)\end{array}\right)$$
$$\wedge$$
$$(tr' > tr) \wedge (tr_M, ref_M) = \text{msg} \circ \text{last}\,(tr')$$

$$= \neg\, ok_M \lor \qquad\qquad\qquad\qquad\qquad\text{[predicate calculus]}$$
$$\left(\begin{array}{l}(\exists\, e, n', r' \bullet G[node(G)(\langle\rangle)/n])\\ \quad \lhd tr_M = \langle\rangle \rhd\\ \quad\left(\begin{array}{l}\exists\, r' \bullet ref_M \in r'\, \wedge\\ \quad G[node(G)(\text{front } tr_M), \text{last } tr_M, node(G)(tr_M)/n, e, n']\end{array}\right)\end{array}\right)$$
$$\wedge$$
$$\exists\, tr, tr' \bullet (tr' > tr) \wedge (tr_M, ref_M) = \text{msg} \circ \text{last}\,(tr')$$

$$= \neg\, ok_M \lor \qquad\qquad\qquad\qquad\qquad\text{[predicate calculus]}$$
$$\left(\begin{array}{l}(\exists\, e, n', r' \bullet G[node(G)(\langle\rangle)/n])\\ \quad \lhd tr_M = \langle\rangle \rhd\\ \quad\left(\begin{array}{l}\exists\, r' \bullet ref_M \in r'\, \wedge\\ \quad G[node(G)(\text{front } tr_M), \text{last } tr_M, node(G)(tr_M)/n, e, n']\end{array}\right)\end{array}\right)$$

$$= \neg\, ok_M \lor \qquad\qquad\qquad\qquad\qquad\text{[predicate calculus]}$$
$$\left(\begin{array}{l}(\exists\, n, e, n', r' \bullet G \wedge n = node(G)(\langle\rangle))\\ \quad \lhd tr_M = \langle\rangle \rhd\\ \quad\left(\begin{array}{l}\exists\, n, e, n, r' \bullet ref_M \in r'\, \wedge\\ \quad G \wedge n = node(G)(\text{front } tr_M) \wedge e = \text{last } tr_M \wedge n' = node(G)(tr_M)\end{array}\right)\end{array}\right)$$

$$= \exists\, n, e, n', r' \bullet ok_M \Rightarrow G\, \wedge \qquad\qquad\text{[property of conditional]}$$
$$\left(\begin{array}{l}(n = node(G)(\langle\rangle))\\ \quad \lhd tr_M = \langle\rangle \rhd\\ \quad\left(\begin{array}{l}ref_M \in r'\, \wedge\\ \quad n = node(G)(\text{front } tr_M) \wedge e = \text{last } tr_M \wedge n' = node(G)(tr_M)\end{array}\right)\end{array}\right)$$

$$= \eta(G) \qquad\qquad\qquad\qquad\qquad\qquad\qquad\text{[definition of } \eta]$$

$$\square$$

This establishes that our transformations preserve traces and maximal refusals. So, to check satisfaction for a graph G, we can use $\theta \circ \nu(G)$, instead of $\eta(G)$.

5.2   Satisfaction for Kripke Structures

The function $\zeta(G)$ defines a Galois connection between the theory of Kripke structures and the theory of traces and maximal refusals.

Definition 7.

$$\zeta(K) \,\widehat{=}\, \exists\, pc, pc', t_k, t_k', ref_k, ref_k' \bullet K\, \wedge$$
$$ok_M = (pc \in \{0, 1\}) \wedge (ok_M \Rightarrow tr_M = t_k' \wedge ref_M = ref_k')$$

The traces tr_M and refusals ref_M that are captured are those of the target states.

Theorem 9. $\zeta(K)$ *defines a Galois connection.*

Proof. Direct consequence of Theorem 3.   □

Using ζ, we can use refinement in the theory of traces and maximal refusals to define satisfaction for Kripke structures as shown below.

Definition 8. *For a property ϕ and a Kripke structure K, we define*

$$K \text{ sat } (pc' = 1 \Rightarrow \phi) \mathrel{\widehat=} \phi \sqsubseteq \zeta(K \wedge pc' = 1)$$

The Kripke structures $\zeta \circ \kappa(P)$ have the same traces and maximal refusals as the reactive process P that they characterise.

Theorem 10. $\theta(P) = \zeta(\kappa(P) \wedge pc' = 1)$

Proof.

$\zeta(\kappa(P) \wedge pc' = 1)$

$$= \zeta \left(\begin{array}{l} \exists\, wait', tr, tr', ref, ref' \bullet Post(P) \wedge \\ \left(\begin{array}{l} \#(tr' - tr) = 1 \wedge pc = 0\ \vee \\ \#(tr' - tr) > 1 \wedge pc = 1 \wedge (t_k, ref_k) = msg \circ last\,(front\ tr')\ \vee \\ tr' = tr \wedge pc = 2 \end{array} \right) \wedge \\ pc' = 1 \wedge (t'_k, ref'_k) = msg \circ last\,(tr') \end{array} \right)$$

$\hfill$ [definition of $\kappa(P)$ and predicate calculus]

$$= \left(\begin{array}{l} \exists\, pc, pc', t_k, t'_k, ref_k, ref'_k \bullet \\ \left(\begin{array}{l} \exists\, wait', tr, tr', ref, ref' \bullet Post(P) \wedge \\ \left(\begin{array}{l} tr = tr' \wedge pc = 2\ \vee \\ \#(tr' - tr) = 1 \wedge pc = 0\ \vee \\ \#(tr' - tr) > 1 \wedge pc = 1 \wedge (t_k, ref_k) = msg \circ last\,(front\ tr') \end{array} \right) \\ \wedge \\ pc' = 1 \wedge (t'_k, ref'_k) = msg \circ last\,(tr') \end{array} \right) \\ \wedge \\ ok_M = (pc \in \{0, 1\}) \wedge (ok_M \Rightarrow tr_M = t'_k \wedge ref_M = ref'_k) \end{array} \right)$$

$\hfill$ [definition of ζ]

$$= \left(\begin{array}{l} \exists\, wait', tr, tr', ref, ref' \bullet Post(P) \wedge \\ \exists\, pc, t'_k, ref'_k \bullet \\ \left(\begin{array}{l} tr = tr' \wedge pc = 2\ \vee \\ \#(tr' - tr) = 1 \wedge pc = 0\ \vee \#(tr' - tr) > 1 \wedge pc = 1 \end{array} \right) \wedge \\ (t'_k, ref'_k) = msg \circ last\,(tr') \wedge \\ ok_M = (pc \in \{0, 1\}) \wedge (ok_M \Rightarrow tr_M = t'_k \wedge ref_M = ref'_k) \end{array} \right)$$

$\hfill$ [predicate calculus]

$$
= \left(\begin{array}{l} \exists\, wait', tr, tr', ref, ref' \bullet Post(P) \wedge \\ \left(\begin{array}{l} \exists\, pc \bullet \\ \quad tr = tr' \wedge pc = 2 \vee \\ \quad \#(tr' - tr) = 1 \wedge pc = 0 \vee \#(tr' - tr) > 1 \wedge pc = 1 \\ ok_M = (pc \in \{0,1\}) \wedge (ok_M \Rightarrow (t_M, ref_M) = \mathrm{msg} \circ \mathrm{last}\,(tr')) \end{array} \right) \wedge \end{array} \right)
$$

[predicate calculus]

$$
= \left(\begin{array}{l} \exists\, wait', tr, tr', ref, ref' \bullet Post(P) \wedge \\ ok_M = tr' > tr \wedge (ok_M \Rightarrow (t_M, ref_M) = \mathrm{msg} \circ \mathrm{last}\,(tr')) \end{array} \right)
$$

[predicate calculus]

$$
= \theta(P)
$$

[definition of $\theta(P)$]

□

This establishes the semantic preservation of our transformation.

As a consequence, it is direct that, to check satisfaction for a graph G, we can use $\kappa \circ \nu(G)$, instead of $\eta(G)$, as shown below.

Theorem 11. G **sat** $\phi \Leftrightarrow \kappa \circ \nu(G)$ **sat** $(pc' = 1 \Rightarrow \phi)$

Proof.

G **sat** ϕ

$= \phi \sqsubseteq \eta(G)$ [definition of sat]

$= \phi \sqsubseteq \theta \circ \nu(G)$ [Theorem 8]

$= \phi \sqsubseteq \zeta(\kappa \circ \nu(G) \wedge pc' = 1)$ [Theorem 10]

$= \kappa \circ \nu(G)$ **sat** $(pc' = 1 \Rightarrow \phi)$ [definition of sat]

□

This is the main result of this paper.

6 Conclusions

We have previously developed a monitor for runtime verification of sequential nondeterministic programs with Kripke-structure semantics. It can check the program's execution behaviour against a subset of LTL safety formulas.

In this paper, we have presented novel UTP theories for normalised graphs and Kripke structures that model these nondeterministic programs. They are complete lattices under the UTP refinement order. Our relation of interest, however, is satisfaction, which we have defined for graphs and Kripke structures. Using this framework, we can justify the soundness of the translation of CSP models via normalised graphs into Kripke structures. This induces a concrete translation from CSP processes to sequential nondeterministic programs.

The framework also indicates how to translate a subset of safety formulas into CSP specifications on traces and refusals. These formulas belong to the formula class handled by the runtime monitor. The framework guarantees that an execution of some CSP process P satisfies a specification a given specification if, and only if, P's translation into a sequential nondeterministic program does.

Temporal model checking of UTP designs (pre and postcondition pairs) based on Kripke structures is discussed in [1]. Like we do, [1] defines satisfaction as an extra relation in a lattice ordered by refinement. Satisfaction is defined for states, and temporal logic operators are modelled as fixed-point operators. We adopt a similar notion of state as variable valuations, but do not formalise temporal operators. On the other hand, we define explicitly a theory of Kripke structures, rather than encode them as designs. Moreover, we capture the relationship between Kripke structures and failure models: directly to action systems encoded as reactive processes and indirectly to normalised graphs. As far as we know, we give here the first account of automata-based theories in the UTP.

An issue we have not covered is the relationship of our theories with the existing UTP CSP theory [7,12]. That amounts to formalising the operational semantics of CSP and the normalisation algorithm of FDR3. Since maximal refusals cannot be deduced from the denotational semantics of CSP [19, p. 124], we do not expect an isomorphism between the theories.

An important property of normalised graphs and Kripke structures that is not captured by our healthiness conditions is connectivity. The definition of a monotonic idempotent that captures this property is left as future work. For Kripke structures, we also do not capture the fact that there are no intermediate stuck states. If we consider that every assignment of values to v is a valid state, then this can be captured by the function $\mathsf{HK6}(K) = (K;\ true) \Rightarrow K$.

Acknowledgements. The work of Ana Cavalcanti and Jim Woodcock is funded by the EPSRC grant EP/H017461/1 and the EU INTO-CPS. No new primary data were created during this study. The work of Wen-ling Huang and Jan Peleska is funded by the grant *ITTCPS – Implementable Testing Theory for Cyber-physical Systems* as part of the German Universities Excellence Initiative.

References

1. Anderson, H., Ciobanu, G., Freitas, L.: UTP and temporal logic model checking. In: Butterfield, A. (ed.) UTP 2008. LNCS, vol. 5713, pp. 22–41. Springer, Heidelberg (2010)
2. Back, R.J., Kurki-Suonio, R.: Distributed cooperation with action systems. ACM Trans. Program. Lang. Syst. **10**(4), 513–554 (1988)
3. Burdy, L., et al.: An overview of JML tools and applications. STTT **7**(3), 212–232 (2005)
4. Butterfield, A.: A denotational semantics for Handel-C. FACJ **23**(2), 153–170 (2011)
5. Cavalcanti, A.L.C., Clayton, P., O'Halloran, C.: From control law diagrams to Ada via *Circus*. FACJ **23**(4), 465–512 (2011)

6. Cavalcanti, A.L.C., Huang, W.L., Peleska, J., Woodcock, J.C.P.: Unified Runtime Verification for CSP - Extended version. Technical report, University of York, Department of Computer Science, York, UK (2015). www.cs.york.ac.uk/circus/hijac/publication.html

7. Cavalcanti, A., Woodcock, J.: A tutorial introduction to CSP in unifying theories of programming. In: Cavalcanti, A., Sampaio, A., Woodcock, J. (eds.) PSSE 2004. LNCS, vol. 3167, pp. 220–268. Springer, Heidelberg (2006)

8. Cavalcanti, A.L.C., Zeyda, F., Wellings, A., Woodcock, J.C.P., Wei, K.: Safety-critical Java programs from *Circus* models. RTS **49**(5), 614–667 (2013)

9. Gibson-Robinson, T., Armstrong, P., Boulgakov, A., Roscoe, A.W.: FDR3 — a modern refinement checker for CSP. In: Ábrahám, E., Havelund, K. (eds.) TACAS 2014 (ETAPS). LNCS, vol. 8413, pp. 187–201. Springer, Heidelberg (2014)

10. Haddad, G., Hussain, F., Leavens, G.T.: The design of SafeJML, a specification language for SCJ with support for WCET specification. In: JTRES. ACM (2010)

11. Harwood, W.T., Cavalcanti, A., Woodcock, J.: A theory of pointers for the UTP. In: Fitzgerald, J.S., Haxthausen, A.E., Yenigun, H. (eds.) ICTAC 2008. LNCS, vol. 5160, pp. 141–155. Springer, Heidelberg (2008)

12. Hoare, C.A.R., Jifeng, H.: Unifying Theories of Programming. Prentice-Hall, Upper Saddle River (1998)

13. Huang, W.L., Peleska, J., Schulze, U.: Contract Support for Evolving SoS. Public Document D34.3, COMPASS (2014)

14. Liu, Z., Jifeng, H., Li, X.: rCOS: refinement of component and object systems. In: de Boer, F.S., Bonsangue, M.M., Graf, S., de Roever, W.-P. (eds.) FMCO 2004. LNCS, vol. 3657, pp. 183–221. Springer, Heidelberg (2005)

15. Miyazawa, A., Lima, L., Cavalcanti, A.: Formal models of SysML blocks. In: Groves, L., Sun, J. (eds.) ICFEM 2013. LNCS, vol. 8144, pp. 249–264. Springer, Heidelberg (2013)

16. Oliveira, M.V.M., Cavalcanti, A.L.C., Woodcock, J.C.P.: A UTP semantics for *Circus*. FACJ **21**(1–2), 3–32 (2009)

17. Peleska, J.: Translating testing theories for concurrent systems. In: Correct System Design, Essays Dedicated to Ernst-Rüdiger Olderog on the Occasion of his 60th Birthday, LNCS. Springer (2015)

18. Roscoe, A.W. (ed.): A Classical Mind: Essays in Honour of C. A. R. Hoare. Prentice Hall International (UK) Ltd., Hertfordshire (1994)

19. Roscoe, A.W.: Understanding Concurrent Systems. Texts in Computer Science. Springer, London (2011)

20. Sherif, A., Cavalcanti, A.L.C., He, J., Sampaio, A.C.A.: A process algebraic framework for specification and validation of real-time systems. FACJ **22**(2), 153–191 (2010)

21. Zeyda, F., Santos, T., Cavalcanti, A., Sampaio, A.: A modular theory of object orientation in higher-order UTP. In: Jones, C., Pihlajasaari, P., Sun, J. (eds.) FM 2014. LNCS, vol. 8442, pp. 627–642. Springer, Heidelberg (2014)

22. Zhu, H., He, J., Qin, S., Brooke, P.: Denotational semantics and its algebraic derivation for an event-driven system-level language. FACJ **27**(1), 133–166 (2015)

Specifying and Analyzing the Kademlia Protocol in Maude

Isabel Pita[✉] and Adrián Riesco

Facultad de Informática, Universidad Complutense de Madrid, Madrid, Spain
ipandreu@sip.ucm.es, ariesco@fdi.ucm.es

Abstract. Kademlia is the most popular peer-to-peer distributed hash table (DHT) currently in use. It offers a number of desirable features that result from the use of a notion of *distance* between objects based on the bitwise exclusive *or* of n-bit quantities that represent both nodes and files. Nodes keep information about files *close* or *near* to them in the key space and the search algorithm is based on looking for the *closest* node to the file key. The structure of the routing table defined in each peer guarantees that the lookup algorithm takes no longer than O(log(n)) steps, where n is the number of nodes in the network.

This paper presents a formal specification of a P2P network that uses the Kademlia DHT in the Maude language. We use sockets to connect different Maude instances and create a P2P network where the Kademlia protocol can be used, hence providing an implementation of the protocol which is correct by design. Then, we show how to abstract this system in order to analyze it by using *Real-Time Maude*. The model is fully parameterized regarding the time taken by the different actions to facilitate the analysis of various scenarios. Finally, we use time-bounded model-checking and exhaustive search to prove properties of the protocol over different scenarios.

Keywords: Kademlia · Distributed specification · Formal analysis · Maude · Real-Time Maude

1 Introduction

Kademlia based distributed hash tables (DHTs) [11] are an essential factor in the implementation of P2P networks since the Kad DHT was incorporated in the eMule client [5]. Among the large number of DHTs studied through theoretical simulations and analysis, such as Chord [25], CAN [21], or Pastry [24], Kademlia is the one that has been chosen for implementation of file sharing systems over large networks due to its relative simplicity. Some of its advantages are that there is only one routing algorithm from start to finish; it prevents a number of attacks by preferring long-standing nodes over new ones in the routing tables; and it allows nodes to learn about the network simply by participating in it.

Research supported by MICINN Spanish project *StrongSoft* (TIN2012-39391-C04-04) and Comunidad de Madrid program *N-GREENS Software-CM* (S2013/ICE-2731).

M. Leucker et al. (Eds.): ICTAC 2015, LNCS 9399, pp. 524–541, 2015.
DOI: 10.1007/978-3-319-25150-9_30

DHTs are mainly used for file sharing applications and decentralized storage systems, due to its lack of security. On the one hand, the large number of users involved in the systems and the absence of a central authority certifying the trust of the participants suggests that the system must be able to operate even if some of them are malicious. On the other hand, the dynamics of the system, mainly the arrival and departure of nodes in P2P networks, and the continuous upload of new data, requires a precise information which is challenging to acquire. Most of the existing studies evaluate these problems experimentally. There is a lack of formal descriptions, even though they have obtained good results in the analysis of other distributed networks and protocols.

This paper presents a distributed specification in Maude [6], a formal specification language based on rewriting logic, of the behavior of a P2P network that uses the Kademlia DHT. Rewriting logic [13] is a unified model for concurrency in which several well-known models of concurrent and distributed systems can be represented. The specification language Maude supports both equational and rewriting logic computations. It can be used to specify in a natural way a wide range of software models and systems and, since (most of) the specifications are directly executable, Maude can be used to prototype those systems. Moreover, the Maude system includes a series of tools for formally analyzing the specifications. Since version 2.2 Maude supports communication with external objects by means of TCP sockets, which allows for the implementation of real distributed applications. Real-Time Maude [19] is a natural extension of the Maude language that supports the specification and analysis of real-time systems, including object-oriented distributed ones. It supports a wide spectrum of formal methods, including: executable specification, symbolic simulation, breadth-first search for failures of safety properties in infinite-state systems, and linear temporal logic model checking of time-bounded LTL formulas. Real-Time Maude strengthens that analyzing power by allowing to specify sometimes crucial timing aspects. It has been used, for example, to specify the Enhanced Interior Gateway Routing Protocol (EIGRP) [22], embedded systems [17], and the AER/NCA active network protocol [16]. Moreover, analysis of real-time systems using Maude sockets, and thus requiring a special treatment for them, has been studied [1,26]. While the algebraic representation of the distribution used in these works follows, as well as our work, the approach presented in [22], the way used to relate logical and physical time allows for a more precise and formal analysis than the one used here, allowing the system to synchronize only when needed.

Our distributed specification of the Kademlia protocol has been implemented on top of the routing protocol described in [22] and uses an external Java clock. Since we formally specify the semantics of the protocol, we obtain a correct *by design* application. Moreover, this distributed system can be simulated and analyzed in Maude if a "centralized" version is provided. This version is obtained by using: (i) an algebraic specification of the sockets provided by Maude; (ii) an abstraction of the underlying routing protocol, which allows the analysis tools to focus on the properties; and (iii) Real-time Maude, as explained above. That is, we abstract some implementation details but leave the protocol implementation unmodified, which allows us to use the centralized protocol to prove properties that must also hold in the distributed version. The analyses that can be performed on the protocol include the simulation of the system to study, for example, how its properties change when its parameters, like the redundancy constant, are modified; examine the reaction of the system to different attacks; and check properties such as that any published file can be found or that files remain accessible even if their publishing peers become offline. Actually, we present different levels of abstraction, which allows us to focus on the properties we want to prove while discarding the unnecessary implementation details.

Our specification is, to the best of our knowledge, the first formal description of a Kademlia DHT. The use of formal methods to describe the behavior of the Kademlia DHT may help to understand the informal description of the protocol and the algorithm given in [11], and to identify areas that are not covered in the description and are being resolved in different ways in different implementations. In particular, the Maude language gives us the opportunity of executing the distributed specification taking into account the time aspects of the protocol in order to detect weak points in the protocol that would be interesting to study. Then using the centralized model that mirrors the distributed one, we can analyze all possible executions of the system, either by searching in the execution tree (which is in fact represented as a graph for efficiency reasons) or by using model checking techniques.

The rest of the paper is structured as follows: Sect. 2 presents the Kademlia protocol and how to specify generic distributed systems in Maude, as well as some related work. Section 3 describes the distributed specification in Maude of this protocol. Section 4 shows how the distributed system can be represented in one single term, while Sect. 5 describes how to simulate and analyze it. Finally, Sect. 6 concludes and presents some future work.

2   Preliminaries and Related Work

We present in this section the basic notions about Maude and Kademlia and the related work.

2.1   Maude

In Maude [6] the state of a system is formally specified as an algebraic data type by means of an equational specification. In this kind of specification we can define new types (by means of keyword sort(s)); subtype relations between types (subsort); operators (op) for building values of these types; and equations (eq) that identify terms built with these operators. We can distinguish between Core Maude [6, PartI], which is implemented in C++ and provides the basic features, and Full Maude [6, PartII], an extension of Maude implemented in Maude itself and used as basis for further extensions, as we explain below.

The *dynamic* behavior of such a distributed system is then specified by rewrite rules of the form $t \longrightarrow t'$ *if* C, that describe the local, concurrent transitions of the system. That is, when a part of a system matches the pattern t and satisfies the condition C, it can be transformed into the corresponding instance of the pattern t'.

In object-oriented specifications, *classes* are declared with the syntax class $C \mid a_1 : S_1, \ldots, a_n : S_n$, where C is the class name, a_i is an attribute identifier, and S_i is the sort of the values this attribute can have, for $1 \leq i \leq n$. An *object* is represented as a term $< O : C \mid a_1 : v_1, \ldots, a_n : v_n >$ where O is the object's name, belonging to a set Oid of object identifiers, and the v_i's are the current values of its attributes. *Messages* are defined by the user for each application (introduced with syntax msg).

In Maude, the state of a concurrent, object-oriented system is called a *configuration*. It has the structure of a multiset made up of objects and messages that evolves by concurrent rewriting. The rewrite rules specify the behavior associated with the messages. By convention, the only object attributes made explicit in a rule are those relevant for that rule. We use Full Maude's object-oriented notation and conventions [6] throughout the whole paper; however, only the centralized specification is specified in Full Maude (which is required by Real-Time Maude), while the actual implementation of the distributed protocol is in Core Maude because Full Maude does not support external objects. The complete Maude code can be found at http://maude.sip.ucm.es/kademlia.

In [22], we described a methodology to implement distributed applications in such a way that the distributed behavior remains transparent to the user by using a routing protocol, the Enhanced Interior Gateway Routing Protocol (EIGRP). Figure 1 presents the architecture proposed in that paper, where the lower layer provides mechanisms to translate Maude messages from and to String (Maude sockets can only transmit Strings); to do so, the user must instantiate a theory requiring a (meta-represented) module with the syntax of all the transmitted messages. The intermediate layer, EIGRP, provides a message of the form to_:_, with the first argument an object identifier (the addressee of the message) and the second one a term of sort TravelingContents, that must be defined in each specific application. We have slightly modified this layer to share the tick! message obtained from the Java server in charge of dealing with time.[1] This layer provides a fault-tolerant and dynamic architecture where nodes may join and leave any moment, and where nodes are always reached by using the shortest path, thus allowing us to implement realistic systems. Finally, the upper layer is the application one, which in our case corresponds to Kademlia. It relies on the lower layers to deliver the messages and focus on its specific tasks, just like the real Kademlia protocol.

---

[1] In the standard implementation, tick! messages are introduced into the configuration each second. However, the time can be customized to get these messages in the time span defined by the user.

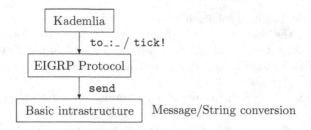

Fig. 1. Layers for distributed applications

2.2 Kademlia

Kademlia is a peer-to-peer (P2P) distributed hash table used by the peers to access files shared by other peers. In Kademlia both peers and files are identified with n-bit quantities, computed by a hash function. Information of shared files is kept in the peers with an identifier *close* to the file identifier, where the notion of distance between two identifiers is defined as the bitwise exclusive or of the n-bit quantities. Then, the lookup algorithm which is based on locating successively *closer* nodes to any desired key has $\mathcal{O}(\log n)$ complexity, where n is the number of nodes in the network.

In Kademlia, every node keeps the following contact information: IP address, UDP port, and node identifier, for nodes of distance between 2^i and 2^{i+1} from itself, for $i = 0, \ldots, n$ and n the identifier length. In the Kademlia paper [11] these lists, called k-buckets, have at most k elements, where k is chosen such that any given k nodes are very unlikely to fail within an hour of each other. k-buckets are kept sorted by the time they were last seen. When a node receives any message (request or reply) from another node, it updates the appropriate k-bucket for the sender's node identifier. If the sender node exists, it is moved to the tail of the list. If it does not exist and there is free space in the appropriate k-bucket it is inserted at the tail of the list. Otherwise, the k-bucket has not free space, the node at the head of the list is contacted and if it fails to respond it is removed from the list and the new contact is added at the tail. In the case the node at the head of the list responds, it is moved to the tail, and the new node is discarded. This policy gives preference to old contacts, since the longer a node has been up, the more likely it is to remain up another hour and also prevents attacks by preferring long-standing nodes.

k-buckets are organized in a binary tree called the routing table. Each k-bucket is identified by the common prefix of the identifiers it contains. Internal tree nodes are the common prefix of the k-buckets, while the leaves are the k-buckets. Thus, each k-bucket covers some range of the identifier space, and together the k-buckets cover the entire identifier space with no overlap. Figure 2 shows a routing table for node 00000000 and a k-bucket of length 5. Identifiers have 8 bits.

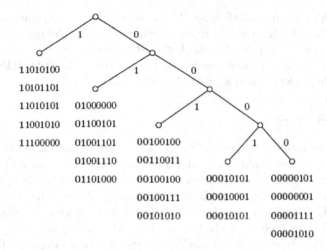

Fig. 2. A routing table example for node 00000000

The Kademlia protocol consists of four Remote Procedure Calls (RPCs):

- PING checks whether a node is online.
- STORE instructs a node to store a file identifier together with the contact of the node that shares the file to publish it to other nodes.
- FIND-NODE takes an identifier as argument and the recipient returns the contacts of the k nodes it knows that are closest to the target identifier.
- FIND-VALUE takes an identifier as argument. If the recipient has information about the argument, it returns the contact of the node that shares the file; otherwise, it returns a list of the k contacts it knows that are closest to the target.

In the following we summarize the dynamics of looking for a value and publishing a shared file from the Kademlia paper [11].

Looking for a Value. To find a file identifier, a node starts by performing a lookup to find the k nodes with the closest identifiers to the file identifier. First, the node sends a FIND-VALUE RPC to the α nodes it knows with an identifier closer to the file identifier, where α is a system concurrency parameter. As nodes reply, the initiator sends new FIND-VALUE RPCs to nodes it has learned about from previous RPCs, maintaining α active RPCs. Nodes that fail to respond quickly are not considered. If a round of FIND-VALUE RPCs fails to return a node any closer than the closest one already seen, the initiator resends the FIND-VALUE to all of the k closest nodes it has not queried yet. The process terminates when any node returns the value or when the peer that started the query has obtained the responses from its k closest nodes.

Publishing a Shared File. Publishing is performed automatically whenever a file needs it. To maintain persistence of the data, files are published by the node that shares them from time to time. Those nodes that have information about the whereabouts of a file publish it more frequently than the node sharing it.

To share a file, a peer locates the k closest nodes to the key, as it is done in the *looking for a value* process, although it uses the FIND-NODE RPC. Once it has located the k closest nodes, it sends them a STORE RPC.

2.3   Related Work

One of the first proposals about using formal methods for analyzing the DHT behavior is due to Borgströn et al., who prove in [4] the correctness of the lookup operation of the DHT-based DKS system, developed in the context of the EU-project [9], for a static model of the network using value-passing CCS. Besides, Bakhshi and Gurov give in [3] a formal verification of Chord's stabilization algorithm using the π-calculus. Lately Lu, Merz, and Weidenbach [10,12] have modeled Pastry's core routing algorithms in the specification language TLA$^+$. The model has been validated using the TLC model checker and they have proved the *CorrectDelivery* safety property stating that there can be only one node responsible for any key at any time using the interactive theorem prover TLAPS of TLA. A different approach is used by P. Zave in [27] to analyze correctness of the Chord DHT protocol. She uses the Alloy language and checks properties with the Alloy analyzer. Properties are expressed as invariants of the system and proved by an exhaustive enumeration of instances over a bounded domain. The analysis revealed some flaws in the original description of the algorithm, which allowed the author to propose some improvements.

Regarding the Kademlia DHT there is a previous work of the first author [20] focused on the Kademlia and the Kad routing tables. The paper highlights the main differences between the Kademlia proposal [11] and its first real implementation in the eMule network. Both routing tables were specified in the Maude formal specification language. The network specification presented in this paper uses the Kademlia routing table. The specification is designed in a modular way to support other *Kademlia style* routing tables, like the one from Kad. This will allow us to compare the behavior of different systems only changing the routing table specification.

3   Protocol Specification

We present in this section the main details of the distributed implementation of the Kademlia protocol. The Kademlia network is modeled as a Maude configuration of objects and messages, where the objects represent the network peers and the messages represent the protocol RPCs.

3.1   Peers

Peers in our specification are objects of class Peer, defined as follows:

```
class Peer | RT : RoutingTable, Files : TFileTable,
             Publish : TPublishFile, SearchFiles : TSearchFile,
             SearchList : TemporaryList .
```

which indicates that the class Peer has the attributes RT, of sort RoutingTable; Files, of sort TFileTable; Publish, of sort TPublishFile; SearchFiles, of sort TSearchFile; and SearchList, of sort TemporaryList These attributes are defined as follows:

- RT is a list that keeps the information of the routing table.
- Files is a table that keeps the information of the files the peer is responsible for publishing. It includes the file identifier, the identification of the peer that shares the file, a time for republishing the file and keep it alive, and a time to remove the file from the table.
- Publish is a table that keeps the information of the files shared by the peer. The information includes the file identifier, the file location in the peer, and a time for republishing the file. This time is greater than the time for republishing of the Files table and prevents the information in the Files table from being removed.
- SearchFiles is a table that keeps the files a peer is looking for. The information includes the file identifier and a waiting time to proceed with the search. This time is used when the file is not found and it should be researched later.
- SearchList is an auxiliary list used in the search and publish processes to keep the information of the nodes that have been already contacted by the searcher/publisher and the state in which the searching/publishing process is. As the searcher/publisher finds out new *closer* nodes to the file identifier, it stores them in this file, and starts sending them messages.

Following is an example of an object of class Peer. We identify the peers by 6-bit quantities, represented by its decimal value to improve readability in the examples presented in this paper. This size provides us with enough nodes for our example network. However the specification may use any n-bit quantity or the complete Kademlia contact information, since it is parameterized with respect to the peers identification.

```
< peer(c(48)): Peer |
  RT : (empty-bucket ! c(14)! c(0)! c(16))!! (empty-bucket ! c(33))!!
       (empty-bucket ! c(60)! c(58)! c(56))!!
       (empty-bucket ! c(50)) +
     c(8)+ PING(c(48),c(14),5,1)+ c(60)c(50) + 4,
  Files : < 32 & c(48);; 19 > # < 38 & c(0);; 4 > # < 54 & c(0);; 8 >,
  Publish : < 22 &"File4"@ 25 > # < 32 &"File5"@ 26 >,
  SearchFiles : < 12 &"File7"; 1 >,
  SearchList : temp-empty >
```

The routing table has four buckets, the first one with the contacts which have its first bit set to 0 (values between 0 and 31), the second with the contacts with its first two bits set to 10 (values between 32 and 47), and so on. The bucket dimension, which is also a parameter of the specification, is set to 3 in our example. We observe that the first and the third buckets are full. In the snapshot shown in the example, the peer has had knowledge of a new contact c(8), but since it should be located in the first bucket, which is full, the peer has sent a PING message to the first contact in the bucket, to verify whether it is still alive. The peer will wait four more time units for the reply before it decides that the contact is not alive and removes the PING message. Meanwhile it has had knowledge of two more contacts c(60) and c(50), which are waiting to be processed. See [20] for a detailed explanation of the routing table specification.

The peer keeps information about three files which has been published by other peers in the Files attribute. These are file 32, which has been published by peer c(48), and files 38 and 54, published by the peer c(0). These files are kept in this peer because it is one of the closest nodes to the file identification. We can choose the redundancy parameter in our specification, in the example it is set to three. The time parameter that appears at the end of each file represents the time remaining for republishing the files to keep them up to date. Each time a peer receives a STORE message for a node that it is already keeping it, updates this time value to the time chosen in the specification for republishing. In this way Kademlia prevents all the redundant peers that keep a file from republishing it at the same time.

The Publish attribute presents two files upload to the network by the peer. The first parameter is the file identifier, next is the file name, and the last parameter is the time remaining for republishing the files to keep them alive.

We observe in the SearchFiles attribute the identifier of a file that the peer wants to search for. Again the last parameter is the time remaining for the process to take place. The searched files are removed from the list when the search process succeed, if it fails the search process is repeated after some time.

Finally, the searchList attribute is a list of contacts used in the publish and look-for processes to find the closest nodes to the file identification. In our example, it is empty since the peer is not performing any of these tasks.

3.2 RPCs

There is a Maude message for each RPC defined in the Kademlia protocol. For example, the FIND-VALUE RPC and its two possible replys are defined as follows:

```
op FIND-VALUE : MyContact BitString -> TravelingContents [ctor] .
op FIND-VALUE-REPLY1 : MyContact BitString
        Set{vCONTACT}{vContact-BitString} -> TravelingContents [ctor] .
op FIND-VALUE-REPLY2 : MyContact BitString MyContact [ctor] .
```

Note that terms of this form will be used to form messages with the operator to_:_ described in Sect. 2.1, where the first parameter is the identifier of the addressee. The first parameter of these operators identifies the peer sending the message, while the second one represents the key the sender is looking for. The reply has also an additional parameter that keeps a set of the k nodes the peer knows that are the closest ones to the target, where k is the bucket dimension or the contact of the owner of the file.

For example, a message from node c(14) to node c(48) requesting for information about file 54 has the form:

```
to peer(c(48)) : FIND-VALUE(c(14),54) .
```

this message will be in the Maude configuration of nodes and messages.

The reply message sended by node c(48) to node c(14) if the node does not have information about file 54 in its Files attribute list, which is not the case in our example, will be:

```
to peer(c(14)) : FIND-VALUE-REPLY1(c(48),54,c(56),c(58),c(50)) .
```

since the closest nodes to file 54 in the routing table of node c(48) are the nodes c(56), c(58) and c(50). The order in which these contacts are returned is not important, since they will be ordered by their distance to the given file in the node that ask for them.

Since node c(48) is one of the closest nodes in the network to file 54 in our example, and it has this file in its File attribute list, the message that it will return contains the contact of the owner of the file, which is node c(0).

```
to peer(c(14)) : FIND-VALUE-REPLY2(c(48),54,c(0)) .
```

3.3   Process Specification in Maude

The specification of the different processes follows their definition. For example, the searching process starts automatically when there are identifiers in the SearchFiles attribute of some connected peer with time for searching equal to one. A greater value indicates that the file has already been searched for, it was not found, and now it is waiting for repeating the search. When the search starts, the auxiliary list SearchList is filled with the *closest* nodes the searcher has in its routing table, and the time of this file in the searchFiles table is set to INF. It will remain with this value until the search process ends. The number of closest nodes used to initialize the auxiliary list is a parameter of the specification. The original Kademlia paper [11] indicates that *it is a system wide concurrency parameter, such as 3*. Notice that in the implementation the file is ordered by the distance of the contact to the file identification. In our example, when node c(48) starts searching file 12 we have:

```
< peer(c(48)): Peer |
  RT : (empty-bucket ! c(14)! c(0)! c(16)) !! (empty-bucket ! c(33)) !!
       (empty-bucket ! c(60)! c(58)! c(56))!! (empty-bucket ! c(50)) +
       c(8)+ PING(c(48),c(14),5,1)+ c(60)c(50) + 4,
  Files : < 32 & c(48) ;;19 > # < 38 & c(0) ;;4 > # < 54 & c(0) ;;8 >,
  Publish : < 22 &"File4"@ 25 > # < 32 &"File5"@ 26 >,
  SearchFiles : < 12 &"File7"; INF >,
  SearchList : < c(14),2,20,0 > < c(16),4,20,0 > < c(0),12,20,0 >
```

where the first value of the nodes in the SearchList is the contact, the second value is the distance from the contact to the searched file, the third is the time that the node will remain in the list if no response is received from a sended

RPC, and the fourth value is a flag that indicates if the contact has already sent the FIND-VALUE RPC, has received the response, or has sent a STORE message and is waiting a response.

The searching process continues by sending FIND-VALUE RPCs to the first nodes in the list to find *closer* nodes to the file identifier. The process is controlled by the rewrite rule:

```
crl [lookfor-file21] :
      < peer(SENDER) : Peer | SearchFiles : < I1 & (S1 ; INF) > # SF,
                              SearchList : SL >
 => < peer(SENDER) : Peer | SearchFiles : < I1 & (S1 ; INF) > # SF,
                            SearchList : set-flag(Tr,SrchListRmve,SL) >
    to peer(Tr) : FIND-VALUE(SENDER, I1)
 if not all-sended(SL) /\ Tr := first-not-send(SL) /\
    messages-in-process(SL) < ParallelSearchRPC /\
    number-nodes-reply(SL) < kSearched .
```

which states that the RPC is only sent if the number of parallel messages is less than the given constant, ParallelSearchRPC, the peer in charge of the search has not received response yet from a certain number of peers given by the kSearched constant, and there are nodes in the search list that have not been contacted yet. Once the RPC is sent, a flag is activated in the search list that marks this node as *in process* with set-flag.

Following our example, when peer c(48) sends the first RPCs, the configuration will have among other peers and messages the following:

```
to peer(c(14)) : FIND-VALUE(c(48),12)
to peer(c(16)) : FIND-VALUE(c(48),12)
to peer(c(0)) : FIND-VALUE(c(48),12)
 < peer(c(48)) : Peer |
   RT :(empty-bucket ! c(14) ! c(0) ! c(16)) !! (empty-bucket ! c(33))!!
       (empty-bucket ! c(60) ! c(58) ! c(56)) !!
       (empty-bucket ! c(50)) +
        c(8) + PING(c(48), c(14),5,1) + c(60) c(50) + 4,
   Files : < 32 & c(48);; 19 > # < 38 & c(0);; 4 > # < 54 & c(0);; 8 >,
   Publish : < 22 &"File4"@ 25 > # < 32 &"File5"@ 26 >,
   SearchFiles : < 12 &"File7" ; INF >,
   SearchList : < c(14), 2, 20, 1 > < c(16), 4, 20, 1 >
                < c(0), 12, 20, 1 >
```

The receivers of the FIND-VALUE messages may find the file the searcher is looking for in its table or it may return the closest nodes it knows about. In the first case, it sends a FIND-VALUE-REPLY2 message to the searcher including the node identifier of the peer that shares the file. When the searcher receives this reply the process finishes by sending a FILE-FOUND message and the file is removed from its searching table. The FILE-FOUND message is a *ghost* message that remains in the configuration to show the files that have been searched and found, hence easing the description of some properties, that just check whether this message appears in the configuration. In the second case, the receiver sends a FIND-VALUE-REPLY1 message to the searcher including the closest nodes to the file identifier it knows about. When the searcher receives this message it changes its search list, adding the nodes ordered by the distance to the objective.

Only nodes closer than the one which proposes them are added. When the full list is traversed, a flag is activated to mark this node as done in the search list. Additionally, the searcher routing table is updated with the `move-to-tail` operation that puts the identifier of the message sender first in the list, so that it will not be removed from the routing table, as it is the last peer the searcher knows it is alive. The searching process continues by sending new `FIND-VALUE` messages to the new nodes in the `SearchList` that have not been asked yet, and are closer to the searched file identifier than the nodes that have already answer the RPC. If the process does not find the file in any of the contacted nodes, it does not remove the file from the `SearchFile` table and initializes its time for a new search.

4 Centralized Simulation

We use Real-Time Maude [19] to analyze our system. Real-Time Maude is an extension of Maude that allows to perform time-bound analyses such as breadth-first search or model checking. However, Real-Time Maude does not support distributed applications, so in order to use it we need to "centralize" our configuration. We discuss below how to achieve this and how to improve (or *abstract*) this representation. In this case, we distinguish between "architecture abstractions," which simplify the state by removing the transitions not related to the properties we want to verify, and "formal abstractions," which refers to established techniques that allow to improve the proofs by different means. It is worth noting that this transformation does not introduce an important overhead on the complexity of the specification: while the distributed implementation of the protocol has around 3100 lines of code, the centralized one has 3300 lines of code, approximately.

4.1 First Architecture Abstraction

As explained above, in order to use the analysis features provided by Real-Time Maude, we need to represent the distributed configuration described in the previous section as a single term. This centralized specification must fulfill the following requirements:

- The underlying architecture must be simulated. This simulation includes not only redirecting the messages, but also possible delays and errors.
- Nodes can connect and disconnect during the process.

In order to solve the first issue, we provide a class `Process` with a single attribute `conf` that keeps the configurations in different locations[2] separated from each other:

```
class Process | conf : Configuration .
```

---

[2] We will use the word *location* to denote the different Maude instances appearing in the distributed system.

Besides "separating" the processes, we must provide an algebraic specification of the built-in sockets. In our case, we use an object of class `Socket` for each two connected locations in the distributed (real) protocol. This class has attributes `sideA` and `sideB`, indicating the two sides of the socket; `delay`, which stores the delay associated to this socket; and `listA` and `listB`, the lists of `DelayedMsg` (pairs of messages and time) sent to `sideA` and `sideB`, respectively:

```
class Socket | sideA : Oid, sideB : Oid, delay : Time,
              listA : List{DelayedMsg}, listB : List{DelayedMsg} .
```

In this way, we can simulate the delay due to the network and specify the architecture with only four rules, two for moving messages into the socket and two more for putting the messages into the target configuration, depending of the side of the socket. For example, the rule moving a message from the list to the side of the socket indicated by `sideA` is specified as follows, where it is important to note that the time of the element being moved has reached 0:

```
rl [receive1] :
   < S : Socket | sideA : O, listA : dl(to O' : TC, O) DML >
   < O : Process | conf : CONF >
=> < S : Socket | listA : DML >
   < O : Process | conf : (to O' : TC CONF) > .
```

In order to simulate errors and disconnections in the peers we have added two attributes to the `Peer` class: `Life` and `Reconnect`, containing values of sort `TimeInf`. Basically, when the `Life` attribute reaches the value 0, it is set to `INF`, the peer cannot receive nor send messages, and the `Reconnect` attribute is set to a random value. Similarly, when `Reconnect` reaches 0, it is set to `INF`, `Life` is set to a random time, and the peer works again.

4.2   Second Architecture Abstraction

Note that the abstraction in the previous section provides an exact correspondence between the distributed system and the centralized one. However, it is possible that most of the properties are either independent of the underlying architecture, independent of the disconnections from the peer, or both. For this reason, we can define more refined abstractions that omit some of these aspects. To abstract the architecture we just use a multiset of peers and messages, so messages sent by a peer reach the addressee immediately; to abstract the connections and disconnections we just remove the `Life` and `Reconnect` attributes introduced above and the associated rules, hence preventing the nodes from unwanted disconnections. In this way we obtain two main advantages: (i) the analysis is optimized, since the number of reachable states is greatly reduced; and (ii) it is easier to understand the results and trace back the causes.

4.3   Formal Abstractions

Beyond simplifying the state with the abstractions above, we can also apply other techniques to improve our proofs. The state space reduction technique in [8] allows us to turn rules (which generates transitions and hence new states during the search and model checking processes) into equations given they fulfill some properties: the specification thus obtained is still a correct executable Maude specification (that is, it is terminating, confluent, and coherent; see [6] for details) and the property is *invisible* for the rules transformed into equations. This invisibility concept informally requires the rules to preserve the satisfiability of the atomic predicates involved in the formulas being proved, and is also the basis for our own abstractions. Another interesting way of reducing the state space can be found in [18].

Regarding infinite systems, an important abstraction can be found in [14]. This abstraction turns an infinite-state system into a finite one by collapsing states by means of equations. This kind of abstraction was not necessary in our case, since our system becomes finite by setting a bound in the execution time.

5   Analysis

We can use now Real-Time Maude in two different ways: to execute the centralized specification and to verify different properties. The former is achieved by using the Maude commands `trew` and `tfrew`, that execute the system (the second one applies the rules in a *fair* way) given a bound in the time; with `find earliest` and `find latest`, that allow the user to check the paths that lead to the first and last (in terms of time) state fulfilling a given property; and with `tsearch`, that checks whether a given state is reachable in the given time. The latter is accomplished by using the `tsearch` command to check that an invariant holds; by looking for the negation of the invariant we can examine whether there is a reachable state that violates it. The specification can also be analyzed by using timed model checking with the command `mc`, that allows the user to state linear temporal logic formulas with a bound in the time.

Note that, before starting the analysis, we need to relate "real-time," as defined by our external Java clock in the distributed specification, and the "real-time" defined by Real-Time Maude. Our distributed specification contains a number of timeouts, defined by natural numbers, and we ask to the Java server to wait this number of seconds. We just mimic this strategy in the centralized specification, using natural numbers (or a constant `INF` standing for infinite time) and we ask Real-Time Maude to wait the maximal possible amount.

We have verified our system with networks form 6 to 20 nodes. We abstract the concrete connections and assume total network connectivity. The life time of each node is randomly chosen, although we use an upper bound life constant to control the ratio of alive nodes. We change the peers that share and search files, as well as the number and time of published and searched files. The analysis of networks with hundreds of nodes using a model checker requires the use of some of the abstraction techniques explained in Sect. 4.3 and it is left as future work.

We can simulate how different attacks may affect a network. For example, in the *node insertion* attack, an attacking peer intercepts a search requests for a file, which are answered with bogus information [15]. The attacking peer creates its own identifier such that it matches the hash value of the file. Then the search requests are routed to the attacking peer, that may return its own file instead of routing the search to the original one. Since the Kademlia network sends the request not only to the closest peer the searcher may find the original file. The `find earliest` command can be used to study different network parameters and check whether this attack is effective. We study if a file may be found in a node that is not the closest one to the file identifier, with the following command:

```
Maude> (find earliest init =>* {< O : Process | conf :
                 (to O' : FILE-FOUND(SENDER, N2) CONF) > CONF'} .)
```

Note that, since the `FILE-FOUND` message returns in its first parameter the peer that is publishing the file, we only need to check whether the peer identifier is the closest to the file identifier.

From the model-checking point of view, there are several properties that can be proved over this protocol. The basic property all P2P networks should fulfill is that if a peer looks for a file that is published somewhere, the peer eventually finds it. We define three propositions (of sort `Prop`, imported from the `TIMED-MODEL-CHECKER` module defined in Real-Time Maude) over the configuration expressing that a peer publishes a file; a peer is looking for that file; and the peer that searches the file finds it. Note that, as in the command above, all the properties are defined taking into account that the configurations are wrapped into objects of class `Process`, that may contain other objects and messages on the `conf` attribute (hence the `CONF` variable used there) and that other processes may also appear in the initial configuration (hence the `CONF'` variable used at the `Process` level):

```
op PublishAFile : Nat -> Prop [ctor] .
eq {< O : Process | conf : (< O' : Peer | Publish :
< I1 & (S1 @ TC4) > # PF > CONF) > CONF'} |= PublishAFile(I1) = true .

op SearchAFile : MyContact Nat -> Prop [ctor] .
eq {< O : Process | conf : (< peer(Searcher) : Peer | SearchFiles :
< I1 & (S1 ; TC3) > # SF > CONF) > CONF'} |=
                              SearchAFile(Searcher,I1) = true .

op FindAFile : MyContact Nat -> Prop [ctor] .
eq {< O : Process | conf : (to peer(Searcher) : FILE-FOUND(I2,I1)
   CONF) > CONF'} |= FindAFile(Searcher,I1) = true .
```

Assuming an initial configuration where a peer publishes the file 54, that is searched by `peer(c(33))`, we can use the following command to check that the property holds:

```
Maude> (mc init' |=t PublishAFile(54) /\ SearchAFile(c(33),54) =>
                    <> FindAFile(c(33),54) in time < 20 .)
Result Bool : true
```

Another basic property is that once a file is published it remains published in some peers unless the publisher is disconnected. We can define the properties `FilePublished`, stating that a peer publishes a file, and `PeerOffline`, indicating that a peer is offline, similarly to the properties above and use the following command to check the property:

```
Maude> (mc init |=t (<> [] (FilePublished(53,c(0))) U PeerOffline(c(0))
         in time < 40 .)
Result ModelCheckResult : counterexample(...)
```

In a network where `peer(c(0))` has published file **53**. Notice that the model checker finds a counterexample. The reason is that all the peers that share the file may be offline at the same time. The property should be reformulated, stating that if the file is published it will always be published again or the publisher will be disconnected:

```
Maude> (mc init |=t ([] <> (FilePublished(53,c(0)) \/ PeerOffline(c(0)))
         in time < 40 .)
Result Bool : true
```

6   Conclusion and Ongoing Work

We have presented in this paper a distributed implementation of the Kademlia protocol in Maude. This distributed system uses sockets to connect different Maude instances and, moreover, to connect each one of these instances to a Java server that takes care of the time. It can be used to share files (only text files in the current specification) using this protocol, allowing peers to connect and disconnect in a dynamic way, adding and searching for new files. Moreover, we also provide a centralized specification of the system, which abstracts most of the details of the underlying architecture to focus on the Kademlia protocol. This centralized specification allows us to simulate and analyze the system using Real-Time Maude to represent the real time implemented in Java in the distributed implementation of the protocol. This centralized implementation of the protocol just mapped real-time to natural numbers. Although this "time sampling" is usual, the relation between physical time and logic time can be refined further. For example, the paper [1], which describes a theory for the orchestration of service-oriented solutions, or [26], which presents a theory for medical devices, provide a much more refined relation, taking into account small deviations due to hardware.

As future work we plan to use the narrowing techniques implemented in Maude [7] to analyze the Kademlia DHT protocol. In this way, we could apply the analyses described in recent works (see e.g. [2,23]) to our system and check whether an error state is reachable from a generic state (a state with variables).

Another line of research is to compare the performance of the routing tables under different parameters, like the bucket dimension or the concurrency parameters used in the node lookup procedure. The comparison can also be done for the different variants of the routing tables taking advantage of the fact that the specification is parametric on the routing table. We also plan to study more complex properties that could apply under other scenarios.

References

1. AlTurki, M., Meseguer, J.: Executable rewriting logic semantics of Orc and formal analysis of Orc programs. J. Logic. Algebraic Meth. Program. **84**(4), 505–533 (2015)
2. Bae, K., Escobar, S., Meseguer, J.: Abstract logical model checking of infinite-state systems using narrowing. In: van Raamsdonk, F. (ed.) 24th International Conference on Rewriting Techniques and Applications, RTA 2013, LIPIcs 21, pp. 81–96. Schloss Dagstuhl-Leibniz-Zentrum fuer Informatik (2013)

3. Bakhshi, R., Gurov, D.: Verification of peer-to-peer algorithms: a case study. In: Combined Proceedings of the 2nd International Workshop on Coordination and Organization, CoOrg 2006, and the Second International Workshop on Methods and Tools for Coordinating Concurrent, Distributed and Mobile Systems, MTCoord 2006, ENTCS, vol. 181, pp. 35–47. Elsevier (2007)

4. Borgström, J., Nestmann, U., Onana, L., Gurov, D.: Verifying a structured peer-to-peer overlay network: the static case. In: Priami, C., Quaglia, P. (eds.) GC 2004. LNCS, vol. 3267, pp. 250–265. Springer, Heidelberg (2005)

5. Breitkreuz, H.: The eMule project. http://www.emule-project.net

6. Clavel, M., Durán, F., Eker, S., Lincoln, P., Martí-Oliet, N., Meseguer, J., Talcott, C.: All About Maude - A High-Performance Logical Framework. LNCS, vol. 4350. Springer, Heidelberg (2007)

7. Clavel, M., Durán, F., Eker, S., Lincoln, P., Martí-Oliet, N., Meseguer, J., Talcott, C.: Maude Manual, version 2.6. http://maude.cs.uiuc.edu/maude2-manual

8. Farzan, A., Meseguer, J.: State space reduction of rewrite theories using invisible transitions. In: Johnson, M., Vene, V. (eds.) AMAST 2006. LNCS, vol. 4019, pp. 142–157. Springer, Heidelberg (2006)

9. Haridi, S.: EU-project PEPITO IST-2001-33234. Project funded by EU IST FET Global Computing (GC) (2002). http://www.sics.se/pepito/

10. Lu, T.: Formal Verification of the Pastry Protocol. Doctoral dissertation, Universität des Saarlandes, December 2013

11. Maymounkov, P., Mazières, D.: Kademlia: a peer-to-peer information system based on the XOR metric. In: Druschel, P., Kaashoek, M.F., Rowstron, A. (eds.) IPTPS 2002. LNCS, vol. 2429, pp. 53–65. Springer, Heidelberg (2002)

12. Lu, T., Merz, S., Weidenbach, C.: Towards verification of the pastry protocol using TLA+. In: Bruni, R., Dingel, J. (eds.) FORTE 2011 and FMOODS 2011. LNCS, vol. 6722, pp. 244–258. Springer, Heidelberg (2011)

13. Meseguer, J.: Conditional rewriting logic as a unified model of concurrency. Theo. Comput. Sci. **96**(1), 73–155 (1992)

14. Meseguer, J., Palomino, M., Martí-Oliet, N.: Equational abstractions. Theo. Comput. Sci. **403**(23), 239–264 (2008)

15. Mysicka, D.: eMule attacks and measurements. Master's thesis, Swiss Federal Institute of Technology (ETH) Zurich (2007)

16. Ölveczky, P., Meseguer, J., Talcott, C.: Specification and analysis of the AER/NCA active network protocol suite in Real-Time Maude. Form. Meth. Syst. Des. **29**, 253–293 (2006)

17. Ölveczky, P.C.: Formal model engineering for embedded systems using Real-Time Maude. In: Durán, F., Rusu, V., (eds.) Proceedings of the 2nd International Workshop on Algebraic Methods in Model-based Software Engineering, AMMSE 2011, EPTCS, vol. 56, pp. 3–13 (2011)

18. Ölveczky, P.C., Meseguer, J.: Abstraction and completeness for Real-Time Maude. In: Proceedings of the 6th International Workshop on Rewriting Logic and its Applications, WRLA 2006, ENTCS, vol. 176(4), pp. 5–27 (2007)

19. Ölveczky, P.C., Meseguer, J.: Semantics and pragmatics of Real-Time Maude. High. Ord. Symbolic Comput. **20**, 161–196 (2007)

20. Pita, I., Fernández-Camacho, M.I.: Formal specification of the Kademlia and the Kad routing tables in Maude. In: Martí-Oliet, N., Palomino, M. (eds.) WADT 2012. LNCS, vol. 7841, pp. 231–247. Springer, Heidelberg (2013)

21. Ratnasamy, S., Francis, P., Handley, M., Karp, R., Shenker, S.: A scalable content-addressable network. In: ACM SIGCOMM Computer Communication Review - Proceedings of the 2001 SIGCOMM Conference, vol. 31, pp. 161–172, October 2001

22. Riesco, A., Verdejo, A.: Implementing and analyzing in Maude the enhanced interior gateway routing protocol. In: Roşu, G. (ed.) Proceedings of the 7th International Workshop on Rewriting Logic and its Applications, WRLA 2008. ENTCS, vol. 238(3), pp. 249–266. Elsevier (2009)

23. Rocha, C., Meseguer, J., Muñoz, C.: Rewriting modulo SMT and open system analysis. In: Escobar, S. (ed.) WRLA 2014. LNCS, vol. 8663, pp. 247–262. Springer, Heidelberg (2014)

24. Rowstron, A., Druschel, P.: Pastry: scalable, decentralized object location, and routing for large-scale peer-to-peer systems. In: Guerraoui, R. (ed.) Middleware 2001. LNCS, vol. 2218, pp. 329–350. Springer, Heidelberg (2001)

25. Stoica, I., Morris, R., Karger, D., Kaashoek, M.F., Balakrishnan, H.: Chord: a scalable peer-to-peer lookup service for internet applications. ACM SIGCOMM Comput. Commun. Rev. **31**, 149–160 (2001)

26. Sun, M., Meseguer, J.: Distributed real-time emulation of formally-defined patterns for safe medical device control. In: Ölveczky, P.C. (ed.) Proceedings of the 1st International Workshop on Rewriting Techniques for Real-Time Systems, RTRTS 2010, EPTCS, vol. 36, pp. 158–177 (2010)

27. Zave, P.: Using lightweight modeling to understand Chord. SIGCOMM Comput. Commun. Rev. **42**(2), 49–57 (2012)

Enforcement of (Timed) Properties
with Uncontrollable Events

Matthieu Renard[1], Yliès Falcone[2]([⊠]), Antoine Rollet[1], Srinivas Pinisetty[3],
Thierry Jéron[4], and Hervé Marchand[4]

[1] LaBRI, Université Bordeaux, Bordeaux, France
{matthieu.renard,antoine.rollet}@labri.fr
[2] Université Grenoble Alpes, Inria, LIG, F-38000 Grenoble, France
ylies.falcone@imag.fr
[3] Aalto University, Espoo, Finland
srinivas.pinisetty@aalto.fi
[4] Inria Rennes Bretagne-Atlantique, Rennes, France
{thierry.jeron,herve.marchand}@inria.fr

Abstract. This paper deals with runtime enforcement of untimed and timed properties with uncontrollable events. Runtime enforcement consists in modifying the executions of a running system to ensure their correctness with respect to a desired property. We introduce a framework that takes as input any regular (timed) property over an alphabet of events, with some of these events being uncontrollable. An uncontrollable event cannot be delayed nor intercepted by an enforcement mechanism. Enforcement mechanisms satisfy important properties, namely soundness and compliance - meaning that enforcement mechanisms output correct executions that are close to the input execution. We discuss the conditions for a property to be enforceable with uncontrollable events, and we define enforcement mechanisms that modify executions to obtain a correct output, as soon as possible. Moreover, we synthesise sound and compliant descriptions of runtime enforcement mechanisms at two levels of abstraction to facilitate their design and implementation.

1 Introduction

Verifying a user-provided specification at runtime consists in running a mechanism that assigns verdicts to a sequence of events produced by an instrumented system w.r.t. a property formalizing the specification. This paper focuses on *runtime enforcement* (cf. [3,12,13,15,21]) which goes beyond pure verification at runtime and studies how to react to a violation of specifications. In runtime enforcement, an enforcement mechanism (EM) takes a (possibly incorrect) execution sequence as input, and outputs a new sequence. Enforcement mechanisms should be *sound* and *transparent*, meaning that the output should satisfy the property under consideration and should be as close as possible to the input. When dealing with timed

© Springer International Publishing Switzerland 2015
M. Leucker et al. (Eds.): ICTAC 2015, LNCS 9399, pp. 542–560, 2015.
DOI: 10.1007/978-3-319-25150-9_31

properties, EMs can act as *delayers* over the input sequence of events [17–19]. That is, EMs buffer input events for some time and then release them in such a way that the output sequence of events satisfies the property.

Motivations. In this paper, we focus on online enforcement of properties with uncontrollable events. Introducing uncontrollable events is a step towards more realistic runtime enforcement. As a matter of fact, uncontrollable events naturally occur in many applications scenarios where the EM has no control over certain input events. For instance, certain events from the environment may be out of the scope of the mechanism at hand. This situation arises for instance in avionic systems where a command of the pilot has consequences on a specific component. In this domain, it is usual to add control mechanisms to check the validity of an event on particular points according to observations. For instance, the "spoiler activation"[1] command decided by the pilot is sent by the panel to a control flight system, and this leads finally to a specific event on the spoilers. Placing an EM directly on the spoiler permits to avoid incoherent events, according to the pilot commands (which are events out of the scope of the EM). In the timed setting, uncontrollable events may be urgent messages that cannot be delayed by an enforcement mechanism. Similarly, when a data-dependency exists between two events (e.g., between a *write* event that displays a value obtained from a previous *read* event), the first *read* event is somehow uncontrollable as it cannot be delayed by the enforcement mechanism without preventing the *write* event to occur in the monitored program.

Challenges. Considering uncontrollable events in the timed setting induces new challenges. Indeed, enforcement mechanisms may now receive events that cannot be buffered and have to be released immediately in output. Since they influence the satisfaction of the property under scrutiny, delays of controllable events stored in memory have to be recomputed upon each uncontrollable event. Moreover, the occurrence of such events has to be anticipated, meaning that all possible sequences of uncontrollable events have to be considered by the enforcement mechanism. Thus, new enforcement strategies are necessary for both untimed and timed properties.

Contributions. We introduce a framework for runtime enforcement of regular untimed and timed properties with uncontrollable events. It turns out that the usual notion of transparency has to be weakened. As we shall see, the initial order between uncontrollable and controllable events can change in output, contrary to what is prescribed by transparency. Thus, we propose to replace transparency with a new notion, namely *compliance*, ensuring that the order of controllable events is maintained while uncontrollable events are output as soon as they are received. We define a property to be enforceable with uncontrollable events when it is possible to obtain a sound and compliant enforcement mechanism for any input sequence. It turns out that a property may not be enforceable

---

[1] The spoiler is a device used to reduce the lift of an aircraft.

because of certain input sequences. Intuitively, enforceability issues arise because some sequences of uncontrollable events that lead the property to be violated cannot be avoided. We give a condition, represented by a property, that indicates whether soundness is guaranteed by the enforcement monitor or not, depending on the input given so far. We describe enforcement mechanisms at two levels of abstraction. The synthesised enforcement mechanisms are sound and compliant whenever the previously mentioned condition holds.

Outline. Section 2 introduces preliminaries and notations. Sections 3 and 4 present the enforcement framework with uncontrollable events in the untimed and timed settings, respectively, where enforcement mechanisms are defined at two levels of abstraction. Section 5 discusses related work. Section 6 presents conclusions and perspectives.

2   Preliminaries and Notation

Untimed notions. An *alphabet* is a finite set of symbols. A *word* over an alphabet Σ is a sequence over Σ. The set of finite words over Σ is denoted Σ^*. The *length* of a finite word w is noted $|w|$, and the *empty word* is noted ϵ. Σ^+ stands for $\Sigma^* \setminus \{\epsilon\}$. A *language* over Σ is any subset $\mathcal{L} \subseteq \Sigma^*$. The concatenation of two words w and w' is noted $w.w'$ (the dot could sometimes be omitted). A word w' is a *prefix* of a word w, noted $w' \preccurlyeq w$ if there exists a word w'' such that $w = w'.w''$. The word w'' is called the *residual* of w after reading the prefix w', noted $w'' = w'^{-1}.w$. The word w'' is then called a *suffix* of w. Note that $w'.w'' = w'.w'^{-1}.w = w$. These standard definitions are extended to languages in the natural way. Given a word w and an integer i such that $1 \leq i \leq |w|$, we note $w(i)$ the i-th element of w. Given a tuple $e = (e_1, e_2, \ldots, e_n)$ of size n, for an integer i such that $1 \leq i \leq n$, we note Π_i the projection on the i-th coordinate, i.e. $\Pi_i(e) = e_i$. Given a word $w \in \Sigma^*$ and $\Sigma' \subseteq \Sigma$, we define the *restriction* of w to Σ', noted $w_{|\Sigma'}$, as the word $w' \in \Sigma'$ whose letters are the letters of w belonging to Σ' in the same order. Formally, $\epsilon_{|\Sigma'} = \epsilon$ and $\forall \sigma \in \Sigma^*, \forall a \in \Sigma$, $(w.a)_{|\Sigma'} = w_{|\Sigma'}.a$ if $a \in \Sigma'$, and $(w.a)_{|\Sigma'} = w_{|\Sigma'}$ otherwise.

Automata. An *automaton* is a tuple $\langle Q, q_0, \Sigma, \rightarrow, F \rangle$, where Q is the set of states, called *locations*, $q_0 \in Q$ is the initial location, Σ is the alphabet, $\rightarrow \subseteq Q \times \Sigma \times Q$ is the transition relation and $F \subseteq Q$ is the set of accepting locations. Any location in F is called accepting. Whenever there exists $(q, a, q') \in \rightarrow$, we note it $q \xrightarrow{a} q'$. Relation $\rightarrow$ is extended to words $\sigma \in \Sigma^*$ by noting $q \xrightarrow{\sigma.a} q'$ whenever there exists q'' such that $q \xrightarrow{\sigma} q''$ and $q'' \xrightarrow{a} q'$. Moreover, for $q \in Q$, $q \xrightarrow{\epsilon} q$ always holds. An automaton $\mathcal{A} = \langle Q, q_0, \Sigma, \rightarrow, F \rangle$ is *deterministic* if $\forall q \in Q, \forall a \in \Sigma, (q \xrightarrow{a} q' \wedge q \xrightarrow{a} q'') \implies q' = q''$. $\mathcal{A}$ is *complete* if $\forall q \in Q, \forall a \in \Sigma, \exists q' \in Q, q \xrightarrow{a} q'$. A word w is *accepted* by $\mathcal{A}$ if there exists $q \in F$ such that $q_0 \xrightarrow{w} q$. The language (i.e. set of all words) accepted by $\mathcal{A}$ is noted $\mathcal{L}(\mathcal{A})$. A *property* is a language over an alphabet Σ. A regular property is a language accepted by an automaton. In the sequel, we shall assume that a property φ is represented by a deterministic and complete automaton $\mathcal{A}_\varphi$.

Timed languages. Let $\mathbb{R}_{\geq 0}$ be the set of non-negative real numbers, and Σ a finite alphabet of actions. An event is a pair $(t, a) \in \mathbb{R}_{\geq 0} \times \Sigma$. We define $\text{date}(t, a) = t$ and $\text{act}(t, a) = a$ the projections of events on dates and actions respectively. A *timed word* over Σ is a word over $\mathbb{R}_{\geq 0} \times \Sigma$ whose real parts are ascending, i.e. σ is a timed word if $\sigma \in (\mathbb{R}_{\geq 0} \times \Sigma)^*$ and $\forall i \in [1; |\sigma| - 1], \text{date}(w(i)) \leq \text{date}(w(i+1))$. $\text{tw}(\Sigma)$ denotes the set of timed words over Σ. For a timed word $\sigma = (t_1, a_1).(t_2, a_2) \ldots (t_n, a_n)$ and an integer i such that $1 \leq i \leq |\sigma|$, t_i is the time elapsed before action a_i occurs. We naturally extend the notions of *prefix* and *residual* to timed words. We note $\text{time}(\sigma) = \text{date}(\sigma(|\sigma|))$, and define the *observation* of σ at time t as the timed word $\text{obs}(\sigma, t) = \max_{\preccurlyeq}(\{\sigma' \mid \sigma' \preccurlyeq \sigma \wedge \text{time}(\sigma') \leq t\})$. We also define the remainder of the observation of σ at time t as $\text{nobs}(\sigma, t) = (\text{obs}(\sigma, t))^{-1}.\sigma$. The *untimed projection* of σ is $\Pi_\Sigma(\sigma) = a_1.a_2 \ldots a_n$, it is the sequence of actions of σ with dates ignored. σ *delayed by* $t \in \mathbb{R}_{\geq 0}$ is the word noted $\sigma +_t t$ such that t is added to all dates: $\sigma +_t t = (t_1 + t, a_1).(t_2 + t, a_2) \ldots (t_{|\sigma|} + t, a_{|\sigma|})$. We also extend the definition of the restriction of σ to $\Sigma' \subseteq \Sigma$ to timed words, such that $\epsilon_{|\Sigma'} = \epsilon$, and for $\sigma \in \text{tw}(\Sigma)$ and (t, a) such that $\sigma.(t, a) \in \text{tw}(\Sigma)$, $(\sigma.(t, a))_{|\Sigma'} = \sigma_{|\Sigma'}.(t, a)$ if $a \in \Sigma'$, and $(\sigma.(t, a))_{|\Sigma'} = \sigma_{|\Sigma'}$ otherwise. A *timed language* is any subset of $\text{tw}(\Sigma)$. Moreover, we define an order on timed words: we say that σ' *delays* σ, noted $\sigma \preccurlyeq_d \sigma'$, whenever $\Pi_\Sigma(\sigma') \preccurlyeq \Pi_\Sigma(\sigma)$ and $\forall i \in [1; |\sigma| - 1], \text{date}(\sigma(i)) \leq \text{date}(\sigma'(i))$. Note that the order is not the same in the different constraints: σ' is a prefix of σ, but dates in σ' exceed dates in σ. We also define a *lexical order* $\preccurlyeq_{\text{lex}}$ on timed words with identical untimed projections, such that $\epsilon \preccurlyeq_{\text{lex}} \epsilon$, and for two words σ and σ' such that $\Pi_\Sigma(\sigma) = \Pi_\Sigma(\sigma')$, and two events (t, a) and (t', a), $(t', a).\sigma' \preccurlyeq_{\text{lex}} (t, a).\sigma$ if $t' < t \vee (t = t' \wedge \sigma' \preccurlyeq_{\text{lex}} \sigma)$.

Consider for example the timed word $\sigma = (1, a).(2, b).(3, c).(4, a)$ over the alphabet $\Sigma = \{a, b, c\}$. Then, $\Pi_\Sigma(\sigma) = a.b.c.a$, $\text{obs}(\sigma, 3) = (1, a).(2, b).(3, c)$, $\text{nobs}(\sigma, 3) = (4, a)$, and if $\Sigma' = \{b, c\}$, $\sigma_{|\Sigma'} = (2, b).(3, c)$, and for instance $\sigma \preccurlyeq_d (1, a).(2, b).(4, c)$, and $\sigma \preccurlyeq_{\text{lex}} (1, a).(3, b).(3, c).(3, a)$.

Timed automata. Let $X = \{X_1, X_2, \ldots, X_n\}$ be a finite set of *clocks*. A *clock valuation* is a function ν from X to $\mathbb{R}_{\geq 0}$. The set of clock valuations for the set of clocks X is noted $\mathcal{V}(X)$, i.e., $\mathcal{V}(X) = \{\nu \mid \nu : X \to \mathbb{R}_{\geq 0}\}$. We consider the following operations on valuations: for any valuation ν, $\nu + \delta$ is the valuation assigning $\nu(X_i) + \delta$ to every clock $X_i \in X$; for any subset $X' \subseteq X$, $\nu[X' \leftarrow 0]$ is the valuation assigning 0 to each clock in X', and $\nu(X_i)$ to any other clock X_i not in X'. $\mathcal{G}(X)$ denotes the set of guards consisting of boolean combinations of simple constraints of the form $X_i \bowtie c$ with $X_i \in X$, $c \in \mathbb{N}$, and $\bowtie \in \{<, \leq, =, \geq, >\}$. Given $g \in \mathcal{G}(X)$ and a valuation ν, we write $\nu \models g$ when for every simple constraint $X_i \bowtie c$ in g, $\nu(X_i) \bowtie c \equiv \text{true}$.

Definition 1 (Timed Automaton [1]). *A timed automaton (TA) is a tuple $\mathcal{A} = \langle L, l_0, X, \Sigma, \Delta, G \rangle$, such that L is a set of locations, $l_0 \in L$ is the initial location, X is a set of clocks, Σ is a finite set of events, $\Delta \subseteq L \times \mathcal{G}(X) \times \Sigma \times 2^X \times L$ is the transition relation, and $G \subseteq L$ is a set of accepting locations. A transition $(l, g, a, X', l') \in \Delta$ is a transition from l to l', labelled with event a, with guard defined by g, and with the clocks in X' to be reset.*

The semantics of a timed automaton $\mathcal{A}$ *is a timed transition system* $[\![\mathcal{A}]\!] = \langle Q, q_0, \Gamma, \to, F_G \rangle$ *where* $Q = L \times \mathcal{V}(X)$ *is the (infinite) set of states,* $q_0 = (l_0, \nu_0)$ *is the initial state, with* $\nu_0 = \nu[X \leftarrow 0]$, $F_G = G \times \mathcal{V}(X)$ *is the set of accepting states,* $\Gamma = \mathbb{R}_{\geq 0} \times \Sigma$ *is the set of transition labels, each one composed of a delay and an action. The transition relation* $\to \subseteq Q \times \Gamma \times Q$ *is a set of transitions of the form* $(l, \nu) \xrightarrow{(\delta, a)} (l', \nu')$ *with* $\nu' = (\nu + \delta)[Y \leftarrow 0]$ *whenever there is a transition* $(l, g, a, Y, l') \in \Delta$ *such that* $\nu + \delta \models g$, *for* $\delta \geq 0$.

A timed automaton $\mathcal{A} = \langle L, l_0, X, \Sigma, \Delta, G \rangle$ is *deterministic* if for any (l, g_1, a, Y_1, l_1') and (l, g_2, a, Y_2, l_2') in Δ, $g_1 \wedge g_2$ is unsatisfiable, meaning that only one transition can be fired at any time. $\mathcal{A}$ is *complete* if for any $l \in L$ and any $a \in \Sigma$, the disjunction of the guards of all the transitions leaving l and labeled by a is valid (i.e., it evaluates to true for any clock valuation).

A *run* ρ from $q \in Q$ is a valid sequence of transitions in $[\![\mathcal{A}]\!]$ starting from q, of the form $\rho = q \xrightarrow{(\delta_1, a_1)} q_1 \xrightarrow{(\delta_2, a_2)} q_2 \ldots \xrightarrow{(\delta_n, a_n)} q_n$. The set of runs from q_0 is noted $\mathrm{Run}(\mathcal{A})$ and $\mathrm{Run}_{F_G}(\mathcal{A})$ denotes the subset of runs accepted by $\mathcal{A}$, i.e. ending in a state in F_G. The *trace* of the run ρ previously defined is the timed word $(t_1, a_1).(t_2, a_2) \ldots (t_n, a_n)$, with, for $1 \leq i \leq n$, $t_i = \sum_{k=1}^{i} \delta_k$. Thus, given the trace $\sigma = (t_1, a_1).(t_2, a_2) \ldots (t_n, a_n)$ of a run ρ from a state $q \in Q$ to $q' \in Q$, we can define $w = (\delta_1, a_1).(\delta_2, a_2) \ldots (\delta_n, a_n)$, with $\delta_1 = t_1$, and $\forall i \in [2; n], \delta_i = t_i - t_{i-1}$, and then $q \xrightarrow{w} q'$. To ease the notation, we will only consider traces and note $q \xrightarrow{\sigma} q'$ whenever $q \xrightarrow{w} q'$ for the previously defined w. Note that to concatenate two traces σ_1 and σ_2, it is needed to delay σ_2: the concatenation σ of σ_1 and σ_2 is the trace defined as $\sigma = \sigma_1.(\sigma_2 +_t \mathrm{time}(\sigma_1))$. Thus, if $q \xrightarrow{\sigma_1} q' \xrightarrow{\sigma_2} q''$, then $q \xrightarrow{\sigma} q''$.

Timed properties. A *regular timed property* is a timed language $\varphi \subseteq \mathrm{tw}(\Sigma)$ that is accepted by a timed automaton. For a timed word σ, we say that σ *satisfies* φ, noted $\sigma \models \varphi$ whenever $\sigma \in \varphi$. A *regular timed property* is a timed language accepted by a timed automaton. We only consider deterministic and complete regular timed properties.

Given an automaton $\mathcal{A}$ such that Q is the set of states of $[\![\mathcal{A}]\!]$ and $\to$ its transition relation, and a word σ, we note $q \text{ after } \sigma = \{q' \in Q \mid q \xrightarrow{\sigma} q'\}$ for $q \in Q$. We note $\mathrm{Reach}(\sigma) = q_0 \text{ after } \sigma$. These definitions are valid in both the untimed and timed cases. We extend these definitions to languages: if L is a language, $q \text{ after } L = \bigcup_{\sigma \in L} q \text{ after } \sigma$ and $\mathrm{Reach}(L) = q_0 \text{ after } L$.

3   Enforcement Monitoring of Untimed Properties

In this section, φ is a regular property defined by a complete and deterministic automaton $\mathcal{A}_\varphi = \langle Q, q_0, \Sigma, \to, F \rangle$. The purpose of an *enforcement mechanism* (EM) for φ is to modify the executions of a running system, represented by words so as to satisfy φ. It takes as input a word, representing an execution, and outputs a word, i.e. an execution. We consider uncontrollable events in the set $\Sigma_u \subseteq \Sigma$. These events cannot be modified by the EM, i.e. they cannot be suppressed nor buffered, so they must be emitted by the EM whenever they are received. Let us note $\Sigma_c = \Sigma \setminus \Sigma_u$ the set of controllable events, which are on the scope of the EM. The EM can decide to buffer them to delay their emission, but it cannot suppress them (nevertheless, it can delay them endlessly, keeping their order unchanged). Thus, the EM may interleave controllable and uncontrollable events.

3.1 Enforcement Functions and their Requirements

An enforcement function is a description of the input/output behaviour of an EM. Formally, we define *enforcement functions* as follows:

Definition 2 (Enforcement Function). *Given an alphabet of actions Σ, an enforcement function is a function $E : \Sigma^* \to \Sigma^*$, i.e. a function that modifies an execution.*

As stated previously, the usual purpose of an EM is to ensure that the executions of a running system satisfy a property, thus its enforcement function has to be *sound*, meaning that its output always satisfies the property:

Definition 3 (Soundness). *E is sound with respect to φ if $\forall \sigma \in \Sigma^*, E(\sigma) \models \varphi$.*

The usual notion of *transparency* in enforcement monitoring [15,21] states that the output of an enforcement function is the longest prefix of the input satisfying the property. The name "transparency" stems from the fact that correct executions are left unchanged. However, because of uncontrollable events, events may be released in a different order from the one they are received. Therefore, transparency can not be ensured, and we define the weaker notion of *compliance*.

Definition 4 (Compliance). *E is compliant w.r.t. Σ_u and Σ_c, noted compliant(E, Σ_c, Σ_u), if $\forall \sigma \in \Sigma^*, E(\sigma)_{|\Sigma_c} \preccurlyeq \sigma_{|\Sigma_c} \wedge \forall u \in \Sigma_u, E(\sigma).u \preccurlyeq E(\sigma.u)$.*

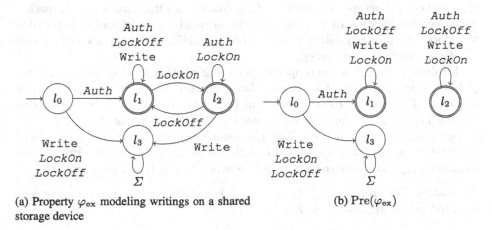

(a) Property φ_{ex} modeling writings on a shared storage device

(b) Pre(φ_{ex})

Fig. 1. A property and its corresponding precondition property of enforceability.

Intuitively, compliance states that the EM does not change the order of the controllable events and emits uncontrollable events immediately upon their reception, possibly followed by stored controllable events. When clear from the context, the partition is not mentioned: E is said to be compliant.

We say that a property is *enforceable* whenever there exists a compliant function that is sound with respect to that property.

Example 1. Figure 1a depicts property φ_{ex} which states that writing to a shared storage device should be authenticated and is authorized only when the device is not locked. φ_{ex} is not enforceable if the uncontrollable alphabet is $\{LockOn, LockOff\}^2$ since reading the word $LockOn$ from l_0 leads to l_3, which is not an accepting location. However, the existence of such a word does not imply that it is impossible to enforce this property for some other input words. If word $Auth$ is read, then location l_1 is reached, and from this location, it is possible to enforce φ_{ex} by emitting $\mathtt{Write}$ only when in location l_1.

3.2   Synthesising Enforcement Functions

Example 1 shows that some input words cannot be corrected by the EM, because of uncontrollable events. This leads us to define another property that captures the sequences that can be input to an EM while ensuring soundness.

Definition 5 (Pre, Enf)

- $\mathrm{Pre}(\varphi) = \langle Q, q_0, \Sigma, \to', Q_{enf} \rangle$, *with*

$$\to' = (\to \cap\, Q_{\overline{enf}} \times \Sigma \times Q) \cup \{(q, a, q) \mid q \in Q_{enf} \wedge a \in \Sigma\}$$

- $\mathrm{Enf}(\varphi) = \langle Q, q_0, \Sigma, \to, Q_{enf} \rangle$.

where $Q_{enf} = \{q \in F \mid (q\ \mathrm{after}\ \Sigma_u^*) \subseteq F\}$, $Q_{\overline{enf}} = Q \setminus Q_{enf}$.

Q_{enf} is the set of accepting locations of $\mathcal{A}_\varphi$ from which it is impossible to reach a non-accepting location by reading only uncontrollable events, and thus possible to enforce the property (since it is possible to indefinitely delay all controllable events to ensure the property).

Intuitively, $\mathrm{Pre}(\varphi)$ is the property specifying whether it is possible to enforce φ from a location that has already been reached by triggering the output sequence of the enforcement mechanism (i.e. a location reached by a prefix of the output) or not. Thus, it can be used to know if soundness is guaranteed or not (i.e. if a location from Q_{enf} has been reached). Since the enforcement mechanism ensures that soundness is satisfied as soon as possible, $\mathrm{Pre}(\varphi)$ is a co-safety property, because once Q_{enf} is reached, φ can be ensured from then.

Example 2. For property φ_{ex}, $Q_{enf} = \{l_1, l_2\}$, and $\mathrm{Pre}(\varphi_{ex})$ is the property represented by the automaton in Fig. 1b.

---

[2] Uncontrollable events are emphasised in italics.

Since it is not possible to enforce φ from locations in $Q \setminus Q_{enf}$, (because uncontrollable events could lead to a location in $Q \setminus F$ trapped with uncontrollable events), an enforcement function should try to always be able to reach locations in Q_{enf} to ensure soundness. Thus, property $\mathrm{Enf}(\varphi)$ holds on a sequence whenever Q_{enf} is reached in $\mathcal{A}_\varphi$ with this sequence. Since $Q_{enf} \subseteq F$, satisfying $\mathrm{Enf}(\varphi)$ is sufficient to satisfy φ. Thus, we shall enforce $\mathrm{Enf}(\varphi)$.

Based on the above definition and the enforcement limitation illustrated in Example 1, we synthesise an enforcement function for φ that is compliant, and sound w.r.t. a property that is as close as possible to φ (see later Propositions 1 and 2).

Definition 6 (store$_\varphi$, E_φ).[3] *Function* store$_\varphi$: $\Sigma^* \to \Sigma^* \times \Sigma^*$ *is defined as follows:*

- store$_\varphi(\epsilon) = (\epsilon, \epsilon)$;
- *for* $\sigma \in \Sigma^*$ *and* $a \in \Sigma$, *let* $(\sigma_s, \sigma_c) = $ store$_\varphi(\sigma)$, *then:*

$$\text{store}_\varphi(\sigma.a) = \begin{cases} (\sigma_s.a.\sigma_s', \sigma_c') & \text{if } a \in \Sigma_u \\ (\sigma_s.\sigma_s'', \sigma_c'') & \text{if } a \in \Sigma_c \end{cases}, \text{ where:}$$

$$\sigma_s' = \max_{\preccurlyeq}(\{w \preccurlyeq \sigma_c \mid \sigma_s.a.w \models \mathrm{Enf}(\varphi)\}),$$
$$\sigma_c' = \sigma_s'^{-1}.\sigma_c,$$
$$\sigma_s'' = \begin{cases} \epsilon & \text{if } \sigma_s.\sigma_c.a \not\models \mathrm{Enf}(\varphi), \\ \sigma_c.a & \text{otherwise,} \end{cases}$$
$$\sigma_c'' = \sigma_s''^{-1}.(\sigma_c.a).$$

The enforcement function E_φ : $\Sigma^* \to \Sigma^*$ *is s.t. for* $\sigma \in \Sigma^*$, $\mathrm{E}_\varphi(\sigma) = \Pi_1(\text{store}_\varphi(\sigma))$.

Intuitively, σ_s is the word that can be released as output, whereas σ_c is the buffer containing the events that are already read/received, but cannot be released as output yet because they lead to an unsafe location from which it would be possible to violate the property reading only uncontrollable events. If σ_s satisfies $\mathrm{Pre}(\varphi)$, then the output will always satisfy the property afterwards.

Upon receiving a new action a, it is output if it belongs to Σ_u, followed by the longest prefix of σ_c that leads to Q_{enf}. If the a is controllable, $\sigma_c.a$ is output if it leads to Q_{enf}, else a is added to the buffer. Property $\mathrm{Enf}(\varphi)$ is used instead of φ to ensure that the output of the enforcement function always leads to locations in Q_{enf}, so that the property will still be satisfied (if it was) upon receiving uncontrollable events.

Enforcement functions as per Definition 6 are sound and compliant.

Proposition 1. E_φ *is sound with respect to* $\mathrm{Pre}(\varphi) \implies \varphi$, *as per Definition 3.*

Proposition 2. E_φ *is compliant, as per Definition 4.*

Notice that for some properties, blocking all controllable events may still satisfy soundness and compliance. However, for any given input σ, $\mathrm{E}_\varphi(\sigma)$ is the longest possible word that ensures to reach Q_{enf}. Controllable events are blocked only when it is not certain that Q_{enf} will be reached.

---

[3] E_φ and store$_\varphi$ depend on Σ_u and Σ_c, but we did not add them in order to lighten the notations.

3.3 Enforcement Monitors

Enforcement monitors are operational descriptions of enforcement mechanisms. Here, we give a representation of the previous enforcement function as a transition system whose output should be exactly the output of the enforcement function defined in Sects. 3.1 and 3.2. The purpose is to ease the implementation, since this representation is closer to the real behaviour of a monitor.

Definition 7 (Enforcement Monitor). *An enforcement monitor $\mathcal{E}$ for φ is a transition system $\langle C^{\mathcal{E}}, c_0^{\mathcal{E}}, \Gamma^{\mathcal{E}}, \hookrightarrow_{\mathcal{E}} \rangle$ such that:*

- $C^{\mathcal{E}} = Q \times \Sigma^*$ *is the set of configurations.*
- $c_0^{\mathcal{E}} = \langle q_0, \epsilon \rangle$ *is the initial configuration.*
- $\Gamma^{\mathcal{E}} = \Sigma^* \times \{\text{dump}(.), \text{pass-uncont}(.), \text{store-cont}(.)\} \times \Sigma^*$ *is the alphabet, where the first, second, and third members are an input, an operation, and an output, respectively.*
- $\hookrightarrow_{\mathcal{E}} \subseteq C^{\mathcal{E}} \times \Gamma^{\mathcal{E}} \times C^{\mathcal{E}}$ *is the transition relation, defined as the smallest relation obtained by applying the following rules in order (where $w/ \bowtie /w'$ stands for $(w, \bowtie, w') \in \Gamma^{\mathcal{E}}$):*

  - **dump:** $\langle q, \sigma.\sigma_c \rangle \xrightarrow{\epsilon/ \text{dump}(\sigma)/\sigma} \langle q', \sigma_c \rangle$, *with* $q \xrightarrow{\sigma} q'$, *and* $q' \in Q_{\text{enf}}$,
  - **pass-uncont:** $\langle q, \sigma_c \rangle \xrightarrow{a/ \text{pass-uncont}(a)/a} \langle q', \sigma_c \rangle$, *with* $a \in \Sigma_u$ *and* $q \xrightarrow{a} q'$,
  - **store-cont:** $\langle q, \sigma_c \rangle \xrightarrow{a/ \text{store-cont}(a)/\epsilon} \langle q, \sigma_c.a \rangle$.

Rule **dump** outputs a prefix of the word in memory (the buffer) whenever it is possible to ensure soundness afterwards. Rule **pass-uncont** releases an uncontrollable event as soon as it is received. Rule **store-cont** simply adds a controllable event at the end of the buffer. Compared to Sect. 3.2, the second member of the configuration represents buffer σ_c in the definition of store$_\varphi$, whereas σ_s is here represented by location q which is the first member of the configuration, such that $q = \text{Reach}(\sigma_s)$.

Proposition 3. *The output of the enforcement monitor $\mathcal{E}$ for input σ is $\mathrm{E}_\varphi(\sigma)$.*

Remark 1. Enforcement monitors as per Definition 7 are somewhat similar to the ones in [13], except that we choose to explicitly keep the memory as part of the configuration and get uniform definitions in the untimed and timed settings (see Sect. 4). Hence, enforcement monitors as per Definition 7 can also equivalently be defined using a finite-state machine, extending the definition in [13].

4 Enforcement Monitoring of Timed Properties

In this section, we extend the framework presented in Sect. 3 to enforce timed properties. Enforcement mechanisms and their properties should be redefined. Enforcement functions need an extra parameter representing the date at which the output is observed. Soundness has to be adapted because, at any time instant, one has to allow the property not to hold, provided that it will hold in the future.

Considering uncontrollable events with timed properties raises several difficulties. First, the order of the events might be modified. Thus, previous definitions of transparency [19], stating that the output of an enforcement function will eventually be a delayed prefix of the input, can not be used in this situation. Moreover, when delaying some events to have the property satisfied in the future, one must consider the fact that some uncontrollable events could occur at any moment (and cannot be delayed). Finally, some properties enforceable in [18] cannot be enforced using uncontrollable events, meaning that it is impossible to ensure the soundness of our enforcement mechanisms, as shown in Example 3. It could be possible to use the same definition of soundness as in Sect. 3, where the output always satisfies the property, but then soundness would have been ensured for less properties (i.e. only for safety properties). Weakening soundness allows to enforce more properties, and to let enforcement mechanisms produce longer outputs.

In this section, φ is a timed property defined by a deterministic and complete timed automaton $\mathcal{A}_\varphi = \langle L, l_0, X, \Sigma, \Delta, G \rangle$ with semantics $[\![\mathcal{A}_\varphi]\!] = \langle Q, q_0, \Gamma, \rightarrow, F_G \rangle$.

4.1   Enforcement Functions and their Properties

An enforcement function takes a timed word and the current time as input, and outputs a timed word:

Definition 8 (Enforcement Function). *An enforcement function is a function from* $\mathrm{tw}(\Sigma) \times \mathbb{R}_{\geq 0}$ *to* $\mathrm{tw}(\Sigma)$.

As for the untimed case, we define the notions of *soundness* and *compliance*.

Definition 9 (Soundness). *An enforcement function E is* sound *w.r.t. φ if* $\forall \sigma \in \mathrm{tw}(\Sigma), \forall t \in \mathbb{R}_{\geq 0}, \exists t' \geq t, E(\sigma, t') \models \varphi$.

Definition 10 (Compliance). *An enforcement function E is* compliant *if* $\forall \sigma \in \mathrm{tw}(\Sigma), \forall t \in \mathbb{R}_{\geq 0}, \exists t' \geq t, E(\sigma, t')_{|\Sigma_u} = \sigma_{|\Sigma_u} \wedge \sigma_{|\Sigma_c} \preccurlyeq_d E(\sigma, t')_{|\Sigma_c}$.

An enforcement function is sound if for any input timed word, at any time instant, the value of the enforcement function satisfies the property in the future. Compliance is similar to the untimed setting but there are noteworthy differences. First, controllable events can be delayed. Moreover, since timing information is attached to events, it is not necessary to consider an event of Σ_u. Indeed, the dates of uncontrollable events are the same in the input and in the output, meaning that they are emitted immediately upon their reception. Compliance states that controllable events can be delayed, but their order must be preserved by the enforcement mechanism (i.e. when considering the projections on controllable events, the output should be a delayed prefix of the input). Regarding uncontrollable events, any uncontrollable event is released immediately when received (i.e. when considering the projections on uncontrollable events, the output should be equal to the input).

As in the untimed setting, we say that a property is *enforceable* whenever there exists a sound and compliant enforcement function for this property.

4.2   Synthesising an Enforcement Function

Example 3 (Non enforceable property). Consider the property defined by the automaton in Fig. 2 with alphabet $\Sigma = \{a, b\}$. If all actions are controllable $(\Sigma_u = \emptyset)$, the property is enforceable because one needs to delay events until clock x exceeds 2. Otherwise, the property is not enforceable. For instance, if $\Sigma_u = \{a\}$, word $(1, a)$ cannot be corrected.

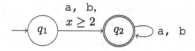

Fig. 2. A timed property enforceable if $\Sigma_u = \emptyset$.

We define a property $\mathrm{Pre}(\varphi)$, indicating whether it is possible to enforce property φ.

Definition 11 (Pre). *Property* $\mathrm{Pre}(\varphi)$ *is defined as the timed property which semantics is* $\langle Q, q_0, \Gamma, \rightarrow', Q_{\mathrm{enf}} \rangle$ *where:*

- $Q_{\mathrm{enf}}(\varphi) = \{q \in F_G \mid (q \text{ after } \mathrm{tw}(\Sigma_u)) \subseteq F_G\}$,
- $Q_{\overline{\mathrm{enf}}}(\varphi) = Q \setminus Q_{\mathrm{enf}}(\varphi)$,
- $\rightarrow' = (\rightarrow \cap\, Q_{\overline{\mathrm{enf}}} \times \Gamma \times Q) \cup \{(q, \gamma, q) \mid q \in Q_{\mathrm{enf}} \wedge \gamma \in \Gamma\}$.

$Q_{\mathrm{enf}}(\varphi)$ is the set of states of $[\![\mathcal{A}_\varphi]\!]$ from which it is impossible to reach a bad state reading only uncontrollable events. Thus, it corresponds to the set of states from which it is possible to enforce the property under consideration. $Q_{\overline{\mathrm{enf}}}$ is the set of states of the semantics of φ from which it is not possible to enforce the property, because there is a timed word containing only uncontrollable events (which cannot be modified nor suppressed) leading to a state that is not accepting. In the following, $Q_{\overline{\mathrm{enf}}}(\varphi)$ is noted $Q_{\overline{\mathrm{enf}}}$ and $Q_{\mathrm{enf}}(\varphi)$ is noted Q_{enf} to ease the notation.

$\mathrm{Pre}(\varphi)$ is a property indicating whether it is possible to enforce φ from the state of the semantics reached after reading a timed word (i.e. every possible continuation leads to Q_{enf}). Note that once Q_{enf} is reached, enforcement becomes effective, then the property will always be satisfied in the future, which explains why $\mathrm{Pre}(\varphi)$ is a co-safety property.

Note that, unlike in the untimed case, Q_{enf}, $Q_{\overline{\mathrm{enf}}}$ and $\mathrm{Pre}(\varphi)$ are defined on the semantics of the automaton representing the property and not on the automaton itself. Indeed, the set of states in the semantics in the untimed setting is the same as the set of locations of the property, thus the use of the semantics is not necessary.

We also define function Safe which, given a state, returns the set of sequences of controllable events that can be emitted safely. Function Safe is then extended to words:

Definition 12 (Safe (states)).

- *Given a state q of the semantics of a timed automaton,*
$$\text{Safe}(q) = \{\sigma \in \Sigma_c^* \mid \forall w \in \text{tw}(\Sigma), \Pi_\Sigma(w_{|\Sigma_c}) \preccurlyeq \sigma \implies \exists w' \in \text{tw}(\Sigma), w \preccurlyeq$$
$$w' \wedge \Pi_\Sigma(w'_{|\Sigma_c}) \preccurlyeq \sigma \wedge \exists q' \in Q_{\text{enf}}, q \xrightarrow{w'} q'\}.$$
- *Given a word $\sigma \in \text{tw}(\Sigma)$, $\text{Safe}(\sigma) = \text{Safe}(q)$, with $q = \text{Reach}(\sigma)$.*

Intuitively, $\text{Safe}(q)$ is the set of sequences of controllable events for which it is always possible to compute dates to reach Q_{enf}, even if any uncontrollable event occurs at any time. Safe shall be used to determine if the enforcement mechanism can release some previously-received controllable events. Contrary to the untimed case, some delay between two consecutive events may be needed to satisfy the property, thus an uncontrollable event could be received by the enforcement mechanism while the delay elapses. Should this happen, the enforcement mechanism needs to compute again the dates for the events it has not output yet in order to reach Q_{enf} if possible. Safe is used to ensure this, i.e. that Q_{enf} remains reachable with the events that have not been output yet even if some uncontrollable events occur.

Let us now define an enforcement function for a timed property φ, denoted as E_φ.

Definition 13 (E_φ).   *Let* store_φ *be the function defined inductively by* $\text{store}_\varphi(\epsilon) = (\epsilon, \epsilon, \epsilon)$, *and for* $\sigma \in \text{tw}(\Sigma)$, *and* (t, a) *such that* $\sigma.(t, a) \in \text{tw}(\Sigma)$, *if* $(\sigma_s, \sigma_b, \sigma_c) = \text{store}_\varphi(\sigma)$, *then*

$$\text{store}_\varphi(\sigma.(t, a)) = \begin{cases} (\sigma_s', \sigma_b', \sigma_c') & \text{if } a \in \Sigma_u, \\ (\sigma_s.\text{obs}(\sigma_b, t), \sigma_b'', \sigma_c'') & \text{if } a \in \Sigma_c, \end{cases}$$

with:

$\kappa_\varphi(\sigma_1, \sigma_2, t) = \min_{\text{lex}}(\max_{\preccurlyeq}(\{w \mid \Pi_\Sigma(w) \preccurlyeq \sigma_2 \wedge \text{date}(w(1)) + \text{time}(\sigma_1) \geq t \wedge \Pi_\Sigma(w) \in \text{Safe}(\sigma_1) \wedge \exists q_i \in \text{Reach}(\sigma_1), q_i \xrightarrow{w} q_{\text{enf}} \in Q_{\text{enf}}\}))$.

$\sigma_s' = \sigma_s.\text{obs}(\sigma_b, t).(t, a)$,

$\sigma_{b_1} = \text{nobs}(\sigma_b, t)$,

$\sigma_b' = \kappa_\varphi(\sigma_s', \Pi_\Sigma(\sigma_{b_1}).\sigma_c, t) +_t \text{time}(\sigma_s')$,

$\sigma_c' = \Pi_\Sigma(\sigma_b')^{-1}.(\Pi_\Sigma(\sigma_{b_1}).\sigma_c)$,

$\sigma_b'' = \kappa_\varphi(\sigma_s.\text{obs}(\sigma_b, t), \Pi_\Sigma(\sigma_{b_1}).\sigma_c.a, t) +_t \text{time}(\sigma_s.\text{obs}(\sigma_b, t))$,

$\sigma_c'' = \Pi_\Sigma(\sigma_b'')^{-1}.(\Pi_\Sigma(\sigma_{b_1}).\sigma_c.a)$,

For $\sigma \in \text{tw}(\Sigma)$, we define $E_\varphi(\sigma, t) = \text{obs}(\Pi_1(s_\sigma).\Pi_2(s_\sigma), t)$, with $s_\sigma = \text{store}_\varphi(\sigma)$.

In the definition of store_φ, the actions of the input belong to one of the three words σ_s, σ_b and σ_c. Word σ_s represents what has already been output and cannot be modified anymore. The timed word σ_b contains the controllable events that are about to be output, such that if σ_b is concatenated to σ_s, the concatenation satisfies φ. The untimed word σ_c contains the controllable actions that remain, meaning that, whatever dates are associated to these actions, it is not sure that Q_{enf} would be reached if it was emitted after σ_b. Yet, the actions of σ_c might be released later (because of the occurrence of an uncontrollable event for instance). Thus it is used to compute the new values for σ_b and σ_c when needed. Note that only the events of σ_c are stored with no dates. Indeed, σ_c is used only when recomputing dates, thus there is not any date to associate to the events in σ_c. κ_φ is computable: even though the number of words satisfying $\Pi_\Sigma(w) \preccurlyeq \sigma_2$ is infinite, since there is an infinite number of possible dates, it is possible to consider only a finite number of words, by considering only words that lead to different regions [1] of the automaton. Moreover, checking if $\Pi_\Sigma(w) \in \mathrm{Safe}(\sigma_1)$ is also computable, because it is a reachability issue, that is computable in the region automaton.

Roughly speaking, the enforcement mechanism described in the previous definition waits for the controllable events of the input to belong to the safe words of the current state, reached with the uncontrollable events (i.e. in $\sigma_c \in \mathrm{Safe}(q)$ if q is the current state), and then starts to emit as many controllable events as possible, with minimum dates greater than the current time. Since the input is safe, Q_{enf} will be reached at some point in the future, and then the enforcement mechanism starts again to wait for the input to be safe for the state reached so far, and goes on like previously.

Proposition 4. E_φ *is sound with respect to* $\mathrm{Pre}(\varphi) \implies \varphi$, *as per Definition 9.*

Proposition 5. E_φ *is compliant, as per Definition 10.*

4.3   Enforcement Monitors

As in the untimed setting, we give here an operational description of an enforcement mechanism whose output is exactly the output of E_φ, as defined in Definition 13.

Definition 14. *An enforcement monitor $\mathcal{E}$ for φ is a transition system $\langle C^{\mathcal{E}}, c_0^{\mathcal{E}}, \Gamma^{\mathcal{E}}, \hookrightarrow_{\mathcal{E}} \rangle$ such that:*

- $C^{\mathcal{E}} = \mathrm{tw}(\Sigma) \times \Sigma_c^* \times Q \times \mathbb{R}_{\geq 0}$ *is the set of configurations.*
- $c_0^{\mathcal{E}} = \langle \epsilon, \epsilon, q_0, 0 \rangle \in C^{\mathcal{E}}$ *is the initial configuration.*
- $\Gamma^{\mathcal{E}} = ((\mathbb{R}_{\geq 0} \times \Sigma) \cup \{\epsilon\}) \times Op \times ((\mathbb{R}_{\geq 0} \times \Sigma) \cup \{\epsilon\})$ *is the alphabet, composed of an optional input, an operation and an optional output. The set of operations is* $\{\mathrm{dump}(.), \mathrm{pass\text{-}uncont}(.), \mathrm{store\text{-}cont}(.), \mathrm{delay}(.)\}$. *Whenever $(\sigma, \bowtie, \sigma') \in \Gamma^{\mathcal{E}}$, it is noted $\sigma / \bowtie / \sigma'$.*
- $\hookrightarrow_{\mathcal{E}}$ *is the transition relation defined as the smallest relation obtained by applying the following rules given by their priority order:*
  - **Dump:** $\langle (t_b, a).\sigma_b, \sigma_c, q, t \rangle \xrightarrow{\epsilon / \mathrm{dump}(t_b, a) / (t_b, a)} \langle \sigma_b, \sigma_c, q', t \rangle$ *if $t_b = t$ with* $q \xrightarrow{(t_b, a)} q'$,
  - **Pass-uncont:** $\langle \sigma_b, \sigma_c, q, t \rangle \xrightarrow{(t, a) / \mathrm{pass\text{-}uncont}(t, a) / (t, a)} \langle \sigma_b', \sigma_c', q', t \rangle$, *with* $q \xrightarrow{(t, a)} q'$ *and $(\sigma_b', \sigma_c') = \mathrm{update}(q', \sigma_b, \sigma_c, t)$,*
  - **Store-cont:** $\langle \sigma_b, \sigma_c, q, t \rangle \xrightarrow{(t, c) / \mathrm{store\text{-}cont}((t, c)) / \epsilon} \langle \sigma_b', \sigma_c', q, t \rangle$, *with* $(\sigma_b', \sigma_c') = \mathrm{update}(q, \sigma_b, \sigma_c.c, t)$,

- **Delay:** $\langle \sigma_b, \sigma_c(l, v), t \rangle \xrightarrow{\epsilon/\ \mathrm{delay}(\delta)/\epsilon} \langle \sigma_b, \sigma_c, (l, v + \delta), t + \delta \rangle,$

where update *is a function computing* κ_φ *from previous definition of* E_φ.

In a configuration $\langle \sigma_b, \sigma_c, q, t \rangle$, σ_b is the word to be output as time elapses; σ_c is what is left from the input; q is the state of the semantics reached after reading what has already been output; t is the current time instant, i.e. the time elapsed since the beginning of the run.

Sequences σ_b and σ_c are as in the definition of store_φ, whereas q represents σ_s, such that $q = \mathrm{Reach}(\sigma_s)$. Function update computes the values of σ_b and σ_c to ensure soundness. Function update represents the computation of function κ_φ in the definition of store_φ.

Proposition 6. *The output of* $\mathcal{E}$ *for input* σ *is* $E_\varphi(\sigma)$.

4.4   Example

Figure 3 depicts a property modeling the use of some shared writable device. One can get the status of a lock through the uncontrollable events *LockOn* and *LockOff* indicating that the lock has been acquired, and that it is available, respectively. The uncontrollable event *Auth* is sent by the device to authorize writings. Once event *Auth* is received, the controllable event *Write* can be sent after having waited a little bit for synchronization. Each time the lock is acquired and released, we must also wait before issuing a new *Write* order. The sets of events are: $\Sigma_c = \{\mathtt{Write}\}$ and $\Sigma_u = \{\mathit{Auth}, \mathit{LockOff}, \mathit{LockOn}\}$.

Let us follow the output of function store_φ over time with the following input word: $\sigma = (1, \mathit{Auth}). (2, \mathit{LockOn}). (4, \mathtt{Write}). (5, \mathit{LockOff}). (6, \mathit{LockOn}). (7, \mathtt{Write}). (8, \mathit{LockOff})$. Let $(\sigma_s, \sigma_b, \sigma_c) = \mathrm{store}_\varphi(\mathrm{obs}(\sigma, t))$. Table 1 gives the values taken by σ_s, σ_b, and σ_c over time. To compute them, first notice that

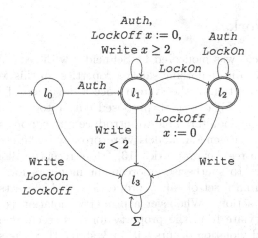

Fig. 3. Property modeling writes on a shared storage device.

Table 1. Values of $(\sigma_s, \sigma_b, \sigma_c) = \text{store}_\varphi((1, Auth). (2, LockOn). (4, \text{Write}). (5, LockOff).$ $(6, LockOn). (7, \text{Write}). (8, LockOff))$ over time.

t	σ_s	σ_b	σ_c
1	$(1, Auth)$	ϵ	ϵ
2	$(1, Auth).(2, LockOn)$	ϵ	ϵ
4	$(1, Auth).(2, LockOn)$	ϵ	Write
5	$(1, Auth).(2, LockOn). (5, LockOff)$	$(7, \text{Write})$	ϵ
6	$(1, Auth).(2, LockOn).(5, LockOff). (6, LockOn)$	ϵ	Write
7	$(1, Auth).(2, LockOn).(5, LockOff). (6, LockOn)$	ϵ	Write . Write
8	$(1, Auth).(2, LockOn).(5, LockOff).$ $(6, LockOn).(8, LockOff)$	$(10, \text{Write}).(10, \text{Write})$	ϵ
10	$(1, Auth).(2, LockOn).(5, LockOff).(6, LockOn).$ $(8, LockOff). (10, \text{Write}).(10, \text{Write})$	ϵ	ϵ

$Q_{\text{enf}} = \{l_1, l_2\} \times \mathcal{V}(\{x\})$. Moreover, we can see that $\text{Write} \in \text{Safe}(l_1)$ because it is always possible to delay the Write event in such a way that the current state remains in Q_{enf}, whatever are the uncontrollable events. Consequently, whenever σ_s leads to l_1, σ_b is empty (because Write is the only controllable event, thus σ_b must start with a Write event).

Figure 4 shows the execution of the enforcement monitor with input $(1, Auth)$. $(2, LockOn). (4, \text{Write}). (5, LockOff). (6, LockOn). (7, \text{Write}).(8, LockOff)$. In a configuration, the input is on the right, the output on the left, and the middle is the current configuration of the enforcement monitor. Variable t keeps track of global time. A valuation is represented as an integer – the value of the (unique) clock x. Observe that the final output at $t = 10$ is the same as the one of the enforcement function: $(1, Auth).(2, LockOn).(5, LockOff).$ $(6, LockOn).(8, LockOff).(10, \text{Write}). (10, \text{Write})$.

5   Related Work

Runtime enforcement was pioneered by Schneider with security automata [21], a runtime mechanism for enforcing safety properties. In this work monitors are able to stop the execution of the system once a deviation of the property has been detected. Later, Ligatti et al. proposed edit-automata, a more powerful model of enforcement monitors able to introduce and suppress events from the execution. Later, more general models were proposed where the monitors can be synthesised from regular properties [13]. More recently, Bloem et al. [6] presented a framework to synthesise enforcement monitors for reactive systems, called as *shields*, from a set of safety properties. A shield acts instantaneously and cannot buffer actions. Whenever a property violation is unavoidable, the shield allows to deviate from the property for k consecutive steps (as in [7]). Whenever a second violation occurs within k steps, then the shield enters into a *fail-safe* mode, where it ensures only correctness. Another recent approach

$t = 0$ $\epsilon/(\epsilon, \epsilon, (l_0, 0), 0)/(1, Auth).(2, on).(4, w).(5, off).(6, on).(7, w).(8, off)$
$\downarrow$ delay(1)

$t = 1$ $\epsilon/(\epsilon, \epsilon, (l_0, 1), 1)/(1, Auth).(2, on).(4, w).(5, off).(6, on).(7, w).(8, off)$
$\downarrow$ pass-uncont$((1, Auth))$

$t = 1$ $(1, Auth)/(\epsilon, \epsilon, (l_1, 1), 1)/(2, on).(4, w).(5, off).(6, on).(7, w).(8, off)$
$\downarrow$ delay(1)

$t = 2$ $(1, Auth)/(\epsilon, \epsilon, (l_1, 2), 2)/(2, on).(4, w).(5, off).(6, on).(7, w).(8, off)$
$\downarrow$ pass-uncont$((2, on))$

$t = 2$ $(1, Auth).(2, on)/(\epsilon, \epsilon, (l_2, 2), 2)/(4, w).(5, off).(6, on).(7, w).(8, off)$
$\downarrow$ delay(2)

$t = 4$ $(1, Auth).(2, on)/(\epsilon, \epsilon, (l_2, 4), 4)/(4, w).(5, off).(6, on).(7, w).(8, off)$
$\downarrow$ store-cont$((4, w))$

$t = 4$ $(1, Auth).(2, on)/(\epsilon, (4, w), (l_2, 4), 4)/(5, off).(6, on).(7, w).(8, off)$
$\downarrow$ delay(1)

$t = 5$ $(1, Auth).(2, on)/(\epsilon, (4, w), (l_2, 5), 5)/(5, off).(6, on).(7, w).(8, off)$
$\downarrow$ pass-uncont$((5, off))$

$t = 5$ $(1, Auth).(2, on).(5, off)/((7, w), \epsilon, (l_1, 0), 5)/(6, on).(7, w).(8, off)$
$\downarrow$ delay(1)

$t = 6$ $(1, Auth).(2, on).(5, off)/((7, w), \epsilon, (l_1, 1), 6)/(6, on).(7, w).(8, off)$
$\downarrow$ pass-uncont$((6, on))$

$t = 6$ $(1, Auth).(2, on).(5, off).(6, on)/(\epsilon, (7, w), (l_2, 1), 6)/(7, w).(8, off)$
$\downarrow$ delay(1)

$t = 7$ $(1, Auth).(2, on).(5, off).(6, on)/(\epsilon, (7, w), (l_2, 2), 7)/(7, w).(8, off)$
$\downarrow$ store-cont$((7, w))$

$t = 7$ $(1, Auth).(2, on).(5, off).(6, on)/(\epsilon, (7, w).(7, w), (l_2, 2), 7)/(8, off)$
$\downarrow$ delay(1)

$t = 8$ $(1, Auth).(2, on).(5, off).(6, on)/(\epsilon, (7, w).(7, w), (l_2, 3), 8)/(8, off)$
$\downarrow$ pass-uncont$((8, off))$

$t = 8$ $(1, Auth).(2, on).(5, off).(6, on).(8, off)/((10, w).(10, w), \epsilon, (l_1, 0), 8)/\epsilon$
$\downarrow$ delay(2)

$t = 10$ $(1, Auth).(2, on).(5, off).(6, on).(8, off)/((10, w).(10, w), \epsilon, (l_1, 2), 10)/\epsilon$
$\downarrow$ dump$((10, w))$

$t = 10$ $(1, Auth).(2, on).(5, off).(6, on).(8, off).(10, w)/((10, w), \epsilon, (l_1, 2), 10)/\epsilon$
$\downarrow$ dump$((10, w))$

$t = 10$ $(1, Auth).(2, on).(5, off).(6, on).(8, off).(10, w).(10, w)/(\epsilon, \epsilon, (l_1, 2), 10)/\epsilon$

Fig. 4. Execution of an enforcement monitor with input $(1, Auth)$. $(2, LockOn)$. $(4, Write)$. $(5, LockOff)$. $(6, LockOn)$. $(7, Write)$. $(8, LockOff)$. $LockOff$ is abbreviated as off, $LockOn$ as on, and Write as w.

by Dolzehnko et al. [11] introduces Mandatory Result Automata (MRAs). MRAs extend edit-automata by refining the input-output relationship of an enforcement mechanism and thus allowing a more precise description of the enforcement abilities of an enforcement mechanism in concrete application scenarios. All the previously mentioned approaches considered untimed specifications, and do not consider uncontrollable events.

In the timed setting, several monitoring tools for timed specifications have been proposed. RT-Mac [20] permits to verify at runtime timeliness and reliability correctness. LARVA [8,9] takes as input safety properties expressed with DATEs (Dynamic Automata with Times and Events), a timed model similar to timed automata.

In previous work, we introduced *runtime enforcement for timed properties* [19] specified by timed automata [1]. We proposed a model of enforcement monitors that work as *delayers*, that is, mechanisms that are able to delay the input sequence of timed events to correct it. While [19] proposed synthesis techniques only for safety and co-safety properties, we then generalized the framework to synthesise an enforcement monitor for any regular timed property [17,18]. In [16], we considered parametric timed properties, that is timed properties with data-events containing information from the execution of the monitored system. None of our previous research endeavors considered uncontrollable events. Considering uncontrollable events entailed us to revisit and redefine all the notions related to enforcement mechanisms (soundness, transparency weaken into compliance, enforcement function, and enforcement monitor). Moreover, we define an enforcement condition as a property, that allows to determine when an enforcement mechanism can safely output controllable events, independently of the uncontrollable events that can be received by the enforcement mechanism.

Basin et al. [4] introduced uncontrollable events in safety properties enforced with security automata [21]. More recently, Basin et al. proposed a more general approach [3] related to enforcement of security policies with controllable and uncontrollable events. Basin et al. presented several complexity results and showed how to synthesise enforcement mechanisms. In case of violation of the property, the system stops the execution. The approaches in [3,4] only handle discrete time, and clock ticks are considered as uncontrollable events. In our framework, we consider dense time using the expressiveness of timed automata. We synthesise enforcement mechanisms for any regular property, and not just safety properties. Moreover, our monitor are more flexible since they block the input word only when delaying events cannot prevent the violation of the desired property, thus offering the possibility to correct violations due to the timing of events.

6   Conclusion and Future Work

This paper extends the research endeavors on runtime enforcement and focuses on the use of uncontrollable events. An enforcement mechanism can only observe uncontrollable events, but cannot delay nor suppress them. Considering uncontrollable events entails to change the order between controllable and uncontrollable events, and to adapt the usual requirements on enforcement mechanisms. Therefore, we weaken transparency into compliance. We define enforcement mechanisms at two levels of abstraction (enforcement functions and enforcement monitors), for regular properties and regular timed properties. Since not all properties can be enforced, we also give a condition, depending on the property and the input word, indicating whether or not the enforcement mechanism can be sound with respect to the property under scrutiny. An enforcement mechanism outputs all received uncontrollable events, and stores the controllable ones, until soundness can be guaranteed. Then, it outputs events only when it can ensure that soundness will be satisfied.

Several extensions of this work are possible. A first extension is to consider more risky strategies regarding uncontrollable events, outputting events even if some uncontrollable events could lead to a bad state. Following such strategies could be guided by an additional probabilistic model on the occurrence of input events. A second extension is to implement the enforcement mechanisms using UPPAAL libraries [14]. A third extension is to use runtime enforcement to modify at runtime the parameters of a system with stochastic behaviour, as done offline in [2]. A fourth extension is to define a theory of runtime enforcement for distributed systems where enforcement monitors are decentralised on the components of the verified system, as is the case with verification monitors in [5]. A fifth extension is to distinguish inputs from outputs in properties, and consider for instance timed i/o automata [10] to formalise properties.

References

1. Alur, R., Dill, D.: The theory of timed automata. In: de Bakker, J., Huizing, C., de Roever, W., Rozenberg, G. (eds.) Real-Time: Theory in Practice. LNCS, vol. 600, pp. 45–73. Springer, Heidelberg (1992)
2. Bartocci, E., Bortolussi, L., Nenzi, L., Sanguinetti, G.: System design of stochastic models using robustness of temporal properties. Theor. Comput. Sci. **587**, 3–25 (2015)
3. Basin, D., Jugé, V., Klaedtke, F., Zălinescu, E.: Enforceable security policies revisited. ACM Trans. Inf. Syst. Secur. **16**(1), 3:1–3:26 (2013)
4. Basin, D., Klaedtke, F., Zălinescu, E.: Algorithms for monitoring real-time properties. In: Khurshid, S., Sen, K. (eds.) RV 2011. LNCS, vol. 7186, pp. 260–275. Springer, Heidelberg (2012)
5. Bauer, A., Falcone, Y.: Decentralised LTL monitoring. In: Giannakopoulou, D., Méry, D. (eds.) FM 2012. LNCS, vol. 7436, pp. 85–100. Springer, Heidelberg (2012)
6. Bloem, R., Könighofer, B., Könighofer, R., Wang, C.: Shield synthesis: runtime enforcement for reactive systems. In: Baier, C., Tinelli, C. (eds.) TACAS 2015. LNCS, vol. 9035, pp. 533–548. Springer, Heidelberg (2015)
7. Charafeddine, H., El-Harake, K., Falcone, Y., Jaber, M.: Runtime enforcement for component-based systems. In: Proceedings of the 30th Annual ACM Symposium on Applied Computing, pp. 1789–1796 (2015)
8. Colombo, C., Pace, G.J., Schneider, G.: LARVA – safer monitoring of real-time Java programs (tool paper). In: Hung, D.V., Krishnan, P. (eds.) Proceedings of the 7th IEEE International Conference on Software Engineering and Formal Methods (SEFM 2009), pp. 33–37. IEEE Computer Society (2009)
9. Colombo, C., Pace, G.J., Schneider, G.: Safe runtime verification of real-time properties. In: Ouaknine, J., Vaandrager, F.W. (eds.) FORMATS 2009. LNCS, vol. 5813, pp. 103–117. Springer, Heidelberg (2009)
10. David, A., Larsen, K.G., Legay, A., Nyman, U., Wasowski, A.: Timed I/O automata: a complete specification theory for real-time systems. In: Proceedings of the 13th ACM International Conference on Hybrid Systems: Computation and Control, pp. 91–100, HSCC 2010. ACM (2010)

11. Dolzhenko, E., Ligatti, J., Reddy, S.: Modeling runtime enforcement with mandatory results automata. Int. J. Inf. Secur. **14**(1), 47–60 (2015)
12. Falcone, Y.: You should better enforce than verify. In: Barringer, H., Falcone, Y., Finkbeiner, B., Havelund, K., Lee, I., Pace, G., Roşu, G., Sokolsky, O., Tillmann, N. (eds.) RV 2010. LNCS, vol. 6418, pp. 89–105. Springer, Heidelberg (2010)
13. Falcone, Y., Mounier, L., Fernandez, J., Richier, J.: Runtime enforcement monitors: composition, synthesis, and enforcement abilities. Form. Methods Syst. Des. **38**(3), 223–262 (2011)
14. Larsen, K.G., Pettersson, P., Yi, W.: Uppaal in a nutshell. Int. J. Softw. Tools Technol. Transf. **1**, 134–152 (1997)
15. Ligatti, J., Bauer, L., Walker, D.: Run-time enforcement of nonsafety policies. ACM Trans. Inf. Syst. Secur. **12**(3), 19:1–19:41 (2009)
16. Pinisetty, S., Falcone, Y., Jéron, T., Marchand, H.: Runtime enforcement of parametric timed properties with practical applications. In: Lesage, J., Faure, J., Cury, J.E.R., Lennartson, B. (eds.) 12th International Workshop on Discrete Event Systems, WODES 2014, pp. 420–427. International Federation of Automatic Control (2014)
17. Pinisetty, S., Falcone, Y., Jéron, T., Marchand, H.: Runtime enforcement of regular timed properties. In: Cho, Y., Shin, S.Y., Kim, S., Hung, C., Hong, J. (eds.) Symposium on Applied Computing, SAC, 2014, pp. 1279–1286. ACM (2014)
18. Pinisetty, S., Falcone, Y., Jéron, T., Marchand, H., Rollet, A., Nguena-Timo, O.: Runtime enforcement of timed properties revisited. Form. Methods Syst. Des. **45**(3), 381–422 (2014)
19. Pinisetty, S., Falcone, Y., Jéron, T., Marchand, H., Rollet, A., Nguena Timo, O.L.: Runtime enforcement of timed properties. In: Qadeer, S., Tasiran, S. (eds.) RV 2012. LNCS, vol. 7687, pp. 229–244. Springer, Heidelberg (2013)
20. Sammapun, U., Lee, I., Sokolsky, O.: RT-MaC: runtime monitoring and checking of quantitative and probabilistic properties. In: 2013 IEEE 19th International Conference on Embedded and Real-Time Computing Systems and Applications, pp. 147–153 (2005)
21. Schneider, F.B.: Enforceable security policies. ACM Trans. Inf. Syst. Secur. **3**(1), 30–50 (2000)

Tool Papers

A Tool Prototype for Model-Based Testing
of Cyber-Physical Systems

Arend Aerts[1], Mohammad Reza Mousavi[2], and Michel Reniers[1(✉)]

[1] Control Systems Technology Group, Eindhoven University of Technology,
Eindhoven, The Netherlands
`a.aerts@student.tue.nl`, `m.a.reniers@tue.nl`
[2] Center for Research on Embedded Systems,
Halmstad University, Halmstad, Sweden
`m.r.mousavi@hh.se`

Abstract. We report on a tool prototype for model-based testing of
cyber-physical systems. Our starting point is a hybrid-system model
specified in a domain-specific language called Acumen. Our prototype
tool is implemented in Matlab and covers three stages of model-based
testing, namely, test-case generation, test-case execution, and conformance analysis. We have applied our implementation to a number of
typical examples of cyber-physical systems in order to analyze its applicability. In this paper, we report on the result of applying the prototype
tool on a DC-DC boost converter.

Keywords: Model-based testing · Conformance testing · Cyber-physical
systems · Hybrid systems · Acumen · Matlab

1 Introduction

Cyber-physical systems have been the focus of much research in the past few
years: their structure and behavior are complex in nature and they often involve
critical applications. Correctness of such systems is a major concern and, hence,
rigorous validation and verification techniques are to be developed to ensure
their correctness. Model-based testing [6] is a rigorous verification technique
that is used to established that the behavior of an implementation conforms to
the specified behavior of a model.

There are some proposals for extending the theory of model-based testing to
the domain of cyber-physical systems [3,4,7,8,11,13]. In this paper, we report
on a prototype model-based testing tool for cyber-physical systems, based on
a variant of the theory presented in [3,4]. We extend the theory of [3,4] by
an offline test-case generation algorithm. Subsequently, in a prototype tool,

M.R. Mousavi has been partially supported by the Swedish Research Council (Veten-
skapsrådet) with award number 621-2014-5057 (Effective Model-Based Testing of
Parallel Systems) and the Swedish Knowledge Foundation (Stiftelsen för Kunskaps-
och Kompetensutveckling) with award number 20140302 (AUTO-CAAS).

M. Leucker et al. (Eds.): ICTAC 2015, LNCS 9399, pp. 563–572, 2015.
DOI: 10.1007/978-3-319-25150-9_32

we implement the three steps of our model-based testing trajectory, namely, test-case generation, test-case execution, and conformance analysis.

Our prototype tool is implemented in Matlab and starts off with a hybrid-system model in a domain-specific language called Acumen Modeling Language [14]. Our choice of Acumen is motivated by the local knowledge and expertise in this particular language. However, the principles described in this paper are defined generically for hybrid-timed state sequences and hybrid automata and, hence, are applicable to a wide set of languages. Based on a model in Acumen, we generate offline test cases that are robust (up to a given threshold) with respect to minor deviations between the model and its implementation. Subsequently, we implement a test-case execution module that interfaces Matlab with the system under test. In our case, we interfaced Matlab with the Acumen simulator which simulates a model of the system under test.

In order to evaluate its applicability, we applied our tool to a few typical examples of cyber-physical systems. In this paper, we focus on one such example, namely the DC-DC boost converter to illustrate the functionality of the tool.

Organization. In Sect. 2, we review our variant of the conformance theory based on the approach of [3,4]. Then, we describe our test-case generation technique for this theory of conformance. In Sect. 3, we describe the general architecture of the tooling. In Sect. 4, we report on the application of our tool to the DC-DC boost converter case study. In Sect. 5, we conclude the paper and present the directions of our future research and implementation activities.

This paper is based on previous work reported in [5].

2   Theory

In this section, we explain the underlying theory of our tool implementation based on and extending the theory of [3,4].

2.1   Semantic Domain

In order to have a model of hybrid-systems behavior, we need to model the input and output trajectories of the system dynamics. In [3,4], it is decided to take a discretized sampling of these trajectories as the basic starting point for conformance testing. The following notion of timed state sequences is defined to this end.

Definition 1 (Hybrid-Timed State Sequence (TSS) [3]). *Consider a sample size $N \in \mathbb{N}$, a dense time domain $\mathbb{T} = \mathbb{R}_{\geq 0}$, and a set of variables V. A hybrid-timed state sequence (TSS) is defined as pair (x,t), where $x \in Val(V)^N$, $t \in \mathbb{T}^N$, and $Val(V) : V \to \mathbb{R}$. The i'th element of a TSS (x,t) is denoted by (x_i, t_i), where $x_i \in Val(V)$ and $t_i \in \mathbb{T}$. Also, we denote the set of all TSSs defined over the set of variables V, considering a specific $N \in \mathbb{N}$, by $\mathbb{TSS}(V, N)$.*

A hybrid system according to [3], defined below, is a mapping from the initial condition and timed sequences of input variables to timed sequences of output variables.

Definition 2 (Hybrid System [3]). *Hybrid system $\mathcal{H}$ with initial condition $H \subset 2^{Val(V)}$, sample size N and input and output variables, respectively, V_I and V_O is modeled as a mapping: $\mathcal{H} : H \times TSS(V_I, N) \mapsto TSS(V_O, N)$. We write $y_{\mathcal{H}}(h_0, (u, t_u))$ to denote the output TSS to which the pair (u, t_u) is mapped by $\mathcal{H}$, considering h_0 as the initial condition.*

2.2 Conformance

The conformance notion [3,4], presented below, compares the output reaction of the model and the system under test to the same input stimuli. The system under test is said to conform to the model, if the output behavior is "similar", i.e., they differ temporally or in signal values not more than the pre-defined τ and ϵ threshold, respectively.

Definition 3 ((τ, ϵ) -Conformance). *Consider a test duration $T \in \mathbb{T}$ and $\tau, \epsilon > 0$; then TSS (y, t) (τ, ϵ)-conforms to TSS (y', t') (both with sample size N and defined on the set V of variables), denoted by $(y, t) \approx_{\tau, \epsilon, V} (y', t')$, if and only if*

1. *for all $i \in [1, N]$ such that $t_i \leq T$, there exists $k \in [1, N]$ such that $t_k \leq T$, $|t_i - t_k| < \tau$ and for each $v \in V$, $\|y_i(v) - y'_k(v)\| < \epsilon$, and*
2. *for all $i \in [1, N]$ such that $t'_i \leq T$, there exists $k \in [1, N]$ such that $t_k < T$, $|t'_i - t_k| < \tau$ and for each $v \in V$, $\|y'_i(v) - y_k(v)\| < \epsilon$.*

A hybrid system $\mathcal{H}$ (τ, ϵ)-conforms to a hybrid system $\mathcal{H}'$ (both with the same sample size and sets of input and output variables), denoted by $\mathcal{H} \approx_{\tau, \epsilon} \mathcal{H}'$, when for each initial condition h_0 and each TSS (u, t_u) on the common input variables V_I, $y_{\mathcal{H}}(h_0, (u, t_u)) \approx_{\tau, \epsilon, V_O} y_{\mathcal{H}'}(h_0, (u, t_u))$.

Choosing the right conformance value for τ and ϵ is application dependent and is left to the user. However, in order to give some insight about the degree of conformance between a specification and a system under test, one may fix a value for τ and determine the minimal value of ϵ for which (τ, ϵ)-conformance holds. The following definition formalizes this concept.

Definition 4 (Conformance Degree). *If $\mathcal{H}_1$ and $\mathcal{H}_2$ are two hybrid systems, given a predefined τ, the conformance degree of $\mathcal{H}_1$ to $\mathcal{H}_2$, denoted by $CD_\tau(\mathcal{H}_1, \mathcal{H}_2)$, is defined as $CD_\tau(\mathcal{H}_1, \mathcal{H}_2) := \inf \{\epsilon : \mathcal{H}_1 \approx_{\tau, \epsilon} \mathcal{H}_2\}$.*

Note that our notion of conformance (degree) simplifies that of [3,4] in a couple of ways: firstly, in our notion the number of discrete jumps is immaterial for our notion of conformance; secondly, we take the sample size of the specification and the implementation to be the same; finally, we simplified the super-dense time domain into a dense time domain. All of these are for the sake of simplicity in presentation (while keeping the definitions still applicable to our practical settings). Generalization to the original setting of [3,4] is straightforward.

2.3   Test-Case Generation

In order to check conformance, we need to stimulate both the model and the system under test and then compare their outputs. To this end, we need to make sure that the inputs fed into the system are valid. Validity in our context has two aspects: firstly our Acumen input models (as well as other typical models of cyber-physical systems) feature state-dependent behavior. In other words, not all combinations of input valuations are valid for system specification. (This aspect is not addressed in the proposal of [3,4] where models are assumed to be input-enabled.) Moreover, since the notion of conformance allows for some deviation between the model and the implementation, the inputs should not be too close to the boundaries of specification states (closer than the specified thresholds τ and ϵ in time and values, respectively); otherwise, the generated test cases may cease to be applicable in the course of test-case execution.

In order to give a generic exposition of our approach, we formulate it using the notion of hybrid automata, quoted below.

Definition 5 (Hybrid Automata [9]). A hybrid automaton is defined as a tuple $(Loc, V, (l_0, v_0), \to, I, F)$, where

- Loc is a finite set of locations;
- $V = V_I \uplus V_O$ is the set of continuous variables;
- l_0 denotes the initial location and v_0 is an initial valuation of V;
- $\to \subseteq Loc \times \mathcal{B}(V) \times Reset(V) \times Loc$ is the set of jumps where:
  - $\mathcal{B}(V) \subseteq Val(V)$ indicates the guards under which the switch may be performed, and
  - $Reset(V) \subseteq Val(V)^2$ is the set of all value assignments to all or a subset of the variables V;
- $I : Loc \to \mathcal{B}(V)$ determines the allowed valuation of variables in each location (called the invariant of the location);
- $F : Loc \to \mathcal{B}\left(V \cup \dot{V}\right)$ describes some constraints on variables and their derivatives and specifies the allowed continuous behavior in each location.

In order to generate test cases for a hybrid automaton, we take two issues into account: validity of inputs in each location and the distance of the values from the location boundaries. These two aspects are summarized in the following notion of "sound and robust" test case. This notion is inspired by the notion of solution of hybrid automata [12].

Definition 6 (Solution). *A solution to the hybrid automaton* $\mathcal{HA} = (Loc, V, (l_0, v_0), \to, I, F)$ *is a function* $s : [1, J] \to \mathbb{T} \to Loc \times Val(V)$ *for some J, where for each* $1 \leq j \leq J$*:* $\mathrm{dom}(s(j)) = [t_j, t_{j+1}]$ *for some* $t_j, t_{j+1} \in \mathbb{T}$*,* $t_1 = 0$*, and*

- *$s(1)(0) = (l_0, v_0)$;*
- *for each* $1 \leq j \leq J$ *and* $t \in [t_j, t_{j+1}]$*:* x *satisfies* $I(l)$ *and* $F(l)$*, where* $(l, x) = s(j)(t)$*; and*
- *for each* $1 \leq j < J$*: there exists* $l \xrightarrow{g,r} l'$ *such that* x *satisfies* g *and* (x, x') *satisfies* r*, where* $(l, x) = s(j)(t_{j+1})$ *and* $(l', x') = s(j+1)(t_{j+1})$*.*

Definition 7 (Sound and Robust Test Case). *A sound and (τ, ϵ)-robust test case of size N for a hybrid automaton is a TSS (y, t) with sample size N on the set V_I of variables if and only if there exists a solution s of the hybrid automaton such that*

1. *for each $i \leq N$, there exists a $j \in \operatorname{dom}(s)$, $t \in \operatorname{dom}(s(j))$, $y_i = s(j)(t) \downarrow V_I$ (soundness),*
   - *$t - \tau \in \operatorname{dom}(s(j))$ and $t + \tau \in \operatorname{dom}(s(j))$ (τ-robustness), and*
   - *for each $\epsilon' \leq \epsilon$, there exists a $t' \in \operatorname{dom}(s(j))$ such that for each variable $v \in V_O$ it holds that $\|val(s(j)(t))(v) - val(s(j)(t'))(v)\| = \epsilon'$ (ϵ-robustness).*

When τ and ϵ are known from the context, we simply use the term "sound and robust test case".

3   Tool

In this section, the implementation of the conformance method in the tooling is discussed. In Fig. 1, an architectural view of our tool is presented. The grey area corresponds to the Graphical User Interface (GUI) which interacts with the tool functionality. The tooling is created in the Matlab R2013b environment. This environment was preferred to keep the implementation generic and also to be able to use the Java compatibility of Matlab in order to interface with various modeling and implementation frameworks.

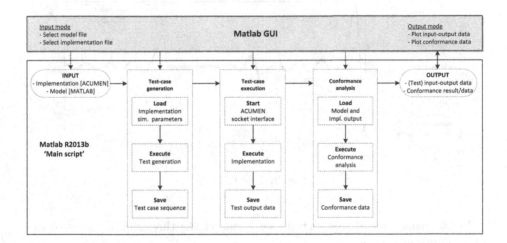

Fig. 1. Tool architecture overview

The three main steps of the conformance method are test-case generation, test-case execution, and conformance analysis and they can easily be recognized in the architecture of the tool depicted in Fig. 1. The application of test-case generation and execution methods results in generating input-output data for both the model and the implementation under test. Application of the conformance analysis, subsequently, results in a conformance judgment possibly accompanied with an additional witness for conformance violation, which is fed into the GUI for visualization purposes.

As depicted in Fig. 1, there is a clear separation between the "Main script" module and the GUI. This division provides us with two builds of the tool, namely the Script Build and the GUI Build. The Script Build contains the full functionality of the tooling which is implemented using Matlab scripting methods (.m files), and is controllable from a command-line interface. The GUI Build contains selected functionality of the tool and offers a GUI for intuitive and easy use, especially for non-expert users. In Fig. 2, a preview of the GUI Build is provided.[1]

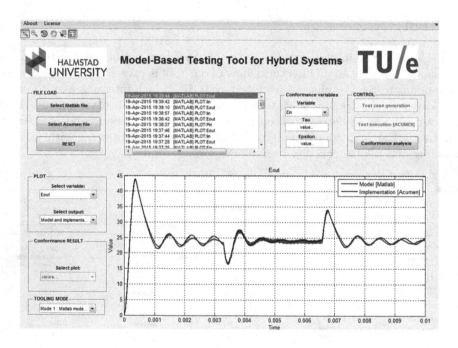

Fig. 2. Tool GUI

In the remainder of this section, we focus on the three main phases of the conformance method.

Test-case Generation. In Fig. 1, before the test-case generation algorithm is executed, the simulation parameters as specified in the Acumen file are loaded into Matlab. This process is performed by an Acumen file parser which extracts the specified simulation parameters and all model variables from the implementation

---

[1] The prototype tool can obtained from http://ceres.hh.se/mediawiki/Tool_Prototype_for_Conformance_Testing_of_CyPhy_Systems.

modeled in Acumen. Definition 7 is then implemented in order to generate sound and robust test cases [5]. We made a slight simplification, by focusing on a subset of hybrid systems in which, firstly, the guards are not time-dependent and secondly, the invariants are only specified as intervals of input variable valuations. This simplified the implementation of the soundness and robustness checks.

Test-case Execution. The test-case execution refers to the application of generated test cases on the implementation modeled in Acumen. In this step, a combination of Java and Matlab code is used in order to execute test cases / inputs on an Acumen (hybrid-system) model. This process involves communication between the Matlab tooling and the Acumen runtime environment. Note that further use of Acumen refers to the Acumen runtime environment.

The (simulator) data that is transferred between Matlab and Acumen uses the JSON-format for information exchange of the Acumen simulator state; see Fig. 3. Since existing JSON parsers failed to unwrap the simulation data from Acumen correctly, a custom Matlab JSON parser was designed and implemented for this purpose. This custom Matlab JSON parser uses the preloaded model variables of the Acumen file parser to extract all model variables of the implementation.

To initiate the communication between the Matlab tooling and Acumen, a command line interface is used (from within Matlab) to start up Acumen in the background. Moreover, Acumen automatically loads a pre-specified Acumen file, in this case the implementation, and starts the simulation of this model. Since this start-up sequence is performed with Acumen in server mode, it creates a socket connection

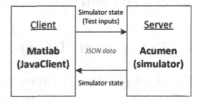

Fig. 3. Socket connection

to execute a co-simulation. Hence, when the simulation of the implementation is automatically started, Acumen waits for a valid socket connection or client, in this case the Matlab tooling. In Fig. 1, this start-up process of the socket connection between Matlab and Acumen is shown as the first step of the test-case execution.

As soon as the Matlab tooling initiates the socket connection by running its embedded Javaclient, the (initial) simulator state is send over to Matlab. When Matlab returns the simulator state to Acumen, one simulation step of the implementation (in Acumen) is performed. This process repeats itself for every timestep of the full simulation duration as specified in the Acumen model file.

Conformance Analysis. The conformance analysis is an implementation of Definition 3. In addition to providing a yes/no answer, the tool provides a visualization of the counter-example in case the conformance relation does not hold. This is achieved by plotting both the specification and the implementation trajectories and depicting the case of violation by a (τ, ϵ) box around the specification point which does not find a counterpart in the implementation (or the other way around). Additionally, our implementation automatically calculates the conformance degree based on Definition 4.

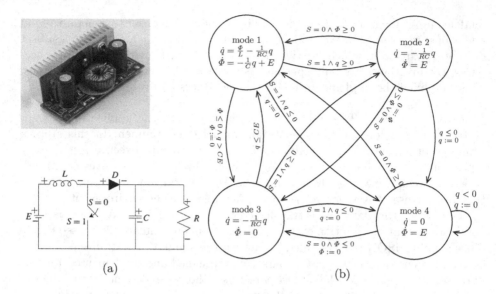

Fig. 4. (a) a PCB [2], its schematic [1], and (b) its hybrid automaton model [1]

4   Experiment: DC-DC Boost Converter

The developed tool has been experimented with on several classical hybrid-system examples, such as the bouncing ball, the thermostat, and the DC-DC boost converter. The DC-DC boost converter example is discussed below. The DC-DC boost converter is a hybrid-system example (see [10]) that originates from the field of electrical engineering and is used to "boost" an input DC voltage to an increased output value. In Fig. 4a, such a boost converter is shown together with a schematic that shows the principle of operation. The boost of the DC voltage is a consequence the combined physical properties of the inductor L and capacitor C, which are controlled by the switch S and diode D. This process transforms the input voltage E to an increased output voltage that is applied to the resistive load R. Note that the control elements of the boost converter transform the otherwise continuous system into a hybrid system. Finally, the system is made input dependent by tuning the resistive load R which results in internal stabilizing behavior of the boost converter.

In Fig. 4b, the hybrid-automaton model of the DC-DC boost converter is shown. The four discrete states of the system are solely dependent on the position of the switch S and the mode of the diode D (conducting/blocking). In addition, the physical properties of the system are modelled by the electric charge q of the capacitor and the magnetic flux ϕ of the inductor. For further understanding of the specified dynamics, state guards and reset maps see [10].

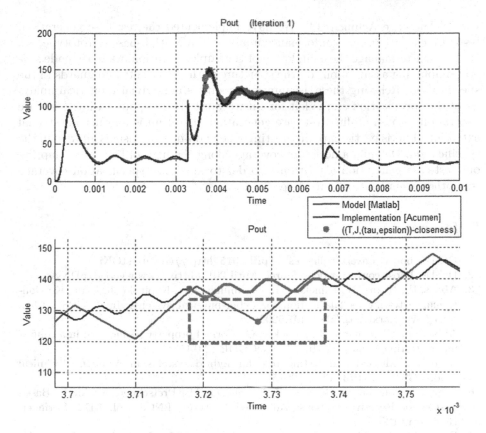

Fig. 5. DC-DC boost converter conformance analysis

In Fig. 5, the output power of a specific boost converter is shown. The blue and black lines indicate the response of the model (in Matlab) and implementation (Acumen) respectively, which are visibly diverging. Hence, conformance analysis is needed in order to evaluate the conformance (degree) of the implementation with respect to the model. Non-conformance is detected and is indicated in red. In the lower sub-plot, an automatic zoom of the non-conformance area is performed in order to provide visual feedback of the τ-ϵ area around the corresponding data point. The following values are used in the conformance analysis of Fig. 5: $\tau = 0.00001$, $\epsilon = 7$.

5   Conclusions and Future Work

In this paper, we reported on an implementation of a conformance testing theory for cyber-physical systems, based on the conformance notion of [3,4]. To this end, we have developed the notion of sound and robust test cases. We have used this notion to generate off-line test cases from a hybrid-system model in the domain specific language Acumen [14]. We have implemented the test-case generation, test-case execution, and conformance analysis in a Matlab-based prototype.

In order to manage the complexity of the implementation, we have made several simplifying assumptions on the structure of the invariants and guards in the specification. Relaxing these assumptions requires non-trivial numerical analysis of the specification and is left for future work. Turning our off-line test-case generation into an on-line test-case generation algorithm is another non-trivial extension. This is particularly interesting when non-determinism is allowed in the specification. Defining a notion of coverage along the lines of [7,8] and adapting our test-case generation algorithm in order to maximize specification coverage is another avenue for our future research.

References

1. DC-DC boost converter [figures], April 2015. http://goo.gl/rstOKi
2. DC-DC boost converter pcb [figure], April 2015. http://www.goo.gl/pDNyw3
3. Abbas, H., Hoxha, B., Fainekos, G., Deshmukh, J.V., Kapinski, J., Ueda, K.: Conformance testing as falsification for cyber-physical systems. In: ICCPS (2014). http://www.arxiv.org/abs/1401.5200
4. Abbas, H., Mittelmann, H., Fainekos, G.: Formal property verification in a conformance testing framework. In: MEMOCODE (2014)
5. Aerts, A.: Model-based testing tool for hybrid systems in Acumen. Technical report. CST 2015.073, TU/e (2015)
6. Broy, M., Jonsson, B., Katoen, J.P., Leucker, M., Pretschner, A.: Model-Based Testing of Reactive Systems: Advanced Lectures. LNCS, vol. 3472. Springer, Heidelberg (2005)
7. Dang, T.: Model-based testing of hybrid systems. In: Model-based Testing for Embedded Systems. CRC Press (2011)
8. Dang, T., Nahhal, T.: Coverage-guided test generation for continuous and hybrid systems. Form. Methods Syst. Des. **34**(2), 183–213 (2009)
9. Goebel, R., Sanfelice, R., Teel, A.: Hybrid dynamical systems. IEEE Control Syst. Mag. **29**(2), 28–93 (2009)
10. Heemels, W.P.M.H., de Schutter, B.: Modeling and control of hybrid dynamical systems. TU/e, Lecture notes course 4K160 (2013)
11. Julius, A.A., Fainekos, G.E., Anand, M., Lee, I., Pappas, G.J.: Robust test generation and coverage for hybrid systems. In: Bemporad, A., Bicchi, A., Buttazzo, G. (eds.) HSCC 2007. LNCS, vol. 4416, pp. 329–342. Springer, Heidelberg (2007)
12. Lemmon, M.D.: On the existence of solutions to controlled hybrid automata. In: Lynch, N.A., Krogh, B.H. (eds.) HSCC 2000. LNCS, vol. 1790, p. 229. Springer, Heidelberg (2000)
13. van Osch, M.: Hybrid input-output conformance and test generation. In: Havelund, K., Núñez, M., Roşu, G., Wolff, B. (eds.) FATES 2006 and RV 2006. LNCS, vol. 4262, pp. 70–84. Springer, Heidelberg (2006)
14. Taha, W., Brauner, P., Zeng, Y., Cartwright, R., Gaspes, V., Ames, A., Chapoutot, A.: A core language for executable models of cyber-physical systems (preliminary report). In: ICDCS (2012)

CAAL: Concurrency Workbench, Aalborg Edition

Jesper R. Andersen, Nicklas Andersen, Søren Enevoldsen, Mathias M. Hansen,
Kim G. Larsen, Simon R. Olesen, Jiří Srba$^{(\boxtimes)}$, and Jacob K. Wortmann

Department of Computer Science, Aalborg University,
Selma Lagerlöfs Vej 300, 9220 Aalborg Øst, Denmark
srba@cs.aau.dk

Abstract. We present the first official release of CAAL, a web-based
tool for modelling and verification of concurrent processes. The tool is
primarily designed for educational purposes and it supports the classical process algebra CCS together with its timed extension TCCS. It
allows to compare processes with respect to a range of strong/weak and
timed/untimed equivalences and preorders (bisimulation, simulation and
traces) and supports model checking of CCS/TCCS processes against
recursively defined formulae of Hennessy-Milner logic. The tool offers a
graphical visualizer for displaying labelled transition systems, including
their minimization up to strong/weak bisimulation, and process behaviour can be examined by playing (bi)simulation and model checking
games or via the generation of distinguishing formulae for non-equivalent
processes. We describe the modelling and analysis features of CAAL, discuss the underlying verification algorithms and show a typical example
of a use in the classroom environment.

1 Introduction

Concurrency is a classical topic taught at many universities as a bachelor or
master degree course in Computer Science. For an introductory course in concurrency, the typical content includes the use of a simple language for the description
of parallel processes (e.g. CCS, CSP, ACP or Petri nets) that is used for modelling
concurrent systems and for explaining the key concepts of equivalence checking
and model checking. At Aalborg University, we offer such an introductory course
called *Semantics and Verification* to the 6th semester software engineering and
computer science students. The course is based on our Reactive Systems book [1]
that, among others, introduces the CCS process algebra (Calculus of Communicating Systems [14]) and bisimulation/model checking approach, including the
corresponding game characterization. In order to motivate the students to study
and appreciate the theoretical concepts in concurrency, we engage them in a
few medium-size modelling exercises. This hands-on modelling experience makes
them realize that designing even small concurrent systems is difficult and that
support by an adequate tool can be very useful. For this purpose, we introduce
the open-source tool CAAL (standing for Concurrency workbench developed at

© Springer International Publishing Switzerland 2015
M. Leucker et al. (Eds.): ICTAC 2015, LNCS 9399, pp. 573–582, 2015.
DOI: 10.1007/978-3-319-25150-9_33

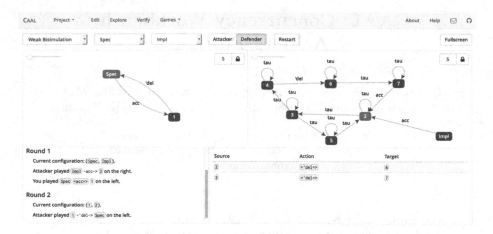

Fig. 1. Game module in CAAL

AALborg university) that supports CCS and TCCS as the input language. CAAL is programmed in TypeScript, a typed superset of JavaScript that compiles into plain JavaScript. The input language of CAAL is an extension of the well-known Concurrency Workbench (CWB) [5] input syntax, so existing CWB projects can be opened in CAAL. The tool is hosted at

http://caal.cs.aau.dk

and it runs in any modern browser but a stand-alone installation is possible too.

CAAL offers an editor with online syntax correction, an explorer for the visualization of the generated labelled transition systems, including different minimizations w.r.t. to strong and weak bisimulation as well as the display of strong/weak and timed/untimed transitions. The explorer module enables an interactive exploration of the state-space via a predefined depth of the view horizon (suitable for exploring large state-spaces), automatic layout with the possibility to lock and rearrange the position of nodes, zoom functionality, simplification w.r.t. structural congruence and export as a raster graphics image. The verification module of CAAL allows to formulate equivalence and model checking queries and verify them either individually or collectively. It is possible to generate distinguishing formulae for non-equivalent processes or enter the game module (see Fig. 1) and interactively play (bi)simulation and model checking games.

Related Work. Concurrency WorkBench (CWB) [5] and its continuation Concurrency WorkBench of the New Century (CWB-NC) [6] are perhaps the best known tools for modelling and analysing CCS processes. Throughout a number of years, CWB has been the tool of choice in courses on concurrency at Aalborg University. Unfortunately, both CWB and CWB-NC are no longer in active development and the latest binaries are from 1999 (CWB) and 2000 (CWB-NC). The download links on the CWB-NC homepage do not work any more and it has become more and more difficult to acquire and install CWB as it relies on an outdated compiler (as a consequence e.g. Mac OS X binaries are not available). Moreover, CWB is only a command line tool and despite the fast verification algorithms it implements, the graphical interface is lacking. Apart from the fact that CAAL provides a modern user interface, integrated process editor and the possibility to visualize the processes and play a variety of (bi)simulation and model checking games, the verification approach also differs. While CWB is using global partitioning algorithms for checking equivalences, we use local on-the-fly approach based on dependency graphs.

Recently, there have been efforts to provide graphical add-ons to CWB as e.g. the Bisimulation Game-Game project [15], but there is no support for model checking games and the tool relies on transition graphs generated by CWB. The tool pseuCo [7] allows to compile an educational Java-based concurrent language into CCS and visualize/minimize the resulting transition systems. TAPAs [4] is another educational tool for specifying and analyzing concurrent processes described in CCSP (CCS plus additional CSP operators). It has a nice GUI but it does not consider bisimulation/model checking games and timed process algebra. Other tools supporting CCS language are more on the experimental level (command line input) and are not targeted towards educational purposes. Let us name here e.g. implementation of CCS in Maude [16] or in Haskell [3].

Finally, there exist mature tools with modern designs like FDR3 [9], CADP [8] and mCRL2 [10], with expressive input languages and efficient analysis methods. Our tool does not aim to compete with them in terms of performance, we are instead focusing on the educational aspects.

2 Modelling Features

CAAL supports the CCS and timed CCS (TCCS) input syntax. Let $\mathcal{A}$ be a finite set of channels, let $\overline{\mathcal{A}} = \{\overline{a} \mid a \in \mathcal{A}\}$ be the set of dual channels[1] and let $Act = \mathcal{A} \cup \overline{\mathcal{A}} \cup \{\tau\}$ be the set of actions. Let $\mathcal{K}$ be a finite set of process names. The collection of CCS expressions is given by the abstract syntax

$$P, Q := K \mid \alpha.P \mid P + Q \mid P \mid Q \mid P[f] \mid P \setminus L \mid 0$$

where $K \in \mathcal{K}$, $\alpha \in Act$, $L \subseteq \mathcal{A}$ and $f : Act \longrightarrow Act$ is the relabelling function satisfying $f(\tau) = \tau$ and $f(\overline{a}) = \overline{f(a)}$ for every $a \in Act$. By convention $\overline{\tau} = \tau$. The behaviour of each process name $K \in \mathcal{K}$ is given by its defining equation $K \overset{\text{def}}{=} P$. The syntax of TCCS is further extended with the delay prefix operator such that for every nonnegative integer d and a process expression P, we have that $d.P$ is also a process expression.

The SOS rules for the CCS and TCCS operators are given in Table 1. For TCCS we support at the moment the discrete time semantics that is defined

---

[1] In CAAL dual channels are prefixed with an apostrophe and the output bar is displayed automatically by the editor.

Table 1. SOS rules for CCS and TCCS

$$\text{ACT}\dfrac{}{\alpha.P \xrightarrow{\alpha} P} \qquad \text{SUM1}\dfrac{P \xrightarrow{\alpha} P'}{P+Q \xrightarrow{\alpha} P'} \qquad \text{SUM2}\dfrac{Q \xrightarrow{\alpha} Q'}{P+Q \xrightarrow{\alpha} Q'}$$

$$\text{COM1}\dfrac{P \xrightarrow{\alpha} P'}{P \mid Q \xrightarrow{\alpha} P' \mid Q} \qquad \text{COM2}\dfrac{Q \xrightarrow{\alpha} Q'}{P \mid Q \xrightarrow{\alpha} P \mid Q'} \qquad \text{COM3}\dfrac{P \xrightarrow{a} P' \quad Q \xrightarrow{\overline{a}} Q'}{P \mid Q \xrightarrow{\tau} P' \mid Q'}$$

$$\text{CON}\dfrac{P \xrightarrow{\alpha} P'}{K \xrightarrow{\alpha} P'} \; K \overset{\text{def}}{=} P \quad \text{REL}\dfrac{P \xrightarrow{\alpha} P'}{P[f] \xrightarrow{f(\alpha)} P'[f]} \quad \text{RES}\dfrac{P \xrightarrow{\alpha} P'}{P \setminus L \xrightarrow{\alpha} P' \setminus L} \; \alpha, \overline{\alpha} \notin L$$

Table 2. SOS rules for unit delays in TCCS (d ranges over nonnegative integers)

$$\text{ONE}\dfrac{}{d.P \xrightarrow{1} (d-1).P} d \geq 1 \quad \text{ACT}\dfrac{}{\alpha.P \xrightarrow{1} \alpha.P} \alpha \neq \tau \quad \text{REL}\dfrac{P \xrightarrow{1} P'}{P[f] \xrightarrow{1} P'[f]}$$

$$\text{SUM}\dfrac{P \xrightarrow{1} P' \quad Q \xrightarrow{1} Q'}{P+Q \xrightarrow{1} P'+Q'} \quad \text{CON}\dfrac{P \xrightarrow{1} P'}{K \xrightarrow{1} P'} \; K \overset{\text{def}}{=} P \quad \text{RES}\dfrac{P \xrightarrow{1} P'}{P \setminus L \xrightarrow{1} P' \setminus L}$$

$$\text{COM}\dfrac{P \xrightarrow{1} P' \quad Q \xrightarrow{1} Q'}{P \mid Q \xrightarrow{1} P' \mid Q'} \; \text{if } P \mid Q \xrightarrow{\tau}\!\!\!\!/$$

in Table 2. The semantics is for simplicity given for single-unit time delays as longer delays are just a syntactic sugar for a series of one time unit delays.

We assume the classical definitions of weak/strong and timed/untimed equivalences and preorders like simulation and trace preorder/equivalence, and bisimilarity (see e.g. [1]) that are supported in CAAL, including their game characterization via two-player games between attacker (trying to disprove the validity of the equivalence/preorder) and defender (supporting its validity).

As for model checking, the tool supports a subset of the modal μ-calculus [12] with recursively defined fixed points given by the syntax:

$$\phi ::= tt \mid f\!f \mid \phi_1 \wedge \phi_2 \mid \phi_1 \vee \phi_2 \mid \langle \alpha \rangle \phi \mid [\alpha]\phi \mid \langle\!\langle \alpha \rangle\!\rangle \phi \mid [[\alpha]]\phi \mid X$$

where $\alpha \in \mathcal{A}ct$ and where X is a variable from a finite set of variables such that every variable has exactly one declaration of the form $X \overset{\text{min}}{=} \phi$ (minimum fixed point) or $X \overset{\text{max}}{=} \phi$ (maximum fixed point). Here the modal operators are available in their strong variants $\langle \alpha \rangle$ (there is an α-successor) and $[\alpha]$ (for all α-successors), as well as the weak ones $\langle\!\langle \alpha \rangle\!\rangle$ and $[[\alpha]]$ that abstract away from τ-actions. We use the abbreviations $\langle A \rangle \phi$ and $[A]\phi$ for a set of actions $A \subseteq \mathcal{A}ct$, standing for $\vee_{\alpha \in A} \langle \alpha \rangle \phi$ and $\wedge_{\alpha \in A} [\alpha]\phi$, respectively. By $\langle - \rangle \phi$ we understand $\langle \mathcal{A}ct \rangle \phi$ and similarly for $[-]\phi$. The same conventions are used for the weak modalities.

For most practical applications it is enough to consider formulae where the recursively defined variables do not contain cyclic references (hence a variable X can refer to itself and/or to another variable Y, but Y may not refer back to X, neither directly or indirectly via other variables). This restriction, adopted by CAAL, allows for faster implementation of the verification engine and makes the interpretation of model checking games between defender (claiming that a process satisfies a given formula) and attacker (claiming that it does not satisfy the formula) a lot easier as we can always uniquely determine whether we are in the context of a minimum or a maximum fixed point. Defender is then the winner of any infinite play whenever we are in the context of maximum fixed point and attacker is the winner if we are in the minimum fixed-point context.

For guarded CCS processes (where every occurrence of a process name is within the scope of action prefixing) we know that two processes are bisimilar if and only if they satisfy exactly the same set of formulae of the Hennessy-Milner logic [11]. In case of strong bisimilarity we allow only the strong modalities in the formulae, and in case of weak bisimilarity we consider only the weak modalities. The theorem implies that if two processes are not bisimilar, we can find the so-called *distinguishing formula* that is satisfied in one of the processes but not in the other one. This can be useful when debugging CCS processes.

Finally, the tool supports also the extension of the logic with time modalities so that we can have formulae of the form $\langle d \rangle \phi$, $[d]\phi$, $\langle\langle d \rangle\rangle \phi$ and $[[d]]\phi$ where d is a nonnegative integer. The modality $\langle d \rangle \phi$ requires that it is possible to delay d time units and then satisfy ϕ, while the modality $[d]\phi$ expresses that whenever it is possible to delay d time units then ϕ must be satisfied. Even though the future after a given time delay is always deterministic, there is a difference between the two operators as if a process cannot delay d time units (due to some enabled τ actions that are urgent in the TCCS semantics) then $[d]\phi$ will be always satisfied while $\langle d \rangle \phi$ will never be satisfied. The weak time delay modalities allow us to interleave the single-unit time delays with arbitrary many τ-actions. The modalities are in CAAL further extended with time intervals such that $\langle d_1, d_2 \rangle \phi$ with $d_1 \leq d_2$ is the abbreviation for $\langle d_1 \rangle \phi \vee \langle d_1 + 1 \rangle \phi \vee \langle d_1 + 2 \rangle \phi \vee \ldots \vee \langle d_2 \rangle \phi$, and similarly $[d_1, d_2]$ stands for $[d_1]\phi \wedge [d_1 + 1]\phi \wedge [d_1 + 2]\phi \wedge \ldots \wedge [d_2]\phi$. The intervals in weak modalities are defined analogously.

3 Verification Engine

The verification algorithms for both equivalence/preorder checking and model checking are based on a fixed-point computation over a structure called *dependency graph* [13]. Such graphs, for the verification problems in question, can be generated on-the-fly and there exist efficient local algorithms by Liu and Smolka [13] for computing the fixed points.

A *dependency graph* is a pair $G = (V, E)$ where V is a finite set of nodes and $E \subseteq V \times 2^V$ is a finite set of hyperedges of the form (v, T) where $v \in V$ is the source node and the nodes in $T \subseteq V$ are called the target nodes. An *assignment* on G is a function $A : V \rightarrow \{0, 1\}$. We define a function F from assignments to assignments as follows: $F(A)(v) = 1$ if and only if there is $(v, T) \in E$ such

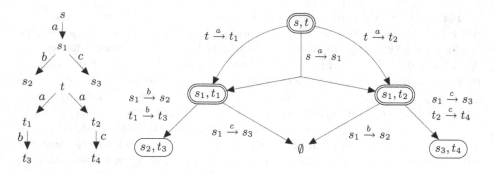

Fig. 2. Two processes s and t (left) and the constructed dependency graph (right)

that $A(v') = 1$ for all $v' \in T$. As all assignments form a complete lattice w.r.t. to the natural point-wise ordering and the function F is monotonic, there is by Knaster-Tarski theorem a unique minimum and maximum fixed point of the function F, denoted by A_{min} resp. A_{max}. The fixed points for G can be computed in linear-time by the use of local on-the-fly Algorithms [13].

We shall now hint at how the verification questions for CCS/TCCS can be encoded in fixed-point computations on dependency graphs. The idea, depicted in Fig. 2 for strong bisimulation, is that nodes in the dependency graph are pairs of processes and for any transition from one of the two processes, we create a new hyperedge with targets that correspond to all possible transitions under the same label from the other process. The hyperedges in the dependency graph are annotated with the transitions that initiated their creation. If for some pair of states there is a transition for which the other process does not have any answer, the resulting set of target nodes is empty and the created hyperedge ensures that the pair will get the value 1 in the minimum fixed-point assignment (denoted in our example by a double circle around the pair). One can prove that for any pair of nodes (s', t') in the dependency graph it holds that $s' \sim t'$ (the states are strongly bisimilar) if and only if $A_{min}((s', t')) = 1$. In order to establish that $A_{min}((s, t)) = 1$, it is enough to explore only a fraction of the dependency graph (e.g. constructing only two hyperedges from (s, t) to (s_1, t_1) and from (s_1, t_1) to the emptyset is sufficient). As the construction of the complete dependency graph can often be avoided by using on-the-fly algorithms, it is sometimes possible to show nonequivalence even for processes with infinitely many reachable states, a situation where the traditional partitioning algorithms will never terminate.

We can also use the computed fixed point on the dependency graph to derive a distinguishing formula for the processes s and t in Fig. 2. First, for every node that has a hyperedge with an empty set of targets, we can directly find such a formula, like for the node (s_1, t_1) where $s_1 \models \langle c \rangle tt$ while $t_1 \not\models \langle c \rangle tt$. From this formula we can now inductively construct a distinguishing formula $[a]\langle c \rangle tt$ for the root node (s, t). Note that this is not the only distinguishing formula, if we e.g. use instead the hyperedge from (s, t) with the two target nodes, we derive the formula $\langle a \rangle (\langle c \rangle tt \wedge \langle b \rangle tt)$ that is arguably more complex than the formula $[a]\langle c \rangle tt$. The problem of finding the simplest distinguishing formula is

---

Algorithm 1. Simple Communication Protocol (CCS) in CAAL

```
 1: Send = acc.Sending;
 2: Sending = 'send.Wait;
 3: Wait = ack.Send + error.Sending + 'send.Wait;

 4: Rec = trans.Del;
 5: Del = 'del.Ack;
 6: Ack = 'ack.Rec;

 7: Med = send.Med';
 8: Med' = 'trans.Med + tau.Err + tau.Med;
 9: Err = 'error.Med;

10: set L = {send, trans, ack, error};
11: Impl = (Send | Med | Rec) \ L;

12: Spec = acc.'del.Spec;
```

---

nontrivial and CAAL uses a greedy heuristic approach to report reasonably small distinguishing formulae.

The approach via dependency graphs is used also for trace-like equivalences and the corresponding dependency graphs are described in master theses available at the tool's homepage. For recursive formulae the construction of dependency graphs requires several copies of the graphs, one for each fixed-point definition, but the same uniform approach is also used here. Finally, the dependency graphs are used for guiding the tool in bisimulation and model checking games.

4   Case Study

We shall now present a simplified version of a communication protocol, where a sender is supposed to forward messages through unreliable medium to a receiver, who then acknowledges it via a direct handshake after which the protocol is again ready to accept another message. A more sophisticated variant of such a protocol (e.g. the Alternating Bit Protocol [2]) is a typical mini-project exercise that we use in our Semantics and Verification course.

The CCS processes describing the protocol are given in Algorithm 1. The sender, defined at lines 1–3, receives a message **acc** from the environment, forwards the message via the internal channel **send** to the medium and then waits for the acknowledgment, an error message from the medium, or tries to resend the message. The receiver, defined at lines 4–6, can receive the message through the medium via the internal channel **trans**, deliver the message to its environment via the output action **'del** and then acknowledge this to the sender. The medium, defined at lines 7–9, communicates with the sender/receiver via the channels **send** and **trans** but can also enter an error state and inform the sender about this (line 9) or silently lose the message and enter its initial state. The implementation of the protocol (line 11) is a parallel composition of the three components described above where all channels except for **acc** and **'del** are

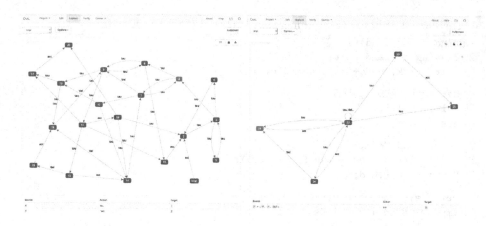

(a) Before weak bisimulation collapse   (b) After weak bisimulation collapse

Fig. 3. Reachable state-space for the process `Impl`

restricted, enforcing a handshake synchronization over these channels. Finally, at line 12, we can see the abstract specification of the protocol. We have deliberately introduced some design errors in order to demonstrate the typical mistakes the students make when modelling more advanced variants of communication protocols. In the rest of this section, we shall demonstrate the debugging options that CAAL offers for analysing and correcting such mistakes.

By entering the verification module of CAAL, we can promptly find out that the processes `Impl` and `Spec` are not weakly bisimilar. We can now enter the explorer module in order to visualize and interactively explore the transition system of the process `Impl` as depicted in Fig. 3a, however, even for this small example, the system is already too large. We can choose to visualize the collapsed transition system where all weakly bisimilar states are merged together as shown in Fig. 3b and here we can already see some design issues. We can e.g. observe that the implementation contains a deadlock (the right-most state).

In general, the labelled transition systems (even after the bisimulation collapse) are often too large to analyse manually. Hence, if two processes are not weakly bisimilar, a natural question to ask is whether they provide the same weak traces (sequences of visible actions). It appears that this is not the case for our example and CAAL informs the user that the process `Impl` can perform the sequence of visible actions `acc`, `'del`, `'del` and such a trace is not possible in the specification. By analysing the trace in the game module, we can see that the problem is at line 3 where the sender has the option to resubmit the message unboundedly many times. Hence we may decide to remove this resubmission option and modify the CCS definition as `Wait = ack.Send + error.Sending`. After this fix, we can now verify that the implementation and the specification are weakly trace equivalent, while they are still not bisimilar. We can ask CAAL to generate a distinguishing formula that holds in `Impl` but not in `Spec` and the tool returns the formula `<<acc>>[['del]]F` that states that the implementation

Algorithm 2. Time Annotated Communication Protocol (TCCS) in CAAL

```
1: Send = acc.2.'send.1.ack.2.Send;
2: Rec = trans.1.'del.2.'ack.8.Rec;
3: Med = send.(3.'trans.Med + 5.tau.Med);
4: Impl = (Send | Med | Rec) \ {send, trans, ack};
5: Spec = acc.'del.Spec;
```

can perform the visible action acc, possibly with some additional τ-transitions before and after, such that after this it is not possible to perform the action 'del (not even preceded by some τ-actions). By entering the game module, the user can play a game against the computer (playing defender) that will reveal to the user (playing attacker) that the formula holds in the state Impl. The game will in fact reveal the presence of a deadlock configuration that we already observed in the explorer.

Alternatively, we can directly formulate the deadlock property as a recursive HML formula X min= [-]ff or <->X. CAAL confirms that the implementation satisfies the property X and in the game module, the computer can convince the user that a deadlock is indeed reachable. The analysis of the discovered deadlock points to the fact that medium should not be allowed to silently discard messages. After changing the definition at line 8 into Med' = tau.Err + 'trans.Med we finally achieve a correct implementation (weakly bisimilar to its specification).

We may also ask if the protocol contains a reachable livelock (an infinite sequence of τ-actions that can be executed in a row). CAAL allows to formulate this property using two recursively defined variables Y min= Z or <->Y (claiming the reachability of livelock) and Z max= <tau>Z (expressing the existence of an infinite τ-sequence). The implementation indeed contains a livelock and the game provides a convincing argument for this fact.

CAAL moreover allows to model TCCS processes. A variant of the communication protocol is given in Algorithm 2, where both the sender, receiver and medium have been annotated with delays such that e.g. the medium needs 3 time units to deliver the message but it will lose it after 5 time units. CAAL will show that the processes Impl and Spec are weakly untimed bisimilar. However, if the receiver process gets just little bit slower and the delay prefix 8 at line 2 is replaced with delay 9 then the weak untimed bisimulation equivalence does not hold anymore as it is now possible that the medium loses a message.

We can also verify that the TCCS process Impl satisfies the formula

$$X \text{ max= [acc] <<0,6>> <'del> tt and [-]X;}$$

expressing the invariant that whenever the action acc is performed then the message can be delivered within 6 time units. The property X actually does not hold but if we ask instead whether the message can be delivered withing 7 time units then it is satisfied.

5   Conclusion

We presented CAAL, an educational tool for modelling and analysing CCS processes. The tool runs in a browser with limited computational resources but it benefits from the efficient on-the-fly algorithms. This is clearly sufficient for the typical student exercises and mini-projects. At the moment, we are exploring the parallelization of the fixed point computation and outsourcing this work to a super-computer via the approach "verification as a web-service". Our Reactive System book [1] is now used at more than 18 universities around the world and we expect that CAAL, once publicly announced, becomes a natural supplement to the concurrency courses based on the CCS formalism.

References

1. Aceto, L., Ingolfsdottir, A., Larsen, K.G., Srba, J.: Reactive Systems: Modelling. Cambridge University Press, Specification and Verification (2007)
2. Bartlett, K.A., Scantlebury, R.A., Wilkinson, P.T.: A note on reliable full-duplex transmission over half-duplex links. Commun. ACM **12**(5), 260–261 (1969)
3. Birgisson, A.: CCS model checker in Haskell (2009). https://github.com/arnar/ccs-searching. Accessed on 03 August 2015
4. Calzolai, F., De Nicola, R., Loreti, M., Tiezzi, F.: TAPAs: a tool for the analysis of process algebras. Trans. Petri Nets Other Models Concurr. **5100**, 54–70 (2008)
5. Cleaveland, R., Parrow, J., Steffen, B.: The concurrency workbench: a semantics-based tool for the verification of concurrent systems. ACM Trans. Program. Lang. Syst. **15**(1), 36–72 (1993)
6. Cleaveland, R., Sims, S.: The NCSU concurrency workbench. In: Alur, R., Henzinger, T.A. (eds.) CAV 1996. LNCS, vol. 1102, pp. 394–397. Springer, Heidelberg (1996)
7. Freiberger, F., Biewer, S., Held, P.: PseuCo (2014). http://pseuco.com. Accessed on 03 August 2015
8. Garavel, H., Lang, F., Mateescu, R., Serwe, W.: CADP 2011: a toolbox for the construction and analysis of distributed processes. Int. J. Softw. Tools Technol. Transfer **15**(2), 89–107 (2013)
9. Gibson-Robinson, T., Armstrong, P., Boulgakov, A., Roscoe, A.W.: FDR3 — a modern refinement checker for CSP. In: Ábrahám, E., Havelund, K. (eds.) TACAS 2014 (ETAPS). LNCS, vol. 8413, pp. 187–201. Springer, Heidelberg (2014)
10. Groote, J.F., Mousavi, M.R.: Modeling and Analysis of Communicating Systems. The MIT Press, Cambridge (2014)
11. Hennessy, M., Milner, R.: Algebraic laws for nondeterminism and concurrency. J. Assoc. Comput. Mach. **32**(1), 137–161 (1985)
12. Kozen, D.: Results on the propositional μ-calculus. Theoretical Computer Science **27**, 333–354 (1983)
13. Liu, X., Smolka, S.A.: Simple linear-time algorithms for minimal fixed points. In: Larsen, K.G., Skyum, S., Winskel, G. (eds.) ICALP 1998. LNCS, vol. 1443, p. 53. Springer, Heidelberg (1998)
14. Milner, R.: A Calculus of Communicating Systems. LNCS, vol. 92. Springer, Berlin (1980)
15. Mosegaard, M., Brabrand, C.: The bisimulation game (2006). http://www.brics.dk/bisim/. Accessed on 03 August 2015
16. Verdejo, A., Marti-Oliet, N.: Executing and verifying CCS in Maude. Technical report, Dpto. Sistemas Informaticos y Programacion, Universidad Complutense de (2002). http://maude.cs.uiuc.edu/maude1/casestudies/ccs/

A Tool for the Automated Verification of Nash Equilibria in Concurrent Games

Alexis Toumi, Julian Gutierrez[✉], and Michael Wooldridge

Department of Computer Science, University of Oxford, Oxford, UK
Julian.Gutierrez@cs.ox.ac.uk

Abstract. Reactive Modules is a high-level specification language for concurrent and multi-agent systems, used in a number of practical model checking tools. Reactive Modules Games is a game-theoretic extension of Reactive Modules, in which concurrent agents in the system are assumed to act strategically in an attempt to satisfy a temporal logic formula representing their individual goal. The basic analytical concept for Reactive Modules Games is Nash equilibrium. In this paper, we describe a tool through which we can automatically verify Nash equilibrium strategies for Reactive Modules Games. Our tool takes as input a system, specified in the Reactive Modules language, a representation of players' goals (expressed as CTL formulae), and a representation of players strategies; it then checks whether these strategies form a Nash equilibrium of the Reactive Modules Game passed as input. The tool makes extensive use of conventional temporal logic satisfiability and model checking techniques. We first give an overview of the theory underpinning the tool, briefly describe its structure and implementation, and conclude by presenting a worked example analysed using the tool.

1 Introduction

Model checking is the best-known and most successful technique for automated formal verification, and is focussed on the problem of checking whether a (computer) system S satisfies a property φ, where typically φ is represented as a temporal logic formula. Model checking has proved to be a very successful technique for systems where S is a complete and monolithic description of the state space of the system. In this case, S is usually called a *closed* system. However, in many situations, especially when dealing with concurrent and distributed multi-agent systems, S can be better represented as a collection of local and inter-dependent processes. In this modelling framework, it is common to understand such processes as *modules*, that is, as being *open* rather than closed systems, in which the behaviour of each process/module may depend on the behaviour of other processes, which constitute its environment, cf., [2,12].

We are interested in the verification of concurrent and multi-agent systems where (computer) processes are modelled as open systems. In particular, we are interested in systems modelled using a game-theoretic approach. In this setting, a system is modelled as a *game*, system components are modelled as *players*

© Springer International Publishing Switzerland 2015
M. Leucker et al. (Eds.): ICTAC 2015, LNCS 9399, pp. 583–594, 2015.
DOI: 10.1007/978-3-319-25150-9_34

(each choosing and then following a given *strategy*), possible computation runs are the *plays* of the game, and the desired or expected behaviour of the system is specified with the *goals* that the players of the game wish to see satisfied. In many cases, for instance when considering reactive systems, such goals can be naturally expressed using temporal logic formulae.

However, because now one is following a game-theoretic approach, it is only natural to ask whether the system has a stable behaviour from a game-theoretic point of view, that is, whether the strategies used by the players modelling the system are in equilibrium [15]. Then, in this case, we talk about *equilibrium checking* rather than model checking. In fact, model checking is a simpler instance of equilibrium checking where either players are forced to cooperate or the whole system is modelled as a one-player game. However, in general, these may not be the best representations of the system.

A way to model the kind of systems just described (open systems) is using the Reactive Modules Language (RML [2]). This is a high-level specification language for reactive, concurrent, and multi-agent systems, which is used in model checking tools such as MOCHA [1] and Prism [13]. However, RML is used to specify general open systems rather than concurrent games. Recently [11], a subset of RML, called the Simple Reactive Modules Language (SRML [19]) was given a game-theoretic interpretation, which provides a game semantics for reactive and concurrent systems written in SRML, and which can be used to perform an equilibrium analysis of open systems modelled as SRML specifications. Indeed, with SRML, one can analyse systems using a language that is much closer to real-world programming and system modelling languages.

In this paper, we present a tool for the automated verification of Nash equilibria in concurrent and reactive systems modelled as concurrent games succinctly represented using the SRML specification language. More specifically, we develop a Python implementation of the above theory of games that, in particular, can be used to solve the *equilibrium checking* problem for this kind of concurrent games/systems. Since the tool, which we call EAGLE ("**E**quilibrium **A**nalyser for **G**ame-**L**ike **E**nvironments"), can be used to automatically check whether a set of strategies forms a *Nash Equilibrium* in a given game-like concurrent system, its analytical power goes beyond model checking.

Related Work. Reactive Modules [2] is used as a specification language in verification tools such as MOCHA [1] and Prism [13]. In each case, open systems modelled as concurrent games can also be specified. However, these tools do not have explicit support for equilibrium analysis. Instead, it is model checking with respect to logics such as PCTL and ATL that these tools allow. MCMAS [14] is another tool for the specification and verification of open systems, modelled as multi-agent systems. In MCMAS, systems are described using the Interpreted Systems Programming Language and properties are described using ATL* and strategy logic—see [5]. Similar to MOCHA and Prism, in MCMAS the analysis of systems focuses on the model checking problem for the logics just mentioned. Because strategy logic can express the existence of Nash equilibria in a concurrent and multi-agent game, in principle, it is possible to analyse some equilibrium properties of MCMAS systems. However, this has to be manually crafted.

Closer to EAGLE is PRALINE [4], a tool for computing Nash equilibria in concurrent games played on graphs. Whereas PRALINE focuses on the synthesis problem (constructing strategies in equilibrium), EAGLE focuses on the verification problem (checking that a given profile of strategies is in equilibrium). There are many other tools available online which either use game techniques for design and verification or allow the analysis of winning strategies in games. For instance, see [3,6,9] for a few references. However, as just said, these tools focus on the study of winning strategies in such games rather than the equilibrium analysis of these systems/games.

2   Preliminaries

Logic. In this paper we will be dealing with logics that extend classical propositional logic. Thus, these logics are based on a finite set Φ of Boolean variables. A *valuation* for propositional logic is a set $v \subseteq \Phi$, with the intended interpretation that $p \in v$ means that p is true under valuation v, while $p \notin v$ means that p is false under v. For formulae φ we write $v \models \varphi$ to mean that φ is satisfied by v. Let $V(\Phi) = 2^\Phi$ be the set of all valuations for variables Φ; where Φ is clear, we omit reference to it and simply write V.

Kripke Structures. We use *Kripke structures* to model the dynamics of our systems. A Kripke structure K over Φ is given by $K = (S, S^0, R, \pi)$, where $S = \{s_0, \ldots\}$ is a finite non-empty set of *states*, $R \subseteq S \times S$ is a total *transition relation* on S, $S^0 \subseteq S$ is the set of *initial states*, and $\pi : S \to V$ is a valuation function, assigning a valuation $\pi(s)$ to every $s \in S$. Where $K = (S, S^0, R, \pi)$ is a Kripke structure over Φ, and $\Psi \subseteq \Phi$, then we denote the *restriction of K to Ψ* by $K|_\Psi$, where $K|_\Psi = (S, S^0, R, \pi|_\Psi)$ is the same as K except that the valuation function $\pi|_\Psi$ is defined as follows: $\pi|_\Psi(s) = \pi(s) \cap \Psi$.

Runs. A *run of K* is a sequence $\rho = s_0, s_1, s_2, \ldots$ where for all $t \in \mathbb{N}$ we have $(s_t, s_{t+1}) \in R$. Using square brackets around parameters referring to time points, we let $\rho[t]$ denote the state assigned to time point t by run ρ. We say ρ is an *s-run* if $\rho[0] = s$. A run ρ of K where $\rho[0] \in S^0$ is referred to as an *initial* run. Let *runs(K, s)* be the set of s-runs of K, and let *runs(K)* be the set of initial runs of K. Notice that a run $\rho \in runs(K)$ induces an infinite sequence $\boldsymbol{\rho} \in V^\omega$ of propositional valuations, viz., $\boldsymbol{\rho} = \pi(\rho[0]), \pi(\rho[1]), \pi(\rho[2]), \ldots$. The set of these sequences, we denote by **runs**(K). Given $\Psi \subseteq \Phi$ and a run $\boldsymbol{\rho}: \mathbb{N} \to V(\Phi)$, we denote the restriction of $\boldsymbol{\rho}$ to Ψ by $\boldsymbol{\rho}|_\Psi$, that is, $\boldsymbol{\rho}|_\Psi[t] = \boldsymbol{\rho}[t] \cap \Psi$ for each $t \in \mathbb{N}$. We can extend the notation for restriction of runs to sets of runs. In particular, we write **runs**$(K)|_\Psi$ for the set $\{\boldsymbol{\rho}|_\Psi : \boldsymbol{\rho} \in \textbf{runs}(K)\}$.

Trees. By a *tree* we here understand a non-empty set $T \subseteq \mathbb{N}_0^*$, such that (*i*) T is closed under prefixes, i.e., for every $u \in T$, also $(u) \subseteq T$, and (*ii*) $u \in T$ implies $ux \in T$ for some $x \in \mathbb{N}_0$. For $s \in S$, a *state-tree* for a Kripke structure $K = (S, S^0, R, \pi)$ is a function $\kappa: T \to S$, where $T \subseteq \mathbb{N}_0^*$ is a tree, $\kappa(\epsilon) \in S^0$, and, for every $u \in \mathbb{N}_0^*$ and $x, y \in \mathbb{N}_0$ such that $ux, uy \in T$, (i) $\kappa(u)$ R $\kappa(ux)$, and (ii) $\kappa(ux) = \kappa(uy)$ implies $x = y$. By $trees(K)$ we denote the state-trees for the Kripke structure K. By a *computation tree* we understand a function $\kappa \colon T \to V(\Phi)$, where T is a tree. For $\Psi \subseteq \Phi$ we write $\kappa|_\Psi$ for the restriction of κ to Ψ, i.e., for every $u \in T$, $\kappa|_\Psi(u) = \kappa(u) \cap \Psi$. Notice that every state-tree $\kappa \colon T \to S$ induces a computation tree $\kappa \colon T \to V(\Phi)$ such that for every $u \in T$ we have that $\kappa[u] = \pi(\kappa(u))$. In such a case κ is said to be a computation tree for K. The set of computation trees for K we denote by $\mathbf{trees}(K)$. We can extend the notation for restrictions of computation trees to sets of computation trees as done for runs, that is, we write $\mathbf{trees}(K)|_\Psi$ for the set $\{\kappa|_\Psi : \kappa \in \mathbf{trees}(K)\}$.

3   Reactive Modules Games

Reactive Modules. The objects used to define agents in SRML are known as *modules*. An SRML module consists of: *(i)* an *interface*, which defines the name of the module and lists the Boolean variables under the *control* of the module; and *(ii)* a number of *guarded commands*, which define the choices available to the module at every state.

Guarded commands are of two kinds: those used for *initialising* the variables under the module's control (**init** guarded commands), and those for *updating* these variables subsequently (**update** guarded commands). A guarded command has two parts: a condition part (the "guard") and an action part, which defines how to update the value of (some of) the variables under the control of a module. The intuitive reading of a guarded command $\varphi \rightsquigarrow \alpha$ is "if the condition φ is satisfied, then *one of the choices available to the module is to execute the action* α". We note that the truth of the guard φ does not mean that α *will* be executed: only that it is *enabled* for execution—it *may be chosen*.

Formally, a guarded command g over a set of Boolean variables Φ is an expression

$$\varphi \rightsquigarrow x_1' := \psi_1; \cdots ; x_k' := \psi_k$$

where φ (the guard) is a propositional formula over Φ, each x_i is a member of Φ and each ψ_i is a propositional logic formula over Φ. Let $guard(g)$ denote the guard of g. Thus, in the above rule, $guard(g) = \varphi$. We require that no variable appears on the left hand side of two assignment statements in the same guarded command. We say that $x_1, \ldots, x_k$ are the *controlled variables* of g, and denote this set by $ctr(g)$. If no guarded command of a module is enabled, the values of all variables in $ctr(g)$ are left unchanged.

Formally, an SRML module, m_i, is defined as a triple $m_i = (\Phi_i, I_i, U_i)$, where: $\Phi_i \subseteq \Phi$ is the (finite) set of variables controlled by m_i; I_i is a (finite) set of *initialisation* guarded commands, such that for all $g \in I_i$, we have $ctr(g) \subseteq \Phi_i$; and U_i is a (finite) set of *update* guarded commands, such that for all $g \in U_i$, we have $ctr(g) \subseteq \Phi_i$.

Moreover, an SRML *arena*, A, is defined to be an $(n+2)$-tuple

$$A = (N, \Phi, m_1, \ldots, m_n)$$

where $N = \{1, \ldots, n\}$ is a set of agents, Φ is a set of Boolean variables, and for each $i \in N$, $m_i = (\Phi_i, I_i, U_i)$ is an SRML module over Φ that defines the choices available to agent i. We require that $\{\Phi_1, \ldots, \Phi_n\}$ forms a partition of Φ (so every variable in Φ is controlled by some agent, and no variable is controlled by more than one agent).

The behaviour of an SRML arena is obtained by executing guarded commands, one for each module, in a synchronous and concurrent way. The execution of an SRML arena proceeds in rounds, where in each round every module $m_i = (\Phi_i, I_i, U_i)$ produces a valuation v_i for the variables in Φ_i on the basis of a current valuation v. For each SRML arena A, the execution of guarded commands induces a unique Kripke structure K_A, which formally defines the semantics of A. Based on K_A, one can define the sets of runs and computation trees allowed in A, namely, those associated with the Kripke structure K; we write $\mathbf{runs}(A)$ and $\mathbf{trees}(A)$ for such sets. Indeed, one can show that for every A there is a K_A such that $\mathbf{runs}(A) = \mathbf{runs}(K_A)|_\Phi$ and $\mathbf{trees}(A) = \mathbf{trees}(K_A)|_\Phi$, that is, with the same runs and computation trees when restricted to Φ. Likewise, for every K there is an SRML module whose runs and computation trees are those of K. In this paper, we provide, amongst others, a Python implementation of all these constructions.

Games. The model of games we consider has two components. The first component is an *arena*: this defines the players, some variables they control, and the choices available to them in every game state. Preferences are specified by the second component of the game: every player i is associated with a *goal* γ_i, which will be a logic formula. The idea, as in several models of strategic behaviour, is that players desire to see their goal satisfied by the outcome of the game. Formally, a game is given by a structure:

$$G = (A, \gamma_1 \ldots, \gamma_n)$$

where $A = (N, \Phi, m_1, \ldots, m_n)$ is an arena with player set N, Boolean variable set Φ, and m_i an SRML module defining the choices available to each player i; moreover, for each $i \in N$, the temporal logic formula γ_i represents the *goal* that i aims to satisfy.[1] Games are played by each player i selecting a *strategy* σ_i that will define how to make choices over time. Given an SRML arena $A = (N, \Phi, m_1, \ldots, m_n)$, a *strategy* for module $m_i = (\Phi_i, I_i, U_i)$ is a structure $\sigma_i = (Q_i, q_i^0, \delta_i, \tau_i)$, where Q_i is a finite and non-empty set of *states*, $q_i^0 \in Q_i$ is the *initial* state, $\delta_i : Q_i \times V_{-i} \to 2^{Q_i} \setminus \{\emptyset\}$ is a *transition function*, and $\tau_i : Q_i \to V_i$ is an *output function*. Note that not all strategies for a module may comply with that module's specification. For instance, if the only guarded update command of a module m_i has the form $\top \leadsto x' := \bot$, then a strategy for m_i should not prescribe m_i to set x to true under any contingency. Strategies that comply with

---

[1] Goals can be given by any logic with a Kripke structure semantics. Although we will consider CTL goals here, due to generality, at this point all definitions will be made leaving this open. Indeed, one could extend our implementation to SRML games with CTL* or μ-calculus goals.

the module's specification are called consistent. Let Σ_i be the set of consistent strategies for m_i. A strategy σ_i can be represented by an SRML module (of polynomial size in $|\sigma_i|$) with variable set $\Phi_i \cup Q_i$. We write m_{σ_i} for such a (strategy) module specification.

Once every player i has selected a strategy σ_i, a *strategy profile* $\vec{\sigma} = (\sigma_1, \ldots, \sigma_n)$ results and the game has an *outcome*, which we will denote by $[\![\vec{\sigma}]\!]$. The outcome $[\![\vec{\sigma}]\!]$ of a game with SRML arena $A = (N, \Phi, m_1, \ldots, m_n)$ is defined to be the Kripke structure associated with the SRML arena $A_{\vec{\sigma}} = (N, \Phi \cup \bigcup_{i \in N} Q_i, m_{\sigma_1}, \ldots, m_{\sigma_n})$ restricted to valuations with respect to Φ, that is, the Kripke structure $K_{A_{\vec{\sigma}}}|_\Phi$. The outcome of a game will determine whether or not each player's goal is or is not satisfied. Because outcomes are Kripke structures, in general, goals can be given by any logic with a well defined Kripke structure semantics. Assuming the existence of such a satisfaction relation, which we denote by "$\models$", we can say that a goal γ_i is satisfied by an outcome $[\![\vec{\sigma}]\!]$ if and only if $[\![\vec{\sigma}]\!] \models \gamma_i$; in order to simplify notations, we may simply write $\vec{\sigma} \models \gamma_i$.

We are now in a position to define a preference relation $\succsim_i$ over outcomes for each player i with goal γ_i. For strategy profiles $\vec{\sigma}$ and $\vec{\sigma}'$, we say that

$$\vec{\sigma} \succsim_i \vec{\sigma}' \text{ if and only if } \vec{\sigma}' \models \gamma_i \text{ implies } \vec{\sigma} \models \gamma_i.$$

On this basis, we can also define the standard solution concept of Nash equilibrium [15]: given a game $G = (A, \gamma_1, \ldots, \gamma_n)$, a strategy profile $\vec{\sigma}$ is said to be a *Nash equilibrium* of G if for all players i and all strategies σ_i' in the game, we have

$$\vec{\sigma} \succsim_i (\vec{\sigma}_{-i}, \sigma_i'),$$

where $(\vec{\sigma}_{-i}, \sigma_i')$ denotes the strategy profile $(\sigma_1, \ldots, \sigma_{i-1}, \sigma_i', \sigma_{i+1}, \ldots, \sigma_n)$. Hereafter, let $NE(G)$ be the set of (pure strategy) Nash equilibria of game G.

4 Reactive Modules Games in Python

Our main contribution is EAGLE, *a Python implementation of the theory of games* described in the previous sections. In particular, EAGLE allows a simple high-level Python description of games specified in SRML, where players are assumed to have branching-time (CTL) goals and strategies can be described as SRML modules. More importantly, EAGLE allows the automated verification of solutions of such games, that is, checking whether a particular profile of strategies is or is not a Nash equilibrium of a given RM game—a problem called *equilibrium checking*. From a systems analysis point of view, this is the game-theoretic equivalent to the *model checking* problem in formal verification. A short description of our verification tool is given next.

Our tool expects as input an RM game $G = (A = (N, \Phi, m_0, \ldots, m_n), (\gamma_i)_{i \in N})$ and a strategy profile $\vec{\sigma}$. Because strategies are modelled as finite state machines with output (which are known as transducers), they can easily be described, uniformly, using SRML. Goals, on the other hand, are written using the syntax for CTL formulae in [7]. For ease of use, a simple command-line interface can be used to input text files with the specification of games. An concrete example will be given later, but all implementation details can be found in [18]. Moreover, EAGLE implements an algorithm—which uses two external libraries for CTL satisfiability and model checking—that automatically solves these multi-player games, that is, their (Nash) equilibrium problem.

More precisely, on input $(G, \vec{\sigma})$, the tool outputs **True** if and only if $\vec{\sigma} \in NE(G)$. We have also implemented, using the command-line interface, a "verbose" mode in which a detailed account of the running process of the algorithm is given. For instance, apart from checking solutions of a given game, the tool reports whether or not players get their goal achieved, and in the case they do not, whether they could benefit from changing the strategy they are currently using. We should note that because in a Nash equilibrium strategy profile no player can benefit from unilaterally changing its strategy, it is the case that if the tool reports that $\vec{\sigma} \notin NE(G)$, then there is some player who does not get its goal achieved, but can change to a different strategy that achieves its goal. On the contrary, if the tool reports that $\vec{\sigma} \in NE(G)$, then no player can benefit from changing its strategy, in particular, those who do not get their goal achieved.

Throughout, we made the following assumptions, which define what a correct input is. In some cases, the assumptions are about the games themselves (1 & 2), and in other cases about the input files (3). In particular, we have made the following assumptions:

1. That the modules, both for the arena and for the strategy profile, respect the specification of SRML. In particular, we require: (a) that no variable is assigned twice in the same guarded command; (b) that the guards to **init** commands are "⊤"; (c) that in the assignment statements $x := \psi$ in **init** commands, ψ is a Boolean constant, ⊤ or ⊥; (d) that for every module $m_i = (\Phi_i, I_i, U_i)$, both I_i and U_i are sets instead of bags, i.e. that they contain only pairwise distinct elements; (e) that for every module $m_i = (\Phi_i, I_i, U_i)$ and for every command $g \in I_i \cup U_i$ we have that $ctr(g) \subseteq \Phi_i$.
2. That the strategy profile is consistent with the arena, as required by the game model.
3. That the input strings for goals are syntactically correct CTL formulae, in particular that they respect the alternation between path quantifiers and tense operators.

To make this concrete, we will, later and in the next section, present some examples.

CTL Satisfiability and Model Checking. In order to solve the equilibrium problem for Reactive Modules games we used a CTL variant of the algorithms first introduced in [10] to check whether a strategy profile is or is not a Nash equilibrium. The technique developed in [10] relies on the existence of two oracles, one for model checking and one for satisfiability of the temporal logic at hand. In the case of this paper, such oracles are for CTL, and can be obtained using any "off-the-shelf" open source external libraries for CTL satisfiability (CTL SAT) and CTL model checking (CTL MC). Specifically, we decided to use the Python CTL model checker MR. WAFFLES [17] and the CTL satisfiability checker in [16], both open source libraries available online.

For CTL MC, the MR.WAFFLES library implements Kripke structures with a class `PredicatedGraph` which extends the `networkx` library for finite graphs with a `predicate` attribute for every node: a list of the propositional variables (represented as strings) that are true at this node. It then provides a `check` method that takes a string representing a CTL formula (in prefix notation) and outputs a list of the states at which the formula is satisfied. Hence, checking whether a Kripke structure satisfies a CTL formula amounts to checking that all the initial states are in this list. For CTL SAT, we use a command-line interface to access an external program that inputs CTL formulae as strings (in infix notation), which is wrapped using a Python subprocess instance.

Concrete Data Structures. We represent propositional variables as ints, and propositional valuations as lists of ints. We implemented a Python class for propositional logic, which we used to store the guards and the Boolean values of guarded commands. There is one subclass for each case in the grammar and two special instances, `T` and `F`, to represent $\top$ and $\bot$. Also, we implemented assignment statements as Python named tuples (`var`, `b`) where `var` is an int and `b` is an instance of the propositional logic class. Guarded commands are implemented as named tuples (`guard`, `action`) where `guard` is an instance of the propositional logic class and `action` is a list of assignment statements. Reactive modules were also implemented as named tuples (`ctrl`, `init`, `update`) where `ctrl` is a list of ints representing the variables the module controls, `init` and `update` are lists of guarded commands.[2]

Input Format. As expected we use Python files, which we then parse using the Python `eval` function. The input to the equilibrium checking algorithm is represented as a Python `dict` with three keys: *(i)* `modules` is a list of reactive modules representing the SRML arena, *(ii)* `goals` is a list of CTL formulae represented as strings in MR.WAFFLES notation, and *(iii)* `strategies` is a list of reactive modules representing the strategy profile. More specifically, we represent modules as Python dictionaries, following the same structure as the named tuples for modules described before. The guards and the Boolean values in guarded commands are expressed using MR.WAFFLES prefix notation, and the propositional variable represented by the int n is simply denoted by `xn`. At this point it is worth noting that using our Python assistant any finite-state strategy can be represented, including non-deterministic ones, by extending the set of controlled variables to represent strategy states without affecting the outcome of the game (of course, as long as the strategy is consistent with its module).

System Architecture. Our system has five Python modules, as follows: 1. A module that implements the command-line interface and the main algorithm; it also implements the verbose mode and prints some running time measurements.

---

[2] EAGLE is being improved and updated frequently. The implementation details in this paper constitute the main design decisions at the moment of submission to ICTAC (in June 2015).

2. A module that implements the propositional logic class. 3. A module that implements the concrete data structures described before, as well as the parsing of input modules and guarded commands. 4. A module that implements the algorithm to translate an arena to its induced Kripke structure, represented as a MR. WAFFLES PredicatedGraph instance. 5. A module that implements a construction to translate an arena, given as a list of modules, into a single CTL formula (used with the CTL SAT command-line interface) representing the branching behaviour of the arena; this module is also responsible for wrapping the CTL SAT command-line interface, using a Python subprocess instance.

Evaluation. EAGLE was tested with a number of systems taken from the literature, and the results are reported in [18]. The running time measures show that its performance is greatly driven by the CTL satisfiability solver, which is used to check whether an alternative player's strategy could be constructed whenever a strategy profile does not satisfy some player's goal. Details can be found in [18]. These experimental results go from two-player games that required hours to be analysed (CTL SAT used) to multi-player games whose equilibrium analysis took a few seconds (only CTL MC used). It was clear, in all cases, that the bottleneck was in the CTL satisfiability subroutine. In the future, we would like to compare EAGLE with PRALINE [4], the only other tool we are aware of that is focused on the equilibrium analysis of concurrent games.

Example. This example illustrates the concrete syntax used for modules in SRML as well as its translation to the concrete syntax in our Python implementation. The SRML module depicted below (on the left), named *toggle*, controls two variables x and y. It has two **init** guarded commands and two **update** guarded commands. The **init** commands define two choices for the initialisation of the pair (x, y): assign it the value $(\top, \bot)$ or the value $(\bot, \top)$. The first **update** command says that if (x, y) has the value $(\top, \bot)$ then the corresponding choice is to assign it the value $(\bot, \top)$, while the second command says that if the pair (x, y) has the value $(\bot, \top)$, we can assign it the value (y, x) in the next state. Note that the two **update** commands define essentially the same choice, but in the first command the action mentions Boolean constants directly, whereas the second command mentions the values of the variables at the current state, and requires to evaluate those to assign the values for the next state. In other words, the module *toggle* first *non-deterministically* picks an initial pair in $\{(\top, \bot), (\bot, \top)\}$, then at each round it *deterministically* toggles between these two pairs. This SRML module is written in our Python assistant for equilibrium checking as shown below (on the right):

```
module toggle controls x, y              { # module "toggle"
   init                                    'ctrl ': [0, 1],
   :: ⊤ ⤳ x' := ⊤, y' := ⊥                'init ': [
   :: ⊤ ⤳ x' := ⊥, y' := ⊤                    "T -> x0' := T, x1' := F",
                                                "T -> x0' := F, x1' := T"],
   update                                  'update ': [
   :: (x ∧ ¬y) ⤳ x' := ⊥, y' := ⊤             "(and x0 !x1) -> x0' := F, x1' := T",
   :: (¬x ∧ y) ⤳ x' := y, y' := x             "(and !x0 x1) -> x0' := x1, x1' := x0"] }
```

5   Case Study: A Peer-to-Peer Communication Protocol

To understand better the usefulness of an equilibrium checking tool, we now present a case study based on the system presented in [8]. Consider a peer-to-peer network with two agents (the extension to $n > 2$ agents is straightforward—we restrict to two agents only due to space and ease of presentation). At each time step, each agent either tries to download or to upload. In order for one agent to download successfully, the other must be uploading at the same time, and both are interested in downloading infinitely often.

While [8] considers an iBG model [10], where there are no constraints on the values that players choose for the variables under their control, we will consider a modified version of the communication protocol: using guarded commands, we require that an agent cannot both download and upload at the same time. This is a simple example of a system which cannot be specified as an iBG, but which has an SRML representation.

We can specify the game modelling the above communication protocol as a game with two players, 0 and 1, where each player $i \in \{0, 1\}$ controls two variables u_i ("Player i tries to upload") and d_i ("Player i tries to download"); Player i downloads successfully if $(d_i \wedge u_{i-1})$. Formally, we define a game $G = (A, \gamma_0, \gamma_1)$, where $A = (\{0, 1\}, \Phi, m_0, m_1)$, $\Phi = \{u_0, u_1, d_0, d_1\}$, and m_0, m_1 are defined as follows:

module m_0 controls u_0, d_0
init
$$:: \top \rightsquigarrow u_0' := \top, \ d_0' := \bot$$
$$:: \top \rightsquigarrow u_0' := \bot, \ d_0' := \top$$
update
$$:: \top \rightsquigarrow u_0' := \top, \ d_0' := \bot$$
$$:: \top \rightsquigarrow u_0' := \bot, \ d_0' := \top$$

module m_1 controls u_1, d_1
init
$$:: \top \rightsquigarrow u_1' := \top, \ d_1' := \bot$$
$$:: \top \rightsquigarrow u_1' := \bot, \ d_1' := \top$$
update
$$:: \top \rightsquigarrow u_1' := \top, \ d_1' := \bot$$
$$:: \top \rightsquigarrow u_1' := \bot, \ d_1' := \top$$

Players' goals can be easily specified in CTL: the informal *"infinitely often"* requirement can be expressed in CTL as *"From all system states, on all paths, eventually"*. Hence, for $i \in \{0, 1\}$, we define the goals as follows: $\gamma_i = \mathbf{AGAF}(d_i \wedge u_{1-i})$.

This is clearly a very simple system/game: only two players and four controlled variables. Yet, checking the Nash equilibria of the game associated with this system is a hard problem. One can show—and formally verify using EAGLE—that this game has at least two different kinds of Nash equilibria (one where no player gets its goal achieved, and another one, which is Pareto optimal, where both players get their goal achieved). In general, the game has infinitely many Nash equilibria, but they all fall within the above two categories. Based on the SRML specifications of players' strategies given below, which can be seen to be consistent with modules m_0 and m_1, we can verify that both $(StPlayer(0), StPlayer(1)) \notin NE(G)$ and $(OnlyUp(0), OnlyUp(1)) \in NE(G)$.

module *StPlayer*(i) **controls** u_i, d_i module *OnlyUp*(i) **controls** u_i, d_i

init   **init**

:: $\top \rightsquigarrow u_i' := \top,\ d_i' := \bot$ :: $\top \rightsquigarrow u_i' := \bot,\ d_i' := \top$

update   **update**

:: $\top \rightsquigarrow u_i' := d_i,\ d_i' := u_i$ :: $\top \rightsquigarrow u_i' := \bot,\ d_i' := \top$

6 Future Work

We see a number of ways in which EAGLE can be improved: From a theoretical point of view, there is no reason to restrict to CTL goals. More powerful temporal logics could be considered. Also, our tool solves games with respect to the most widely used solution concept in game theory: Nash equilibrium. However, other solution concepts could be considered. It would also be useful to support, *e.g.*, quantitative/probabilistic reasoning or epistemic specifications so that more general agent's preference relations or beliefs can be modelled. Finally, even though our verification system is quite easy to use, we could implement a more user-friendly interface to input temporal logic goals. At present, our main limitations are given by the syntax used by the two external libraries we use to solve the underlying CTL satisfiability and model checking problems.

Acknowledgment. EAGLE was implemented by Toumi as part of his final Computer Science project [18] at Oxford. Both EAGLE and [18] can be obtained from him. (To obtain EAGLE or [18], please, send an email to Alexis.Toumi at gmail.com). We also acknowledge the support of the ERC Research Grant 291528 ("RACE") at Oxford.

References

1. Alur, R., Henzinger, T.A., Mang, F., Qadeer, S., Rajamani, S., Tasiran, S.: MOCHA: modularity in model checking. In: Hu, A.J., Vardi, M.Y. (eds.) CAV 1998. LNCS, vol. 1427, pp. 521–525. Springer, Heidelberg (1998)
2. Alur, R., Henzinger, T.A.: Reactive modules. Form. Meth. Syst. Des. **15**(1), 7–48 (1999)
3. Berwanger, D., Chatterjee, K., De Wulf, M., Doyen, L., Henzinger, T.A.: Alpaga: a tool for solving parity games with imperfect information. In: Kowalewski, S., Philippou, A. (eds.) TACAS 2009. LNCS, vol. 5505, pp. 58–61. Springer, Heidelberg (2009)
4. Brenguier, R.: PRALINE: a tool for computing nash equilibria in concurrent games. In: Sharygina, N., Veith, H. (eds.) CAV 2013. LNCS, vol. 8044, pp. 890–895. Springer, Heidelberg (2013)
5. Čermák, P., Lomuscio, A., Mogavero, F., Murano, A.: MCMAS-SLK: a model checker for the verification of strategy logic specifications. In: Biere, A., Bloem, R. (eds.) CAV 2014. LNCS, vol. 8559, pp. 525–532. Springer, Heidelberg (2014)
6. David, A., Jensen, P.G., Larsen, K.G., Mikučionis, M., Taankvist, J.H.: Uppaal Stratego. In: Baier, C., Tinelli, C. (eds.) TACAS 2015. LNCS, vol. 9035, pp. 206–211. Springer, Heidelberg (2015)

7. Emerson, E.A.: Temporal and modal logic. In: van Leeuwen, J. (ed.) Handbook of Theoretical Computer Science Volume B: Formal Models and Semantics, pp. 996–1072. Elsevier, Amsterdam (1990)
8. Fisman, D., Kupferman, O., Lustig, Y.: Rational synthesis. In: Esparza, J., Majumdar, R. (eds.) TACAS 2010. LNCS, vol. 6015, pp. 190–204. Springer, Heidelberg (2010)
9. Friedmann, O., Lange, M.: Solving parity games in practice. In: Liu, Z., Ravn, A.P. (eds.) ATVA 2009. LNCS, vol. 5799, pp. 182–196. Springer, Heidelberg (2009)
10. Gutierrez, J., Harrenstein, P., Wooldridge, M.: Iterated boolean games. In: IJCAI, IJCAI/AAAI (2013)
11. Gutierrez, J., Harrenstein, P., Wooldridge, M.: Verification of temporal equilibrium properties of games on Reactive Modules. Technical report, University of Oxford (2015)
12. Kupferman, O., Vardi, M., Wolper, P.: Module checking. Inf. Comput. **164**(2), 322–344 (2001)
13. Kwiatkowska, M., Norman, G., Parker, D.: PRISM 4.0: verification of probabilistic real-time systems. In: Gopalakrishnan, G., Qadeer, S. (eds.) CAV 2011. LNCS, vol. 6806, pp. 585–591. Springer, Heidelberg (2011)
14. Lomuscio, A., Qu, H., Raimondi, F.: MCMAS: a model checker for the verification of multi-agent systems. In: Bouajjani, A., Maler, O. (eds.) CAV 2009. LNCS, vol. 5643, pp. 682–688. Springer, Heidelberg (2009)
15. Osborne, M.J., Rubinstein, A.: A Course in Game Theory. MIT Press, Cambridge (1994)
16. Prezza, N.: CTLSAT (2015). https://github.com/nicolaprezza/CTLSAT
17. Reynaud, D., Mr. Waffles: (2015). http://mrwaffles.gforge.inria.fr
18. Toumi, A.: Equilibrium checking in Reactive Modules games. Technical report, Department of Computer Science, University of Oxford (2015)
19. van der Hoek, W., Lomuscio, A., Wooldridge, M.: On the complexity of practical ATL model checking. In: AAMAS, pp. 201–208. ACM (2006)

Short Papers

A Mathematical Game Semantics
of Concurrency and Nondeterminism

Julian Gutierrez[✉]

Department of Computer Science, University of Oxford, Oxford, UK
julian.gutierrez@cs.ox.ac.uk

Abstract. Concurrent games as event structures form a partial order model of concurrency where concurrent behaviour is captured by nondeterministic concurrent strategies—a class of maps of event structures. Extended with winning conditions, the model is also able to give semantics to logics of various kinds. An interesting subclass of this game model is the one considering deterministic strategies only, where the induced model of strategies can be fully characterised by closure operators. The model based on closure operators exposes many interesting mathematical properties and allows one to define connections with many other semantic models where closure operators are also used. However, such a closure operator semantics has not been investigated in the more general nondeterministic case. Here we do so, and show that some nondeterministic concurrent strategies can be characterised by a new definition of nondeterministic closure operators which agrees with the standard game model for event structures and with its extension with winning conditions.

Keywords: Concurrent games · Event structures · Closure operators

1 Introduction

Event structures [13] are a canonical model of concurrency within which the partial order behaviour of nondeterministic concurrent systems can be represented. In event structures, the behaviour of a system is modelled via a partial order of events which are used to explicitly model the causal dependencies between the events that a computing system performs. Following this approach, in the model of event structures, the interplay between *concurrency* (independence of events) and *nondeterminism* (conflicts between events) can be naturally captured.

Event structures have a simple two-player game-theoretic interpretation [16]. Within this framework, games are represented by event structures with polarities, and a strategy on a game is a polarity-preserving map of event structures satisfying some behaviour-preserving properties. In [16], concurrent games were presented as event structures and proposed as a new, alternative basis for the semantics of concurrent systems and programming languages. The definition of strategies as presented in [16] was given using *spans of event structures*—a family of maps of event structures. This definition has been both generalised and specialised to better understand particular classes of systems/games.

© Springer International Publishing Switzerland 2015
M. Leucker et al. (Eds.): ICTAC 2015, LNCS 9399, pp. 597–607, 2015.
DOI: 10.1007/978-3-319-25150-9_35

For instance, in [20] the original definition of strategies was given a characterisation based on *profunctors*, and related sheaves and factorisation systems, a more abstract presentation that can provide links with other models of concurrency based on games. In the other direction, in [19], Winskel studied a subclass of concurrent systems corresponding to deterministic games. In this simpler setting, concurrent strategies were shown to correspond to *closure operators*.

In this paper, we will investigate a model of strategies that is intermediate between the representations based on closure operators (which correspond to *deterministic strategies*) and profunctors (which correspond to the general model of *nondeterministic strategies*). In particular, we provide a mathematical model, which builds on closure operators and has a simple game-theoretic interpretation, where some forms of concurrency and nondeterminism are allowed to coexist.

Semantic frameworks based on closure operators are not new. In fact, they have been used in various settings as a semantic basis, amongst other reasons, because they can provide a mathematically elegant model of concurrent behaviour—see, *e.g.*, [3,7,14,17,19], for some examples. In particular, semantics based on closure operators provide an intuitively simple *operational* reading of their behaviour. However, such a simplicity comes at a price: the interplay between concurrency and nondeterminism must be severely restricted.

The model we provide here inherits many of the desirable features of systems with closure operator semantics, but also some of its limitations. In particular, it can be used to represent concurrent systems/games represented as event structures having a property called *race-freedom*, a structural condition on event structure games which ensures that no player can interfere with the moves available to the other. Our main results are significant since most known applications of games as event structures fall within the scope of the class of race-free games (cf. Sect. 6). The various models of games and strategies we have described above can be organised, in terms of *expressive* power, as shown in Fig. 1.

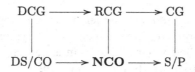

Fig. 1. The following abbreviations are used: Deterministic Concurrent Games (DCG); Race-free Concurrent Games (RCG); General Concurrent Games (CG); Deterministic Strategies (DS); Closure Operators (CO); Nondeterministic CO (NCO); Strategies as spans of event structures (S) and profunctors (P). The model of strategies in bold (**NCO**) is the one investigated in this paper.

Structure of the Paper. The rest of the paper is organised as follows. Section 2 presents some background material on concurrent games as event structures and Sect. 3 introduces nondeterministic closure operators. Section 4 describes when and how concurrent strategies can be characterised as nondeterministic closure operators and Sect. 5 extends such a characterisation to games with winning conditions. Section 6 concludes, describes some relevant related work, and puts forward a number of potential interesting application domains.

2   Concurrent Games as Event Structures

An *event structure* comprises $(E, \leq, \mathrm{Con})$, consisting of a set E, of *events* which are partially ordered by $\leq$, the *causal dependency relation*, and a nonempty *consistency relation* Con consisting of finite subsets of E, which satisfy axioms:

$$\{e' \mid e' \leq e\} \text{ is finite for all } e \in E,$$
$$\{e\} \in \mathrm{Con} \text{ for all } e \in E,$$
$$Y \subseteq X \in \mathrm{Con} \implies Y \in \mathrm{Con}, \text{ and}$$
$$X \in \mathrm{Con} \ \& \ e \leq e' \in X \implies X \cup \{e\} \in \mathrm{Con}.$$

The *configurations* of E consist of those subsets $x \subseteq E$ which are

Consistent: $\forall X \subseteq x.\ X$ is finite $\implies X \in \mathrm{Con}$, and
Down-closed: $\forall e, e'.\ e' \leq e \in x \implies e' \in x.$

We write $\mathcal{C}(E)$ for the set of configurations of E. We say that an event structure is well-founded if all its configurations are finite. We only consider well-founded event structures. Two events e_1, e_2 which are both consistent and incomparable with respect to causal dependency in an event structure are regarded as *concurrent*, written $e_1 \ co \ e_2$. In games the relation of *immediate* dependency $e \rightarrow e'$, meaning e and e' are distinct with $e \leq e'$ and no event in between plays an important role. For $X \subseteq E$ we write $[X]$ for $\{e \in E \mid \exists e' \in X.\ e \leq e'\}$, the down-closure of X; note that if $X \in \mathrm{Con}$ then $[X] \in \mathrm{Con}$. We use $x {-}\!\!\subset y$ to mean y covers x in $\mathcal{C}(E)$, i.e., $x \subset y$ with nothing in between, and $x \stackrel{e}{-}\!\!\subset y$ to mean $x \cup \{e\} = y$ for $x, y \in \mathcal{C}(E)$ and event $e \notin x$. We use $x \stackrel{e}{-}\!\!\subset$, expressing that event e is enabled at configuration x, when $x \stackrel{e}{-}\!\!\subset y$ for some configuration y.

Let E and E' be event structures. A *map* of event structures is a partial function on events $f : E \rightarrow E'$ such that for all $x \in \mathcal{C}(E)$ its direct image $fx \in \mathcal{C}(E')$ and if $e_1, e_2 \in x$ and $f(e_1) = f(e_2)$ (with both defined) then $e_1 = e_2$. The map expresses how the occurrence of an event e in E induces the coincident occurrence of the event $f(e)$ in E' whenever it is defined. Maps of event structures compose as partial functions, with identity maps given by identity functions. Thus, we say that the map is *total* if the function f is total.

The category of event structures is rich in useful constructions on processes. In particular, *pullbacks* are used to define the composition of *strategies*, while *restriction* (a form of equalizer) and the *defined part* of maps will be used in defining strategies. Any map of event structures $f : E \rightarrow E'$, which may be a partially defined on events, has a *defined part* the total map $f_0 : E_0 \rightarrow E'$, in which the event structure E_0 has events those of E at which f is defined, with causal dependency and consistency inherited from E, and where f_0 is simply f restricted to its domain of definition. Given an event structure E and a subset $R \subseteq E$ of its events, the *restriction* $E {\restriction} R$ is the event structure comprising events

$\{e \in E \mid [e] \subseteq R\}$ with causal dependency and consistency inherited from E; we sometimes write $E \setminus S$ for $E \restriction (E \setminus S)$, where $S \subseteq E$.

Event Structures with Polarity. Both a game and a strategy in a game are represented with event structures with polarity, comprising an event structure E together with a polarity function $pol : E \to \{+, -\}$ ascribing a polarity $+$ (Player) or $-$ (Opponent) to its events; the events correspond to moves. Maps of event structures with polarity, are maps of event structures which preserve polarities. An event structure with polarity E is *deterministic* iff

$$\forall X \subseteq_{\text{fin}} E.\ Neg[X] \in Con_E \implies X \in Con_E,$$

where $Neg[X] =_{\text{def}} \{e' \in E \mid pol(e') = -\ \&\ \exists e \in X.\ e' \leq e\}$. We write $Pos[X]$ if $pol(e') = +$. The *dual*, $E^\perp$, of an event structure with polarity E comprises the same underlying event structure E but with a reversal of polarities.

Given two sets of events x and y, we write $x \subset^+ y$ to express that $x \subset y$ and $pol(y \setminus x) = \{+\}$; similarly, we write $x \subset^- y$ iff $x \subset y$ and $pol(y \setminus x) = \{-\}$.

Games and Strategies. Let A be an event structure with polarity—a game; its events stand for the possible moves of Player and Opponent and its causal dependency and consistency relations the constraints imposed by the game.

A *strategy (for Player)* in A is a total map $\sigma : S \to A$ from an event structure with polarity S, which is both *receptive* and *innocent*. Receptivity ensures an openness to all possible moves of Opponent. Innocence, on the other hand, restricts the behaviour of Player; Player may only introduce new relations of immediate causality of the form $\ominus \rightarrowtail \oplus$ beyond those imposed by the game.

Receptivity: A map σ is *receptive* iff
$$\sigma x \xrightarrow{a} \subset\ \&\ pol_A(a) = -\ \implies\ \exists! s \in S.\ x \xrightarrow{s} \subset\ \&\ \sigma(s) = a.$$

Innocence: A map σ is *innocent* iff
$$s \rightarrowtail s'\ \&\ (pol(s) = +\ \text{or}\ pol(s') = -)\ \implies\ \sigma(s) \rightarrowtail \sigma(s').$$
Say a strategy $\sigma : S \to A$ is *deterministic* if S is deterministic.

Composing Strategies. Suppose that $\sigma : S \to A$ is a strategy in a game A. A counter-strategy is a strategy of Opponent, so a strategy $\tau : T \to A^\perp$ in the dual game. The effect of playing-off a strategy σ against a counter-strategy τ is described via a pullback. Ignoring polarities, we have total maps of event structures $\sigma : S \to A$ and $\tau : T \to A$. Form their pullback,

$$
\begin{array}{ccc}
P & \xrightarrow{\ \Pi_2\ } & T \\
{\scriptstyle \Pi_1}\downarrow & \lrcorner & \downarrow{\scriptstyle \tau} \\
S & \xrightarrow[\ \sigma\ ]{} & A.
\end{array}
$$

The event structure P describes the play resulting from playing-off σ against τ. Because σ or τ may be nondeterministic there can be more than one maximal configuration z in $\mathcal{C}(P)$. A maximal z images to a configuration $\sigma\Pi_1 z = \tau\Pi_2 z$ in $\mathcal{C}(A)$. Define the set of *results* of playing-off σ against τ to be

$$\langle\sigma, \tau\rangle =_{\text{def}} \{\sigma\Pi_1 z \mid z \text{ is maximal in } \mathcal{C}(P)\}.$$

Winning Conditions. A *game with winning conditions* comprises $G = (A, W)$ where A is an event structure with polarity and the set $W \subseteq \mathcal{C}(A)$ consists of the *winning configurations (for Player)*. Define the *losing conditions (for Player)* to be $L = \mathcal{C}(A) \setminus W$. The dual $G^\perp$ of a game with winning conditions $G = (A, W)$ is defined to be $G^\perp = (A^\perp, L)$, a game where the roles of Player and Opponent are reversed, as are correspondingly the roles of winning and losing conditions.

A strategy in G is a strategy in A. A strategy in G is regarded as *winning* if it always prescribes moves for Player to end up in a winning configuration, no matter what the activity or inactivity of Opponent. Formally, a strategy $\sigma : S \to A$ in G is *winning (for Player)* if $\sigma x \in W$ for all $\oplus$-maximal configurations $x \in \mathcal{C}(S)$—a configuration x is $\oplus$-maximal if whenever $x \overset{s}{-\!\!\!-\!\!\subset}$ then the event s has $-$ve polarity. Equivalently, a strategy σ for Player is winning if when played against any counter-strategy τ of Opponent, the final result is a win for Player; precisely, it can be shown [5] that a strategy σ is a winning for Player iff all the results $\langle\sigma, \tau\rangle$ lie within W, for any counter-strategy τ of Opponent.

3   Nondeterministic Closure Operators

It is often useful to think "operationally" of a strategy $\sigma : S \to A$ as an function that associates to a configuration of A another configuration of A that, potentially, can be played next. Since, in general, a concurrent strategy can be nondeterministic then such a function may not be between configurations of A, but rather a function from $\mathcal{C}(A)$ to the powerset of $\mathcal{C}(A)$, denoted by $\wp(\mathcal{C}(A))$. In particular, for *race-free* concurrent games—those games which satisfy a structural condition called race-freedom, to be defined in the following section—given a strategy $\sigma : S \to A$, we define $\sigma_\mu : \mathcal{C}(A) \to \wp(\mathcal{C}(A))$ with respect to σ as follows:

$$y' \in \sigma_\mu(y) \text{ iff } \exists x, x' \in \mathcal{C}(S). \ \sigma x = y \ \& \ x' \in f_\mu^{\rightarrow}(x) \ \& \ \sigma x' = y'$$

for some operator $f_\mu^{\rightarrow} : \mathcal{C}(S) \to \wp(\mathcal{C}(S))$, also defined with respect to $\sigma : S \to A$, as a *nondeterministic closure operator* $f^{\rightarrow} : \mathcal{C}(S) \to \wp(\mathcal{C}(S))$, that is, as an operator from $\mathcal{C}(S)$ to $\wp(\mathcal{C}(S))$ that satisfies the following properties:

1. $\forall x' \in f^{\rightarrow}(x). \ x \subseteq^+ x'$,
2. $\forall x' \in f^{\rightarrow}(x). \ \{x'\} = f^{\rightarrow}(x')$,
3. $x_1 \subseteq^- x_2 \implies f^{\rightarrow}(x_1) \subseteq f^{\rightarrow}(x_2)$

In fact (for 3):

$$\forall x_1' \in f^{\rightarrow}(x_1). \ \exists \ x_2' \in f^{\rightarrow}(x_2). \ x_1' \subseteq x_2' \text{ and}$$
$$\forall x_2' \in f^{\rightarrow}(x_2). \ \exists! x_1' \in f^{\rightarrow}(x_1). \ x_1' \subseteq x_2'.$$

That is, such that for some x, x' in $\mathcal{C}(S)$ and $f_\mu^\rightarrow$, the diagram below commutes:

$$
\begin{array}{ccc}
x & \xrightarrow{f_\mu^\rightarrow} & x' \\
\sigma \downarrow & & \downarrow \sigma \\
y & \xrightarrow{\sigma_\mu} & y'
\end{array}
$$

Remark. *If $f_\mu^\rightarrow$ is deterministic in the sense that the image of $f_\mu^\rightarrow(x)$ is a singleton set, for every $x \in \mathcal{C}(S)$, then $f_\mu^\rightarrow$ can be regarded as a usual closure operator on the configurations of S, with the order given by set inclusion. To see this, simply let x', $\{x'\}$, and $f_\mu^\rightarrow(x)$ be $\mathrm{cl}^\rightarrow(x)$, where $\mathrm{cl}^\rightarrow(x) = \bigcup f_\mu^\rightarrow(x)$, and eliminate quantifiers as they are no longer needed. Moreover, the condition that $Pos[x_1] = Pos[x_2]$ (given by $x_1 \subseteq^- x_2$ in 3) can be eliminated too as no positive event of $\bigcup f_\mu^\rightarrow(x_1)$ is inconsistent with a positive event of x_2. And since $f_\mu^\rightarrow(x)$ is the set of maximal configurations in $\{x' \in \mathcal{C}(S) \mid x \subseteq^+ x'\}$ we know that $f_\mu^\rightarrow$ preserves negative events; then we can also omit all references to polarities so as to yield the following presentation: 1. $x \subseteq \mathrm{cl}^\rightarrow(x)$; 2. $\mathrm{cl}^\rightarrow(x) = \mathrm{cl}^\rightarrow(\mathrm{cl}^\rightarrow(x))$; 3. $x_1 \subseteq x_2 \implies \mathrm{cl}^\rightarrow(x_1) \subseteq \mathrm{cl}^\rightarrow(x_2)$. These facts are formally presented below.*

Proposition 1 (Deterministic Games). *Let A be a game and $\sigma : S \to A$ a concurrent strategy. If S is deterministic, then $f_\mu^\rightarrow$ is a closure operator.*

4   Strategies as Nondeterministic Closure Operators

In [5] it was shown that in order to build a bicategory of concurrent games, where the objects are event structures and the morphisms are concurrent strategies (that is, innocent and receptive maps of event structures), a structural property called *race-freedom* had to be satisfied by the 'copy-cat' strategy in order to behave as an identity in such a bicategory. Race-freedom proved again to be a fundamental structural property of games as event structures when studying games with winning conditions: it was, in [5], shown to be a necessary and sufficient condition for the existence of winning strategies in well-founded games.

Race-freedom, formally defined below, is satisfied by all concurrent games we are aware of. Informally, race-freedom is a condition that prevents one player from interfering with the moves available to the other player. Formally, a game A is *race-free* if and only if for all configurations $y \in \mathcal{C}(A)$ the following holds:

$$
y \xrightarrow{a} \subset\ \&\ y \xrightarrow{a'} \subset\ \&\ pol(a) \neq pol(a') \implies y \cup \{a, a'\} \in \mathcal{C}(A).
$$

Race-freedom proves to be useful again. It is shown to be a necessary and sufficient condition characterising strategies as nondeterministic closure operators. To see that race-freedom is necessary, consider the following simple example.

Example 2 (Race-Freedom). Let A be the game depicted below. The wiggly line means conflict, that is, that the set of events $\{\ominus, \oplus\}$ is not a configuration of A.

$$\ominus \sim\!\!\sim\!\!\sim \oplus$$

This game is not race-free. Moreover, there is a strategy for Player that cannot be represented as a nondeterministic closure operator, namely, the strategy $\sigma : S \rightarrow A$ that plays $\oplus$. To see that this is the case, consider condition 3 of nondeterministic closure operators (the other two conditions are satisfied). Let $f^{\rightarrow} : \mathcal{C}(S) \rightarrow \wp(\mathcal{C}(S))$ be a candidate nondeterministic closure operator to represent σ. Observe that even though $\emptyset \subseteq \{\ominus\}$, it is not the case that $f^{\rightarrow}(\emptyset) \subseteq f^{\rightarrow}(\{\ominus\})$; indeed, $f^{\rightarrow}(\emptyset) = \{\{\oplus\}\}$ and $f^{\rightarrow}(\{\ominus\}) = \{\{\ominus\}\}$.   □

Proposition 3. *Let A be a concurrent game that is not race-free. Then, there is a nondeterministic strategy $\sigma : S \rightarrow A$ for Player that do not determine a nondeterministic closure operator on $\mathcal{C}(S)$—and similarly for Opponent.*

Then, if one wants to build a model where *every* strategy has a nondeterministic closure operator representation for *every* game, race-freedom will be a *necessary* condition. This is not a surprising result since, as mentioned before, copy-cat strategies, which can be represented as conventional closure operators, require this condition. What is, therefore, much more interesting is that race-freedom is in fact a *sufficient* condition too, as shown by the result below.

Theorem 4 (Closure Operator Characterisation). *Let $\sigma : S \rightarrow A$ be a nondeterministic concurrent strategy in a race-free concurrent game A. Then, the strategy σ determines a nondeterministic closure operator on $\mathcal{C}(S)$.*

Proof (Sketch). Since A is race-free then S is race-free (because S cannot introduce inconsistencies between events of opposite polarity). Then, $f_\mu^{\rightarrow}(x)$ is the set of $\oplus$-maximal configurations that cover x, namely $f_\mu^{\rightarrow}$ is the nondeterministic closure operator determined by σ, as shown next.

Suppose $\sigma x = y$ & $\sigma x' = y'$ & $y' \in \sigma_\mu(y)$. Then (for 1) $x \subseteq x'$ and $Neg[x] = Neg[x']$, for every $x' \in f_\mu^{\rightarrow}(x)$. And, (for 2) as every x' is $\oplus$-maximal, then it cannot be extended positively by any configuration; hence, $f_\mu^{\rightarrow}(x') = \{x'\}$. Now, (for 3) suppose $\sigma x_1 = y_1$ & $\sigma x_2 = y_2$ & $y_2 \in \sigma_\mu(y_1)$, with $x_1 \subseteq^- x_2$ (and therefore $y_1 \subseteq^- y_2$). Thus, since

- $f_\mu^{\rightarrow}(x_1)$ is the set of $\oplus$-maximal configurations that cover x_1, and
- $Pos[x_1] = Pos[x_2]$, and
- S is race-free,

then $f_\mu^{\rightarrow}(x_1) \subseteq f_\mu^{\rightarrow}(x_2)$, because x_2 enables at least as many $\oplus$-events as x_1; recall that $x_1 \subseteq^- x_2$ means that $Neg[x_1] \subseteq Neg[x_2]$ and $Pos[x_1] = Pos[x_2]$.   □

Informally, what Theorem 4 shows is that whereas in the deterministic case, a strategy $\sigma : S \rightarrow A$ can be seen as a partial function between the configurations of A which satisfies the axioms of a closure operator, in the nondeterministic race-free setting, a strategy can be seen as a partial function from $\mathcal{C}(A)$ to $\wp(\mathcal{C}(A))$ which satisfies the axioms of a nondeterministic closure operator. This, we believe, gives a more operational view of strategies than the one given by strategies as maps of event structures [16] or as certain fibrations and profunctors [20].

Race-free/Probabilistic Games. Because our nondeterministic closure operator characterisation of strategies only applies to race-free games, a natural question is whether race-freedom is either a mild or a severe modelling restriction. (We already know that race-freedom is not a real restriction with respect to sequential systems, but it is a restriction with respect to concurrent ones.) Even though we do not address such a question in this paper, we would like to note that a possible way to relax the race-freedom structural condition is by moving to a quantitative setting where races were allowed but only in a probabilistic manner, that is, to a setting where players' choices are associated with a probability distribution.

5 Characterising Winning Strategies

Theorem 4 provides a key closure operator (game semantic) characterisation of the model of nondeterministic concurrent strategies in games as event structures. It relies, in particular, in the fact that the games are race-free. Under the same conditions, other general theorems for games with winning conditions can also be given with respect to the new closure operator game semantics. In particular, we extend the characterisation of strategies as nondeterministic closure operators to games with winning conditions. We start by providing the following result.

Theorem 5. *Let A be a race-free game. A strategy $\sigma : S \to A$ in (A, W) is winning iff $\sigma_\mu(y) \subseteq W$ for all $y \in \mathcal{C}(A)$ under σ.*

Based on Theorem 5, which relates the standard definition of strategies as maps of event structures with strategies as nondeterministic closure operators, known techniques to characterise winning strategies can be used so that such concurrent strategies can be characterised, instead, with respect to the existence of nondeterministic closure operators. First, let us define the set of results of a concurrent game via nondeterministic closure operators.

Given A and two nondeterministic closure operators σ_μ and τ_μ for Player and Opponent, their *one-step composition* at $y \in \mathcal{C}(A)$, denoted by $(\sigma_\mu \bowtie \tau_\mu)(y)$, induces the following set of configurations: $\{\sigma_\mu(y') \subseteq \mathcal{C}(A) \mid y' \in \tau_\mu(y)\}$. Now, let the set R be the *partial results* of playing-off σ_μ against τ_μ, which is inductively defined as follows: $(\sigma_\mu \bowtie \tau_\mu)(\emptyset) \subseteq R$ and if $y \in R$ then $(\sigma_\mu \bowtie \tau_\mu)(y) \subseteq R$. Finally, similar to the case where the results of a concurrent game are computed using a pullback construction, we define the set of *results* of the game as the maximal elements of R, which we simply denote by $\sigma_\mu \bowtie \tau_\mu$. Using these definitions one can show that $\bowtie$ is a commutative operator, that is, that the following holds.

Proposition 6. *Let σ and τ be two strategies for Player and Opponent. Then*

$$\sigma_\mu \bowtie \tau_\mu = \tau_\mu \bowtie \sigma_\mu$$

The equivalence relation given by Proposition 6 ensures that the two strategies can be played in *parallel* while preserving the same set of results—a property of the composition of strategies in the model of games as event structures.

Based on the above results, one can also show that winning strategies, when represented as nondeterministic closure operators, can be characterised with respect to the sets of results obtained when composing them with every deterministic strategy, represented as closure operators, for the other player. Finally, the following result fully captures the notion of winning in race-free games.

Theorem 7 (Winning Strategies). *Let A be a race-free concurrent game. The nondeterministic closure operator σ_μ is winning for Player if and only if $(\sigma_\mu \bowtie \tau_\mu) \subseteq W$, for all closure operators τ_μ for Opponent.*

Theorem 7 follows from results about winning strategies [5], and the fact that not only every strategy $\sigma : S \to A$ determines a unique (partial) nondeterministic closure operator $\sigma_\mu : \mathcal{C}(A) \to \wp(\mathcal{C}(A))$, but also every operator σ_μ is determined by some (total) nondeterministic closure operator $f^{\to} : \mathcal{C}(S) \to \wp(\mathcal{C}(S))$.

6 Conclusions, Application Domains, and Related Work

In this paper, we studied a mathematical model, which builds on closure operators and has a game-theoretic interpretation, where some forms of concurrency and nondeterminism are allowed to coexist. In particular, the model extends those based on deterministic games—and hence on closure operators too.

Indeed, deterministic games/strategies are already important in the model of games as event structures. Strategies in this kind of games can represent stable spans and stable functions [18], Berry's dI-domains [4], closure operator models of CCP [17], models of fragments of Linear Logic [1,3], and innocent strategies in simple games [8], which underlie Hyland–Ong [9] and AJM [2] games. Strategies in deterministic games are also equivalent to those in Melliès and Mimram [11] model of asynchronous games with receptive ingenious strategies.

However, none of the models above mentioned allow a free interplay of nondeterminism and concurrency: either nondeterminism is allowed in a sequential setting, or concurrency is studied in a deterministic setting. Still, nondeterminism is needed in certain scenarios, or may be a desirable property. We would like to mention three prominent cases: concurrent game models of logical systems [5], formal languages with nondeterministic behaviour [12,15], and concurrent systems with partial order behaviour—also called 'true-concurrency' systems [13].

Logical Systems. In order to give a concurrent game semantics of logical systems such as classical or modal logics, the power to express nondeterministic choices is needed, in particular, in order to be able to interpret disjunctions in a concurrent way—a "parallel or" operator. Deterministic strategies—and hence conventional closure operators—are unable to do this in a full and faithful way.

Formal Languages. Another example where nondeterminism is allowed is within formal languages such as ntcc [12], a nondeterministic extension of CCP, and in simple programming languages with nondeterminism as the one initially studied by Plotkin using powerdomains [15]. Whereas in the former case no game theoretic model has been studied, in the latter case no closure operator semantics has been investigated. Indeed, to the best of our knowledge, no game theoretic characterisation of powerdomains has been defined so far. An interesting potential application would be a (nondeterministic closure operator) game characterisation of Kahn–Plotkin concrete domains [10] given the simpler structure of nondeterministic choices allowed in such a denotational model.

True-concurrency. In concurrent systems with partial order behaviour, such as Petri nets or asynchronous transition systems, both concurrency and nondeterminism are allowed at the same time, which prevents the use of conventional closure operators as the basis for the definition of a fully abstract model. In all of these cases, the model of concurrent games as event structures could be used as an underlying semantic framework, and in particular our nondeterministic closure operator characterisation/semantics when restricted to race-free systems. A good starting point would be to consider free-choice nets [6], since in this case race-freedom can be easily imposed by associating it with conflicts in the net.

Acknowledgment. I thank Paul Harrenstein, Glynn Winskel, and Michael Wooldridge for their comments and support. Also, I acknowledge with gratitude the support of the ERC Advanced Research Grant 291528 ("RACE") at Oxford.

References

1. Abramsky, S.: Sequentiality vs. concurrency in games and logic. Math. Struct. Comput. Sci. **13**(4), 531–565 (2003)
2. Abramsky, S., Jagadeesan, R., Malacaria, P.: Full abstraction for PCF. Inf. Comput. **163**(2), 409–470 (2000)
3. Abramsky, S., Melliès, P.: Concurrent games and full completeness. In: LICS, pp. 431–442. IEEE Computer Society (1999)
4. Berry, G.: Modèles complètement adéquats et stables des lambda-calculs typés. Ph.D. thesis, University of Paris VII (1979)
5. Clairambault, P., Gutierrez, J., Winskel, G.: The winning ways of concurrent games. In: LICS, pp. 235–244. IEEE Computer Society (2012)
6. Desel, J., Esparza, J.: ree Choice Petri Nets, Cambridge Tracts in Theoretical Computer Science, vol. 40. Cambridge University Press, Cambridge (1995)
7. Gutierrez, J.: Concurrent logic games on partial orders. In: Beklemishev, L.D., de Queiroz, R. (eds.) WoLLIC 2011. LNCS, vol. 6642, pp. 146–160. Springer, Heidelberg (2011)
8. Harmer, R., Hyland, M., Melliès, P.: Categorical combinatorics for innocent strategies. In: LICS, pp. 379–388. IEEE Computer Society (2007)
9. Hyland, J.M.E., Ong, C.L.: On full abstraction for PCF: I, II, and III. Inf. Comput. **163**(2), 285–408 (2000)
10. Kahn, G., Plotkin, G.: Concrete domains. Technical report, INRIA (1993)
11. Melliès, P.-A., Mimram, S.: Asynchronous games: innocence without alternation. In: Caires, L., Vasconcelos, V.T. (eds.) CONCUR 2007. LNCS, vol. 4703, pp. 395–411. Springer, Heidelberg (2007)
12. Nielsen, M., Palamidessi, C., Valencia, F.D.: Temporal concurrent constraint programming: Denotation, logic and applications. Nord. J. Comput. **9**(1), 145–188 (2002)

13. Nielsen, M., Winskel, G.: Models for concurrency. In: Handbook of Logic in Computer Science, pp. 1–148. Oxford University Press (1995)
14. Olarte, C., Valencia, F.D.: Universal concurrent constraint programing: symbolic semantics and applications to security. In: SAC, pp. 145–150. ACM (2008)
15. Plotkin, G.D.: A powerdomain construction. SIAM J. Comput. **5**(3), 452–487 (1976)
16. Rideau, S., Winskel, G.: Concurrent strategies. In: LICS, pp. 409–418. IEEE Computer Society (2011)
17. Saraswat, V.A., Rinard, M.C., Panangaden, P.: Semantic foundations of concurrent constraint programming. In: POPL, pp. 333–352. ACM Press (1991)
18. Saunders-Evans, L., Winskel, G.: Event structure spans for nondeterministic dataflow. Electron. Notes Theoret. Comput. Sci. **175**(3), 109–129 (2007)
19. Winskel, G.: Deterministic concurrent strategies. Formal Aspects Comput. **24**(4–6), 647–660 (2012)
20. Winskel, G.: Strategies as profunctors. In: Pfenning, F. (ed.) FOSSACS 2013 (ETAPS 2013). LNCS, vol. 7794, pp. 418–433. Springer, Heidelberg (2013)

First Steps Towards Cumulative Inductive Types in CIC

Amin Timany[(✉)] and Bart Jacobs

iMinds-DistriNet, KU Leuven, Leuven, Belgium
{amin.timany,bart.jacobs}@cs.kuleuven.be

Abstract. Having the type of all types in a type system results in paradoxes like Russel's paradox. Therefore type theories like predicative calculus of inductive constructions (pCIC) – the logic of the Coq proof assistant – have a hierarchy of types $\mathtt{Type}_0$, $\mathtt{Type}_1$, $\mathtt{Type}_2$, ..., where $\mathtt{Type}_0 : \mathtt{Type}_1$, $\mathtt{Type}_1 : \mathtt{Type}_2$, ... In a cumulative type system, e.g., pCIC, for a term $\mathtt{t}$ such that $\mathtt{t}: \mathtt{Type}_i$ we also have that $\mathtt{t}: \mathtt{Type}_{i+1}$. The system pCIC has recently been extended to support universe polymorphism, i.e., definitions can be parametrized by universe levels. This extension does not support cumulativity for inductive types. For example, we do not have that a pair of types at levels i and j is also considered a pair of types at levels $i + 1$ and $j + 1$.

In this paper, we discuss our on-going research on making inductive types cumulative in the pCIC. Having inductive types be cumulative alleviates some problems that occur while working with large inductive types, e.g., the category of small categories, in pCIC.

We present the pCuIC system which adds cumulativity for inductive types to pCIC and briefly discuss some of its properties and possible extensions. We, in addition, give a justification for the introduced cumulativity relation for inductive types.

1 Introduction

The type system of the proof assistant Coq, a variant of the predicative calculus of inductive constructions (pCIC) (see [3] for details), has recently been extended to support universe polymorphism [6]. There the calculus is extended with support for universe polymorphic definitions and inductive types are treated by considering copies of them at different universe levels – so long as levels satisfy constraints imposed by the inductive type and the environment. In this system the simple definition for a category, `Class Category: Type := {O: Type; H: O →O→Type; ...}` defines a type $\mathtt{Category}_{ij}$ where i is the universe level for objects and j is the universe level for homomorphisms. This allows a straightforward definition of the category of small[1] categories, `Definition Cat: Category := {|O:= Category; H:= Functor;...|}` which defines a term of type category, $\mathtt{Cat}_{ijkl}:\mathtt{Category}_{ij}$, with object type $\mathtt{Category}_{kl}$[2]. However, inductive types such as `Category` not being

---

[1] Here, smallness and largeness are to be understood as relative to universe levels.
[2] Subject to side constraints on universe levels, e.g., $k, l < i$.

© Springer International Publishing Switzerland 2015
M. Leucker et al. (Eds.): ICTAC 2015, LNCS 9399, pp. 608–617, 2015.
DOI: 10.1007/978-3-319-25150-9_36

cumulative implies that having a term t such that t: $\mathtt{Category}_{kl}$ and
t : $\mathtt{Category}_{k'l'}$ is possible if and only if $k = k'$ and $l = l'$.

This side condition, however, has undesirable consequences. First and foremost, the term Cat above is not the category of all small categories, rather all categories at *some particular* lower universe level. Furthermore, statements about Cat impose restrictions on its universe levels. That is, only those copies of Cat that conform to the restrictions imposed are subject to the stated fact. For instance, showing that the trivial category (a category with a single object and its identity arrow) with object type unit: $\mathtt{Type}_0$ is the terminal object of $\mathtt{Cat}_{ijkl}$, implies $k = 0$. Also, showing that $\mathtt{Cat}_{ijkl}$ has exponentials (functor categories) implies $j = k = l$. The latter restriction is inconsistent with the restriction $n < m$ on TypeCat: $\mathtt{Category}_{mn}$, the category of types and functions in Coq: Definition Type_Cat: Category := {|O:= Type; H:= fun A B ⇒ A → B;...|}. Note

that here m is the level of *type of* O:=Type@{k} for some k while n is the level of type of A → B, i.e., Type@{k}, hence $n = k$. This means, a copy of Cat cannot both have exponentials and a copy of TypeCat in its objects. Furthermore, having $\mathtt{Cat}_{ijkl}$ cartesian closed restricts it so that $j = k = l = 0$. For further details of using universe levels to represent smallness/largeness in category theory see [7]. There, in addition to the issues mentioned above, we shortly discuss how this representation works intuitively and as expected.

It is, furthermore, noteworthy that such issues are not particular to category theory and are rather prevalent in any case incorporating large inductive types. Take the well-known definition of sets in type theory with inductive types: Inductive Ens: Type :=ens : Π(A: Type),(A →Ens) →Ens. In this case, $\mathtt{Ens}_i$: $\mathtt{Type}_{i+1}$ has constructor $\mathtt{ens}_i$: Π(A: $\mathtt{Type}_i$),(A →$\mathtt{Ens}_i$) →$\mathtt{Ens}_i$. As a result, the ensemble of small ensembles, ens Ens (λ(x: Ens). x), can't be formed as x in the body of the lambda-term is at a strictly lower universe level than the result ensemble.

To solve these problems, explicit lifting functions, e.g., Lift_Ens: $\mathtt{Ens}_i$ →$\mathtt{Ens}_j$ with $i \leq j$, could be used. They allow formation of terms such as the ensemble of *lifted* small ensembles. However, we can't prove, or even specify, Π(t: T), t = Lift_T t. As a result, working with such lifted values and types depending on them in particular is very complicated.

The rest of this paper is structured as follows: in Sect. 2 we present an extension of pCIC with cumulative inductive types and discuss its properties. In Sect. 3, we introduce lpCuIC; a subsystem of pCuIC for which we can prove soundness by reducing it to the soundness of pCIC using lifter terms. These lifters will in addition provide an intuitive reason why the cumulativity relation introduced in pCuIC is suitable. In Sect. 4, we conclude with discussing possible extensions to the presented system.

2   pCIC with Cumulative Inductive Types (pCuIC)

In this section we present the predicative calculus of cumulative inductive constructions (pCuIC for short), an extension of the predicative calculus of inductive constructions (pCIC) which additionally supports cumulativity for inductive types. The definition of pCuIC is identical to that of pCIC, except for the cumulativity rule C-IND of Fig. 2. The rules for typing judgements of this system are presented in Fig. 1. In the sequel, we use $x, y, z \ldots X, Y, Z \ldots$ to denote variables, $m, n, \ldots, M, N, \ldots$ for terms, $i, j, \ldots$ for natural numbers and s to stand for a sort, i.e., Prop or Type_i.

$$(\text{EMPTY}) \quad \frac{}{\cdot \vdash}$$

$$(\text{DECL}) \quad \frac{\Gamma \vdash T : s \quad x \notin \Gamma}{\Gamma, x : T \vdash}$$

$$(\text{TYPE}) \quad \frac{\Gamma \vdash}{\Gamma \vdash \text{Type}_i : \text{Type}_{i+1}}$$

$$(\text{PROP}) \quad \frac{\Gamma \vdash}{\Gamma \vdash \text{Prop} : \text{Type}_i}$$

$$(\text{VAR}) \quad \frac{\Gamma \vdash \quad (x : T) \in \Gamma}{\Gamma \vdash x : T}$$

$$\frac{\Gamma \vdash t : (\Pi x : A.B) \quad \Gamma \vdash t' : A}{\Gamma \vdash (t\ t') : B[t'/x]} \quad (\text{APP})$$

$$\frac{\Gamma \vdash A : s \quad \Gamma, x : A \vdash B : s' \quad (s, s', s'') \in R_\Pi}{\Gamma \vdash \Pi x : A.\ B : s''} \quad (\text{PROD})$$

$$(\text{LAM}) \quad \frac{\Gamma, x : A \vdash t : B}{\Gamma \vdash (\lambda x : A.\ t) : (\Pi x : A.\ B)}$$

$$\frac{\Gamma \vdash t : A \quad \Gamma \vdash B : s \quad A \preceq B}{\Gamma \vdash t : B} \quad (\text{CONV})$$

$$\frac{A \in Ar(s) \quad \Gamma \vdash A : s' \quad (\Gamma, X : A \vdash C_i : s \quad C_i \in Co(X) \quad \forall 1 \leq i \leq n)}{\Gamma \vdash \text{Ind}(X : A)\{C_1, \ldots, C_n\} : A} \quad (\text{IND})$$

$$\frac{I \equiv \text{Ind}(X : A)\{C_1, \ldots, C_n\} \quad \Gamma \vdash I : A \quad 1 \leq i \leq n}{\Gamma \vdash \text{Constr}(i, I) : C_i[I/X]} \quad (\text{CONSTR})$$

$$\frac{I \equiv \text{Ind}(X : \Pi \vec{x} : \vec{A}.\ s)\{C_1, \ldots, C_n\} \quad \Gamma \vdash \vec{a} : \vec{A} \quad (s, s') \in R_\xi \quad \Gamma \vdash c : (I\vec{a})}{\Gamma \vdash Q : (\Pi \vec{x} : \vec{A}.(I\ \vec{x}) \to s') \quad (\Gamma \vdash f_i : \xi(I, Q, \text{Constr}(i, I), C_i) \quad \forall 1 \leq i \leq n)}{\Gamma \vdash \text{Elim}(c, Q)\{f_1, \ldots, f_n\} : (Q\ \vec{a})\ c} \quad (\text{ELIM})$$

Fig. 1. Typing judgements

2.1 Typing Rules

Figure 1 contains typing rules of PCuIC where the conversion/cumulativity relation, $\preceq$, of rule CONV is defined in Fig. 2.

The relation R_Π governs the level of products formed in the system and is given by $R_\Pi = \{(\_, \text{Prop}, \text{Prop}), (\text{Type}_i, \text{Type}_j, \text{Type}_{max(i,j)})\}$. In other words, Prop is impredicative while Type is predicative. The relation R_ξ governs formation of eliminations, $R_\xi = \{(\text{Prop}, \text{Prop}), (\text{Type}_i, \text{Type}_j), (\text{Type}_i, \text{Prop})\}$ That is, we do not allow terms that are not proofs to be constructed by case analysis on a proof. The judgement $\Gamma \vdash$ expresses validity of context Γ and judgement $\Gamma \vdash t : A$ expresses the fact that term t has type A under context Γ. In case x does not appear freely in B, we abbreviate $\Pi x : A.\ B$ as $A \to B$.

Rules IND, CONSTR and ELIM, respectively, concern formation of inductive types, their constructor terms and their elimination. For further reading on inductive types in calculus of constructions refer to [4,5]. Here, arity for a sort s, $Ar(s)$, types strictly positive in X, $Pos(X)$ and types of constructors, $Co(X)$, are as follows:

$$Ar(s) := \Pi \vec{x} : \vec{M}. \ s \qquad Pos(X) := \Pi \vec{x} : \vec{M}. \ X \vec{m}$$
$$Co(X) := X \vec{m} \mid Pos(X) \to Co(X) \mid \Pi \vec{x} : \vec{M}. \ Co(X)$$

provided that in $Pos(X)$ and $Co(X)$, above, X does not appear in $\vec{m}$ or $\vec{M}$. This is to ensure that X appears in constructors only strictly positively. In Fig. 1, ξ is the type for eliminators defined below.

Definition 1 (Eliminator Type). *Let C be a type of constructor for X and let Q and c be two terms. Then, the type of eliminator for C, $\xi(I, Q, c, C) \equiv (\xi_X(Q, c, C))[I/X]$ is defined as follows:*

$$\xi_X(Q, c, P \to N) \quad = \Pi p : P. \ (\Pi \vec{x} : \vec{M}. \ (Q \ \vec{m} \ (p \ \vec{x}))) \to \xi_X(Q, (c \ p), N)$$
$$\text{for } P \equiv \Pi \vec{x} : \vec{M}. \ (X \ \vec{m})$$
$$\xi_X(Q, c, \Pi x : M. \ N) = \Pi x : M. \ \xi_X(Q, (c \ x), N)$$
$$\xi_X(Q, c, X \ \vec{a}) \quad = (Q \ \vec{a} \ c)$$

∎

In this system, variable x in terms $\lambda x : A. \ B$ and $\Pi x : A. \ B$ are bound in B and variable X in $Ind(X : A)\{C_1, \ldots, C_n\}$ is bound in $C_1, \ldots, C_n$. We consider terms equal up to renaming of bound variables, α-conversion. Additionally, we assume that before any substitution, if necessary, α-conversion is performed so as to prevent any variable capture.

2.2 Reduction Rules

The computational rule corresponding to inductive types, expectedly, corresponds to induction/recursion. The elimination of a term of an inductive type should perform a case analysis on its input and apply the corresponding provided elimination for that case by recursively eliminating any argument of the constructor that is of the inductive type. This will be made more clear later. For now let us consider recursors for constructors. A recursor for a constructor, as the name suggests, takes the arguments of a constructor and performs the provided elimination by recursively eliminating sub-terms. The recursor $\mu(I, F, f, C)$ for a constructor C of an inductive type I takes two terms f and F. The term f is the term that performs elimination for constructor C while term F corresponds to recursive elimination of sub-terms.

Definition 2 (Recursor). *Let C be a type of constructor for X and F and f be terms. Then, recursor $\mu(I, F, f, C) = (\mu_X(F, f, C))[I/X]$ is defined as follows:*

$$\mu_X(F, f, P \to N) \quad = \lambda p : P. \mu_X(F, (f \ p \ (\lambda \vec{x} : \vec{M}. \ (F \ \vec{m} \ (p \ \vec{x})))), N)$$
$$\text{for } P \equiv \Pi \vec{x} : \vec{M}.(X \ \vec{m})$$
$$\mu_X(F, f, \Pi x : M. \ N) = \lambda x : M. \ \mu_X(F, (f \ x), N)$$
$$\mu_X(F, f, X \ \vec{a}) \quad = f$$

∎

We consider two computation rules for pCuIC, β, for function application, $(\lambda x : A. t)t' \to_\beta t[t'/x]$ and ι for elimination of inductive types,

$$\mathsf{Elim}((\mathsf{Constr}(i, I)\ \vec{m}), Q)\{f_1, \ldots, f_n\} \to_\iota (\mu(I, F_{elim}(I, Q, f_1, \ldots, f_n), f_i, C_i)\ \vec{m})$$

for $I \equiv \mathsf{Ind}(X : A)\{C_1, \ldots, C_n\}$, $A \equiv \Pi\vec{x} : \vec{A}.s$ where $F_{elim}(I, Q, f_1, \ldots, f_n) \equiv \lambda\vec{x} : \vec{A}.\ \lambda c : (I\ \vec{x}).\ \mathsf{Elim}(c, Q)\{f_1, \ldots, f_n\}$. In the sequel, we write $\simeq$ to denote definitional equality, i.e., $\alpha\beta\iota\eta$-conversion. For proofs of why eliminator types and recursors above are well-typed, refer to [4,5].

As an example of inductive types and their elimination, let us define in pCuIC the prime example of inductive types, natural numbers, $nat \equiv \mathsf{Ind}(X : \mathsf{Type}_0)\{X, X \to X\}$. Let us use $Zero \equiv \mathsf{Constr}(1, nat) : nat$ and $Succ \equiv \mathsf{Constr}(2, nat) : nat \to nat$ to refer to the zero and successor constructors of the natural numbers. We construct the eliminator for type nat as follows.

$$\cdot \vdash \left(\lambda Q : (nat \to s).\lambda f_1 : (Q\ Zero).\right.$$
$$\left.\lambda f_2 : (\Pi p : nat.\ (Q\ p) \to Q\ (Succ\ p)).\ \lambda n : nat.\ \mathsf{Elim}(n, Q)\{f_1, f_2\}\right) :$$
$$\left(\Pi Q : (nat \to s).\ (Q\ Zero) \to (\Pi p : nat.\ (Q\ p) \to Q\ (Succ\ p)) \to \Pi n : nat.\ Q\ n\right)$$

Which is precisely the induction (in case $s =\mathsf{Prop}$) and recursion principle for natural numbers.

As another example of an inductive type consider the *even* predicate defined inductively. $even \equiv \mathsf{Ind}(X : nat \to \mathsf{Prop})\{X\ Zero, \Pi n : nat.\ X\ n \to X\ (Succ\ (Succ\ n))\}$ This type has two constructors. The first constructor constructs a proof that $Zero$ is an even number. The second constructor, takes a natural number n and a proof that n is even and produces a proof that $(Succ\ (Succ\ n))$ is even.

2.3 Cumulativity

The relation $\preceq$ in rule CONV reflects both convertibility and cumulativity. Rules for this relation are depicted in Fig. 2. Rule C-IND corresponds to cumulativity of inductive types. Intuitively, rule C-IND establishes relation $I\ \vec{m} \preceq I'\ \vec{m}$, if every arity type and constructor parameter type of I is a subtype of the corresponding type in I'. As a condition of C-IND we have $\forall i.\ X\ \vec{m_i} \simeq X\ \vec{m_i'}$. This means that the ith constructor of I and I' if applied to the same terms must produce instances of I and I' with the same values for arities.

As an example, consider the type of categories which in pCuIC is of the form: $Category_{i,j} \equiv \mathsf{Ind}(X : \mathsf{Type}_{max(i+1,j+1)})\{\Pi o : \mathsf{Type}_i.\Pi h : o \to o \to \mathsf{Type}_j.N\}$ for $i, j \in \mathbb{N}$ where i and j don't appear in N. Clearly, we can use C-IND to derive $Category_{i,j} \preceq Category_{k,l}$ given that $i \le k$ and $j \le l$. A similar argument can show that $Ens_i \preceq Ens_j$ given that $i \le j$. Hence, pCuIC doesn't suffer from the problems mentioned earlier regarding the category of small categories and the ensemble of small ensembles.

$$\frac{}{\text{Prop} \preceq \text{Type}_i} \text{(C-Prop)} \qquad \frac{i \leq j}{\text{Type}_i \preceq \text{Type}_j} \text{(C-Type)} \qquad \frac{A \simeq A' \quad B \preceq B'}{\Pi x : A. \, B \preceq \Pi x : A'. \, B'} \text{(C-Prod)} \qquad \frac{A \simeq B}{A \preceq B} \text{(C-Conv)}$$

$$\frac{A \simeq A' \quad A' \preceq B' \quad B \simeq B'}{A \preceq B} \text{(C-Congr)}$$

$$I \equiv (\text{Ind}(X : \Pi \vec{x} : \vec{N}. \, s)\{\Pi \vec{x_1} : \vec{M_1}. \, X \, \vec{m_1}, \ldots, \Pi \vec{x_n} : \vec{M_n}. \, X \, \vec{m_n}\}$$
$$I' \equiv (\text{Ind}(X : \Pi \vec{x} : \vec{N'}. \, s')\{\Pi \vec{x_1} : \vec{M_1'}. \, X \, \vec{m_1'}, \ldots, \Pi \vec{x_n} : \vec{M_n'}. \, X \, \vec{m_n'}\}$$

$$\frac{s \preceq s' \quad \forall i. \, N_i \preceq N'_i \quad \forall i,j. \, (M_i)_j \preceq (M_i')_j}{\qquad length(\vec{m}) = length(\vec{x}) \quad \forall i. \, X \, \vec{m_i} \simeq X \, \vec{m_i'} \qquad}{I \, \vec{m} \preceq I' \, \vec{m}} \text{(C-Ind)}$$

Fig. 2. Conversion/cumulativity relation

2.4 Properties

Although we do not provide any proof, we believe that the following two conjectures, stating properties of pCuIC and relation $\preceq$, respectively, hold and can be proven in a way akin to their counterparts in [2] or [4].

Conjecture 1. pCuIC has the following properties:

1. Church-Rosser property for $\beta\iota$-reduction   *(Church-Rosser)*
2. $\beta\iota$ strong normalization   *(Strong Normalization)*
3. Every derivation $\Gamma \vdash t : A$ has a sub-derivation that derives $\Gamma \vdash$ and every derivation $\Gamma, x : T, \Gamma' \vdash$ has a sub-derivation that derives $\Gamma \vdash T : s$ for some sort s *(Context-Validity)*
4. if $\Gamma \vdash t : A$, then there is a sort s such that $\Gamma \vdash A : s$ *(Typing-Validity)*
5. if $\Gamma \vdash t : A$ and $t \rightarrow^*_{\beta\iota} t'$ then $\Gamma \vdash t' : A$ *(Subject Reduction)* ∎

Conjecture 2. Properties of $\preceq$:

1. $\preceq$ is a partial order relation over $\simeq$: $\dfrac{}{t \preceq t} \quad \dfrac{t \preceq t' \quad t' \preceq t''}{t \preceq t''} \quad \dfrac{t \preceq t' \quad t' \preceq t}{t \simeq t'}$
2. The relation $\preceq$ is well-founded, i.e., there is no infinite decreasing chain $A_0 \succ A_1 \succ \ldots$, where $t \prec t'$ if $t \preceq t'$ and $t \not\simeq t'$ *(Well-Founded)*
3. if $\Gamma \vdash t : A$, then there exists B such that $\Gamma \vdash t : B$ and for any C such that $\Gamma \vdash t : C$ we have $B \preceq C$ *(Principal Type)* ∎

The system presented in this paper, pCuIC, has a strictly richer type system compared to pCIC. In other words, $\Gamma \vdash_{\text{pCIC}} t : A$ implies $\Gamma \vdash_{\text{pCuIC}} t : A$ but the converse does not hold. Consider the instance of the ensemble of small ensembles expressed in pCuIC as $EE_i \equiv (\text{Constr}(1, Ens_{i+1}) \, Ens_i \, (\lambda x : Ens_i. \, x)) \cdot \vdash EE_i : Ens_{i+1}$ is derivable in pCuIC but not in pCIC. The inductive type Ens_i is defined as: $Ens_i \equiv \text{Ind}(X : \text{Type}_{i+1})\{\Pi A : \text{Type}_i. \, (A \rightarrow X) \rightarrow X\}$.

In pCuIC, Π types are considered invariant in their domain type. However, we believe that results similar to those discussed in this paper hold for the case with full contravariance for the domain type of Π types. Note that points 2 and 3 of Conjecture 2 don't hold in the version of pCuIC with full contravariance. Although, we believe they do hold in the subsystem $pCuIC^n$, a subsystem of PCuIC in which the universe levels are squashed such that Type_n is the type of all types (note that Type_n itself has no type in $pCuIC^n$). This treatment is similar to that of ECC^n in the proof of quasi normalization of ECC in [2].

For systems such as pCuIC, the strong normalization and subject reduction properties (stated in Conjecture 1 for pCuIC) imply (see [2]) the soundness and decidability of type checking. However, as of writing of this paper we have not yet proven the conjectures above. Another approach to proving soundness of pCuIC is by reducing it to the soundness of pCIC. That is, the soundness of pCuIC follows from the following conjecture:

Conjecture 3. Let $\Gamma \vdash_{\mathsf{pCIC}} T : s$ be a pCIC type such that $\Gamma \vdash_{\mathsf{pCuIC}} t : T$. Then there exists a term t' such that $\Gamma \vdash_{\mathsf{pCIC}} t' : T$. ∎

In other words, every pCIC type that is inhabited in pCuIC is also inhabited in pCIC. We can use this conjecture to prove the soundness of pCuIC as follows. Let $False \equiv \mathsf{Ind}(X : \mathsf{Prop})\{\}$ be the inductive type with no constructors. According to the Conjecture 3, if there is a term t such that $\cdot \vdash_{\mathsf{pCuIC}} t : False$ then there is a term t' such that $\cdot \vdash_{\mathsf{pCIC}} t' : False$ which implies unsoundness of pCIC which is a contradiction. The main difficulty in proving this conjecture though is the fact that there are types in pCuIC that are not valid types in pCIC. As an example consider any type that involves the term EE_i (ensemble of small ensembles) above. Lifting such a term results in a type where the dependent argument is ensemble of lifted small ensembles (and not ensemble of small ensembles). Note that such terms and types can be part of a term which has a pCIC type. The situation is particularly complicated with the functions whose domain type is a type that is not a valid pCIC type.

3   Lesser pCuIC

In this section we introduce the lesser pCuIC (lpCuIC for short) which is a subsystem of pCuIC for which we can prove soundness by reducing it to the soundness of pCIC. This furthermore gives an intuition why the cumulativity relation introduced in this paper for inductive types is suitable.

Definition 3 (lpCuIC). *The system lpCuIC is the system pCuIC where rules* C-IND *and* APP *are replace respectively by:*

$$I \equiv (\mathsf{Ind}(X : \Pi \vec{x} : \vec{N}.\ s)\{\Pi(\vec{x_1} : \vec{M_1}.\ X\ \vec{m_1}, \ldots, \Pi \vec{x_n} : \vec{M_n}.\ X\ \vec{m_n}\}$$

$$I' \equiv (\mathsf{Ind}(X : \Pi \vec{x}' : \vec{N}'.\ s')\{\Pi \vec{x_1}' : \vec{M_1}'.\ X\ \vec{m_1}', \ldots, \Pi \vec{x_n}' : \vec{M_n}'.\ X\ \vec{m_n}'\}$$

$$s \preceq s' \quad \forall i.\ N_i \preceq_{\mathsf{pCIC}} N'_i \quad \forall i,j.\ (M_i)_j \preceq_{\mathsf{pCIC}} (M'_i)_j$$

$$\frac{length(\vec{m}) = length(\vec{x}) \quad \forall i.\ X\ \vec{m_i} \simeq X\ \vec{m_i}'}{I\ \vec{m} \preceq I'\ \vec{m}} \quad \text{(C-IND')}$$

$$(\text{App'}) \quad \frac{\Gamma \vdash t : (\Pi x : A.\ B) \quad \Gamma \vdash t' : A}{\Gamma \vdash (t\ t') : B[t'/x]} \quad (\Gamma \vdash_{\text{pClC}} t' : A \text{ or } x \notin FV(B))$$

We furthermore impose the restrictions that in any derivation of $\Gamma \vdash_{\text{lpCulC}} t : T$, *for any sub-derivation of* $\Gamma' \vdash_{\text{lpCulC}} t' : T'$ *we have* $\Gamma' \vdash_{\text{pClC}} T' : s$ *and for any sub-derivation of* $\Gamma' \vdash_{\text{lpCulC}} T' : \Pi \vec{x} : \vec{A}.s$ *we have* $\Gamma' \vdash_{\text{pClC}} T' : \Pi \vec{x} : \vec{A}.s$ ∎

In other words, lpCuIC is a subsystem of pCuIC in which every valid type is also a valid type in pCIC and any functions whose output type depend on their input can't be applied to terms that are not of the appropriate type in pCIC.

In lpCuIC, we define the following lifters for the cumulativity relation. These lifters are then used to show that any type inhabited in lpCuIC is also inhabited in pCIC. This will give us a soundness proof for lpCuIC.

Definition 4 (Lifters). *Let* T *and* T' *be two terms such that* $T \preceq_{\text{lpCulC}} T'$. *Then, we define the lifter* $\Upsilon_{T \preceq_{\text{lpCulC}} T'}$ *recursively on derivation of* $T \preceq_{\text{lpCulC}} T'$. *If the last rule used to derive* $T \preceq_{\text{lpCulC}} T'$ *is:*

$$\text{C-Prop, C-Type or C-Conv} \quad then \quad \Upsilon_{T \preceq_{\text{lpCulC}} T'} = \lambda x : T.\ x$$

$$\text{C-Prod} \quad then \quad \Upsilon_{\Pi x : A.\ B \preceq_{\text{lpCulC}} \Pi x : A'.\ B'} = \lambda f : \Pi x : A.\ B.\lambda x : A'.$$
$$\Upsilon_{B \preceq_{\text{lpCulC}} B'}\ (f\ x)$$

$$\text{C-Congr} \quad then \quad \Upsilon_{A \preceq_{\text{lpCulC}} B} = \Upsilon_{A' \preceq_{\text{lpCulC}} B'}$$

$$\text{C-Ind} \quad then \quad \Upsilon_{I\ \vec{t} \preceq_{\text{lpCulC}} I'\ \vec{t}} = \lambda x : I\ \vec{t}.\text{Elim}(x, Q)\{\phi_1, \ldots, \phi_n\} \ for:$$

$$Q \equiv \lambda \vec{y} : \vec{M}_A.\ \lambda z : I\ \vec{y}.I'\ \vec{y} \quad \phi_i = \upsilon(I, Q, \text{Constr}(i, I), C_i, \text{Constr}(i, I'), C_i')$$

$$I' \equiv \text{Ind}(X : \Pi(\vec{x} : \vec{M}_A').\ s')\{C_1', \ldots, C_n'\}$$

$$I \equiv \text{Ind}(X : \Pi(\vec{x} : \vec{M}_A).\ s)\{C_1, \ldots, C_n\}$$

Here, the constructor lifter for C, $\upsilon(I, Q, c, C, f, C') = \upsilon_X(Q, c, C, f, C')[I/X]$ *is defined as follows:*

$$\upsilon_X(Q, c, P \to N, f, P' \to N') = \lambda p : P.\ \lambda z : (\Pi \vec{x} : \vec{M}.Q\ \vec{t}\ (p\ \vec{x})).$$
$$\upsilon_X(Q, (c\ p), N, (f\ z), N')$$
$$for\ P \equiv \Pi \vec{x} : \vec{M}.\ X\ \vec{t}$$

$$\upsilon_X(Q, c, \Pi x : M.\ N, f, \Pi x : M'.\ N') = \lambda x : M.\ \upsilon_X(Q, (c\ x), N, (f\ x), N')$$
$$\upsilon_X(Q, c, X\ \vec{t}, f, X\ \vec{t}) \qquad = f$$

∎

Lemma 1 (Type Correctness of Lifters). *Let* T *and* T' *be two terms such that* $T \preceq_{\text{lpCulC}} T'$ *and* $\Gamma \vdash_{\text{lpCulC}} T : s$ *and* $\Gamma \vdash_{\text{lpCulC}} T' : s'$. *Then,* $\Gamma \vdash_{\text{pClC}}$ $\Upsilon_{T \preceq_{\text{lpCulC}} T'} : T \to T'$. ∎

Theorem 1 (Inhabitants in lpCuIC). *Let* t *and* T *be terms such that* $\Gamma \vdash_{\text{lpCulC}} t : T$. *Then there exists* t' *such that* $\Gamma \vdash_{\text{pClC}} t' : T$. ∎

Corollary 1 (Soundness of lpCuIC). $\cdot \vdash_{\mathsf{lpCuIC}} t : False$ *implies that there exists* t' *such that* $\cdot \vdash_{\mathsf{pCIC}} t' : False$. ∎

For proofs of the above lemma, theorem and corollary refer to [8].

The ensemble of ensembles EE_i is a valid term in lpCuIC, i.e., $\cdot \vdash_{\mathsf{lpCuIC}} EE_i : Ens_{i+1}$. For the cumulativity relation $Ens_i \preceq_{\mathsf{lpCuIC}} Ens_{i+1}$, we have the lifter:

$$\Upsilon_{Ens_i \preceq_{\mathsf{lpCuIC}} Ens_{i+1}} = \lambda x : Ens_i.\mathsf{Elim}(x, \lambda z : Ens_i.Ens_{i+1})\{\phi\}$$

for $\phi \equiv \lambda A : \mathtt{Type}_i.\lambda p : (A \to Ens_i).\lambda z : (A \to Ens_{i+1}).\mathsf{Constr}(1, Ens_{i+1})\ A\ z$. As an example, Theorem 1 gives the following term in pCIC for EE_i:

$$\mathsf{Constr}(1, Ens_{i+1})\ Ens_i\ (\lambda x : Ens_i.\Upsilon_{Ens_i \preceq_{\mathsf{lpCuIC}} Ens_{i+1}}\ x)$$

Which is the ensemble of lifted small ensembles and a valid term in pCIC.

The whole purpose of lpCuIC is to demonstrate an intuition of the workings of pCuIC and why we believe it has the properties discussed earlier. Note that although lpCuIC can express terms like ensemble of ensembles, it does not provide us with a flexible enough working environment. As an example, the type $eq\ Ens_{i+1}\ EE_i\ EE_i$ is not a valid lpCuIC type when $eq\ T$ is the equality for type T. This is due to the fact that EE_i and hence $eq\ Ens_{i+1}\ EE_i\ EE_i$ is not a valid type in pCIC. Here the inductive equality type is defined as: $eq \equiv \lambda A : \mathtt{Type}_i.\ \lambda x : A.\ \mathsf{Ind}(X : A \to \mathtt{Prop})\{X\ x\}$

4   Discussion and Conclusion

We presented pCuIC which extends pCIC with cumulativity for inductive types and discussed issues that this treatment helps mitigate. We furthermore justified the cumulativity relation for inductive types that we introduced by showing that there is a sub-system of pCuIC, lpCuIC, in which any such cumulativity relation has a corresponding lifting in pCIC. This, in addition, allowed us to reduce soundness of lpCuIC to the soundness of pCIC.

Inductive types considered lack parameters and mutual inductive types (see [5] for details). Parameters can be considered as variables in the context while an inductive type is being defined. For instance, consider the type of equality eq defined above. There, A and x are parameters of the inductive type eq. In general, the values of parameters can influence the variance of types involving them in an inductive definition. Consider $F : \mathtt{Type}_i \to \mathtt{Type}_j \vdash \mathsf{Ind}(X : \mathtt{Type}_l)\{\Pi A : \mathtt{Type}_k.(F\ A) \to X\}$. In this case we can't determine, e.g., whether $F\ A \preceq F\ B$ for $A \preceq B$. Hence separate analysis of different instances of inductive types with different parameters can help make the cumulativity results more fine-grained. In a different approach, we could add support for variables in the context, e.g., F above, to specify variance of their result with respect to their input, if appropriate, in addition to their type.

On the other hand, mutually inductive types are restricted to only appear strictly positively in one another. Therefore, although it is subject to further

research, it seems natural that the approach presented here can be straightforwardly extended to the case of mutual inductive types.

Another interesting case is when we have x: Type_i in an inductive type. We
have not considered variance of x in our relation. Doing so will result in having,
e.g., $list\ A \preceq list\ B$ for $A \preceq B$. Such cumulativity relations can be very useful
in practice, lessening the need of explicit conversions.

We believe that the typical ambiguity and also elaboration and unification
algorithms presented in [6] can be directly extended to this system. However,
as higher order unification is undecidable in general, lifting functions can be
used as hints to facilitate unification when necessary. Note that these liftings
are not based on case analysis on the input anymore and are hence free of the
aforementioned problems.

Acknowledgements. This work was funded by EU FP7 FET-Open project ADVENT
under grant number 308830.

References

1. Coquand, T., Paulin, C.: Inductively defined types. In: COLOG 1988, Proceedings of
   International Conference on Computer Logic, Tallinn, USSR, pp. 50–66, December
   1988
2. Luo, Z.: An Extended Calculus of Constructions. Ph.D. thesis, University of Edin-
   brugh, Department of Computer Science, June 1990
3. The Coq development team: Coq 8.2 Reference Manual. Inria (2008)
4. Paulin-Mohring, C.: Inductive definitions in the system Coq - rules and properties.
   In: Proceedings of International Conference on Typed Lambda Calculi and Applications, TLCA 1993, Utrecht, The Netherlands, pp. 328–345, March 1993
5. Paulin-Mohring, C.: Introduction to the Calculus of Inductive Constructions,
   November 2014. https://hal.inria.fr/hal-01094195
6. Sozeau, M., Tabareau, N.: Universe polymorphism in Coq. In: Klein, G., Gamboa,
   R. (eds.) ITP 2014. LNCS, vol. 8558, pp. 499–514. Springer, Heidelberg (2014)
7. Timany, A., Jacobs, B.: Category theory in Coq 8.5. CoRR abs/1505.06430 (2015),
   http://www.arxiv.org/abs/1505.06430, accepted for a presentation at the 7th Coq
   workshop, Sophia Antipolis, France on 26 June 2015
8. Timany, A., Jacobs, B.: First Steps Towards Cumulative Inductive Types in CIC:
   Extended Version. Technical Report CW684, iMinds-Distrinet, KU Leuven, March
   2015. http://www2.cs.kuleuven.be/publicaties/rapporten/cw/CW684.abs.html

Author Index

Printed in the United States
By Bookmasters